International Conference on Design, Automation, and Control (ICDAC 2020)

IOP Conference Series: Materials Science and Engineering
Volume 1123

Vellore, India
6-8 January 2020

Editors:

R. Vasudevan
D. Malligarjuna Reddy
E. Rajkumar
T. Narendiranath Babu

ISBN: 978-1-7138-4608-6
ISSN: 1757-8981

Printed from e-media with permission by:

Curran Associates, Inc.
57 Morehouse Lane
Red Hook, NY 12571

Some format issues inherent in the e-media version may also appear in this print version.

Institute of Physics
Dirac House, Temple Back
Bristol BS1 6BE UK

Phone: 44 1 17 929 7481
Fax: 44 1 17 920 0979

techtracking@iop.org

Additional copies of this publication are available from:

Curran Associates, Inc.
57 Morehouse Lane
Red Hook, NY 12571 USA
Phone: 845-758-0400
Fax: 845-758-2633
Email: curran@proceedings.com
Web: www.proceedings.com

TABLE OF CONTENTS

IOP Conf. Series: Materials Science and Engineering 1123 (2021) 011001 doi:10.1088/1757-899X/1123/1/011001

Proceedings of the International Conference on Design, Automation and Control

The international conference on Design, Automation and control (ICDAC) is organized by the Department of Design and Automation, School of Mechanical Engineering, VIT Vellore, during January 6-8, 2020.

The school has been frequently organizing events of this stature for the benefit of the students, faculty members, research scholars and industry experts. The focus of this conference is to share the advancements in Design and Automation domain.

Recent trends in manufacturing such as Cloud-based automation, Cyber-physical production systems, Nano composites and micromechanics, additive manufacturing are the major topics, which are deliberated in this conference. Further the recent developments in automated fault detection, diagnostics, and prognostics, modelling and simulation, shape memory alloys, functionally graded materials and more topics will be addressed in this conference.

I hope this conference will provide the platform for the researchers, academicians and industry experts to share their knowledge in the respective fields.

I wish the participants a technically fruitful stay in VIT and wish the event a grand success.

Dr.R.Vasudevan

The Conference chair

IOP Conf. Series: Materials Science and Engineering 1123 (2021) 011001 doi:10.1088/1757-899X/1123/1/011001

List of commitees

S.No	Committees	Faculty Coordinators	Faculty Team Members
1	**Food committee**	Prof. S. Devendiran	Prof. Ramesh Babu Vemuluri Prof. Senthilnathan N Prof. Kanish T.C Prof. P. M. Anil
2	**Accommodation and Transportation committee**	Prof. Vinoth Jebaraj A	Prof. Seenuvasaperumal P Prof. Bala Murali Prof. Jayaprakash N Prof. S. Senthil Kumaran
3	**Reception and Registration committee**	Prof. Darius Gnanaraj S	Prof. Rajay Vedharaj I S Prof. Bharanidaran R Prof. Sharan Chandran M Prof. Deepa Prof. Fathima Patham
4	**Session management committee**	Prof. Sugavaneswaran M	Prof. Venkatesan K Prof. Anandavel K Prof. Deepak Kumar Biswal Prof. Arulvel S
5	**Souvenir committee**	Prof. Manikandan M	Prof. Renold Elsen S Prof. Rajasekhara Reddy M Prof. Niranjan Behra Prof. Rajesh. M Prof. Anoop A M Prof. Rahul Singh Sikarwar
6	**Guest Care committee**	Prof. Venkatesan.S	Prof. Bikash Routh Prof. Arivarasu Prof. Vinod Kumar Sharma Prof. Senthur Prabhu
7	**Printing and Purchase committee**	Prof. Kalaiarasan. G	Prof. Sathish Paulraj Gundupalli Prof. Velu M Prof. Giriraj M Prof. M. Natarajan Prof. P. Baskar
8	**Honorarium committee**	Prof. Arun Tom Mathew	Prof. Ramakrishnan R Prof. Chinmaya Prasad Mohanthy

IOP Conf. Series: Materials Science and Engineering 1123 (2021) 011001 doi:10.1088/1757-899X/1123/1/011001

			Prof. S. Senthil Kumar Prof. Ramakrishnan R Prof. Rijesh M
9	**Programme Schedule committee**	Prof. Bendict Thomas	Prof. Saurabh Gupta Prof. Jaya krishna Prof. Arivarasu M
10	**Hall arrangement committee**	Prof. Raghuraman DRS	Prof. Sovan Sundar Dasgupta Prof. Suraj Kushe Shekar Prof. Anuj Kumarm
11	**Paper Evaluation committee**	Prof. Akash Mohanty	**Thermal** Prof. Devendra Kumar Patel Prof. Ashok.K Prof.Bibhiti bhusan sahoo Prof.Nataraj.G Prof.Nithin Batula Prof.Ravi verma **Automotive** Prof. K. Ravi Prof. Govindarasu N **Manufacturing** Prof. A.K. Jeevanantham Prof. S.K.Ariful Rehman Prof. Dega Nagaraju Prof. Jayaveera pandiarajan Prof. T. Sampath Kumar **Design and Mechatronics** Prof. Mallikarjuna Reddy Prof. Edwin Sudhagar P Prof. Krishna S Prof. Boopathi. M

			Prof. Ramakrishnan R Prof. Bharanidaran.R Prof. Sudhir Raj

Editors:

Name + Title(s)	*Address (work +*	*Email + URL*
Prof.Vasudevan.R	Professor, Dean, Conference chair, School of Mechanical Engineering, Vellore Institute of Technology, Vellore, Tamilnadu, 632014 Ph:9994097854	Email:vasudevan.r@vit.ac.in URL: www.vit.ac.in
Prof. D.Malligarjuna Reddy,	Associate Professor, Conference co-chair, Head of the department, Department of Design and Automation, School of Mechanical Engineering, Vellore Institute of Technology, Vellore, Tamilnadu, 632014 Ph:9600845926	Email:dmreddy@vit.ac.in URL: www.vit.ac.in
Prof. Rajkumar.E	Associate Professor, Convener, Department of Design and Automation, School of Mechanical Engineering, Vellore Institute of Technology, Vellore, Tamilnadu, 632014 Ph:9944354944	Email: rajkumar.e@vit.ac.in; URL: www.vit.ac.in
Prof.Narendiranath babu.T	Associate Professor, Convener, Department of Design and Automation,	Email:narendiranathbabu.t@vit.ac.in;

	School of Mechanical Engineering, Vellore Institute of Technology, Vellore, Tamilnadu, 632014 Ph: 9176127206	URL: www.vit.ac.in

Peer review declaration

All papers published in this volume of *IOP Conference Series: Materials Science and Engineering* have been peer reviewed through processes administered by the Editors. Reviews were conducted by expert referees to the professional and scientific standards expected of a proceedings journal published by IOP Publishing.

- **Type of peer review: Single blind**
- **Conference submission management system:**
  By email: icdac2020@vit.ac.in
- **Number of submissions received: 123**
- **Number of submissions sent for review:123**
- **Number of submissions accepted:70**
- **Acceptance Rate (Number of Submissions Accepted / Number of Submissions Received X 100): 56.69**
- **Average number of reviews per paper: 2**
- **Total number of reviewers involved: 60**
- **Any additional info on review process:**
- **Contact person for queries: Dr.Rajkumar.E** rajkumar.e@vit.ac.in
-

IOP Conf. Series: Materials Science and Engineering 1123 (2021) 012001 doi:10.1088/1757-899X/1123/1/012001

A FLUID-STRUCTURE INTERACTION BASED SIMULATION STUDY OF PIEZOELECTRIC MICROPUMP FOR DRUG DELIVERY APPLICATION

Abhijeet B. Vante [1]***, R. Ramakrishnan*** [2]

[1,2] Department of Design and Automation, School of Mechanical Engineering, Vellore Institute of Technology, Vellore, India,

[1] abhijeetbhupal.vante@vit.ac.in

[2] ramakrishnan.rc@vit.ac.in

Abstract. Micropump technology has garnered the attention of researchers because of its diversity in numerous applications. Fast response and low power consumption make the piezoelectric micropump an obvious choice for controlled drug delivery applications. It requires fluid flow with minimum flow pulsations and backpressure. Because of the complex structure of the valved micropump, an accurate analytical flow solution is difficult to attain. Hence, a numerical simulation was performed on the model of piezoelectric micropump with microvalve using a finite element solver. A 3D Two-way Fluid-Structure Interaction (FSI) study was carried out. This illustrates the pressure exerted by fluid onto the solid and the change in fluid flow as the resultant. The solution was obtained for the fluid flow in a continuously deforming geometry. The analysis of results gives stresses developed on the microvalve, change in flow discharge and actuator displacement with respect to input frequency and voltage. The experimental results are presented to validate the simulation. This study is useful to establish a precise and controlled drug delivery using micropump in biomedical applications.

1. INTRODUCTION

A drug delivery system (DDS) consists of a method or a device that introduces the drug into the body. To improve the therapeutic index of the drug, it is very crucial for DDS to perform with optimum efficiency. The performance of DDS depends on the place, time and rate of the drug delivery. The drug delivery system is broadly classified into a sustained/ controlled drug delivery system and periodic/manual drug delivery. To achieve controlled or sustained drug delivery, it is important that flow should be with minimum flow pulsation and backpressure. A piezoelectric micropump is used in this study for simulation and experiment as shown in Figure 1.

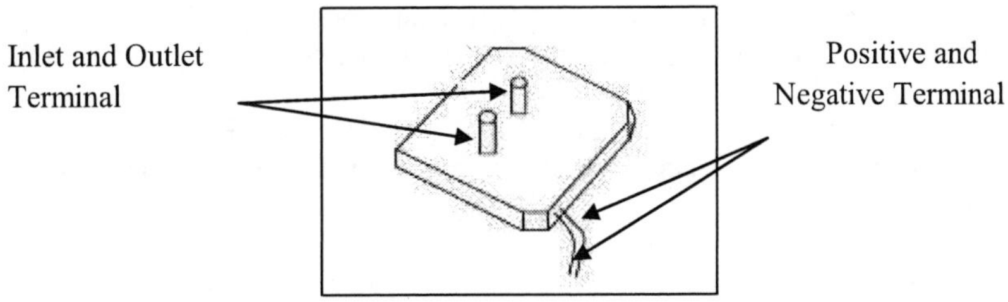

Figure 1. A Piezoelectric Micropump

Large actuation time and rapid response make the piezoelectric micropump a vital component for drug delivery system. Specifications of the micropump components are tabulated in Table 1.

Table 1. Specifications of the Piezoelectric Micropump

Piezoelectric Micropump	Values
Fluid flow rate	20ml/min
Drive frequency	10-60Hz
Drive voltage	60-250Vp-p
Maximum pressure	35KPa
Physical dimensions	33×33×5.5mm

As compared to valveless micropump, a micropump with valves has the advantage in reducing the backpressure of the fluid. The pump houses 3 layers of material- PZT(Lead Zirconate Titanate) actuator, a diaphragm, and a metal plate.

Moreover, Traditional fluid flow simulation does not consider the FSI(Fluid-Structure Interaction) phenomenon. FSI is the multiphysics coupling between fluid mechanics and structural mechanics. It takes place between the moving structure and the surrounding fluid flow. If the deformations of the structure are large, the velocity and pressure field of fluid will change as a result and the problem becomes bi-directionally coupled FSI. In this, the fluid field and pressure field affect the structural deformation and structural deformation affects the flow and pressure. In the micropump, fluid mesh near structure undergoes large deformation and becomes distorted. (ALE formulation)Algebraic Lagrangian-Eulerian (ALE) provides undistorted mesh for the fluid around corners.

FSI is an important phenomenon to analyse the flow behaviour of the micropump. Alongside fluid mechanics and structural mechanics, flow simulations of piezoelectric micropump also consist of electromechanics. The working performance of the micropump can be improved by demonstrating the indicative parameters such as backpressure, pressure behaviour inside the system, flow rate and actuation pressure. [1] FSI is further classified as one-way coupling and two-way coupling. Two-way coupling is difficult to attain prior to its complex geometry, intricate physics of fluids and complicated fluid-structure interaction. [2] But two-way FSI is also helpful in quantifying and parameterizing the flow and pressure quantities for any real-time problems. [3] So the objective of using FSI changes with the change in application. In the case of piezoelectric micropump, FSI is being used in simulation study of both valveless and microvalve pumps. [4] FSI can also be used to confirm the experimental results. [5] Simulation study to understand the flow pulsation and backpressure from the human body are not well explored. A simulation model on 3D piezoelectric micropump is a challenging task.

An attempt has been made in this paper to simulate and control a piezoelectric micropump to obtain a solution for fluid flow in a continuously deforming geometry. 3D two-way fluid-structure interaction is carried out to obtain stresses developed on the microvalve, change in flow discharge and actuator displacement with respect to the given input frequency and voltage. In this paper a simulation study based on 3D two-way FSI of piezoelectric micropump with microvalve is conducted. The analysis of backpressure, flow characteristics and the pressure parameters are presented for the developed 3D simulation of FSI model.

2. 3D MODELING

2.1 Structure of the Piezoelectric Micropump

The Piezoelectric micropump selected for the study has a single chamber with inlet and outlet. The structure of the pump consists of a piezoelectric actuator disc. Also, the pump houses a 3 layer structure with a diaphragm and a metal plate placed underneath the piezo actuator. The piezoelectric pump is a unimorph with just a single layer of piezo actuator. This micropump contains 2 microvalves to curb the amount of backpressure. Figure 2 depicts the structure of the micropump-

IOP Conf. Series: Materials Science and Engineering 1123 (2021) 012001 doi:10.1088/1757-899X/1123/1/012001

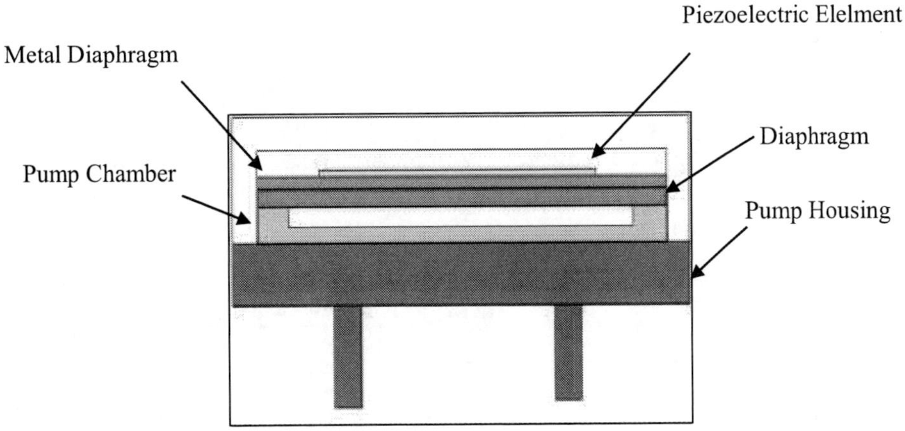

Figure 2. Structure of Piezoelectric Micropump

2.2 3D Model of the Piezoelectric Micropump

The 3D solid model of the piezoelectric micropump (figure 2) is built in the COMSOL Multiphysics package. Piezoelectric actuator is made up of PZT (Lead Zirconate Titanate) material. The pump diaphragm and pump housing uses COC material. The inlet and outlet microvalves are flexible. The microvalves were made from EPDM(ethylene propylene diene monomer). These valves are deflected by the velocity and pressure exerted by the fluid. Simulation is performed using this 3D model of piezoelectric micropump.

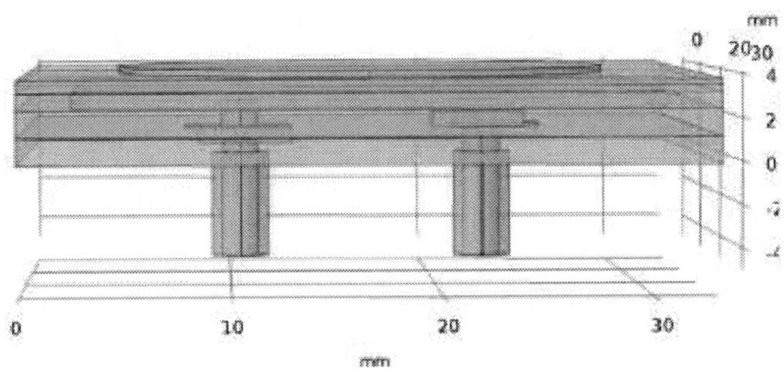

Figure 3. 3D solid model of Piezoelectric Micropump

3 Dimensional two-way FSI was carried on the piezoelectric micropump to obtain the simulation results for fluid flow and von mises stress results.

3. RESULT AND DISCUSSION

3.1 Simulation of Piezoelectric Micropump

The micropump selected for the study works on piezoelectric actuation. Piezoelectric pumps are simple in structure, lightweight and it can handle small quantity of water. These attributes makes it an obvious choice. Simulation of piezoelectric micropump involves the multiphyics approach wherein different physics are combined together to perform the simulation. This study involves the electric-fluid-structure interaction approach. Structural mechanics, electrostatics and fluid dynamics are combined together to get the results for flow rate and stress concentration of the pump.Table 2 shows the change in flow rate with respect to input voltage.

IOP Conf. Series: Materials Science and Engineering 1123 (2021) 012001 doi:10.1088/1757-899X/1123/1/012001

Table 2. Fluid flow rate vs Voltage input

Piezoelectric Pump				
Voltage	**Frequency**	**Volume**	**Time**	**Flowrate**
in V	**in Hz**	**in ml**	**Min**	**ml/min*10-5**
100			1.35	2.25
150	**20**	10	**47.94sec**	**1.95**
200			28.86sec	1.5
250			23.93sec	1.2

The simulation results foe flow rate indicate that it is in proportion with change in input voltage and frequency.

3.2 Simulation of Piezoelectric actuator

Piezoelectric actuator used in the micropump is unimorph i.e. the pump uses only one layer of the piezoelectric material. The piezoelectric layer uses PZT (Lead Zirconate Titanate)material. The piezoelectric material is bonded to a diaphragm and a metal plate. The structure of the piezoelectric actuator is shown in figure 4. The actuation of the piezoelectric material is carried out by applying the potential to the top of the piezoelectric material. The piezoelectric material deforms along with the metallic plate and the diaphragm.

Figure 4. Simulation of the piezoelectric actuator

The actuation of the piezoelectric disc will inititate the suction of the working fluid in the micropump. Piezoelectric actuator was actuated by using the electric potential.(1) Electric potential is applied at the top of the plate and bottom side of the actuator is marked as ground. Equation (1) gives the actuation for an input voltage and frequency for 150V and 20Hz respectively.

$$V_0 = 150 * \sin(20 * t[1/s]) \quad (1)$$

Where t stands for time in seconds and s stands for the frequency in Hertz.The effect of variation of input voltage on the deflection of the diaphragm can be seen in Figure 4-

IOP Conf. Series: Materials Science and Engineering 1123 (2021) 012001 doi:10.1088/1757-899X/1123/1/012001

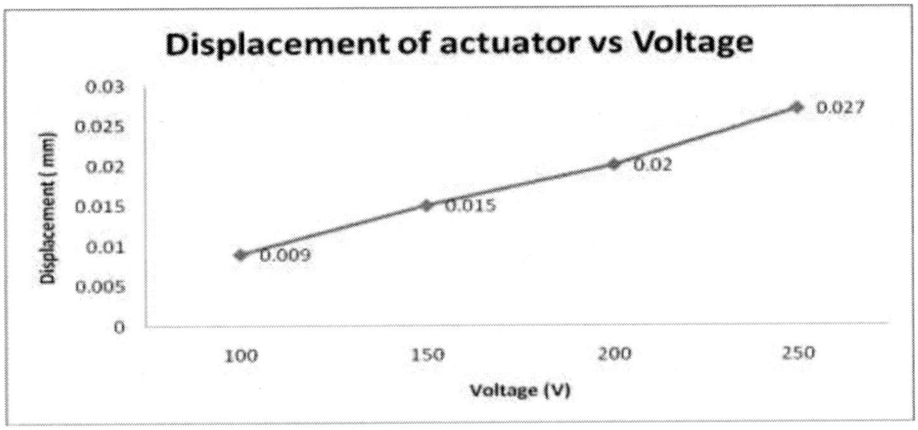

Figure 5. Displacement of Piezoelectric Actuator

3.3 Simulation of Microvalve

The micropump contains two microvalves in the structure. Figure 6 gives the von mises stress distribution in the microvalves. These valves are placed to minimize the backpressure. The valves are placed at the inlet and the outlet of the pump. As per the structure of the pump, the inlet and outlet of the pump are predefined. The position of the valves is inverted to restrict the flow of the water in only one direction. As compared to the valveless micropump, this valved structure aids in better backpressure control and ensures the unidirectional flow.

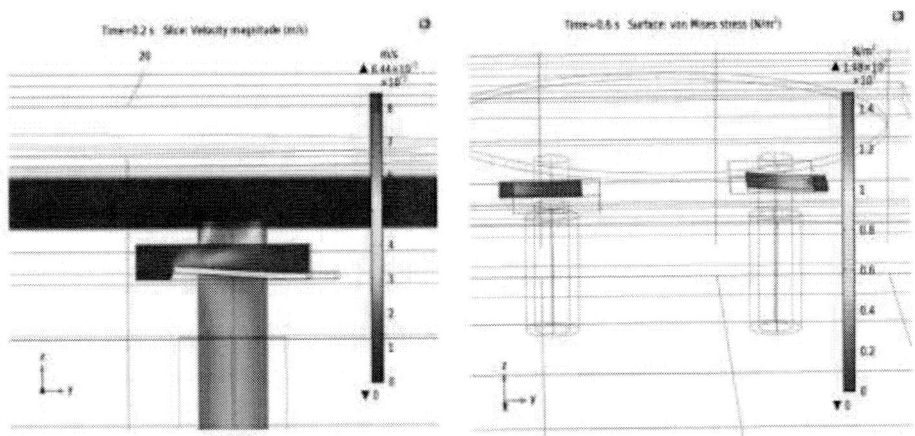

Figure 6. Simulation of the microvalve(a)deflection of the microvalve(b)Two microvalves in position.

3.4 Experimentation

For the validation of the simulation results, experimentation is carried out. Experimental setup involves the piezoelectric micropump, a controller to vary the voltage input to the pump, the input-output reservoir and tubing to connect the pump to the reservoir. Input voltage is varied from 60Volts to 250Volts and frequency is kept constant at 20Hz to match the simulation scenario. The flow rates are calculated for pumping 10ml of water. Table 3 shows the flow rate variation for different sets of voltage and frequency.

Table 3. Flow rate variation for a different set of voltage and frequency

Piezoelectric Pump				
Voltage in V	**Frequency in Hz**	**Volume in ml**	**Time Min**	**Flowrate ml/min**
100	20	10	1.35	7.4
150			**47.94sec**	**12.5**
200			28.86sec	20.79
250			23.93sec	25.07

4. CONCLUSION

A 3 dimensional solid model for piezoelectric micrpump is modeled in COMSOL Multiphysics package. Simualtion is carried out on a finite element model. Simualtion of micropump contains electro-fluid-structure interaction study. This study is a bidirectional electro-fluid-structural interaction. This gives the flow rate and backpressure distribuation in various parts of the micropump. Also, the results indicate that flow rate, actuation of piezo element and backpressure developed are directly proportional to the voltage and frequency input. This study will be helpful in designing the various types of external and implantable micropumps for drug delivery application. The future scope of this research includes further optimization and finetuning of the model to get the accurate flow output with minimum backpressure.

REFERENCES

[1] Johari J and Majlis B Y 2008 Flow Behavior of Fluid-Structure Interaction (FSI) System of Piezoelectric Actuated Valveless Micropump (PAVM) pp. 216–220

[2] Benra F, Dohmen H J, Pei J, Schuster S, and Wan B A 2011 Comparison of One-Way and Two-Way Coupling Methods for Numerical Analysis of Fluid-Structure Interactions vol. 2011

[3] Takaddus A T, Gautam P, and Chandy A J A fluid-structure interaction (FSI)- based numerical investigation of peristalsis in an obstructed human ureter pp. 1–21.

[4] Podder P K, Samajdar D P, Mallick D, and Bhattacharyya A 2012 Design and simulation of micro-pump, micro-valve and micro-needle for biomedical applications *CODEC 2012 - 5th Int. Conf. Comput. Devices Commun* vol 3 pp 1–4

[5] Li Y, Le V T, Goo N S, Kim T H, and Lee C S 2017 High actuation force of piezoelectric hybrid actuator with multiple piezoelectric pump design.

AN ENHANCED MATHEMATICAL MODEL FOR CLOUD BASED DATA ORIENTED JOB ANALYSIS

Manishankar S[1], Santosh Anand[2]

Department of Computer Science, Amrita School of Arts & Sciences, Mysuru
Amrita Vishwa Vidyapeetham, India
smanishankar@asas.mysor.amrita,edu[1],santoshanand.jha@gmail.com[2]

Abstract. Data processing or data analytics is the common functionality that is attached to most of the real world applications. The amount of processing required for data oriented tasks or jobs are quiet high. To resolve the processing issue the most common approach deployed is by using a high performance cluster. Setting a cluster over real time infrastructure leads to a very expensive solution. A Cloud based infrastructure remains as an ideal support for setting up a cluster. Managing the cluster over a Cloud is a challenging task as allocation of infrastructure based on the task schedule is a critical parameter. The proposed mathematical model introduces a strategy to allocate the infrastructure and manage the load of the cluster based on Queuing model. The experimental setup is made on top of private cloud and Hadoop based data processing jobs are tested. The proposed data oriented resource optimizer enhances the performance of the cluster by balancing the increased load due to data processing jobs. The result shows the enhanced improvement in performance compared to default resource manager.

Keywords: *Cloud Computing, Hadoop Cluster, Queuing model, Resource optimization.*

1. INTRODUCTION

Storage of real time generated data from information system is a major challenge in most of the data management system and data warehouses. Storing data is not only the concern it is also important how the data can be processed and handled. Different methods are used to support exploratory, statistical, and predictive which can analyze data with different views that help in interpret and visualizing data according to the various research perspectives [1]. So there arises a requirement for an efficient and structurally strong data processing system like Hadoop to support the data processing.

The research work carried on job oriented data analysis introduces the concept of having multiple jobs run on a cluster, the analysis is termed as job oriented because the scheduler allocates the job based on the type of operation. A queuing model based on response time of cluster is setup to provide a support for the cluster. The elastic cloud helps in reallocating or reutilizing the resource based on the system use. The importance of setting a real time data based analysis in various fields like sensor network can be incorporated where the sensor inputs can be analyzed with cluster [1]. The performance results are compared with the YARN based scheduler. The system is an efficient cloud based solution for jobs of varying sizes and types. The queuing model enhances the performance of the mediation node which is key in making the system faster as it takes in inputs [2].

Real time application like sensor load balancing to optimize energy and enhance the life time of wireless sensor network. The protocol proposed in the research work to design the path which is similar queuing model for making the process execute [3]The major requirement in future storage system is the need for setting up an integrated multi cluster environment which can execute parallel data processing and analytical operations. A kind of system that will optimize the resources involved in storage and data processing system [5] and which has job-oriented view for each of the operation performed on data. Various Thus, the research introduces a efficient and deeper approach involved in Big Data processing and analytics.

2. RELATED WORK

Data Oriented job Analysis in a Distributed Cluster Environment is a field of research were lot of works have been carried on because all real time applications are connected with data storage and retrieval . Survey of recent works in the research field are depicted below [3] Considers the FBC (frequency-based center) and proposes a new function for performing K-means clustering on a categorical dataset. [4] The work uses the FBC concept to find the new center of the categorical data set. Has developed an algorithmic framework for solving the projected clustering problem, and has tested its performance on synthetic data.[5] SHadoop is implemented. An enhanced and fully attuned version of Hadoop, is intended to shorten the performance time, cost of MapReduce work, particularly for short term work.[6] There are 3 alternates solutions of the K-Means algorithm for incorporating binary data streams. These variations contain the K-means which is extensible online, the , and the K-means incremental approach, which is a anticipated variant that can achieve a higher quality solution in a shorter time. The proposed algorithm can be used to monitor transactions.[7] proposed the work by using the greedy method to select the random point to solve the problem. Using the idea of cluster center initialization. [8] The clustering of the numerical data set is the distance from the Manhattan distance, not the Euclidean distance. In [9] the use of association rule mining (ADMLCAR) anomaly detection Multi-tag classification is used to solve MLC problems. In overall, multi-label classification difficulties are mostly answered by one of two approaches: problem conversion, algorithm adaptation. [10]A compacted over turned directory data structure that might help in creeping for words in dictionary direction such that all the directories built for millions of documents need not be administered has been projected. [11] has implemented PCA-based graphical similarity measurements and generate different Laplacian spectra for spectral clustering. [12]Lecturing the problem of k-medoid clustering by MapReduce agenda for distributed computing on commodity machines to assess its competence. [13], further motivated by the Map-Reduce has revisited the two-stage flow shop problem and have given a dynamic program for minimizing the total flow time especially when all the jobs arrive at the same time. The paper also proposes a scheduling model that captures the challenges in Map-Reduce scheduling. [14], they have proposed simplified and effective strategies which use ratio-slot between maps and reduce tasks which act as an effective knob.

The experimental results show the working correctness and compilation of schemes under both simple, complex and mixed workloads.[15] had studied the Map Reduce job performance in HadoopV1 and YARN with different resource configuration. This made them to propose Map Reduce, which can dynamically manage a number of corresponding tasks running on each node. [16], has proposed in the case study for Smart City Corporation that the data in today's world is generated rapidly from various sources such as mobile phones, machines, etc and these data generated has to be handled in order to serve a variety of purposes. [17] has studied Map Reduce which is a prevalent time for large-scale data dispensation in cloud computing. Though, the slot-based Map Reduce system (e.g., Hadoop MRv1) may possibly perform poor due to resource allocation in an un-optimized manner.[18] Explains they range real time cluster scheduling approach to explanation for the two-phase calculation style of Map Reduce. Also progress measures for scheduling jobs built on user quantified limit limitations.[19] A new method to scheduling jobs using genetic algorithms (GA) on a service grid. The fitness function to minimize the average execution time of M ($\leq$ N) machines that schedule N jobs onto the grid. Two models were proposed to predict the execution time of a single job or multiple jobs on each machine with a different system load. A single service type model is used to schedule jobs for a single service to a machine, while multiple service type models schedule jobs for multiple services to a machine. The predicted execution time from these models is used as input to the genetic algorithm to schedule N jobs to M machines on the grid. Experiments on a small grid of four machines show that the new job scheduling method significantly reduces the average execution time [20] an algorithm has developed to list heterogeneous tasks on heterogeneous processors in a circulated system. The scheduler runs in an environment changing resources and adapts variable system resources. It uses genetic algorithms to minimize the execution time. Comparing the scheduler with six other schedulers, three batch modes, and three immediate mode schedulers. Experiments show that the algorithm is superior to other algorithms and can achieve near-optimal efficiency.

IOP Conf. Series: Materials Science and Engineering 1123 (2021) 012002 doi:10.1088/1757-899X/1123/1/012002

3. PROPOSED MODEL OF CLOUD-BASED CLUSTER

Real time data tasks and operations related to data processing requires a high performance, high available and reliable computing environment. Cluster based systems are found to support computing environment efficiently. A cloud based computing system is very much desirable as virtualizing the infrastructure helps in a better way to setup a computing cluster with less cost. The proposed architecture is a Hadoop based cluster setup over a private OpenNebula Cloud. Hadoop cluster are responsible for the execution of the Big data processing job, whereas Cloud provides the elastic infrastructure to support the executions.

The proposed model is having multiple cluster which is setup on top of the private Cloud. The Cloud is configured with Virtual machines as depicted in Figure 1 to manage cluster.

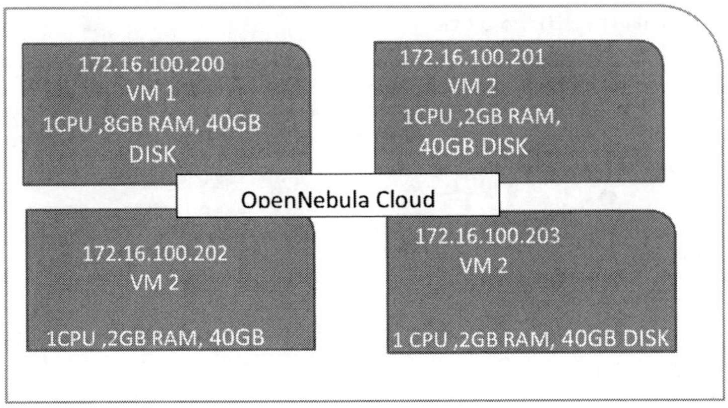

Figure 1. Architecture of the Cloud-based cluster

The cluster has a Name node, Data node and processing nodes. Important task is to monitor the cluster when jobs are executed. A performance monitor is formed with help of a monitoring tool Ganglia. Various types of data processing jobs are tested on top of the cluster. Running parallel jobs of varying size of cluster seems to increase the load of the cluster by a large extend. As data processing jobs have many parallel executing jobs there arises a need for a model to allocate the data oriented jobs in the cluster.

Hadoop Cluster has many type of schedulers, YARN (Yet Another Resource Negotiator) in Hadoop plays a vital role in managing the resources in the cluster. YARN ensures the availability of resources within the Hadoop. To monitor the performance of the system we require a ganglia to be installed. Ganglia being a cluster monitor records the readings from processing environment which can be utilized by the proposed schedule in the research to work efficiently, once the cluster is set up and running it is important to try out various jobs in the cluster and record the performance using ganglia.

The Hadoop cluster is run with a variety of jobs such as Word Count, Number Sum and transaction management. Map Reduce operation on YARN is compared to give a faster result than traditional MapReduce. YARN components are Client who submits a MapReduce Job of any one category above mentioned. Resource Manager controls the usage of resources in the Hadoop cluster. Node Manager is responsible for running the jobs assigned, it creates containers to hold the job and also monitors the usage of the job container. Application Master coordinates MapReduce jobs and schedule task by negotiating with the resource manager. Hadoop Distributed Filesystem shares the jobs between YARN components. The java interface we created runs a code necessary for the job. The job object is identified and a new application is submitted to the resource manager.

The resource manager checks the availability and forwards the job to the node manager. Node manager creates the container for the job and starts the application master. The application manager creates a YARN child required to run the job with the help of a node manager YARN will execute Map as well as reduce task and acquire the job resources from the HDFS. Once the job is submitted an automatic application Id is retrieved from the resource manager. Distributed task allocation and management is done by Application Master if job size is small the Application Master executes both Map and Reduce task in single JVM. Condiitons for the job to run on single JVM is

- Less than 10 Mappers in job
- Single reducer task
- Input data size is less than one HDFS block

If the tasks are not satisfying these condition then the scheduler has to check the availability of the resource and see where to place the task. Node manager allocates the containers that has Map and Reduce job present in them. Once the containers are scheduled the total sum of all job containers should not execeed the assigned threshold.

Calculating the physical memory limit of a job is also essential the container memory usage will be termed by a calculation where

CM-Container Memory

HS-Heap size of Virtual machine

NL-Native Library

MU-Memory used by other spawned process. JP- java VM per Region.

$$CM=\sum [\![HS + NL + MU + JP]\!] \quad (3.1)$$

Equation 3.1 depicts that the commuted memory is a total of all virtual machine memory partitions. As it seen even though the default schedulers allocate maximum and minimum values for the memory allocation. For example a Capacity scheduler has a minimum memory of 1024MB to maximum memory of 10240MB, so any request that is between Min and Max will be processed by the scheduler. The memory model in such a way that in allocated memory some part is assigned to Map and other is assigned to reduce based on the number of Mapper and Reducer function and configuration in Yarn site.XML. The Node manager in configuration of Yarn.site.XML has a 10 Mappers or 6 Reducers may be a combination of the above. So the node manager Yarn scheduler capacity of minimum allocation will be given a client request. Based on the availability Application master assigns the Node manager to allocate the container based on number of Map and Reduce present in the Job assigned. For each task Node manager starts a Yarn child, which is executed in the allocated JVM. It also copies resource locally as configurations or Jar files. Then Yarn executes the Map and Reduce tasks. The report of the running job and task is polled by the Application Master. The task continues until the application master is confirmed that job is done. Client can monitor the status of job by polling the application master with help of progress monitor. Once the cluster is set and the performance is observed, there arises a requirement for model for assigning jobs to the YARN so the optimization of the load of cluster is possible

4. Mathematical Model for data-oriented job analysis

To assign P Process to N node in C cluster we require queuing model in which a mediation server MS is introduced which is the entry point of the user request to run job arrives. The jobs are forwarded to the Cluster nodes CNi for processing the jobs. Where i=1,2,3....N are the nodes of the cluster. The scheduler or load balancer is represented as M/M/1 Queue which works with probability and arrival rates are calculated by Poisson distribution. The arrival rate of jobs are represented as λ to the load balancer as a Poisson process. A service rate is fixed for the cluster nodes, if λ is less than the service rate S, the schedule is fixed and the task is executed.

Each cluster Node access the datacenter DCi with a probability β, the rate in which the server will be processing the I/O operation depends on the service rate of the model M/M/1 Queue. For calculating the cluster performance the measurement of response time of the mediation node, cluster node are evaluated.

The total response time T of the Cloud will be a sum of response time of the queuing servers.

$$T=T_{MS}+T_{CN} \quad (3.2)$$

The response time of the MS will be obtained based on response time of the M/M/1 queue, which is calculated a

$$T_{MS}=\frac{1/s}{1-\lambda/s} \quad (3.3)$$

Where S is the service and λ is the arrival rate

Processing response time of cluster also plays a vital role in modeling the server. The N nodes of the cluster which is in the queuing model, the response time can be calculated as

$$T_{CN}=\frac{1}{\alpha}+\frac{P(n,\rho)}{n\alpha-\lambda} \quad (3.4)$$

Where n is the number of clusters and α and λ service rate and arrival rate of jobs . The P(n,ρ) is the probability of job getting added to the queuing model

$$P(n,\rho)=\frac{(\frac{cp}{c})X(\frac{1}{1-p})}{\sum_{k=0}^{c-1}\frac{cp^{n}}{k!}+(\frac{1}{1-p})} \quad (3.5)$$

Once the response time and memory is calculated the server takes performance values monitored by the ganglia and apply to the queuing model to get the response time and the queuing delay

5. IMPLEMENTATION AND RESULTS

In order to conduct the proposed research a Hadoop cluster environment is setup with help of Hadoop. The Hadoop cluster is monitored for the performing using ganglia observations collected for various predefined data and jobs are recorded. The Hadoop cluster has component like Name Node, Resource Manager with YARN,File System of distributed nature and Hadoop Map reduce Ecosystem and YARN based parallel processing of dataset. There will be a name node who act as a master and who will have the job tracker and job analyzer which we introduced in the proposed architecture.
The slave node has task trackers which are coordinated by adaptive scheduler and each one has a data node and does the map reduce operation required. YARN with adaptive scheduler enables dynamic resource utilizations and job oriented computations which can run various applications without having any worry about increase the load the resource manager who manages and allocate cluster resources feeds the data to the adaptive scheduler the node manager who enforces node resource allocation and management is connected to resource pool and job table. The application manager manages the life cycle of the application in co-ordination with the proposed job oriented analytic engine.

The results from the performance monitor shows that the cluster is heavily loaded and the utilization initially is degraded. The performance degradation of YARN occurs when the jobs are of varying sizes as depicted in Figure 2.

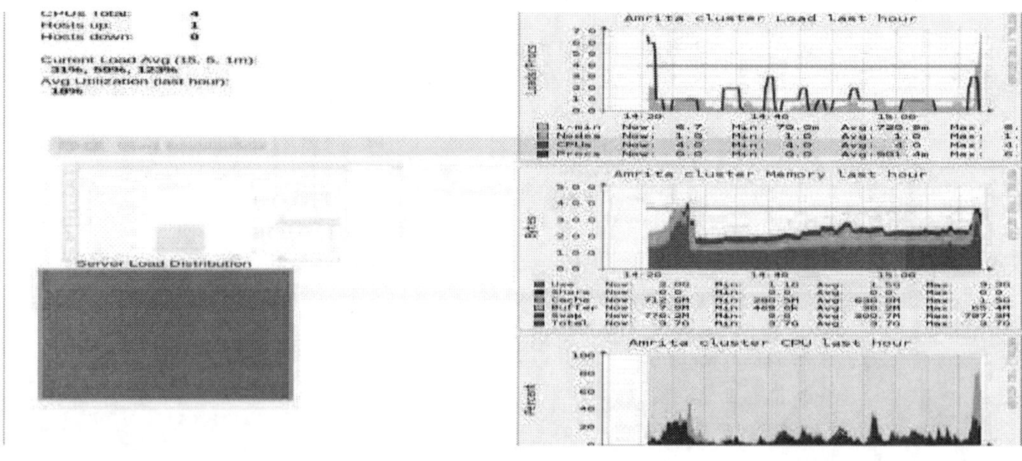

Figure 2. Performance degradation without Yarn optimization.

The Table 1 depicted above show the cluster node response data .Data which gives job id denotes the number assigned to the incoming jobs. Rid depicted as Resource id represents allocated resources. Job type which is categorized based on the data by job analyzer puts the jobs into three categories High, Medium, low. Execution time represents the time taken for the execution of the job in job tracker noted from ganglia. Memory denotes the usage of the memory, R status depicted as resource status denotes whether the job is assigned to a resource or not. CPU depicts CPU utilization.

The overall load of the cluster while a particular job is running is noted as load. The time taken for map and reduce jobs inside the slave node are recorded with help of task tracker and listed as map time and reduce time. Job analyzer which collects the input data and decides on job type based on the mathematical model and decision algorithm.

IOP Conf. Series: Materials Science and Engineering 1123 (2021) 012002 doi:10.1088/1757-899X/1123/1/012002

Table 1. Cluster performance data obtained from monitor.

Job id	Rid	Job type	Exe time	Memory
1	1	H	1.454	1971053909
2	2	L	0.2973333	25346068
3	1	L	0.9513333	25023829
4	1	H	1.989	27986542
5	3	L	0.3433332	26789979
6	1	L	0.98	27000019
7	4	H	2.01	30023126
8	2	L	0.4532234	27651239
9	1	L	1.02	27609862
10	5	H	1.39	438457132
11	1	M	1.46	773632000
12	3	L	1.04	496620612
13	4	H	1.54	446523433
14	1	M	1.07	7865425548
15	2	L	1.06	499932168

Once the response time and memory are calculated the model helps in categorizing the job based on arrival rate

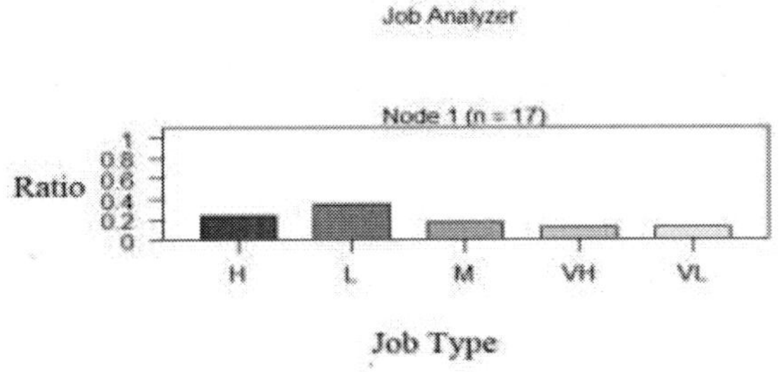

Figure 3. Categorization of jobs based on response time

The figure 3 depicts job analyzer categorizes the job into various level sets based on the load execution time, CPU utilization and memory. The data obtained from ganglia used to do the categorization and with the help of a decision tree algorithm job analyzer divides the job as depicted in the diagram above. The jobs are categorized into high, low ,medium, very high, very low. training data set consist of 17 various types of jobs based on different inputs.

The YARN with the adaptive scheduler makes critical decisions on allocating jobs to resource of the cluster. Our modified scheduler receives input from the job analyzer and makes mapping of jobs to the table and searches in the resources for the best available resource to execute the job. The use of cross mutation genetic algorithm helps in correlating and mapping jobs to the resource. The algorithm creates the fitness function which can evaluate the job based on parameter like execution time, cpu utilization, memory utilization, load etc and generations of job in accordance to resource created.

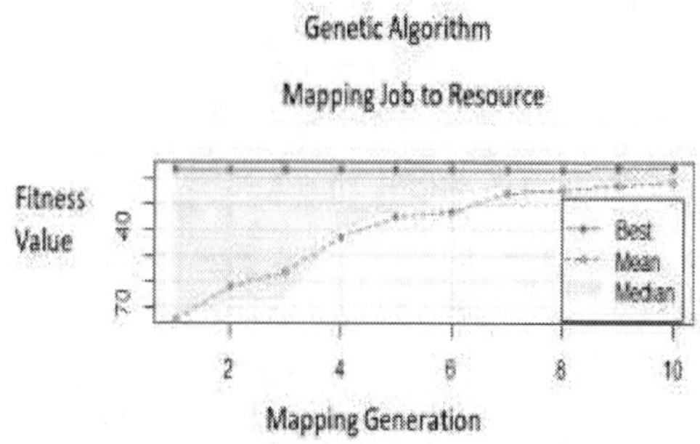

Figure 4. Genetic algorithm plot

The figure 4 above depicts the plot of genetic algorithm mapping resources and jobs. The plot depicts the generations in x-axis and fitness in y-axis. The best mean, median are depicted in plot.The mapping of resources to job helps in resource optimization. The most appropriate resource is identified by the job analyser and the genetic algorithm supporting it. Based on the fitness function and the cross mapping function an exact match of the job to resource is obtained. The results of the comparison of the YARN scheduler and the proposed Queuing model based schedule

Table 1. Execution time results

Number of Jobs	Number of Resources	Execution Time(YARN)	Execution Time(Adaptive Scheduler)
5	3	5.035	2.034
10	5	12.34	8.34
15	7	17.05	13.04
20	8	21.35	16.05

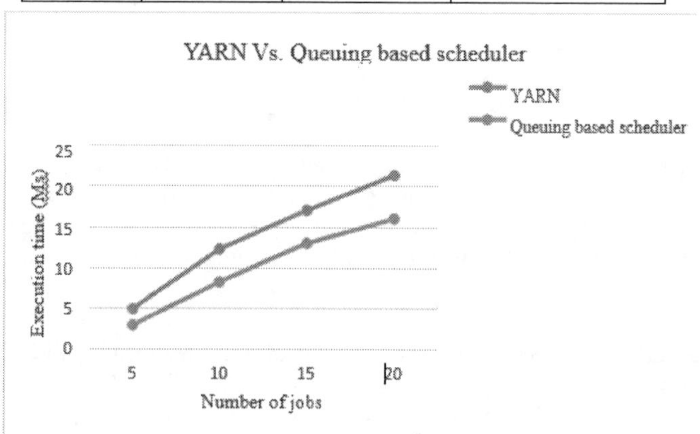

Figure 6: Comparison of Execution time of YARN and Queuing based scheduler

The graph represented in Figure 5 represents the comparison of execution time with presence of queuing model and default YARN scheduler in the Hadoop cluster.

6. CONCLUSION

Distributed cluster environment that does computation like data processing and analytics require a proper model that can support the scheduler in handling heterogeneous data processing jobs. So, this work aims to provide a platform which can handle the heterogeneous data operation with accessible infrastructure and resources by including few classifications of that data during preprocessing stage and then applying various analytics with the help of queuing based scheduler for the multi clustered environment. The classification of jobs and resources are based on the mathematical model. The mathematical distribution model is applied to make selection of resource and job based on the input data. A scheduler supervises the correlation between job and resources allocation which enhances the performance of cluster environment and thus providing an efficient job-oriented computation and analysis.

REFERENCES

[1] Gonzalez, T F 1985 Clustering to minimize the maximum intercluster distance *Journal Theoretical Computer Science* volume 38,page 293-306

[2] Aggarwal C C, Wolf J L, Yu P S, Procopiuc C and Park J S 1999 Fast algorithms for projected clustering.mIn *ACM SIGMoD Record* Vol. 28, No. 2, pp. 61-72 ACM.

[3] Santosh Anand Pillai and Atulya Radhakrishna A Protocol for the Effective Utilization of Energy in Wireless Sensor Network with Load Balancing Pages: 82-86, DOI: 10.14419/ijet.v7i3.3.14491

[4] Gu R, Yang X, Yan J, Sun Y, Wang B, Yuan C and Huang Y 2014 SHadoop: Improving MapReduce performance by optimizing job execution mechanism in Hadoop clusters *Journal of parallel and distributed computing* 74(3) 2166-2179

[5] Ordonez, C 2003 June Clustering binary data streams with K-means. In Proceedings of the 8th *ACM SIGMOD workshop on Research issues in data mining and knowledge discovery*(pp. 12-19). ACM

[6] Li Q and Liu X 2014 November A K-medoids clustering algorithm with initial centers optimized by a P system. In *International Conference on Human Centered Computing* (pp. 488-500). *Springer Cham*

[7] Legany Csaba, Juhas., Sandor Babos and Attila 2006 Cluster Validity Measurement Techniques In proceedings of the *5th WSEAS International Conference on Artificial Intelligence, Knowledge Engineering and Data Bases* pp:388-393 WSEAS

[8]. `Santosh Anand and Akarsha R R 2018 A Protocol for the Effective Utilization of Energy in Wireless Sensor Network, Participated and presented paper in ICTEMPS-2018 conducted by Geethanjali College of Engineering and Technolog Pages: 9398, DOI: 10.14419/ijet.v7i3.3.14495,

[9] Prathibhamol C P, Amala G S and Kapadia M 2016 September Anomaly detection based multi label classification using Association Rule Mining (ADMLCAR). *In Advances in Computing, Communications and Informatics (ICACCI), 2016 International Conference IEEE* pp. 2703-2707

[10] Velusamy K, Venkitaramanan, D Vijayaraju N, Suresh G and Madhu D 2013 Inverted indexing in big data using hadoop multiple node cluster *International Journal of Advanced Computer Science and Applications (IJACSA)*, 4(11)

[11] Kavitha K R, Sandeep S, and Praveen P R 2016 December Improved spectral clustering using PCA based similarity measure on different Laplacian graphs. In *Computational Intelligence and Computing Research (ICCIC) IEEE International Conference* on pp. 1-6)

[12] Harikumar S and Thaha S S 2016 August MapReduce model for k-medoid clustering. In *Data Science and Engineering (ICDSE) International Conference* IEEE pp. 1-7

[13] Moseley B, Dasgupta A, Kumar R, and Sarlós T 2011 June On scheduling in map-reduce and flow-shops. In Proceedings of the twenty-third annual *ACM symposium on Parallelism in algorithms and architectures* pp. 289-298

[14] Yao Y, Wang J, Sheng B, Tan C C and Mi N 2017 Self-adjusting slot configurations for homogeneous and heterogeneous hadoop clusters *IEEE Transactions on Cloud Computing* 5(2) 344-357

[15] Liang F, and Lau F C 2015 May SMapReduce: Optimising resource allocation by managing working slots at runtime. In *Parallel and Distributed Processing Symposium (IPDPS) IEEE International* pp. 281-290

[16] Barbieru C, and Pop F 2016 March Soft real-time hadoop scheduler for big data processing in smart cities. In Advanced Information Networking and Applications (AINA) *IEEE 30th International Conference* pp. 863-870

[17] Tang S, Lee B S and He B 2014 Dynamicmr: A dynamic slot allocation optimization framework for mapreduce clusters *IEEE Transactions on Cloud Computing* 2(3) 333-347

[18] Kc K and Anyanwu K 2010 November Scheduling hadoop jobs to meet deadlines. In *Cloud Computing Technology and Science (CloudCom) IEEE Second International Conferenc*e pp. 388-392

[19] Gao Y, Rong H, and Huang J Z 2005 Adaptive grid job scheduling with genetic algorithms. Future Generation Computer Systems 21(1) 151-161

[20] Page A J and Naughton T J 2005 Framework for task scheduling in heterogeneous distributed computing using genetic algorithms. Artificial Intelligence Review 24(3-4) 415-429.

AN OVER-VIEW OF THE APPLICATIONS OF DFA (DESIGN FOR ASSEMBLY) TECHNIQUES ON AUTOMOBILE COMPONENTS FOR REDUCING ASSEMBLY TIME AND COST

Shrey Arvind Jaiswal [1], Darius Gnanaraj S [2]

[1] UG student, SMEC VIT, Vellore, India,
shrey.jaiswal2017@vitstudent.ac.in
[2] Department of Design and Automation, SMEC, VIT, Vellore, India,
dariusgnanaraj.s@vit.ac.in

Abstract. Design for Manufacturing and Assembly (DFMA) techniques are useful for minimizing the cost of products through design and process improvements. DFMA is a combination of Design for Manufacturing (DFM) and Design for Assembly (DFA). This paper focuses case studies on the successful application of DFA techniques on Automobile components and reports the reduction in assembly time and cost of the product successfully achieved by different R&D works carried out by researchers around the World, collected through literature survey. Information available related to the following topics are collected and analyzed: minimize part count, design parts with self-locating features, design parts with self-fastening features, minimize reorientation of parts during assembly, standardization of parts, minimum use of fasteners, modular design, design for a base part to locate other components and design for component symmetry for insertion. The findings reported in this over- view are useful to practicing design engineers.

1. INTRODUCTION

The automobile industry has grown at a faster rate during the last 5 years. It is one of the largest worldwide industries with more than 4 million cars and commercial vehicle produced in India, one of Asia's largest markets. Competition has increased among industries due to the entry of global players through joint ventures, fully owned units, collaborations with local industries and as a result, economic conditions have changed. This action forces the industries to reduce the prices, which starts straight away from the lower levels of manufacturing supply base. Due to this pressure and constraints on the pricing, various techniques are to be used at the basic design levels of a product. DFMA (Design for Manufacturing and Assembly) is one such technique which reduces the price for a manufactured component. It is a technique which is used by majority of the industries. It enables the company to not only reduce the price but also the time for production with improved quality.

2. STEPS FOLLOWED IN DFMA

DFMA is a combination of DFM (Design for Manufacturing) and DFA (Design for Assembly). "Design for Manufacture" (or DFM) means the design for ease of manufacture of the collection of parts that will form the product after assembly and "Design for assembly" (or DFA) means the design of the product for ease of assembly [1]. This process was developed by Boothroyd Dewhurst in 1960 in US. There are various processes in the market which increase the productivity of an organization. But DFA stands out to be one of the major processes. It tries to reduce the number of parts by replacing then by a single assembly, which in turn not only reduces the number of parts but also the time of assembly and cost of the product which shoots up the productivity. The Flowchart given below shows the basic steps to be followed in DFMA:

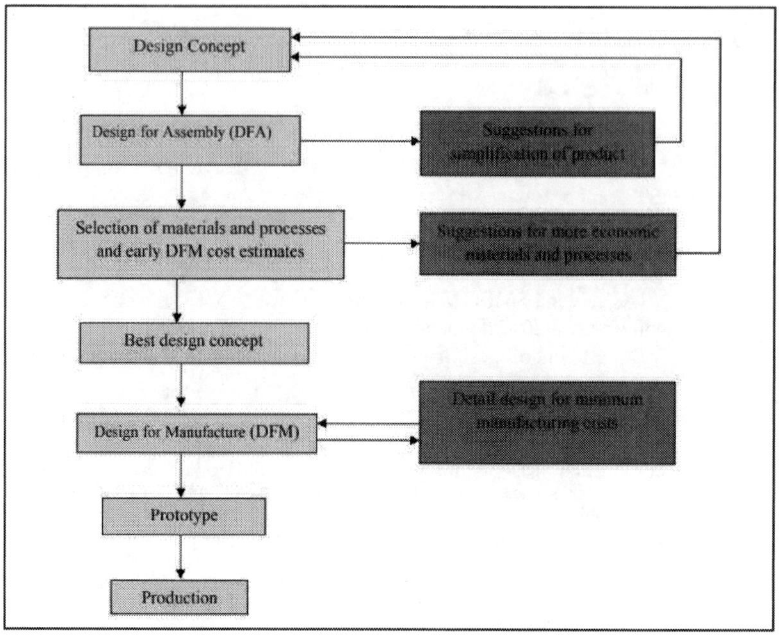

Figure 1. Steps followed in DFMA

2.1 Guidelines for DFA

Guidelines for DFA are given below :

- Minimizing the part count
- Designing for ease in insertion
- Standardizing the parts
- Designing for the process capabilities
- Awareness about the alternatives for process capabilities
- Making sure there is some free space for handing
- Making sure about pre--positioning
- Ensuring easy testing access
- Providing component structure which is modular
- Differentiating the variants at end processes
- Checking if the material is compatible
- Preferred flow of parts

2.2 The Design efficiency

The design efficiency can be calculated by an index called DFA index. It tells how easily a component can be assembled.

$$\text{DFA index} = \frac{N_m * T_a}{T_{ma}} \quad (1)$$

N_m= Minimum no. of parts (Theoretical) T_a= Minimum assembly time per part (sec) T_{ma} = Estimated total assembly time.

3. DFA ANALYSIS

The DFA analyses of various automobile parts done were studied and the current progress in the world of DFMA is shown in the following sub section:

3.1 Gear Box

DFA analysis was done on a gearbox by Shete et al. [2]. The gearbox analyzed was a 4-stage gearbox with helical gears and was assembled manually. The DFA index for the existing assembly was found to be 11.73 (as shown in Table 1) and the time for total assembly was found to be 940 seconds. As we know that design efficiency can be improved by analyzing it by DFMA software, the gearbox was then analyzed and the complete gearbox was divided into four sub-assemblies since it reduced the assembly time. After a few changes in the design like use of standard fasteners and reducing the parts, the DFA index was increased drastically and found to be 23.40 (as shown in table 1) and the assembly time was reduced to 630 seconds.

Table 1. DFA analysis of a gear box [2]

	Entries Including Repeats	**Original Design**	**Re-Design**	**Saving s**
1	Parts meet minimum criteria	30	19	11
2.	Parts are candidates for elimination	53	41	12
3.	Analyzed subassemblies	5	4	1
4.	Separate assembly operations	0	0	0
5.	Total entries	88	64	24
	Assembly labor time, s			
1.	Parts meet minimum part criteria	297.5 2	185.88	111.64
2.	Parts are candidate for elimination	622.1 5	415.95	206.2
3.	Insertion of analysed subassemblies	17.7 0	28.16	-10.46
4.	Separate assembly operation	0	0	0
5.	Total assembly labor time, s	937.3 7	629.99	307.38 s
	Design Efficiency			
1.	DFA index	11.7 3	23.40	11.67

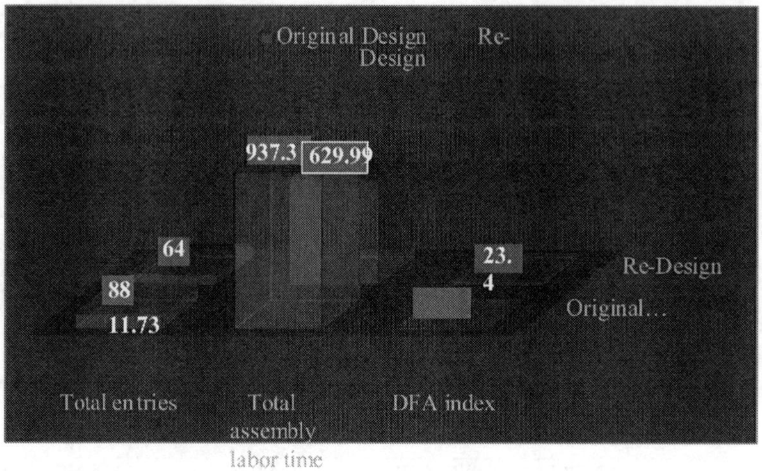

Figure 2. Comparison between Original and Re-design of Gearbox

3.2 Bumper middle pole assembly

A DFMA analysis was done on a middle pole assembly of a bumper by Beijing Automotive Technology Center [3]. Middle pole's main function is to provide points for mounting front grill and compound horn. Apart from this it also has to have sufficient stiffness to withstand impact force while closing of the bonnet cover.

The original design had many fasteners and number of parts. DFA was applied and the count of bolts and fasteners were reduced and the time for assembly was reduced from 552 seconds to 131 seconds. The DFA index was augmented by 70%.

Table 2. DFA Analysis of middle pole Assembly [3]

Sl. No	Item	Original Design	Re-Design	Savings
1	Parts	30	10	20
2.	Fasteners	20	6	14
3.	Actual Time, s	552 s	131 s	421 s
4.	Manufacturing Costs ₹	₹ 432.31	₹301.94	₹130.38

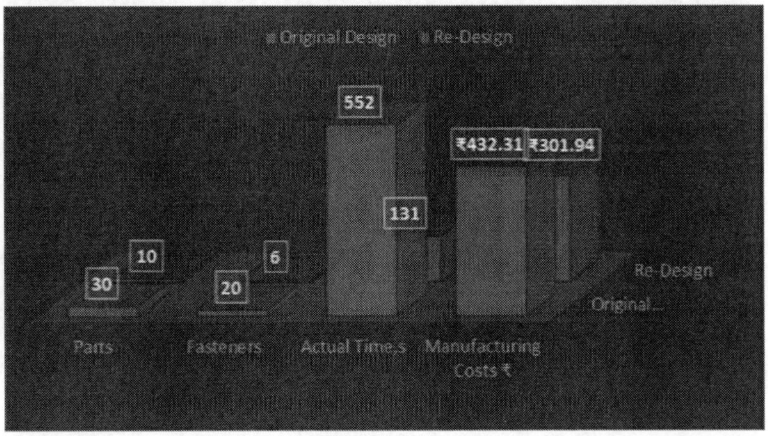

Figure 3. Comparison between Original and Re-design of bumper middle pole assembly

3.3 Instrument Panel

Peter Dewhurst [4] conducted DFA analysis on the instrumental panel of a model truck to reduce the cost of manufacturing and assembly. The original design had a total of 482 steps in assembly and a total of 123 fasteners and the total time for assembly was 3780 sec which on applying DFA was reduced to 121 steps in assembly, 23 fasteners and the time was reduced to 894 sec. The original design had a major of 4 injection molding structures which in proposed design was brought down to 2 injection moldings structures. A noticeable change was witnessed in the total cost.

Table 3. DFA Analysis of an Instrumental Panel [4]

Sl. No	Item / processes	Original Design	Re-Design	Savings
1	Assembly steps	482	121	361
2.	Fasteners	123	23	100
3.	Items for elimination	173	39	134
4.	Assemblie	130	15	115
5.	Time, s	3780 s	894 s	2886 s
6.	Cost/IP	₹ 3149.14	₹ 608.12	₹ 2541.02

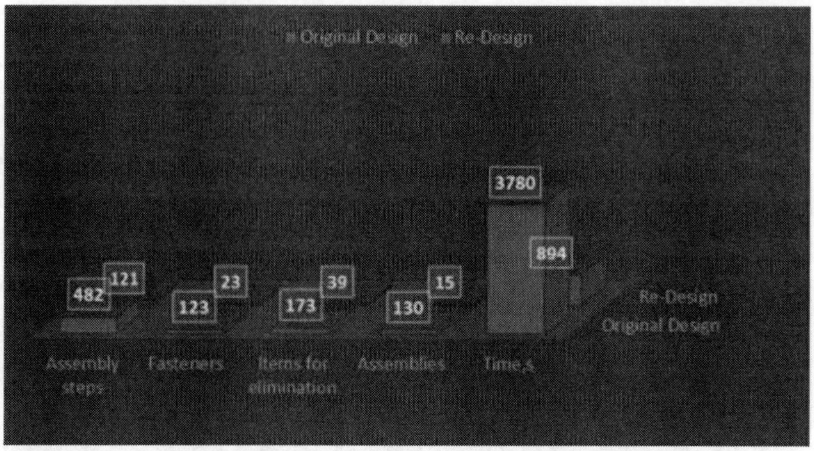

Figure 4. Comparison between Original and Re-design for Instrumental Panel

3.4 Jump Seat

Richard F. Johnson carried out DFA analysis on a jump seat. Jump seat is a seat which can be folded up and is mostly used in cars, taxis, busses and airplanes. The DFA analysis was done by "Magna Seating Group" with the aim of reducing the materials, sub-assemblies and tooling costs. On performing DFA analysis on the original design the number of separate parts was found to be 105 which were made up of 4 different types of materials. There were 6 major subassemblies and the total assembly time as found to be 1440 sec. which on Redesigning the part count was reduced to 19, the sub-assemblies were brought down to 5 and the total assembly time was found to be 258 seconds

Table 4 DFA analysis of a Jump Seat

Sl. No	Item	Original Design	Re-Design	Savings
1	Parts	105	19	86
2.	Sub- assemblies	6	5	1
3.	Actual Time,s	1440 s	258 s	1182 s

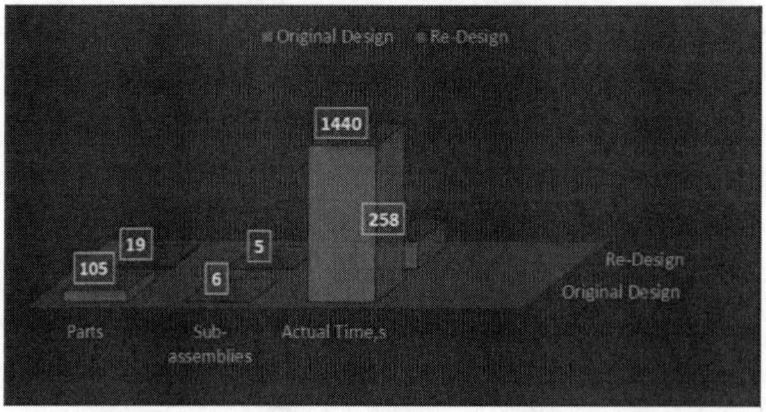

Figure 5. Comparison between Original and Re-design of a Jump Seat

3.5 Diesel Engine

Carla and Marcelo [6] performed the DFMA analysis on the cylindrical block of diesel Engine. Casings were not there in the cylinder block; open refrigeration system incorporated and it is connected directly to the gear box. When the analysis was done, the bearing cover bores were found to have potential for elimination. The number of parts found before DFMA analysis were 29, the assembly time was found to be 381 sec and a DFA index was 0.23. The non-pass bored and milling the assembly were found to be unnecessary and apart from this the block painting and the serial number plate of the engine were eliminated. After the re-design, the minimum number of parts was 29 and the assembly time was found to be 301 and DFA index was 0.29. This resulted in a reduction of price by 0.41%.

Table 5. DFA analysis of Diesel engine (cylindrical block) [6]

Sl. No	Item	Original Design	Re-Design	Savings
1	Parts	29	29	0
2.	Operation time, s	381	301	80
3.	DFA	0.23	0.29	0.06

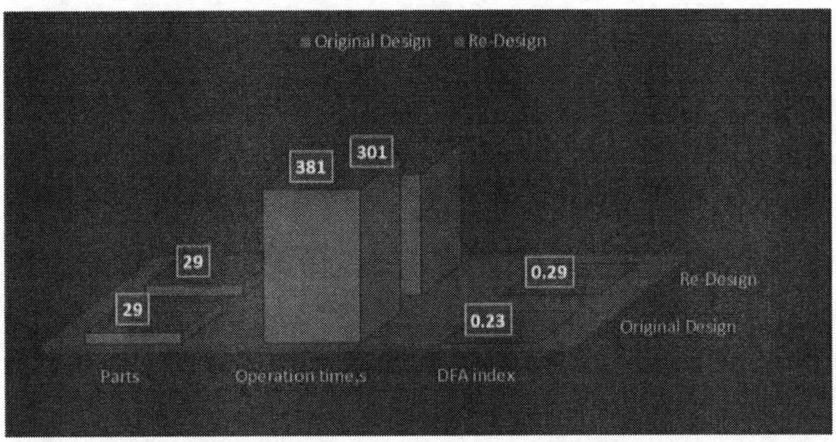

Figure 6. Comparison between Original and Re-design of cylindrical block

3.6 Burring Tool

It is designed for performing flanging operation in components used in automobiles It is an automotive tool assembly. Nilesh and Rajesh[7] have done the DFMA analysis on burring tool . The product was completely studied and the CAD model was fed into the DFA software to analyse the tool Tthe total part number was found to be 315 with a assembly time of 8709.26 seconds and the DFA index was found to be 2.68 .On applying DFA, few redesign changes were implemented like standardization of fasteners, redesigning of lifting hooks, guide post and lower cap. The part number was reduced to 241, the assembly time was reduced to 6883.58 sec and DFA index was brought down to 3.39. The cost was reduced to 23.5% by applying DFMA.

Table 6. DFA analysis of Burring Tool [7]

Entries Including Repeats		Original Design	Re-Design	Savings
1	Parts meet minimum criteria	30	30	0
2.	Parts for elimination	185	147	38
3.	Analyzed subassemblies	4	4	0
4.	Separate assembly operations	96	60	36
5.	Total entries	315	241	74
Assembly labor				
1.	Parts meet minimum criteria	836.70	836.70	0 S
2.	Parts for elimination	6326.83	4857.95	1468.88 s
3.	Insertion of analysed subassemblies	734.33	734.33	0 s
4.	Separate assembly operation	811.40	454.60	356.8
5.	Total assembly labor time	8709.26	6883.58	1825.68 s
Design Efficiency				
1.	DFA index	2.68	3.39	0.71

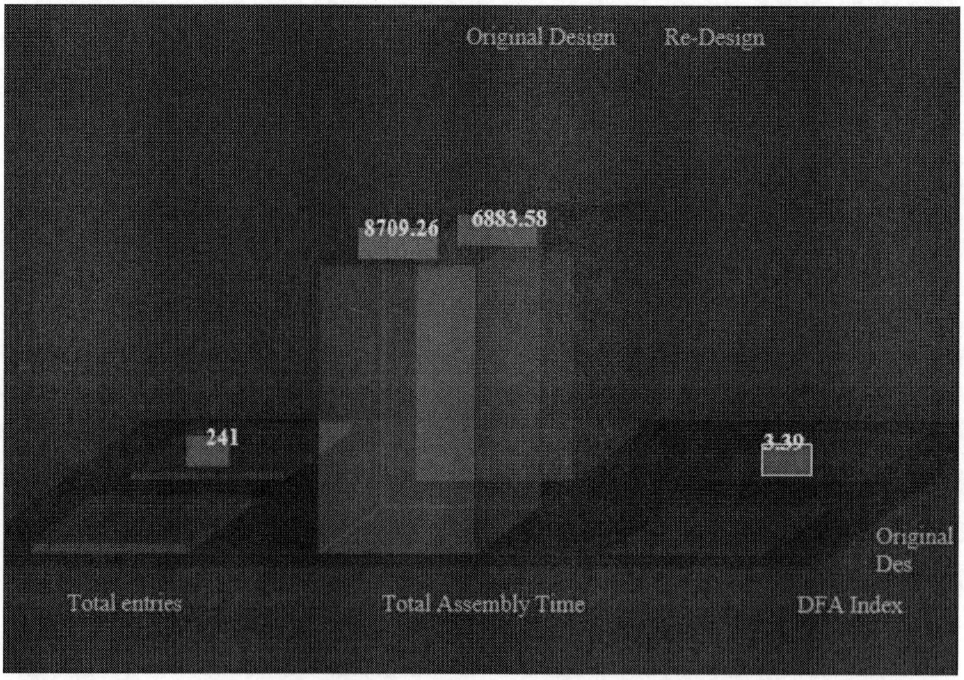

Figure 7. Comparison between Original and Re-design of Burring Tool

IOP Conf. Series: Materials Science and Engineering 1123 (2021) 012003 doi:10.1088/1757-899X/1123/1/012003

4.0 Summary

An overview of applications of DFA analysis of various automobile components is presented in this paper. DFA has a great potential to reduce the price of the component and the time for assembling the product. It follows a basic rule of eliminating all those parts which do not contribute to product's functionality and properties. It has been shown through many examples that DFA provides a concrete support for reducing the assembly time. It is evident in many cases that the design efficiency can be improved substantially by applying DFA concept. While developing a new product, if DFA is applied at the early stages of design it can yield much better results. Therefore it can be inferred that DFA can be a great asset to product development and design process.

References

[1] Geoffrey Boothroyd, Peter Dewhurst, and Winston Knight, *Product Design for Manufacture and Assembly, 2nd edition, Revised and expanded,* CRC Press, 2001

[2] Shete T S, Kulkarni P R and Pawar P V 2019 Design for Manufacturing and Assembly Technique
Applicable For a Gearbox. *8th National Conference on "Recent Developments in Mechanical Engineering [RDME-2019],* M.E.S. College of Engineering, Pune, Maharashtra, India, *IOSR Journal of Engineering* (IOSR JEN) 2019 pp. 14-19

[3] Wan Ni 2019 Re-design on Vehicle Front Support Structure Using DFMA. Beijing Automotive Technology Center Co., Ltd. https://www.dfma.com/pdf/wanpaper.pdf

[4] Peter Dewhurst Evaluation of Instrument Panel Designs for Cost of Manufacture
and Environmental Impacted. University of Rhode Island. Kingston

[5] Richard F. Johnson 2019 Using DFA to enhance Value Engineering/ Value Analysis workshop outcomes. Magna Interior Systems- Seating group.*12th Annual International Forum on DFMA*. 1997. Newport, RI. https://www.dfma.com/resources/johnson.asp [Accessed on 4 November 2019]

[6] Carla Estorilio and Marcelo César Simião. Cost reduction of a diesel engine using the DFMA method, *Product: Management and Development* 4(2) PP 95

[7] Nilesh Kailas More, Rajesh B. Buktar, Syed Mubasheer Ali, Sanjay Samant
Design for Manufacturing and Assembly (DFMA) Analysis of Burring Tool Assembly
Int. Research J of Engineering and Technology (IRJET) 2(6) pp 845 e ISSN: 2395 -0056 .

IOP Conf. Series: Materials Science and Engineering 1123 (2021) 012004 doi:10.1088/1757-899X/1123/1/012004

ANALYSIS OF DISC BRAKE

Nishant Barik[1], Sayyad Abdul Razak Khadari [2]
[1,2]B. Tech Mechanical (specialization in Automotive) Engineering,
Vellore institute of technology

Abstract. This paper deals with the structural analysis of the rotor disc of disc brake. The rotor disc of a disc brake goes through many forces that are applied on it to stop the wheel. The deterioration of the disc brake is a major factor which should be kept in mind during a designing and choosing of the material of disc brake. The material properties are taken in context with the type of forces that act on the brake. In this paper all the disc brakes from past year few years have been analyzed and on the basis of that new model has been designed. The materials taken are compared and to see which shows the best result in the new model disc. The disc brake modelling in done on SOLIDWORKS followed by analysis in ANSYS WORKBENCH 18.1. The dimensions are taken with respect to the MOTOGP brakes.

1.INTRODUCTION

Braking system is a very important part in a vehicle as it helps in slow down or stop the vehicle. Brakes are mechanical devices which absorbs the energy of a moving wheel to inhibit its motion. The kinetic energy of the wheels is converted to heat energy due to friction because of the surface of pads and the rotor disc. Greater the force applied on the disc more will be the heat deliberation and better will be the stopping tendency. The pressure applied on the disc brake also depend on inertia of the car. Braking system not is just help during motion it also help during parking of the car. If braking system are classified on the basis of friction contact basis, we have drum brakes and disc brakes.

In drum breaks the drum (housing of the brake shoes) and the wheel hub are attached to each other. So, when the wheel moves the outer part of the drum rotate and the inner part which consist of the brake shoes are held constant. When we apply the brake the brake shoe comes in contact with the friction surface of the outer surface. This contact with the rotating drum deaccelerates the wheel.

In disc brake in place of drum we have disc rotor which is attached with the wheels is shown in figure 1. The rotor is fixed between two calipers. The calipers had brake shoes which comes in contact with the friction part when brakes are applied.

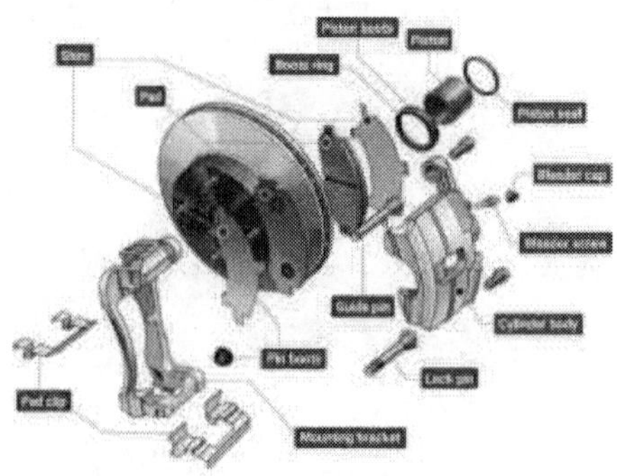

Figure 1. Disassembly Disc Rotor

A disc brake consists of:

1. Brake pads
2. Caliper (piston in fitted in it)
3. Rotor

Mechanism of disc brake:

When the brake pedal which is connected to the master cylinder is pressed. The brake fluid (filled in master cylinder) is pressurized and creates a pressure (pascal's law) in the piston inside the caliper. This moves the brake pads to the friction surface of the rotating disc. This help in stopping the car. Nowadays we also have ABS (anti-lock braking system) fitted with the assembly.

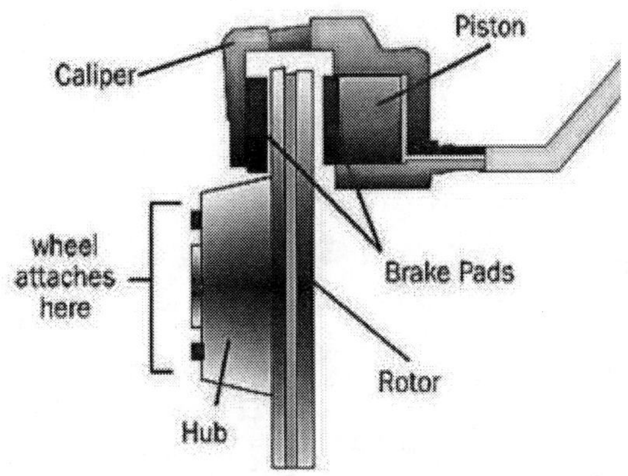

Figure 2. Assembly Disc Rotor

The rotor disc is the part of the disc brake which has the friction surface. This part come in contact with the brake pads when brake pedal is pushed. If the material used in not sufficient enough to handle the stress then this will lead to rotor disc breakdown. Also, the size and the holes for cooling of the disc brake is very important for proper functioning of the disc brake. Generally, manufactures try to keep the rotor disc as small and as light as possible for better efficiency. The rotor discs are commonly made by grey iron steel but nowadays new metals and alloys are made which give better performance than grey iron steel is shown in figure 2.

2. METHODOLOGY

SOLIDWORKS are used for making the models of various disc brakes. After the disc are made the disc thermal analysis has been done in ANSYS workbench 18.1. Three model , 1990, 2000, 2014 model has been made and analyzed on the basis of the negative and positive factors new 2019 model has been made.

The dimensions and the design of old model of the disc brake are taken from *MOTOGP BRAKE INFOGRAPICS of BREBO* special edition. Also, some assumptions are taken with respect to the design. But the final design (2019) is made on the report study of the old models. Basic dimension is assumed. After this the models are exported to *the ANSYS* workbench 18.1.

IOP Conf. Series: Materials Science and Engineering 1123 (2021) 012004 doi:10.1088/1757-899X/1123/1/012004

During analysis *in ANSYS* we have taken few assumptions that the material is a isotopic and the stresses are applied on all the wheels only if the brake are applied and there is a contact between the friction pad and the rotor disc.

FACTORS:

The factors what was kept in mind during modelling of the disc brake are:

1. Light weight
2. Better cooling
3. Environment friendly
4. Cost efficient
5. Availability of the material
6. Size and structure with reference to the brake pads.

3. RESULTS AND DISCUSSION

1990:

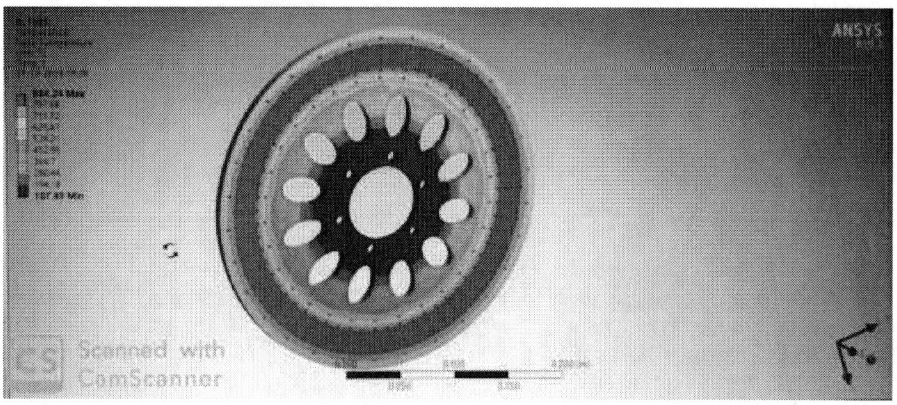

Figure 3. Analysis of 1990's Disc Rotor

Result

- This is the rotor model used in 1990.
- These kind of rotor does not have holes at the area of contact of friction pads.
- They have holes just after the contact patch, this helps in little cooling.
- Since there is no air flow at the contact, so temperature increase that in turn decrease the friction.
- Due to less cooling ability there is maximum detrition on the friction pads shown in figure 3.
- In 80's the discs are not at the outside or at the edge of the wheel they are a bit inside of the wheel, which makes it even harder to the air to get inside and cool the discs.

2000:

Figure 4. Analysis of 2000's Disc Rotor

Result

- This is a disc rotor used in 2000.
- These disc rotors have slits instead of holes.
- This increases the air flow at the area of contact, which decreases the temperature shown in figure 4.
- This has more friction than 1990 model
- 2000 is the year when people started thinking about heating of break discs.
- The slit formation in the break disc increased the rate of cooling but it decreased the strength and life of break disc.
- The slits decrease the structural integrity of the break disc.

2014:

Figure 5. Analysis of 2014's Disc Rotor

Result

- This is disc rotor model used in 2014
- This model is the combination of 2000 and 1990.

➢ This model has long and wide slits along with holes, and a thin structural design at the center.

➢ In 2014 technology and power of engine has developed very much, which also increases the complication in breaking shown in figure 5.

➢ This intern effects the cooling of the of the break disc.

➢ We have to apply more effective system of cooling which could suppress the upraise of the power of engine.

Figure 6. Analysis of 2019 Disc Rotor

Result:

➢ In 2019 model we added fins to the disc
➢ These fins are design similar to a diffuser
➢ Area of intake is smaller than output
➢ Which causes drop in pressure between the fins.
➢ To cover the low-pressure air flows through the holes in the break contact patch
➢ which causes increase in the rate of cooling shown in figure 6.
➢ This disc brake has more volume than other which causes to decrease the hear per unit volume

Table 1. Material Properties

Parameters	**Grey cast Iron**	**Aluminum metal matrix composites**	**Stainless steel**	**Carbon ceramic**
Density (kg/m^3)	7100	2820.6	7750	1800
Poison's	0.25	0.35	0.3	0.31
Young's modulus	125	113.7	190	195
Specific heat	586	828.43	500	755
Thermal conductivit y	54	147.95	26	40

Grey cast iron:

Grey cast iron type of iron with grey color due to the presence of graphite fractures in it. it has fine flakes which give extra strength to the material. It has 2.5-4% carbon and 1-3% silicon. It is used to make gears, automotive suspension etc. it has high thermal conductivity that mean the has easy heat flow movement.

Aluminum metal matrix composites:

This material has gain popularity in automobile aerospace etc. this composite has high strength, low density and good resistance when compared to other metals. This metal is a reinforcement of material like fly ash, silicon, graphite, organic material etc.

Stainless steel:

This material contains 16-26% chromium, 6-22% nickel (Ni) and low carbon. It has excellent corrosion resistance, ductile and readily formable, good performance at high temperature.

Carbon ceramic:

Majority application in aerospace derived technology and powerful cars. The component main matrix is silicon carbide and silicon element. Light weight and better thermal resistance. This is a carbon fiber reinforced silicon carbide composite. Give extra stress and shear resistance.

4. CONCLUSION:

In this paper we have done the analysis of disc of disc brake. After the whole analysis it was proven that the 2019 model had better design than the rest of the disc brake. it has small holes in the place where there will be the contact of the fiction pads also fins to suck in air so that it can give better cooling than rest of the model. Also due to more area of contact with the rotor and the friction pad it will give better friction and braking efficiency will increase. On the basis of material Carbon composite is the best material as is 50 % less weight than grey cast iron. It also gives better braking response and higher stability due to high Young's Modulus than other material. Also due to less wear resistance it will give longer life of the brakes. And high thermal stability giving better control.

5. REFERENCE:

[1]. Toshikazu Okamura and Masanori Imasaki, Increasing thermal strength of brake discs by improving material homogeneity, Kiriu corporation,2009.
[2]. Ashesh shah, Innovative concept of front disc break module with weight reduction and cost optimization, Tata motors.Ltd,2014.
[3]. Toshikazu Okamura, Experimental Study of Effect of Brake-Disc Surface Texture on Friction and Wear through Small-Scale Tests on Tribotester,2013.
[4]. Ashesh Anil, Conversion of Drum Brake System to Disc Brake with CAE and CFD: Resulted in Optimized Brake Rotor Design and Improved Performance,2017.
[5]. Andrew C.smith, A New, High Torque Brake Design Using Sliding Discs, Delphi corporation,2003.
[6]. Dragan Alekendric, Dynamic Control of Disc Brake Performance, University of Belgrade, Serbia,2012.
[7]. Nikhil Mule, Analysis and correlation of thermomechanical deformation of a disc brake rotor, chassis breakes international,2015
[8]. Toshikazu Okamura, Interactive effects of thermal defamation and wear on lateral runout and thickness variation of brake disc rotors,2016
[9]. R.C. Parker, The Performance and Characteristics of the Disc Brake, Ferodo Ltd,2000.
[10]. Toshikazu Okamura, Fundamental Study on Thermal Behavior of Brake Discs,2006
[11]. Masaaki Nishiwaki, Study on Disc Brake Squeal,2000
[12] Rudolf Limpert, Brake Design and Safety, Third Edition R-398, Published by SAE International with a Product Code of R-398, ISBN of 978-0-7680-3438-7.

BIO-COMPOSITES: A STUDY ON BEHAVIOR OF OIL PALM MESOCARP FIBER REINFORCED KGG

V.Srikanth1, Dr.K.Raja Narender Reddy2, Dr.D.K.Nageshwar Rao3
[1]Department of Mechanical Engineering, KITS, Warangal, India,srikanthvallala345@gmail.com
[2]Department of Mechanical Engineering, KITS, Warangal, India,krajanr@rediffmail.com
[3]Department of Mechanical Engineering, B.I.T., Bahir Dar University, Ethopia,
dr.dknrao@gmail.com

Abstract

The use of petroleum based non-degradable polymers and the effect of petroleum products directly impacting lives. Therefore, eco-friendly biodegradable composites are attracting more and more researchers. Biodegradable composites are prepared by using hand layup process using Kondagogu gum(KGG) resin with oil palm mesocarp fiber(OPMF) as reinforcement. Mechanical properties such as tensile, flexural, impact and Thermogravimetric (TG/DTA) of the composites are evaluated. The effect of fiber loading on the mechanical and thermal properties of the composite is also studied.

Keywords: Kondagogu Gum(KGG), Oil Palm Mesocarp(OPMF), NaoH, Mechanical & Thermal Properties.

1. Introduction

The day to day increased awareness of environmental issues based in natural fibers reinforced with plastic polymer composite has been the major priority in many industrial sectors. The ever-growing numbers of biocomposite researchers in each year are well aware of environmental issues. Every researcher is eager to find a solution and substitute traditional materials that can harm our world. Bio-Composites are good flexibility, highest stiffness and low-cost. Bio-composites are the top choice made by consumers and the limited supply of petroleum makes bio composites more popular. Various among natural plant proteins, natural gum, such as kondagogu gum, gum karaya, rosin,soy protein and wheat protein isolated(WPI) all are growing attention in scientific research and industry due to light access and low cost. India has major source of forest gums, mamely, gum kondagogu (Cocholospermum gossypium), gum karaya (Sterculia urens roxb). Gum kondagogu belongs to the family of Bixaceae[1]. Substitute natural fibers in replacement of synthetic fibers, such as glass, and carbon are more recent much attention has been paid to tackle this environment issue [2]. Composites are mainly controlled by fiber/matrix interface and ability to transfer loads from matrix to fiber for find a mechanical properties. But the major tendency to apply a natural fibers, it absorbs moisture (hydrophilic) which is less interfacial bonding between matrix and fibre. According to them, Researchers suggest that surface modification techniques improve the interfacial bond between fiber & matrix through a chemical treatment process [3]. As for engineering applications biocomposites find an important place in environmental protection and developing less mass composites for automobiles, structural beams, packaging etc....[4,5].

V.T.P. Vinod & R.B. Sashidhar Experimental work in the laboratory on this gum has resulted in a special identification of this gum KGG, compared to the well-established and commercially exploited gum karaya and KGG has unique properties as balanced to different gum trees [6]. Dani Jagadeesh et al. Scrutinize the bestmethod conditions for preparing films from Wheat protein isolated (WPI) to Glycerol content was determined by changing 10 to 25% and glutaraldehyde from 0 to 40%. Various factors affecting the structure of these films have been examined by measuring their mechanical (tensile strength and extension), spectral and thermal properties and moisture content[7]. Xue Li et al. and K.obi Reddy et al. In this review, various chemical modification processes are reported to be used to develop compounds such as alkali, silane, acetylation, benzoylation, acrylation, maleated coupling agents, isocyanates, permanganate and other therapies. The interfacial bonding between the fiber surface and matrix may not only modify fibre surface and also improve the strength of fibre. Composites are less water absorption and developed their mechanical properties in the aim of chemical treatment[8,9]. V. K. Thakur et al. research work is focused on the effective use of lignocellulosic natural fibers as reinforcement using polymer resin as a novel matrix. Green composites were prepared using the compression molding technique with different fiber contents. The physicomechanical and thermal characteristics of the different composite samples were investigated as a function of fiber contents [10]. K Raja Narender Reddy et al. obtained the results the effect of treated and untreated woven century fiber composites are investigated for evaluating their mechanical & thermal properties. Higher tensile, hardness, higher glass transition and thermal decomposition temperatures, and low water absorption properties are reported in alkaline-treated composites[11]. V.T.P Vinod et al. discussed Gum kondagogu (Cochlospermum gossypium), a natural biopolymer, was investigated to assess its solution native gum in aqueous condition, indicating a predominantly liquid- to solid-like behavior and conformational properties of the gum were determined by deacetylated gum was thermally more stable than native gumand and various physico-chemical properties of this gum are yet to be characterized like scanning electron microscopy (SEM), differential scanning calorimetry (DSC), static light scattering (SLS) [12,13,14]. Jagadeesh Dani et al. biocomposites were prepared using wheat protein isolate (WPI) was cross-linked with glutaraldehyde in the presence of glycerol plasticizer. These polymer composites were subsequently subjected to evaluation for their mechanical, morphological (SEM), thermal (TGA/DMA) properties[15].

Come to the current study, natural gum was extracted from a plenty plants available in India forest, Kondagogu Gum. Kondagogu gum, scientifically known as “Cochlospermum gossypium”, is a newly-identified gum. Similarly natural fibre also extracted from locally-available plant, Oil Palm Mesocarp fibre (OPMF). The OPMF reinforced composites were prepared using KGG as the matrix and investigated their mechanical & thermal properties of the composite thus formed were evaluated as a function of fibre content and matrix in nature. These fibers were collected using simple manual water rewriting procedure prior to alkaline treatment and characterized by tensile, flexural, impact and Thermogravimetric (TG/DTA). The obtained results were then analyzed and compared with other prevalent natural fiber-based composites.

2. Experimental procedure

2.1 Materials

KGG powder (Grade-I) was procured from Nutriroma Pvt.Ltd, Hyderabad,India. Oil palm mesocarp fiber were obtained from 3F oil plam industries located at yernagudem, Devarapalle, West Godavari, Andhra Pradesh. V.T.P Vinod was reported by Analytical profile of gum Kondagogu, 6.1% Crude Protein (g %), 7.3% of ash (g %), Water binding capacity (ml/g) 35.1, pH-4.9 to 5.0[6]. Glutaraldehyde minimum assay is 25% was purchased from Narmada chemical industries,Kukatpally, Hyderbad. Glycerol about 98% purified and ammonia solution 25%, analytical grade sodium hydroxide (NaoH), HCL, deinozed water were purchased from Rekha chemicals, Hanamkonda, Warangal, India.

2.2 Alkaline (NaoH) treatment

The chemical treatment of fiber is aimed at improving the adhesion between the fiber surface and the polymer matrix may not only modify the fiber surface but also increase fiber strength. Alkaline treatment or mercerization is one of the most used chemical treatment of natural fibers when used to reinforce thermoplastics and thermosets. The important modification done by alkaline treatment is the disruption of hydrogen bonding in the network structure, thereby increasing surface roughness. This treatment removes a certain amount of lignin, wax and oils covering the external surface of the fiber cell wall, depolymerizes cellulose and exposes the short length crystallites. For treatment process water by volume is taken along with 2% of NaOH mixed with water. The fibers are soaked in the water for 24 hours, figure1. (a) the red oil represents after removing of oil palm mesocarp fibre, then the fibers are washed thoroughly with distilled water to remove the final residues of alkali. Good bonding is expected due to improved wetting of fibers with the matrix. In order to develop composites with better mechanical properties and good environmental performance, it is necessary to impart fibers by chemical treatments.

(a) (b) (c)

Figure 1. Alkali treatment of oil palm mesocarp fiber (a)Red oil of mesocarp fibe (b) Treated fiber (c) Untreated or raw fiber

2.3 Preparation of resin

KGG gum was set up by utilizing a throwing technique we depicted beneath. To prepare a resin, the desired amounts of KGG powder was accurately weighed and dispensed into a clean glass beaker containing one liter (by wt. of 2.5 times KGG) of deionzed water. Then entire solution was stirred upto 20-30 minutes with help of magnetic stirrer at the room temperature, and then desired amount of 1M NaoH & 1M HCl was added to neutralize the solution of gum. Contiguous incubation for 20 min, Glycerol & Glutaraldehyde weighed separately, mixed with 25-30% times (by wt. of KGG powder) to a final pH of 7.0-8.0 in figure 2. Different quantities glutaraldehyde was added as a crosslinkinig agent to increase the properties of tensile and thermal stability. To decrease the brittleness of the resin, glycerol was added in different proportions.

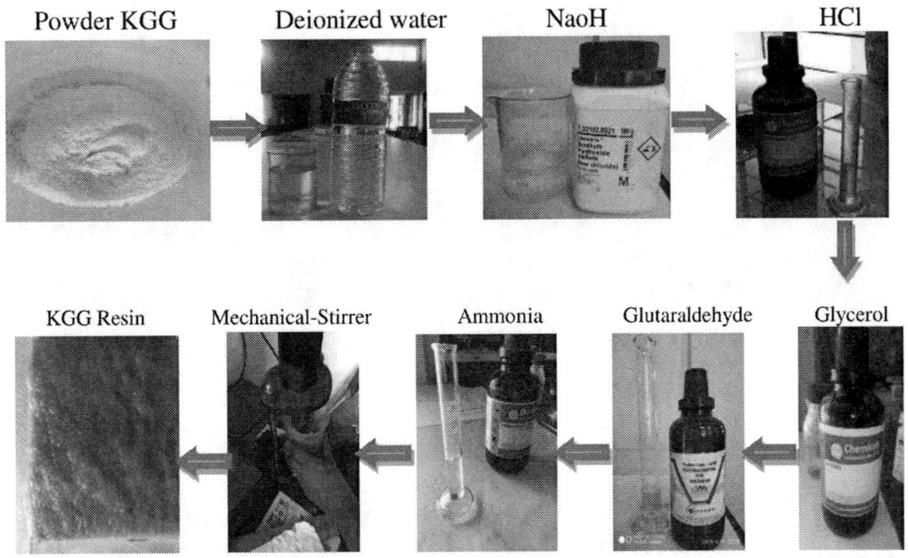

Figure 2. Preparation of KGG Resin

2.4 Fabrication of Composite

The bio-composite laminates were prepared at Kakatiya Institute of Technology & Science (KITSW) facility at Yerragattu gutta, Waarangal,India. Prior to the laminating process, extracted raw fibre of oil palm mesocarp to remove the excess of moisture and to prepare the random short fibre-reinforced composites, the OPMF were cut into lengths of 4mm, 8mm & 12mm as shown in Fig. 3 (a).The hand layup process is used to preapared the composites, OPMF were subsequently added to the mixture which was evenly mixed before being placed on the mould as shown in Fig. 3(b). The fibres and resins are allowed to cure with in 24 hrs. The process was repeated for OPMF with length ranging from 4mm, 8mm & 12mm. The OPMF/KGG composite laminates were produced with fibre weight of 05, 10, 15% as shown in Fig. 3(c).

(a) (b) (c)

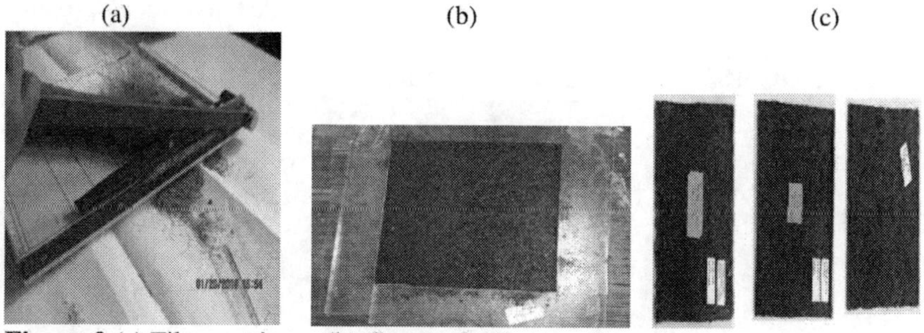

Figure 3.(a) Fiber cutting (b) Composite Preparation (c)OPMF/KGGComposites with different weight fractions

2.5 Analysis

2.5.1 Tensile test of the bio-composites

The tensile test specimens were prepared according to ASTM D638 standards. The tensile properties of KGG/OPMF randomly oriented bio-composites were studied an Zwick / Roell Z010 from the fig 4.(a) universal testing machine with a load of 10KN used for testing at a crosshead speed of 3mm/min maintaining a gauge length of 50mm.In each laminate three specimens were tested and the average values are reported for figure 4.(b) &(c), we determined the parameters of maximum stress, modulus, & % elongation at break.

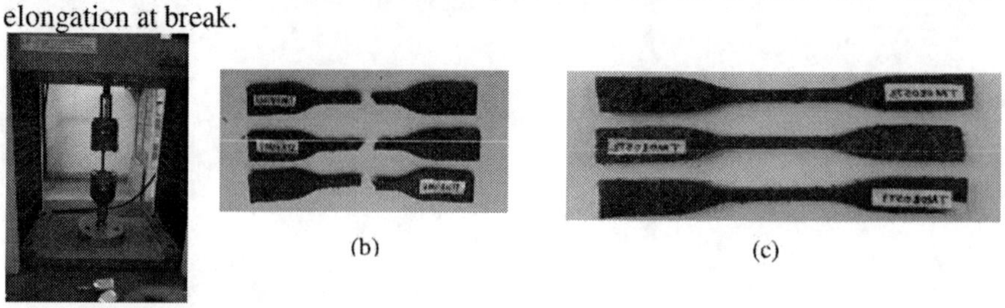

(b) (c)

(a)

Figure 4.(a)Universal Testing Machine (b) After test (c) Before test

IOP Conf. Series: Materials Science and Engineering 1123 (2021) 012005 doi:10.1088/1757-899X/1123/1/012005

2.5.2 Flexural test of the bio-composites

Flexural properties were characterized using a rectangular bar samples in an Zwick / Roell Z010 machine with a three point bending geometry at a crosshead speed of 5mm/min.The force was easured by a load cell of 10KN and the displacement determined from the movement of the crosshead. Test specimens were prepared and stamped in accordance to ASTM D790.The average values were calculated from three for each sample.

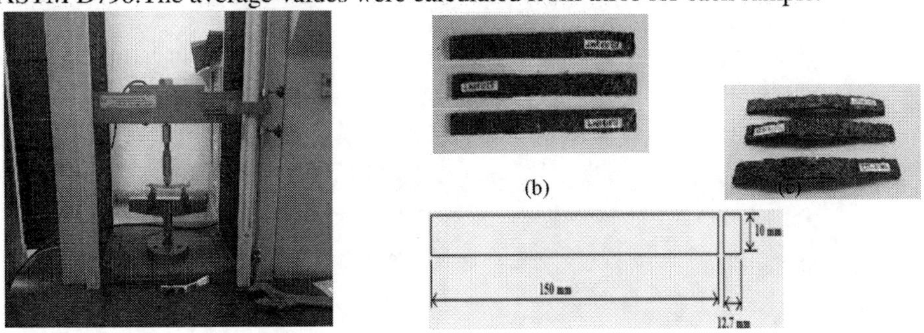

(a) (d)

Figure 5.(a)Three point bending moment (b)Before test (c)After test (d) Dimensions of specimen

2.5.3 Impact test of the bio-composites

The Izod impact test was carried out according to ASTM D256 and supplier is Deepak poly plastic Pvt ltd, impact tester (IZOD/Charpy Impact Tester) used to determine the impact strength of specimens. The sample size is 127 X 12.7 X 3mm^3. The average of three values were calculated from each sample.

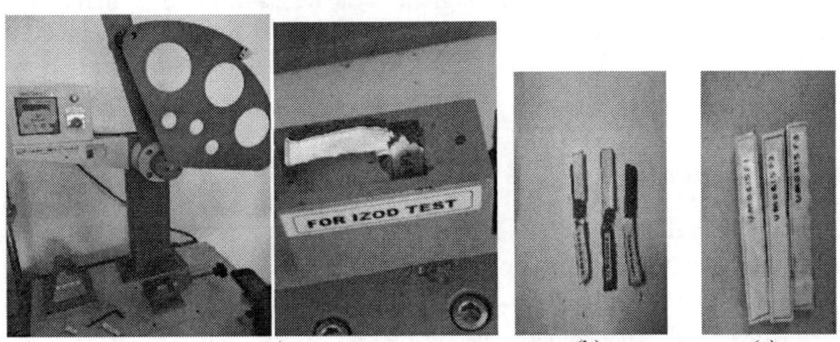

(a) (b) (c)

Figure 6. (a)Specimens placed on Impact test machine (b) breaking specimens (c) Before test

2.5.4 Thermo-Gravimetric Analysis

Thermogravimetric analysis (TGA) is to understanding the thermal stability of composite materials. It also offers a more precise control of heating condition such as variable temperature range and accurate heating rate and needs only a small quantity of sample for analysis. Data are recorded as TGA curve of weight or weight percentage as a function temperature or time. The instrument used for this analysis is model **NETZSCH, STA 2500 Regulus** in NIT,Warangal. The samples were heated from 25 – 300 °C, at heating rate 10°C/min. For analysis purpose were done under flowing nitrogen at a constant flow rate of 20mL/min.

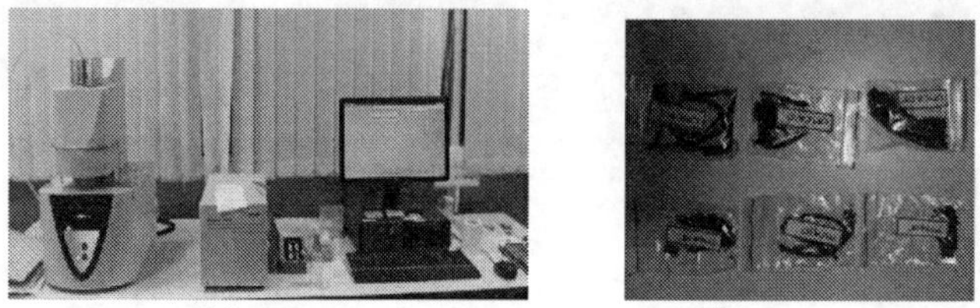

(a) (b)

Figure 7.(a) TGA (b) Specimens of TGA

3.Results & Discussions

3.1 Tensile test

The strength of the fiber reinforced composites depends on the type of fiber, matrix, fiber length, fiber content and bonding strength between the fiber and matrix. Tensile specimens were tested in the FIE universal testing machine with a crosshead speed of 10 mm/min and the results are tabulated in Table. Yield force is the force where the material begins its permanent deformation. All the laminates have shown increase in the yield force, where mostly the alkaline treated laminates have more yield force when compared to their counter untreated laminates, this decrease in yield force for untreated fiber may be due to the presence of red oil and wax on the fiber which restricts the proper binding of the fiber with the resin, the yield force can even be increased by trying different available chemical treatments. From the figure 8.(a) shows among all the laminates the treated fiber of length 4mm with 10% of weight ratio laminate have more yield force, it can be observed that the yield force has decreased with the increase in the fiber length, as the composites are made of randomly oriented fiber there may be voids created in the laminates with the increase of the fiber length. All the laminates with 10% weight ratio of fiber have increased yield force when compared to 5% and 15% weight ratio of fibers

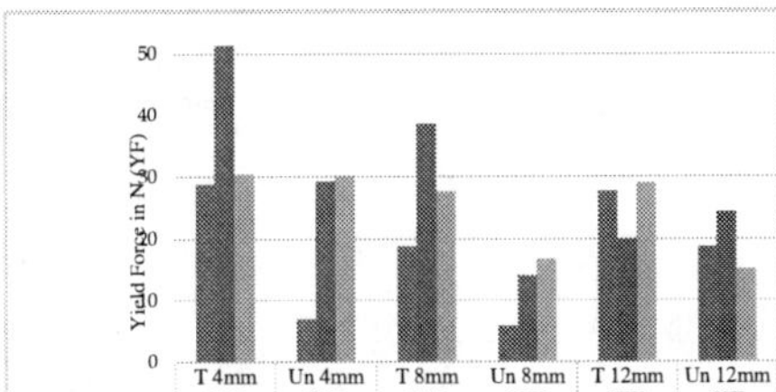

	T 4mm (YF)	Un 4mm (YF)	T 8mm (YF)	Un 8mm (YF)	T 12mm (YF)	Un 12mm (YF)
5% fiber	28.89	7	18.79	5.88	27.7	18.67
10% fiber	51.42	29.34	38.67	14.07	20.03	24.37
15% fiber	30.43	30.3	27.7	16.77	29.03	14.98

Figure 8.(a) Yield force vs % fiber

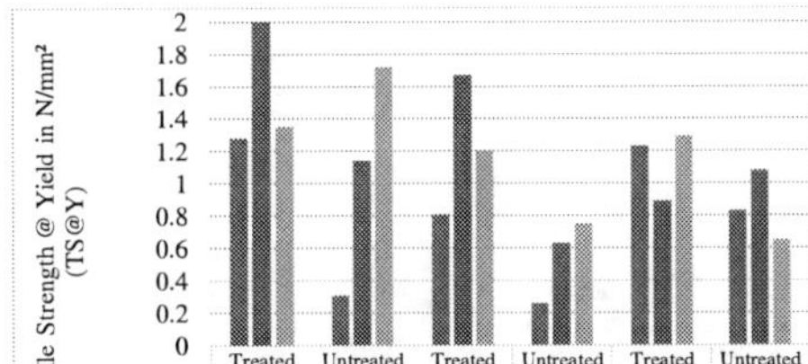

	Treated 4mm (TS@Y)	Untreated 4mm (TS@Y)	Treated 8mm (TS@Y)	Untreated 8mm (TS@Y)	Treated 12mm (TS@Y)	Untreated 12mm (TS@Y)
5% of fiber	1.28	0.31	0.81	0.26	1.23	0.83
10% of fiber	2.15	1.14	1.67	0.63	0.89	1.08
15% of fiber	1.35	1.72	1.2	0.75	1.29	0.65

b)Tensile strength at treated and untreated fiber

The below graph figure 8.(c) is indicates a tensile strength of Oil Palm Mesocarp fiber at different lengths and weights of treated and untreated fiber. The above chart all the laminates have shown increase in the tensile strength, where mostly the alkaline treated laminates have more tensile strength when compared to their counter untreated laminates, this decrease in tensile strength for untreated fiber may be due to the presence of red oil and wax on the fiber which restricts the proper binding of the fiber with the resin, the tensile strength can even be increased by trying different available chemical treatments. Among all the laminates the treated fiber of length 4mm with 10% of weight ratio laminate has more tensile strength, it can be observed that the strength has decreased with the increase in the fiber length, as the composites are made of randomly oriented fiber there may be voids created in the laminates with the increase of the fiber length

Mostly all the laminates with 10% weight ratio of fiber have increased tensile strength when compared to 5% and 15% weight ratio of fibers except some laminates this may be due to the improper lay-up of laminate or defect in the cutting of the specimen from the laminate which can be overseen as the 4mm length 10% treated fiber has more tensile strength. %Elongation of a material is the measure of its ductility. In composites with the increase in tensile strength there is decrease in % elongation both can be increased with some correction in the cross-linking agent, All the laminates have shown % elongation, where mostly the alkaline treated laminates have less % elongation when compared to their counter untreated laminates in most of the laminates. Among all the laminates the treated fiber of length 4mm with 10% of weight ratio laminate has very less % elongation, it can be observed that the % elongation has increased with the increase in the fiber length except for 8mm length fiber, from the tensile strength graph it is seen that 4mm length fiber has more strength and here it has less % elongation these both can be increased with the increase in the amount of cross-linking agent or by using other cross-linking agents.

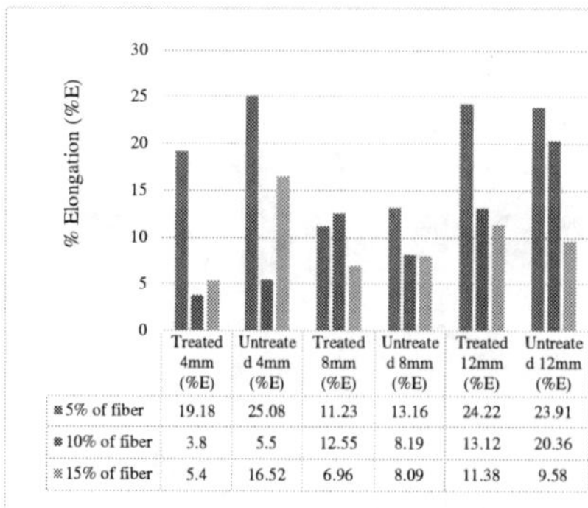

	Treated 4mm (%E)	Untreated 4mm (%E)	Treated 8mm (%E)	Untreated 8mm (%E)	Treated 12mm (%E)	Untreated 12mm (%E)
5% of fiber	19.18	25.08	11.23	13.16	24.22	23.91
10% of fiber	3.8	5.5	12.55	8.19	13.12	20.36
15% of fiber	5.4	16.52	6.96	8.09	11.38	9.58

Figure 8.(c)% elongation vs % fiber

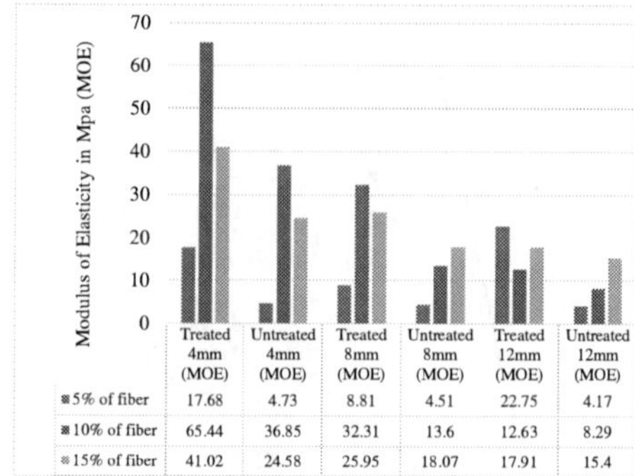

	Treated 4mm (MOE)	Untreated 4mm (MOE)	Treated 8mm (MOE)	Untreated 8mm (MOE)	Treated 12mm (MOE)	Untreated 12mm (MOE)
5% of fiber	17.68	4.73	8.81	4.51	22.75	4.17
10% of fiber	65.44	36.85	32.31	13.6	12.63	8.29
15% of fiber	41.02	24.58	25.95	18.07	17.91	15.4

(d) Modulus of elasticity vs % fiber

From the figure chart 8.(d) shows a Modulus of elasticity measures the stiffness of material. All the laminates have shown increase in the modulus of elasticity, where mostly the alkaline treated laminates have more modulus of elasticity when compared to their counter untreated laminates, this decrease in elasticity for untreated fiber may be due to the presence of red oil and wax on the fiber which restricts the proper binding of the fiber with the resin, the modulus can even be increased by trying different available chemical treatments. Among all the laminates the treated fiber of length 4mm with 10% of weight ratio laminate has more modulus of elasticity, it can be observed that the modulus of elasticity has decreased with the increase in the fiber length, as the composites are made of randomly oriented fiber there may be voids created in the laminates with the increase of the fiber length. mostly all the laminates with 10% weight ratio of fiber have increased modulus of elasticity when compared to 5% and 15% weight ratio of fibers except some laminates this may be due to the improper lay-up of laminate or defect in the cutting of the specimen from the laminate which can be overseen as the 4mm length 10% treated fiber has very high modulus of elasticity.

After analyzing all the tensile testing results of the specimens it can be concluded that the 4mm treated 10% fiber has good tensile properties like Modulus of Elasticity is 65.44 N/mm², Tensile Strength@ Break is 2.19 N/mm², % elongation is lowest of all which is 3.8%, tensile strength at yield is 2.15 N/mm², yield force is 51.42N.

Note: **UM0405**: UM: untreated oil palm mesocarp fiber

TM0405: TM:Treated Oil palm mesocarp fiber

04 : Length of the fiber
05: Percentage of the fiber

04 : Length of the fiber
05: Percentage of the fiber

IOP Conf. Series: Materials Science and Engineering 1123 (2021) 012005 doi:10.1088/1757-899X/1123/1/012005

3.2 Flexural Test

Flexural strength is the bending strength of the material. All the laminates have shown increase in the flexural strength, where mostly the alkaline treated laminates have more flexural strength when compared to their counter untreated laminates, this decrease in strength for untreated fiber may be due to the presence of red oil and wax on the fiber which restricts the proper binding of the fiber with the resin. Among all the laminates, the treated fiber of length 4mm with 5% of weight ratio laminate has more flexural strength, it can be observed that the strength has decreased with the increase in the fiber length, as the composites are made of randomly oriented fiber there may be voids created in the laminates with the increase of the fiber length. mostly all the laminates with 5% weight ratio of fiber have increased strength when compared to 10% and 15% weight ratio of fibers except some laminates this may be due to the improper lay-up of laminate or defect in the cutting of the specimen from the laminate which can be overseen as the 4mm length 5% treated fiber has more flexural strength and it may be because of the presence of the high amount of resin which increases the hardness.

Stiffness at the initial step of the bending process is known as the flexural modulus, flexural modulus and tensile modulus are calculated individually for composites as they are nonhomogeneous materials. All the laminates have shown increase in the flexural modulus, where mostly the alkaline treated laminates have more flexural modulus when compared to their counter untreated laminates, this decrease in flexural modulus for untreated fiber may be due to the presence of red oil and wax on the fiber which restricts the proper binding of the fiber with the resin.

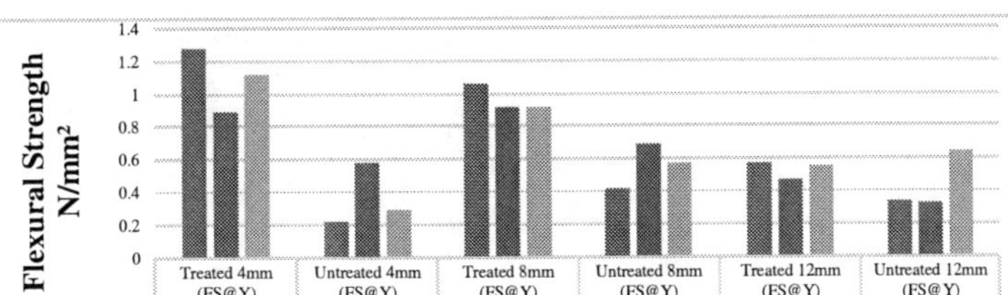

	Treated 4mm (FS@Y)	Untreated 4mm (FS@Y)	Treated 8mm (FS@Y)	Untreated 8mm (FS@Y)	Treated 12mm (FS@Y)	Untreated 12mm (FS@Y)
5% of fiber	1.28	0.22	1.06	0.42	0.57	0.34
10% of fiber	0.89	0.58	0.92	0.69	0.47	0.33
15% of fiber	1.12	0.29	0.92	0.57	0.55	0.64

Figure.8(e) flexural Strength vs % fiber

Among all the laminates the treated fiber of length 4mm with 5% of weight ratio laminate has more flexural modulus, it can be observed that the flexural modulus has decreased with the increase in the fiber length, as the composites are made of randomly oriented fiber there may be voids created in the laminates with the increase of the fiber length. mostly all the laminates with 5% weight ratio of fiber have increased modulus when compared to 10% and 15% weight ratio of fibers except some laminates this may be due to the improper lay-up of laminate or defect in the cutting of the specimen from the laminate which can be overseen as the 4mm length 5% treated fiber has very high flexural modulus.After analysing all the flexural testing results of the specimens it can be concluded that the 4mm treated 5% fiber has good flexural strength which is 1.28 N/mm².

3.3 Impact Strength

The impact strength of the OMPF/KGG bio-composite laminate is tested using impact testing machine. Izod impact test is carried out to analyze the impact energy observed by the OPMF reinforced with KGG bio-composite laminates. The impact testing machine gives the energy loss capability of the material to withstand suddenly applied load. The experiments are conducted with different length and wt% of fiber loading for OPMF reinforced natural fiber composite (see Figure 8(f)). The figure indicates that the increase in length of fibers increases the impact energy which leads to the impact strength. The figure also indicates that the increase is observed only up to 12mm treated length and 5% weight of the laminates, further increase in wt% does not show any improvement in the impact strength of OPMF/KGG bio-composite laminates. From the analysis of the above, it has been asserted that the treated 12mm and 5wt% of composite laminates is performing, better than the other Untreated and treated different length and weight percentage considered. The impact energy normally increases with the increase in length of the treated fibers. It is found that the impact force is uniformly distributed in the unidirectional fiber reinforced composites. In 12mm treated length and 5% weight of composites, the increases in length of the fiber content with sufficient distribution of the matrix leads to the proper crack path and absorb maximum impact energy. Based on the experimental analysis, we are observed treated bio-composites 12mm length fiber and 5% fiber have highest impact strength (i.e 834.64 J/m²) then compared to a overall untreated fibers and the least impact strength 4mm 15% fiber (i.e 135.17 J/m²).

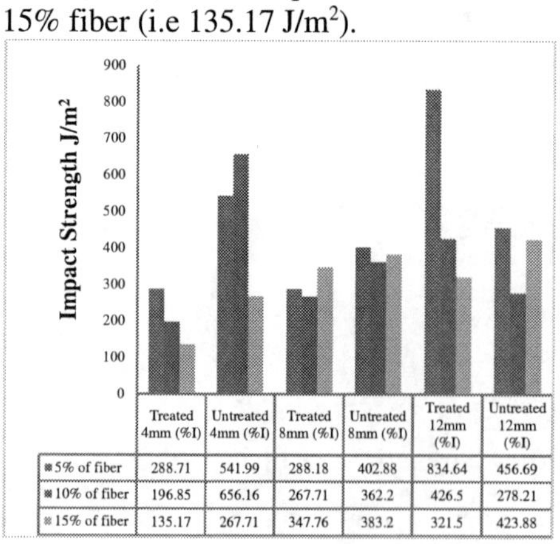

	Treated 4mm (%I)	Untreated 4mm (%I)	Treated 8mm (%I)	Untreated 8mm (%I)	Treated 12mm (%I)	Untreated 12mm (%I)
5% of fiber	288.71	541.99	288.18	402.88	834.64	456.69
10% of fiber	196.85	656.16	267.71	362.2	426.5	278.21
15% of fiber	135.17	267.71	347.76	383.2	321.5	423.88

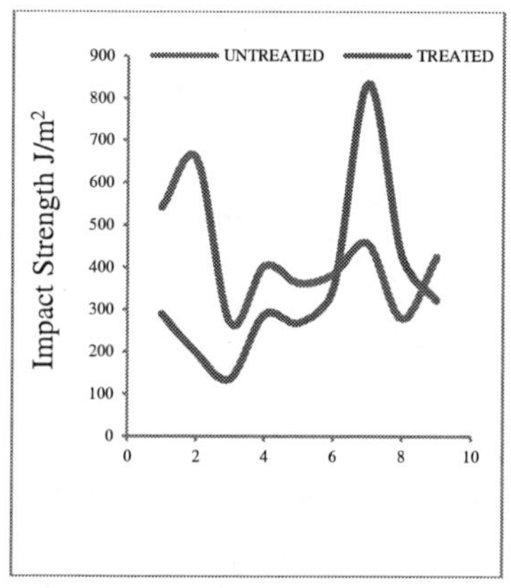

Figure.8 (f) Impact strength comparison for different length and weight fractions of OPMF/KGG bio-composites.

3.4 Thermogravimetric Analysis (TGA)

Thermogravimetric (TGA) analysis curves as a function of the temperature of treated & untreated Oil palm mesocarp fiber(OPMF) reinforced Kondagogu gum resin (KGG) with the variation of fiber length and wt% are given in Figure 8(g). Thermal decomposition of the composites takes place in the temperature range 25 °C to 300 °C. The percentage of mass temperature curves are shown in Figure 8(g), which shows that increased treated 4mm length (TM0410)fiber content and decreasing remaining fiber, there is also a reduction in the mass loss as a function of temperature.

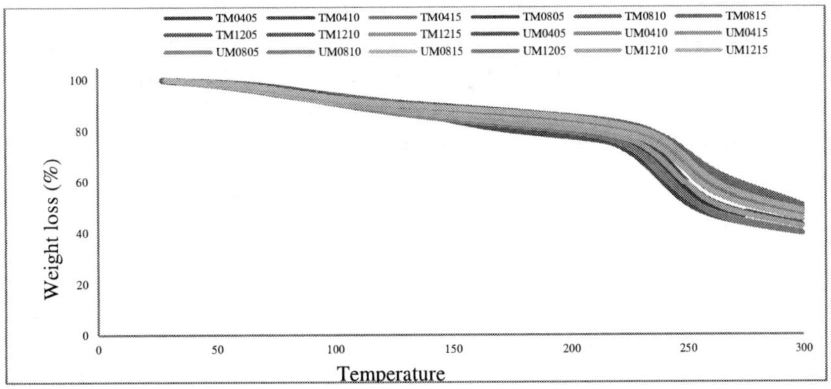

Figure.8 (g) TGA analysis in treated and untreated fibers of different length and wt% OPMF/KGG bio-composites

The TGA result in weight loss with respect to the temperature of treated and untreated fiber comparison in different length and wt% was illustrated in Figure 8(g). From the experimental analysis treated 8mm length 10% weight of laminate(TM0810) is shows the best results compared to all increased in length and wt% of the treated and untreated laminates. The better results it shows from the treated laminate(TM0810) temperature of 60°C the composite loses only 1.8% of the initial weight, with 120 °C the weight loss is already 8.1 % of mass that corresponds to the removal of solvent KGG in the matrix. Between 150°C and 250 °C the weight loss was approximately 33.63 % due to degradation and volatilization with OPMfibers present in the biocomposites. After that, the composite maintains a linear mass loss up to 300°C, where the final residue is only 51.76% of the original mass. The thermal degradation for all composites display a three-step degradation process, where the first stage is attributed to the decomposition of KGG resin, the second stage is attributed to the decomposition of OPMF fiber and the third step is attributed to the decomposition of Additives(Glutaraldehyde and glycerol).

According to the derivative thermogravimetric (DTG) curves in Figure 8(h),the bettered results are shown from the graph the treated 12mm length and 05% wt of laminate(TM1205) is shows a very less derivative weight loss in comparison to all untreated and treated (TM0415) 4mm length and 15% weight of the fiber is shows the highest derivative weight loss (4mm, 8mm & 12mm length of fibers). Let us explaining (TM0415)in the treated 4mm length and 15% wt of the laminate of small degradation peaks were detected between 213 °C to 230 °C due to decomposition of the organic elements in the composite. The second stage around 240 °C to 267 °C due to decomposition of the hemicellulose constituents of the OPMF content, while the third step degradation between 270 °C to 300 °C shows the decomposition of the bio composite components. From the DTG curve displayed a similar trend for Tm0405 to um1215 around this temperature due to more amount of fiber content.

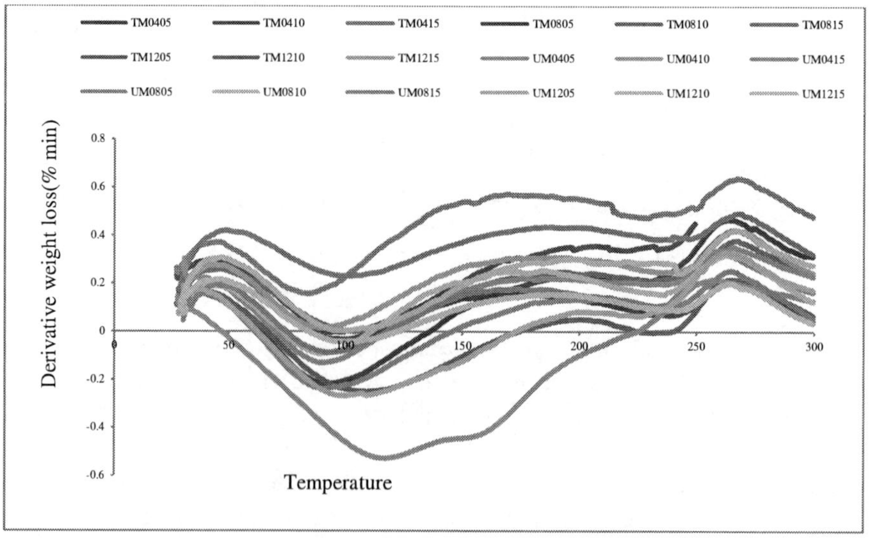

Figure.8 (f) DTG analysis in treated and untreated OPMF/KGG bio-composites

4.Conclusion

This study concerns the characterization of the mechanical and thermal properties of treated and untreated OPMF/KGG biocomposites. The conclusions we obtained from this investigation are as follows:

1. The alkaline-treated fibres are superior strength compared to that of the untreated fibres; tensile testing results of the specimens it can be concluded that the 4mm treated 10% fiber has good tensile properties like

Modulus of Elasticity is 65.44 N/mm², Tensile Strength@ Break is 2.19 N/mm², % elongation is lowest of all which is 3.8%, tensile strength at yield is 2.15 N/mm², yield force is 51.42N. The tensile modulus for 4mm untreated fiber increased and decreased with the increasing weight ratio and the same behaviour is exhibited by the treated 4mm fiber. this may be due to the impoper distribution of the fiber.The treated 4mm fiber has increased values when compared to untreated 4mm fiber due to the decrease of hydroxyl and carboxyl in the treated fiber which increases the binding capacity and the load transfer capacity of the fiber. The 8mm untreated fiber has decrease in values when compared its treated one here to it showed an increase and decrease behaviour and the overall values are less when compared to 4mm fiber, the behaviour of 12mm fiber is also same as the above fiber and its overall values are less than 4mm and 8mm length fibers.

2. The flexural strength decreases linearly with increase in length and wt% of fiber. The overall result analysis in treated and untreated analysing all the flexural testing results of the specimens it can be concluded that the 4mm treated 5% fiber has higher strain rate is observed and good flexural strength which is 1.28N/mm². The flexural modulus for 4mm untreated fiber increased and decreased with its increasing length and its values are lesser than the treated 4mm fiber and in it the modulus decreased and increased with increasing length, the same behaviour is exhibited by 8mm and 12mm treated fiber.Oil palm mesocarp is a lignocellulose material, hydroxyl and carboxyl groups are present in cellulose and hemicellulose of the fiber due to which it shows poor adhesion and decrease in load transfer. So, the chemical treatment of the fiber increases the binding and load transfer capacity of the composite

3. The impact energy increases linearly with increase in length and wt% of fiber. When all the bio composite specimens are observed treated mesocarp fibre of 12mm length and 5% fibre have highest impact strength (ie 834.64 J/m2) and 4mm 15% fiber have least impact strength (ie 135.17 J/m2). Among all the untreated mesocarp specimens 4mm length and 10% of fibre specimen have the highest impact strength (ie 656.16 J/m2) and 4mm length 15% fibre have the least impact strength (267.71 J/m2).In the comparison of treated bio-composites 12mm length fiber and 5% fiber have highest impact strength (i.e 834.64 J/m2) then compared to a overall untreated fibers and the least impact strength 4mm 15% fiber (i.e 135.17 J/m2).

4. From the TGA analysis we revealed that treated laminate is TM0810 very less weight loss of mass degraded to comparison of all untreated laminates, similarly differential thermal analysis (DTA) demonstrated that the differential weight loss as a function of time is 0.53% at 149°C. upto 172°C, the sample exhibited an endothermic stage. From the analysis, it can be concluded that the crystallisation temperature of the OPMF/KGG bio composite is 213°C. The lowest peak at 228°C indicates the melting temperature of bicomposites. After that the fibre exhibits an exothermic in nature as it will release the energy of heat and decrease to a maximum peak level.

REFERENCES

Journal articles

[1]. Sudheer Kumar,Sushanta K. Samal,Smita Mohanty &Sanjay K. Nayak, Recent Development of Biobased Epoxy Resins: A Review, *Journal Polymer-Plastics Technology and Engineering,* Volume 57, 2017 - Issue ,PP 133-155.

[2]. B.C. Mitra, Environment Friendly Composite Materials: Biocomposites and Green Composites:Review, *Defence Science Journal*, Vol. 64, No. 3, May 2014, pp. 244-261.

[3]. Huang Gu, Tensile behaviours of the coir fibre and related composites after NaOH treatment, *Materials and Design* ,volume 30 (2009) 3931–3934.

[4]. Georgios Koronis, Arlindo Silva, Mihail Fontul, Green composites: A review of adequate materials for automotive applications, Composites: Part B 44 (2013) 120–127.

[5]. Maya Jacob John, Sabu Thomas, Biofibres and biocomposites: Review, *Carbohydrate Polymers*, 71 (2008) 343–364.

[6]. V.T.P. Vinod, R.B. Sashidhar, Surface morphology, chemical and structural assignment of gum Kondagogu (Cochlospermum gossypium DC.): An exudate tree gum of India, *Indian Journal of Natural Products and Resources*, Vol. 1 (2), June 2010, pp. 181-192.

[7]. Dani Jagadeesh, D. Jeevan Prasad Reddy, A. Varada Rajulu, Preparation and Properties of Biodegradable Films from Wheat Protein Isolate, *Journal of Polymers and Environment* (2011) 19:248–253.

[8]. Xue Li,Lope G. Tabil, Satyanarayan Panigrahi, Chemical Treatments of Natural Fiber for Use in Natural Fiber-Reinforced Composites: A Review, *Journal of Polymers and Environment*, (2007) 15:25–33.

[9]. K. Obi reddy, K. Raja Narender Reddy, Jun Zhang, Jinming Zhang, and A. Varada rajulu, Effect of Alkali Treatment on the Properties of Century Fiber, Journal of Natural Fibers, 10:282–296, 2013.

[10]. V. K. Thakur, A. S. Singha, and M. K. Thakur, Ecofriendly Biocomposites from Natural Fibers: Mechanical and Weathering Study, *International Journal of Polymer Analysis and Characterization.,2013* Volume 18:1, PP 64–72.

[11]. K Raja Narender Reddy, DKN Rao, K Gopala Kishan Rao and Kamal K Kar, Studies on woven century fiber polyester composites, *Journal of Composite Materials*, 2012, volume 46(23), PP: 2919–293.

[12]. V.T.P. Vinoda. R.B. Sashidharb, K.I. Suresh, B. Rama Rao, U.V.R. Vijaya Saradhi, T. Prabhakar Rao, Morphological, physico-chemical and structural characterization of gum kondagogu (Cochlospermum gossypium): A tree gum from India, *Food Hydrocolloids,* volume *22* (2008) 899-915.

[13]. V. T. P. Vinod, R. B. Sashidhar, V. U. M. Sarma, and U. V. R. Vijaya Saradhi, Compositional Analysis and Rheological Properties of Gum Kondagogu ***(Cochlospermum gossypium***): A Tree Gum from India, *Journal of agricultural and food chemistry*, 2008, volume56, PP: 2199-2207.

[14]. V.T.P. Vinod, R.B. Sashidhar, Solution and conformational properties of gum kondagogu *(Cochlospermum gossypium)* - A natural product with immense potential as a food additive, *Food Chemistry*, (2009), Volume 116, PP: 686-692.

[15]. Jagadeesh Dani, Jeevan Pradsad Reddy D., Varada Rajulu A.,Li R, Green Composites From Wheat Protein Isolate and Hildegardia Populifolia Natural Fabric,*Polymer Composites,* 2011, Volume32, Issue3Pages 398-406.

IOP Conf. Series: Materials Science and Engineering 1123 (2021) 012006 doi:10.1088/1757-899X/1123/1/012006

Characterization of AMCs Produced by Stir Casting Technique

Saleem.M[1], Nishant Kumar raj[2], Yogesh kumar[3], Pranav Kumar[4]

[1]Aeronautical department, Veltech Rangarajan Dr. Sagunthala R&D Institute of Science and technology, Chennai, India, Strokesaleem1@gmail.com

[2]Aeronautical department, Veltech Rangarajan Dr. Sagunthala R&D Institute of Science and technology, Chennai, India, rajnishant3627@gmail.com

[3]Aeronautical department, Veltech Rangarajan Dr. Sagunthala R&D Institute of Science and technology, Chennai, India, Yogeshdewangan684@gmail.com

[4]Aeronautical department, Veltech Rangarajan Dr. Sagunthala R&D Institute of Science and technology, Chennai, India, Bharadwaj.ts.ts.7713@gmail.com

Abstract. Stir casting is the simplest and cheapest method to manufacture composite materials. The objective of the research work is to produce Aluminum Matrix Composites with LM5 as matrix material and Fly Ash as reinforcement material at 0, 5, 10 % by weight, using stir casting technique and to study the mechanical properties such as tensile test, impact test, hardness test. The wettability between all these materials is assured by stir casting process. The microstructure of the composite show that the reinforcement material is homogeneously distributed in the matrix materials. Tensile Strength decreases with increase in weight percentage of Fly Ash. The weight percentage of Fly Ash has no effect on the impact strength of the composite. The hardness increases with the increase in weight percentage of fly ash and hence the fly ash can be used for increasing the wear resistance of the composites, where hardness plays a major role in industries which can turn industrial waste into industrial wealth. ***Key words:*** *AMCs, Stir Casting, Microstructure, characterization.*

1.Introduction

From past many years need for advanced materials in engineering is increasing tremendously day by day. Conventional materials has limitations in fulfilling our needs for stiffness, toughness, lightness and density. For satisfying our need of better material with more advantages metal matrix composites are best for industries. Constituents for making MMCs are taken such that it should give better results according to our need. MMCs are widely used in many manufacturing companies such as medical, aerospace, automotive, civil, marine and electronics industries for their superior properties like high module, low thermal expansions, high strength, low weight and low ductility.

There are many method available for fabrication of composites materials but stir casting method seems to be best among all because it gives better result as compared to others and it is simple also. For mass production of materials such as MMCs stir casting method seems to be more useful. In Aluminium metal matrix composites Aluminium is taken as matrix phase and the component or material which is enclosed in it is taken as reinforcement phase. The materials taken in matrix phase have a continuous character and is more ductile from the reinforcement phase component while component or material taken in reinforcement phase have a discontinuous character and is usually stronger than the matrix phase.

Fly ash is widely used as a reinforcement phase in thermal power plants because of its low density and inexpensiveness. Normally Fly ash includes some amount of calcium oxide (Cao), aluminium oxide (Al2o3) as well as Silicon dioxide (Sio2). It has very low density so when it is mixed with molten metal overall weight and density of the material is reduced. Apart from MMCs it has many more applications in different area such as in bricks, Asphalt concrete, Geopolymers and can also be used as catalyst for converting polyethylene to crude oil.

In this research work Aluminium (LM5) is reinforced with fly ash to give a strong, light weight and high stiffness material. For effortless and economical method, stir casting method has been used. Many tests were also conducted to the new material in the lab and many results were found out.

2.MATERIAL SELECTION

LM5 has been used as a matrix phase and fly ash has been taken as reinforced phase. For satisfying the need of strong and better material many procedure has to be done for selecting the material. Because of good corrosion resistance and high strength LM5 is being used in many manufacturing industries. For knowing better constituent of material number of constituents have been taken which is shown in table1.

Table 1. Constituents of materials for reinforcement

Sl.No.	Aluminium (%)	Fly ash (%)
1	95	5
2	90	10
3	85	15
4	80	20
5	75	25

3. TESTING and RESULTS:

After Reinforcing Aluminium and Fly ash a new material has been formed which has been tested with different methods to know about its strength as well as its hardness value. Five types of specimens was created with different proportion of Aluminium and Fly ash and was tested. Figure 1 and 2 shows front view and top view of two samples created from stir casting technique.

Fig 1. Front view of samples created from stir casting techniques

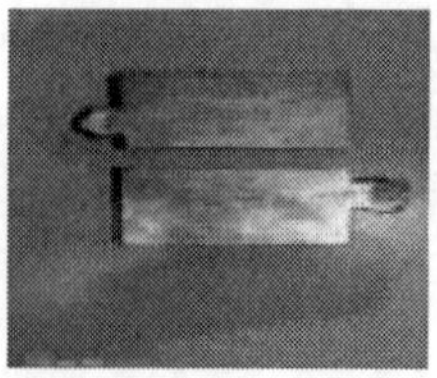

Fig 2. Top view of samples created from stir casting techniques

Figure 3 shows Rockwell hardness testing machine in which samples has been tested. Many more testing also has been done in Tensile as well as charpy test.

Fig 3. Rockwell hardness testing machine

Many tabulation has been formed from the results taken from different types of testing. Table 2 shows ultimate strength values for all the different specimens. All the specimens have been tested in the testing machine present in our laboratory.

Table 2. Ultimate strength values for specimen

S.N	Aluminium %	Fly ash%	Ultimate strength(Mpa)
1	95	5	678
2	90	10	698
3	85	15	725
4	80	20	744
5	75	25	756

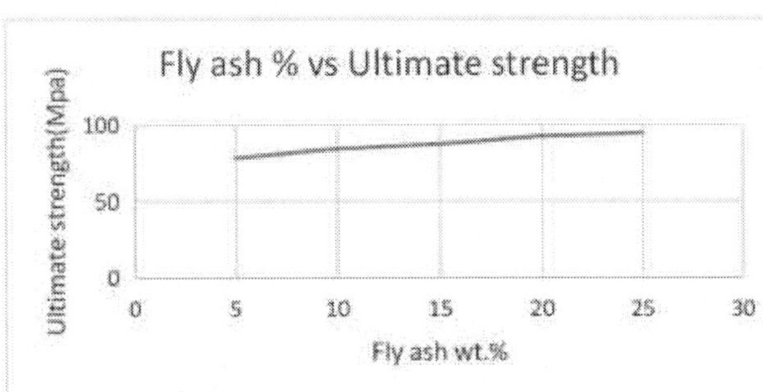

Fig 4. Graph of fly ash percentage vs ultimate strength

Figure 4 shows the graph plotted between Ultimate strength and Fly ash Weight percentage. From the graph it has been found that as increasing the fly ash weight percentage in our material ultimate strength is also increasing. From this we can conclude that Fly ash weight percentage is directly proportional to Ultimate strength. Table 3 shows the value of hardness which has been calculated from hardness testing machine.

Table 3. Value of hardness from hardness testing machine

S.N0.	Aluminium%	Fly ash %	Hardness(HRA)
1	95	5	78
2	90	10	84
3	85	15	87
4	80	20	92
5	75	25	94.4

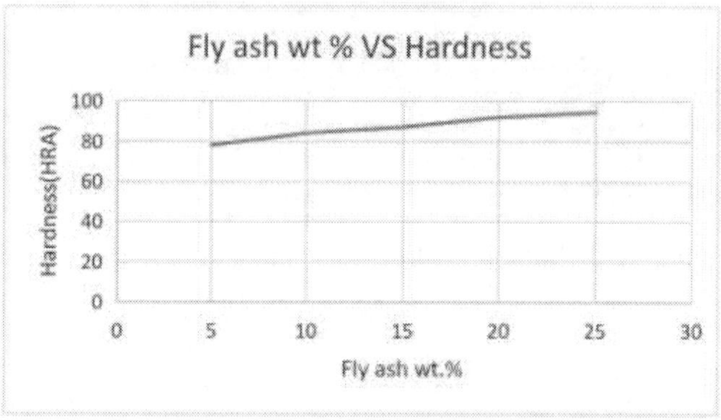

Fig 5. Graph of Hardness vs Fly ash weight percentage

Figure 5 shows the graph plotted between Fly ash weight percentage and Hardness. Similarly, like previous one in this also we have seen that while increasing Fly ash weight percentage Hardness of the material is also increasing. After hardness test toughness test has also been done to the material. Charpy test has been done on the material to determine its toughness. Table 4 shows the value of toughness that has been calculated by charpy test to each specimen.

Table 4. value of toughness for individual element from charpy test

S.NO.	Aluminium %	Fly ash %	Toughness(j/m)^3
1	95	5	71.2
			74
			78
			71
			81
2	90	10	130
			127
			122
			118
			116

From all the above results we have found that as we increase the amount of Fly ash in Aluminium metal matrix composites, Ultimate strength, Hardness as well as Toughness of the material also increases. So we can say that increasing of percentage of fly ash leads to improve the material quality. Different graphs have also been plotted to show the variation of values of our results.

4. Conclusion

The concern research work is done for the production of aluminium matrix composites with LM5 as matrix material and fly ash as reinforcement material at 0,5,10% by weight using stir casting materals. Here we have selected the materials and analyse uts constituents and reinforced them. The materials formed by reinforcment of different materials constituent was tested and strength and hardness are analysed. Rockwell hardness testing machines are used for the testings and ultimate strength values are determinesd for individual materials. Some more hardness testing and strength testing are done by charpy test. From all the reinforcements , testings and determined and nanalysed data of hardness , toughness and strength we came to the conclusion that increasing amount of fly ash in aluminium metal matrix composites lead to improve the material quality.

5. References

[1] Ranjan T P, Pillai M R, Pal B C, Satyanarayana K G and Rohtagi P K 2001 *Proc. of national conf. on Recent Advances in materials processing (RAMP- 2001)* India pp 327-324

[2] Sudipta kumar and Ananda Theertan 2008 Production and characterization of aluminium -fly ash composite using stir casting method, NIT , Rourkela

[3] Suman kant and Ajay Singh Verma stir casting process in particulate aluminium metal matrix composite : A review, *Int. J. of Mechanics and solids*

[4] Surappa M K 2003 Aluminium matrix composites challenges and opportunities Sadhana 28(1-2),319-334.

[5] Jims G, John Wessley, Saidu Srinivas and Azgar Ali M D 2019 Preparation and characterization of an aluminium 6061 alloy based metal matrix composite

[6] Sachin Mohal Microstructural investigation of aluminium silicon carbide particulate metal matrix composite Fabricated by stir casting

[7]https://www.sciencedirect.com/science/article/pii/S2214785315005271

[8] Rajsekhar Gangaram Bhandare and Parshuram M. sonawane 2014 Preparation of aluminium matrix composite by using stir casting method & its characterization.

CRASH TEST ANALYSIS OF BUMPERS OFAUTOMOBILES USING LS-DYNA

Aman Chaure [1], Gaurav Mathur [2], Narendiranath Babu T[3]

[1] School of Mechanical Engineering, Vellore Institute of Technology, Vellore, India, amanchaure@gmail.com
[2] School of Mechanical Engineering, Vellore Institute of Technology, Vellore, India, gauravmathgm@gmail.com
[3] School of Mechanical Engineering, Vellore Institute of Technology, Vellore, India, narendiranathbabu.t@vit.ac.in

Abstract. This article deals with calculations of deformation and failure of the bumpers of vehicles at different speeds against different obstacles. The main aim of bumpers is to absorb the impact energy in the form of stain energy and efficiently transfer it to the chassis to avoid permanent deformation of the whole frame of any motor vehicle. Different materials have been tried out with and without supporting structures (Ribs) which are known to increase the overall strength of the bumpers while keeping the weight to a minimum. In this experiment, 3 metals namely Sheet Iron, Aluminum and Titanium of varying thickness. Using LD-DYNA 108 cases were implemented. If the component passes the test, the Factor of Safety, Max Stress and the Trends in Variation of these parameters were calculated, the Max Energy Absorbed, Transmitted, the Time required for Failure are extrapolated from the data. Comparisons are drawn between failure, with and without support structures. In the end, overall effectiveness of the support structures in enhancing the load transfer, absorbing strain energy and increasing factor of safety while keeping the mass in check. *Keywords: Car Bumpers, Crash Analysis, LS-Dyna.*

1. INTRODUCTION

Automotive industries have constantly focused on weight reduction to improve the mile to fuel efficiency of the vehicles. This method causes serious damage to the safety of the passengers in the events of a crash. In this study, we have considered three materials models namely Grade 5 Titanium (Ti-6Al-4V) [1], Stainless Steel (SS-304)[2],[4],[7] and Al 99.96 percentage (Pure Annealed)[11]. Using LS-Dyna, multiple simulations were done at different speeds (40, 50, 60 Kmph). Considering strength and bending stiffness, the bumper with various thickness was designed. The bumper was tested against two different obstacles (Wall and Pole). Simulation of a total of 54 cases yielded enough data to redesign the bumper and repeat the same experiment on the second iteration, the results are discussed below. In the recent years, Bumpers are being made with plastic to reduce the weight.

2. METHODOLOGY

2.1 Experiment Design

Bumpers are amongst the first components to fail in an event of a full-frontal crash. Thinner metal sheets with extra supporting structures in place, can replace plastic sheet bumpers. In this study, we have used three different materials as a replacement of the plastic bumpers. The materials used were SS-304[2],[4],[7], Al 99.96 percentage [11], Ti-6Al-4V.[6],[10]

Table 1. Properties of the materials used in Simulations

Sr. No	Material Model	Young's Modul	Density	Poisson's	Strength Coefficie nt (GPa)	Hardening Exponen	Plastic Failur
1	Aluminum 99.96% Pure Annealed[11	68.94	2.685e-06	0.33	0.0974	0.33587	0.50
2	Titanium	113.8	4.430e-06	0.342	1.564608	0.18231	0.14

	(Ti6Al4V)[6][10]						
3	Stainless Steel (AISI-304)[2][4][7]	207	7.830e-06	0.3	1.332	0.395	0.70

The bumper is tested against a wall and a pole (both marked as completely rigid in all cases), at three different speeds (40, 50, 60 Kmph). A second iteration is based on the stress acting and the design is not considerably changed except for adding ribs at probable points of failure. After creation of 108 cases, the simulations were conducted to yield various energy plots. Stress, Strain &Deformation Contours.

2.2 Design

The bumper design included all the required features in a standard car. The novelty in our design that to improve the bending stiffness we have made a corrugated design. Corrugations and Ribs are the design additions that tend to improve the strength parameters. A plastic coating to hide corrugations will keep the visuals intact and reduce the weight of the bumpers; we have considered three thickness in our study (5mm, 9mm and 13mm). The Corrugations and the ribs increase the bending stiffness of the design without increasing much weight. Thus overall, the strength to weight ratio is optimised. The locations of the rib were decided on the basis of the stress concentration locations and their respective levels. Figure 1 shows the Bumper design without ribs. The main dimensions of the design are shown in the Figure 2.

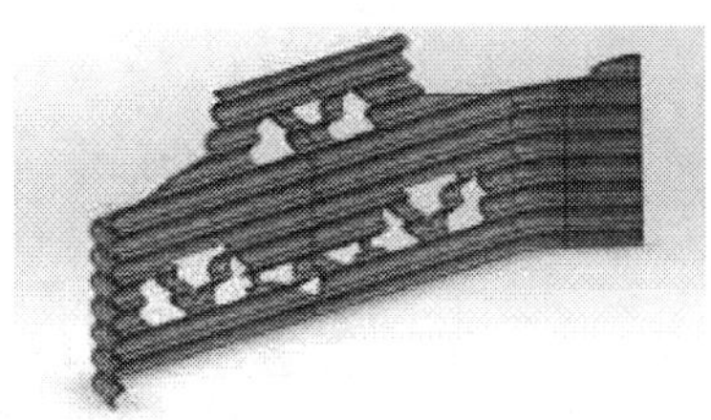

Figure 1. Isometric View of the design without Ribs.

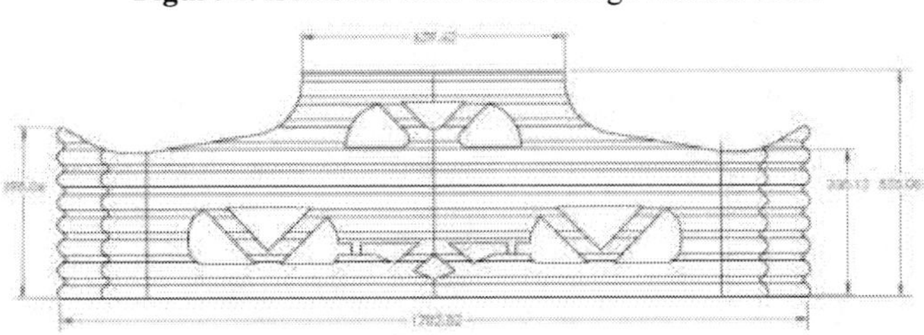

Figure 2. Dimensions of the bumper in mm.

2.3 Iteration 2

After careful examination of the stress distribution and stress propagation, several locations were noted down and at those places; ribs were added to the design along the direction of bending to increase the bending stiffness. Effects due to the changes in the design were studied and plotted. All the calculations were done to evaluate the Strain energy stored in the design when it crashed with the two common kinds of obstacles namely wall and pole. Figure 5 shows the modified design of the bumper along with the ribs. Figure 3 & 4 shows the locations of maximum stress.

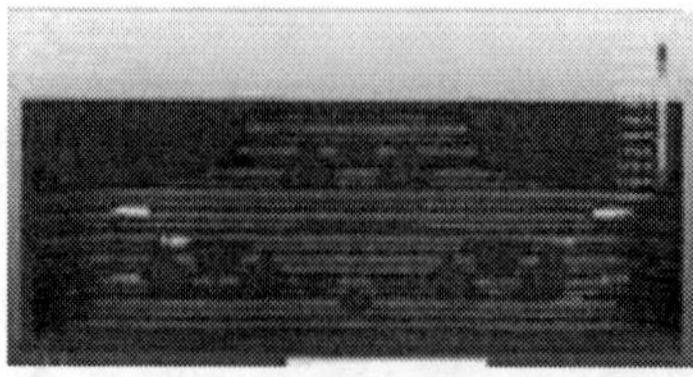

Figure 3. Front View of the Stress Developed

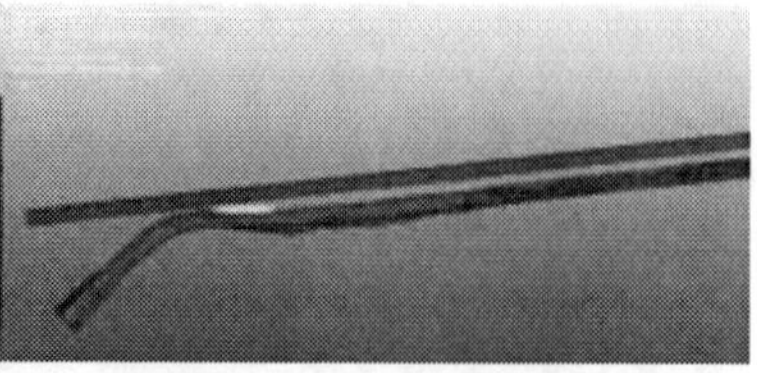

Figure 4. Top View of the Stress Developed

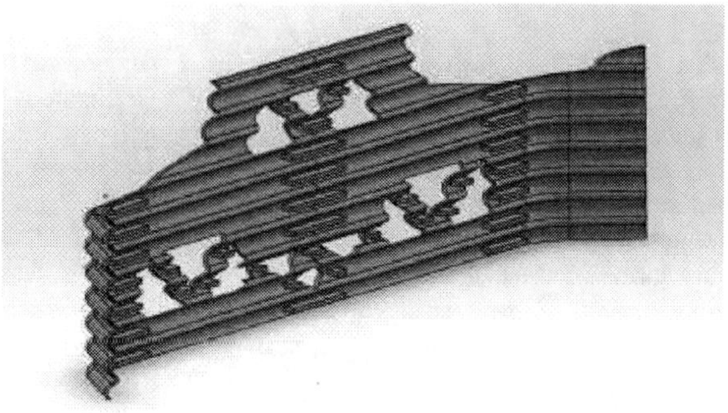

Figure 5. Modifications in the Design to Reduce Stress

2.4 Method of simulation

Firstly, the meshed model was created in the 'HYPERMESH' software, then this model was exported as a '.k' file. The material model was set to MAT-018 (power law plasticity) for bumper in the LS-PrePost Software [9]. All three materials (Aluminum, Stainless Steel, Titanium) were defined under the same category along with the properties given in the Table 1. The Material model MAT-020 (Rigid) [12] was used for the obstacle, since we didn't want to increase the simulation time for calculating the deformations in the obstacle along with same steel material. The section was defined as solid and both Bumper and Obstacle. Then the boundary conditions such velocity & fixed support were defined respectively for Bumper and the obstacle. At last the Termination time was iteratively determined to be around 4-8 milliseconds, for the wide range of simulations. The plots for Nodal Displacements and Forces under 'ASCII-rcforc' & D3PLOT were considered for calculating the work done and strain energy development calculations.

$$\sigma_y = k\varepsilon^n = k\left(\varepsilon_{yp} + (\bar{\varepsilon})^p\right)^n$$

Here, σ_y=yield Stress; ε_{yp}= Elastic strain; ( ) = effective

plastic strain.

2.5 Meshing

The meshing was done in Hyperworks with the given Specification data. The R-triad type was used for getting propert geometric Curvatures into account. It was observed that in the given size for the element the geometry was matched within limit and mesh senstivity was conserved. Table 2 contains the parameters of the Meshing. Figure 6 & 7 shows the meshing of the bumper and the obstacles.

Table 2. Meshing Parameters

Sno	**Part Name**	**Mesh Type**	**2D-Type**	**3D-Type**	**Min Ele**	**Feature Angle**	**Max Element Size**	**No.of Element**	**No of Nodes**
1.	5mm	V-Tetra	R-triad	Tetras	6	30	10	134062	45953
2.	9mm	V-Tetra	R-triad	Tetras	6	30	10	113865	36724
3.	13mm	V-Tetra	R-triad	Tetras	6	30	10	133983	39948
4.	Pole	V-Tetra	R-triad	Tetras	50	30	50	228	105
5.	Wall	V-Tetra	R-triad	Tetras	100	30	100	1068	396

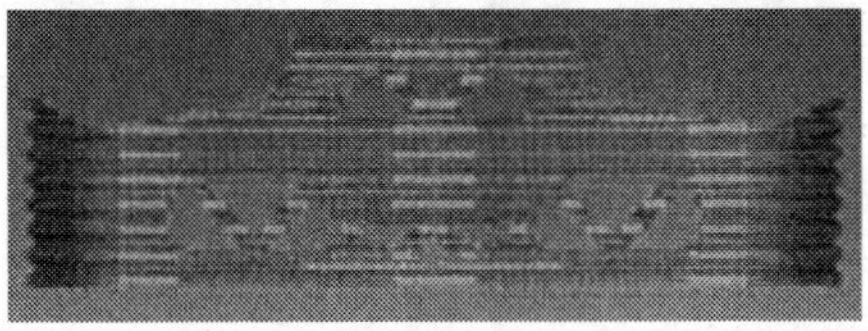

Figure 6. Meshing Image of the Bumper

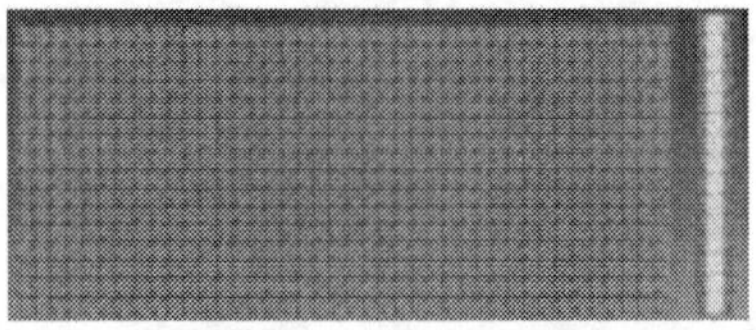

Figure 7. Meshing Image of Obst

2.6 Simulation Parameters

Table 3. Simulation Parameters.

Simulation	Material	Speed	Barrier	Thickness	With	Sim	Without
1	Al 99.96%	40	Wall	5	W	55	W/O
2	SS-304	40	Wall	5	W	56	W/O
3	Ti6Al5V	40	Wall	5	W	57	W/O
4	Al 99.96%	50	Wall	5	W	58	W/O
5	SS-304	50	Wall	5	W	59	W/O
6	Ti6Al5V	50	Wall	5	W	60	W/O
7	Al 99.96%	60	Wall	5	W	61	W/O
8	SS-304	60	Wall	5	W	62	W/O
9	Ti6Al5V	60	Wall	5	W	63	W/O
10	Al 99.96%	40	Pole	5	W	64	W/O
11	SS-304	40	Pole	5	W	65	W/O
12	Ti6Al5V	40	Pole	5	W	66	W/O
13	Al 99.96%	50	Pole	5	W	67	W/O
14	SS-304	50	Pole	5	W	68	W/O
15	Ti6Al5V	50	Pole	5	W	69	W/O
16	Al 99.96%	60	Pole	5	W	70	W/O
17	SS-304	60	Pole	5	W	71	W/O
18	Ti6Al5V	60	Pole	5	W	72	W/O
19	Al 99.96%	40	Wall	9	W	73	W/O
20	SS-304	40	Wall	9	W	74	W/O
21	Ti6Al5V	40	Wall	9	W	75	W/O
22	Al 99.96%	50	Wall	9	W	76	W/O
23	SS-304	50	Wall	9	W	77	W/O
24	Ti6Al5V	50	Wall	9	W	78	W/O
25	Al 99.96%	60	Wall	9	W	79	W/O
26	SS-304	60	Wall	9	W	80	W/O
27	Ti6Al5V	60	Wall	9	W	81	W/O
28	Al 99.96%	40	Pole	9	W	82	W/O
29	SS-304	40	Pole	9	W	83	W/O
30	Ti6Al5V	40	Pole	9	W	84	W/O
31	Al 99.96%	50	Pole	9	W	85	W/O
32	SS-304	50	Pole	9	W	86	W/O
33	Ti6Al5V	50	Pole	9	W	87	W/O
34	Al 99.96%	60	Pole	9	W	88	W/O
35	SS-304	60	Pole	9	W	89	W/O
36	Ti6Al5V	60	Pole	9	W	90	W/O
37	Al 99.96%	40	Wall	13	W	91	W/O
38	SS-304	40	Wall	13	W	92	W/O
39	Ti6Al5V	40	Wall	13	W	93	W/O

40	Al 99.96%	50	Wall	13	W	94	W/O
41	SS-304	50	Wall	13	W	95	W/O
42	Ti6Al5V	50	Wall	13	W	96	W/O
43	Al 99.96%	60	Wall	13	W	97	W/O
44	SS-304	60	Wall	13	W	98	W/O
45	Ti6Al5V	60	Wall	13	W	99	W/O
46	Al 99.96%	40	Pole	13	W	100	W/O
47	SS-304	40	Pole	13	W	101	W/O
48	Ti6Al5V	40	Pole	13	W	102	W/O
49	Al 99.96%	50	Pole	13	W	103	W/O
50	SS-304	50	Pole	13	W	104	W/O
51	Ti6Al5V	50	Pole	13	W	105	W/O
52	Al 99.96%	60	Pole	13	W	106	W/O
53	SS-304	60	Pole	13	W	107	W/O
54	Ti6Al5V	60	Pole	13	W	108	W/O

3. RESULTS AND DISCUSSION

3.1 Strain Energy

Energy absorbed by Aluminum without ribs is much less than the Ribs case. All the materials in all the cases follow this trend. The maximum energy absorbed is 16.73 KJ by a Stainless Steel Bumpers. The cumulative strain energy absorbed by bumper without ribs is 0.450 KJ while the energy absorbed by the bumper is 3.5 KJ. There is almost 600% increase in Strain energy absorbed.

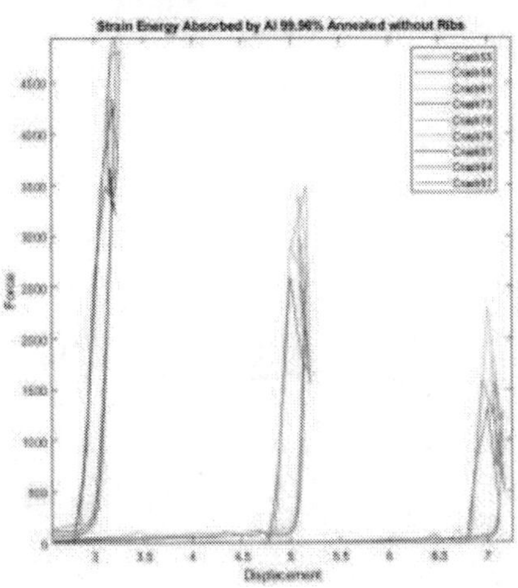

Figure 8. Strain Energy absorbed by Al Al Without ribs (Wall)

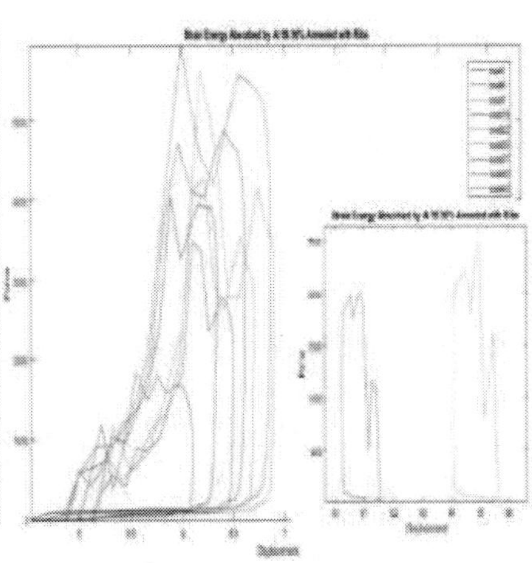

Figure 9. Strain Energy absorbed by With Ribs (Wall)

Figure 8, Figure 9 represent Strain energy absorbed by Al bumper without and with ribs respectively. The maximum force exerted is almost the same in both the cases, but the displacement is higher the case of ribs, thereby increasing strain energy absorbed. The crash number can be referenced from above for particular case. It is a common observation that the common trend is as the speed increases the lesser energy is absorbed and more amount of energy is transmitted. The lower the thickness the lesser energy is absorbed.

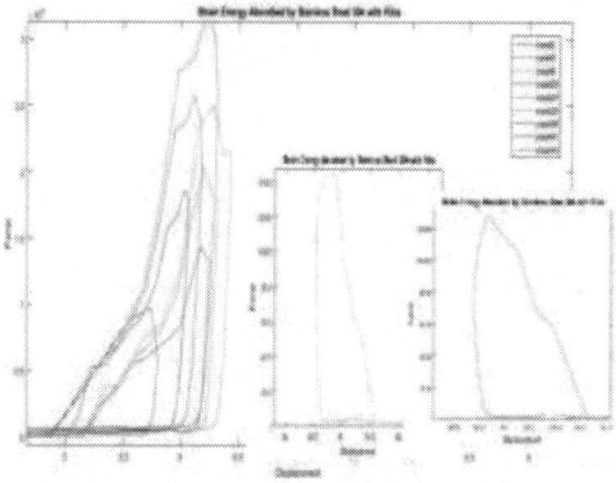

Figure 10. Strain Energy absorbed by SS absorbed Without ribs (Wall)

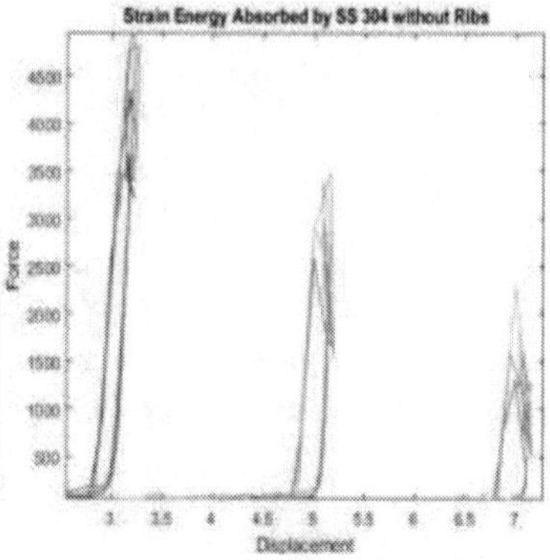

Figure 11. Strain Energy with ribs (Wall)

Figures 10 and 11 represent energy absorbed by SS-304; all the trends followed by aluminum is followed by SS.

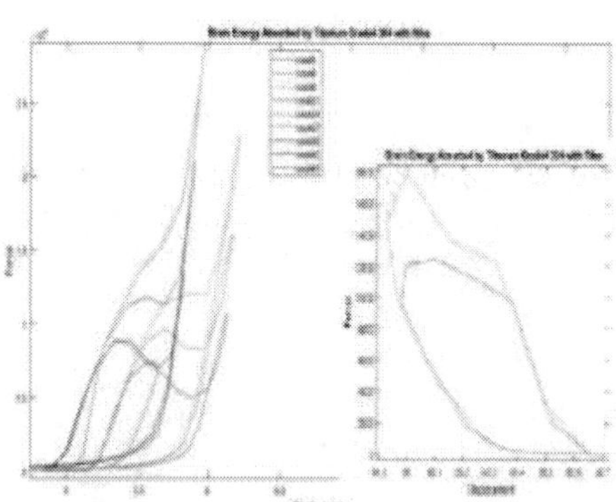

Figure 12. Strain Energy absorbed by Ti absorbed Without ribs (Wall)

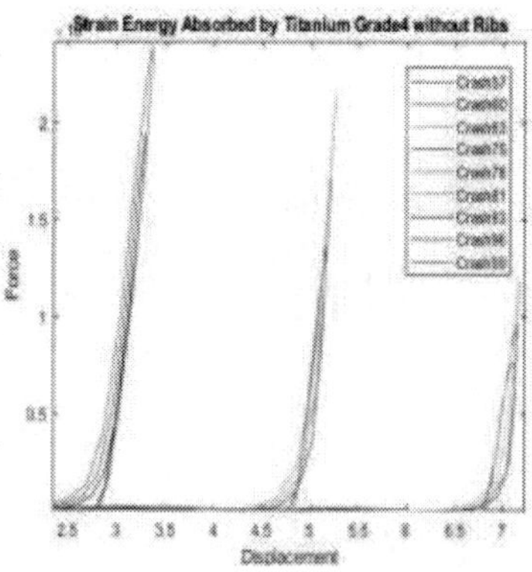

Figure 13. Strain Energy by Ti with ribs (Wall)

Figures 12 and 13 represent the crash by bumpers made of titanium with and without ribs. All the trends followed by Al and SS are followed by Ti only in the case Titanium without ribs. However, Titanium with ribs presents a peculiar case; the thickness does not really matter in this case. The work done is slightly increasing mostly due to the stiffness of the material.

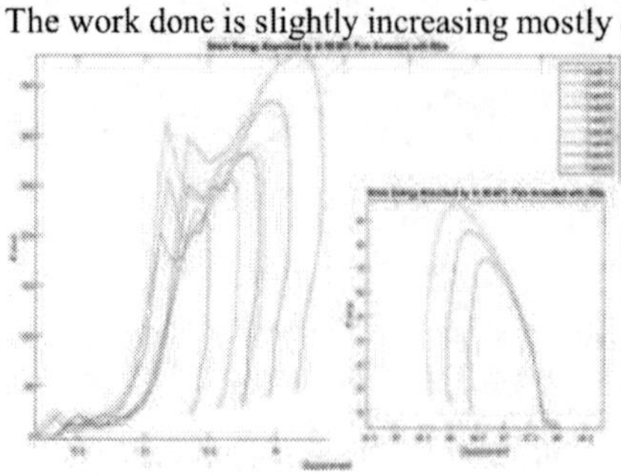

Figure 14. Strain Energy absorbed by Al absorbed Without ribs (Pole)

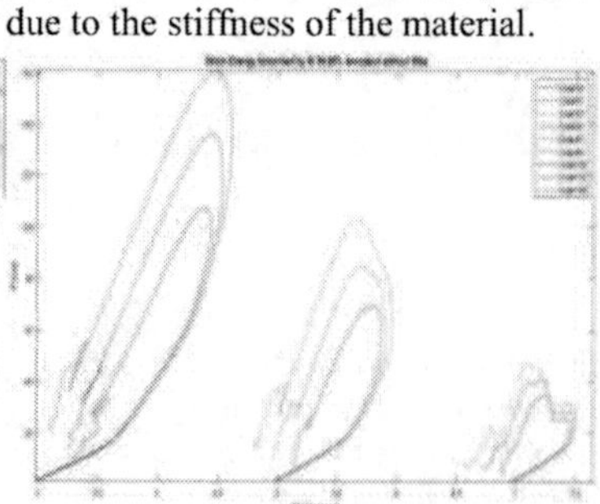

Figure 15. Strain Energy by Al with ribs (Pole)

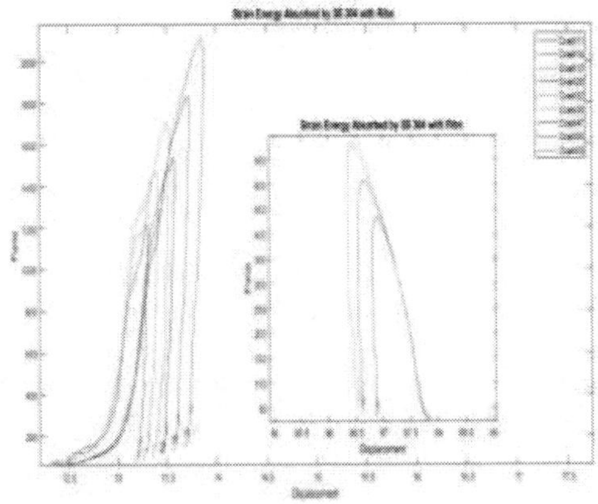

Figure 16. Strain Energy absorbed by SS absorbed Without ribs (Pole)

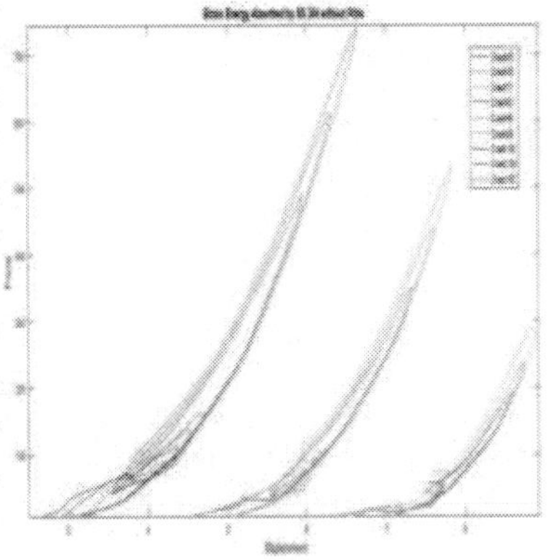

Figure 17. Strain Energy by SS with ribs (Pole)

Figures 14 to 17 represent crashes made by Aluminum and Stainless Steel. All the trends observed in wall cases all followed by the aluminum and Stainless steel.

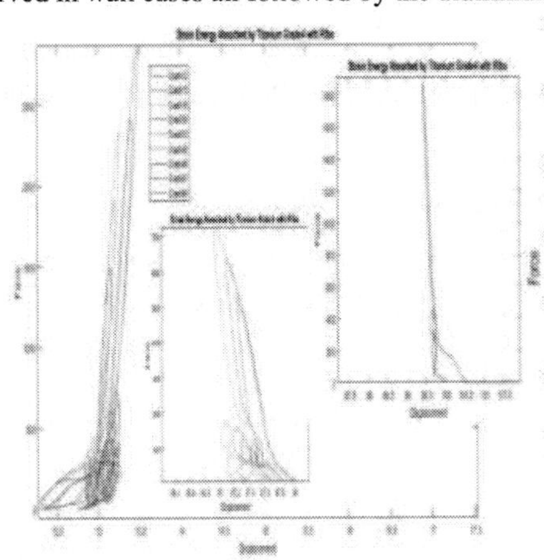

Figure 18. Strain Energy absorbed by Ti absorbed Without ribs (Pole)

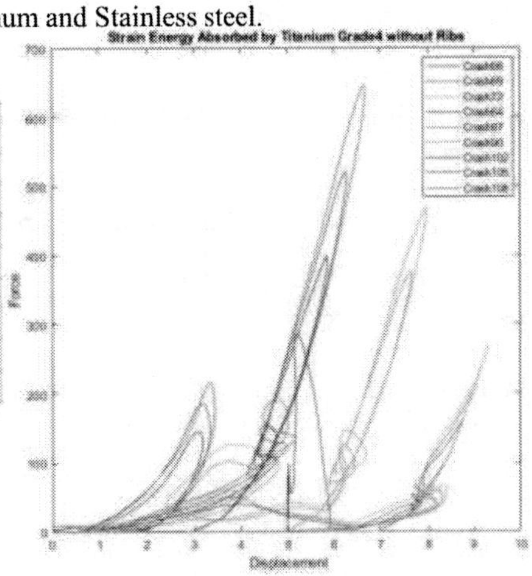

Figure 19. Strain Energy by Ti with ribs (Pole)

Figures 18 and 19 represent Titanium colliding with a pole; the bumper starts to vibrate just as it collides, the loops represent vibrations on the bumper. All the other trends followed by all cases are followed by this case too.

3.2 Stress Contours

4. FUTURE WORK

The study can be expanded to plastics and composites, Composites can be used to reduce the weight, as their stiffness to mass ratio is high. Plastics are currently being used in the industry. Other factors also effect the bumpers, Cost of manufacturing and thickness of the materials should be considered in real time.

5. CONCLUSIONS

Thin Stainless Steel or Aluminum sheets can be used to make bumpers to improve the safety of the cars. Metals are stiffer than plastics and can be used as an efficient alternative to current plastics. Titanium can be used, which can reduce the weight of the bumper considerably and can easily absorb the force by vibration.

6. REFERENCES

[1] OdenbergerB L, Oldenburg M and Thilderkvist P 2011Tool development based on modelling and simulation of hot sheet metal forming of Ti–6Al–4V titanium alloy. *Elsevier J. of Materials Processing Tech.* Vol211, Issue 8, pp 1324-1335

[2] Amit Kumar Gupta, Hansoge Nitin Krishnamurthy and Yashjeet Singh 2013 Development of constitutive models for dynamic strain aging regime in Austenitic stainless steel 304, Materials & Design, March Vol 45, pp 616-627.

[3] DYNAmore GmbH. LS-Dyna Support. 2019. https://www.dynasupport.com/howtos/general/consistent-units [Accessed 11 Nov 2019].

[4. Azom , Azo Materials 2001 https://www.azom.com/properties.aspx?ArticleID=965 [Accessed 11 November 2019]

[5] Callister Jr and William D Fundamentals of Materials Science and Engineering (2nd ed.), United States of America: John Wiley & Sons, pp 199, ISBN 978-0-471-47014-4

[6] Matweb, ASM Aerospace Specification Metals Inc. http://asm.matweb.com/search/SpecificMaterial.asp?bassnum=MTP641[Accessed 11 Nov 2019]

[7] Matweb, ASM Aerospace Specification Metals Inc. http://asm.matweb.com/search/SpecificMaterial.asp?bassnum=mq304a[Accessed 11 Nov 2019]

[8]. Matweb, ASM Aerospace Specification Metals Inc. http://asm.matweb.com/search/GetReference.asp?bassnum=mq304a[Accessed 11 Nov 2019]

[9]. http://www.lstc.com/dynamat/pdfs/mat_018_theory.pdf

[10]. Development of a Titanium Alloy Ti-6Al-4V Material Model Used in LS-DYNA. http://www.tc.faa.gov/its/worldpac/techrpt/tc15-23.pdf

[11]. http://www.varmintal.com/aengr.htm

[12] http://www.lstc.com/dynamat/pdfs/mat_020_theory.pdf

DESIGN AND ANALYSIS OF ALUMINUM MATRIX COMPOSITE PISTON

M. Saleem [1], Shiva Sam Kumar. G [2], A. Ashique Rosh Suresh[3]

[1] Department of Aeronautical Engineering, Veltech Rangarajan Dr.Sagunthala R&D Institute of Science & Technology, Chennai, India

[2] Department of Aeronautical Engineering, Veltech Rangarajan Dr.Sagunthala R&D Institute of Science & Technology, Chennai, India

[3] Department of Aeronautical Engineering, Veltech Rangarajan Dr.Sagunthala R&D Institute of Science & Technology, Chennai, India

Abstract. The Piston is meant to be the maximum load withstanding structure in an internal combustion engine. The Piston of an IC engine is assigned to transform thermal energy into mechanical energy by expanding gases and to sustain for a longer period. To increase the lifetime and thermal resistance of the piston, the material should have higher strength to density ratio and lower coefficient of thermal expansion. Thus, the study of reinforced Aluminum metal (A2024 alloy) as a frame and organic compound (fly ash) as a filler or matrix (metal matrix or hybrid composites), which has higher strength to weight ratio is done and this paper deals with the determination and study of structural and thermal stress distribution on Piston made of Aluminum Matrix Composite. A piston model is prepared in CATIA V5R20 (Modelling software) and the properties is examined using CAE software (ANSYS).**Keywords:** AMCs, A2024, engine, energy, Ansys

1. INTRODUCTION

Piston is the most crucial part in an engine. The advancement and development in the material selection geometry and manufacturing have been drastically increasing from past few years. The purpose of piston in an internal combustion engine is to transfer the maximum amount of force extracted from the expanding of gases in the cylinder to the crank shaft through the connecting rod [3,6]. A piston engine is said to be sustainable and efficient if the piston is capable of transferring the maximum amount of force extracted from expanding of gases in the cylinder to the crank shaft with in a suitable short period of time and do the same for a longer run with minimal wear and tear and minimal deformation. In this research, a piston is made up of Aluminum metal matrix (light weight metal matrix) [1,2,4]. Essential and increasing advancement in the development of light metal composite framework has been attained in the last decades and thus the advancements could be introduced into most of the industrial applications. Metal matrix (MMC's has been utilized monetarily as reinforced piston [7-9].

Thus, the study of reinforced Aluminum metal (A2024 alloy) as a frame and organic compound (fly ash) as a filler or matrix (metal matrix or hybrid composites) which has higher strength- to-weight ratio and high stress withstanding structure [5,10]. This paper deals with the determination and study of temperature distribution, structural and thermal stress distribution on piston made of Aluminum matrix composite.

2. MATERIAL SELECTION

The composite material is fabricated by stir casting method with Aluminum 2024 (A2024) alloy as a reinforcement metal and fly ash (organic compounds) as filler or matrix. Stir casting is the method of fabricating composite material in liquid state, where the dispersed phase is mixed with the matrix metal (molten state) by means of mechanical stirring. Aluminum matrix composites have strength and are highly ductile. The properties of metal matrix composite has been tabulated down in Table 1 .

Table 1. PROPERTIES OF THE MATERIAL USED FOR PISTON

SR. No.	PARAMETER	VALUE
1.	Density	$2.2\frac{g}{(cm)^3}$
2.	Tensile strength	$227\frac{N}{(mm)^2}$
3.	Elongatio	0.80
4.	Poison's	0.22
5.	Young's modulus	185.3Gpa
6.	Thermal conductivity	$156\frac{W^{-1}}{K}$ m

3. DESIGN OF PISTON

A prototype of piston is designed in CATIA V5R20 modelling software as it an easy modelling tool in which files can be easily exported to any analysis software in igs format. Figure 1. shows the front view of the piston with dimensions whereas Figure 2. demonstrates the isometric view of piston model in CATIA V5 R20 modeling software.

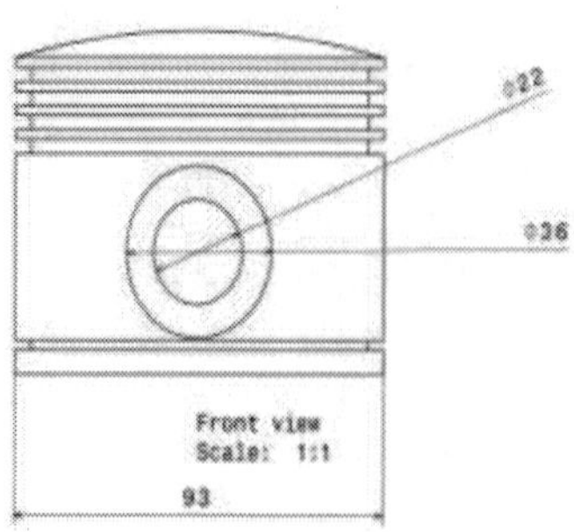

Figure 1. Front view of piston

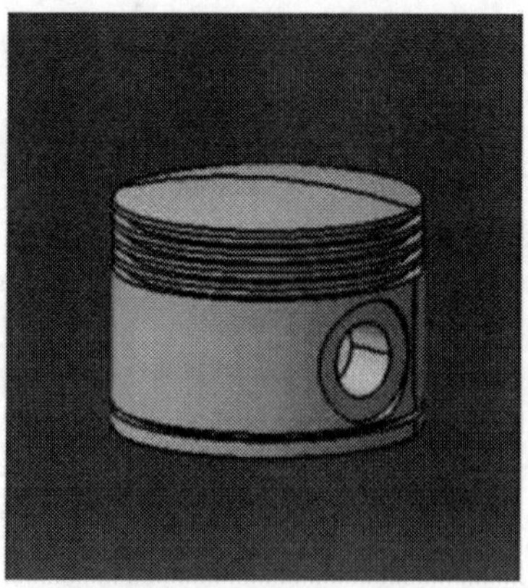

Figure 2. Isometric view

IOP Conf. Series: Materials Science and Engineering 1123 (2021) 012008 doi:10.1088/1757-899X/1123/1/012008

4. ANALYSIS

Analysis of temperature distribution, thermal stress and stresses acting in static condition on piston. Analysis of any material is done by the following process:

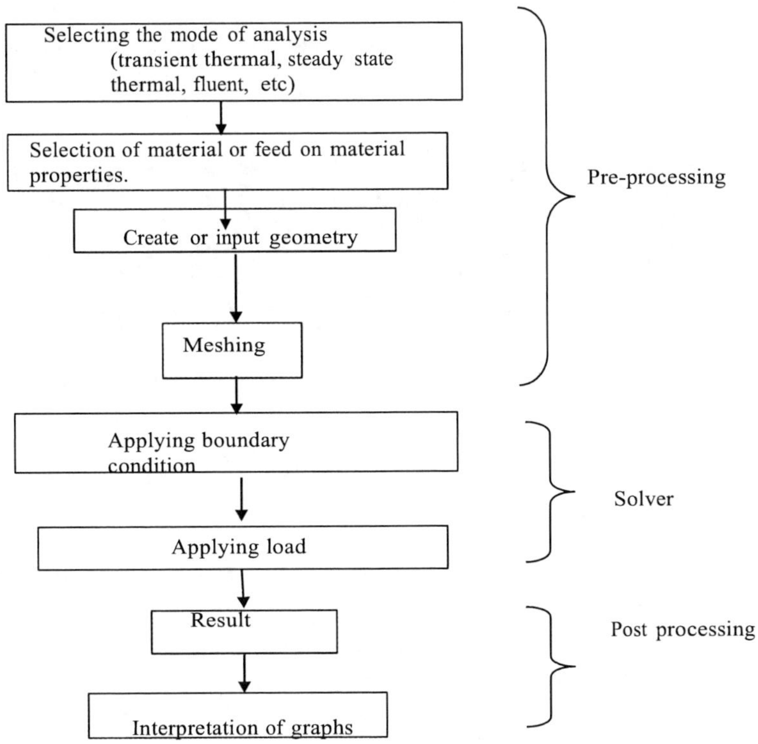

This type of analysis is called as FEA (Finite Element Analysis).

4.1 STATIC STRUCTURAL ANALYSIS

The material properties of the metal matrix as given in the table has been feed to the imported CATIA geometry file and is ready for discretization.

4.2 MESHING

A fine mesh with 100 percent relevance and fast transition has been applied. The piston has been divided into 499529 nodes and 336082 elements as shown in Figure 3.

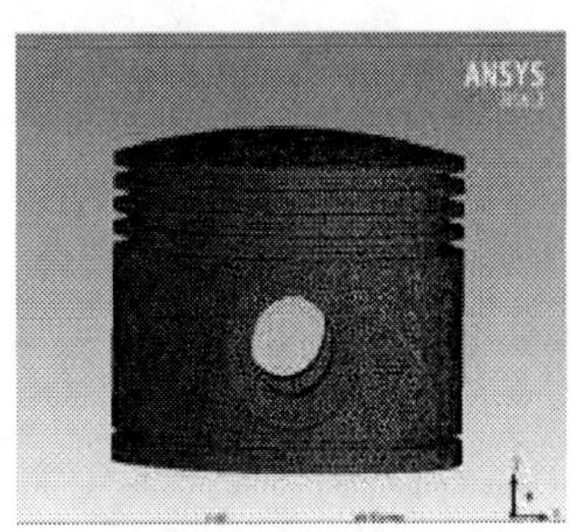

Figure 3. Fine mesh

4.3 BOUNDARY CONDITION

The surface of the piston and piston pin hole are subjected to frictionless support. When a face is subjected to frictionless support, the transitional and rotational motion normal to the face is locked. In this paper, we consider the effect of side thrust face to be negligible but in reality, it has some impact on deformation and stress on the piston surface, but pressure forces are taken in record and uniform temperature is maintained throughout the analysis. 9Mpa of force is applied on the piston crown as shown in the figure 5.

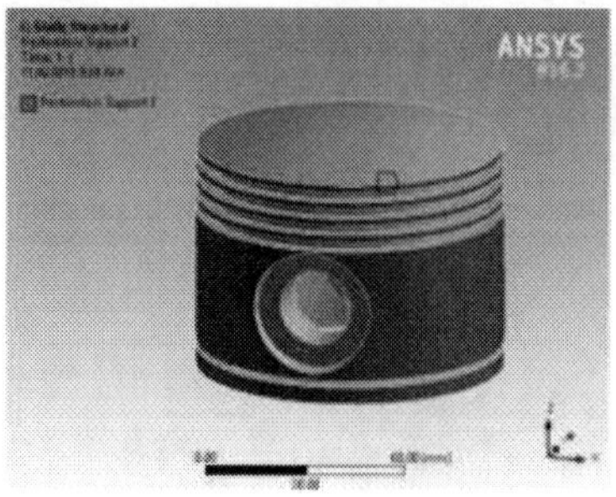

Figure 4. Frictionless support

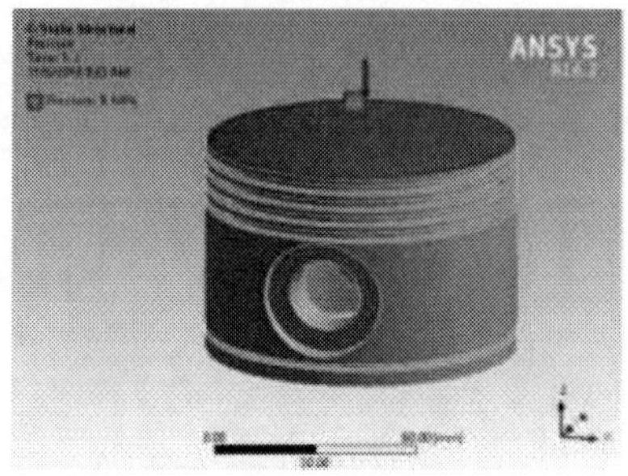

Figure 5. 9 Mpa pressure load

4.4 THERMAL ANALYSIS

Thermal analysis of Aluminum matrix piston is done in ANSYS to find the suitability at maximum temperature. The maximum and minimum temperature values are set as 1500°C and 22°C respectively and convection film coefficient of $1.2 * \frac{10^{-3}W}{(mm)^2}°C$ is applied.

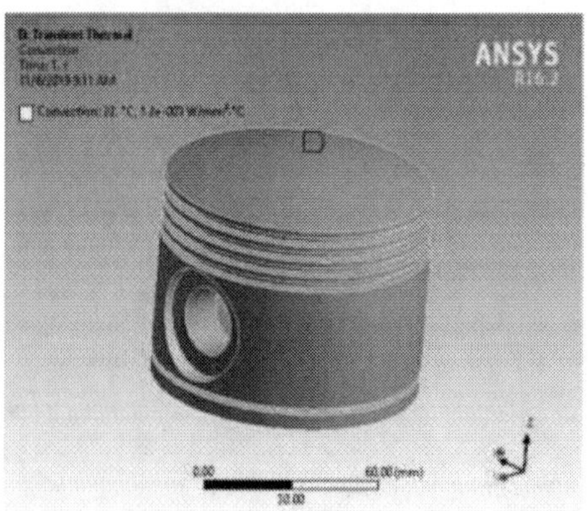

Figure 6. Convection film coefficient

4.5 THERMAL STRESS

A hollow circular cylinder of some cross section as metal matrix composite piston is modelled with 10mm depth. Thermal stress analysis is done in static structural solver by applying the boundary conditions on inner and outer face of cylinder as 200°C and 300°C respectively. The outer face of the cylinder is locked to 6 degrees of freedom and a pressure load of 500Mpa is applied on the inner wall of piston. The figure 8. shows the boundary and loading conditions over the aluminum metal matrix composites.

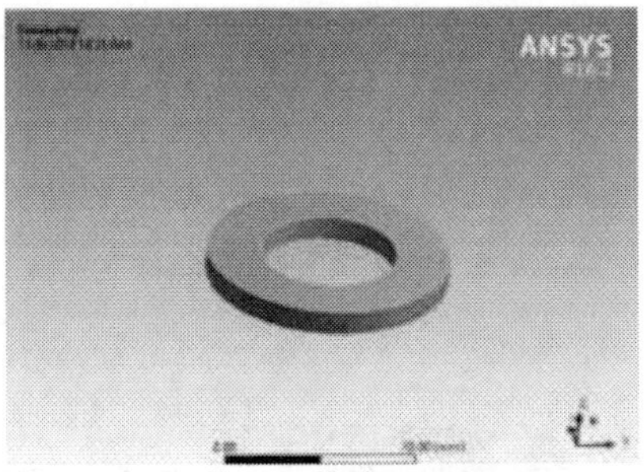

Figure 7. Hollow cylinder geometry

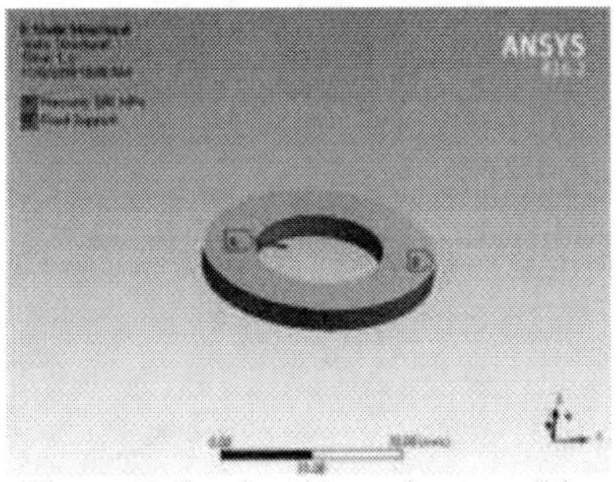

Figure 8. Load and boundary conditions

5. RESULTS AND DISCUSSONS

5.1 STATIC STRUCTURAL ANALYSIS

The static structural analysis is used to find out the total deformation, von misses, stresses and strains. The static analysis of Aluminum matrix piston is done by ANSYS software for the combustion pressure of 9Mpa. The results of total deformation, equivalent stress and elastic strain has been shown from Figure 9-11.

The maximum deflection induced in the piston is at top of the piston. The maximum von misses stress is developed in the inner boss fillet area which is due to the concentrated stress and application of pressure on the top surface of the piston. The maximum stress is visualized at the inner side of the piston pin boss fillet area. The results of total deformation, equivalent stress, equivalent strain has been tabulated in Table 2.

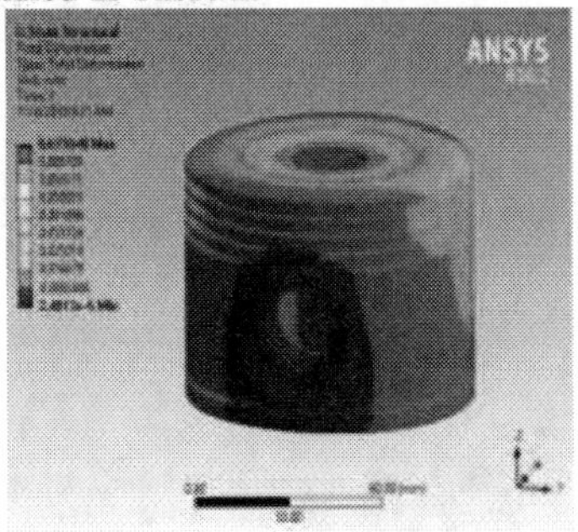

Figure 9. Total deformation

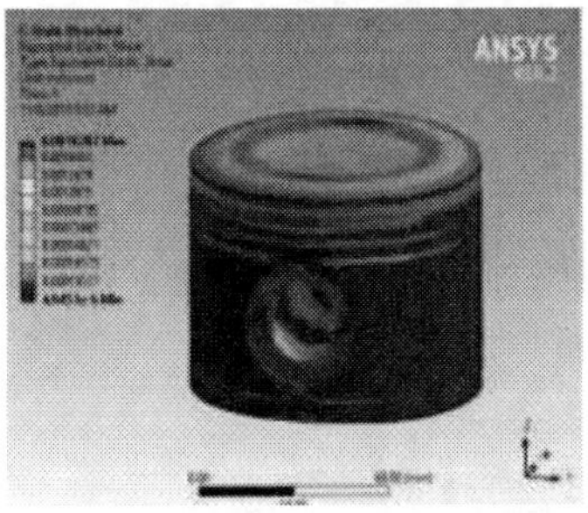

Figure 10. Equivalent elastic strain

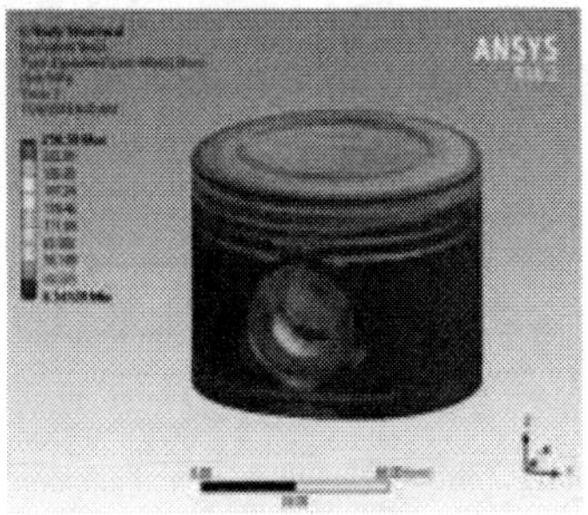

Figure 11. Von-mises stress

IOP Conf. Series: Materials Science and Engineering 1123 (2021) 012008 doi:10.1088/1757-899X/1123/1/012008

5.2 THERMAL ANALYSIS

Transient thermal, steady-state thermal and static structural solver of ANSYS workbench is chosen for thermal analysis of metal matrix composite piston. The Figure 12- 14 displays the result of temperature distribution, thermal stress and total heat flux generated in aluminum metal matrix composite piston.

Figure 13. shows the total heat flux generation on the metal matrix composite piston under the given temperatures. The maximum heat flux region is on the interior region of the piston and is not accessible which shows that the restriction of thermal stress is also not accessible and thus the material is safe. The results of thermal stress, total heat flux has been tabulated in Table 3.

Figure 12. Temperature distribution

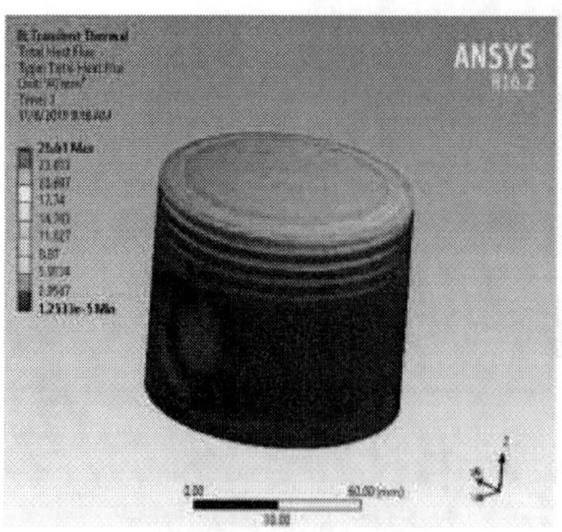

Figure 13. Total heat flux

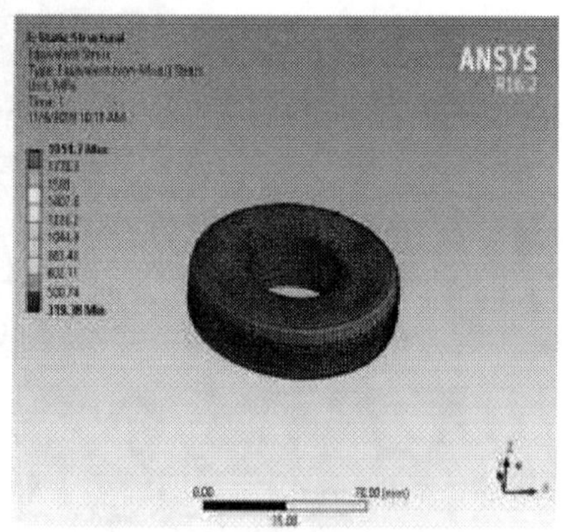

Figure 14. Thermal stress

Table 2. STATIC STRUCTURAL ANALYSIS OF PISTON

MATERIAL	TOTAL DEFORMATION (mm)	VON MISES STRESS (Mpa)	VON MISES STRAIN
Aluminum metal matrix composite piston	0.075048	250.59	0.0016287

Table 3. THERMAL ANALYSIS

MATERIAL	TEMPERATURE	TOTAL HEAT FLUX	THERMAL STRESS (Mpa)
Aluminum metal matrix composite	1500	26.61	1951.7

6. CONCLUSION

Thus, the result of thermal and structural analysis of Aluminum matrix composite piston is interpreted and studied and it is proven that Aluminum matrix composite piston with A2024 as reinforcement metal and fly ash (organic compound) as filler or matrix is light weight structure, has a high tensile strength and also has a high- pressure withstanding structure. Aluminum matrix composite (A2024 + fly ash) piston have a greater strength and the structure is highly ductile, thus the heat energy is easily transformed to mechanical energy. As the maximum heat flux generated in the piston is inaccessible, it says that the piston is safe throughout the combustion and sustainable. The deformation in the piston due to application of heavy pressure load is very minimum and negligible. Hence, in this research paper it is proven that the metal matrix composite can be used in industrial applications especially in automobile industries cause of its unique structural and thermal properties.

REFERENCES

[1] Pulkit Garg, Anbesh Jamwal, Devendra Kumar, Kishor Kumar Sadasivuni and Pallav Gupta 2019 Advance research progresses in aluminum matrix composites: manufacturing & applications. J. *of Materials Research and Tech.* Vol 8 Issue 5 pp 4924-4939,

[2] Venkatesan S and Anthony Xavior M 2008 Analysis of Mechanical Properties of Aluminum Alloy Metal Matrix Composite by Squeeze Casting *A Review Materials Today: Pro.* Vol 5 Issue 5 Part 2 pp 11175-11184

[3] Arunachalam Ramanathan, Pradeep Kumar Krishnan and Rajaraman Murali raja 2019 A review on the production of metal matrix composites through stir casting Furnace design properties challenges, and research opportunities *J. of Manufacturing Processes* Vol 42 Pages 213-245

[4] Avinash Gavel, Suswagata Poria and Prasanta Sahoo 2019 Design of experiments analysis of abrasive friction behavior of Al-TiB2 composites materials Today: *Proc. In press corrected proof* Available online

[5] Jae-Gil Jung, Young- Hee Cho, Jung-Moo Lee, Hyoung-Wook Kim and Kwangjun Euh 2019 Designing the composition and processing route of aluminum alloys using CALPHAD: Case studies Calphad, Vol 64 pp 236-247,

[6] Narender Panwar and Amit Chauhan 2018 Fabrication methods of particulate reinforced Aluminum metal matrix composite- A review Materials Today: Proc Vol 5 Issue 2 Part 1 pp 5933- 5939

[7] 2018 Manufacturing of light automobile engine piston head using sheet metal Procedia Manufacturing, Vol 15 pp 940-948

[8] Siv Ananthan S, Raja Laxman Reddy V, Samson Jerold and Samuel C 2019 Preparation and evaluation of mechanical properties of 6061Al-Al2O3 metal matrix composites by stir casting process. Materials Today: *Pro. In press corrected proof* Available online 3

[9] Smirnov A, Peretyagin P and Bartolomé J F 2019 Processing and mechanical properties of new hierarchical metal-graphene flakes reinforced ceramic matrix composites *J. of the European Ceramic Society* Vol 39 Issue 12 pp 3491-3497

[10] Pradeep Kumar Krishnan, John Victor Christy, Ramanathan Arunachalam, Abdel- Hamid I. Mourad and Majumder Manik Chandra 2019 Production of aluminum alloy-based metal matrix composites using scrap aluminum alloy and waste materials: Influence on microstructure and mechanical properties *J. of Alloys and Comp.* Vol 784 5 pp 1047-1061.

IOP Conf. Series: Materials Science and Engineering 1123 (2021) 012009 doi:10.1088/1757-899X/1123/1/012009

Design and Analysis of Composite Micro Aerial Vehicle

Saleem.M[1], Nishant Kumar Raj[2], Shreshth Gupta[3], Sunny Ramesh Izate[4]

[1]Aeronautical department, Veltech Rangarajan Dr. Sagunthala R&D Institute of Science and technology, Chennai, India,Strokesaleem1@gmail.com
[2]Aeronautical department, Veltech Rangarajan Dr. Sagunthala R&D Institute of Science and technology, Chennai, India,rajnishant3627@gmail.com
[3]Aeronautical department, Veltech Rangarajan Dr. Sagunthala R&D Institute of Science and technology, Chennai, India,shreshthgood@gmail.com
[4]Aeronautical department, Veltech Rangarajan Dr. Sagunthala R&D Institute of Science and technology, Chennai, India,Sunnyizate12@gmail.com

Abstract. Now a day getting materials which suit our requirement is very tough, so today almost all the industries are going to composite materials. Reinforced and matrix combination giving for new material which has high strength and reduction in weight ratio. These materials are called composites. This research work represents Aluminium matrix composite produced by stir casting techniques. Aluminium matrix composites have less weight, low cost and high stiffness. Composites are used in aerospace industries for reduction in weight. Aluminium matrix composites are tested under various mechanical tests and we're carried out the characteristics. In this work micro aerial vehicle is designed by using CATIA. Structural and flow analysis were carried out by computational methods. Keywords: composite, LM5, Structural, MAV, Analysis

1. INTRODUCTION

From past 100 years' growth of composite materials is tremendously increasing in aerospace field. For satisfying the need of light weight and strong material in aerospace industries two materials with different properties are reinforced with some techniques which gives a new material with very unique properties. This unique material is classified as composites material. The new material gives superior properties than the materials which have been combined. Due to its properties like less weight, high fatigue resistance and high impact strength application of composite materials is increasing rapidly in aerospace industries. Rather than aerospace industries these materials are also used invidious other industries such as automotive, construction, glasses, defence, nuclear etc. Designers, engineers have possibilities for bringing their ideas as it provides flexibility in design and it also has very unique material properties.

An Unmanned Aerial Vehicle (UAV) is the type of aircraft which is operated directly by flight control either by computer in the vehicle or by the remote control by the pilot from ground. Micro aerial vehicle is the type of UAV with very small size operating at very low speed and very low Reynolds number. Its Reynolds number is lower than 150,000.
Due to its small dimension it is widely used for Reconnaissance, intelligence, armed attacking and surveillance in military. Instead of military applications it is also used in urban application for monitoring the traffic and for surveillance of the area. MAVs can also tell about the weather condition with the help of their sensors and can also be used for biochemical sensing.
For the optimized use of MAV there has been a procedure for validating the material selection. The material selection will affect the performance, capabilities and endurance of the MAV. For achieving the property of light weight for an MAV, optimized procedure is being followed for selecting material. Composites seem to be better option for MAVs as it gives flexibility and better mechanical response.

2. MATERIAL SELECTION

In order to get better properties, we have to select our material in such a way that it should satisfy the need of the manufacturer. From past few years it is been observed that metal matrix composites are used more than grid composites because of their enhanced properties. During development of the MAV different types of structure were tested with different results. For this MAV LM5 and Fly ash has been reinforced together with stir casting technique. It has been tested with different methods in our laboratory and was found very light weight and strong as compared to other composite materials. We also found that its impact strength and fatigue resistance is also high as compared to other composite materials. Table 1 show different properties of the material which has been known from the test conducted in our laboratory.

Table 1. Properties of Material

S. No	Properties	Results
1	Density	2.8
2	Toughness	71.2
3	Hardness (HRA)	78
4	Ultimate tensile strength (MPa)	678

3. CONCEPTUAL DESIGN OF MAV

The conceptual design is the initial step which is being taken before manufacturing of the MAV. As there are numerous MAV's present now a day there are plenty of MAV with common features. Based on that proper literature survey has been done on different MAV and then based on sweet spot taken from them a new MAV has been designed in CATIA V5. Table 2 shows some of the detailed parameters taken from the literature survey and fig1 Shows the conceptual design of MAV designed from the parameters in CATIA.

Table 2. Dimensions of MAV

S. No	Parameters(m)	Value
1	Maximum chord length	10
2	Total length of fuselage	16.5
3	Wing span	21
4	Wing area	138
5	Aspect Ratio	1.23

Figure 1. Conceptual Design of MAV

Figure 2 shows the fuselage design in which flapping mechanism and motors are to be mounted on the bottom side and wings are attached in the upper side. Same material with same properties has been applied in fuselage also in CATIA. Many slots and holes are made additionally in unwanted area to reduce the weight of the fuselage.

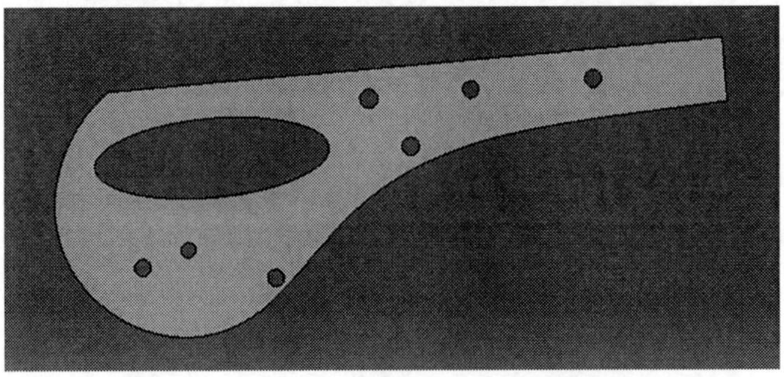

Figure 2. Conceptual Design of Fuselage

4. ANALYSIS

Basic analysis has been done on the fuselage part as well as on the whole MAV with the help of ANSYS. Finite Element Analysis has been done on the fuselage and Von-misses stress and strain has been found. Load has been applied on the upper part of the fuselage because of wing and lift produced. Figure 3 and Figure 4 shows the variation of stress and strain in the fuselage due to load applied.

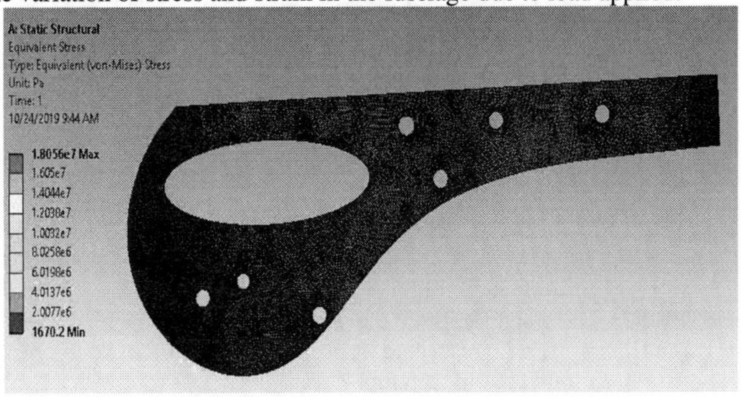

Figure 3. Stress Analysis on Fuselage

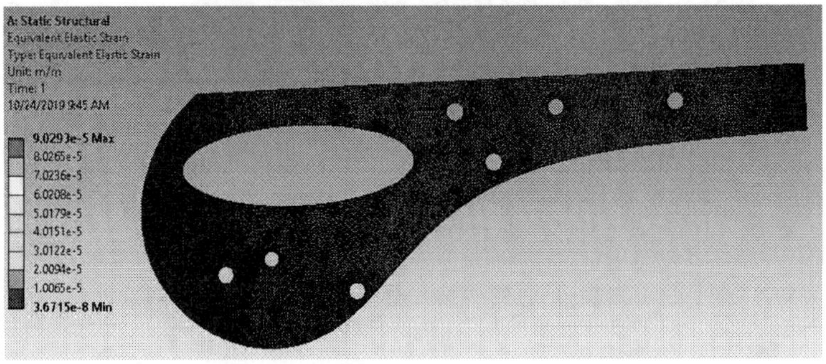

Figure 4. Strain Analysis on Fuselage

Flow visualization has been also done in ANSYS fluent. Flow analysis has been done on outer body of the MAV. It is also designed in CATIA to protect the inner body and the whole mechanism. It has been meshed in hyper mesh software and 313101 nodes and 274492 elements has been generated by cut shell method.

SST K-w and energy equation has been used for visualizing the flow over the surface of MAV. It has been analysed at the velocity of 2m/s with gauge pressure at zero. COUPLED method has been used for analysing the flow. Coefficient of drag and L/D ratio has been also found out from the software. Fig5 and fig6 shows pressure and velocity contour acting over the body of the MAV.

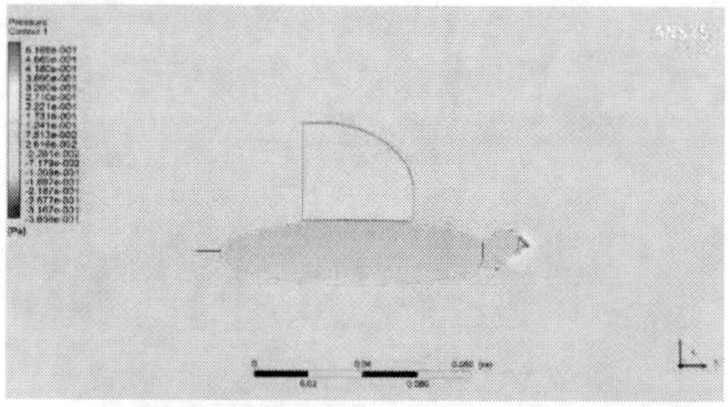

Figure 5. Pressure Contour over MAV

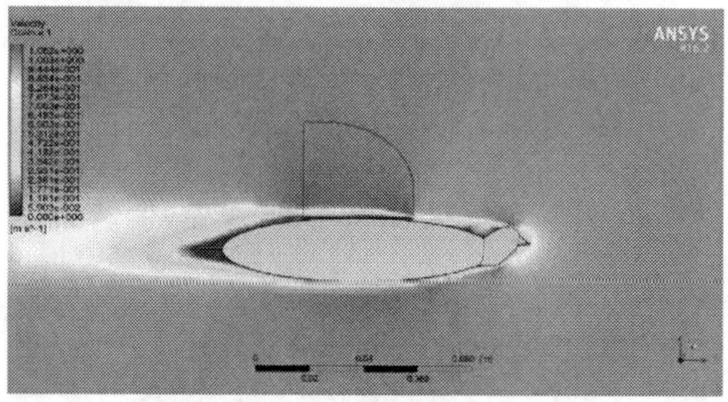

Figure 6. Velocity Contour over MAV

5. GRAPHS

Various graphs have also been validated form the software. Fig7 shows the graph of mesh at X-coordinate with respect to its position. Different points tell the position of the mesh with respect to X-coordinates. Fig8 shows the value of static pressure with respect to its position over the body of the MAV. Value of turbulence kinetic energy has also been found out in the graph.

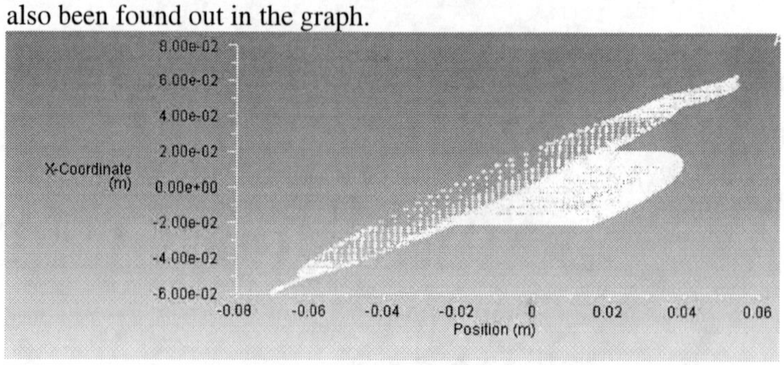

Figure 7. Graph of Mesh at X-coordinate

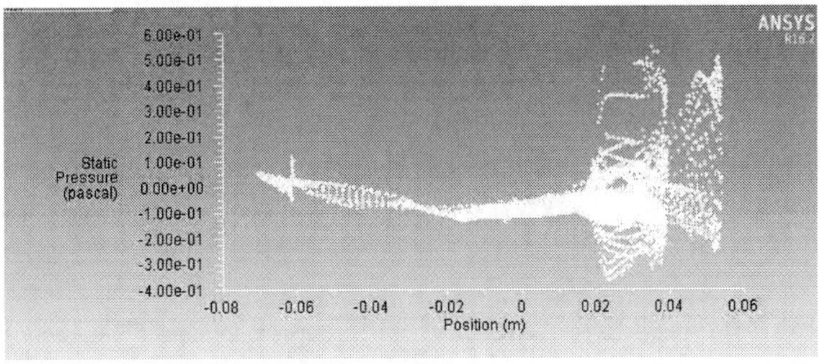

Figure 8. Static Pressure vs. Position Graph

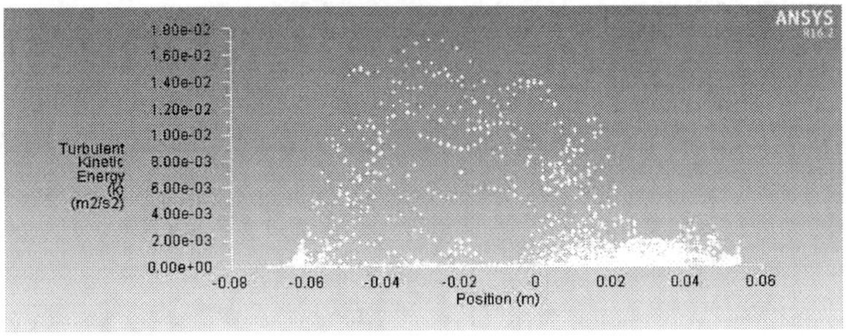

Figure 9. Turbulent Kinetic Energy vs. Position Graph

6. RESULT AND DISCUSSION

Whole model has been designed in CATIA with the dimensions taken from literature survey. The fuselage structure has been made separately for analysis. The fuselage structure has been analysed in FEA and it has been found out that both stress and strain are acting mainly at middle part of the fuselage because of the load of wing and lift produced. Maximum stress was found out to be 1.6×10^{7} Pa at the middle section of the fuselage. Pressure contour has been seen more at the head, starting of the wing and at the end of the tail part. Similarly, it has been seen that flow visualization around body of the MAV was symmetric and smooth. Value of cd has been found from the analysis as 0.024 and L/D value has also been found out. It has been seen that the flow is smooth and symmetric while applying the material as LM5+fly ash. Highest pressure seen over the body is 1.724×10^{-1} Pa which is at the starting of the head and velocity seems to be reduced at the upper body of the fuselage. Minimum velocity found on the body is 5.93×10^{-2} ms^{-1}. Different graphs have also been validated and many outputs has been taken from that.

7. CONCLUSION

With a general analysis of MAV with aluminium metal matrix composite we concluded that material selection plays a major role in reduction of weight as well as for getting high strength. We have seen that as we increase weight percentage of fly ash strength and hardness of the material is also increasing. From analysis we have seen that flow over the whole body is uniform and symmetrical. Our material develops very low coefficient of drag and it also develops sufficient amount of Lift to drag ratio. Also our material can be used for manufacturing various aerodynamic body which need less weight and more strength.

REFERENCES

[1] Sachin Mishra and Sahil Garg 2014 Design and Analysis of Composite Structure for Development of A Flapping Wing Micro Air Vehicle *J. of Basic and Applied Engineering Research* Vol 1, No 4 pp.90-92

[2] Hara R O 2011 The Material Property and Structural Dynamic Characterization of the Manduca Sexta Forewing for Application to Flapping Micro Air Vehicle Design, Dissertation Prospectus Air Force Institute of Technology

[3] DeLeon N E 2011 Manufacturing and Evaluation of a Biologically Inspired Engineered MAV Wing Compared to the Manduca Sexta Wing Under Simulated Flapping Conditions Air Force Institute of Technology Master's Thesis

[4] Xie L, Wu P and Ifju P 2010 Advanced Flapping Wing Structure Fabrication for Biologically-Inspired Hovering Flight *51st AIAA/ASME/ASCE/AHS/ASC Structures, Structural Dynamics, and Materials Converence AIAA*

[5] Michelson R C and Reece S 1998 Update on Flapping Wing Micro Air Vehicle Research in *Int. RPV Conf. on Bristol*

6] Norris A 2008 Investigation of Reinforced Membrane Wings for Use in Micro Air Vehicles Air Force Institute of Technology Dissertation Prospectus AFIT/GA/ENYDSY-09S

[7] Sims T, Palazotto A and Norris A 2010 A Strucural Dynamic Analysis of a Manduca Sexta Forewing *Int. J. of Micro Air Vehicles* vol 2 no 3

[8] Norris A, Palazotto A and Cobb R 2006 Structural Dynamic Characterization of an Insect Wing: Towards the Developement of Bug Sized Flapping Wing Micro Air Vehicles American Institute of Aeronautics and Astronautics

[9] Malik A and Qureshi B 2010 Vibration Analysis of Flapping Wing Micro Air Vehicle Using Finite Element Methods in *Proc of the World Cong. on Engineering* Vol 2 2004.

[10] Masarati P, Morandini M, Quaranta G and Vescovini R 2011 Multibody Analysis of a Micro-Aierial Vehicle Flapping Wing *MULTIBODY DYNAMICS, ECCOMAS Thematic Conf.*

[11] Wood R 2007 Design Fabrication and Analysis of a 3DOF 3cm Flapping Wing MAV in *Conf. on Intelligent Robots and System IEEE*, San Diego CA USA

DESIGN OF AIRFOIL FOR LOW REYNOLDS NUMBER FLIGHTS USING PYTHON AUTOMATION CODE

Somashekar V[1], Immanuel Selwyn Raj A[2]

[1]School of Mechanical Engineering, VIT University, Vellore, Tamil Nadu - 632014, India, shekar.mtech.ph.d@gmail.com

[2]School of Mechanical Engineering, VIT University, Vellore, Tamil Nadu - 632014, India, immanuel.selwynraj@vit.ac.in

Abstract. The design of airfoil represents a facet of aerodynamics. The low-Reynolds number flight vehicle manufacturers/designers are seeking the efficient and effective method of airfoil design code for design and development. This research paper presents an efficient, low- speed airfoil design using the Python automation code. The main objective is to design the airfoil to achieve high aerodynamic performance for future fixed-wing micro aerial vehicle (MAV) applications. In this study, the fixed-wing micro aerial vehicle travel with the speed range between 6-20 m/s and the Reynolds number $2x10^5$ are considered. The Python automation code is developed to achieve the optimum air foil as per the user requirements The designed airfoil is analysed using ANSYS Fluent and is validated with wind tunnel experiments. According to the findings, the designed airfoil has a high lift aerodynamic performance at low Reynolds number and could be ideal for the design and development of micro aerial vehicles. The maximum variation of the coefficients of lift and drag from reference (S1223 airfoil) and designed airfoil numerical results are 13.64% and 16.66%, respectively at 10 angle of attack (AOA). The maximum range and endurance estimated for both reference and designed airfoil. The maximum range 6% increased for designed airfoil as compared to the reference airfoil. It is strongly recommended that any aerodynamic design engineer can use this Python automation code to design the airfoil as per the user requirements.***Keywords:*** *Low-Reynolds Number, Airfoil Design, Python Automation Code, Micro Aerial Vehicle, Low Speed Wind Tunnel.*

1.INTRODUCTION

Airfoil plays a significant role in the aerodynamics of the low Reynolds number flights. The design or selection of airfoil processes is essential for the design and development of low Reynolds number flight vehicles and significant at the earliest low Reynolds number flight vehicle design stage to support design engineers in selecting or designing asuitable airfoil as required by the user. The airfoil aerodynamic characteristics include the coefficient of lift, drag and pitching moment that are essential to evaluate by performing the test at the desired working condition of the airfoil [1].

The concept of low Reynolds number in aerodynamics predominant for civil and military applications. These include high altitude devices, wind turbines, human-powered vehicles, remotely piloted vehicles, sailplanes, unmanned aerial vehicle and micro aerial vehicle[2-5]. In the 21st century, the micro aerial vehicles (MAVs) and unmanned aerial vehicles (UAV's)are becoming increasingly curious, required extensive and in-depth research into 2D airfoils and 3D wings [6]. The Python programming language was created by Guido van Rossum in the late 1980s. In IT industries such as NASA, Industrial Light and Magic, Yahoo, Facebook, and Google, Python is used mainly for software development. The knowledge of the python community can accomplish great things with python, but the beauty of python is that it is open to beginning programmers and allows them to deal with interesting problems faster than many other, more complex languages that have a steeper learning. Sunanda et al. [8] At Re = 4000, the author analysed thirteen different airfoils. This study was carried out by adjusting the geometry of the airfoil, i.e. camber, thickness, and airfoil roughness.Kunz[9] studied the low Reynolds number airfoil geometry effects on performance by varying camber, thickness, and leading and trailing edge shape at Re = 10000. The optimized airfoil achieved a maximum lift-to-drag ratio five percent at Re = 2000 and an average lift-to-drag ratio of about 4 percent at Re = 6000. It was found that the camber line plays animportant role in achieving the maximum lift-to-drag ratio.Mateescu and Abdo[10] Many NACA airfoils were tested at different angles of attack and the Reynolds number ranged from 400 to 6000. The lift-to-drag ratio depends on camber, thickness, and Reynolds number. Pranesh et al.[11] studied NACA 0008 airfoil aerodynamic characteristic at Re =2000 and 6000.Srinath and Mittal[12] studied aerodynamic characteristic of the NACA 0012 airfoil at Re $\leq$ 500 at a different angle of attack. Interesting was the optimized airfoil and different geometry shapes were obtained. It has been concluded that more design space leads to better aerodynamic forms.Kumar et al. [13] proposed a method to optimize the shape with a wide range of design variables.

For Re = 1000 and Re = 10,000 at $\alpha = 4°$, the optimized airfoil was designed. Large ranges of low Reynolds number airfoils reported in the literature, particularly for free flight model aircraft, sailplanes, and heavy-lift radio-controlled aircraft.[14-16]Arvind Santhanakrishnan et al. [17] studied experimental and numerical investigations on NACA 4415 airfoil at chord-based Reynolds number lesser than 100000, by varying static camber the stall angle of attack increased. S. S. Talya et al. [18] multidisciplinary optimization of gas turbine airfoil design was studied, despite various limitations on airfoil geometry. Using the Kreisselmeier-Steinhauser (K- S) function method, the restricted multi-objective optimization problem is solved

In this research work, the author developed a Python automation code to design the optimum airfoil as per the user requirementsfor both civilian and military applications. For a selected S1223 reference airfoil, the author principally interested in maximizing lift-to-drag ratio to increase the overall aerodynamic efficiency of the low Reynolds number flights. The lift-to- drag ratio depends on wing geometry and the flow conditions. The designed airfoil results validated with ANSYS Fluent, experiments results from the literature, and wind tunnel experiments.

2. Methods & Materials

2.1 Python Automation Code

Python is a general-purpose programming language of interpretation developed by Guido van Rossum and first released in 1991.The python program is a language-construct an approach that aims to help programmers write clear, logical code for small and large-scale research projects [7]. The python automation code with XFoil solver generated to design low Reynolds number airfoil for different application based on the user requirements. The python automation code sequence of operation, as shown in Figure 1.

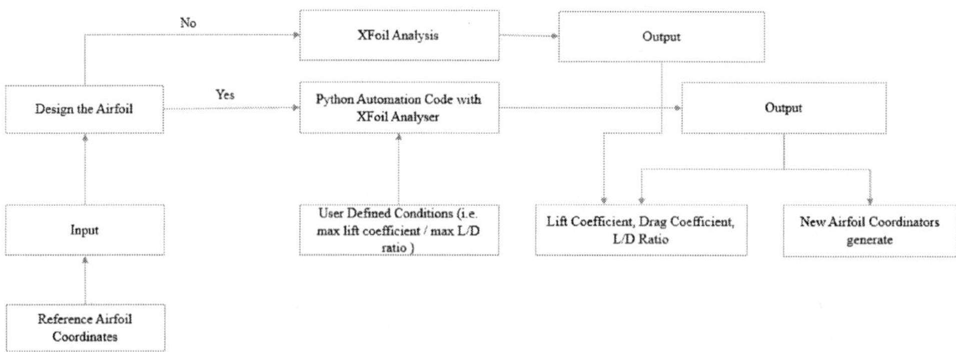

Figure 1. Block diagram of the python automation code for design the airfoil as per the user requirements

The low Reynolds number S1223 airfoil, considered as a reference airfoil and investigated in this research work, has a maximum thickness of 12.1% at 19.8% chord, a maximum camber of 8.1% at 49% chord, and the S1223 airfoil profile is shown in Figure 2. This airfoil exclusively designed and used for low Reynolds number applications [8]. The outcome of the Python automation code generated new airfoil coordinates as per the user requirements. The designed airfoil maximum thickness of 14.7% at 20.1% chord, a maximum camber of 10.8% at 47.7% chord, and the designed airfoil profile is shown in Figure 3.

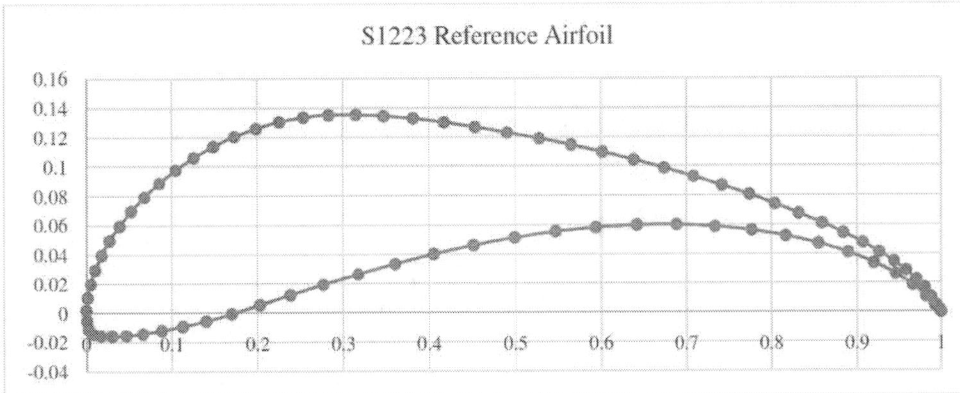

Figure 2. S1223 reference airfoil profile [8]

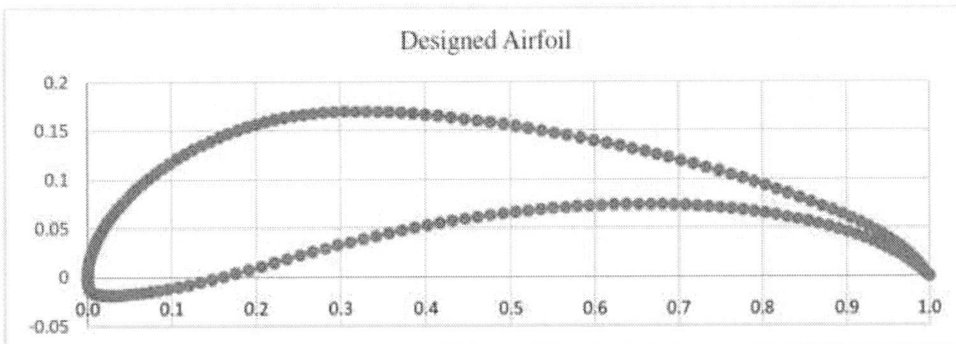

Figure 3. Designed airfoil profile

2.2 Numerical techniques

The numerical simulations of the flow around S1223 reference airfoil and designed airfoil were carried out using the commercial CFD software, ANSYS 15. Using ICEM-CFD, the C- type domain generated with pressure-far-field boundary conditions, as shown in Figure 4. The total number of elements found that 34740 with 0.03 mm first spacing. Figure 4 shows the grid generation over the airfoil.

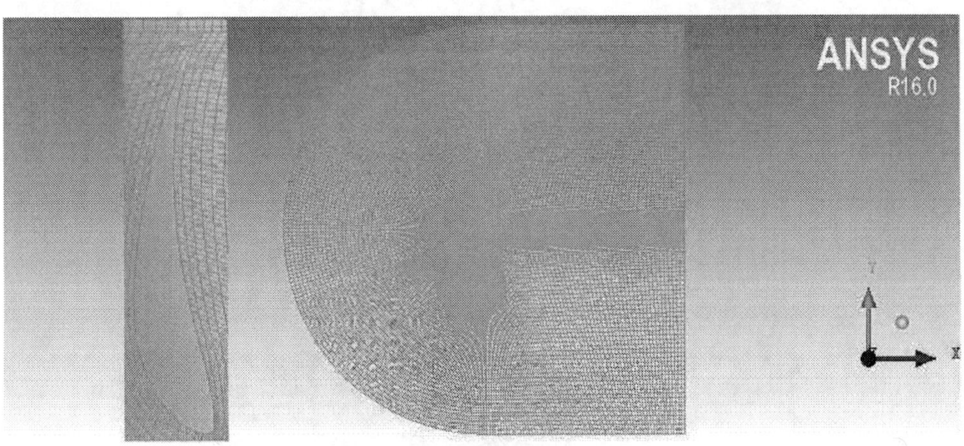

Figure 4. C-type Grid using ICEM-CFD

Table 1. Detailed information on the setup of CFD solvers for both reference and optimized airfoil used in this study

Mesh	2-D structured	Reference values	
First cell thickness	0.03 % chord	Area (m)	1
Total mesh elements	34740	Density (kg/m^3)	1.225
Solver	Pressure based	Length (m)	0.3048
Viscous model	laminar	Temperature (k)	288.16
Pressure	101325	Velocity (m/s)	9.63 m/s
Fluid	Air	Viscosity (kg/m-s)	1.79E-05
Angle of Attack	-4 to +14 Degrees (for every 2° of intervals)	The ratio of Specific Heat	1.4
Boundary conditions	Pressure-far-field	Reynolds Number	200000

Table 1 shows detailed information on the setup of CFD solvers for both reference and optimized airfoil used in this study. The reference and optimized airfoil are simulated for the above-mentioned conditions and extracted aerodynamic characteristics.

2.3 Wind Tunnel Experimental Setup

The experiments were conducted in a low Reynolds number wind tunnel shown in Figure 5. It isopen circuit suction type, made up of sheet metal, consists of test cross-section 21 × 13 × 15 cm,air-speed up to 15 m/s and contraction ratio 9:1. The overall size of the low Reynolds number windtunnel 127 cm (L) × 36 cm (B) × 36 cm (H). Experimental data were captured at Reynolds number of were captured at Reynolds number of 2 × 105 , to compute the coefficients of lift and drag. The pressure readings on designed airfoil model were recorded by using multi-tube manometer.

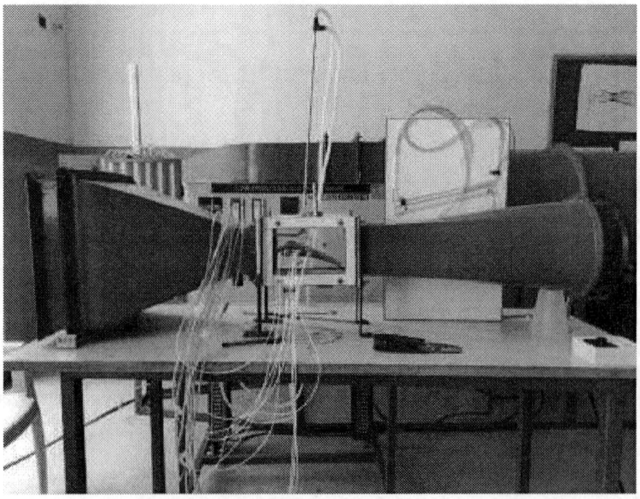

Figure 5. Wind tunnel experimental setup to analyse aerodynamic coefficients over a designed airfoil

IOP Conf. Series: Materials Science and Engineering 1123 (2021) 012010 doi:10.1088/1757-899X/1123/1/012010

3. Results and Discussion:

3.1 Python Automation Code Results

The lift and drag coefficients of the designed airfoil at a various angle of attack were calculated from the automated python code using Xfoil solver. Figure 6(a) shows the profiles of the lift and drag coefficients vs. angle of attack and lift-to-drag polar plot. The lift and drag coefficients of the designed airfoil as a function of AOA and lift-to-drag ratio plot are shown. These graphs are the result of the integration of surface pressure distribution around designed airfoil computed in Xfoil with the help of python automation code. At Reynolds number 2×10^5, coefficient of lift smoothly increases linearly till 12 AOA. It describes, the laminar boundary layer remains attached to the designed airfoil surface and leaves at the trailing edge to satisfy the Kutta condition. The recorded maximum coefficient of lift value is 2.466, and the corresponding coefficient of drag is 0.04922. Further increasing the AOA, it will lead to a stall, the coefficient of drag increases and decreases the coefficient of lift due to the adverse effect. Similarly, it was observed that by increasing the AOA (-4 - 14) the drag coefficient also increases.
The Figure 6(b) shows the lift-to-drag vs. angle of attack. From the Xfoil results, the recorded maximum lift-to-drag ratio of 67.40 at 4°angle of attack. The L/D ratio increases drastically at 4° angle of attack. It is due to the effect of adverse pressure gradient is insignificant over the upper surface of the airfoil. Further increasing the AOA, the L/D ratio gradually decreases when the adverse pressure gradient is significant.

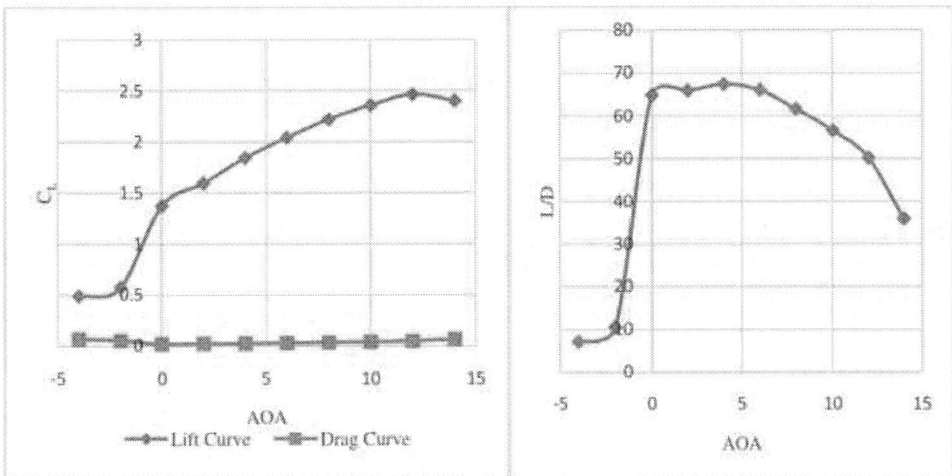

Figure (a) Figure (b)

Figure 6. The designed airfoil Xfoil results; (a) Lift and Drag coefficients vs. angle of attack; and (b) lift-to-drag vs. angle of attack

3.2 Numerical Results

Lift and Drag Coefficients: The lift and drag coefficient characteristics for the reference and designed airfoil is presented in Figure7. The aerodynamic parameters, lift and drag coefficients for a different angle of attack ranging from -4° to 14° at Re = 0.2 x 10 were computed. From the graph, it is demonstrable that the maximum coefficient of lift for the designed airfoil is higher than that for the reference airfoil. The corresponding values reported for the reference, and designed airfoil is 2.0177 and 2.3364 at an angle of attack10°, respectively.

Further increasing the AOA it will lead to a stall, due to adverse effect, the coefficient of drag increases and decreases the coefficient of lift for the case of low Reynolds number conditions. Similarly, the corresponding drag coefficient values reported for the reference and designed airfoil are 0.0588 and 0.0636 at an angle of attack 10 , respectively. Because, at higher AOA, the coefficient of drag rapidly increases due to increases of frontal area and also the increases of boundary layer thickness.

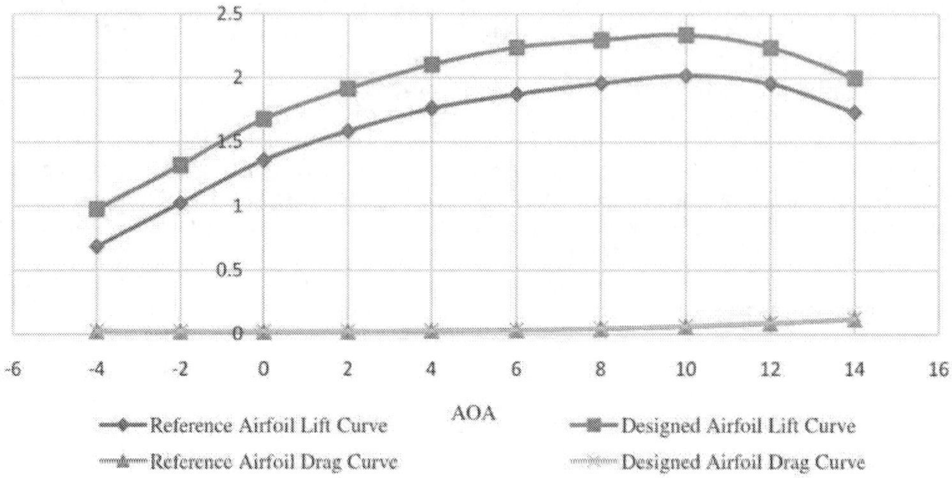

Figure 7. Lift and drag coefficient vs angle of attack for designed airfoil CFD results

Lift-to-Drag ratio: The lift-to-drag ratio is conferred as L/D at a different angle of attack (AOA) for both reference and designed airfoil in Figure8. From the graph, it is evident that the lift-to-drag ratio for the designed airfoil is higher than that for the reference airfoil. The corresponding values reported for the reference, and designedairfoil is 76.99 and 81.18 at an angle of attack 2 , respectively. Beyond this angle of attack, the lift-to-drag ratio is substantially reduced due to drag penalty as the reference and optimizeddesigned airfoil approaches stall angle of attack. The average increment of lift-to-drag ratio compared to the reference and designed airfoil is 3.2% for a different angle of attack.

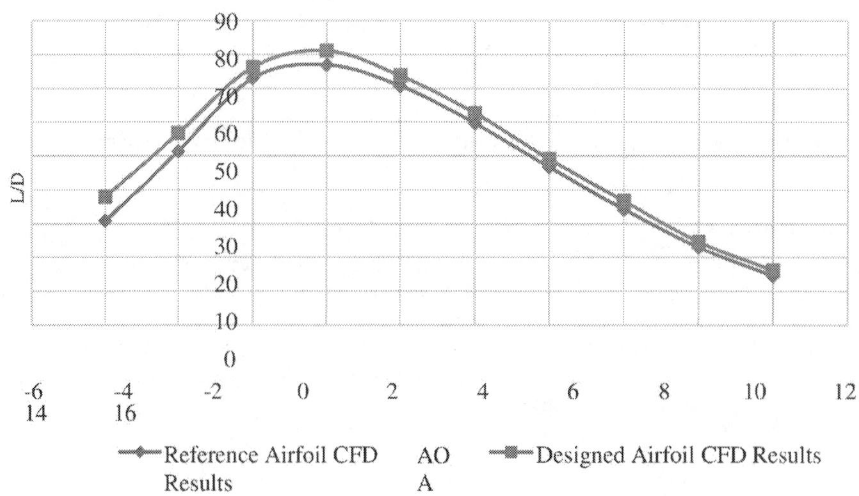

Figure 8. Lift-to-drag ratio vs.angle of attack for designed airfoil CFD results

IOP Conf. Series: Materials Science and Engineering 1123 (2021) 012010 doi:10.1088/1757-899X/1123/1/012010

3.3 Wind Tunnel Experimental Results

In this work, investigated aerodynamic performance characteristics over a designed airfoil using low Reynolds number wind tunnel experiments. The wind tunnel experiments were conducted for different flight conditions, i.e. angle of attack 0° to 20° degrees at Reynolds number 2×10^5 Figure 9 shows the profiles of the lift and drag coefficients vs. AOA and lift-drag polar plot.

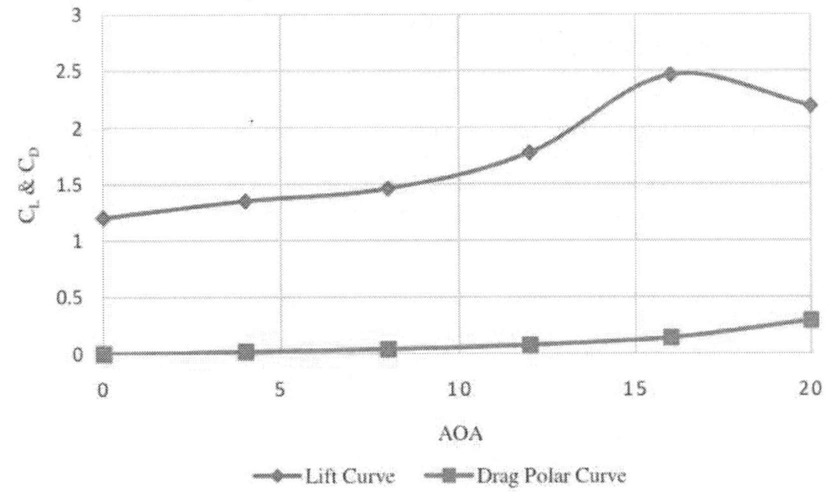

Figure (a)

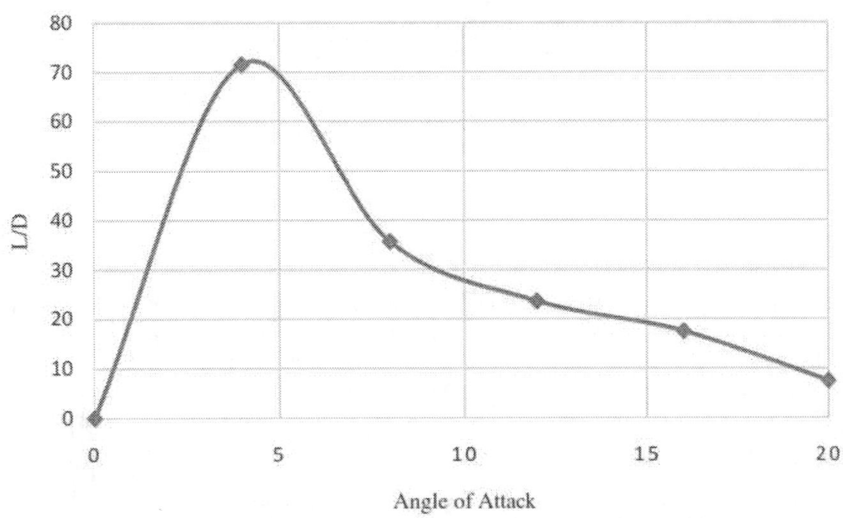

Figure (b)

Figure 9. The designed airfoil lift and drag coefficients; *(a)* airfoil Lift and Drag coefficients vs. angle of attack; and (b) lift-drag polar curve

From the experimental results, the recorded maximum lift and drag coefficients is 2.46 and 0.1412 at Reynolds number 2×10^5 , respectively. The lift coefficient increases linearly by increasing the AOA (0° to 16°) and it reaches the maximum at 16 ° AOA. Further increasing the AOA it will lead to stall. Similarly, it was observed that by increasing the AOA (8° - 20°) the drag coefficient also increases.

The Figure 9(b) shows the lift-to-drag vs. angle of attack. From the experimental results, the recorded maximum lift-to-drag ratio of 71.49 at 4°angle of attack. The L/D ratio increases drastically at 4° angle of attack. It is due to the effect of adverse pressure gradient is insignificant over the upper surface of the airfoil. Further increasing the AOA, the L/D ratio gradually decreases when the adverse pressure gradient is significant.

In this research work Barnhart, et al.'s MAV 2 battery-powered model [20] considered to validate the aerodynamic performance characteristics i.e. maximum endurance and range for both reference and designed airfoil. The following empirical relation used to calculate the maximum Endurance and Range. [21] In these experiments, the maximum endurance and range estimated for the reference airfoil to be 1.924 hours and 31.92 km. The maximum endurance and range estimated for the designed airfoil to be 1.938 hours and 33.96 km.

$$E_{max} = R_t^{1-n} \left(\frac{\eta_{tot} V \times C}{\frac{2}{\sqrt{\rho S}} C_{DO}^{\frac{1}{4}} \left(2W \sqrt{\frac{k}{3}} \right)^{\frac{3}{2}}} \right)^n \; Hour$$

$$R_{max} = R_t^{1-n} \left(\frac{\eta_{tot} V \times C}{\frac{1}{\sqrt{\rho S}} C_{DO}^{\frac{1}{4}} \left(2W \sqrt{\frac{k}{3}} \right)^{\frac{3}{2}}} \right)^n \sqrt{\frac{2W}{\rho S} \sqrt{\frac{k}{C_{DO}}}} \times 3.6 \quad KM$$

4. Validation:

The reference and designed airfoil aerodynamic characteristics were numerically investigated at various AOA and aerodynamic performances were validated with wind tunnel experiments and literature. Figure 10 shows the lift coefficient versus angle of attack for the reference and designed airfoil data at= 0.2×10^6 respectively.In general, the present data agree with their measurements, thus providing validation for the present research results. The lift and drag coefficients validation process carried out for different conditions as follows.

Case 1: The maximum variation of the coefficients of lift and drag from reference and designed airfoil CFD results are 13.64% and16.66% at 10°AOA.

Case 2: The maximum variation of the coefficients of lift and drag from Xfoil solver and CFD solver fordesigned airfoil are 1% and 33% at 10°AOA.

Case 3: The maximum variation of the coefficients of lift and drag from Xfoil solver and experimental results from the literature are 1% and 0% at 10°AOA.

Case 4: The maximum variation of the coefficients of lift and drag from CFD solver and experimental results from the literature is 22.95% and 33.33% at 10°AOA.

90

Case 5: The maximum variation of the coefficient of lift from Reference airfoil (literature) and wind tunnel experiments results is 14.63% at 16°AOA

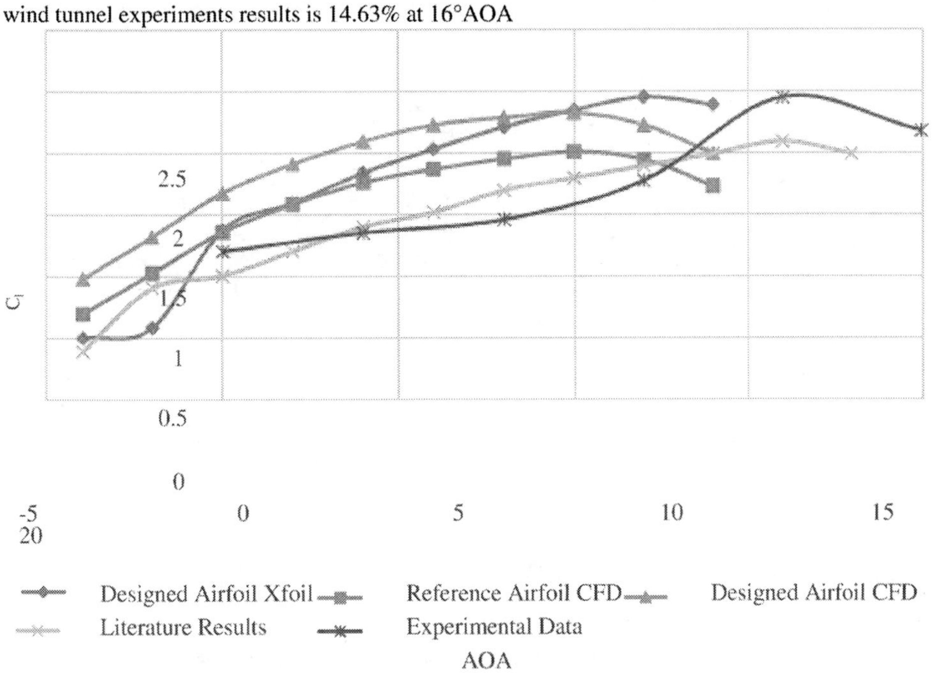

Figure 10. Lift coefficient vs.angle of attack for the designed airfoil

Lift-to-Drag Ratio: The lift-to-drag ratio is presented as L/D at a different angle of attack (AOA) for both reference airfoil and designed airfoil in Figure11. From the graph, it is evident that the lift-to-drag ratio for the designed air foil is higher than that for the reference airfoil.

The lift-to-drag ratio validation process carried out for different conditions as follows.

Case 1: The maximum variation of the lift-to-drag ratio from reference and designed airfoil CFD results are 5.16% at 2° AOA.

Case 2: The maximum variation of the lift-to-drag ratio from Xfoil solver and CFD solver for designed airfoil are 18.18% at 2° AOA.

Case 3: The maximum variation of the lift-to-drag ratio from Xfoil solver and experimental results from the literature are 23.07% at 2°AOA.

Case 4: The maximum variation of the lift-to-drag ratio from CFD solver and experimental results from the literature are 38.27% at 2° AOA.

Case 5: The maximum variation of the lift-to-drag ratio from Reference airfoil (literature) and wind tunnel experiments results is 24.46% at 4° AOA.

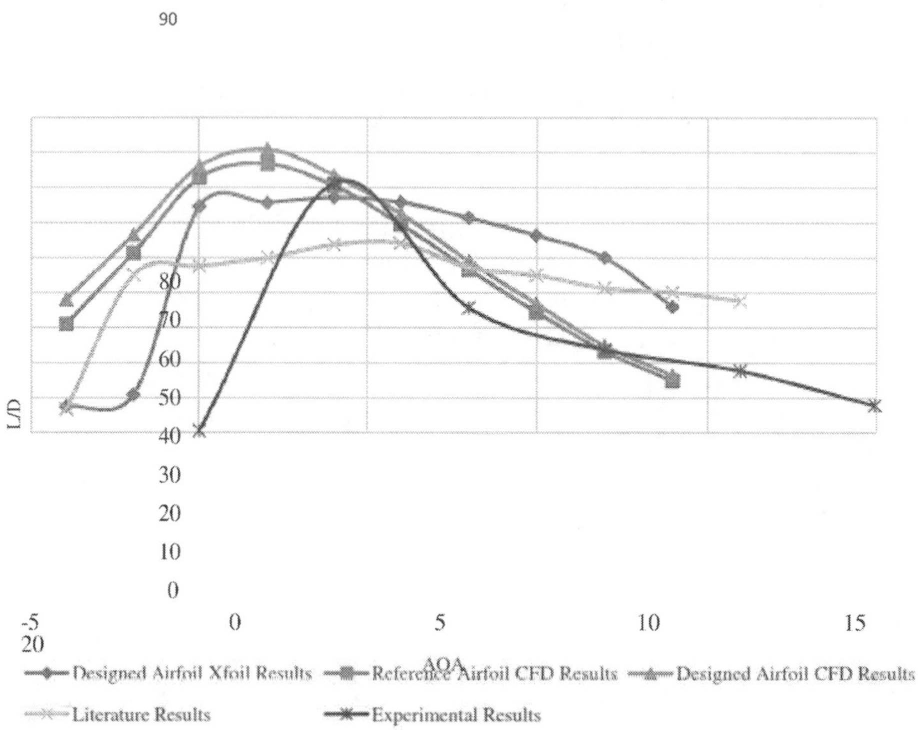

Figure 11. Lift-to-Drag ratio vs.angle of attack for the designed airfoil

Conclusions

In low Reynolds number airfoil design problems, based on the applications and user requirements, different constraints may be considered to the airfoil geometry. In this research, developed the Python automation code interface with Xfoil solver to design the low Reynolds number airfoil geometry for both conventional and commercial applications as per the user requirements.Based on the developed code, the user can constraints two parameters, i.e. (i) maximize the lift coefficient (ii) maximize the lift-to-drag ratio. In this paper, the author constrained to maximize the lift-to-drag ratio parameter at $R = 0.2 \times 10^6$ for low Reynolds number flight applications. Based on the results, the following major conclusion made.

1. The developed python automation code interface with Xfoil solver tested based on the user-defined input, the designed airfoil achieved as per the user requirements.
2. The maximum variation of the lift and drag coefficients from CFD solver and experimental results from the literature are 38.27% at 2° AOA.
3. The designed airfoil lift-to-drag ratio (airfoil efficiency) increases by 38.27%, low Reynolds number flight with the higher lift-to-drag ratio are more efficient than those with lower lift-to-drag ratios.
4. High lift-to-drag ratio airfoils are suitable for low Reynolds number applications < 10 [19]
5. In these experiments, the maximum endurance and range estimated for referenceairfoil to be 1.924 hours and 31.92 km. The maximum endurance and range estimated for designed airfoil to be 1.938 hours and 33.96 km.
6. The maximum range and endurance estimated for reference airfoil to be 1.924 hours and 31.92 km. The maximum endurance and range estimated for the designed airfoil to be 1.938 hours and 33.96 km. The maximum range 6% increased for designed airfoil as compared to the reference airfoil.
7. The designed high lift-to-drag ratio airfoil is strongly recommended for low Reynolds number flight applications.

REFERENCE

[1] AnhBao DINH, Khanh Hieu NGO and Nhu Van NGUYEN 2015 An Efficient Low-Speed Airfoil Design Optimization Process Using Multi-Fidelity Analysis for UAV Flying Wing the *3rd Vietnam Conf. on Control and Automation VACC*
[2]]Yang Z F, Fred L H and Hu H 2007 An experimental investigation on the flow separation on a low-Reynolds-number airfoil *45th AIAA aerospace sciences meeting and exhibit*
[3] Mueller T J and DeLaurier J D 2003 Aerodynamics of small vehicles *Annu Rev Fluid Mech* Vol 35 Issue 1 pp 89–111.
[4] Carmichael B H 1981 Low Reynolds number airfoil survey California Low Energy Transport systems; Report No NASA-CR-165803-Vol1.
[5] Shirzad Hosseinverdia,and Hermann Fasela F 2015 Laminar-turbulent transition in a laminar separation bubble in the presence of free-stream turbulence *IUTAM_ABCM Symposium on Laminar Turbulent Transition, Procedia IUTAM 14* pp 570 – 579 *Elsevier*
[6] Lei Juanmian , Guo Feng and Huang Can 2012 Numerical study of separation on the trailing edge of a symmetrical airfoil at a low Reynolds number *Chinese J.l of Aeronautics* vol 26 Issue 4 pp 918–925.
[7] Kuhlman, Dave 2012 A Python Book Beginning Python, Advanced Python, and Python Exercises Section 1.1 Archived from the original
[8] Sunada, S Sakaguchi A and Kawachi K 1997, Airfoil Section Characteristics at a Low Reynolds Number *J. Fluids Eng.* Vol 119 Issue 1 pp 129–135.
[9] Kunz P J 2003 Aerodynamics and design for ultra-low Reynolds number flight Ph. D. Thesis *Stanford University* Stanford.
[10] Mateescu D and Abdo M 2010 Analysis of flows past airfoils at very low Reynolds numbers *Proc. Inst. Mech. Eng* Part G *J. Aerosp. Eng* Vol 224 Issue7 pp 757–775.
[11] Pranesh C, Sivapragasam M and Deshpande M D 2014 Aerodynamic Characteristics of Flow Past Naca 0008 Airfoil At Very Low Reynolds *Numbers J. Aerosp. Sci. Technol* Vol 66 Issue 4 pp 247–266.
[12] Srinath D N and Mittal S 2009 Optimal airfoil shapes for low Reynolds number flows *Int. J. Numer. Methods Fluids* Vol 61 Isuue 4 pp 355–381.
[13] Kumar N, Diwakar A, Attree S K and Mittal S 2013 A method to carry out shape optimization
[14] Selig, M S, Guglielmo J J, Broeren A P and Giguère P 1995 Summary of low-speed airfoil data *SOARTECH publications* USA Vol 1.
[15] Selig M S, Lyon C A, Giguère P, Ninham C P and Guglielmo J J 1996 Summary of low-speed airfoil data *SOARTECH publications* USA, 1996, Vol 2.
[16] Lyon C A Broeren A P, Giguère, P, Gopalarathnam and Selig M S Summary of low-speed airfoil data, *SOARTECH publications USA* Vol 3
[17] Arvind Santhanakrishnan , Nan Pern J and Jamey Jacob D Optimization and Validation of a Variable Camber Airfoil 2005 Fluid Mechanics Lab Mechanical Engineering Dept Uni. Of Kentucky, Lexington, KY 40506 USA *13th AIAA/ASME/AHS Adaptive Structures Conf.* pp 18–21
[18] Talya S S, Rajadas J N and Chattopadhy A 2000 Multidisciplinary optimization for gas turbine airfoil design, Inverse Problems in Engineering 8:3 pp 283-308 DOI: 10.1080/174159700088027731
[19] Evangelista R, Pfenninger W, Mangalam S M et al 1987 Design and wind tunnel test of a high-performance low Reynolds number airfoil AIAA 87-2349
[20] Dilek FundaKurtulus 2016 Introduction to micro air vehicles Concepts Design and Applications https://www.researchgate.net/publication/304490025
[21] Lance Traub W 2011 Range and Endurance Estimates for Battery Powered Aircraft J. OF AIRCRAFT Vol. 48 No 2 2011http://www.smartgauge.co.uk/peukert2.html.

DESIGN OPTIMISATION OF SPUR GEAR USING GENETIC ALGORITHM

Ajay Pillai K[1], Aritra Guha Ray[2], Shivam Kaul[3], Narendiranath Babu.T[4]

[1]Undergraduate Student, Mechanical, VIT Vellore, Tamil Nadu
[2] Undergraduate Student, Mechanical, VIT Vellore, Tamil Nadu
[3] Undergraduate Student, Mechanical, VIT Vellore, Tamil Nadu
[4] Associate Professor, Mechanical, VIT Vellore, Tamil Nadu

Abstract. Gears are one of the most commonly used power transmission mechanisms in most types of machinery and. The design of gears is highly complicated involving the satisfaction of many constraints such as strength, pitting resistance, bending stress, scoring wear, and interference in involute gears. In addition, using conventional or traditional optimization techniques to solve this problem could not give optimum results. In this study, a spur gear pair was modelled and was subjected to static structural analyses for varying gear material. Stress analyses and deformation analyses were performed for each material and the optimum design for reduced weight structural stability was chosen. The design was then subjected to optimisation by Genetic Algorithm. A stochastic approach as a Genetic Algorithm (GA) is applied in this paper to find the optimal combination of design parameters for minimum weight of spur gears. The purpose of this study is minimizing the weight and the centre distance of one pair of spur gears. This objective was accomplished by the means of the GA under some constraint such as bending strength, a contact stress and each dimension conditions of gears, which must be satisfied. The results are calculated by using MATLAB tools of Genetic algorithm with three type of materials, which are alloy steel, cast iron, and epoxy glass composites. Keywords. Genetic Algorithm , MATLAB, epoxy glass composites.

1. INTRODUCTION

Spur gears are a type of cylindrical gear, with shafts that are parallel and coplanar, and teeth that are straight and oriented parallel to the shafts

A lot of research has been done in the past in the area of spur gear dimensioning and its characteristics, material used and stresses it is subjected to. But design optimisation requires further fine tuning in terms of weight reduction and several other parameters. Optimization design of gear reducer includes the continuous variables and discrete variables. Many studies indicate that the genetic algorithm has strong ability of general optimization, which is very effective in treating optimization problem containing continuous and discrete variables.

1.1 Literature Gap

It was observed that even though a thorough analysis was done on the material properties as well as the subsequent parameters that determine the life of the gear, such as stress and wear off rate, a final optimised design was never obtained.

IOP Conf. Series: Materials Science and Engineering 1123 (2021) 012011 doi:10.1088/1757-899X/1123/1/012011

2. OBJECTIVE FUNCTION

2.1 Centre Distance

$$a = \frac{m(z_1+z_2)}{2} \quad (1)$$

Thus, the objective function for minimizing centre distance of spur gear train is expressed as:

F1-objective = F (Z1, Z2, m) (2)

Where *a* is the centre distance, *m* is the module, *Z1* and *Z2* is the number of teeth on pinion and on gear respectively.

2. Gear Weight

A weight reduction of gears improves performance of nonstationary systems and also it saves the material, which leads to cost reduction and easy assembling. The following is the equation of the weight of the gear set

$$\frac{\pi}{4}\rho m^2 b(Z_1^2 + Z_2^2) \quad (3)$$

So, the objective function for minimizing weight is presented as:

F2-objective = F (Z1, m, b) (4)

Where *b* represents the tooth width and ρ the density of the material.

2.2 Constraints

The geometry of gears train is very specific. Therefore, they are many constraints of assembly, dimension, contact stress, and bending stress must be satisfied and introduced in the objective function for the best selection of optimum results. All the constraints are formulated as shown belo

The maximum and minimum tooth width:

$$g_1(x) = \frac{b}{m} - 12 \leq 0 \quad (5)$$

$$g_2(x) = 6 - \frac{b}{m} \leq 0 \quad (6)$$

The maximum and minimum number of teeth:

$$g_3(x) = z_1 - 30 \leq 0 \quad (7)$$

$$g_4(x) = 18 - z_1 \leq 0 \quad (8)$$

The constraint of design bending stress:

$$g_5(x) = \frac{i+1}{amby}[M_t] - [\sigma_b] \leq 0 \quad (9)$$

The constraint of design contact stress:

$$g_6(x) = \frac{i+1}{a}\sqrt{\frac{i+1}{ib}E[M_t]} - [\sigma_c] \leq 0 \quad (10)$$

3. METHODOLOGY

3.1 Methodology

This project attempts to achieve the most suitable design in terms of functioning and weight reduction, following the genetic algorithm.

A spur gear prototype was designed with the help of **SolidWorks** software. Its properties would be compared by assigning three different materials to it. Upon analysis of these properties, the material that shows the least static stress shall be chosen as the best among the given lot of materials.

After this, the spur gear shall be subjected to design optimisation in terms of weight reduction using the Genetic Algorithm and the obtained result shall be compared with that of the gear before it was subjected to the algorithm.

3.2 Modelling

The modelling of the gear parts was done on SolidWorks^{TM} 2016. Gear Parameters were chosen for the design considering the power requirements of a windmill. The gear parameters are shown in Table 2.

Table 1. Gear Design Parameters (Default Cast Iron)

Inputs			
Parameters	Symbol	Unit	Value
Power Output	P	kW	25
Gear Ratio	i	-	2.5
Pinion Speed	N1	rpm	1100
Working Life	N	cycles	1.32×109
Calculated			
Centre Distance	a	mm	157.500
Module	m	mm	5
Pinion teeth	Z1	-	18
Face width	b	mm	47.250
Revised Torque	[Mt]	kN-m	312.960
Design Bending Stress	σ_b	N/mm2	88.984
Design Contact Stress	σ_c	N/mm2	855.855

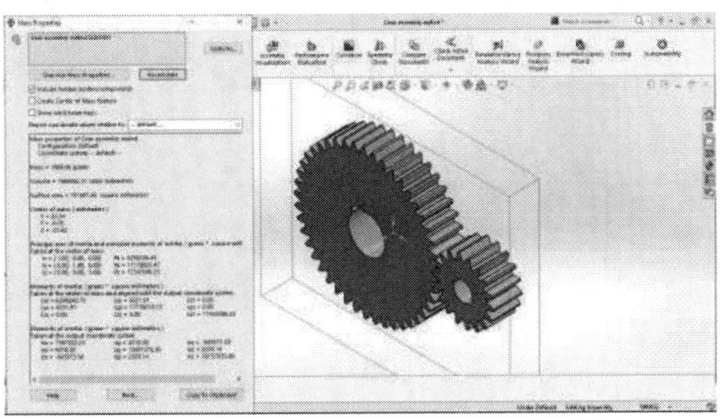

Figure 1. SolidWorks Model of Spur Gear

Table 2: Mass properties of Gear assembly as measured in SolidWorks (Configuration: Default)

Mass	1969.06 grams
Volume	1969062.01 cubic millimetres

3.3 Static Structural Analysis

The spur gear assembly was subjected to structural analysis in Ansys^{TM} 2016, Static Structural Software. The pinion and the gear wheel were subjected to a frictionless support on the interior surface of the hole where the shafts sit. The pinion was then applied with a turning moment (Figure 2).

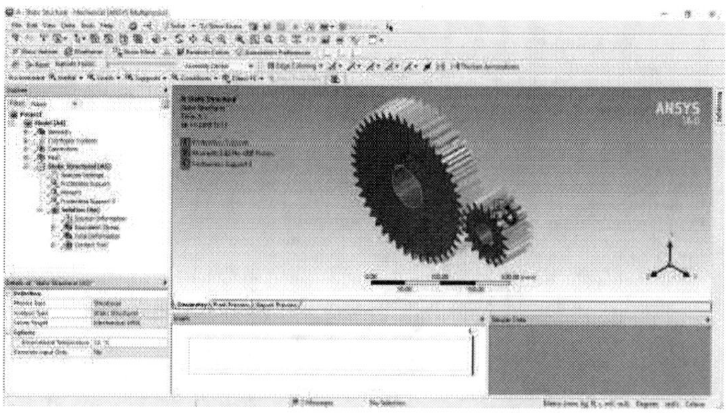

Figure 2. Static Structural Elements

The equivalent stress and total deformation were then observed for three different materials viz. Alloy Steel, Cast Iron, Epoxy Glass Composites. Results of stress analysis on Alloy Steel is shown in Figure 3.

Figure 3. Equivalent Stress in Alloy Steel

Total deformation analysis on Alloy Steel is shown in Figure 4.

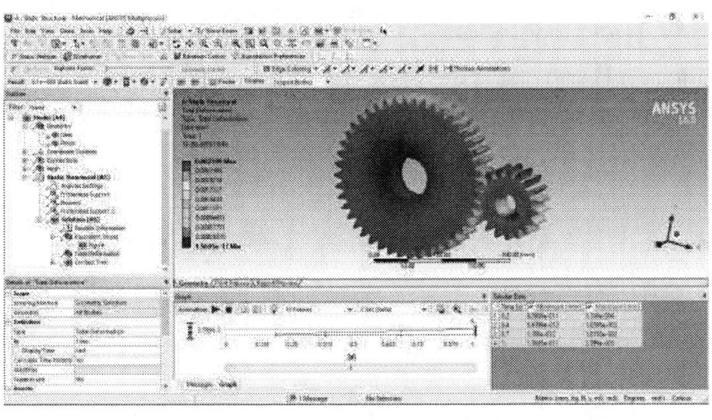

Figure 4. Total Deformation in Alloy Steel

3.4 Genetic Algorithm Implementation

The Optimization of weight and centre distance is carried out by using GA with help of a program developed on *MATLAB* platform. Many Factors are considered in this optimization as such as torque, material, width of tooth, gear ratio and input power as defined in the table below (Table 3).

Table 3. Design Variables and Parameters

Parameters	Symbol	Unit	Value
Material Density			
Alloy Steel	ρ	kg/m3	7850
Cast Iron	ρ	kg/m3	7200
Epoxy Glass Composite	ρ	kg/m3	1900
Young's Modulus			
Alloy Steel	E	MPa	2.1×105
Cast Iron	E	MPa	1.4×105
Epoxy Glass Composite	E	MPa	0.72×105
Number of Pinion Teeth	Z1	-	[18 30]
Module	m	mm	[1.25 6]
Face width	b	mm	[9 72]

B. Constraints:

The geometry of gears train is very specific. Therefore, they are many constraints of assembly, dimension, contact stress, and bending stress must be satisfied and introduced in the objective function for the best selection of optimum results. All the constraints are formulated as shown below:

4. RESULTS AND DISCUSSION

4.1 Structural Analysis Results:

Table 4. Structural Analysis Results

Material	**Equivalent Stress (MPa)**	**Total Deformation (mm)**	**Mass (kg)**
Alloy Steel	23.991	0.002599	15.457
Cast Iron	24.077	0.004605	14.177
Epoxy Glass Composite	28.86	0.029556	3.741

The analysis of the given materials shows that the stress developed and also the deformation was least in alloy steel. However, the mass of the gear made with this material was more as compared with cast iron and glass composite.

IOP Conf. Series: Materials Science and Engineering 1123 (2021) 012011 doi:10.1088/1757-899X/1123/1/012011

4.2 Genetic Algorithm Results:

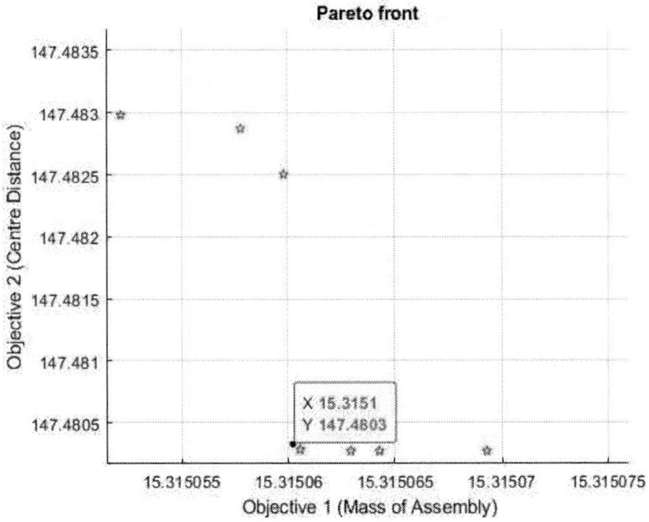

Figure 5. Optimization of Cast Iron

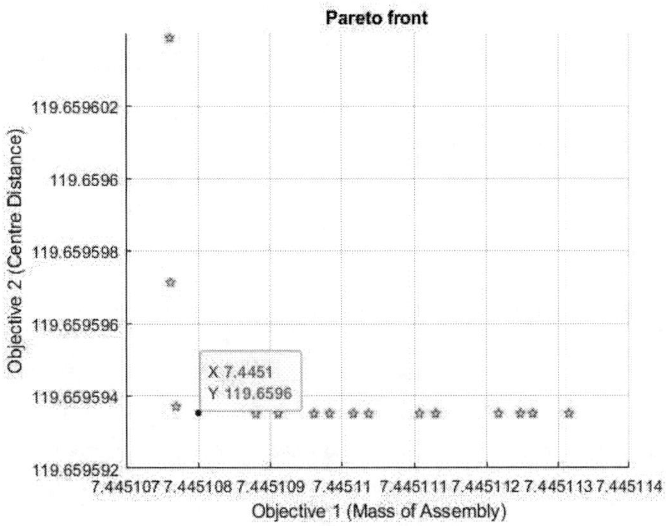

Figure 6. Optimization of Alloy Steel

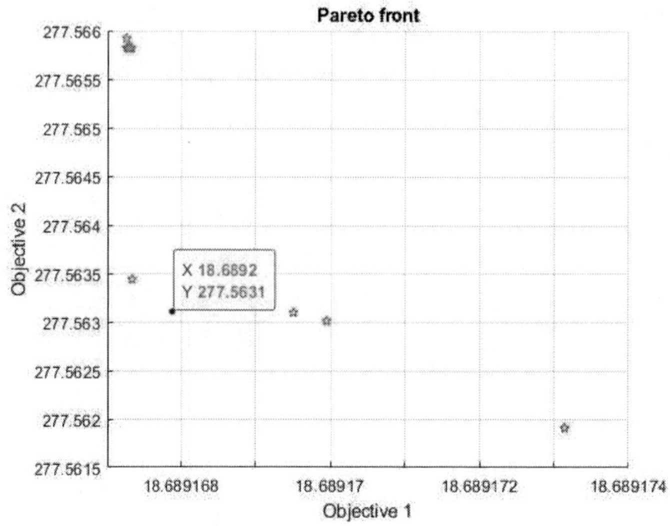

Figure 7. optimization of Epoxy Glass Composite

Table 5. Genetic Algorithm optimization Results

Materials	Gen	Z1	m (mm)	b (mm)	Centre Distance (mm)	Weight (kg)
Cast Iron	218	18	4.682	52.591	147.49	15.315
Alloy Steel	122	20	3.418	35.625	119.66	7.445
Epoxy Glass Composite	329	27.24	5.822	68.668	277.563	18.689

The table shows that alloy steel presents the best fitness of the weight of assembly and centre distance with a value of 7.445 kg and 119.66 mm respectively. The model of gears train with the epoxy glass composite represent the high weight and centre distance with an average in the tooth width value. This is because of the low design bending tress and design contact stress values for epoxy material. Thus, for the safety of design and to meet all the constraints the epoxy assembly needs to be much larger and bulker in size compared to the metallic assemblies. The optimal result of the module was found to be equal to 3.418 mm with the Alloy Steel material. The variation of the best fitness of the weight and centre distance are presented in the figures 8 to 9.

IOP Conf. Series: Materials Science and Engineering 1123 (2021) 012011 doi:10.1088/1757-899X/1123/1/012011

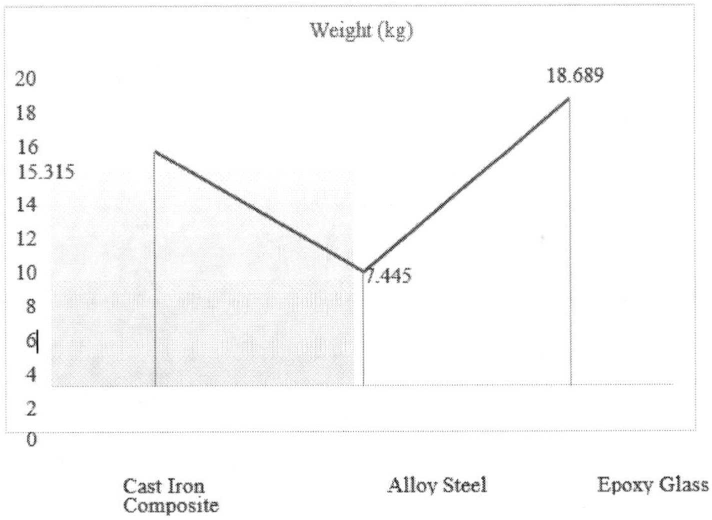

Figure 8. Weight vs Material

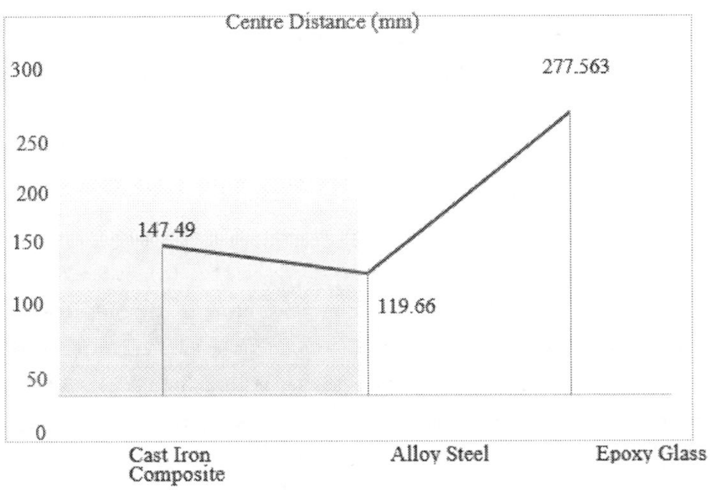

Figure 9. Centre Distance vs Material

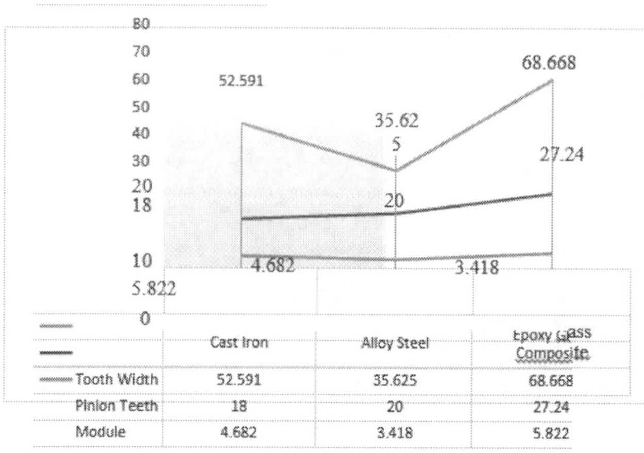

Figure 10. Optimum value of Design Variables for Different Materials

In figure 10, the optimal value of the number of teeth is presented in the case of cast iron. For the module and the tooth width, the alloy steel gives the best value. Therefore, the choice of materials is an exercise considering both the Genetic Algorithm results as well as the stress and deformation values obtained from the Structural Analysis. It gives us a vision on the concentrations of stress, allowable deformation, design modification possibilities and to estimate the lifecycle of the system.

5. CONCLUSION

Design optimization was performed on the spur gears as designed on SOLIDWORKS. Initially, three different materials viz. Alloy Steel, Cast Iron and Epoxy Glass Composite, were analyzed in terms of their equivalent stress and mass. Based on the acquired result, Alloy Steel was chosen as a suitable material in comparison with the others. Upon doing so, Genetic algorithm was incorporated to further enhance the design of the Alloy Steel spur gear. It was seen that alloy steel presents the best fitness of the weight of assembly and center distance with a value of 7.445 kg and 119.66 mm respectively. The optimal result of the module was found to be equal to 3.418 mm with the Alloy Steel material.

REFERENCE

[1] Peruboina Nagavasavi and Govindaiah T 2015 The difference in properties of various materials used to make the spur gear using finite element analysis Vol 4 Issue No 12

[2] Srenatthan P L and Vignesh P 2016 Analyse fatigue wear and weight reduction in spur gears used in heavy duty vehicles Vol 1 Issue 1

[3] Madhusudan G and Vijayasimha C R To obtain alternate designs for a given set of primary input data under the influence of varying secondary parameters

[4] Kaoutar Daoudi and El Mostapha Boudi Genetic algorithm approach for spur gears design optimization

[5] Pawar P B and Abhay Utpat A Replacement of conventional metallic gears with suitable composite material in order to improve performance

[6] Houck C R, Joines J and MG Kay 1995 A genetic algorithm for function optimization: a Matlab implementation Ncsu-ie tr *igi.tugraz.at Ncsu-ie tr* igi.tugraz.at

[7] Chipperfield A J and Fleming P J 1995 IET The MATLAB genetic algorithm toolbox.

[8] Yokota T, Taguchi T and Gen M 1998 A solution method for optimal weight design problem of the gear using genetic algorithms *Computers & industrial engineering Elsevier* Vol 35 Issues 3–4 pp 523-526

[9] Wang H and Wang H 1994 Optimal engineering design of spur gear sets Mechanism and machine theory *Elsevier* Vol 29 Issue 7 pp 1071-1080

[10] Mohan Y M and Seshaiah T 2012 Spur gear optimization by using genetic algorithm- *Int. J. of Eng. Research Citeseer* Vol 2 Issue 1 pp 311-318

Design Optimization of Universal Joints for All-Terrain Vehicles

Sagar Yanda[1], Gunji Bala Murali [2*], Dr. Siva Shankar raju[3], Bill Jason Duckworth [4], Chandra Sen [5], Sidakdeep Singh Chadha [6]
[2, 4, 5, 6] Department of Design and Automation, Vellore Institute of Technology, Vellore, Tamilnadu, India 632014
[1, 3] Department of Mechanical Engineering, Aditya Institute of Technology and Management, Tekkali, India 532201
[1]sagaryanda@gmail.com, [2]balamurali.g@vit.ac.in, [3]sivaraju80@gmail.com, [4]billjason.duckworth2016@vitstudent.ac.in, [5]chandra.sen2016@vitstudent.ac.in, [6]sidakdeepsingh.2016@vitstudent.ac.in.

Abstract. Universal joint mechanism widely used in the automobile power transmission systems. This joint mechanism desired to permit some angular deviations along the axis of rotation. The cardan joint comprises of the three main parts including the input drive shaft, the output drive shaft and the cross like piece. Two points of the cross piece connect to the input drive shaft and two other points connect to the output drive shaft. Connections are provided by the needle bearings. It is an important aspect of these bearings that while in action they never go through complete cycles. In other words, each of these bearings revolves only a few degrees around its axis before returning to its original position. The paper aims to develop a UV joint that is suitable for All-Terrain Vehicles (ATVs). These types of vehicles have to traverse rough terrain and hence require taller suspension set up. Conventional universal joints limit this ability as the allowable articulation angle is very less. Due to this, they are prone to wear and malfunction and need to be replaced in comparatively short intervals of time. Naturally, it means that such parts have limited life span. We have hence developed a UV joint design which allows for better articulation angle while maintaining strength and reducing weight at the same time. **Keywords**: Universal joint, All-Terrain Vehicles, Computer Aided Design, Ansys.

1. Introduction

The idea of a mechanism that can provide slight inclination to the direction of the rotating shafts has long been thought of by the engineers. Failure of the universal joints can end up with serious consequences and it can be very costly. It causes sudden disruption in the supply of power between its source and the consumer device. Therefore, many studies are performed in order to recognize the nature of the corresponding forces and the failure within these mechanisms. All such studies concentrated on the failure in the universal joints or its connecting rods, in a general sense. There is hardly any research report specifically concerned with the optimization of the design of a universal joint to prevent failure while improving the articulation angle and the weight.

Sheikh, S. M used Pro/E software for the design and HyperMesh for analysis of UV joints. Basically the author focused on modification of UV joint geometry using topology optimization concept and free size optimization concept to reduce the mass of UV [1]. Neve et al. [2] uses different epoxy materials of composites for driving shaft. Out of those, Kevlar/Epoxy composite has given maximum strength compared to the others composite materials. During the study, 90° fibre angle is giving better fundamental frequency compared to other angles. Vasekar et al. [3] is concerned with the failure analysis and weight optimization of a universal joint yoke of an automobile power transmission. The weight optimization has been carried out with help of hyper mesh to reduce the weight from 1.11 kg to 0.967kg (12.8% reduction in weight).

Gagandeep Singh & Rajbir Singh [4] have proposed optimised universal joint using two materials to reduce the stress and mass of UV joint. The author has done failure analysis also to know at portion the UV joint fails. Vesali, Farzad et.al. [5] have analysed the dynamics of universal joint and failures for the improvement of the UV joint. This is achieved by the deriving the different motions associated with the UV joint.

Kang, and Weidan Wang [6] investigated faults associated with the double cross universal joint, including looseness, wear, and intermediate shaft torsional deformation. First, the authors propose a transmission model and a new fault model for this joint. Then, the model parameters of the optimal transmission performance are analyzed. Finally, the main contribution of this article is in discussing the transmission performance as related to looseness, wear, and the intermediate shaft torsional deformation.

The paper is organised as follows: section-1 deals with introduction and literature of the ASP problem. Section-2 explains about the improvement made for the fruit fly algorithm and how it is implemented to solve ASP problem. Section-3 deals with the results obtained from the proposed methodology. Section-4 explains about conclusion made from the proposed method in the paper.

2. Methodology

In this section, testing of universal joints made by AISI4340 and Al7075 for factor of safety, minimum deformation, stress analysis and weight optimization. Finite element analysis method is used for stress analysis to determine the stress condition at failed section.

The main aim is to design a universal joint that provides maximum articulation angle while maintaining strength of the joint and also reducing the weight. The cup was designed keeping in mind an Original Equipment Manufacturer (OEM) part by DANA SPICER used for TATA Ace vehicle. This helped us to arrive at a baseline for designing the part. After performing many iteration and were analyzed at each stage to arrive at the final part.

After obtaining the basic final design, different parts of the driveshaft such as the spindle and flange were created. The shaft was designed to be a hollow shaft as it provides much better torsional rigidity as compared to a solid shaft.

The designs were analyzed for both aluminum 7075 and steel 4340. Aluminum is a lighter metal and proves to be more weight effective. But at the same time, it compromises on strength. Due to this, certain sections need to be made thicker in order to preserve the strength of the joint.

Aluminum is a very desirable metal because it is more malleable and elastic than steel. Aluminum can go places and create shapes that steel cannot, often forming deeper or more intricate spinning's. Especially for parts with deep and straight walls, aluminum is the material of choice. Steel is a very tough and resilient metal but cannot generally be pushed to the same extreme dimensional limits as aluminum without cracking or ripping during the spinning process.

While malleability is very important for manufacturing, aluminum's greatest attribute is that it is corrosion resistant without any further treatment after it is spun. Aluminum doesn't rust. With aluminum there is no paint or coating to wear or scratch off. Steel or "carbon steel" in the metals world usually needs painted or treated after spinning to protect it from rust and corrosion, especially if the steel part will be at work in a moist, damp or abrasive environment.

Even with the possibility of corrosion, steel is harder than aluminum. Most spinnable tempers and alloys of aluminum dent ding or scratch more easily as compared to steel. Steel is strong and less likely to warp, deform or bend underweight, force or heat. Nevertheless the strength of steel's tradeoff is that steel is much heavier /much denser than aluminum. Steel is typically 2.5 times denser than aluminum.

Thus, aluminum is a more suitable material as compared to steel as we get weight reduction along with good strength.

2.1 Modeling

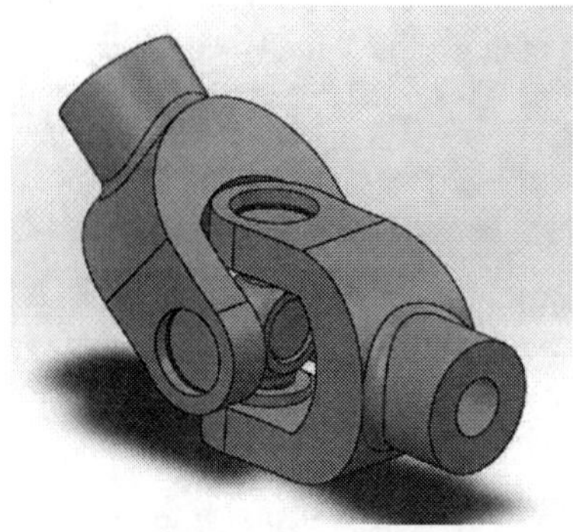

Figure 1. Existing Design (based on the Polaris RANGER)

Figure -1 represents the existing model of UV joint used in ALL-Terrain Vehicles (Polar Ranger) designed in Solid Works. Keeping in mind the inherent design inconstancies of the conventional design used in ATVs, which were discussed in introduction section, a new designed is proposed by changing the articulation angle. The modified design with exploded view is shown in Figure 2 designed in Solid Works.

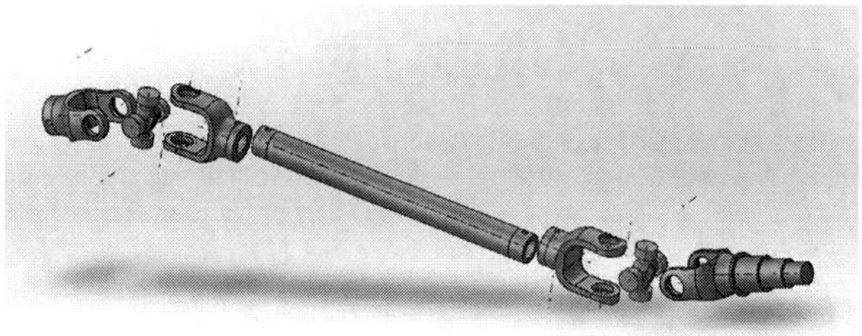

Figure 2. Represents the exploded assembly view

All these parts of the back transmission were created keeping in mind the primary goal of increasing the ride height of the ATV while reducing weight and preserving the strength of the joint. The detail specification about all the parts is shown in the table-1.

Table 1. Represents the specifications of transmission system parts

Specification	Old design	Optimized design
Shape	Straight cup	Tapered cup
Shaft OD	30mm	35mm
Flange thickness	4mm	6mm
Cup diameter	65mm	67mm

2.1 Simulation in ANSYS

Finite Element Analysis (FEA) was carried out on the individual modules of the joint in ANSYS Structural. Using these simulations, we optimized the individual modules for stress, deformation and FOS. Simulation was conducted for both steel and aluminum joints.
The reason for analyzing individual modules instead of the entire assembly was that it helps to reduce the computation time required and also helps to reinstate that each part will be able to suffice to the requirement.

The parts were meshed with a 1mm mesh in automatic mode to uniformly mesh the entire part. The torque was applied to either the "ears" of the yoke or the internal region of the yoke depending on its position in the assembly. The torque was applied to either the "ears" of the yoke or the internal region of the yoke depending on its position in the assembly.

The torque was calculated as follows from the previous literature data:

Engine Torque =19 Nm

CVT reduction ratio= 3.9:1

Gearbox reduction ratio-=7.2:1

Overall reduction ratio= CVT reduction ratio x Gearbox reduction ratio (1)

=28.08

Driveshaft Torque=Overall ratio x Engine Torque (2)

=533.33 Nm ~600 Nm (approx.)

These values were taken into account while designing the joint and analyzing it. These values represent the maximum possible value of torque that may be experienced by the joint at any given point of time.

Steel AISI 4340

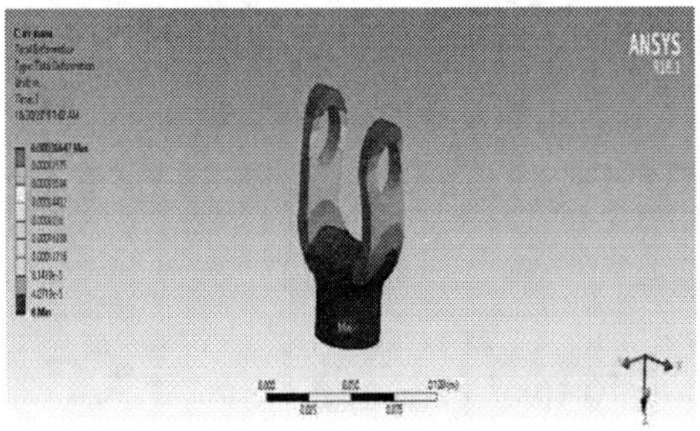

Figure 3. Represents the yoke deformation

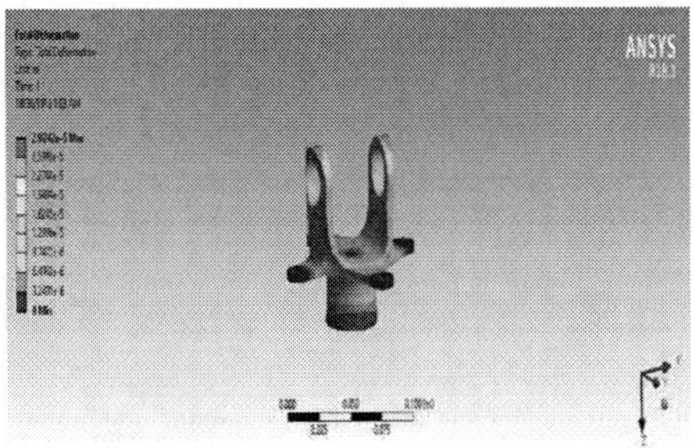

Figure 4. Represents the Flange deformation

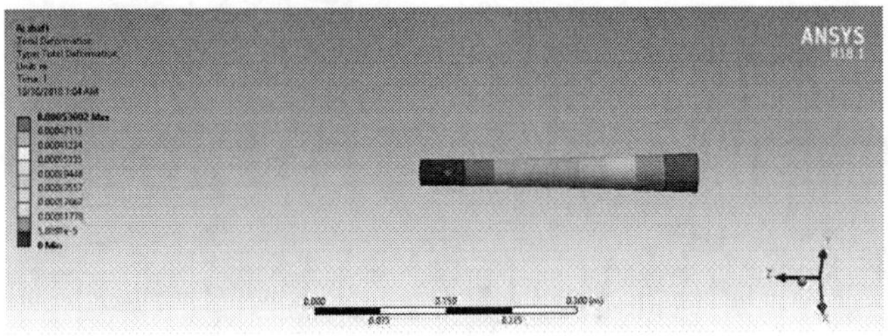

Figure 5 Represents the shaft deformation

In Figures 3, 4 & 5 represents the deformation of yoke, flange and shaft of back transmission system made of steel AISI 4340 for ALL-Terrain Vehicle.

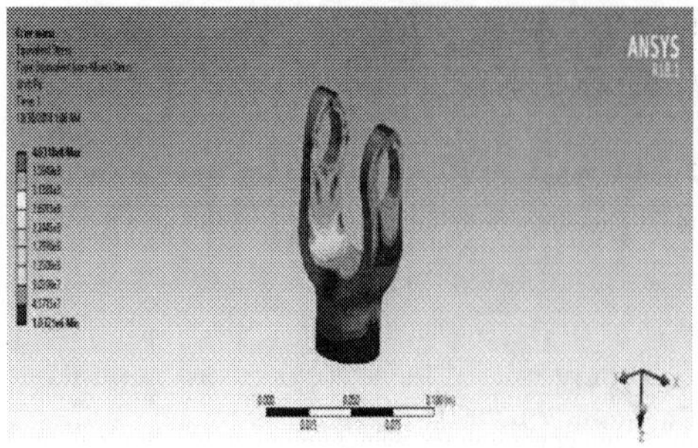

Figure 6. Represents the stress of **yoke**

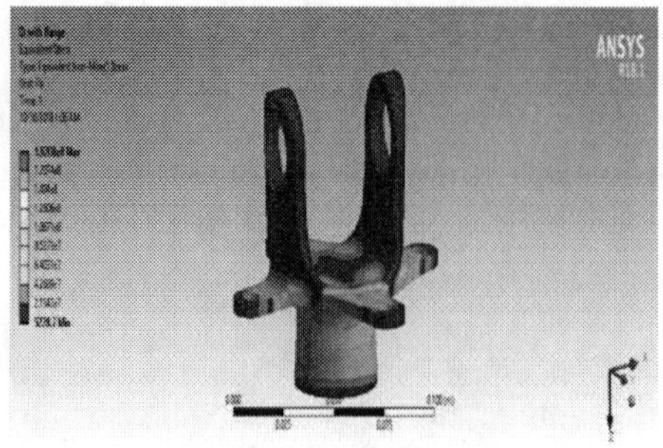

Figure 7. Represents the stress of flange

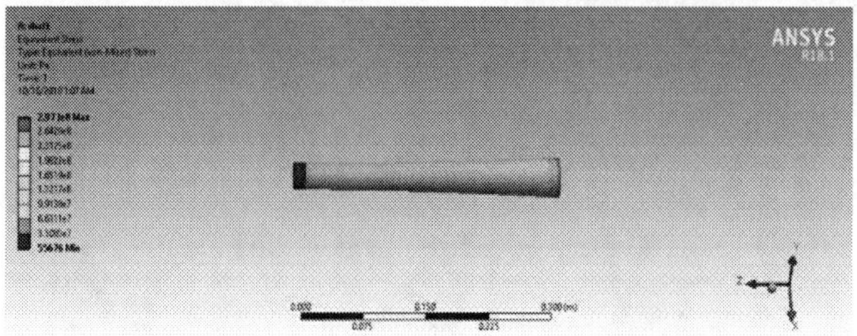

Figure 8. Represents the stress of shaft

In Figures 6, 7 & 8 represents the stress of yoke, flange and shaft of back transmission system made of steel AISI 4340 for ALL-Terrain Vehicle.

Aluminum 7075 (T6)

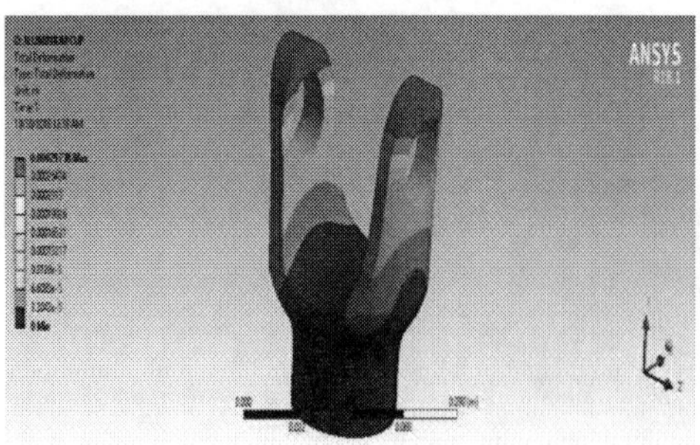

Figure 9. Represents the deformation of yoke

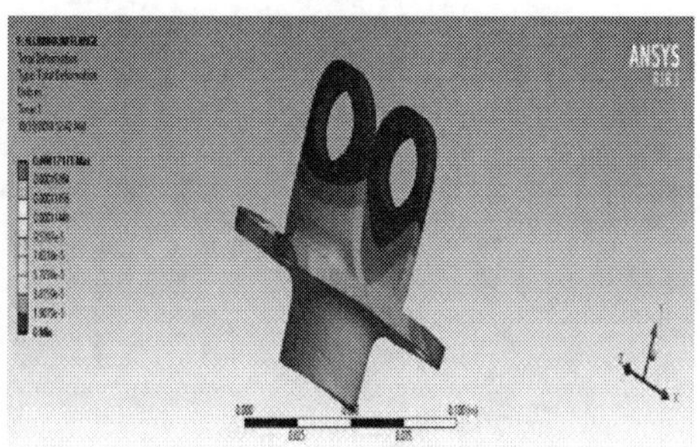

Figure 10. Represents the deformation of flange

In Figures 9, 10 & 11 represents the deformation of yoke, flange and shaft of back transmission system made of aluminum 7075 for ALL-Terrain Vehicle.

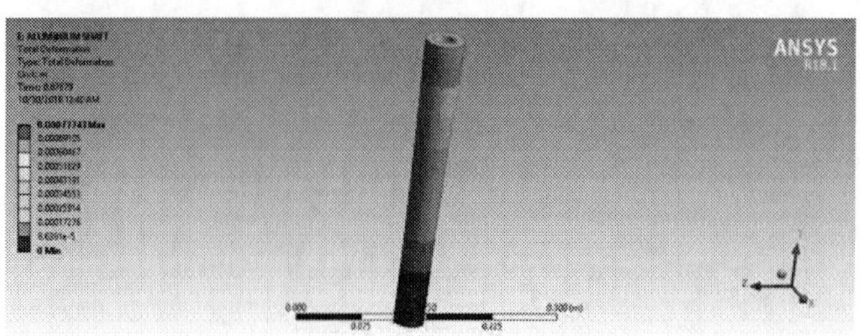

Figure 11. Represents the deformation of shaft

IOP Conf. Series: Materials Science and Engineering 1123 (2021) 012012 doi:10.1088/1757-899X/1123/1/012012

In Figures 12, 13 & 14 represents the stress of yoke, flange and shaft of back transmission system made of aluminum 7075 for ALL-Terrain Vehicle.

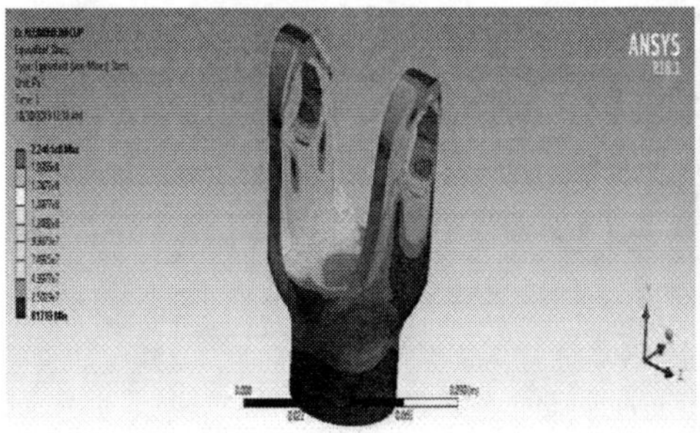

Figure 12. Represents the stress of yoke

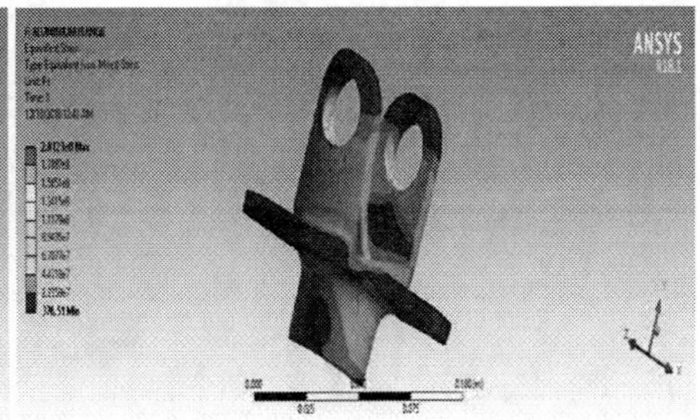

Figure 13. Represents the stress of flange

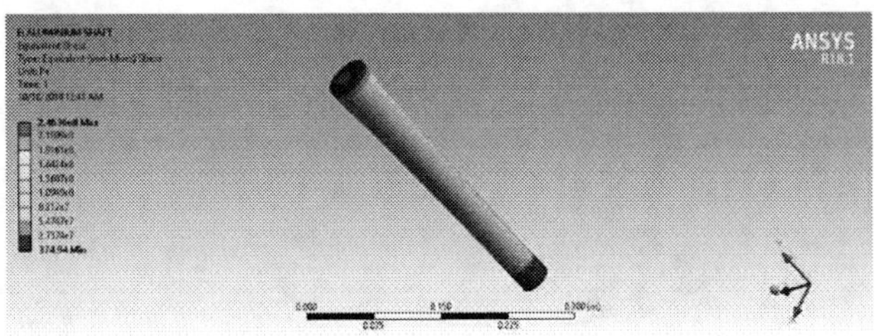

Figure 14. Represents the stress of shaft

3. RESULTS AND DISCUSSION

From the results obtained from the analysis, AISI 4340 steel UV and Al 7075 UV have close resemblances in their respective results. This shows that both the materials will be suitable for the universal joint. But, Aluminum will be preferred over Steel as it is much lighter and is providing us with similar results in both cases. As can be seen, the values of stress and deformation are well in the acceptable limits of both the materials. The Factor of Safety is satisfactory for the required application of off-roading.

Our primary motive of increasing the ride height of the vehicle has been achieved as the max permissible angle is 65 degrees (approx.) as compared to 37.45 degrees for the conventional design. We also see a 78% reduction in weight for the Aluminum design as compared to the conventional design

Table-2: Represents the different analysis values of steel and aluminum

Type of analysis	**Steel UV**	**Aluminum UV**
	Yoke	
Deformation	0.36mm	0.29mm
Stress	403 MPa	224 MPa
	Shaft	
Deformation	0.53mm	0.77mm
Stress	231 MPa	210 MPa
	Flange	
Deformation	0.23mm	0.17mm
Stress	192 MPa	201 MPa

Table-2 represents the various analysis values of steel and aluminum material of UV joint used in ALL-Terrain Vehicles. Form the table Al is having better results than the Steel.

Figure 15. Angle before Optimization (37.45 ◦)

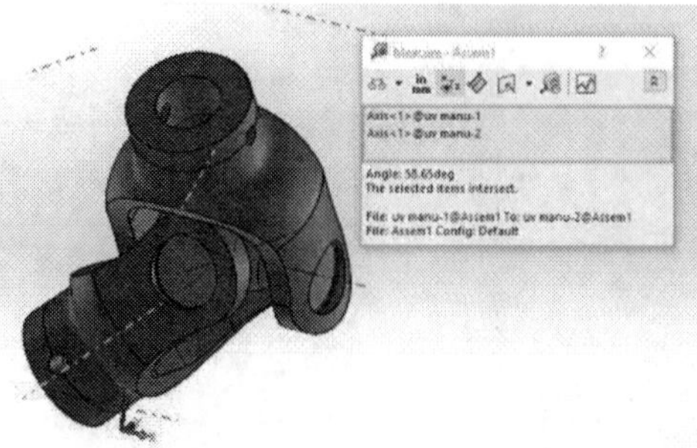

Figure 16. Angle after Optimization (58.65°)

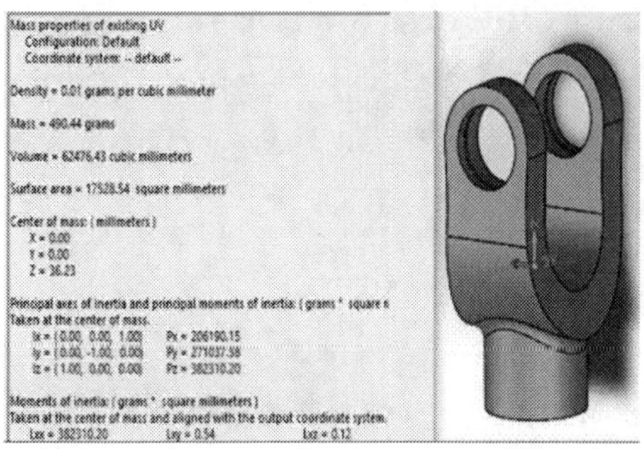

Figure 17. Weight of original design (490.44 grams)

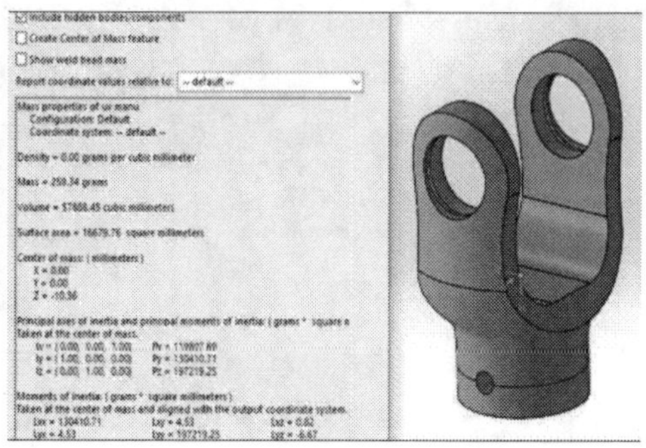

Figure 18. Weight of optimized design (259.34 grams)

Figures 15, 16, 17 &18 represents the original and modified articulation angle of UV joint and weight reduction of yoke respectively.

4. Conclusion

In this paper, optimization of UV joint design by changing the articulated angle is carried out for use in higher ride height vehicles such as ATVs. The weight has also been lowered while maintaining strength of the joint. This can help to improve the fuel efficiency of the vehicle in the long run. To optimize the articulated angle, two types of materials has been considered: steel and aluminum and different analysis like deformation, stress and FOS has been done to generate modified shape of UV joint with reduction in weight.

References

[1] Sheikh, Siraj Mohammad Ali 2012 Analysis of universal coupling under different torque Condition." *Int. J. Eng. Sci. Adv. Technol* Vol 2 Issue 3 pp 690-694

[2] Maheta, Virendra V, Anup Patel B, Virendra Maheta V and Anup Patel B 2015 Design, Analysis and optimization in Automobile Drive shaft Int. J. Vol 1 pp 432-439.

[3] Vasekar, Avinash C, and Ranjitsinha Gidde R 2015 Failure analysis and optimization of universal joint yoke subjected by torsion and shear." Int. research J. of Eng. and Tech. Vol 2

[4] Singh, Gagandeep, and Rajbir Singh 2016 FEM Based Optimization of Universal Joint Using Two Materials Imperial J. of interdisciplinary Research Vol 2 Issue 11

[5] Vesali, Farzad, Mohammad Ali Rezvani and Mohammad Kashfi 2012 Dynamics of universal joints, its failures and some propositions for practically improving its performance and life expectancy J. of mechanical science and Tech. Vol 26 Issue 8 pp 2439-2449.

[6] An, Kang and Weidan Wang 2017 Transmission performance and fault analysis of a vehicle universal *joint Advances in Mechanical Engineering* Vol 9 Issue 5 1687814017707478.

Design, Analysis, Control and Manufacturing of a Head Massaging Machine

[1]Aman Chaure, [2]Rajkumar.E

[1]Under Graduate student, SMEC, VIT, Vellore, India,
amanchaure@gmail.com,

[2]Associate professor, Department of Design and Automation, SMEC, VIT, Vellore, India

rajkumar.e@vit.ac.in

Abstract. Nowadays, a lot of people are suffering from unexpected Headaches at odd hours of the day due to their hectic schedule and busy lifestyle. Many devices for massage therapy have already been introduced in the market for the same but none of them are actually worth the price as they fail to deliver the quick necessary relief. For overcoming this problem, we have come up with a new design for head massaging which mainly works on the ideal concept of massage therapy by pressing the optimum pressure points located at the extremes of the forehead. The novelty of this paper lies in the mechanism involved that will help us to achieve the desired output. We will be presenting the complete design, control and manufacturing for the product by a series of simulations which will includes Linear Static Stress analysis for evaluating the strength of the design which can produce the right amount of force, Vibrational/Modal Frequency response for analysing the stability while functioning, Fatigue Test for evaluating the maximum stress the design can bear for a good life and finally comes the Thermal Analysis for evaluating the heat dissipation from the electronics. The Computer Aided Design will be developed on SOLIDWORKS Software and the Finite Element Simulations will be conducted on ANSYS for optimization. Some post simulations will also be conducted such as drop test for evaluating the chances of sudden failure under the circumstances of impact loading. At last there will be a complete manufacturing planning for the entire design along with its DFMA planning.
Keywords. Optimization, Computer Aided Design, Finite Element Simulations

I. Introduction

Massage therapy the most important part of physiotherapy), is one of the most useful way of relieving ourselves from mental stress. For this a head massaging machining has been designed with actuators in the most promising location for the quick relief operation. A rocker arm mechanism has been modified and designed to serve the purpose. The push rod of the rocker arm along with the cam shaft is replaced with the Linear actuators for serving the same purpose. The effective output of work maintained by the designed control system for optimum functionality

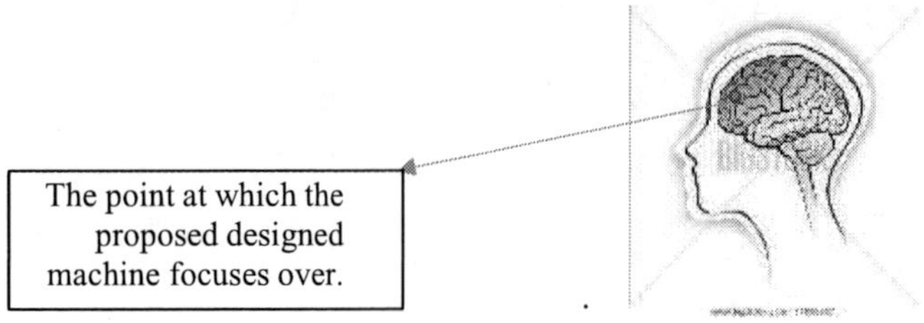

Figure 1. The point of Action of Force for massage

IOP Conf. Series: Materials Science and Engineering 1123 (2021) 012013 doi:10.1088/1757-899X/1123/1/012013

2. Literature Review

Juvenile Rheumatoid Arthritis: Benefits from Massage Theraphy [1], this paper focuses on the effect of head massaging over the children for decrease in the stress levels and the anxiety, Head massage machine, a currently existing Patent in the same field [2]. Hand type electric massage machine, another consisting patent [3], The short-term effects of myofascial trigger point massage therapy on cardiac autonomic tone in healthy subjects [4], this paper contains the study of point massage therapy for stress relieving effects over the body, The effect of massage in patients with chronic tension headache [5].

3. Problem description

A lot of designs are currently existing under the head massaging division or category. But our design mainly focuses over the two points each on either side of the forehead serving only a single action but having the greatest effect of its little functionality. The main problem is that all the currently existing designs are focusing over multiple points of the head and fore head but are not capable of providing quick relief by the same. For this problem I have come up with a very simple solution by focusing all the energy of a small device over the most promising location of the head.

4. Methodology

The methodology for out design is quite simplistic. We have started with rough design for the same product and then by analysing the same with Triz contradiction matrix followed by a 2nd design. Then Simulating the same with Static Analysis in SOLIDWORKS Software for analysing its functionality. The detailed study is giving in the steps followed here.

Step 1: 1st conceptual Design, containing the details just about the basic idea.
Step 2: Triz Contradiction Matrix with results, for improving the design on the basis of Triz concepts.
Step 3: 2nd iterative design with the changes proposed by the Triz matrix, giving us the improves design.
Step 4: Analysing the same design for its functionality in SOLIDWORKS Simulation package.

Step 5: The manufacturing planning for the same using FLASHPRINT Software for printing the structure in a single go.
Step 6: The Control Planning for the same with Arduino UNO and a Linear Motor.

TRIZ Contradiction Matrix

S No.	Improving Factors\Worsening Factors:	20 Stress	21 Stability	24 Function Efficiency	27 Loss Energy	40 Harmful Effects of Generated System	41 by Manufacturability
1	weight of Moving Object	28, 31, 40, 35	35, 1, 30, 39	1, 2, 28, 3	19, 6, 3, 10	5, 24, 27, 21	1, 5, 36, 25
2	weight of stationary Object	35, 31, 8, 40	15, 5, 17, 30	35, 31, 7, 3	28, 35, 7, 31	17, 37, 31, 3	35, 8, 9, 1
12	Duration of Action of Moving Object	35, 3, 17, 14	35, 24, 40, 13	13, 1, 19, 12	10, 24, 35, 23	21, 28, 15, 24	35, 4, 10, 3

15	Force/Torque	35, 14, 9, 3	35, 10, 24, 19, 35, 6, 13	19, 15, 2,	10, 35, 40, 22	15, 18, 35, 6
16	Energy used in the moving object	19, 35, 5, 9	13, 19, 24, 2, 13, 28, 12	19, 15, 12,	2, 10, 3, 22	28, 5, 30, 10

Outcomes of the Triz Matrix:

• Principle 3. Local quality
• A. Change an object's structure from uniform to non-uniform, change an external environment
• B. Make each part of an object function in conditions most suitable for its operation.
• C. Make each part of an object fulfill a different and useful function.
• Principle 5. Merging
• A. Bring closer together (or merge) identical or similar objects, assemble identical or similar parts toperform parallel operations.
• B. Make operations contiguous or parallel; bring them together in time.
• Principle 6. Universality
• A. Make a part or object perform multiple functions; eliminate the need for other parts.

5. Modelling

The isometric View for the Model is shown from both Front and Back Side in the figures 1 to 2. Then comes the orthographic views from figure 3 to figure 4. In these figures the dimensional views are presented along with all the dimensions in mm.

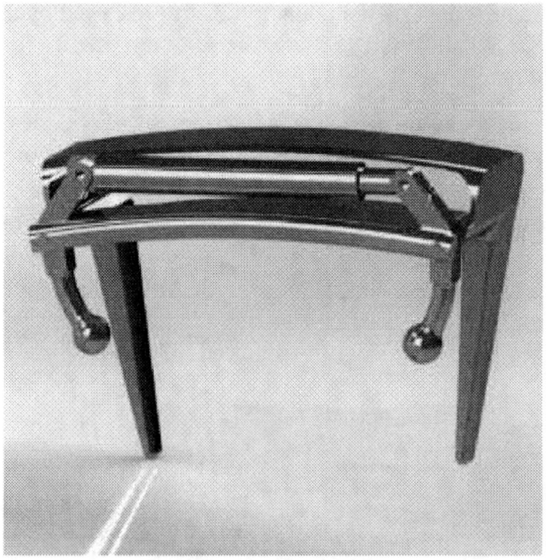

Figure 2. Isometric view of the Head Massaging Machine.

IOP Conf. Series: Materials Science and Engineering 1123 (2021) 012013 doi:10.1088/1757-899X/1123/1/012013

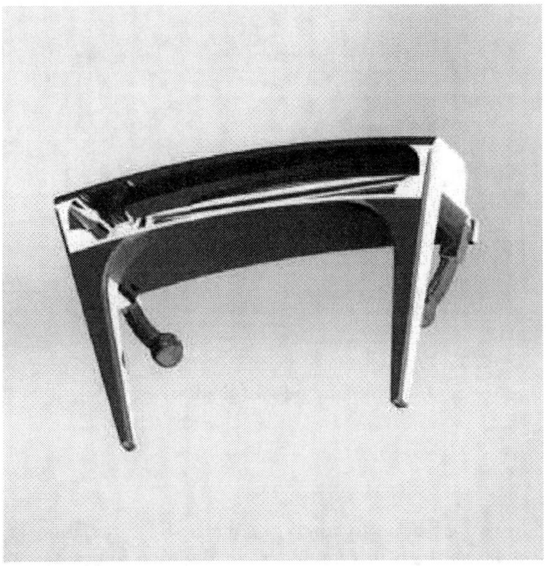

Figure 3. Isometric view of the Head Massaging Machine.

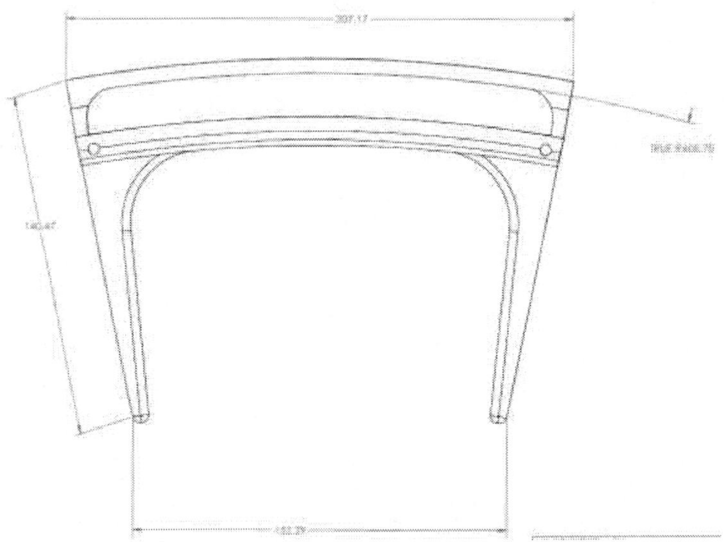

Figure 4. Orthographic view of the top head part.

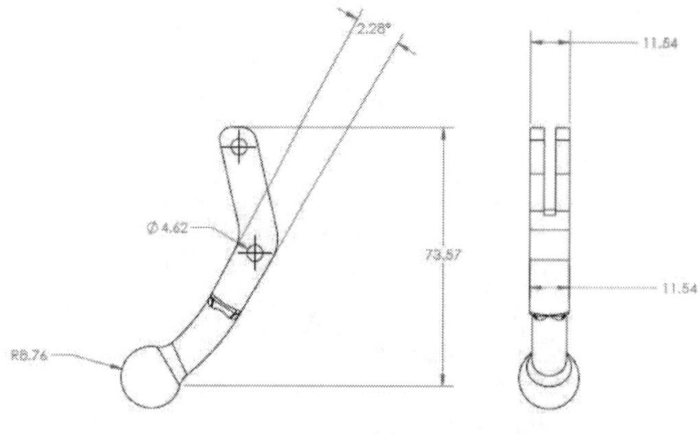

Figure 5. Orthographic view of the top head part.

6. Experimentation

The Finite Element Analysis was done to carry out the Static Simulation to Check the Stress and Deformation and The Strain over the Structure under the Reaction forces over the Structure itself. The Factor of Safety was also Calculated for the same for verifying the overall safety limits for the Functionality.

Table 1. Study Properties

Study name	Static 1
Analysis type	Static
Mesh type	Solid Mesh
Thermal Effect:	On
Thermal option	Include temperature loads
Zero strain temperature	298 Kelvin
Include fluid pressure effects from SOLIDWORKS Flow Simulation	Off
Solver type	FFEPlus
Inplane Effect:	Off
Soft Spring:	Off
Inertial Relief:	Off
Incompatible bonding options	Automatic
Large displacement	Off
Compute free body forces	On
Friction	Off
Use Adaptive Method:	Off

Result folder	SOLIDWORKS document (C:\Users\AMAN CHAURE\Desktop\books to complete\New product development)

7. Results and Discussions

Units

Unit system:	SI (MKS)
Length/Displacement	mm
Temperature	Kelvin
Angular velocity	Rad/sec
Pressure/Stress	N/m^2

Material Properties

Model Reference	Propertie	Component
	Name: **Nylon 6/10** Model type: **Linear Elastic Isotropic** Default failure criterion: **Max von Mises Stress** Yield strength: **1.39043e+08 N/m^2** Tensile strength: **1.42559e+08 N/m^2** Elastic modulus: **8.3e+09 N/m^2** Poisson's ratio: **0.28** Mass density: **1,400 kg/m^3** Shear modulus: **3.2e+09 N/m^2**	**SolidBody 1(Boss-Extrude1)(13 mm pin-1), SolidBody 1(Boss-Extrude1)(13 mm pin-2), SolidBody 1(Boss-Extrude1)(28 mm pin-1), SolidBody 1(Boss-Extrude1)(28 mm pin-2), SolidBody 1(Scale1)(cylinder-1),** **SolidBody 1(Cut-**
Figure 6: Model with Constraints		

Resultant Forces

Reaction forces

Selection set	Units	Sum X	Sum Y	Sum Z	Resultant
Entire Model	N	1.04606e-05	-15.4034	3.45294	15.7856

Reaction Moments

Selection set	Units	Sum X	Sum Y	Sum Z	Resultant
Entire Model	N.m	0	0	0	0

Name	Type	Min	Max
Displacement1	URES: Resultant Displacement	0.000e+00 mm Node: 3046	6.210e-02 mm Node: 3114

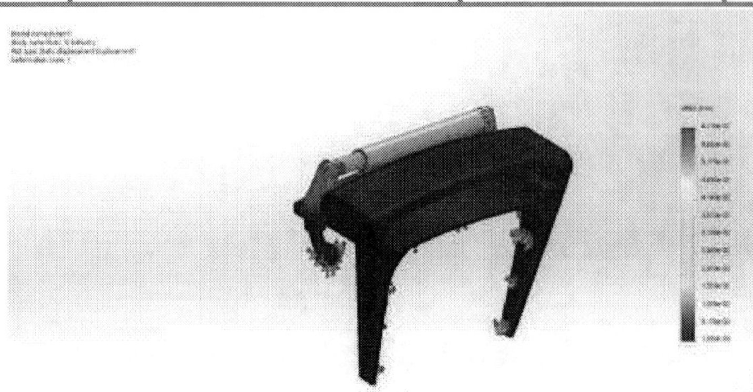

Figure 7. Displacement Contour

Name	Type	Min	Max
Strain1	ESTRN: Equivalent Strain	7.819e-11 Element: 9115	2.921e-04 Element: 2417

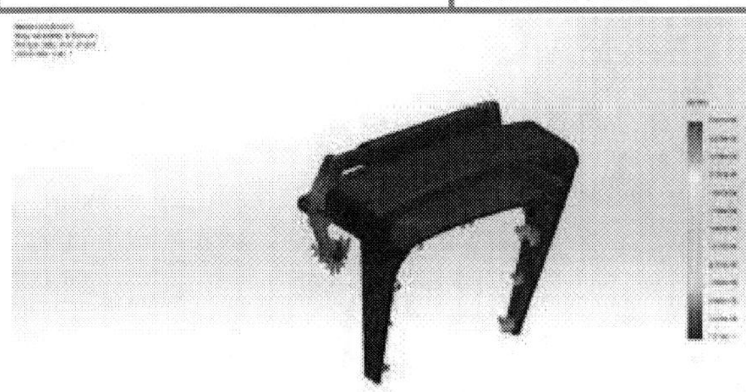

Figure 8. Strain Contour

Name	Type
Displacement1{1}	Deformed shape

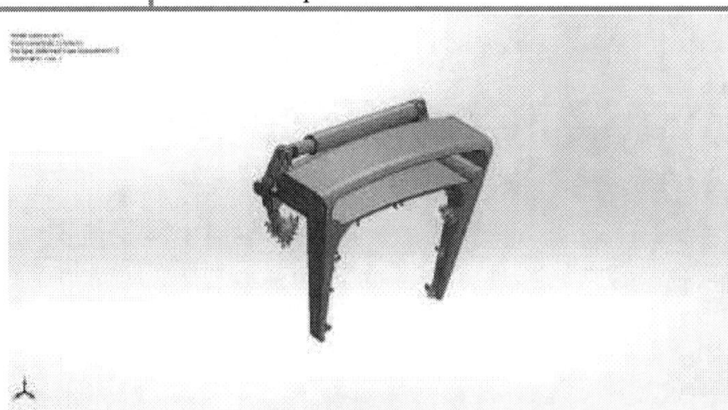

Figure 9. Deforment

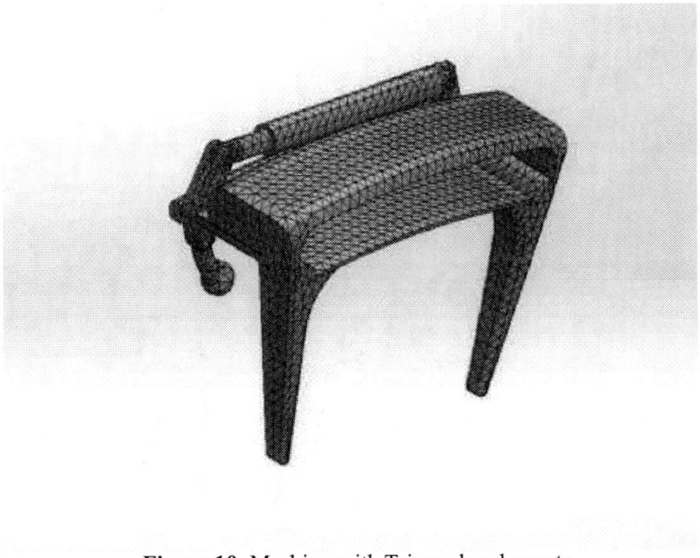

Figure 10. Meshing with Triangular elements.

8. Manufacturing Simulation

The recent Trends in the market are the advancements in the additive manufacturing. With the ease of Manufacturing, additive manufacturing plays a very important role in the new Product Development. Thereby we will be using the 3D printing with Flash Forge Guider II printer for Additively Manufacturing the product with Nylon 6 material. The hinges and pins of Standard sizes can be used for assembly purpose if not preferred for printing.

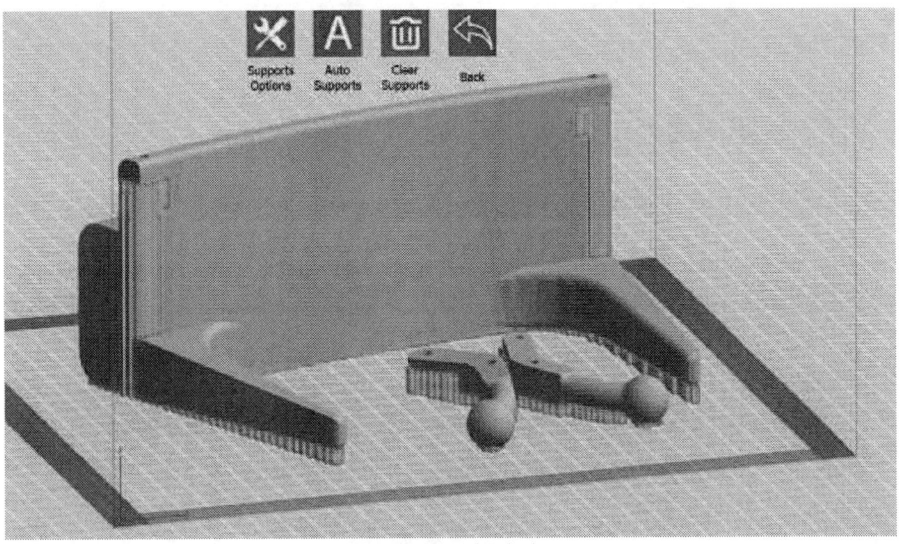

Figure 11. 3d Printing Simulation

Time required for Printing with Flash Forge Guider Printer is Calculated to be around 27 Hours and its analysis can be seen in the below photograph & The Preferred Configuration for Printing Parameters is given below:

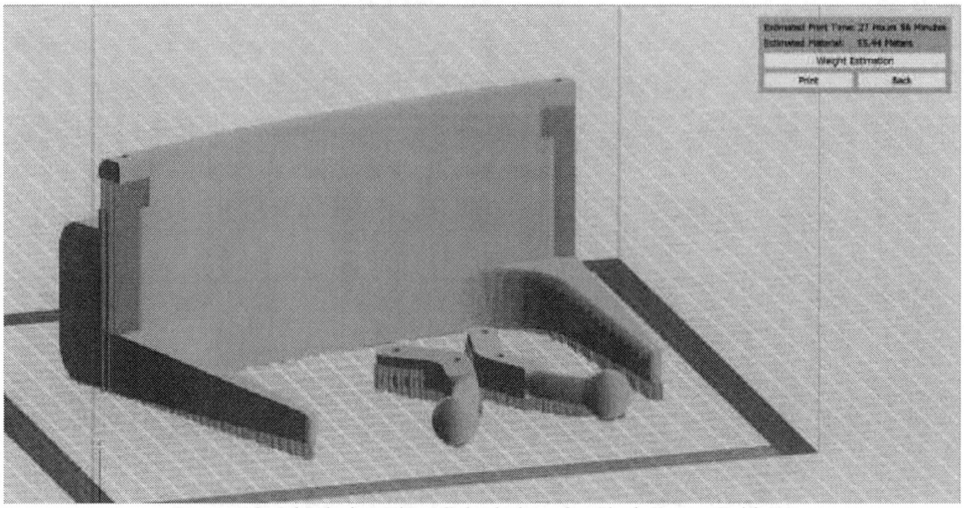

Figure 12. 3d Printing Time Calculations for Flash Forge Guider 2 Printer.

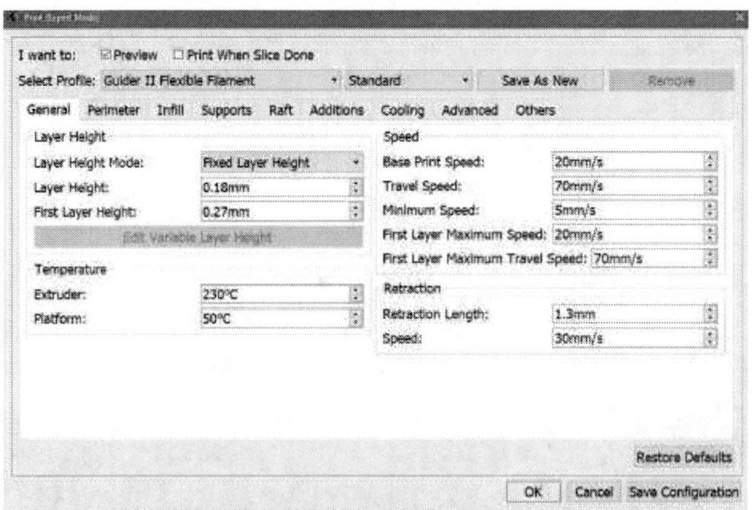

Figure 13. Parameters for Printing Nylon 6/10 With Flash Forge Guider 2 printer.

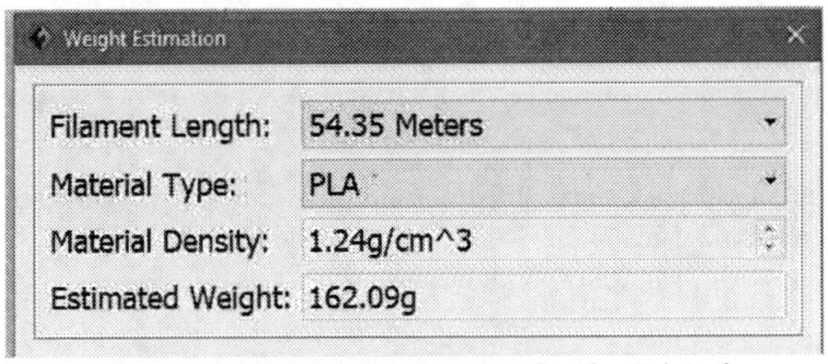

Figure 14. Weight Estimation for the Printed Parts

9. Control System

The Control system has been designed to give pulse load to the rocker arms of the product giving it a too & fro motion along with some delay in the between two cycles. This Control System is powered by an Arduino UNO & Genuino UNO,

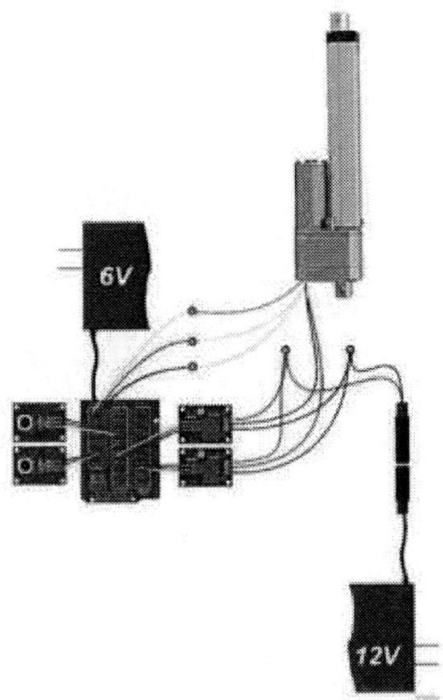

Figure 15. The image for the control system [8], from an open source document.

10. Prototype Development

The Prototype development will be done in future for this product after filing a design patent. The same manufacturing planning will be done for making a functional prototype for the same.

11. Conclusions

By this project, we can conclude that after this planning the machine is ready for the manufacturing and testing in the authorization of a Physician. After testing in the real time and by the affiliation of a doctor, the machine will be ready for production on a large scale.

References

[1] http://www.goodfellow.com/E/Polyamide-Nylon-6.html

[2]http://www.matweb.com/search/datasheet.aspx?MatGUID=8d78f3cfcb6f49d595896ce6ce6a2ef1&ckck=1

[3] http://www.regner.es/product/ra-47x-high-speed-linear-actuator-with-feedback/

[4] https://electronicsforu.com/electronics-projects/controlling-linear-actuator-arduino

[5]https://www.hackster.io/robotgeek-projects-team/control-a-large-linear-actuator-with-arduino-8a3953

[6] https://help.solidworks.com/2018/english/SolidWorks/cworks/t_Performing_Static_Analysis.htm

DESIGN, DEVELOPMENT AND TESTING OF A MULTIFUNCTIONAL JAR AND BOTTLE OPENER

Harshit Kuchhal [1], Rajkumar.E[2]

[1] Under Graduate student, SMEC, VIT, Vellore, India, harsh.kuchhal@gmail.com

[2] Associate Professor, Department of Design and Automation, SMEC, VIT, Vellore, India, rajkumar.e@vit.ac.in

Abstract. Many fields in Engineering Technology today are focused on encouraging and motivating students to implement their educational concepts in the real world to fight with the increasing demands of new technology in the global market. This paper explores how a project interwoven through the Engineering Technology curriculum can be used to create a product to be sold in the today's market to do basic household jobs and make them easier. The main focus of the bottle opener product existing in the market is that they are used on specific type of bottles and if to be used for different needs different product and will reduce the feasibility and portability as a greater number of products will be used to carry out these operations. Therefore, to eliminate this problem the objective of the project is to develop a multifunction bottle opener which can be used to deal with all the types of bottles available in the market. This product will work on the basic principles of mechanism like rack and pinion mechanism, cork plug and stress bending. Many researchers compared the shapes and mechanism in their work. Though there are similar work, but in the present work casts a unique design and has a functionality difference which makes this project exclusive from other projects. The optimized form of a common product that is a multi-functional bottle and jar opener is obtained from this project which can be used to open different styles of cap of different bottles. Keywords. multi-functional, rack and pinion, stress bending,

1. INTRODUCTION

The product is a solution to the needs of the customer. Anything that can satisfy a customer's requirement and complete the demand in the market is called a successful product. A typical product development has various phases. The first phase is understanding the opportunity which deals with the identification of customer needs which define the problem to be addressed various tools like customer survey interviews and interviews are used to understand the customer needs. The next stage requires to develop different rough concept (schematic diagram) based on the requirement of the customer known as the concept development process. A concept is then selected out of the of those designed earlier which has the best characteristics and which has the most optimal features required by the customer. Various process like Pugh matrix, panel selection etc. can be used to select the best concept by rating the criteria's selected for the product and scoring them on the basis of their performance. TRIZ Matrix is applied to the product to provide them with the solutions to the process problems that they face. The features of the product can be improved which are required and some features can be downgraded which doesn't contribute much to the product's performance. The final stage implementing a thought deals with the embodiment designing, design for x etc. It is the phase where the material selection, dimensions & tolerances for the product is finalized. In this research study, the general product development process has been kept in mind while designing of the product according to the market needs. A multi-functional jar and bottle opener has been designed based on the process of new product development.

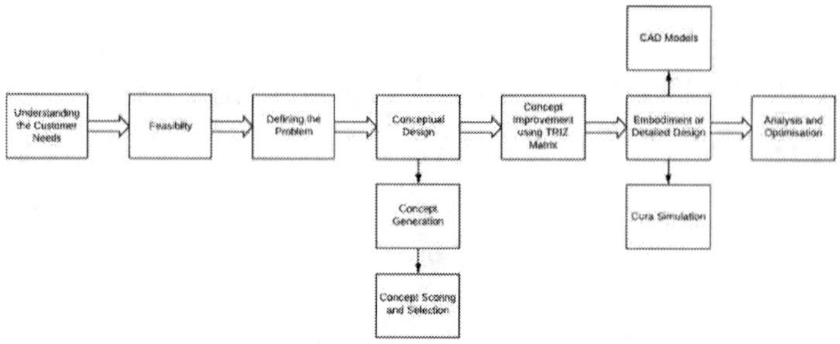

Figure 1. Modern Design Process

2. PROBLEM DESCRIPTION

The need for a multifunctional jar and bottle opener arises from the market and customer analysis which reveals that the normal bottle openers that are available in market doesn't serve the purpose of multi- functionality i.e. they don't have the ability to open the jars. This leads to the commuter to heavily rely on different classes of bottle or openers or the usage of hand which can cause injury while opening bottles with hard fixed caps. Hence it is understood that the multifunctional jar and bottle opener is an alternative to the conventional bottle opener existing in the market. From the field survey it has been found that even though there are multifunctional bottle openers in the market, they are found to be less feasible as they don't serve the purpose of opening jars, even though they are multi-functional. The multi-functional configuration is also found to be not simple which leads to difficulty in manufacturing it. This leads to the requirement of designing and developing a new jar and bottle opener made out of material with high strength that can be operated easily and also very portable. Therefore, the main focus of the bottle opener product existing in the market is that they are used on specific type of bottles and if to be used for different needs, different product and will reduce the feasibility and portability as a greater number of products will be used to carry out these operations. Therefore, to eliminate this problem the objective of the project is to develop a multifunction bottle opener which can be used to deal with all the types of bottles available in the market.

3. FUNCTIONAL MODELLING

The purposeful modelling is nothing however establishing purposeful structure of the merchandise of interest. Solution principles or designs based on conventional methods do not provide required design when latest technologies and inventions are to be used. The solution of the problem should be expressed as a list of the required functions and essential constraints. The created purposeful tree has been developed supported the necessity statement. The need statement has been obtained by the small field survey and literature survey. Fig. 2 and 3 show a simple black box diagram and functional tree of a multifunctional jar and bottle opener respectively. The functional tree has been generated based on the conceptual analysis of the product requirement. The need statement has been obtained by the small field survey and literature survey.

IOP Conf. Series: Materials Science and Engineering 1123 (2021) 012014 doi:10.1088/1757-899X/1123/1/012014

Figure 2. Simple Black Box of the product

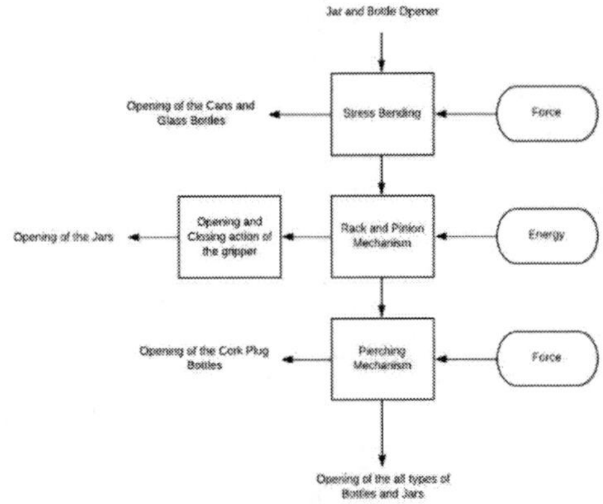

Figure 3. Functional Tree of a Multifunctional Jar and Bottle Opener

4. CONCEPTUAL DESIGN

Conceptual design is one of the important stages of a product design. After functional modelling, various concepts will be removed based on the various concept generation methods like Pugh matrix, conceptual brainstorming etc.

4.1 CONCEPT GENERATION

Concept generation or getting the ideas, is the most critical step in the engineering design process. Starting with a group of client desires and target specifications, the process concludes with an array of product alternatives from which a final design is selected. There are multiple steps concerned within the generic thought generation method, as well as various approaches. This research paper reviews associated critiques these totally different views among the context of with success developing an electronic medical product that's innovative in style and client attractiveness. The concept generation process required the following steps

- Understanding the problem and the customer needs
- Searching externally through the internet and the markets to find out whether a product like this exists or not
- Searching internally by generating ideas which can be used to compensate the needs of the customer.
- Reflecting the results by getting the feedback of people and making changes in the concept design according the survey reviews.

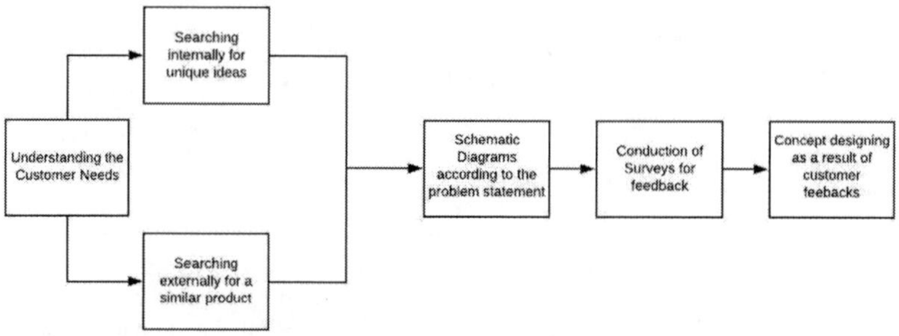

Figure 4. Flowchart showing the concept generation process

Figure 5. Concept 1

Figure 6. Concept 2

Figure 7. Concept 3

Based on the above-mentioned steps, various concepts have been generated as shown in the Fig.5-7

4.2 CONCEPT EVALUATION AND SELECTION

Concept Generation may provide a lot of concepts to the designer that are related to the problem statement, but not all of them are feasible and can be achieved effectively. In this process the designer tries to limit the concepts generated earlier according to the criteria that are relevant for the production of our concept such as Manufacturing Cost, Weight of the product, Ease of Operation etc. In a typical development process, once the various product ideas are generated, the generated ideas need to be evaluated to pick the most effective conception among the varied-out ideas. Pugh matrix is one of the most effective and appropriate method to select the most feasible concept out of other ideas. The first step in the concept evaluation process is to select a list of criteria which have to be kept in mind while designing of the product and these much be beneficiary for the customers. These criteria mainly depend on the type of product and therefore can be unique for each product.

4.2.1 PUGH MATRIX

The table 1 shows Pugh chart that has been developed for a Multifunctional Jar and Bottle opener product. The criteria chosen for this is based on Multifunctional characteristics of the product (i.e. variety of the products it can be used upon), Portability of the product, Ease of Operation, Manufacturing cost, Safety to the customer and strength of the Product. All these criteria are provided some weightage according to the requirement of the customer as each of them cannot be considered to be equally required in the product. Giving weightage to each criteria can provide extra layer to distinguish among the concepts. It can also provide a form of "robustness" assessment. The weightage of each criteria depends on the customisation of the product and hence varies for each product uniquely. All these concepts are evaluated taking the product which is existing in the market as the baseline or the datum value. The concepts are evaluated for each of the above-mentioned criteria. The scoring and weightage are as shown in Table1.

Table 1. Pugh Matrix

Crite	**Weightage**	**Conce**					
		Concept 1		**Concept 2**		**Concept 3**	
		Score	Total	Score	Total	Score	Total
Multifunctionali	30.00	2	0.6	-1	-	0	0
Portabi	10.00	0	0	2	0.2	2	0.2
Ease Of	15.00	1	0.15	1	0.15	1	0.15
Manufacturing	10.00	0	0	0	0	1	0.1
Safety to the	20.00	1	0.2	1	0.2	1	0.2
Streng	15.00	2	0.3	1	0.15	0	0
	Sum of all		1.25		0.7		0.65
	Sum of all		0		-		0
	Total Score		1.25		0.4		0.65
	Continue?		**Ye**		**N**		**N**

The net score is obtained for each design concept based on the weightage and the scoring of each criteria. The concept that has the highest score as compared to other concepts is selected as the best concept by the process of Pugh Matrix Methodology.

5. TRIZ MATRIX

Triz Matrix is a procedure to formalize the process of innovating a product. The TRIZ matrix is applied to get different solutions. The TRIZ is a systematic way to think about issues and get a hint about the most plausible solutions. These solutions are highly plausible because many of these solutions have been successfully implemented in the past and are therefore tried and tested with time. With the assistance of the TRIZ Matrix, one will initial develop multiple solutions. This will help with the rational process of evaluating all the possible solutions and then coming up with the most feasible one. The TRIZ method believes that all technological innovations that have occurred in different industries have been based on a set of "inventive principles". TRIZ method states that every specific problem that an organization faces can be reduced to a general problem that has been faced before. This general problem then has a general solution based on one of the 40 inventive principles that the matrix contains. Thus, the use of the matrix suggests the general solution. This general solution can then be used to come up with a solution to the problem that the organization is facing.

Table 2. Triz Matrix

		Worsening Features		
		Shap	**Ease of Repair D**	**uration of action of the**
Improvin g Features	**Weight of the stationary**	13, 10, 29,	2,27,28,11	-
	Volume of the stationary	7, 2, 35	1	-
	Strength	10,30,35,40	27,11,3	27,3,26
	Ease of	15,34,29,28	12,26,1,32	29, 3, 8, 25

Table 3. Inference Matrix

Improving Features	Worsening Features	Triz Principles
Weight of the stationary object	Shape	**Preliminary Action** Pre arrange objects such that they come into action from the most convenient place.
Volume of the stationary object	Shape	**Segmentation** Dividing an object into different parts.
Strength	Ease of repair	**Local Quality** Make each part of an object fulfil a different useful action
Ease of operation	Duration of action of the moving object	**Local Quality** Make each part of an object fulfil a different useful action

From the results of the inference matrix we can find out the solutions for the product to make it even better as per the demands of the customer.

6. EMBODIMENT DESIGN

Embodiment design is one of the most important part in the process of the modern product development. In this process the main product design is been made with all the specific dimensions and economic criteria as per the requirement of the product to the stage where the detailed design can be sent directly for production purpose. There are various stages carried out as a part of the Embodiment Design. CAD model, Ultimaker modeling and FEA analysis have been carried out for our product.

6.1 FEA ANALYSIS

FEA is used to evaluate the different types of fatigue failure of the proposed in the product design. The location points of the failure and the intensity of different stresses and displacements can be found out through this analysis. The slider of the product is modelled and simulated in the Solidoworks environment. The product is made in a meshed model to make it possible to do the mesh analysis on the product. For the failure analysis of the product the shape of the rim of the bottle opener is kept to be constant as it should not deform under any condition of the loading conditions and the force of 100N and 200N is applied normal to the face of the slider as the average force exerted on the slider by the palm of an average person for opening a bottle. The material selected to manufacture the slider was AISI 316 Annealed Stainless-Steel Bar. Two types of failure test have been conducted on the product i.e. failure under static loading and Fatigue simulation.

6.1.1 STATIC LOADING SIMULATION

The load has been taken as 100N and 200N normal to the face of the slider and the results obtained have been shown in the Fig 8-13 and Table 4

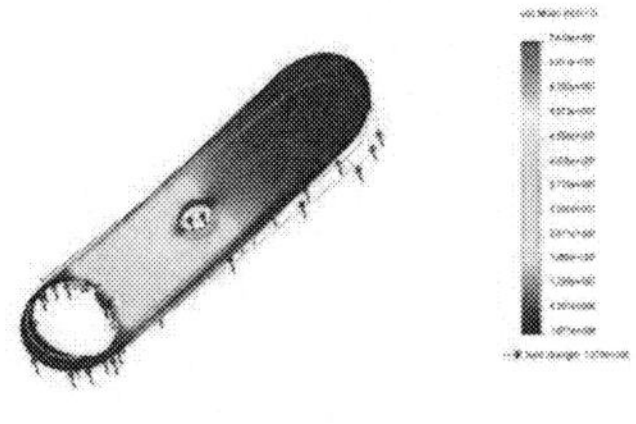

Figure 8. Von Mises Stress under static load 100N

Figure 9. Von Mises stress under static load of 200N

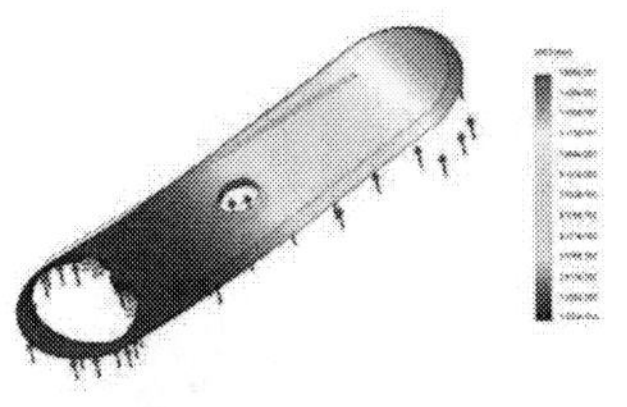

Figure 10. Displacement under static load of 100N

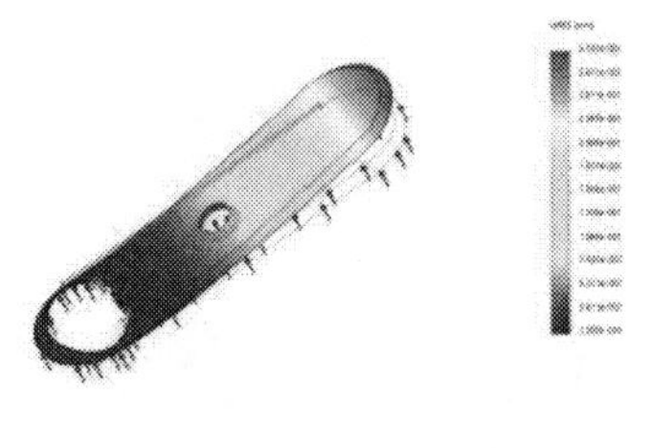

Figure 11. Displacement under static load of 200N

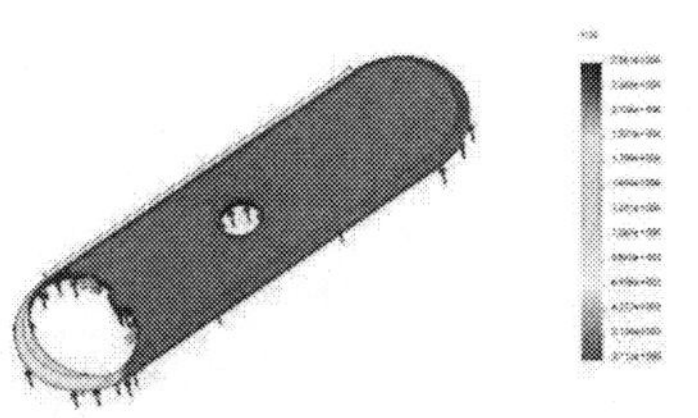

Figure 12. Factor of Safety under static load of 100N

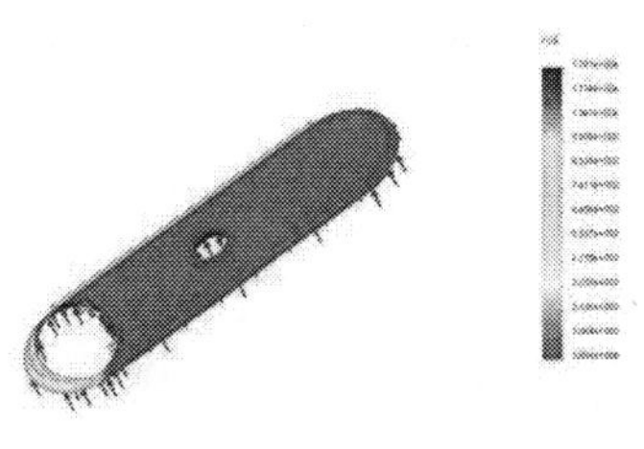

Figure 13. Factor of safety static load of 200N

Table 4. FEA Solution under Static Loading

Loa d	Type of Load	Von-Mises Stress	Total Deflection	Yield Strength	Factor of
100 N	Normal to the plane	37.15	0.156	137	3.71
200 N	Normal to the plane	74.15	0.31	137	1.86

As the proposed design satisfies the criteria for safety, the product is considered safe to use under normal loading conditions.

6.1.2 STATIC LOADING SIMULATION

The cycle frequency has been taken as 20,000 cycles as an average lifetime use of the product. The product is been analysed under the same loading conditions as done previously. The results obtained from the test has been shown in the Fig and table 5.

Figure 14. Damage Percentage in Fatigue Simulation under a load of 100N

Figure 15. Damage Percentage in Fatigue Simulation under a load of 200N

Figure 16. Total Life Cycle in Fatigue Simulation under a load of 100N

Figure 17. Total Life Cycle in Fatigue Simulation under a load of 200N

Figure 18. Load Factor in Fatigue Simulation under a load of 100N

Figure 19. Load Factor in Fatigue Simulation under a load of 200N

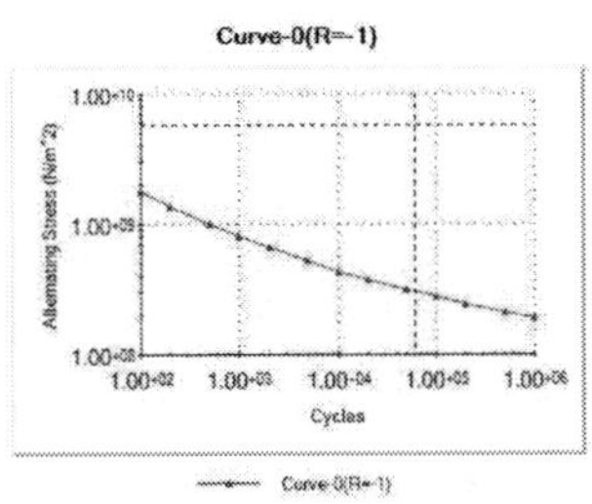

Figure 20. S-N Curve Fatigue Simulation under a load of 100N

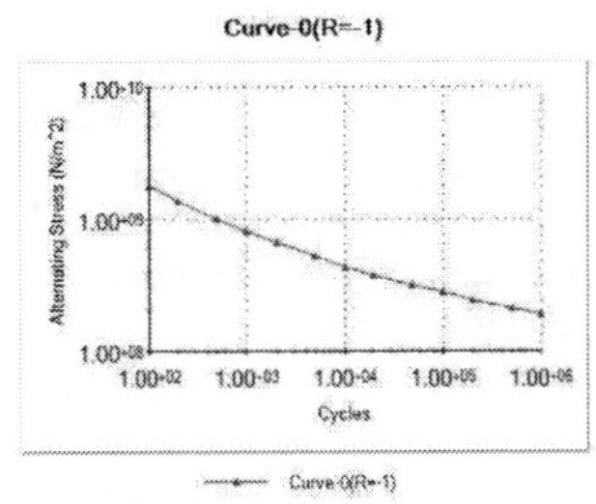

Figure 21. S-N Curve Fatigue Simulation under a load of 200N

Table 5. Fatigue Simulation Solution

Fatigue	Load	Type of	Damage (%)	Life Cycles	Load
20000	100	Normal to the	2	10^	3.372*10^4
20000	200	Normal to the	2	10^	1.523*10^4

As the proposed design satisfies the criteria for safety, the product is considered safe to use under normal loading conditions.

7. CAD MODEL

This is the stage where the concept generated after all the tests gets converted into actual 3D object with all the appropriate dimensions and tolerances. It contains various parts like handle, slider, nut and screw, rack with gripper and piercing tool. The product has been joined in such a manner that it achieves the maximum feasibility in all the requirements of the product.

IOP Conf. Series: Materials Science and Engineering 1123 (2021) 012014 doi:10.1088/1757-899X/1123/1/012014

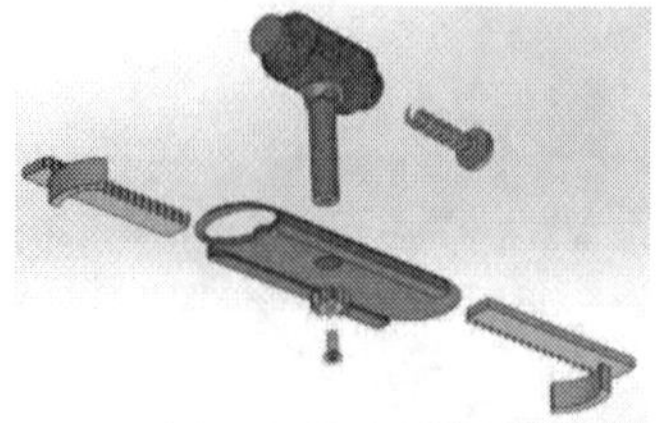

Figure 22. Exploded View of the Multifunctional Jar and Bottle Opener

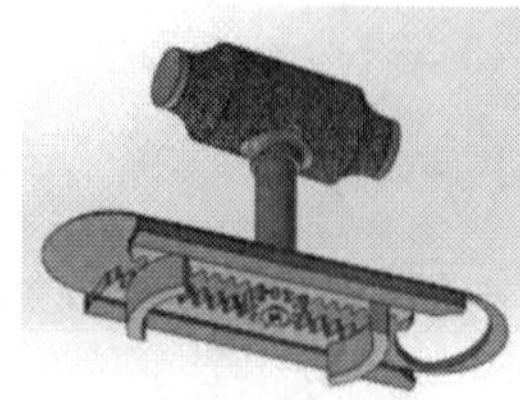

Figure 23. Assembled View of the Multifunctional Jar and Bottle Opener

Fig 22 shows the disassembled view of the multifunctional jar and bottle opener. This view shows the multifunctional character of the proposed design, which aids in the ease for operation and portability of the product. Fig 23 shows assembled view of the product which aids to the final representation of the product when ready to use

8. ULTIMAKER CURA SIMULATION

The recent trends in manufacturing is single component manufacturing by avoiding the joints and fixations leading to better material properties, also by decreasing the chances for failures and defects to take place and serves the function better than ever before. This can be achieved by only known technology "Additive Manufacturing". In additive manufacturing initially, the part is sliced into different layers with lesser the layer thickness for better the surface finish. Then, other constrains shell thickness or wall thickness is specified for higher the value better the surface strength. Then, the fill density of the part model to be manufactured is specified to give the optimal internal strength. These constrains are inputted to the slicing software CURA where it generates the g-codes and m-codes for the part and constrains given designing the tool path to be followed and the amount of material to be used. This data is then forwarded to the machine which is driven by the code to manufacture the component developing layer by layer.

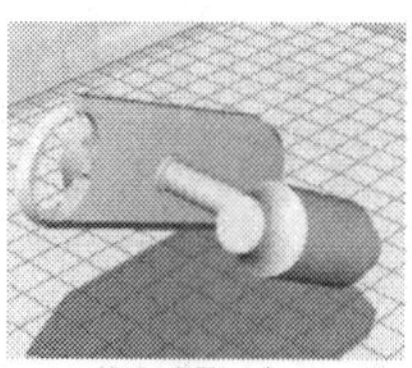

Figure 24. 3D view of product in the CURA bed

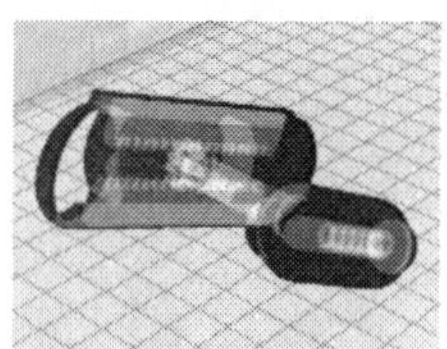

Figure 25. X-Ray view of the product in the CURA bed

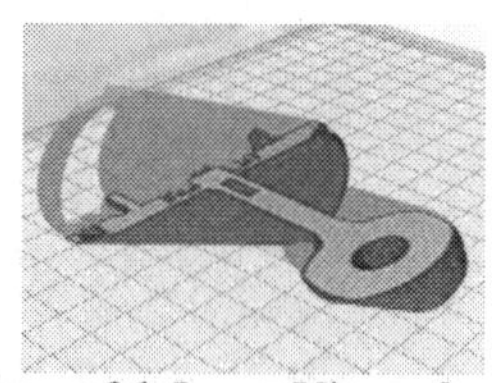

Figure 26. Layer View of the product

9. CONCLUSION

In this study, a novel multifunctional jar and bottle design has been developed. The proposed design has good portability and multifunctional characteristics when it is compared with the existing bottle openers available in the market. The design has been evaluated with existing designs using Pugh chart methods and has been made better with the help of solutions using TRIZ Matrix. The CAD model of the product was made based on the previous results and the CURA simulation of the product was processed. The FEA analysis has been carried out for the product to check the strength and possible failures of the product under dynamic loading conditions.

10. FUTURE WORK

The Current study, deals with a conceptual and embodiment design of a multifunctional jar and bottle opener. The downstream activities within the development method like cost of the material used for the manufacturing of the product, proof of conception needs to be developed.

REFERENCES

[1] Zhiqiang Yu et.al 2013 The Design of a Wine and Beer Bottle Opener *J. of Computational and Theoretical Nanoscience*

[2] Irwin J L 2008 To Build A Better Bottle Opener: Interweaving A Project Through the Engineering Technology Curriculum Article in *ASEE 2008 Annual Conf. & Exposition* At Pittsburgh, Pennsylvania

[3] David Shipstone 1999 New Approaches Mechanics of a Bottle Opener *Article in Physics Education*

DRIVE SHAFT DESIGN AND ANALYSIS FOR MINI BAJA

Gowtham S [1], Raaghul V [2], Sreeharan B N [3], Praveen Kumar R A [4], Mohamed Kasim S [5]

[1]UG Scholar of Third Year Mechanical Engineering, Department of Mechanical Engineering, Kumaraguru College of Technology, Coimbatore, Tamil Nadu, India. gowtham.17me@kct.ac.in

[2] UG Scholar of Third Year Mechanical Engineering, Department of Mechanical Engineering, Kumaraguru College of Technology, Coimbatore, Tamil Nadu, India, raaghul.17me@kct.ac.in

[3] Assistant Professor of Mechanical Engineering, Department of Mechanical Engineering, Kumaraguru College of Technology, Coimbatore, Tamil Nadu, India, sreeharan.b.mec@kct.ac.in

[4, 5] Alumni, Department of Mechanical Engineering, Kumaraguru College of Technology, Coimbatore, Tamil Nadu, India.

Abstract. A wide variety of automobile design and racing events happening across the country to encourage the student community in the design of automobile vehicles. In this regard, the Society of Automotive Engineers (SAE) enables a student to gain real world experience in the fields of design, analysis and manufacturing of the all-terrain vehicle by conducting such events. This paper describes about the design and analysis of drive shaft for mini BAJA vehicle that has participated in one of the national level competitions conducted by SAE. Drive shaft is one of the major components used in BAJA. The drive shaft discussed here is used to transmit the required torque from the differential to the wheels. All the components of the drive shaft were designed analytically, modeled in *SolidWorks* and analyzed using *Ansys* software to ensure its safety. The main objective is to fabricate more reliable and eco- friendly drive shaft by minimizing the cost and increasing the overall performance. Keywords: Design and analysis, Drive shaft, Terrain vehicle, Torque transmission, Safety

1. INTRODUCTION

Half Axle is a device used to transfer the power by means of rotational motion from the differential gearbox to the wheel. One end is engaged with the differential gearbox while another to the wheel hub assembly. The suspension system used in this work is a double-wishbone which is used to make the full-floating axle. The load that came on axle is the only torsional or driving force. From the viewpoint of functionality, the axle should have the rigidity at the lower weight. This work was focused on the BAJA SAE INDIA event as the ATVs must have less weight at the same rigidity is also required. By optimizing the design and upgrading the material it can be achieved. The design is validated on the Ansys software to ensure the safety and it has been used in the BAJA vehicle in the competition.

2. TERRAIN VEHICLE

2.1 Identification of problem

MINI BAJA events are intercollegiate events involving purely students. Their budget is limited that they must build the entire vehicle within the given budget. Having a tight budget must not affect vehicle performance (in this paper the design quality of the driveshaft). A lot of teams use drive shafts from OEMs. Which might help them clear the event but will not gain scores in the design event thus affecting their overall score.

2.2 Objective of the paper
The primary objective of the project is not to use composite materials since they cost much more than typical steel or aluminum alloy drive shafts. Further, no compromises should be made on the build quality of the drive shaft. The project focuses on reducing as much weight possible yet not failing the ATV to satisfy the judges in every way and to survive all road conditions and toughest of tracks.

2.3 All terrain vehicles
An all-terrain vehicle (ATV), additionally referred to as a quad, three-wheeler, four-track, hackney carriage, or curve cycle, as outlined by the Yankee National Standards Institute (ANSI) could be a vehicle that travels on unaggressive tires, with a seat that's straddled by the operator, alongside handlebars for steering management. As the name implies, it's designed to handle a wider kind of piece of ground than most different vehicles. Though it's a street-legal vehicle in some countries, it's not street-legal inside most states and provinces of Australia, the US or Canada.

2.4 Purpose of drive shaft
The torque that is produced from the engine and transmission must be transferred to the rear wheels to push the vehicle forward and reverse. The drive shaft must provide a smooth, uninterrupted flow of power to the wheels. The drive shaft is used to transfer this torque.

2.5 Functions of drive shaft
❖ The primary work is to transmit torque from the transmission to the rear wheels. During the operation, it is necessary to transmit maximum low gear torque developed by the engine.

❖ The drive shaft must be capable of rotating at very fast speeds required by the vehicle

❖ The drive shaft must also operate through at constantly varying angles.

❖ The suspension of the ATV determines the angle through which the drive shaft has to vary.

❖ This change in the angle determines the significance of the need of face width of the spline in the drive shaft.

❖ The input velocity must equal the output velocity to get the maximum performance of the drive shaft. Single cardan drive shafts usually fail to produce that, causing failure of the drive train. Double cardan or CV joint drive shafts overcome this drawback.

2.6 Design considerations
➢ Handling of high torques which is required by the ATV must be taken care.

➢ The diameter and the length of the drive shaft being fixed, the material must be selected.

➢ The shaft is subjected to shear loads and fatigue loads. Hence a highly durable design is needed to sustain all the driving conditions.

2.7 Drive shaft arrangement in ATV
Generally, drive shafts in automobiles are considered as the propeller shaft which transfers power from transmission in front wheel drive car to differential at the back. The shaft which transmits torque from differential to the rear wheels are known as rear axles. In the design of ATV for BAJA the drive axles are known as drive shafts since it is the only component to transmit power from transmission or differential to rear wheels.

3. LITERATURE REVIEW

[1]Jasim M.Salman et. al. [2013] studied to improve properties of 7075-T6 such as impact toughness, thermal age hardening behavior and corrosion resistance in 3.5% NaCl solution by using quenching in 30% polyethylene glycol and addition alloying elements, i.e. boron (B) to this alloy. Results showed improved impact toughness by (30%) when quenching in water, and by (50 %) when quenching in 30% PAG corresponding to the base alloy at aging temperature 150°C.

[2]BHIRUPANKAJPRAKASHet. al.[2014] The graphite/carbon/fiberglass/aluminum driveshaft tube was developed as a direct response to industry demand for greater performance and efficiency in light trucks, vans and high performance automobiles. The main reason for this was significant saving in weight of drive shaft; the results showed that the final composite drive shaft has a mass of about 2.7 kg, while this amount for steel drive shaft is about 10 kg.
[3]P.N. SHARMA et. al. [2016] The common normal at the action between two teeth must always pass through a pitch point in such a way that it divides line joining the centers of the mating gears in the reverse ratio of angular velocity. Increasing pressure angle increases the width of the base of the gear tooth, leading to greater strength• and load carrying capacity.
[4] L.VALTIER et. al. [2002] spline couplings have a reputation of being able to accommodate some angular misalignment, the dangers of tooth fracture, fatigue, and wear that may result should be recognized. Unfortunately, it is exceedingly difficult to achieve perfect alignment in practice; thus, the design process should prudently allow for some minimum misalignment attainable and accept a reasonable penalty.
[5] BASKAR et. al. [2013] Due manufacturing and design fault, raw material faults, maintains faults, material processing faults, to avoid this problem various techniques such as either, modify material or geometry and conduct analysis on the same and finally implemented in the solution or used to find out the best design of Joint with considering the all the factor such as weight, cost, Fatigue life, stress distribution, stiffness.

4. DESIGN CALCULATIONS

4.1 Strength: Maximum shear stress theory

$$\tau_{max} = \frac{16}{\pi d^3}\sqrt{(M_b^2 + M_t^2)}$$

where,
Mb = maximum bending moment in Nm
Mt = maximum twisting moment in Nm
d = diameter of the shaft in m
τ_{max} = maximum shear stress in N/mm^2

Considering bending moment will be maximum at the center since the shaft is symmetric,

$$M_b = \frac{WL^2}{24} = \frac{0.086 \times 0.394^2}{24} = 0.55 \text{ Nm}$$

Where,
L = length of the complete drive shaft from support to support
W = load per unit length, which is 0.086 kN/m
Thus

$$\tau_{max} = \frac{16}{\pi \times 0.035^3}\sqrt{0.55^2 + 700^2} = 83150354.6\ N/m^2$$

τ_{max} = 83.2 N/mm^2

Figure 1. Drive shaft

Shear strength for Al7075T6 is 331 N/mm^2 , and the fatigue strength being 159 N/mm^2 which is greater than the actual shear stress of 83.2 N/mm^2.
Therefore, the design is safe.

4.2 Strength: Angle of twist

$$\theta = \frac{180}{\pi}\frac{M_t}{J}\frac{l}{G}$$

Where,
Mt = maximum twisting moment in N.mm l
= length of the center shaft in mm
J = polar moment of inertia = 1.47∗10^{-7} N/mm^2
G = modulus of rigidity = 25.8∗10^9 N/mm^2
Θ = angle of twist in degrees

$$\theta = \frac{180\times 0.258\times 10^7}{1.47\times 25.8\times 10^9} = 8.5°$$

Θ is minimum and even decreases when the wheel starts to spin.Therefore, the design is safe.

4.3 Whirling of shaft

$$f_n = 3.562\sqrt{\frac{GEI}{WL^4}}$$

Where,
G = Acceleration due to gravity.
E = young's modulus = 71.7*10^9 N/mm^2
I = Area of moment of inertia = 73.7*10^{-9} mm^4
Density of AL7075T6 = 2.81 g/cm^3 = 2.81*10^3 kg/m^3
Volume per unit length = A∗L = 96.2*10^4 mm^3
Mass per unit length = Density∗volume = 2.7 kg
Bending Natural Frequency,

$$f_n\ (\text{revs/sec}) = 3.562\sqrt{\frac{GEI}{WL^4}}$$

$$= 3.562\sqrt{\frac{9.81\times 71.7\times 10^9\times 73.7\times 10^{-9}}{2.7\times 0.2584}}$$

$$= 3.562\sqrt{4333243.1873}$$

= 7290 rev/sec

From Engine specifications and transmissions design, the maximum speed at which the shaft rotates is 560 rev/sec.

So, the design is safe.

4.4 Spline strength: Lewis factor equation

$$\sigma = \frac{W_t p}{f y}$$

Where,
Wt = tangential tooth load. F
= face width of tooth
Y = Lewis form fraction.
P = diametrical pitch.
Maximum Torque, T = 700 Nm
Pitch circle Radius, R = 14.5 mm
Total load, F = 48275.86 N
No. of teeth = 29
Load shared by each tooth,

$$W_t = \frac{48275.86}{29}$$

= 1664.69

N Lewis factor equation,

$$\sigma = \frac{W_t p}{fy}$$

$$= \frac{1664 \times 1}{25 \times 0.43} = 154.8 \frac{N}{mm^2}$$

In general, the yield strength for EN24 is 680 N/mm^2.
From Lewis factor equation, the stress is 154.8 N/mm^2 < 680 N/mm^2. Thus, the design is safe.

5. ANALYSIS AND TESTING

5.1 Static structural analysis

Static structural analysis helps in determining the displacements, stresses, strains, and forces in any structures or components. Here, the structures response to change in the loads expected to happen slowly. Types of loading that can be applied in static structural analysis include:

- Externally applied pressures and forces
- Steady state inertial forces
- Imposed displacements
- Temperatures for thermal strain

This type of analysis can either be linear or non-linear. All types of nonlinearities are allowed - large deformations, stress stiffening, plasticity, contact (gap) elements, hyper elasticity and so on.

5.2 Safety factor

Safety factor (SF) is the ratio of absolute strength of structures to the actual load applied. Reliability of a particular design can be measured using safety factor. It expresses stronger a system needs to be for intended load. Safety factors are often calculated using detailed analysis because comprehensive testing is impractical on many projects. Many systems are intentionally built much stronger than needed for normal usage to allow for emergency situations, unexpected loads, misuse, or degradation.

5.3 Modal analysis

A good solution can be delivered only after the function of the modal being tested modal analysis is performed by using ANSYS software. The figures of the results are shown below:

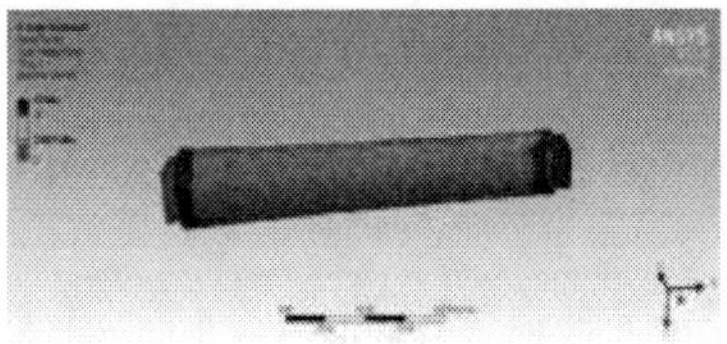

Figure 2. Structural analysis of aluminium -7075T6

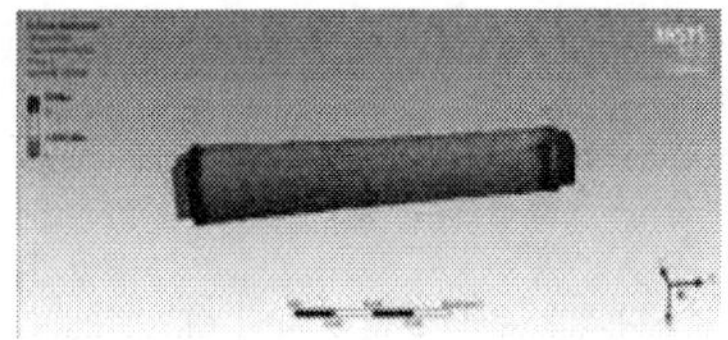

Figure 3. Structural analysis of EN24

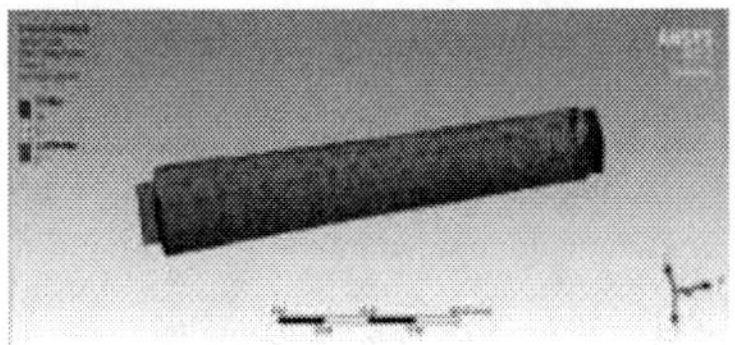

Figure 4. Structural analysis of grade 1 titanium

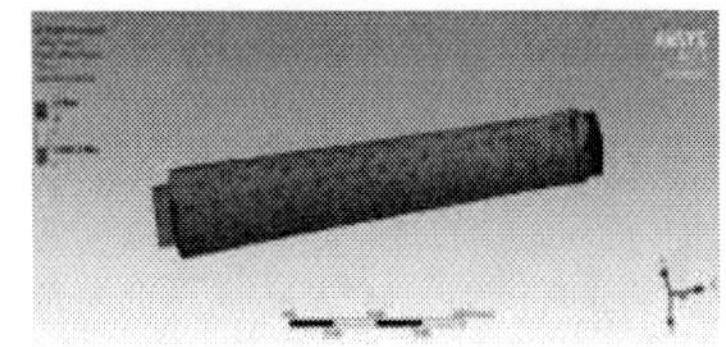

Figure 5. Structural analysis of 304 stainless steel

6. FABRICATION PROCESS

6.1 Lathe

Lathe is a rotating machine which rotates about an axis to perform various operations such as cutting, knurling, sanding, drilling, deformation, facing, and turning. The drive shaft material is purchased around billet. The round billet is then produced to the lathe. The round billet is then turned to the desired diameter and cut at the edges.

6.2 CNC machine

CNC machining is the short for "computer numerical control" machining is a manufacturing process in which computer software is pre-programmed to decide the movement of the machining tool or component in order to achieve the desired output. The process can be used to control a range of complex machinery, from grinders and lathes to mills and routers. With CNC, three dimensional tasks can be achieved in minutes. The components called cups which hold the cross carrying a roller bearing is machined using CNC machine.

The versatility of the CNC machine makes it easy to manufacture the cup within minutes. By importing the model data to CNC, codes are generated to make precise cuts and then the machining takes place. Finally, the desired component that is required for this project work is fabricated.

6.3 Electric discharge machine

Electrical discharge machining (EDM), also known as spark machining, spark eroding, burning, die sinking, wire burning or wire erosion, is a manufacturing process whereby a desired shape is obtained by using electrical discharges (sparks). Material is removed from the work piece by a series of rapidly recurring current discharges between two electrodes, separated by a dielectric liquid and subject to an electric voltage. One of the electrodes is called the tool-electrode, or simply the "tool" or "electrode," while the other is called the work piece-electrode, or "work piece." The process depends upon the tool and work piece not making actual contact. The sparks generated make the cut on the subjected shaft carrying cup at the end. The precision of the cut produced by EDM is unmatched by even many of the current technologies. Spline in the male cup is machined with the help of EDM.

7. MATERIAL SELECTION

The materials have been analyzed in static and structural analysis using ANSYS software and the data have been tabulated below:

Table 1. Numerical simulation results

Material	Safety factor
EN24	1.9563
Aluminium 7075T6	2.0937
Aluminium 6061T6	1.1488
Stainless Steel 304	0.8862
Titanium Grade 1	1.0159

From the above table 1, Aluminium 7075T6 is the best chosen option to manufacture the drive shaft which meets all the design criteria and the project objectives.

8. RESULTS AND DISCUSSION

The design is focused on factors like strength, fatigue and whirling of the solid drive shaft. Primarily a set of materials are taken into consideration which are widely used in the field of automobile industries to manufacture drive shaft.

Figure 6. Baja vehicle in competition using designed and fabricated drive shaft

A SOLIDWORKS model is created with the length and diameter constraints set the by design of other components of the ATV. The maximum amount of stress the shaft has to withstand is calculated as well. With these in hand, all the above-mentioned materials are put to test both analytically and theoretically.

9. ACKNOWLEDGEMENTS

We would like to thank all the people who made their role unavoidable. Without their contribution this paper would have not been come true. Also, we thank our Institution head and Department head for providing facilities without which this paper would not be possible. We would also like to thank the industry for giving the constant support to carryout experimental work in real time practices.

10. CONCLUSION

With the completion of the project, the primary objectives of the project are met with maximum efficiency. Even at high speeds, the drive shaft maintained not to whirl. This contributed highly to the overall performance of the ATV. The selection of material resulted in much lesser manufacturing cost when compared to reinforced composite drive shafts and meeting the other objective of designing and fabricating a budget drive shaft favoring the enthusiastic students to participate in all the ATV design and racing events across the country.

REFERENCES

[1] Sivakandhan C and SureshPrabhu P 2012 CompositeDrive Shaft is a Good Strength and Weight Saving toCompare Conventional Materials Design and Analysis of EGlass/Epoxy Composite Drive Shaft for Automotive Applications *European J. of scientific Research* Vol 76 No 4 2012 pp.595-600.
[2] Rangaswamy T, Vijayarangan S, and Chandrashekar, R.A.&Venkatesh 2002 Optimal design and analysis of automotive composite *driveshaft Int. Symp. of Research Students on Materials Sci. and Engg.* pp 1-9.
[3] Mohammad Reza Khoshravan, AminPaykani and AidinAkbarzadeh 2011 Design and modal analysis of composite drive shaft for automotive application *IJEST* 11, Vol.3 No 4
[4] Mouritz A P and Thomson R S 1999 Compression, flexure and shear properties of a sandwich composite containing defects Composite structures pp 263-278
[5] Potluri P, Kusak E and Reddy T Y 2003 Novel stitch bonded sandwich composite structures", Composite structures pp 251-259.
[6] Jung-Seok Kim, Hyuk-Jin Yoon and Kwang-Bok Shin 2011 A study on crushing behaviors of composite circular tubes with different reinforcing dibres *Int. J. of Impact Engg* Vol 38 pp 198-207
[7] Davoodi M.M, Sapuan S M, Ahmad D, Aidy A, Khalina A and Mehdi Jonoobi 2011 Concept selection of car bumper beam with developed hybrid bio-composite material Materials and Design Vol 32 pp 4857-4865.
[8] Kedar S Pandya, Ch Veerraju and Naik N K 2011 Hybrid composites made of carbon and glasswoven fabrics under quasi static loading *Materials and Design* Vol 32 pp 4094-4099.
[9] Thimmegowda Rangasamy and Sabapathy Vijayarangan 2005 Optimal sizing and stacking sequence of composite drive shafts Materials Science (Medziagotyra) Vol 11 No 2 pp 133-139.
[10] Dudley Bukey D 1899 Horse-*Power U.S. Patent 631,198* .

IOP Conf. Series: Materials Science and Engineering 1123 (2021) 012016 doi:10.1088/1757-899X/1123/1/012016

DYNAMIC ANALYSIS OF MULTI DISC ROTATING STEPPED SHAFT THROUGH SIGNAL PROCESSING

K Babu Rao [1], D. Mallikarjuna Reddy [2]
[1] Research Scholar, Design and Automation Department, School of Mechanical Engineering, VIT University, Vellore, India, kbabu.rao2016@vitstudent.ac.in
[2] Associate Professor, Design and Automation Department, School of Mechanical Engineering, VIT University, Vellore, India, dmreddy@vit.ac.in

Abstract. This study identifies a method for detection of irregularities like open cracks or grooves on a rotating stepped shaft with multiple discs, based on the Wavelet Transforms. Cracks are represented as reduction in diameter of shaft (groove) with small width. Single as well as multiple grooves are considered on stepped shaft at locations of stress concentration. Translational or rotational response curves/mode shapes are extracted from finite element analysis of rotor with and without grooves. Discrete and continuous 1D wavelet transforms applied on resultant response curve or mode shapes. The results show that rotational response curves or mode shapes are more sensitive to shaft cracks and key contributors to identify the location of cracks than translation response curves or mode shapes. Discrete Wavelet Transforms are accurate enough to locate the groove of smaller size. Effectiveness of detection by Wavelets Transforms is analyzed for Single as well as Multiple grooves with increase in grove depth. Increase in groove depth can be quantified by increase in wavelet coefficient and it can be an indicator.

1. Introduction

Identification of damage in the machine elements has been the special interest among the researchers in last few decades. Increase in demands towards more efficient and high speed system demands more robust, reliable design and condition monitoring systems. Best and reliable defect identification techniques are available during and after manufacturing of components. This quality checks cannot stop the defects to generate during operation of machine due to fatigue, operating conditions, lack of maintenance and environment effects. Rotors are no more exemption to it. In fact this type of defects leads to catastrophic failure.

Studying the response of the system can reveal the changes in dynamic behaviour of rotor dynamic system. Many methods are available to identify the changes in the running behaviour of rotor dynamics system such as modal analysis, acoustic analysis, wear debris analysis, mode shapes, transfer functions and response of the system at different locations. The first model parameter used to diagnose the health of the rotor system is natural frequency. Signal process techniques like Fourier transforms along with natural frequency change can only capable of detecting damages and will not give any information about the location and severity of damage. The Sensitivity of this method is very low, possibility of damage identification is not guaranteed because there is very small change in frequency content of signal.

Wavelet Transforms (WT) is showed a better tool for extracting the salient feature from vibration response [8] to identify and locate cracks. WT carry the location information along with frequency and amplitude. Spatial vibration data such as operation deflection shapes (ODS) or mode shapes is reliable tools to detect the location of the cracks or damages [8]. The ODS at a particular frequency with different loading conditions is gaining importance for detection of crack in recent years [9]. Luc

Chouinard et al [1] estimated the reliability of WT in detection of cracks by analysing mode shapes. They defined a criteria based on probability and statics of wavelet coefficients to judge the severity of cracks in beams. S Swamy et al [4] proposed a method based on fundamental mode shapes with continuous wavelet transforms (CWT) to identify and quantify single as well as multiple damages in beam structure. S. Suresha and D. M. Reddy [5] used CWT to identify the sensitivity of mode shapes in detecting the damage location and quantification. They concluded that CWT is promising technology to identify the damages in plate structures [5, 8].

Rims Janeliukstis et al [7] defined a damage estimate reliability parameter based on wavelet scales and isotropic Pet Hat wavelet to validate scales of WT coefficients. Rouhollah Pour et al [2] applied WT on higher order mode shapes to detect the cracks in cantilever beam. Roger Serra and Lautaro Lopez et al [6] used stationary wavelet transforms to de-noise the signal of mode shapes with and without damage in cantilever beam and used CWT on difference between this signals to identify the damage.

Factor dimension of the detailed signals of discrete wavelet transforms (DWT) can be used to locate the cracks accurately [10]. Time synchronised wavelet analysis method was proposed and its effectiveness to detect faults in rotor system [12]. Maosen Cao and Pizhong Qiao [13] presented a method integrated WT to ODS for damage diagnosis. Curvature of mode shape is one the characteristic that can be used to detect the crack [2] [15]. Mode shape curvatures and spatial wavelet transform can be used to identify the location of damage in rotor [11]. Many researchers have sued ODS to identify the crack or damage location in rotor systems using WT. C Bhargav Sai and D Mallikarjuna Reddy [14] proposed CWT on ODS can detect the cracks in the rotor. S. Rathna Prasad and A. S. Sekhar [9] used displacement and strain energy to locate and quantify the damage in noisy signal from ODS of rotor. Banks and colleagues [8], published a papers on crack detection using the wavelet Transforms. In this study we identified the sensitive parameter of ODS to detect, locate and quantify the dame or crack in rotor dynamics systems.

2. WAVELET TRANSFORMS AND ANALYSIS

Wavelet transform provides three outputs, scale (inverse of frequency), time or space and "wavelet coefficient". This wavelet coefficient resembles signal behaviour [15]. Below subsection describe the basic theories of CWT and DWT.

Continuous Wavelet Transform (CWT)

Let ψ (t) is a real or complex valued function in limited time domain. That is ψ (t) has values in a certain time range and zeros elsewhere. The CWT coefficient of mother wavelet $\psi_{u,s}(t)$ which is given as

$$Wf(s,u) = \int_{-\infty}^{\infty} f(t) \frac{1}{\sqrt{s}} \psi^* \left(\frac{t-u}{s}\right) dt \qquad (1)$$

With this transformation, one can map a one dimensional signal f(t) to a two dimensional coefficientsWf(s, u).

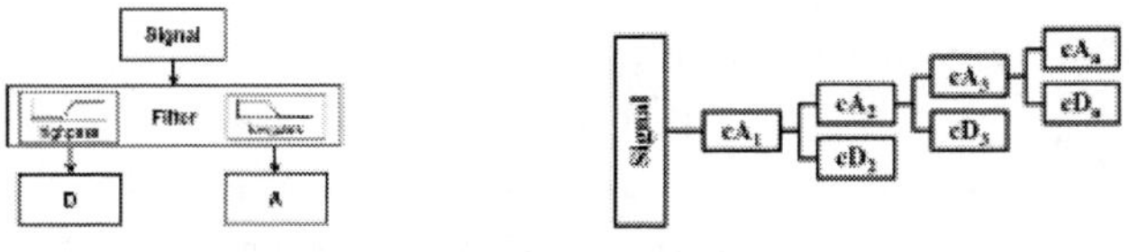

Figure 1. Wavelet Decomposition Tree

Discrete Wavelet Transform (DWT)

DWT process the signals in two channel sub bands approximation and detail. Low frequency content in a signal is referred as Approximation and High frequency content in a signal is referred as Detail. Discrete wavelet approximation and detail coefficients can be calculated as follows. The

*** Formatting Issue - Best Available Paper/Graphic ***

decomposition can be iterative, the filtering process for multilevel decomposition is shown figure (1).

$$W(j_0, k) = \frac{1}{\sqrt{s}} \sum_t f(n)\, \psi_{jo,k}(n) \tag{2}$$

$$W(j, k) = \frac{1}{\sqrt{s}} \sum_t f(n)\, \psi_{j,k}(n) \qquad j \geq j_0 \tag{3}$$

3. METHODOLOGY

The following block diagram as shown in the figure (2) indicates steps flowed in this study. This diagram provides details about the process and techniques used for the current analysis.

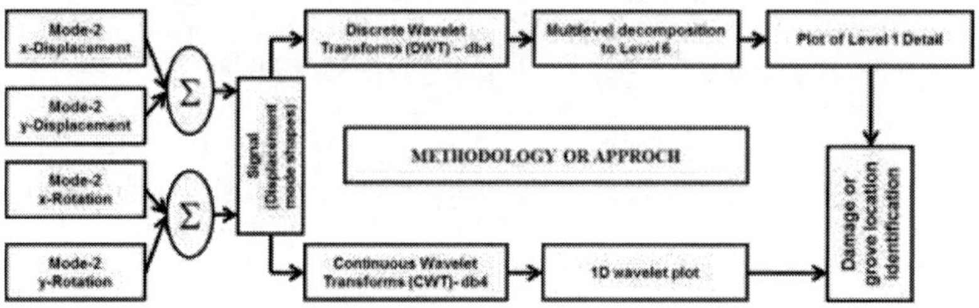

Figure 2. Analysis Block diagram

Nodal displacements at specific speed of rotor are extracted and displacement mode shapes are formed. This displacement mode shape transformed to CWT and DWT domains. Difference between wavelet coefficients of rotor with and without grove is analysed for damage or grove location identification. In this study Daubechies wavelet type 4 (db4) is considered for both DWT and CWT. Db4 wavelet function is an orthonormal wavelet and having highest number of vanishing movements.

4. NUMERICAL MODEL AND EQUATION OF MOTION

In this study finite element analysis of stepped shaft with 3 Disc rotor model with four degree of freedom per node is considered. Rotor shaft is modelled using Euler and Timoshenko beam theory. Shear, rotary inertia and gyroscopic effects are considered in this analysis. The equation of motion of stepped rotor systems in fixed coordinates is given as

$$[M]\ddot{q} + [C + G]\dot{q} + [K]q = [Q] \tag{4}$$

Where [M] is mass matrix, [C+G] gyroscopic and damping matrix and [K] stiffness matrix of shaft, bearings and groove or crack elements. Two translational displacements u and v are along OX and OY axis respectively. Two Rotational Displacements θ and φ are along OX and OY respectively. The rotor rotates clockwise with angular velocity Ω about OZ as shown in figure (1). It is most common to term translation and rotation displacement as bounce and tilt motion, respectively. Bearing are isotropic, stiffness in x and y direction are $K_{xx}=K_{yy}=1e9$ N/m^2.

DESCRIPTION OF STEPPED SHAFT ROTOR MODEL

Uniform diameter shaft are rare in the any physical applications. Steps are necessary for some functionality and cannot be avoided. Shaft with three diametric steps (∅40.0mm, ∅50.0mm and ∅60.0mm) are considered in this study. The total length of the shaft is 1.0m. Figure (3) describes the positional, dimensional and nodal details of stepped rotor, bearings, discs and grove or cracks.

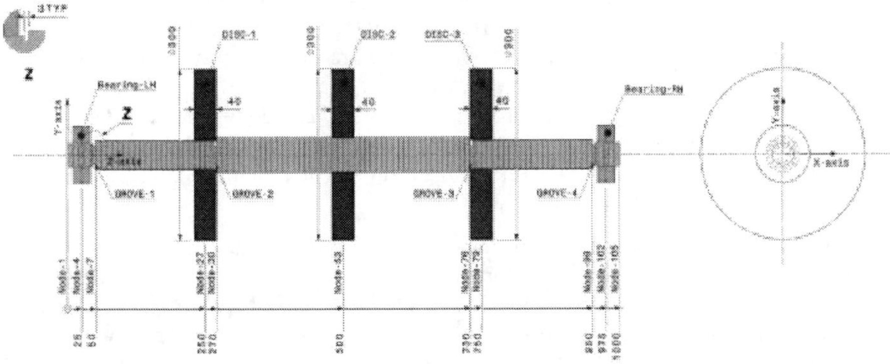

Figure 3. Stepped shaft rotor model with dimensional and nodal information

Discretization with more number of elements can increase the accuracy but need more computational time and data to store. A convergence test based on mode-4 natural frequency reveals that a minimum 80 number of elements required for discretization, however we considered 105 elements for this study, Refer figure (4) for details.

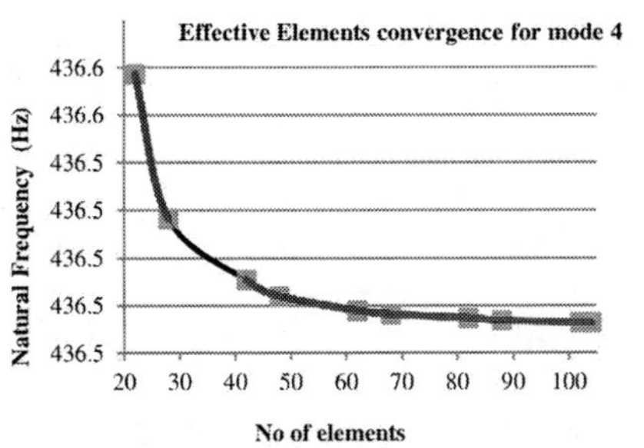

Figure 4. No of elements Convergence plot

Table 1. Natural frequency of different cases

	WG	C1	C2	C3	C4	C5
NF1	22.88	22.88	22.82	22.82	22.75	22.75
NF2	54.39	54.38	54.27	54.29	53.92	54.19
NF3	213.22	213.10	211.62	211.81	209.40	210.40
NF4	436.53	435.03	414.25	414.75	396.00	397.95
NF5	629.00	628.70	604.21	604.60	574.20	575.21

Groove or open cracks are modelled by reduction in the diameter of the shaft element. Groove is provided at highly stress concentrations locations or areas. At these locations the section changes occur which have an effect on modal displacement. It is highly important to clearly distinguish the causes of these singularities. Studying the Effectiveness of the crack or grove detection method at this location is vital.

NUMERICAL SIMULATION

A finite element code is generated in MATLAB for stepped rotor model with different groove location and combinations as follows

A single groove or an open crack between node 6 & 7 (Near bearing LH).

A single groove or an open crack between node 29 & 30 (Near disc-1)

Two groove or an open crack between node 6 & 7, 29 & 30 (Near bearing LH & disc-1)

Three groove or an open crack between node 6 & 7, 29 & 30, 76 &77 (Near bearing LH, disc-1 & disc-2)

Four groove or an open crack between node 6 & 7, 29 & 30, 76 &77, 99 & 100 (Near bearing LH, disc-1, disc-2 & bearing RH)

Model analysis is performed on healthy rotor and rotor with groove or cracks. Table 1 shows the details of first five natural frequencies against healthy rotor (WG) and cracked rotor Case C1 to C5 with 40% reduction in diameter of grove elements. These natural frequencies are taken at a rotor speed of 3100 RPM which is close to second natural frequency. Reductions in natural frequency are evident for presence of groove. Response of all four degrees of freedom at every node is recorded and Operational Deflection Shapes are formed separately for translational and rotational degrees of freedom. Bounce mode or translation mode shape curve is formed by resultant of X-displacement and Y-displacement. Tilt mode or rotational mode shape is formed by resultant of rotational Displacements θ and φ.

Signals of translational and rotational mode shapes are analysed separately and collectively with Continuous and discrete wavelet transforms. Grove depth is varied from 10% to 50% of base diameter for all cases as stated above. Signal of translational and rotational mode shapes are generated by subtracting the nodal displacement without grove from without grove. DWT coefficients are calculated for level 6 decompositions. It is observed that crack or grove location can be identified by plot of detail-1 of the DWT and CWT transforms.

RESULTS AND DISCUSSION

1st Translational ODS and its DWT, CWT with db4 mother wavelet for healthy rotor are shown in the figure (5). It is observed that ODS curve 5(a) is having discontinuities at location where diameter of the shaft changes. Correspondingly we can see the spikes at location of discontinuities in DWT plot 5(b) which is detail at Level-1. CWT plot 5(c) doesn't show these discontinuities. Generally CWT will show higher amplitudes at bearing restrains.

Figure 5. Healthy rotor 1st translational response plots. (a) ODS, (b)DWT details-L1, (c) CWT

Figure 6. Case-1:10% grove depth 1st translational response plots. (a) RODS, (b)DWT, (c) CWT

ODS of rotors with grove or crack are formed by subtracting the deflection without grove from with grove for all mode shapes. We termed this ODS as Resultant ODS (RODS). 1st Translational RODS and its DWT, CWT with 10% grove-1 depth is shown in the figure (6). It is observed that DWT 6(b) and CWT 6(c) plots are similar as healthy rotor in figure (5) with change in magnitude of coefficients. Therefore 1st translation mode is not effective in identifying the location of damage it can only give information about discontinuities in shaft.

1st Rotational RODS and its DWT, CWT with 10% grove-1 depth (Case-1) is shown in the figure (7). It is observed that DWT 7(b) plot showing spikes with higher magnitude at grove-1 location. In and CWT 7(c) plot the higher magnitude (bright colour) at low scales i.e. scale -1 (High frequency) indicates the discontinuity. We observe a small discontinuity at middle of the disc. Therefore rotational modes are more sensitive to the groove then translation modes but other small spikes are also observed. This may be conflating the claim to identify the grove or crack location clearly.

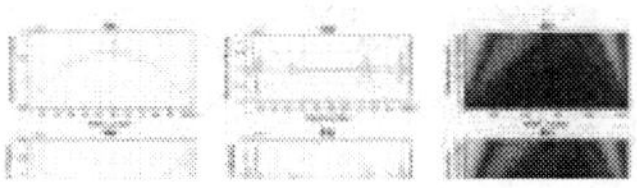

Figure 7. Case-1:10% grove depth 1st rotational response plots. (a) RODS, (b)DWT details-L1, (c) CWT

Figure 8.Case-1:10% grove depth 2nd rotational response plots. (a) RODS, (b)DWT details-L1, (c) CWT

2nd Rotational RODS and its DWT, CWT with 10% grove-1 depth (Case-1) is shown in the figure (8). It is observed that DWT 8(b) plot showing spikes with higher magnitude only at grove-1 location. In and CWT 8(c) plot the higher magnitude (bright colour) at low scales i.e. scale -1 (High frequency) observed only at grove locations. Therefore 2nd rotational modes are more sensitive and reliable to identify the crack or grove location clearly.

Figure 9. Case-2:10% grove depth 2nd rotational response plots. (a) RODS, (b)DWT details-L1, (c) CWT

Figure 10. Case-3:10% grove depth 2nd rotational response plots. (a) RODS, (b)DWT details-L1, (c) CWT

Figure 11. Case-4:10% grove depth 2nd rotational response plots. (a) RODS, (b)DWT details-L1, (c) CWT

*** Formatting Issue - Best Available Paper/Graphic ***

Figure 12. Case-5:10% grove depth 2nd rotational response plots. (a) RODS, (b)DWT details-L1, (c) CWT

2nd Rotational RODS and its DWT, CWT with 10% grove depth for Case-2, Case-3, Case-4 and Case-5 are shown in the figure (9), figure (10), figure (11) and figure (12) respectively. It is observed that DWT and CWT plot showing spikes with higher magnitude only at grove location.

Figure 13 to 16 plots related to increase in grove depth from 20% to 50% for case-2 of this study. Observing figures 9, 13 to 16 it is clear that DWT coefficient magnitude increase as the crack or grove depth increases. 2D-CWT plots cannot revile the increase in magnitude of wavelets. A 3D CWT plots can provide this information. To proceed further, in future crack depth or grove depth can be quantified by analyse the wavelet coefficients for different depths.

Figure 13. Case-2:20% grove depth 2nd rotational response plots. (a) RODS, (b)DWT details-L1, (c) CWT

Figure 14. Case-2:30% grove depth 2nd rotational response plots. (a) RODS, (b)DWT details-L1, (c) CWT

Figure 15 Case-2:40% grove depth 2nd rotational response plots. (a) RODS, (b)DWT details-L1, (c) CWT

Figure 16. Case-2:50% grove depth 2nd rotational response plots. (a) RODS, (b)DWT details-L1, (c) CWT

5. Conclusions

This paper present a non-destructive damage detection and location identification technique based on operational mode shapes along with discrete and continuous wavelet transforms. Daubechies mother wavelet for continuous wavelet transforms and daubechies order 4 mother wavelet with level 6 decomposition is used for discrete wavelet transforms. Finite element method is used to analyse the stepped rotor with 3 discs. Grove or crack was created at stress concentration location by reducing the diameter of the grove element. Operational mode shape data is processed through wavelet transforms to identify the location of damage. 1st Translational mode shapes can reveal the information about discontinuity in the rotor while 2nd rotational mode shapes can accurately locate the grove or crack location. Despite of grove or crack locations are at discontinuities the proposed method can effectively and accurately locate the grove or crack element. This research study proved that wavelet transforms of 2nd operational mode shapes is a promising method to identify and locate the crack in rotor machine condition monitoring.

References

[1] Luc Chouinard, Vahid Shahsavari and Josée Bastien 2019 Reliability of Wavelet Analysis of Mode Shapes for the Early Detection of Damage in Beams *Front. Built Environ* Vol 5 pp 91
[2] Rouhollah Pour, Asgari Marnani J and Tabatabei A A 2018 A Novel Method for Crack Detection in Steel Cantilever Beam Using Wavelet Analysis by Combination Mode Shapes, *Int. J. of Image, Graphics and Signal Processing(IJIGSP)* Vol 10 No 4 pp 1-12.
[3] Muyideen Abdulkarem, Norhisham Bakhary, Mohammadreza Vafaei and Norhazilan Noor 2017 Mode Shape and Mode Shape Difference Evaluation to Damage Location in Plate Structures *Int. J. of Applied Engineering Research* Vol 12 pp 14620-14627.
[4] Shivasharanayya Swamy, Mallikarjuna Reddy D and Jaya Prakash G 2016 Damage detection and identification in beam structure using modal data and wavelets World *J. of Modelling and Simulation* Vol 13 No 1 pp 52-65.
[5] Suresha s and Mallikarjuna Reddy D 207 Study of sensitivity of mode shapes in damage identification using continuous wavelet transform *ARPN J. of Engg and Applied Sciences* Vol 12 No 15 pp 4539-4545.
[6] Roger Serra and Lautaro Lopez 2016 Damage detection methodology on beam-like structures based on combined modal Wavelet Transform strategy Mechanics & Industry Vol 18 pp 807.
[7] Rims Janeliukstisa, Sandris Rucevskisa, Pavel Akishina and Andris Chatea 2016 Wavelet transform based damage detection in a plate structure *World Multidisciplinary Civil Engineering-Architecture-Urban Planning Symp.* , Procedia Engineering Vol 161 Issue 127 – 132.
[8] María Gómez j, Cristina Castejón and Juan C. García-Prada 2016 Review of Recent Advances in the Application of the Wavelet Transform to Diagnose Cracked Rotors, MDPI Algorithms Vol 9 pp 19.
[9] Rathna Prasad and Sekhar A S 2015 Damage detection in rotors using wavelet analysis of operational deflection shapes *ICSV22, Florence (Italy)* pp 1-8.
[10] Yong-Ying Jiang, Bing Li, Zhou-Suo Zhang, and Xue-Feng Chen 2015Identification of Crack Location in Beam Structures Using Wavelet Transform and Fractal Dimension, *Hindawi publishing Corporation Shock and Vibration* Article ID 832763, 10 pages.
[11] Meng Hee Lim and Leong M S 2013 Detection of Early Faults in Rotating Machinery Based on Wavelet Analysis *Hindawi publishing Corporation Advances in Mechanical Engineering*, Article ID 625863, 8 pages.
[12] Aleksandra ziaja and krzysztof mendrok 2011 mode shapes subtraction and wavelet analysis for damage detection, 2011, Diagnostyka - diagnostics and structural health monitoring Vol 2 issue 58 pp 31-38.
[13] Maosen Cao and Pizhong Qiao 2008 Integrated wavelet transforms and its application to vibration mode shapes for the damage detection of beam-type structures *Smart Mater. Struct.* 17 055014 pp 17
[14] Cherukuri Bhargav Sai and Mallikarjuna Reddy D 2016 Dynamic Analysis of Faulty Rotors Through Signal Processing, Applied Mechanics and Materials Vol 852 pp 602-606.
[15] Mallikarjuna Reddy D, Jayaprakash G and Swarnamani S 2008 Detection and localisation of damage in bridge model by spatial wavelet based approach, International Journal of Modelling Identification and Control Vol 3 Issue 2 pp 156-161

FINITE ELEMENT ANALYSIS OF THERMAL FATIGUE LOADING OF NANO MATERIALS COATED TURBINE BLADE FOR CRITICAL APPLICATIONS

Kiran Ashok, Aswin C K, Ben Abraham Samuel, Prof. Akash Mohanty

[1] Mechanical, Vellore Institute of Technology, Vellore, India, aswinck96@gmail.com
[2] Mechanical, Vellore Institute of Technology, Vellore, India, benabraham592@gmail.com
[3] Mechanical, Vellore Institute of Technology, Vellore, India, kiranashok007@gmail.com

Abstract. Thermal barrier coatings (TBC) have structures which are very complex and are employed in high temperature condition. The phenomenon of cracking is usually noticed at both the Top Coat (TC) and the interface between the bond coat and the thermally grown oxide layer under the Thermal Cyclic Fatigue (TCF) loading. In this study, the cracking behavior in Bond Coat (BC) and interface for TBC were thoroughly determined by using the commercially available ABAQUS/CAE 6.14-3 FEA software. This study also analyzed the key aspects causing the cracking behaviors in TBC. Various crack propagation path in the Bond Coat (BC) could be obtained on changing the thermal wave forms. The beginning location of TC (Top Coat) was independent of the roughness in the interface. The thermally grown oxide layer (TGO) which is thinner at the beginning could increase the service life of TBC. From the results it is evident that in order to characterize the service environment, the thermal fatigue load and the working temperatures have to be designed

1. INTRODUCTION

Failure of blade in gas turbine engines could have a harsh strike on the provision of engines and on the management of Engine soundness. Failure of blade could occur due to high rotational speed at elevated temperature. Thermal barrier coatings (TBCs) are high temperature coating material that is applied to metallic surface. These TBCs act as insulators to parts that are exposed to high and continued heat loads by the application of thermal insulators [Fig. 1.1]. Fatigue life is expressed in terms of cycles to failure. Cycles to failure can be stated as the number of loads a material can sustain until a crack starts to appear on the material. For the detailed study Finite Element Analysis (FEA) is used. FEA gives an insight of imperfections that occurred during the manufacturing process. FEA is done with the help of ABAQUS/CAE 6.14-3 software on the turbine blade and the residual stress and the interface temperatures of different coating layers were calculated. Behaviour of the surface of turbine blade was also studied by analysing the surface for crack and creep formation. In this analysis, the nanoparticle coated TBC also considered for the same purpose.

2. SCOPE OF THE WORK

Turbine blades which are under cyclic thermal load has tendency to create thermal residual stress accumulation in the coating leading to early delamination of the multi layered coating. So TBC coated materials came into existence in order to overcome this scenario, but even these advanced coating have their limitation. Most of the work done so far deals with the macroscopic elements. This work deals with Nano- particle property and nanoparticle coated TBC which is a new opening, since nanoparticle has an excellent thermal property, an attempt was made to do such analysis. Stress analysis can be performed to predict the propagation of cracks. Moreover the idea of thermal insulation of the TBC materials also studied by plotting a graph between the temperature vs. thickness.

2.1 Deliverables For Nano

Use of Nano materials in the near future is going to be high because of their high strength, good stiffness and functional properties that are better than conventional materials. The adoption of Nano materials such as YSZ opens up a promising substitute to enhance the performance and structure of a TBC because of its excellent properties which are mainly high micro-hardness, less thermal diffusivity, elastic modulus, smoother roughness etc. We will be seeing a lot of high strength composite along with the incorporation of the Nano materials due to more than satisfactory requirements it meets.

2.2 Limitation

Since we are working on Nano structured materials the calculations may deviate from the tolerance limits that are required. This problem can be solved by generating micrometer and a nanometer meshes wherever necessary.5% is the standard for engineering tolerance used for comparison. But as we are dealing Nano structured materials in the current study, the value deviations for FEM approach can be up to 10-15% (approximately) when we are comparing the results.

3. OBJECTIVE OF THE WORK

- To calculate the residual stress of the traditional thermal barrier coating using FEA and also do the analysis of nanoparticle coated TBC.

- To find the thermal insulation of TBC material and to find out the interface temperature at each surface.

a) Top coat-TGO interfaces

b) TGO – Bond coat interfaces

c) Bond coat- substrate interfaces

IOP Conf. Series: Materials Science and Engineering 1123 (2021) 012017 doi:10.1088/1757-899X/1123/1/012017

4. METHODOLOGY

The following flow chart which is depicted in the Figure 1 represents the workflow that is carried during the FEA analysis in ABAQUS software.

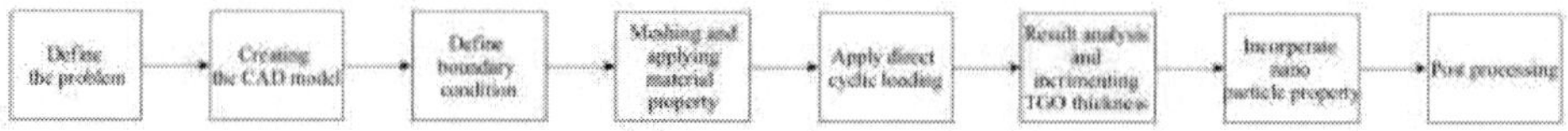

Figure 1. Flowchart showing FEA method

The analysis of the Thermal Barrier Coated surface can be done using the software ABAQUS/CAE16.4-3. This is a graphical and powerful stress analysis tool and mathematical tool to solve boundary condition using Finite Element Analysis methods. An aerofoil CAD model was prepared in order to resemble the turbine blade geometry but for the analysis purpose, a simple circular cylinder was considered with TBC having same thickness dimensions. The analysis is done on a cylinder for reducing the complexity. A schematic illustration of the model is shown in **Figure 2**. The radius of the circular cylinder was taken to be 30 mm. The thicknesses of various layers in the coating for the 1st iteration have been mentioned in Table 4.1. The model is imported to ABAQUS platform in order to do the analysis. The file format which is imported to ABAQUS is .iges.

Figure 2. Schematic representation of coating Specimen

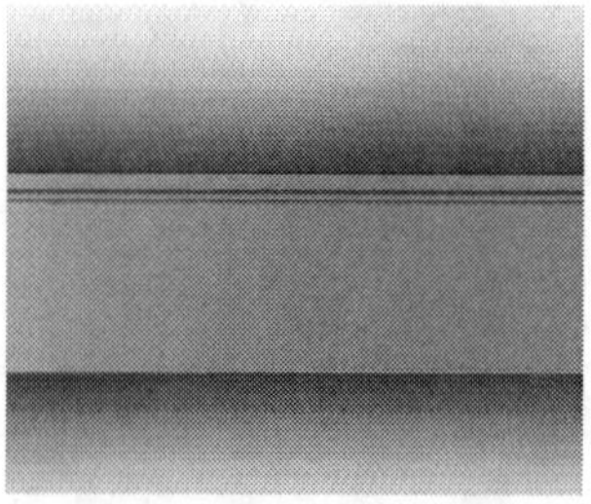

Figure 3. Thickness of different layers

Table 1. Thickness of different layers

SL NO.	LAYERS	THICKNESS (in μm)
1	TOP COAT	300
2	TGO + BOND COAT	5 + 145
		15 + 135
		30 + 120
3	SUBSTRATE	3000

Multiple iterations were performed with varying thickness so as to understand how the Temperature changes with the Thermally Grown Oxide (TGO) thickness. For this purpose, the thickness was considered as 5 μm, 15 μm, and 30 μm. Study of the effect of Nano-particle reinforced material property was also analysed by changing the material property of topcoat from traditional to nanoparticle property. All the iterations were done for 20 cycles which contains a heating, dwell and cooling period. The mesh shape was chosen to be square with varying size. The cell size range was varied from 1 μm to 50 μm. The meshes were very fine at the TGO layer and the substrate layer has more coarse cells comparatively in order to reduce simulation timing. Number of nodes generated 1432154 numbers of cells 514227. Temperature varying property of each layer found experimentally was imported into the software in the form of a table. From the properties it was evident that the Top coat (Ceramic) was having the least thermal conductivity. TGO (Thermally Grown Oxide) layer is necessary for the adhesion of TC (Top Coat) and BC (Bond Coat).

Turbine blades which are under cyclic thermal load has tendency to create thermal residual stress accumulation in the coating leading to early delamination of the multi layered coating. So TBC coated materials came into existence in order to overcome this scenario, but even these advanced coating have their limitation. Most of the work done so far deals with the macroscopic elements. This work deals with Nano- particle property and nanoparticle coated TBC which is a new opening, since nanoparticle has an excellent thermal property, an attempt was made to do such analysis. Stress analysis can be performed to predict the propagation of cracks. Moreover the idea of thermal insulation of the TBC materials also studied by plotting a graph between the temperature vs. thickness.

a) Boundary Constraint

The constraints of symmetric boundary are used because of axisymmetric configuration. The different layers considered are isotropic and homogeneous. For the purpose of structural analysis fix all the edges in the bottom and sides of each face are fixed. The top boundary is set free. A maximum temperature of 1150 ^{o} C is given at the top most point of the Top Coat and a temperature of 27 ^{o} C is given at the bottom most point. The residual stress is induced during the cooling period of the thermal cycles. Phase transformation and creep mechanisms introduction which results in stress-relieving and can effect stress state in TBCs are assumed inactive during the simulation. All the boundaries are fixed since the main focus is on the TGO layer.

b) Cyclic Loading

For the setup of thermal cyclic loading the entire TBC's (Thermal Barrier Coating) is to be kept in a constant temperature field, which varies with time in every Thermal Cycle (TC). The Finite Element (FE) calculation starts at 1150 °C which is a stress-free state followed bycooling down to room temperature and then heat up again up to 1150 °C. 20 thermal cycles continue the same way [11]. Each cycle is heated from 20°C to 1150 °C for 0.15 h heating period. It has a dwell period of 4 h for oxidation at 1150 °C. Each cycle is cooled down from 1150 °C to 20 °C for 0.3 h. During the idle period or the dwell period, a continuous oxidation takes place between the TC-BC interfaces, which in turn thickens the TGO layer at a decreasing rate.

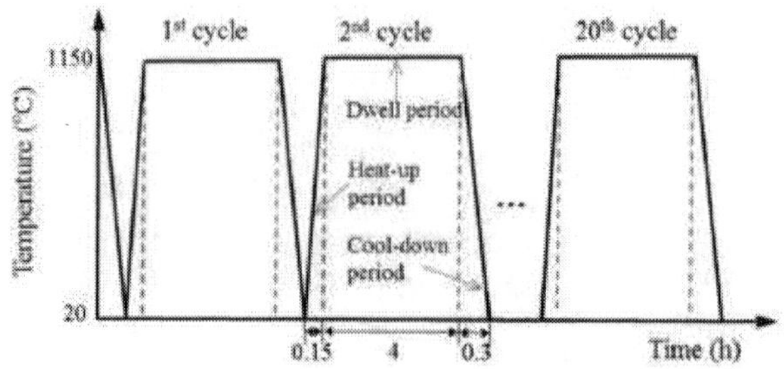

Figure 4. Cyclic loading

The TGO thickness h is achieved at dwell time t by a parabolic equation: h2 = 2kpt, where kp is the parabolic rate constant. The creep strength of TGO layer depends on the grain size. The creep strength is affected by the cat ion impurities. The cat ion impurities are Yttrium or Zirconium. The creep strength of TC (Top Coat) depends on the procession conditions, so there is a wide range of creep parameters for both TC and TGO.

5. Result And Discussion

The materials with traditional and Nano material property was successfully simulated and the graphs shows that the proposed material that is the Nano coated material has a higher efficiency in insulating the temperature, at high temperature working conditions. Stress analysis is presented as a prerequisite for crack behaviour. The results show the effect of TGO when the TBC coated material is subjected to Thermal Cyclic Loading (TCL). Maximum shear stress is used to determine the variation of residual stress at various cycles and is depicted below. Each cycle there was a decrease in maximum principal stress in the TGO layer and Top Coat. The graph below shows the temperature vs. thickness in the case of 5μm. The graph shows that the insulation property of the top coat is larger compared to the TGO and bond coat.

5.1 Temperature flow along the TBC

Figure. 5 Temperature contour in traditional coated TC

Figure. 6 Temperature contour in Nano coated TC

5.2 Comparison of temperature variation along the TBC interface of different TGO thickness

Figure 7 Shows the Temperature vs. Thickness variation for 5μm. The top coat which is ceramic coating made up of Yttrium stabilised Zirconia insulates the major temperature from entering to the substrate and this is the main reason for the protection of the substrate from degrading at a faster rate. From the graph a comparison was done between the Nano YSZ and the traditional YSZ top coat with the same thickness of TGO layer of 5 μm. It was observed that the insulation property of the Nano coated YSZ top coat has more efficient insulating property than the traditional YSZ topcoat and naturally there was a larger temperature drop in the Nano coated, top coat and the drop was almost 25 □C. The temperature that was obtained from the graph at various interfaces for the traditional top coated YSZ for 5μm are as follows 398 □C for TC/TGO, 395 □C for TGO/BC, 302 □C for BC/substrate. In the case of nanomaterial top coated YSZ for 5μm are as follows 374 □C for TC/TGO, 36 □C for TGO/BC, 283 □C for BC/substrate coat.

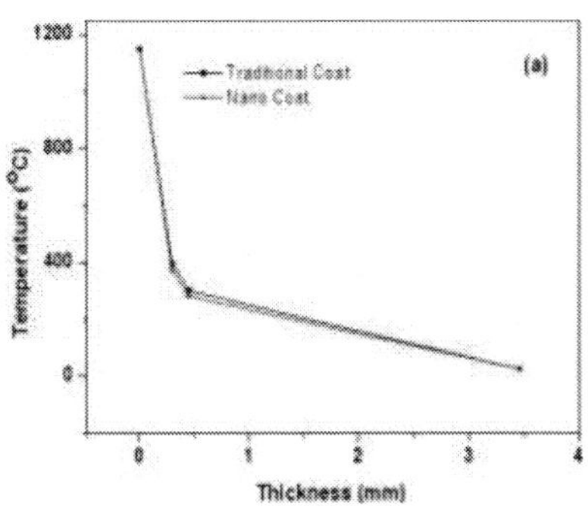

Figure. 7 Temperature vs. Thickness 5μm

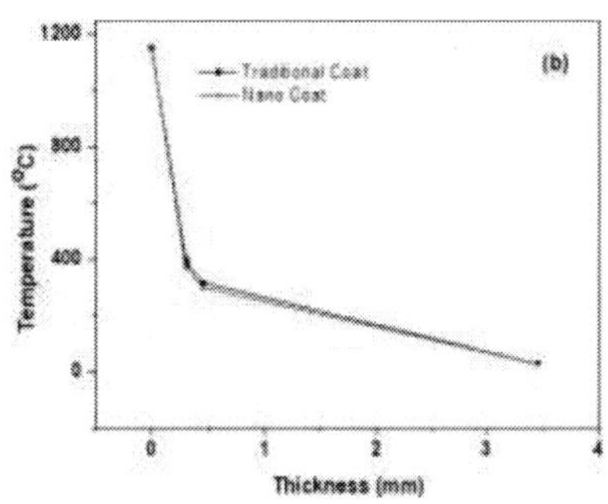

Figure. 8 Temperature vs. Thickness 15μm

Figure 9 shows the temperature vs. thickness variation. There was no change in the thickness of the ceramic coating made up of Ytrititium stabilised Zirconia remain the same but there was an increase in the TGO thickness to 15μm and this also lead to the decrease in the bond coat-135μm as the total thickness of bond coat plus TGO layer should remain 150μm. The TGO layer is formed by the corrosion of the bond coat and the aluminium present in the bond coat gets oxidised more quickly increasing the TGO layer. Again a comparison was made between Nano YSZ and the traditional YSZ top coat but in this case we have considered a different thickness of TGO which is 15μm. A larger temperature drop in the Nano coated top coat and the drop was almost 25 □C. The temperature that was obtained from the graph at various interfaces for the traditional top coated YSZ for 15μm are as follows- 399.5 □C for TC/TGO, 381 □C for TGO/BC, 315 □C for BC/substrate. In the case of nanomaterial top coated YSZ for 15μm are as follows 374 □C for TC/TGO, 368 □C for TGO/BC,

298 □C for BC/substrate coat. The temperature drop at the bond coat was comparatively reduced in comparison with the 5μm thick TGO layer, from this it can be seen that the temperature drop that was created got reduced with increase in TGO thickness.

IOP Conf. Series: Materials Science and Engineering 1123 (2021) 012017 doi:10.1088/1757-899X/1123/1/012017

Figure 10 represents the temperature variation with thickness, In the case of 30µm thick TGO layer. There was no change in the thickness of the ceramic-top coat and only the material property was varied, that was traditional and Nano particle properties are used for this purpose of analysis. There was an increase in the TGO thickness to 30µm and this also leads to the decrease in the bond coat thickness to 120µm as the total thickness of bond coat plus TGO layer should remain 150µm. Temperature drops in TBC material when it moves from top most point of the topcoat to the TGO Topcoat interface. The temperature that was obtained from the graph at various interfaces for the traditional top coated YSZ for 15µm are as follows 402 °C for TC/TGO, 382 °C for TGO/BC, 341 °C for BC/substrate. In the case of nanomaterial top coated YSZ for 15µm are as follows 375 °C for TC/TGO, 361 °C for TGO/BC, 311 °C for BC/substrate coat. The temperature drop at the bond coat was comparatively reduced in comparison with the 5µm and 15µm thick TGO layers, from this it can be seen that the temperature drop that was created got reduced with increase in TGO thickness.

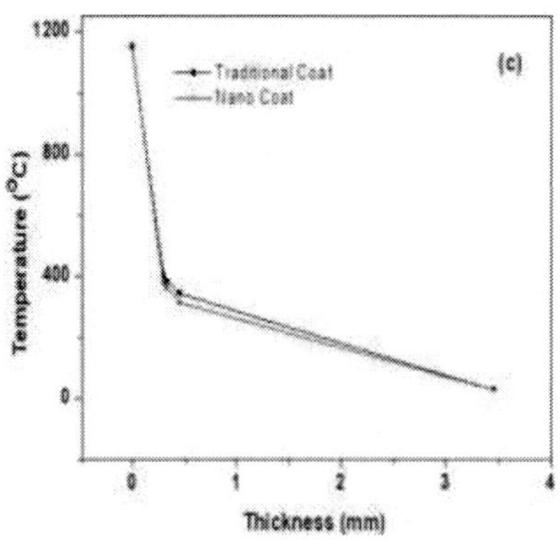

Figure. 9 Temperatures vs. Thickness 15µm

5.3 Temperature vs. TGO thickness

Figure 11 represents the Temperature vs. TGO thickness at the Topcoat/ TGO interface. The graph represents that the temperature ranges from 398 °C to 402 °C. From these graphs the first inference that was obtained is that the thermal insulation doesn't very much in topcoat with change in TGO layer thickness. The major change that was made during the analysis is change in material property of the top coat. The Nano material top coat when compared with the traditional material property put out a very positive result by dropping the temperature difference at the interface of the layers by 250 °C. This way the insulating property of topcoat is enhanced when compared with the earlier used traditional top coat materials.

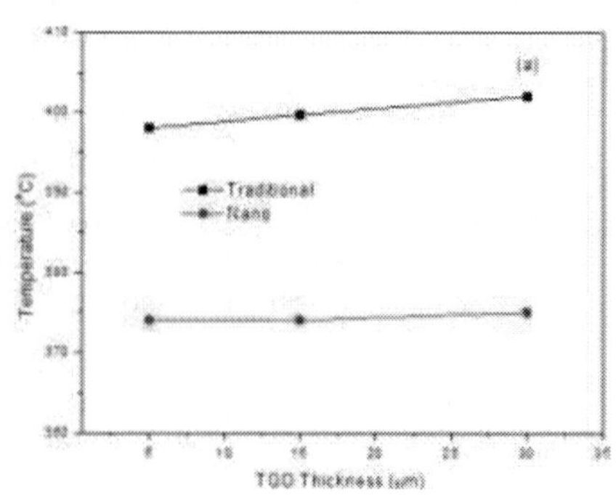

IOP Conf. Series: Materials Science and Engineering 1123 (2021) 012017 doi:10.1088/1757-899X/1123/1/012017

Figure. 10 Temp at TC/TGO interface vs. TGO thickness

Figure 11 shows the temperature variation with thickness, In the case of 15μm thick TGO layer. There was no change in the thickness of the ceramic coating made up of Yttrium stabilised Zirconia remain the same but there was an increase in the TGO thickness to 15μm and this also lead to the decrease in the bond coat-135μm as the total thickness of bond coat plus TGO layer should remain 150μm. The TGO layer is formed by the corrosion of the bond coat and the aluminium present in the bond coat gets oxidised more quickly increasing the TGO layer. Again a comparison was made between Nano YSZ and the traditional YSZ top coat but in this case we have considered a different thickness of TGO which is 15μm. A large temperature drop in the Nano coated top coat and the drop was almost 25 °C. The temperature that was obtained from the graph at various interfaces for the traditional top coated YSZ for 15μm are as follows 399.5 °C for TC/TGO, 381 °C for TGO/BC, 315 °C for BC/substrate. In the case of nanomaterial top coated YSZ for 15μm are as follows 374 °C for TC/TGO, 368 °C for TGO/BC, 298 °C for BC/substrate coat. The temperature drop at the bond coat was comparatively reduced in comparison with the 5μm thick TGO layer, from this it can be seen that the temperature drop that was created got reduced with increase in TGO thickness.

Figure 12 represents the temperature variation with thickness. In the case of 30μm thick TGO layer. There was no change in the thickness of the ceramic-top coat and only the material property was varied, that was traditional and Nano particle properties are used for this purpose of analysis. There was an increase in the TGO thickness to 30 μm and this also leads to the decrease in the bond coat thickness to 120μm as the total thickness of bond coat plus TGO layer should remain 150μm. Temperature drops in TBC material when it moves from top most point of the topcoat to the TGO Topcoat interface. The temperature that was obtained from the graph at various interfaces for the traditional top coated YSZ for 15μm are as follows 402 °C for TC/TGO, 382 °C for TGO/BC, 341 °C for BC/substrate. In the case of nanomaterial top coated YSZ for 15μm are as follows 375 °C for TC/TGO, 361 °C for TGO/BC, 311 °C for BC/substrate coat. The temperature drop at the bond coat was comparatively reduced in comparison with the 5μm and 15μm thick TGO layer. From this it can be seen that the temperature drop that was created got reduced with increase in TGO thickness.

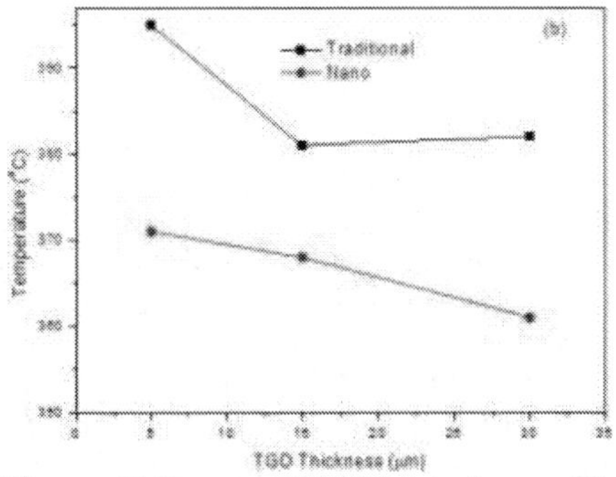

Figure. 11 Temp at BC/TGO interface TGO thickness

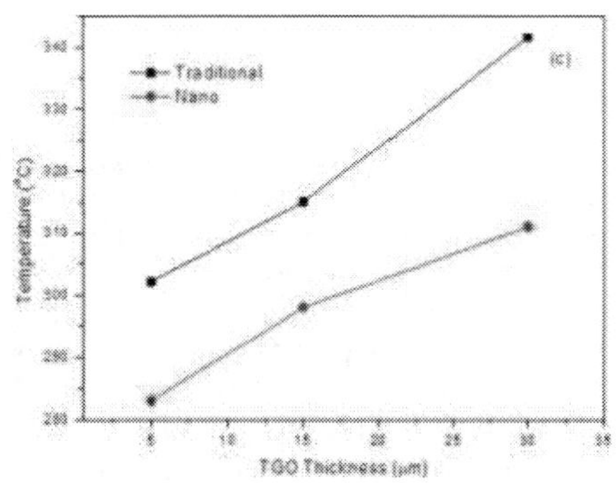

Figure. 12 Temp at BC/substrate vs vs. TGO thickness

IOP Conf. Series: Materials Science and Engineering 1123 (2021) 012017 doi:10.1088/1757-899X/1123/1/012017

5.4 Maximum principal Stress vs. Time

Figure 14 shows the plots of Maximum principal stress vs. time. In the graph which is shown a superimposition of 4 graphs all of them showing the maximum principal stress at various layers in 5μm thick TGO layer. The plot shows the variations in maximum principal stress when passing through the time and has to cross through the heating, dwell and cooling condition. The red and the green line show the variation of the 5μm thick TGO layer maximum principal stress. From the plot the Nano particle coated TGO has lesser stress compared to the traditional coating. In the case maximum principal stress varies from 215.42 MPa to 143.57 MPa and the residual stress varies from 130.43 to 70.49 at end of 20 cycles in the case of TGO for traditional coating. In the case of Nano particle coated TBC the stress variation is comparatively lower and the variation of maximum principal stress varies from 201.42 to 129.57. The variation in the case of residual stress is as follows 116.44 to 56.19.The dark blue and line and the light blue line shows the variation of the 5um thick Top Coat layer maximum principal stress and residual stress. From the plot the Nano particle coated top coat has lesser stress compared to the traditional coating. In the case maximum principal stress varies from 293.42 MPa to 233.57 MPa and the residual stress varies from 210.77 to 135.49 at the end of 20 cycles in the case of TGO for traditional coating. In the case of Nano particle coated TBC the stress variation is comparatively lower and the variation of maximum principal stress varies from 275.99 to 203.57. The variation in the case of residual stress is as follows 190.41 to 130.19. From the plot the Nano particle coated TGO has lesser stress compared to the traditional coating.

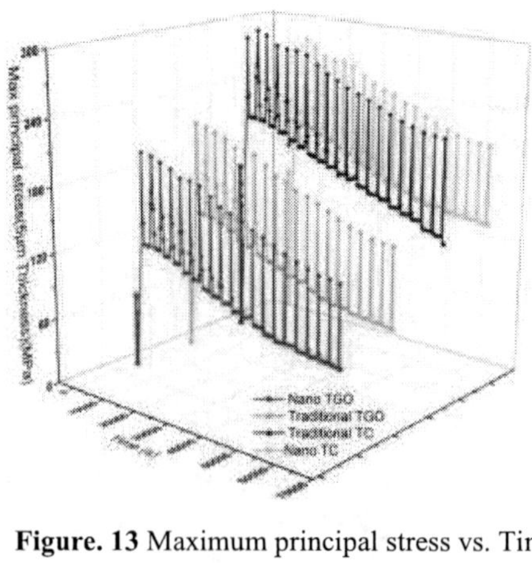

Figure. 13 Maximum principal stress vs. Time (5 μm)

Figure 13 plots the Maximum principal stress vs. time. In the graph which is shown a superimposition of 4 graphs all of them showing the maximum principal stress at various layers in 5μm thick TGO layer. The plot shows the variations in maximum principal stress when passing through the time and has to cross through the heating, dwell and cooling condition. The red and the green line show the variation of the 5μm thick TGO layer maximum principal stress. From the plot the Nano particle coated TGO has lesser stress compared to the traditional coating. In the case maximum principal stress varies from 215.42MPa to 143.57MPa and the residual stress varies from 130.43 to 70.49 at end of 20 cycles in the case of TGO for traditional coating. In the case of Nano particle coated TBC the stress variation is comparatively lower and the variation of maximum principal stress varies from 201.42 to 129.57. The variation in the case of residual stress is as follows 116.44 to

56.19.The dark blue and line and the light blue line shows the variation of the 5μm thick Top Coat layer maximum principal stress and residual stress. From the plot the Nano particle coated top coat has lesser stress compared to the traditional coating. In the case maximum principal stress varies from 293.42MPa to

IOP Conf. Series: Materials Science and Engineering 1123 (2021) 012017 doi:10.1088/1757-899X/1123/1/012017

233.57MPa and the residual stress varies from 210.77 to 135.49 at the end of 20 cycles in the case of TGO for traditional coating. In the case of Nano particle coated TBC the stress variation is comparatively lower and the variation of maximum principal stress varies from 275.99 to 203.57. The variation in the case of residual stress is as follows 190.41 to 130.19.From the plot the Nano particle coated TGO has lesser stress compared to the traditional coating

Figure 14 plots the Maximum principal stress vs. time. In the graph which is shown a superimposition of 4 graphs all of them showing the maximum principal stress at various layers in 30µm thick TGO layer. The plot shows the variations in maximum principal stress when passing through the time and has to cross through the heating, dwell and cooling condition. The red and the green line show the variation of the 30µm thick TGO layer maximum principal stress. From the plot the Nano particle coated TGO has lesser stress compared to the traditional coating. In the case maximum principal stress varies from 250.49 to 178.49 and the residual stress.
varies from 165.49 to 106.34 at end of 20 cycles in the case of TGO for traditional coating. In the case of Nano particle coated TBC the stress variation is comparatively lower and the variation of maximum principal stress varies from 245.42 to 174.57. The variation in the case of residual stress is as follows 116.44 to 111.19.The dark blue and line and the light blue line shows the variation of the 5µm thick Top Coat layer maximum principal stress and residual stress. From the plot the Nano particle coated top coat has lesser stress compared to the traditional coating. In the case maximum principal stress varies from 349.9 to 279.57 and the residual stress varies from 266.47 to 206.97 at the end of 20 cycles in the case of TGO for traditional coating. In the case of Nano particle coated TBC the stress variation is comparatively lower and the variation of maximum principal stress varies from 289.99 to 213.57. The variation in the case of residual stress is as follows 206.41 to
147.19.From the plot the Nano particle coated TGO has lesser stress compared to the traditional coating.

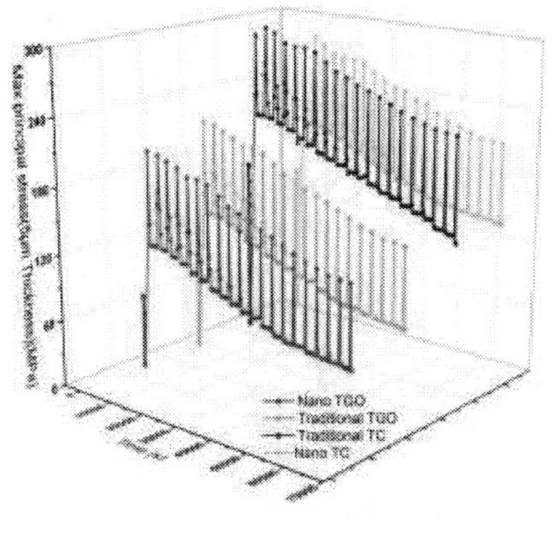

Figure. 13 Maximum principal stress vs. Time vs. Time (15µm)

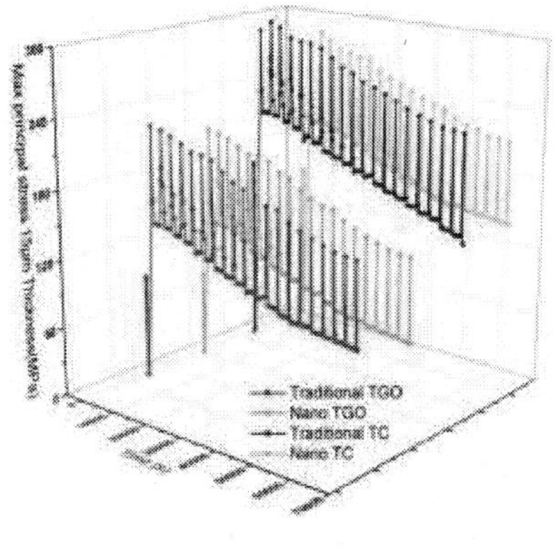

Figure 14. Maximum principal stress (30µm)

5 CONCLUSION

The interface temperature for the TBC material was calculated with varying temperature and the graphs were plotted. The residual stress graph with time was plotted. A comparative study of Nano particle and traditional coated material in terms of the insulation property and the residual stress variation in both the cases. From these results the proposed material which is the Nano coated YSZ can be utilized for the critical applications of the turbine blade as it has higher thermal insulating properties compared to the traditional. So the use of Nano- coated YSZ can increase the life of the turbine blade and hence be economical.

Reference

[1] Biao Li, Xueling Fan, Tiejun Wang and Kun Zhou 2015 Interfacial fracture behaviour of double-ceramic-layer thermal barrier coating system with segmented *structure J. of Engg Fracture Mechanics*

[2] Cen, W Y Qin and Yu Q M 2014 Analysis of interface delamination in thermal barrier coating system with axisymmetric structure based on corresponding normal and Tangential stresses *J. of Surface and Coatings Tech.*

[3] Jianan Song, Shaolin Li, Xiaoguang Yang, Duoqi Shi and Hongyu Qi 2016 Numerical study on the competitive cracking behavior in TC and interface for thermal barrier coatings under thermal cycle fatigue loading, *J. of Surface aCoatings Tech.*

[4] Jiang B S, Xu H D, Wanga and Lu Y U 2017 Modelling functionally graded materials in heat transfer and thermal stress analysis by means of graded finite *elements J. of Applied Mathematical Modelling*

[5] Jishen Jiang, Lingxin Jiang, ZhenweiCai, Weizhe Wang, Xiaofeng Zhao, Yingzheng Liu and Zhaomin Cao 2016 Numerical stress analysis of the TBC-film cooling system under operating conditions considering the effects of thermal gradient and TGO growth *J.of Surface and Coatings Technology*

[6] Jishen Jiang, Weizhe Wang, Xiaofeng Zhao, Yingzheng Liu, Zhaomin Cao and Ping Xiao 2018 Numerical analyses of the residual stress and top coat cracking behavior in thermal barrier coatings under cyclic thermal loading *J. of Engg fracture mechanics*

[7] Marcin,Białas 2008 Finite element analysis of stress distribution in thermal barrier coatings J. *of Surface & Coatings Tech.*

[8] Moridi, Azadi M and Farrahi G H 2012 Thermo-mechanical stress analysis of thermal barrier coating system considering thickness and roughness effects *J. of Surface and Coatings*

[9] Nayebpashaee, Seyedein S H, Aboutalebi M R, Sarpoolaky H and Hadavi S M M 2015 Finite element simulation of residual stress and failure mechanism in plasma sprayed thermal barrier coatings using actual microstructure as the representative volume *J. of Surface and coatings tech.*

[10] Valente C Bartuli M Sebastiani and Casadei F 2004 Finite element analysis of residual stress in plasma-sprayed ceramic *coatings J. of SAGE*

[11] Abbas, Guo H and Ramzan M 2018 Comparitive study on effect of oxide thickness on stress distribution of traditional and Nano structural zirconia coating systems *J. of Ceramics*

[12] Wang, Wanga Y, Zhang W Q, Suna X G, Hea J Q, Pana Z Y and Wanga C H 2011 Finite element simulation of stress distribution and development in 8YSZ and double-ceramic- layer La2Zr2O7/8YSZ thermal barrier coatings during thermal shock *J. of Applied Surface Science*

[13] Youho Lee, Jeong Ik Lee and Hee Cheon NO 2010 Mechanical analysis of surface- coated zircaloy cladding *J. of Surface & Coatings Tech.*

[14] Abbas, Guo H and Ramzan M 2018 Comparitive study on effect of oxide thickness on stress distribution of traditional and Nano structural zirconia coating systems *J. of Ceramics*

[15] Yu, Cen L and Wang Y 2017 Numerical study of residual stress and crack nucleation in thermal barrier coating system with plane model *J. of Ceramics International*

[16] Yu, Zhou H L and Wang L B 2015 Influences of interface morphology and thermally grown oxide thickness on residual stress distribution in thermal barrier coating system *J. of Ceramics Int.*

FMEA ANALYSIS ON OFFSHORE WIND TURBINES USING FUZZY MULTIMOORAMETHOD

Nitin Kotkunde [1]***, Mayank Chauhan*** [2]***, Karthick Chetti*** [3]
[1] Mechanical Engineering, BITS Pilani, Hyderabad, India, nitink@hyderabad.bits-pilani.ac.in
[2] Mechanical Engineering, BITS Pilani, Hyderabad, India, h20181060154@hyderabad.bits-pilani.ac.in
[3] Mechanical Engineering, BITS Pilani, Hyderabad, India, h20181060163@hyderabad.bits-pilani.ac.in

Abstract. Failure mode effective analysis (FMEA) is a quality instrument that is being used to recognize potential failure modes and its related consequences on different complex sub-structures in a system. The tool is effectively used for continuous improvement of efficiency, reliability, and quality. However, the traditional FMEA methods using risk priority number (RPN) values have been criticized for having many immanent limitations, affecting its effectiveness in real-world applications. Many risk priority models are emerging in the field of safety and reliability engineering. Among these models, Multi-criteria decision making (MCDM) methods are amongst the most widespread approaches employed to rank failure modes. The main aim of the present work is to implement FMEA analysis using fuzzy MULTIMOORA method. The proposed model has the competence of representing the imprecise knowledge and uncertainties of FMEA team members. FMEA analysis has been performed on offshore wind turbines consisting of four sub-assemblies for nine potential failure modes. The failure modes and its risk factors have been well-defined and appropriate linguistic variables are assigned to the failure modes according to the knowledge of team members. The responses of the team members have been aggregated and operated by the fuzzy model which prioritizes failure modes. The comparison of traditional FMEA rankings with the proposed model displays that a more precise and rational ranking can be attained by the incorporation of MULTIMOORA method using fuzzy set theory to FMEA. ***Keywords:*** *FMEA, Fuzzy, MULTIMOORA, Reliability, RPN, Safety*

1. INTRODUCTION

Safety is a major concern for program managers, design engineers, and reliability practitioners as the system designs have become more complex with increasing customer requirements. Failure modes and effects analysis (FMEA) initially came into existence in the aerospace industry during the Apollo mission where the influence of complex system failures on system performance, safety and mission success was to be evaluated during the early 1960s[1]. FMEA is usually a very attractive tool popularly used to assist the dependability and safety of products in various industries. Safety should be considered at the initial design stage to identify risks as early as possible which tends to reduce the lifecycle costs [2]. FMEA is applied at a component level and it assesses what can go wrong, and how it can affect the system. The main idea behind application of FMEA is to allow the specialists to set up the failure modes of a design, method, product or system in order to allocate the inadequate assets to the maximum risk items [3]. An effective FMEA provides potential failure preventive measures to assure high reliability, improve quality and reduction of failures [4].

IOP Conf. Series: Materials Science and Engineering 1123 (2021) 012018 doi:10.1088/1757-899X/1123/1/012018

FMEA is a methodology that uses potential failure modes which are aimed at preventing failures of a product by the evaluation of impacts, hence determining a priority of possible actions for reducing the occurrence of such failures. The application of FMEA is done at the system level, functional level or at the service level of a product [5]. The analysis of FMEA distinguishes the possible failure modes, evaluates the cause and effects of dissimilar failure modes and thereby diminishes the chance of high-risk failures. Conventional FMEA makes use of a risk priority number to calculate failure modes that's demarcated as the arithmetic product of risk elements O, S and D, where in O is the possibility of the failure, S is the significance of the failure, and D is the capacity to discover the failure before the effect is finished. For the calculation of the fee of RPN of every failure mode, each of risk elements are gauged on a mathematical scale from 1 to 10. The risk of a product or a system is directly proportional to the RPN of the failure mode. The RPNs are ranked and preventive measures and actions are engaged on the high-jeopardy failure modes [6]. Figure 1 shows the steps involved in the traditional FMEA method.

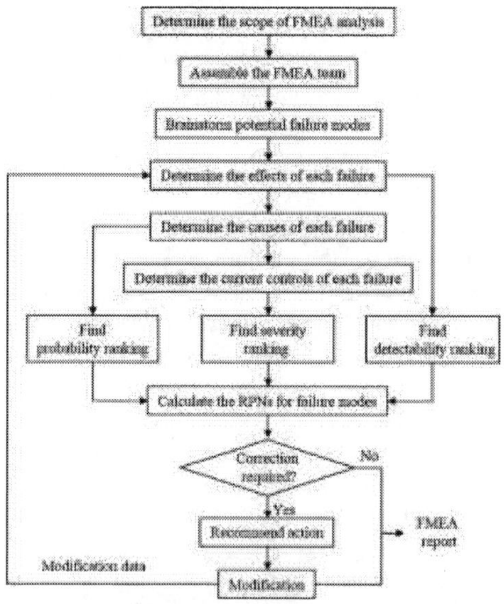

Figure 1. Traditional FMEA Methodology

The conventional FMEA has gone under a scanner for a few decades and is of many intrinsic deficiencies leading to its limit in actual applications. It has been criticized due to various reasons which include its inefficacy to govern the priorities of potential risk failures. Diverse assimilation of risk factors may yield the equal value of RPN however their concealed risk outcomes may be numerous. This might cause a misuse of sources and a few high-hazard failure modes may match unobserved. The team members may not have a right skill set and pleasant degree of knowledge regarding the risk analysis of all failure modes. This arises due to the complexity of design, system or provider which makes the effects of risk evaluation display the characteristics of imprecision, fuzziness and uncertainty. The traditional analysis assumes that the risk factors contain equal weights which might not be applicable in real-life applications. The direct and indirect correlations among failure modes are not taken into consideration. The mathematical formula for calculating RPNs lacks scientific basis and considering the complexity of the systems involved nowadays, it needs to be efficient enough to encompass all the effects of risk factors. The traditional FMEA makes use of only three risk factors and ignores other potential risk factors such as economic aspects. The algebraic formula for evaluating RPN is highly sensitive to deviation in risk factor assessments. Slight deviations in one rating may have significant changes in RPN values, which is undesirable. The quantification of three risk factors are generally difficult for FMEA members as the inputs typically have vague information especially in a linguistic way like 'important',' high',' low'. Hence rating them on a scale from 1 to 10 becomes a challenge to the team members and may cause uncertain judgments on the identified failure modes [7].

Numerous theories have been evolving to overcome vagueness and uncertainty arising in assessment of criticality. These include the fuzzy set, intuitionistic fuzzy set (IFS) and the Dempster–Shafer theory. The multi criteria choice making (MCDM) methods are emerging as the dominant procedures hired to prioritize the failure modes recognized through FMEA. These strategies enhance the effectiveness and experimental validity of consequences acquired from risk evaluation. The MCDM models below uncertain environment has a sequence of precise benefits which differentiate them from other FMEA models related to artificial intelligence. The FMEA approaches are experiencing detailed study as well as a continuous extension in the working envelope of its applications. Decision making tools such as uncertainty theories are used to model the obscurities arising from risk assessments. The MCDM practices have wide opportunities for improvement of the vitality analysis process of FMEA [8]. Figure 2 shows the numerous FMEA methods which are evolving in the field of safety and reliability engineering.

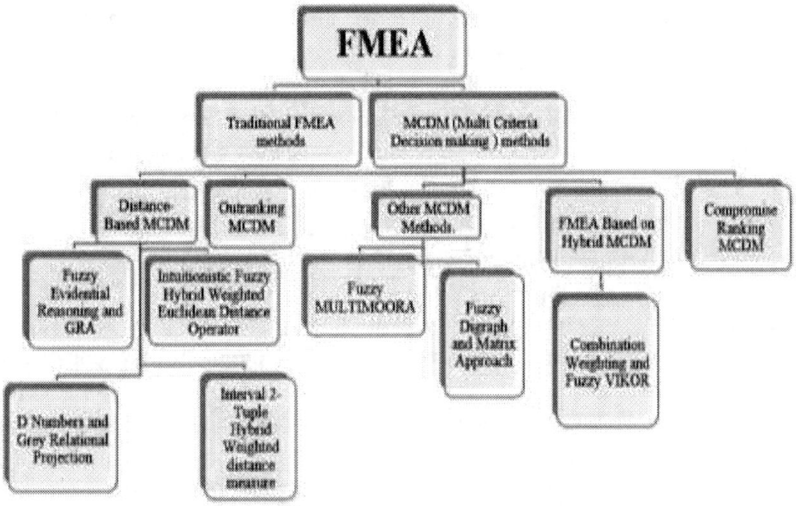

Figure 2. Different FMEA Methods

The fuzzy set theory is a fundamental concept of science. It handles imprecise information by using a certain formulation that manipulates the mathematical terms. The flexible or inaccurately defined boundaries of classes or categories can be represented and managed by the fuzzy set theory. This theory represents the classes by the use of characteristic functions which obtains its values from an ordered set of membership values. The membership function that defines the fuzzy set Z charts the elements of the universe X to the interval [0, 1], $Z:X \rightarrow [0,1]$. The membership function fZ(x) categorizes the fuzzy set Z in X and associates each point X with a real number in the unit interval. The association level of x with the set Z is represented by fZ(x). It is critical to analyze the type of fuzzy set for the criteria of its aptness for handling the succeeding optimization methods. The commonly used groups of the membership functions which are well defined in the real numbers space trapezoidal, triangular and exponential membership functions [9]. The fuzzy set takes care of the uncertainties resulting from human judgements on risk factors. The fuzzy set theory laid foundation for a risk priority model for prioritizing the failure modes and this led to a proposal of

MULTIMOORA method. The fuzzy ratings assessed the risk factors and the relative weights and the ranking of failure modes was accomplished through a comprehensive MULTIMOORA method [10]. The MCDM methods can be used amongst a number of team members or analysts who are experts in their respective areas. In the present study, floating off-shore wind turbines are considered for FMEA evaluation. The necessary data required for evaluation is obtained by the risk assessment on the off-shore wind turbines using correlation FMEA [11]. The present research mainly focuses on the systematic investigation of fuzzy MULTIMOORA method for evaluating on off shore wind turbines.

2. FLOATING OFFSHORE WIND TURBINES

Floating offshore wind turbine (FOWT) is a recent approach towards the demand for renewable sources of energy. Floating offshore wind turbines could be a game-changer as it has a lot of opportunities to offer. They can be installed in deep water sites with dynamic and harsh wind conditions. They provide a lot of environmental benefits compared to traditional wind turbine installation. The major drawbacks are the cost of installation of FOWT along with maintenance and experience of the workers. The location can be a disadvantage due to high lead repair time. An effective risk assessment would significantly reduce the corrective maintenance costs. FMEA is a widely applied methodology on the risk assessment of FOWT. The incorporation of FMEA to FOWT has drawbacks as it has a lot of complex parts and the FMEA is incapable of including these risk parameters [12]. In this study, we compare the nine potential failure modes of FOWT by using the RPN ratings of the failure modes already conducted on the four subassemblies of the offshore wind turbines [11]. The uncertainty theories can be used for risk assessment hence we use the fuzzy MULTIMOORA method for the FMEA analysis.

The FOWT is a multi-functional system. Table 1 describes the various subassemblies and their respective potential failure modes. The application of FMEA calls for the categorization of failure modes rendering to diverse system capabilities which are greater accurate for gauging the risk and reliability. The sub-assemblies are categorized according to the main functions of a driven FOWT. Table 2 relates the failure modes and corresponding reasons with end effects it could have on the system. The severity, occurrence and detection ratings for the failure modes are defined by Table 3, Table 4 and Table 5 respectively. Table 6 shows the RPN rating as the arithmetic multiplication of risk factors.

Table 1. FOWT Function Structure [11]

Subassemblies	Failure Modes
Blade system	The blades culpabilities are connected to structural failures which include fatigue of the fibrous composite materials used.
Generator components	The leading faults in electrical components are due to short-circuits or open-circuit of the winding or due to overheating.
Transmission system (Bearing, Gearbox and Coupling)	The gearbox is susceptible to shock loads triggered by the variability of the wind speeds and the corrosion.
Support system (Nacelle tower, mooring system, floating foundation)	The crucial failure reasons of the support structure are corrosion, hull collision and welding cracking.
Auxiliary system (lightning protection	The primary failure modes are hydraulic pressure error and temperature.

system, cooling system, hydraulic system)	

Table 2. End Effects of failure modes in subassemblies

Subassembly	Failure Modes	End Effects	Cause of Failures
A. Generator	1. Deformation of bearing	Equipment impairment	Over tightening or loosening of bearing shaft
	2. Overheat	FOWT system shutdown	Disproportionate system oscillation
	3. Winding failure	FOWT system shutdown	Winding corrosion
B. Electrical controls	4. Conversion failure	Disconnection from grid	Invert power input error
	5. Transform winding failure	Disconnection from grid	Iron core corrosion
	6. Output voltage error	Disconnection from grid	Friction between rotor and stator
	7. Yaw positioning inaccuracy	Inefficiency	Yaw damper inaccuracy
C. Support system	8. Fracture of mooring line	FOWT system closure	Fatigue damage
D. Auxiliary systems	9. Failure of cooling system	FOWT system closure	Pipeline leakage

Table 3. Severity Ratings

Scale	Description	Criteria
1	Minor	Electricity produced but repair necessary
2	Marginal	Decrease in the capacity to produce electricity
3	Critical	Loss of capacity to produce electricity
4	Catastrophic	Major impairment to the turbine

Table 4. Occurrence Ratings

Scale	Description	Criteria
1 to 2	Extremely Unlikely	Probability is less than 0.0001
3 to 5	Remote	Probability ranges from 0.0001 and 0.0001
6 to 8	Occasional	Probability ranges from 0.0001 and 0.01
9 to 10	Frequent	Probability is more than 0.01

Table 5. Detection Ratings

Scale	Description	Criteria
1 to 2	Practically Firm	Existing checking procedures always spot the failure
3 to 5	High	Good Probability that existing monitoring procedures will spot the failure

6 to 8	Low	Low Probability that existing monitoring procedures will spot the failure
9 to 10	Practically Impossible	No existing monitoring procedures accessible to spot the failure

Table 6. RPN Ratings

Failure Mode	S	O	D	RPN
1. Bearing Deformation	4	6	6	144
2. Overheat	3	7	6	126
3. Winding Failure	4	8	7	224
4. Conversion Failure	3	7	8	168
5.Transformation winding failure	3	8	8	192
6. Output Voltage error	3	7	8	168
7.Yaw positioning inaccuracy	3	7	6	126
8. Mooring line fracture	4	6	8	192
9. Cooling system failure	4	6	8	192

3. FUZZY MULTIMOORA METHOD

Multi-Objective optimization with the aid of ratio analysis (MOORA) is the idea which comes under MULTIMOORA method. MCDM technique makes use of a new threat precedence model where fuzzy set concept and MULTIMOORA technique are implemented for the evaluation of failure modes and rating in FMEA. The chance elements and their relative weights are pondered as fuzzy variables and calculated via the usage of fuzzy linguistic phrases and fuzzy rankings. The risk ranking of the failure modes is described using a complete MULTIMOORA method which aids in taking preventive movements for safety and reliability improvement [12]. Figure 3 indicates the whole method for imposing the MULTIMOORA model.

3.1 Methodology

The following steps are followed to prioritize failure modes using fuzzy MULTIMOORA method.

Step 1 - Objectives of risk evaluation is recognized and FMEA scope is described.
The fundamental step is outlining the objectives of risk assessment. It is significant to give condensed thought to this step as it will govern the whole process of risk assessment and ranking procedure. The scope of the FMEA analysis has been well well-defined to attain the desired rankings.

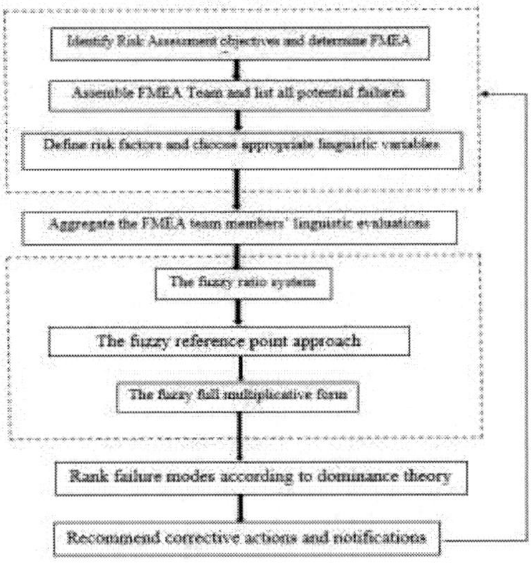

Fig. 3 Proposed methodology using fuzzy multimoora method

Step 2 - FMEA team is assembled and all potential failure modes enlisted.
A team of cross-functional specialists has to be fashioned considering the specific hassle scope and its complete size. All capability failure modes in conjunction with their reasons are enlisted the use of Brainstorming periods and cause-effect illustrations and their ability outcomes for every sub-function are analysed. To investigate the MULTIMOORA method, we have taken two team members having equal weight (0.5). We consider the same failure modes considered for the calculation of RPN according to Table 6. For our convenience, we name it FM1, … FM9.Instead of giving numbers as ratings, the two team members can rate the failure modes and risk factors in linguistic terms which are used in everyday language. The linguistic ratings given by the two team members (TM1, TM2) are represented in Table 7.

Table 7. Team Members Ratings

	TM_1			TM_2		
Failure Modes	**O**	**S**	**D**	**O**	**S**	**D**
FM 1	M-H	V-H	M-H	H	E-H	M-H
FM 2	H	M-H	M-H	M-H	H	M-H
FM 3	V-H	V-H	H	E-H	H	M-H
FM 4	H	M-H	H	M-H	M	V-H
FM 5	V-H	M-H	H	H	H	H
FM 6	H	M-H	H	M-H	H	H
FM 7	H	M-H	M-H	H	M	M-H
FM 8	M-H	V-H	H	M-H	H	H
FM 9	M-H	V-H	H	H	E-H	H

Table 8. Ratings for Weights

Weights	O	S	D
TM_1	M-H	H	E-H
TM_2	H	E-H	V-H

Step 3 - Related risk factors are defined and suitable linguistic variables are selected.

A definite finite set of risk elements and their calculation metrics are established to evaluate the hazard of failure modes. We need to choose suitable linguistic terms for the significance weights of risk factors and the scores of failure modes concerning every threat issue. The linguistic terms for risk factors and weights are transformed into fuzzy numbers with the assistance of their fuzzy equivalents shown in Table 9 and Table 10 respectively. Hence the converted fuzzy values for our ratings are shown in Table 11 and Table 12. The two team members are indicated by TM1 and TM2.

Table 9. Risk Factors Fuzzy Equivalents

Risk factors	
Linguistic ratings	**Fuzzy equivalents**
Extremely Low (E-L)	(0, 0, 0, 0)
Very Low (V-L)	(0, 0, 1, 2)
Low (L)	(1, 2, 2, 3)
Medium Low (M-L)	(2, 3, 4, 5)
Medium (M)	(4, 5, 5, 6)
Medium High (M-H)	(5, 6, 7, 8)
High (H)	(7, 8, 8, 9)
Very High (V-H)	(8, 9, 10, 10)
Extremely High (E-H)	(10, 10, 10, 10)

Table 10. Weights Fuzzy Equivalents

Weights	
Linguistic ratings	**Fuzzy equivalents**
Extremely Low (E-L)	(0, 0, 0, 0)
Very Low (V-L)	(0, 0, 1, 2)
Low (L)	(1, 2, 2, 3)
Medium Low (M-L)	(2, 3, 4, 5)
Medium (M)	(4, 5, 5, 6)
Medium High (M-H)	(5, 6, 7, 8)
High (H)	(7, 8, 8, 9)
Very High (V-H)	(8, 9, 10, 10)
Extremely High (E-H)	(10, 10, 10, 10)

Table 11. Equivalent Fuzzy Ratings for Risk Factors

	TM_1			TM_2		
Failure modes	**O**	**S**	**D**	**O**	**S**	**D**
FM 1	(5,6,7,8)	(8,9,10,10)	(5,6,7,8)	(7,8,8,9)	(10,10,10,10)	(5,6,7,8)
FM 2	(7,8,8,9)	(5,6,7,8)	(5,6,7,8)	(5,6,7,8)	(7,8,8,9)	(5,6,7,8)
FM 3	(8,9,10,10)	(8,9,10,10)	(7,8,8,9)	(10,10,10,10)	(7,8,8,9)	(5,6,7,8)
FM 4	(7,8,8,9)	(5,6,7,8)	(7,8,8,9)	(5,6,7,8)	(4,5,5,6)	(8,9,10,10)
FM 5	(8,9,10,10)	(5,6,7,8)	(7,8,8,9)	(7,8,8,9)	(7,8,8,9)	(7,8,8,9)
FM 6	(7,8,8,9)	(5,6,7,8)	(7,8,8,9)	(5,6,7,8)	(7,8,8,9)	(7,8,8,9)
FM 7	(7,8,8,9)	(5,6,7,8)	(5,6,7,8)	(7,8,8,9)	(4,5,5,6)	(5,6,7,8)
FM 8	(5,6,7,8)	(8,9,10,10)	(7,8,8,9)	(5,6,7,8)	(7,8,8,9)	(7,8,8,9)
FM 9	(5,6,7,8)	(8,9,10,10)	(7,8,8,9)	(7,8,8,9)	(10,10,10,10)	(7,8,8,9)

Table 12. Equivalent Fuzzy Ratings for Weights

Weights	O	S	D
TM_1	(5, 6, 7, 8)	(7, 8, 8, 9)	(10, 10, 10, 10)
TM_2	(7, 8, 8, 9)	(10, 10, 10, 10)	(8, 9, 10, 10)

Step 4 - Assembled the team members' linguistic assessments

The collective fuzzy rankings of failure modes regarding every risk factor can be calculated to develop the fuzzy group assessment matrix after the FMEA crew contributors give their selections on chance elements using linguistic phrases. The aggregation for risk factors given by Table 7 is calculated by equation (1) and for aggregation of weights as shown in Table 8 is calculated by equation (2).

$$r_{ij} = (r_{ij1}, r_{ij2}, r_{ij3}, r_{ij4}) = (\sum_{k=1}^{l} \lambda_k r^k{}_{ij1}, \sum_{k=1}^{l} \lambda_k r^k{}_{ij2}, \sum_{k=1}^{l} \lambda_k r^k{}_{ij3}, \sum_{k=1}^{l} \lambda_k r^k{}_{ij4}) \quad (1)$$

The aggregated fuzzy equivalent of weight of each risk factor w_j is calculated as

$$w = (w_{j1}, w_{j2}, w_{j3}, w_{j4}) = (\sum_{k=1}^{l} \lambda_k w^k{}_{j1}, \sum_{k=1}^{l} \lambda_k w^k{}_{j2}, \sum_{k=1}^{l} \lambda_k w^k{}_{j3}, \sum_{k=1}^{l} \lambda_k w^k{}_{j4}) \quad (2)$$

Table 13. Aggregated Fuzzy Ratings

	O	S	D
FM 1	(6,7,7.5,8.5)	(9,9.5,10,10)	(5,6,7,8)
FM 2	(6,7,7.5,8.5)	(6,7,7.5,8.5)	(5,6,7,8)
FM 3	(9,9.5,10,10)	(7.5,8.5,9,9.5)	(6,7,7.5,8.5)
FM 4	(6,7,7.5,8.5)	(4.5,5.5,7.5,7)	(7.5,8.5,9,9.5)
FM 5	(7.5,8.5,9,9.5)	(6,7,7.5,8.5)	(7,8,8,9)
FM 6	(6,7,7.5,8.5)	(6,7,7.5,8.5)	(7,8,8,9)
FM 7	(7,8,8,9)	(4.5,5.5,6,7)	(5,6,7,8)
FM 8	(5,6,7,8)	(7.5,8.5,9,9.5)	(7,8,8,9)
FM 9	(6,7,7.5,8.5)	(9,9.5,10,10)	(7,8,8,9)
Weights	(0.6,0.7,0.75,0.85)	(0.85,0.95,0.9,0.95)	(0.9,0.95,1,1)
$\hat{r}$	r_1 =26.4	r_2 =26.37	r_3 =24.443

Step 5 - Normalized fuzzy assessment matrix is developed using the fuzzy ratio system which describes normalization of the fuzzy numbers.
The normalization is accomplished with the aid of comparing appropriate values of fuzzy numbers.

The aggregated fuzzy values are calculated by "(3)" and "(4)" is used to calculate $\hat{r}$. The normalized fuzzy assessment matrix is shown in Table 14.

$$x_{ij} = (x_{ij1}, x_{ij2}, x_{ij3}, x_{ij4}) = (w_{j1}\frac{r_{ij1}}{\hat{r}}, w_{j2}\frac{r_{ij2}}{\hat{r}}, w_{j3}\frac{r_{ij3}}{\hat{r}}, w_{j4}\frac{r_{ij4}}{\hat{r}}) \quad (3)$$

$$r_j = \sqrt{\sum_{i=1}^{m} r^2{}_{ij4}} \quad (4)$$

Table 14. Normalized Fuzzy Assessment Matrix

Failure Modes	O	S	D
FM 1	(0.1364,0.1857,0.2131,0.2738)	(0.2901,0.3423,0.3413,0.3603)	(0.184,0.2332,0.286,0.3273)
FM 2	(0.1364,0.1857,0.2131,0.2738)	(0.1934,0.2522,0.256,0.3063)	(0.184,0.2332,0.286,0.3273)
FM 3	(0.2046,0.252,0.2842,0.3221)	(0.2418,0.3063,0.3072,0.3423)	(0.221,0.2721,0.307,0.3478)
FM 4	(0.1364,0.1857,0.2131,0.2738)	(0.1451,0.1982,0.256,0.2522)	(0.276,0.3304,0.368,0.3478)
FM 5	(0.1705,0.2255,0.2558,0.306)	(0.1934,0.2522,0.256,0.3063)	(0.258,0.3109,0.327,0.3682)
FM 6	(0.1364,0.1857,0.2131,0.2738)	(0.1934,0.2522,0.256,0.3063)	(0.258,0.3109,0.327,0.3682)
FM 7	(0.1592,0.1857,0.2274,0.2899)	(0.1451,0.1982,0.2048,0.2522)	(0.184,0.2332,0.286,0.3273)
FM 8	(0.1137,0.1592,0.1989,0.2577)	(0.2418,0.3063,0.3072,0.3423)	(0.258,0.3109,0.327,0.3682)
FM 9	(0.1364,0.1857,0.2131,0.2738)	(0.2901,0.3423,0.3413,0.3603)	(0.258,0.3109,0.327,0.3682)

Step 6 - The fuzzy reference point technique

The fuzzy maximal objective reference point (MORP) vector $r^*=(x_1^*,x_2^*,....x_n^*)$ is attained according to the matrix $X=[x_{ij}]_{mxn}$. The j_{th} coordinate of the reference point be similar to the fuzzy maximum or minimum of the j_{th} risk factor x_j^* and can be calculated by "(5)"

$$x_j^* = \begin{cases} (\max x_{ij1}, \max x_{ij2}, \max x_{ij3}, \max x_{ij4}), j \le g; \\ (\min x_{ij1}, \min x_{ij2}, \min x_{ij3}, \min x_{ij4}), j > g \end{cases} \quad (5)$$

The distance of each failure mode from the fuzzy MORP can be calculated by using "(6)". The ranking orders of all failure modes are defined according to the eccentricity from the reference point.

$$d_i = \max d(x_j^*, x_{ij}) \quad (6)$$

Step 7- The final rankings are obtained from the fuzzy full multiplicative form as indicated in Table 15.

Table 15. Fuzzy Multiplicative Form

Failure Modes	Yi	Y_i	Ranking
FM 1	(0.6106,0.7611,0.8409,.9614)	2.376	5
FM 2	(0.5139,0.6711,0.7555,0.9073)	2.135	8
FM 3	(0.6673,0.8303,0.8982,1.0121)	2.548	2
FM 4	(0.5576,0.7142,0.8373,0.9146	2.258	7
FM 5	(0.6217,0.7886,0.8391,0.9804)	2.418	3
FM 6	(0.5876,0.7488,0.7964,0.9482)	2.309	6
FM 7	(0.4883,0.617,0.7185,0.8694)	2.023	9
FM 8	(0.6132,0.7763,0.8334,0.9682)	2.388	4
FM 9	(0.6843,0.8389,0.8818,1.0023)	2.549	1

Step 8 - Examine the outcomes and develop recommendations to reinforce the system performance.

The contrast of the fuzzy MULTIMOORA with the RPN method is shown in Table 16.

Table 16. Comparison of Rankings

Failure Modes	RPN	MULTIMOORA
FM1	7	5
FM2	9	8
FM3	1	2
FM4	6	7
FM5	4	3
FM6	5	6
FM7	8	9
FM8	2	4
FM9	3	1

4. CONCLUSIONS

-The rankings obtained from the fuzzy MULTIMOORA method is compared from the rankings obtained from the traditional RPN method.
-The fuzzy MULTIMOORA methods encompasses all the uncertainties and vagueness of the data and is expected to provide effective FMEA rankings.
- We have successfully implemented the MULTIMOORA model for the FMEA analysis and obtained comparable priorities of failure modes.
-This model can also be extended to additional team members, failure modes and risk factors depending upon the requirement and complexity of the system to be analyzed.

REFERENCES

[1] Khairul Annuar Abdullah and Mohd Salihin Ngadiman 2005Development of FMEA Information System for Manufacturing Industry, *International Conference on Modeling and Analysis of Semiconductor Manufacturing* Malaysia

[2] Tommaso Sgobba, Firooz Allahdadi A, Isabelle Rongier and Paul Wilde D 2013 *Safety Design for Space Operations* (Third Edition) USA Butterworth-Heinemann publications, 2013 pp 337-469

[3] Hu-Chen Liu 2016 *FMEA Using Uncertainty Theories and MCDM Methods*, Springer Science, Singapore pp 2-9

[4] Liu HC and You J X 2016 Failure mode and effect analysis under uncertainty: an integrated multiple criteria decision-making approach, *IEEE Transactions Reliability* pp 1027-1042

[5] Duane Kritzinger 2017 Failure Modes and Effects Analysis Author links open overlay panel , *Aircraft System Safety Assessments for Initial Airworthiness* Vol 5 pp 101-132

[6] Stamatis DH 2003 *Failure mode and effect analysis: FMEA from theory to execution* New York 2003

[7] Haibin Liu, Xinyang Deng and Wen Jiang 2017 *Risk Evaluation in Failure Mode and Effects Analysis Using Fuzzy Measure and Fuzzy Integral* School of Electronics and Information

[8] Carlson C 2012 *Effective FMEAs: achieving safe, reliable, and economical products and processes using failure mode and effects analysis* New Jersey Wiley

[9] Abdelgawad M and Fayek A R 2010 Risk management in the construction industry using combined fuzzy FMEA and fuzzy AHP *J. of construction Engg and Management* Vol 136 Issue 9 pp 1028–1036

[10] Liu HC and Fan XJ Evaluating the risk of failure modes with extended MULTIMOORA method under fuzzy environment, *Engg. Applications of Artificial Intelligence* Vol 34 pp 168–177

[11]Jichuan Kangb and Liping Sunb 2017 Risk assessment of floating offshore wind turbine based on correlation-FMEA, *Ocean Engineering* Vol 129 pp 382–388

[12] Brauers WKM and Zavadskas EK, 2010 Project management by MULTIMOORA as an instrument for transition economies, *Technological and Economic Development of Economy* Vol 16 Issue 1 pp 5–24

[13] Brauers WKM and 2012 Zavadskas EK, Robustness of MULTIMOORA: a method for multi-objective optimization *Informatica* Vol 23 Issue 1 pp 1–25

[14] Brauers WKM and Zavadskas EK 2011 MULTIMOORA optimization used to decide on a bank loan to buy property, *Technological and Economic Development of Economy* Vol 17 Issue 1 pp 174

FUZZY LOGIC BASED DIGITAL HYDRAULICS CONTROL OF BLADE PITCH ANGLE IN WIND TURBINES

V. Lakshmi Narayanan[1], R. Ramakrishnan[2]
[1, 2] Department of Design and Automation, Vellore Institute of Technology, Vellore, India,
[1] lakshmi.narayan2017@vitstudent.ac.in, [2] ramakrishnan.rc@vit.ac.in

Abstract. Pitch control system is generally employed in wind turbine to mitigate load and maintain uniform power generation at above-rated wind speed regions. Hydraulic system has more power to weight ratio and so it is incorporated in the pitch system of large scale wind turbines. Some of the issues related to the usage of conventional valves in Hydraulic Pitch System (HPS) are: internal leakage, throttling losses, high power loss, less fault-tolerant, requires high manufacturing tolerance and more sensitive to contamination. In order to overcome these issues, digital hydraulics technology should be introduced as Digital Hydraulics Pitch System (DHPS). Commercially, Proportional Integral Derivative (PID) controller is used as pitch controller but these controller performance doesn't hold good when excessive disturbance or change in operating point occurs in the system. So, heuristic-based Fuzzy Logic Controller (FLC) is preferred which has the ability to surpass the PID problems. In this paper, heuristic FLC based control strategy is proposed for a novel DHPS configuration to control the pitching action. Simulation model of DHPS is developed and system simulations are conducted. The comparative study on the effectiveness of FLC for DHPS and conventional valve controlled HPS is conducted. ***Keywords**: Wind turbine, Pitch control, Fuzzy logic controller, Digital hydraulics*

1. INTRODUCTION

The need for renewable energy has started to exponentially increase due to the harmful effects created by the non-renewable energy sources (fossil fuel and coal) over the environment. Among the renewable energy sources, wind energy is the most promising source of electric energy [1, 2, 3]. Since, Wind Turbine (WT) is given more priority than conventional power plant by many countries, the technology related to WT has started to increase. In order to generate optimal power and also to safe guard the WT during high wind speeds, Pitch Control System (PCS) and yaw control system are mostly preferred [4, 5, 6]. Generally, PCS is employed at above rated wind speed condition to maintain uniform power generation and to mitigate erratic blade loads [7,8]. Basically the two types of PCS used in WT are: electro-mechanical and Hydraulic Pitch System (HPS). Electric motors are used as pitch actuator in case of electro-mechanical pitch system where as hydraulic cylinder/motor are used in case of HPS. Since HPS are more robust to disturbance and has more power to weight ratio than its counterpart, HPS is mostly preferred in large scale WT [9, 10]. The advantage of using hydraulic motor as the end actuator in HPS over hydraulic cylinder is that the final pitch angle is directly proportional to displacement of the hydraulic motor [11, 12]. Conventionally, pump control and valve control technique are preferred in HPS [13]. The reason for considering valve controlled HPS to be better because, it has higher bandwidth than the pump control HPS [14]. Conventional valve controlled HPS uses proportional or servo valves to control flow rate and also uses directional valve to control the direction of the fluid as shown in (1).

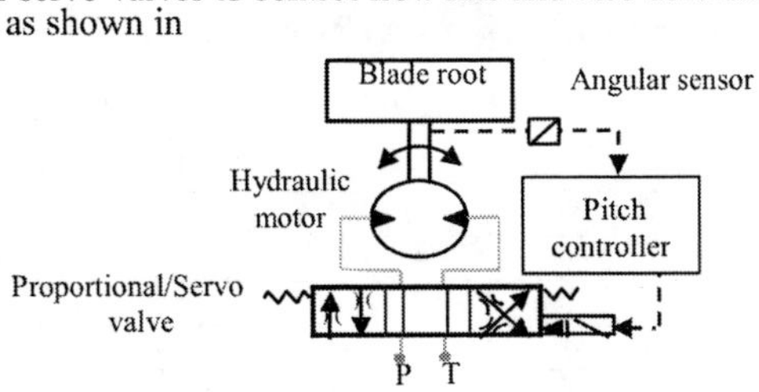

Figure 1. Conventional valve controlled hydraulic pitch system [15]

These valves has few drawbacks like internal leakage, throttling losses, high power loss, less fault tolerant, requires high manufacturing tolerance, high cost and more sensitive to contaminations [16,17,18]. The drawback of the conventional valve controlled HPS can be overcome by implementing Digital Hydraulics (DH) technology in HPS. DH uses 2/2 hydraulic valve arranged in parallel combination along with various sizes of orifice known as Digital Flow Control Unit (DFCU) as shown in. (2). Based on positioning the DFCU in the hydraulic system, DH can be categorised into few types such as: digital displacement motor and pump, switching control and parallel valve technology [19, 20]. Here digital motor concept is implemented. While designing the DFCU, sizing of orifice play a vital role. The different methods utilized for sizing the orifice are to attain the varying flow rates are discussed in [19, 21].

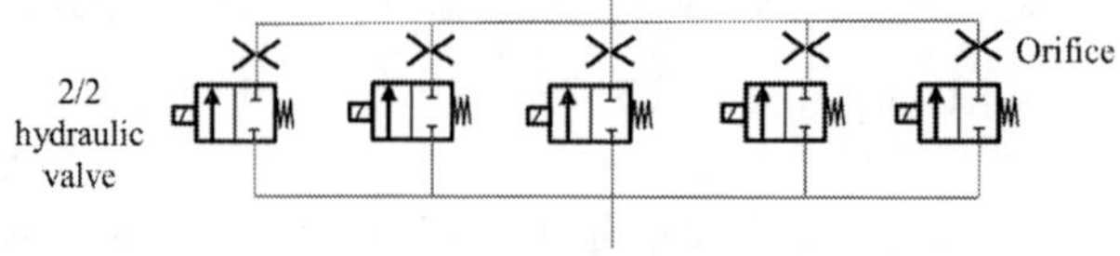

Figure 2 Digital Flow Control Unit

Though there are many advantages in DH, the controllability of DH is still a tedious task [16, 23]. The different valve actuation methods that are implemented in DH system are: Pulse Width Modulation (PWM), Pulse Number Modulation (PNM), Pulse Frequency Control (PFC), Modified Pulse Width Modulation (MPWM) and Pulse Code Modulation (PCM) [19, 21, 24]. P + PID controllers were incorporated in [24] to analyse the robust stability of DH under unknown load mass and the results indicated that the controller has better tracking performance. In [25] a novel fine positioning method was developed where four DFCU were used which resulted in the accurate positioning of the hydraulic system. Zero-Flowrate-Switching (ZFS) control method makes the valve to turn off when the flowrate through the valve is zero [26]. The output of ZFS were found to be effective. Though different control techniques were opted in DH, still DH control has a large scope to improve the controllability due to extreme non-linearity of 2/2 valves. In order to improve the controllability by overcoming the nonlinearities, heuristic based Fuzzy Logic Controller (FLC) is implemented in this paper.

The contribution of this paper involves the development of a novel DHPS for wind turbine and also FLC. The output of FLC is DFCU pair valve state selection which results in varying the flow rate and also controlling the direction of the hydraulic fluid by using DFCU so that the required pitching action takes place at the pitch actuator to achieve the appropriate pitch angle.

This paper is structures as follows: section 2 presents the description of Digital Hydraulic Pitch Control System (DHPCS). Section 4 describes the modelling of DHPCS. Section 5 gives the simulation results and discussion. Finally, conclusion are discussed in section 6.

2. System Description

The Figure 2 illustrate the proposed novel DHPCS which consist of FLC and DHPS. The DHPS consist of fixed displacement pump driven by an electric motor. As discussed earlier the DFCU consist of 2/2 valves and orifice which is attached at the end of the each valves. The sizing of orifice were carried out using the binary series method [2^0, 2^1, 2^2, 2^3....2^n]. Since five valves are used in each DFCU, the DFCU is called 5 bit DFCU and it is possible to achieve 2^n-1 states. In this proposed configuration 2^5-1=31 states with different flowrates for each states can be achieved as shown in table 1. DFCU AT and PA are connected at A port of the hydraulic motor and the DFCU PB and BT are connected at the B port of the hydraulic motor. Based on the required direction of rotation of hydraulic motor either DFCU PA and BT pair or DFCU PB and AT pair is triggered so that clockwise/anticlockwise rotation are achieved. The pitch angle generated at the hydraulic motor sensed by an angular sensor and the data's are fed to the FLC.

Table 1. DFCU pair valve states

V 1	V 2	V 3	V 4	V 5	States	Flow
0	0	0	0	1	1	Q_1
0	0	0	1	0	2	Q_2
0	0	0	1	1	3	Q_3
0	0	1	0	0	4	Q_4
.	.	.	.	.	.	.
.	.	.	.	.	.	.
.	.	.	.	.	.	.
1	1	1	1	1	31	Q_{31}

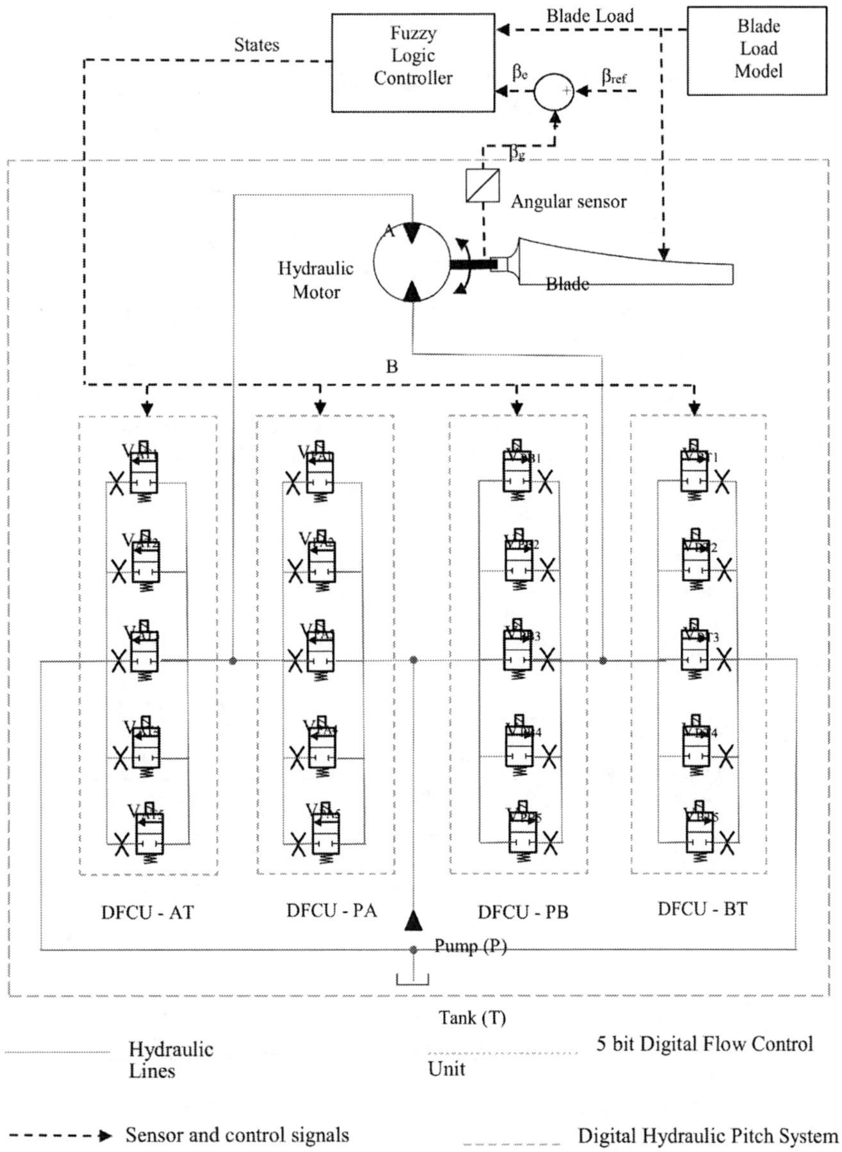

Figure 3. Proposed digital hydraulic pitch control system

3. SYSTEM MODELLING

In this section, the blade load model and FLC are modeled for the proposed DHPCS. The closed loop control strategy of DHPCS is shown in (4). FLC was chosen for the proposed configuration due to its advantage over the Proportional Integral (PI) and Proportional Integral Derivative (PID) [27]. The inputs to the FLC are blade load P*l* and pitch angle error βe where βe is obtained from (1)

$$\beta_e = \beta_{ref} - \beta_g \quad (1)$$

Here βref is pitch angle reference and βg is pitch angle generated. The output of the FLC is states

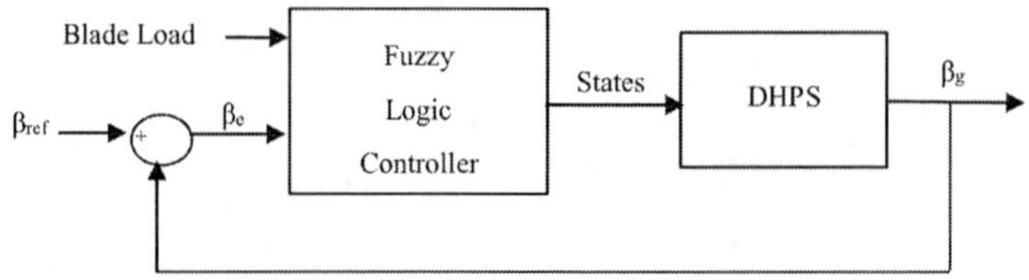

Figure 4. Block diagram of proposed digital hydraulic pitch control system

3.1 Blade load model

The load that is developed over the blades of the wind turbine due to varying wind speed play an important role in the pitch system since the DHPS should overcome the blade load P*l* and has to achieve the required pitch angle. The P*l* is arrived from [28] as shown in (2).

$$P\!l \; 390 \; 170.9 \; 92.96 \; 0.39^{2} \; 0.24 \; 3.73^{2} \qquad (2)$$

The different blade loads for varying wind speed and pitch angle are shown in the table 2 which is obtained by substituting the wind speed and pitch angle values in (2). Blade load model is developed by using (2) in Matlab/SIMULINK and the developed model will generate the loads which the DHPS should overcome.

Table 2. Blade load under different pitch angle and wind speed [28]

Pitch Angle / **Wind speed (m/s)**	**0°**	**5°**	**10°**
11	1428 Nm	867	141
12	1600 Nm	1020 Nm	280
13	1765 Nm	1182 Nm	445
14	1929 Nm	1335 Nm	574
15	2114 Nm	1500 Nm	730
16	2266 Nm	1672 Nm	900
17	2420 Nm	1824 Nm	1035 Nm
18	2554 Nm	2000 Nm	1206 Nm
19	2707 Nm	2138 Nm	1355 Nm
20	2830 Nm	2307 Nm	1547 Nm

3.2 Fuzzy Logic Controller

FLC is based on the rules which are helpful when the dynamics of the system and also the complete non linearities of the system are not known. Similar to human beings making decision, FLC applies reasoning and so the controller rules possess expert knowledge of the system. The main advantage of FLC is that mathematical description of the system which is to be controlled is not required. Fuzzy Logic Toolbox™ which is available in Matlab/Simulink. Generally there are three stages in the FLC and they are fuzzification, fuzzy rules and defuzzification. In fuzzification process the inputs are converted into fuzzy sets using linguistic terms and membership functions. Both the inputs and output uses Triangular Membership Functions (TMF) which is shown in (5). TMF are highly sensitive when variables arrives at zero value [29]. Fuzzy rules are assigned as shown in the table 2, where if and then statements are used to coin the rules like �� �1=�1 ��nd �2=�2 ����N �=�. The last process in the FLC is defuzzification where the fuzzy sets are converted into precise action with real values

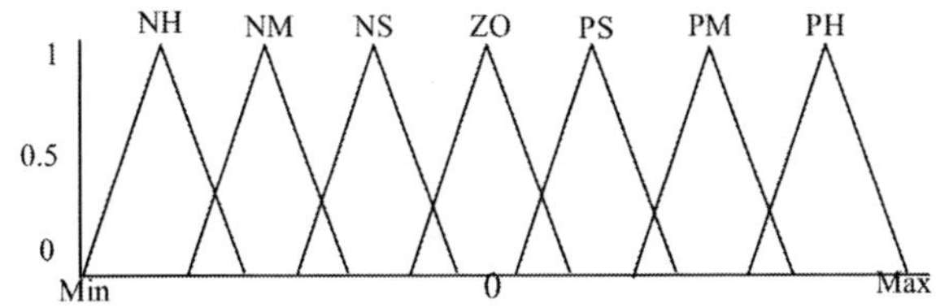

Figure 5 Triangular membership function for inputs and output

Table 3. Fuzzy rules

Pitch angle error / Blade Load	NH	NM	NS	ZO	PS	PM	PH
NH	NH	NH	NM	ZO	PM	PH	PH
NM	NH	NM	NM	ZO	PM	PM	PH
NS	NM	NM	NS	ZO	PS	PM	PM
ZO	NM	NS	NS	ZO	PS	PS	PM
PS	NM	NM	NS	ZO	PS	PM	PM
PM	NH	NM	NM	ZO	PM	PM	PH
PH	NH	NH	NM	ZO	PM	PH	PH

4. SIMULATION RESULTS AND DISCUSSION

The performance of the FLC is tested by implementing it in DHPS. Simulations were conducted from 3 sec to 10 sec (since it took 0-3 sec for the hydraulic system to initiate). Reference pitch angle data's of a 2MW wind turbine was obtained from [30]. A random wind profile (6a) was given as an input to the reference pitch angle model to generate the required βref which the DHPS should track.

Figure 6b shows the comparison of βg to βref. The βg follows the same trend as βref. At the same time, there is a lag in terms of magnitude between βref and βg is due to the nonlinearities exist in the DHPS. The magnitude of error is shown in (6d) and the maximum error observed was 1.009° which is indicated in (6d). The minimum error indicates that the FLC has a better tracking ability which was developed for DHPS. Since the wind speed is drastically changing, FLC plays a vital role in tracking the pitch angle at all conditions. Future work involves in reducing the lag so that the tracking performance can be improved and also to reduce the hydraulic system initiation lag.

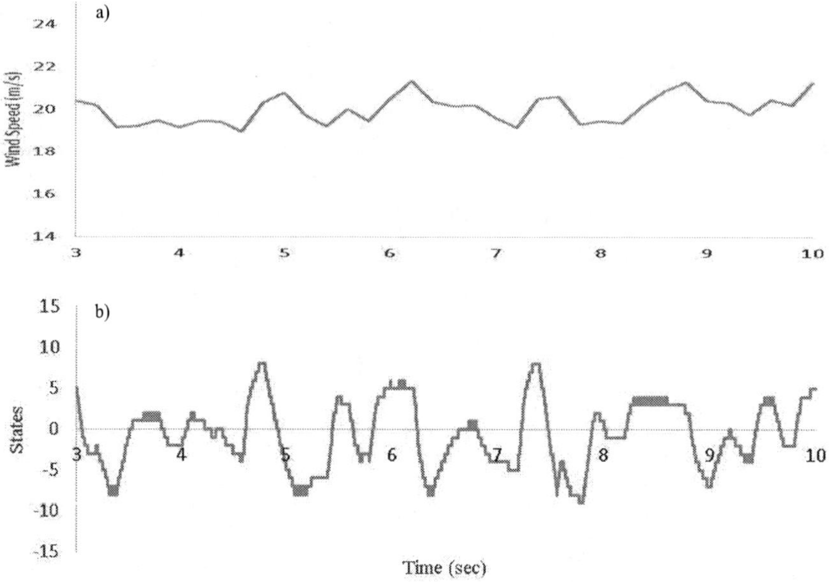
a)
Wind Speed (m/s)
24
22
20
18
16
14
3
4
5
6
7
8
9
10
b)
States
15
10
5
0
-5
-10
-15
Time (sec)

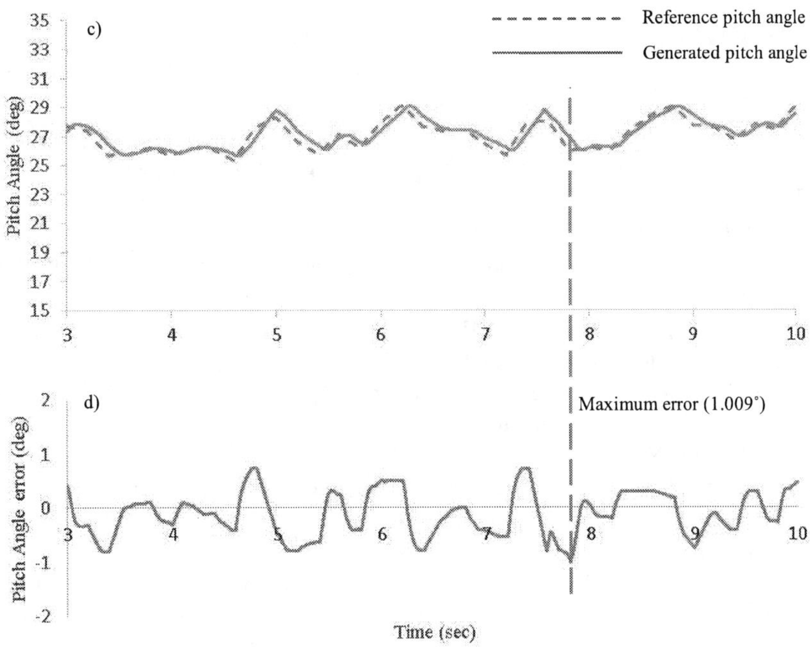

5. CONCLUSION

Hydraulic pitch system delivers high power to weight ratio to mitigate load and maintain uniform power generation. In this paper, Digital hydraulics technology is implemented in hydraulic pitch system to achieve the same performance of proportional or servo valve controlled system. The output of the fuzzy logic controller selects the DFCU pair and states to achieve the required pitch angle. The results shows that the fuzzy logic controller has better tracking ability and the maximum error percentage that was observed to be 3.89%. The results were tested for various wind pattern and the output was found to be effective. So by implementing proposed DHPCS, cost effective, highly robust, fault tolerant and high response pitch system can be established in wind turbines.

REFERENCES

[1] Connor B, Iyer S N, Leithead W E and Grimble M J 1992 Control of a horizontal axis wind turbine using H infinity control., *In Proc. of the First IEEE Conf. on Control Applications* pp 117-122.

[2] Sholapurkar R B and Mahajan Y S 2015 Review of wind energy development and policy in India. *Energy Technology & Policy* Vol *2* Issue 1 pp 122-132

[3] Petersen E L 2017 In search of the wind energy potential. *J. of Renewable and Sustainable Energy* Vol 9 Issue 5 pp 052301

[4] Yuan Y and Tang J 2017 On advanced control methods toward power capture and load mitigation in wind turbines, *Engineering* Vol 3(4) pp 494-503.

[5] Tong W 2010 *Wind power generation and wind turbine design* WIT press.

[6] Njiri, J G and Soeffker D 2016 State-of-the-art in wind turbine control: Trends and challenges, *Renewable and Sustainable Energy Reviews* Vol 60, pp 377-393.

[7] Fernando D B, Hernán D B and Ricardo J M 2006 Wind turbine control systems: principles, modelling and gain scheduling design

[8] Menezes E J N, Araújo A M and da Silva N S B 2018 A review on wind turbine control and its associated methods, *J. of cleaner production* 174 pp 945-953.

[9] Tong, W, *Fundamentals of wind energy,* In Wind power generation and wind turbine design, WIT Press pp 25

[10] Yin X X, Lin Y G, Li W, Gu Y J, Lei P F and Liu H W 2015 Adaptive back-stepping pitch angle control for wind turbine based on a new electro-hydraulic pitch system, *Int. J. of Control* 88(11) pp 2316-2326.

[11] Yin X X, Lin Y G, Li W, Gu Y J, Wang X J and Lei P F 2015 Design, modeling and implementation of a novel pitch angle control system for wind turbine, *Renewable Energy* Vol 81 pp 599-608

[12] Yin X X, Lin Y G, Li W, Liu H W and Gu Y J 2015 Adaptive sliding mode back-stepping pitch angle control of a variable-displacement pump controlled pitch system for wind turbines. *ISA transactions* Vol 58 pp 629-634

[13] Lei X, Ni M, Li D and Ma Y 2011 Study on simulation of digital pump-control cylinder position control system, *Procedia Engineering* Vol 16 pp 729-736.

[14] Turner C W Babbitt G R, Balton C S, Raimao M A and Giordano D D 2004 Design and control of a two-stage electro-hydraulic valve actuation system, *SAE transactions* pp 833-846.

[15] Yin X X, Lin Y G, Li W and Gu Y J 2014 Integrated pitch control for wind turbine based on a novel pitch control system *J. of Renewable and Sustainable Energy* Vol 6 Issue 4 pp 043106

[16] Laamanen M S A and Vilenius M 2003 Is it time for digital hydraulics, *In eighth Scandinavian Int. conf. on fluid power*

[17] Scheidl R, Linjama M and Schmidt S 2012 Is the future of fluid power digital?, *Proc. of the Institution of Mechanical Engineers, Part I: J. of Systems and Control Engineering*, 226(6), pp 721-723

[18] Laamanen A, Siivonen L, Linjama, M. and Vilenius M 2004 Digital flow control unit-an alternative for a proportional valve? *In Bath Workshop on Power Transmission and Motion Control, Professional Engineering Publishing* pp 297

[19] .Linjama M Digital fluid power State of the art 2011 *In 12th Scandinavian International Conf. on Fluid Power, Tampere, Finland* pp. 18-20.

[20] Merrill K, Holland M, Batdorff M and Lumkes Jr 2010 Comparative study of digital hydraulics and digital electronics *Int. J. of fluid power* Vol 11 Issue 3 pp 45-51

[21] Kalaiarassan G and Krishnamurthy K 2019 Digital hydraulic single-link trajectory tracking control through flow-based control *Measurement and Control*

[22] Paloniitty M and Linjama M High-linear digital hydraulic valve control by an equal coded valve system and novel switching schemes. *Proc. of the Institution of Mechanical Engineers, Part I: J. of Systems and Control Engineering* Vol 232(3) pp 258-269.

[23] Pedersen N H, Johansen P and Andersen T O 2018 Challenges with respect to control of digital displacement hydraulic units

[24] Linjama M, Huova M and Huhtala K 2016 Model-based force and position tracking control of an asymmetric cylinder with a digital hydraulic valve *Int. J. of Fluid Power* 17(3) pp 163-172

[25] Huova M, Linjama M, Siivonen L, Deubel T, Försterling H, and Stamm E 2018 Novel Fine Positioning Method for Hydraulic Drives Utilizing On/Off-Valves. *In BATH/ASME Symp. on Fluid Power and Motion Control* pp 46

[26] Peng S 2017 An Zero-Flowrate-Switching (ZFS) Control Method Applied in a Digital Hydraulic System *In Proc. of 15th Scandinavian Int. Conf. on Fluid Power* pp. 172- 177

[27] Eker I and Torun Y 2006 Fuzzy logic control to be conventional method *Energy conversion and management* Vol 47(4) pp 377-394.

[28] Wang C S and Chiang M H, 2016 A novel pitch control system of a large wind turbine using two-degree-of- freedom motion control with feedback linearization control *Energies* Vol 9(10) pp 791

[29] Eltamaly A M and Farh H M 2013 Maximum power extraction from wind energy system based on fuzzy logic control *Electric Power Systems Research* 2013 Vol 97 pp 144-150.

[30] Zhou F and Liu J 2018 Pitch controller design of wind turbine based on nonlinear PI/PD control. *Shock and Vibration*

IMPROVING THE EFFICIENCY OF STATIONARY CONCRETE PUMPS USING WHITWORTH QUICK RETURN MECHANISM

Suraj Shyam[1], Shivam Kaul[2],Johnson Jesudoss[3],Narendiranath Babu.T[4]

[1]School of Mechanical Engineering, Vellore Institute of Technology, Vellore, India, surajshyam@outlook.com

[2]School of Mechanical Engineering, Vellore Institute of Technology, Vellore, India, shivamkaul98@gmail.com

[3]Quality Management, SCHWING Stetter India Pvt Ltd., Chennai, India, johnson.a@schwingstetterindia.com

[4]School of Mechanical Engineering, Vellore Institute of Technology, Vellore, India, narendiranathbabu.t@vit.ac.in

Abstract. The objective of this paper is to study the working of concrete pumps and to establish an alternative mechanism that would enhance their efficiency. A concrete pump is a machine used for transferring semi-solid, ready-mix concrete by pumping. As observed from the analysis of their functioning, it was evident that the transition of energy from mechanical to hydraulic, and back to mechanical during the process was accompanied with significant energy loss, which comprehensively reduced the efficiency of the concrete pump. A great proportion of this energy was lost through heat dissipation. Reducing the number of energy conversions posed to be a promising concept, for which the usage of hydraulic system in the current setup was to be withdrawn. The Whitworth Quick Return mechanism sufficed as a suitable alternative for the required purpose. It was observed from the mathematical analysis of the new mechanism that the theoretical flow rate increased from $45m^3/h$ to $73m^{3/}h$, indicating a higher volume of concrete being pumped at the same input energy supply.

Keywords: *Concrete Pump, Efficiency, Energy Conversion, Hydraulic System, Whitworth Quick Return Mechanism.*

1. INTRODUCTION

A concrete pump is a machine used for transferring semi-solid ready-mix concrete by pumping. There are two types of concrete pumps. The first type of concrete pump is attached to a truck or longer units are on semi-trailers. It is known as a boom concrete pump because it uses a remote-controlled articulating robotic arm (called a boom) to place concrete accurately. Boom pumps are used on most of the larger construction projects as they are capable of pumping at very high volumes and because of the labor saving nature of the placing boom.

The second main type of concrete pump is either mounted on a truck or placed on a trailer, and it is commonly referred to as a line pump or trailer-mounted concrete pump. This pump requires steel or flexible concrete placing hoses to be manually attached to the outlet of the machine. Those hoses are linked together and lead to wherever the concrete needs to be placed. Line pumps normally pump concrete at lower volumes than boom pumps and are used for smaller volume concrete placing applications such as swimming pools, sidewalks, and single family home concrete slabs and most ground slabs. The pump works by one piston drawing concrete into a cylinder from a hopper while the other one simultaneously pushes its concrete out into the discharge pipes. A valve which determines which cylinder is open to the concrete hopper and which one is open to the discharge pipes switches over each time the pistons reach the end point and the process continues with the first cylinder now discharging and the second drawing fresh concrete from the hopper. To be able to perform this, the concrete pumps utilize high amounts of energy which they obtain through simple energy conversions. A diesel engine provides mechanical energy to run the hydraulic unit, which in turn regulates the fluid flow in the cylinders by supplying adequate liquid pressure on the pistons. The pistons then move synchronously and push out the concrete at the necessary rate. This conversion of energy from mechanical (engine) to hydraulic (hydraulic unit) and again to mechanical (pistons) is accompanied by many losses. The losses can be in the form of heat, wearing, and bends which lead to pressure drop. It may thus be inferred that the concrete pumps which make use of the above mechanism do not perform at high efficiencies.

At a time where industrialization and construction is at peak, a lot of investment in monetary terms and in energy is made on concrete pumps. This paper addresses the problem seen in the conventional mechanism and suggests an alternative mechanism to counter the energy losses. A method of approach was to reduce the number of energy conversions that took place. In the process, the Whitworth Quick Return mechanism posed as a suitable opportunity to replace the previously used mechanism. Following this, the hydraulic unit was dropped from the system and replaced by the Whitworth mechanism, providing the means to run the concrete pumps solely on mechanical energy.

2. MODEL INVESTIGATION

2.1 Model Type and Specifications

For the purpose of this paper, the model analyzed was **SP1200D** and all calculations, inferences and results were obtained in regards to the model's specifications. SP1200D is a stationary concrete pump which is manufactured at SCHWING Stetter GmbH. The organization deals with a large variety of concrete pumps, both stationary and boom type. Some examples of their other models include SP 1420, SP 8800, SP 4800 and SP 4507. These models differ on the basis of the volume of concrete they can function with. Due to its simplicity, SP 1200 D was chosen as the specimen for this paper. With their due permission, data pertaining to the specified concrete pump was obtained and investigated. The specifications of the model are mentioned in Table 1:

Table 1. Specifications of SP 1200 D

Properti	**Uni**	**SP 1200**
Versio	-	Diese
Motor/Engine rating	kW	49
Nominal Speed	RP	210

Pumping cylinder Diameter X	mm	180 X
Differential cylinder drive	-	RS
Max. stroke count per	-	24
Max. theoretical output	m^3/h	45
Max. concrete pressure	bar	70

In order to proceed with the Whitworth Quick Return mechanism, it is necessary to determine the efficiency of the concrete pumps that follow the current method and then compare it with the results obtained with the alternative mechanism.

Figure 1. SP 1200 D Stationary Concrete Pump Outlay

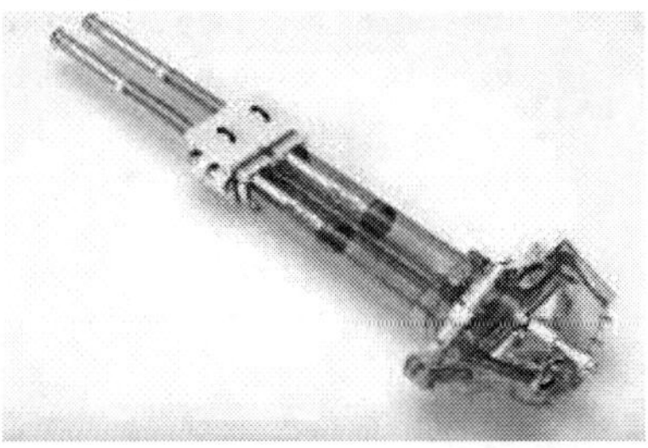

Figure 2. Pumping Mechanism

2.2 Calculation of Efficiency

All the technical data relevant to the calculation of efficiency is listed below (as observed from Table 1):

Engine/Motor Capacity, Pin = 49 kW

Bore Diameter x Stroke Length = 180 mm (D) x 1200 mm (S)

Max. Strokes per minute, N = 24 Claimed

Max. Theoretical Concrete Output = 45 m3 /h

Actual Output Rate of Concrete = Qb

Volumetric Efficiency in Cylinder, ηv = 60%

$$\text{Fluid Power, } FP = Q_b \times P_{b(c)} \tag{1}$$

$$Q_b = \left(\frac{\pi}{4}\right) \times D^2 \times S \times \left(\frac{N}{60}\right) \times \eta_v$$

$$Q_b = (0.785398) \times 0.18^2 \times 1.2 \times \left(\frac{12}{60}\right) \times 0.6$$

$$Q_b = 0.003664\ m^3/s = 0.003664 \times 3600\ m^3/h$$

$$\boxed{Q_b = 13.191673\ m^3/h}$$

$$P_{b(c)} = 7{,}000\ kPa$$

$$FP = 0.003664 \times 7000 = 25.650476\ kW$$

$$\text{Efficiency, } \varepsilon = \left(\frac{\text{Output Power}}{\text{Input Power}}\right) \times 100 = \left(\frac{25.650476}{49}\right) \times 100 \tag{2}$$

$$\boxed{\varepsilon = 52.347909\%}$$

3. PROPOSAL OF ALTERNATIVE PUMPING MECHANISM

3.1 Ideology

The hydraulic system is responsible for the losses of energy in the concrete pumping mechanism as it is apparent from the calculations made in the previous section. In order to improve the efficiency with which concrete is pumped, an idea of approach to the problem is to remove the hydraulic system completely and replace it with a suitable alternative that would produce the same output with a greater efficiency. The reduction in the number of energy conversions occurring during the process would imply a lesser wastage in energy. Therefore, it is necessary that only one energy type is used throughout the entire process of pumping. Since mechanical energy obtained from the diesel engine is the initiator of the entire process, abiding to it throughout seemed ideal. Hence, an alternative mechanism is sought after that would be driven purely by mechanical energy and simultaneously sucks in and pushes out concrete with better efficiency. Not only should this mechanism produce better output, but is should also consume lesser energy for its functioning. This way, the efficiency of the pump would increase significantly. The Whitworth Quick Return Mechanism posed as a substitute for the current mechanism.

3.2 Whitworth Quick Return Mechanism and its Working Principle

Whitworth quick return mechanism is an apparatus that converts circular motion (rotating motion following a circular path) into reciprocating motion (repetitive back-and-forth linear

IOP Conf. Series: Materials Science and Engineering 1123 (2021) 012020 doi:10.1088/1757-899X/1123/1/012020

motion) in presses and shaping machines, which are utilized to shape stocks of metal into flat surfaces. Unlike the crank and slider, the forward reciprocating motion is slower rate than the return stroke. At the bottom of the drive arm, the peg only has to move through a few degrees to sweep the arm from left to right, but it takes the remainder of the revolution to bring the arm back. This is why it is called quick return mechanism. Figure 3 shows a simple Quick Return Mechanism's link arrangement.

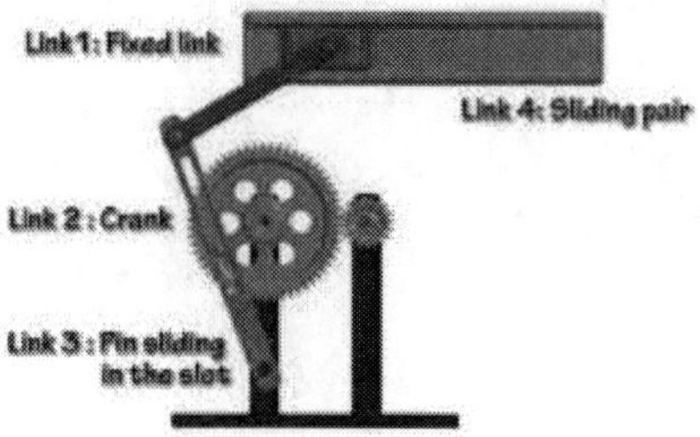

Figure 3. Whitworth Quick Return Mechanism

In the conventional mechanism of pumping, a rocker valve was made use of at the output of the two cylinders. This valve would timely change its position from one cylinder to another. While it closes the opening of one cylinder, the other cylinder pushes out the concrete. Meanwhile, the piston inside the cylinder whose opening was closed sucks in concrete from the hopper with the aid of vacuum pressure. To make the output flow of concrete regulated, the two cylinders as well as its rocker valve were retained from the previous mechanism. However, that would mean that both the cylinders couldn't have the same time periods for the strokes, which would imply that the actuators wouldn't move synchronously. To solve this synchronization problem of the actuators, the requirement of two rams of opposite speeds i.e. one with a quicker returning motion and another with a quicker forward motion was needed.

3.3 Basic Working Principle of Concrete Pump

The objective remains to pump concrete through the cylinders. Two cylinders are being made use of in such a manner that while one pushes the concrete through the outlet, the other sucks concrete through the hopper. The mechanism, however, is initiated without the usage of hydraulics. The power of the engine is directed to only rotate the bull gear at desired RPMs to achieve the exact motion sequence.

The cylinder motions are defined as follows: Forward: Pump out Concrete
Backward: Suck in Concrete

Cylinder 1:

1 second to move forward, 2 seconds to return back

Cylinder 2:

2 seconds to move forward, 1 second to return back

Following this time period allotment, the cylinders always remain in sync with one another despite having different ram speeds. While one pushes out concrete, the other sucks it in and in the same definite amount of time.

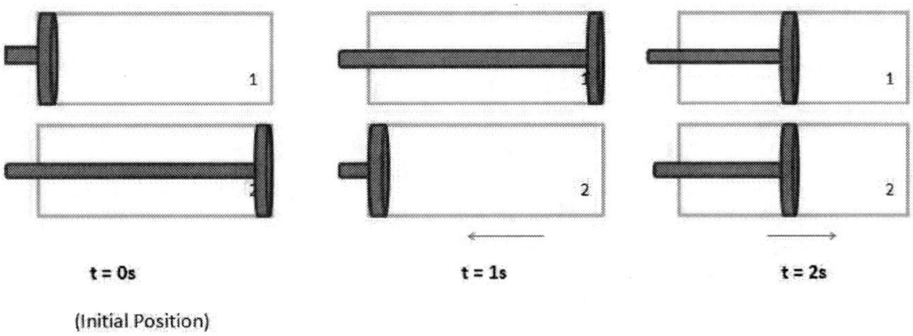

Figure 4. Schematic Diagram of ram movement inside cylinders

3.4 Dimensions of the CAD Model

Table 2. Specifications of CAD Model

Featur	Notation	Uni	Valu
Inner Cylinder Diameter	D_i	mm	180
Outer Cylinder Diameter	D	mm	200
Stroke	L	mm	120
Ram	D_r	mm	180
Width/Thickness of Piston Head	T_r	mm	60
Bull Gear/ Wheel Diameter	D	mm	600
Bull Gear/ Wheel Thickness	T	mm	20
Distance between Cylinder	D_c	mm	100

3.5 CAD Model of Suggested Mechanism

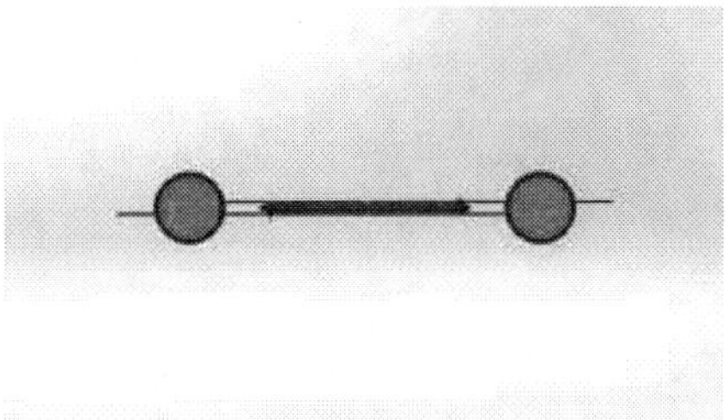

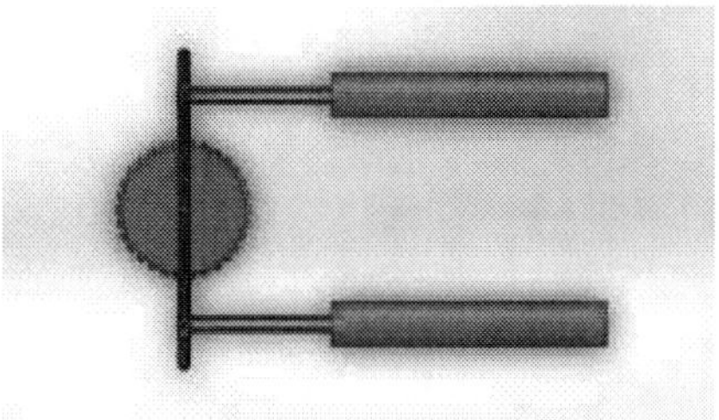

Figure 5. Front View of suggested mechanism mechanism

Figure 6. Top View of suggested

3.6 Calculation of Efficiency

Mass of Ram
Material = Mild Steel (Density, ρ_r = 7000 kg/m^3)
Mass of Ram, $M_r = Density \times Volume$

$$M_r = \rho_r \times \pi \times \left(\frac{D_r^2}{4}\right) \times T_r \quad (3)$$

$$M_r = 7700 \times 3.14 \times \left(\frac{0.18^2}{4}\right) \times 0.06$$

$$\boxed{M_r = 11.756468\ kg}$$

Mass of Concrete
Material = Concrete (Density, ρ_c = 2400 kg/m^3)
(Assumption: 100% of the volume is included during suction.)
Mass of Concrete, $M_c = Density \times Volume$

$$M_c = \rho_c \times \pi \times \left(\frac{D_i^2}{4}\right) \times L \quad (4)$$

$$M_c = 2400 \times 3.14 \times \left(\frac{0.18^2}{4}\right) \times 1.2$$

$$\boxed{M_c = 73.287073\ kg}$$

Acceleration of Ram

The ram uniformly accelerates for half the displacement and then uniformly retards towards the end.
Let us assume:
Displacement of Ram = s
Initial Velocity = u
Time duration = t
Acceleration in forward direction = af
Acceleration in backward direction = ab
Taking Cylinder 1 (one second for forward motion and two seconds for backward motion) into consideration: During Forward Stroke:

$$s = \frac{1}{2} \times stroke = \frac{1}{2} \times 1.2 = 0.6\ m$$
$$u = 0\ m/s$$
$$t = 0.5\ s$$
$$s = ut + \frac{1}{2}at^2 \tag{5}$$
$$0.6 = 0 + \frac{1}{2}(a_{r(f)})\left(\frac{1}{2}\right)^2$$
$$a_{r(f)} = \frac{1.2}{0.25} = 4.8\ m/s^2$$

Similarly, Retardation = 4.8 m/s^2

During Backward Stroke:

$$u = 0\ m/s$$
$$t = 1.0\ s$$
$$s = ut + \frac{1}{2}at^2$$
$$0.6 = 0 + \frac{1}{2}(a_{r(b)})(1.0)^2$$
$$a_{r(b)} = 1.2\ m/s^2$$

Similarly, Retardation = 1.2 m/s^2

For the ram to complete one cycle of motion (i.e.) from one end to another and then back, the bull gear completes one rotation and the time taken to do is 3 seconds.

$$N_{req} = \frac{60}{3} = 20\ RPM$$

Therefore, required RPM for Bull Gear;

Now, in 3 seconds, both the pumps have pushed out Concrete once.

Theoretical Maximum Output Flow Rate of Concrete $= Q_{b(new)}$

Volume of Cylinder $= V_c$

Therefore, $3 \times Q_{b(new)} = 2 \times V_c$

$$Q_{b(new)} = \left(\frac{2}{3}\right)\left(\frac{\pi}{4}\right)D_i^2 L = \left(\frac{2}{3}\right)\left(\frac{\pi}{4}\right) \times 0.18^2 \times 1.2$$
$$Q_{b(new)} = 0.020357\ m^3/s = 0.020357 \times 3600\ m^3/h$$
$$\boxed{Q_{b(new)} = 73.287073\ m^3/h}$$

4. OBSERVATIONS AND RESULTS

Taking into consideration that the volumetric efficiency of the concrete filling the cylinder is around 60% (as observed), the actual concrete output amounts to 13.2 m^3/hr/cylinder, or 26.4 m3/hr for the entire system. Calculating the Fluid Power using this flow rate and the pressure on the piston head on concrete side and using this, an efficiency of 52.35% was obtained.

As observed from the above calculations, the Whitworth Quick Return Mechanism seeks to pump out concrete much quicker, at a theoretical output rate of approximately 73.3 m^3/hr, as opposed to the theoretical output supplied by the existing model, which amounted to approximately 45 m^3/hr.

With only the CAD model in hand and no physical prototype to support the theoretical findings, it is difficult to assess the volumetric efficiency of the Whitworth Mechanism and this is left for the future scope of analysis of the mechanism.

5. CONCLUSION

It is clear that the concrete output rate that the current mechanism offers can be theoretically bettered with the help of the Quick Return Mechanism that was designed. With the notion that an increased output for the same amount of input power results in an improved efficiency, it is easy to assume that the efficiency increases in the new mechanism.

However, mechanical engineering is a field of study where assumptions need to be proven before they can be accepted. It is critical to calculate the pressure on the ram heads, which can be best performed experimentally. Such an experiment would prove to be very capital-intensive and time-consuming. Support obtained from organizations may help in carrying out the execution of the theoretical model and verifying its correctness as obtained on paper.

Further advancements can be achieved, mainly in lines with selection of alternative power sources which could replace the diesel engine. This is desirable, owing to the fact that diesel, like many other fossil fuels, is a depleting entity. Moreover, consumption of diesel contributes to release of greenhouse gases, such as carbon dioxide, which are essentially the main culprits of the alarming rise in the global average temperatures. Taking these into consideration, one can always select a suitable replacement, probably a chemically or an electrically running device, that is not only sustainable for the machine but also for the environment.

The scope of research in this area is limitless. Further enhancements in the model can be achieved, in terms of its basic body outlay, the mechanism and other aspects, which would benefit both economically and functionally.

REFERENCES

[1]. Ye. Koshkarev, Influence of Engine Load on Concrete Pump Operating Parameters, In *International Conference on Industrial Engineering*, Surat, India, 21-23 December 2017. *Procedia Engineering*, 2017, pp.1684-1689.

[2] Chipperfield A J and Fleming P J 1995 IET The MATLAB genetic algorithm toolbox.

[3] Yokota T, Taguchi T and Gen M 1998 A solution method for optimal weight design problem of the gear using genetic algorithms *Computers & industrial engineering Elsevier* Vol 35 Issues 3–4 pp 523-526

[4] Wang H and Wang H 1994 Optimal engineering design of spur gear sets Mechanism and machine theory *Elsevier* Vol 29 Issue 7 pp 1071-1080

[5]. https://www.schwingstetterindia.com/products/stationary-pump/sp-1200

Investigation On Smart Materials For Light Duty Goods Vehicle Chassis Frame Using Fea

Dr. K.G.Saravanan [1], Prof.S.Lakshmana Kumar [2], Prof.S.Vetrivel [3]
[1] Associate professor, Department of Mech Engg, Sona College of Technology, Salem, India, kgsmechanical@gmail.com
[2]Assistant professor (S.G), Department of Mech Engg, Sona College of Technology, Salem, India
[3]Assistant professor, Department of Mech Engg, Sona College of Technology, Salem, India

Abstract. Weight directly affects performance and emission parameters of vehicle. The main contributor of weight in heavy vehicle is chassis frame [1]. The reduction of weight of chassis improves fuel economy and reduces carbon dioxide emission. There are several conventional and non-conventional chassis materials available in the market [2]. Composite materials future looks promising because of their lesser weight and higher performance [3]. Proper usage of such product will help in fuel economy thus, saving the fuel and environment [4]. Also it helps to achieve higher performance readily than other conventional materials [5]. Hence, in this Project, we are going to reduce the weight of TATA ACE chassis using composites after the design and analysis of chassis frame [6]. Different materials are tested in ANSYS software are the material having the best desirable properties is recommended for further application in the automotive industry [7].
Keywords - Aluminum Propylene, principal stress, solid work, CATIA V5, ANSYS workbench v15.

1. Introduction

- Chassis is a French term which is used to denote complete body and structure of the vehicle.
- It contains all the major units necessary to propel the vehicle direct its motion, stop it and allow it to run smoothly over uneven surfaces.
- The chassis is the major weight constituting part of automobile.
- The chassis constitutes about 20% of the automobile weight.

2. Functions of the Chassis Frame

1. To carry load of the passengers or goods carried in the body.
2. To support the load of the body, engine, gear box etc.
3. To withstand the forces caused due to the sudden braking or acceleration.
4. To withstand the stresses caused due to the bad road condition.
5. To withstand centrifugal force while cornering.

3. Types of Chassis Frame

3.1 Conventional Frame

It is non-load carrying frame. The loads of the vehicle are transferred to the suspensions by the frame. This suspension in the main skeleton of the vehicle is supported on the axles through springs. The body is made of flexible material like wood and isolated frame by inserting rubber mountings in between. The frame is made of channel section or tubular section or box section.

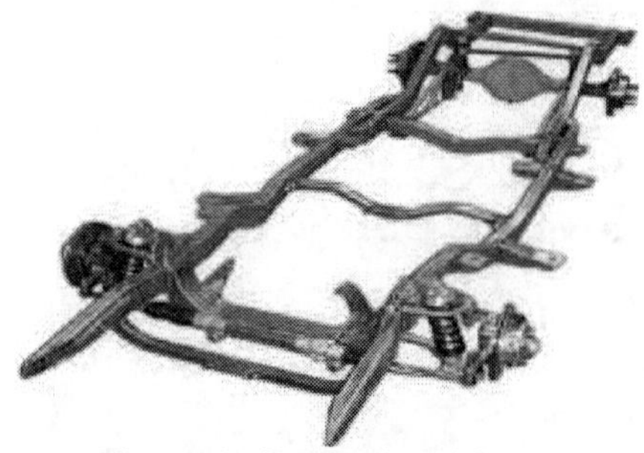

Figure 1. Conventional frame

3.2 Integral Frame

This frame is used now a day in most of the cars. There is no frame and all the assembly units are attached to the body. All the functions of the frame carried out by the body itself. Due to elimination of long frame it is cheaper and due to less weight most economical also. Only disadvantage is that the repairing is difficult.

Figure 2. Integral Frame

3.3 Semi-Integral Frame

In some vehicles half frame is fixed in the front end on which engine gear box and front suspension is mounted. It has the advantage when the vehicle is met with accident the front frame can be taken easily to replace the damaged chassis frame. This type of frame is used in some of the European and American cars.

Figure 3. Semi-integral Frame

4. Types of Frame Sections

The frames are made of following sections.

1. Box section
2. Channel section
3. Tubular section

Box section is rectangular in cross section.

Box section is used in short members of the frame.

Channel section is used in long members of the frame.

Tubular section is used these days in three-wheelers, scooters and matadors, pick-ups frames.

Channel section is good for bending, tubular in torsion and box in bending and torsion.

Figure 4. Various Frame Sections

5. Important Factors to Consider

The material chosen should have lesser weight than previously used material for mini truck frame without compromising strength.

The chassis should be able to withstand various acting forces in the frame.

The newly designed chassis should be safe, efficient and economical.

6. Selection of Materials

The materials are selected by considering density, strength, cost and availability parameters. By studying the various properties of materials these materials are selected for Analysis. They are as follows.

Steel
Aluminum epoxy
Aluminum propylene

6.1 Steel

Steel is existing chassis material of TATA Ace chassis so analysis is done on the reference of steel regarding their weight, cost and strength properties. Carbon composites and its growing use is one of the driving factor for analyzing. Carbon composites are having higher strength as well as good fatigue resistant properties. Aluminum epoxy is having very long pot life (working time) as well as less brittle in nature. The combined weight of aluminum along with the epoxy is lesser in comparison with the other composites.Aluminium propylene is already used in a new material called injector. It has already promised that this material can reduce the weight of chassis by 50% and enhance the performance of automotive vehicle.

6.2 Aluminum Epoxy

Aluminum/polymer composites are attractive materials for a wide range of industrial applications due to the combination of properties such as low density, corrosion resistance, thermal stability, and ease of fabrication. Aluminum/epoxy composites are fabricated by cast molding method using 10 wt. % unmodified and salinized aluminum powders. They are having excellent adhesive properties. Cost of epoxy is three times less than same weight of carbon fiber. Epoxy tends to be stronger than other resins and is less brittle in nature. Density of epoxy is same as polyester resin but viscosity is higher than other resins. They are much stronger in adhesion. They tend to have greater working life.

IOP Conf. Series: Materials Science and Engineering 1123 (2021) 012021 doi:10.1088/1757-899X/1123/1/012021

Table 1. Properties of epoxy

Density	1100 kg/m^3
Young's modulus	6.5 GPa
Ultimate Strength	3450 N/mm^2

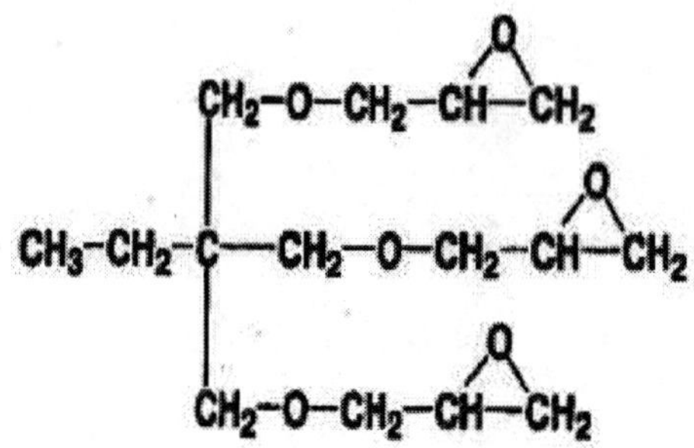

Figure 5. Chemical representation of epoxy

6.3 Aluminum Propylene

Aluminum propylene is recently used as automotive material. Infector has developed a patented lightweight technology which uses aluminum skin bonded with an expanded propylene (EPP) core. Infector can reduce weight from 300kg typical for chassis to 160kg. To add to its advantage, the sandwiched material is 100% recyclable. It is lesser in weight but comparative stronger than other material counterparts.

Table 2. Properties of propylene

Density	80 kg/m^3
Young's modulus	9.162 GPa
Ultimate Strength	321 MPa

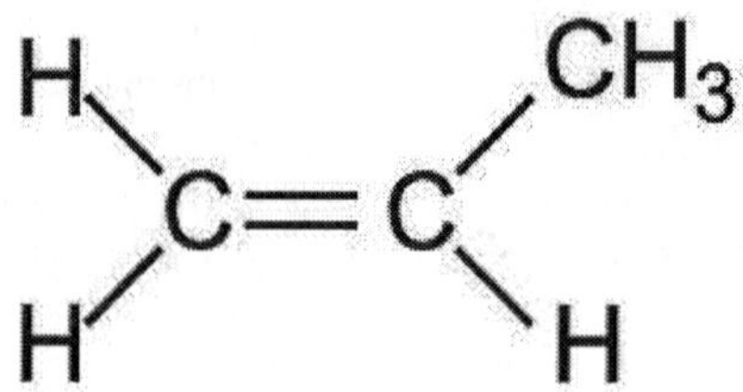

Figure 6. Chemical representation of propylene

7. Frame calculation

7.1 C – Section Chassis

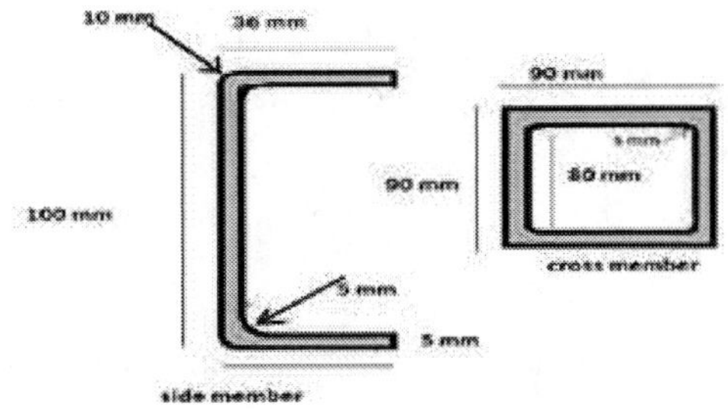

Figure 7. C – Section Chassis

Length of vehicle = 4340 mm
Width of vehicle = 1565 mm
Height of vehicle = 1858 mm
Wheelbase = 2380 mm
Track width = 1320 mm
Length of chassis = 4201 mm
Width of chassis = 808 mm
Dimensions of side bar = 100mm x 36mm x 5 mm
Dimensions of cross bar = 90 mm x 90 mm
Gross vehicle weight = 2180 kg
Kerb weight = 1180 kg

Pressure:

Hence the total area of application of load as calculated from chassis dimensions= 1182600 mm^2.

Total load to be applied = 2150 x 9.81 = 21091.5 N

Pressure to be applied = 1091.5/1182600 = 0.017834 MPa.

Volume Occupied = 2*(90*5*4201+36*5*4201*2) +5*(808*90*5*2+798*80*5*2)
=27133758mm^3

Structural steel

Density = 7850 kg/m^3

Weight of the Steel Frame chassis = Density *Volume Occupied
=7850* 27133758*10^{-9}
=213 kg

For applying composite over the chassis section we are going to modify the design structure of chassis having rectangular box frame instead of C channel frame.

7.2 Rectangular box chassis

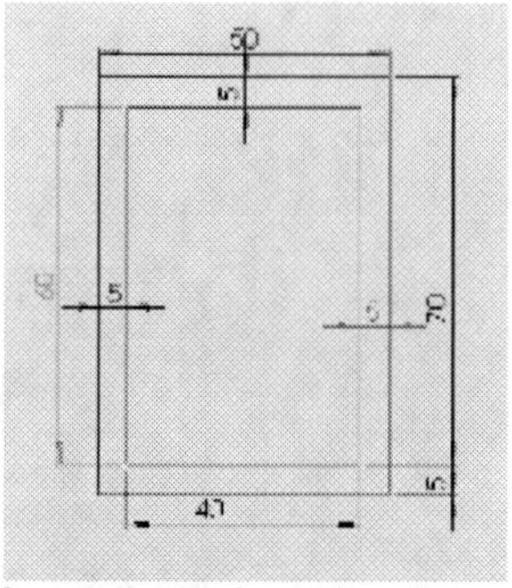

Figure 8. Dimension of rectangular box chassis cross section

(Note: Dimensions of Aluminum epoxy and Aluminum Propylene chassis are same)

Length of vehicle	= 4340 mm
Width of vehicle	= 1565 mm
Height of vehicle	= 1858 mm
Wheelbase	= 2380 mm
Track width	= 1320 mm
Length of chassis	= 4201 mm
Width of chassis	= 808 mm
Dimension of side bar	= 70 X 50 X 5mm
Dimension of Cross bar	= 70 X 50 X 5 mm
Total area of application of load	= 1899850 mm^2
Gross vehicle weight	= 2066 kg
Kerb weight	= 1066 kg
Total load to be applied	= 2050*9.81
	= 20110.5N
Pressure to be applied	= load/Area
	= 20110.5/1899850
	= 0.0105853 Mpa

Total volume Occupied = volume occupied by aluminum+ volume occupied by resin

volume occupied by aluminum = 1300*4201*2+1300*5*708
= 15524600 mm^3

volume occupied by composite = 60*40*4201*2+60*40*5*708
= 28660800 mm^3

Total volume occupied = 15524600+28660800
= 44185400 mm^3

Volume Concentration of Aluminum = 15524600/(15424600+28660800)
= 0.351

Volume Concentration of resin = 28660800/ (15424600+28660800)
= 0.6486

IOP Conf. Series: Materials Science and Engineering 1123 (2021) 012021 doi:10.1088/1757-899X/1123/1/012021

For Aluminum Epoxy,

Weight of Aluminum epoxy chassis = weight of aluminum+ weight of epoxy Weight of aluminum = Density *Volume occupied by aluminum

= 2700*15524600*10^{-9}

= 41.91 kg

Weight of epoxy = Weight of hardener+ Weight of resin

Density of hardener = 940 kg/m^3

Density of resin = 2250 kg/m^3

Mix ratio for epoxy by volume = 7.23/1 (epoxy/hardener)

Weight of hardener =Volume*density

=28660800*10^{-9}*(1/8.23)*940

=3.27 kg

Weight of resin = Volume*density

= 28660800*10^{-9}*7.23/8.23*2250

= 56.65 kg

Weight of epoxy = 3.27+56.65

= 59.92 kg

Total weight of chassis = weight of aluminum+ weight of epoxy

= 41.91+59.92

= 101.83 kg

Young's modulus of epoxy (E_{poxy}) =6.5 Gpa

Young's modulus of aluminum ($E_{aluminum}$) =68.9 Gpa

Equivalent Young's modulus for Aluminum Epoxy = 75.4 GPa

For Aluminum Propylene,

Weight of Aluminum propylene chassis = weight of aluminum+ weight of Poly propylene

Weight of aluminum = Density *Volume occupied by aluminum

= 2700*15524600*10^{-9}

= 41.91 kg

Weight of Polypropylene = Density of Polypropylene*Volume of Polypropylene

=80*28660800*10^{-9}

=2.3088 kg

Total Weight of Aluminum Propylene Chassis

= 41.91+2.8666 = 44.796 kg

Young's Modulus of Aluminum=68.9 GPa

Young's Modulus of Polypropylene = 9.162 Gpa

Equivalent Young's Modulus of Aluminum Propylene = 78.062 GPa

8. Design of Chassis Frame

After calculation of chassis frame, the sample chassis of TATA ACE mini truck frame is designed using SolidWorks software. The design procedure was based on creating a model, viewing it, assembling parts as required, and then generating any drawings which are required. The original chassis of Tata Ace is designed for structural analysis of steel frame. For Analyzing composite section, the chassis design is modified. Instead of original C-channel section ladder frame, rectangular box section type ladder chassis is designed. The dimensions of both side bars and cross bars of modified sections are kept same.

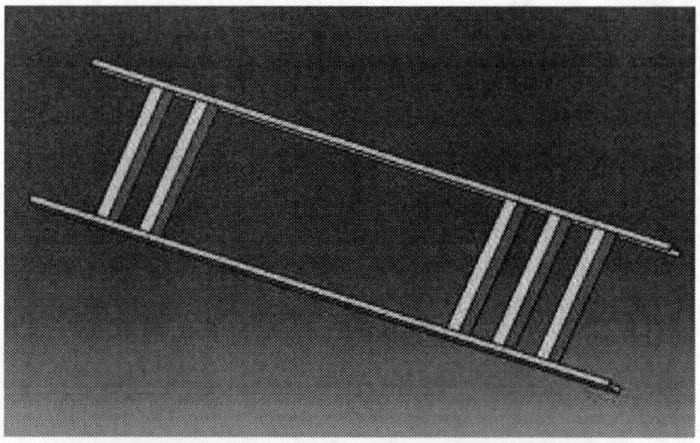

Figure 9. C-Section Original Chassis frame

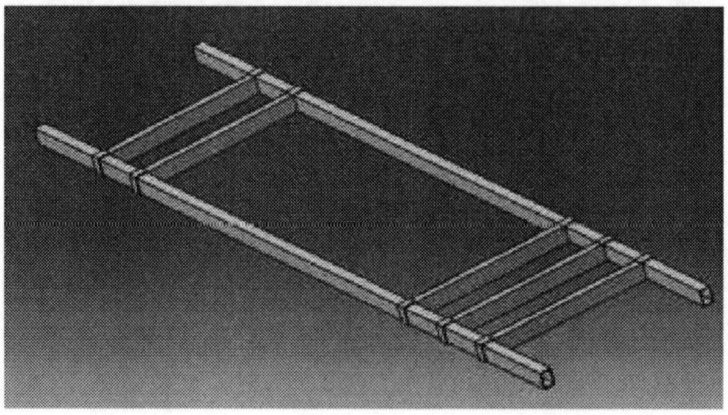

Figure 10. Rectangular Cross Section Modified Chassis frame

9. RESULTS AND DISCUSSIONS

9.1 Analysis of Chassis Frame:

After designing the chassis frame, the frame is subjected to the analysis. The analysis is done in ANSYS software. CAD model of TATA is created and imported into ANSYS for FEA. The design created by SolidWorks is used in ANSYS for structural analysis. The C channel section which is the conventional chassis is used to analyze structural steel. The second modified design is used to analyze Aluminum epoxy and Aluminum propylene respectively. The size of elements is kept as minimum as possible to get the accurate results and at some points the finer meshing is also done to get better results. Static analysis is used to find maximum stress region. After analysis, the obtained results are used to conclude the best material.

The designed model from of chassis using SolidWorks is saved in IGES format which can be directly imported into ANSYS workbench.

9.2 Meshing:

The meshed view as shown in above figure has 28212 nodes and 13331 elements.

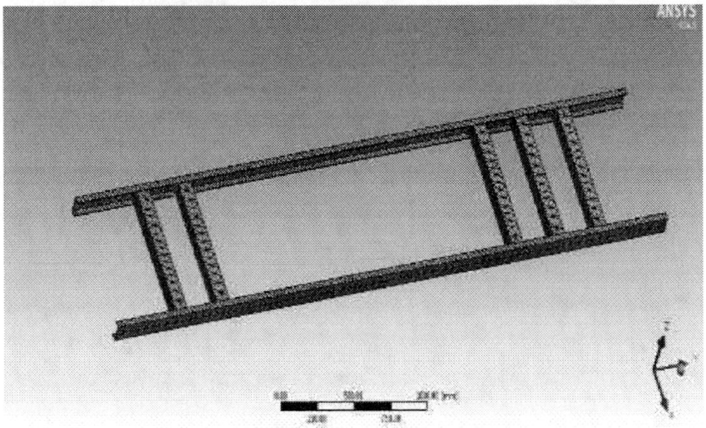

Figure 11. Mesh view of chassis using steel

The meshed view as shown in above figure has 28212 nodes and 13331 elements.

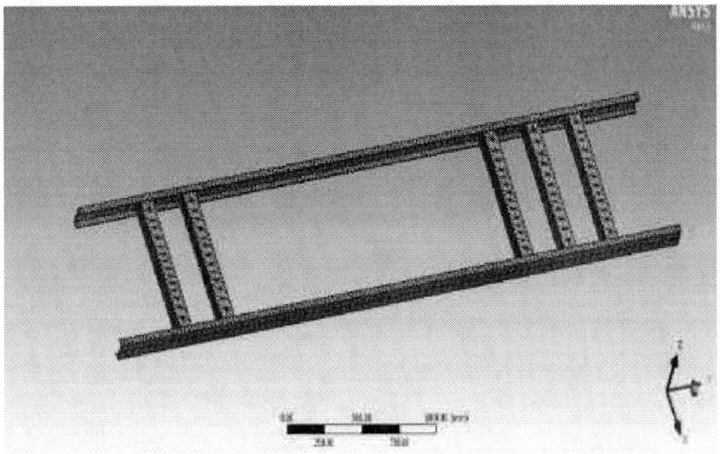

Figure 12. Mesh view of chassis using Aluminum Epoxy

The meshed view as shown in above figure has 28212 nodes and 13331 elements.

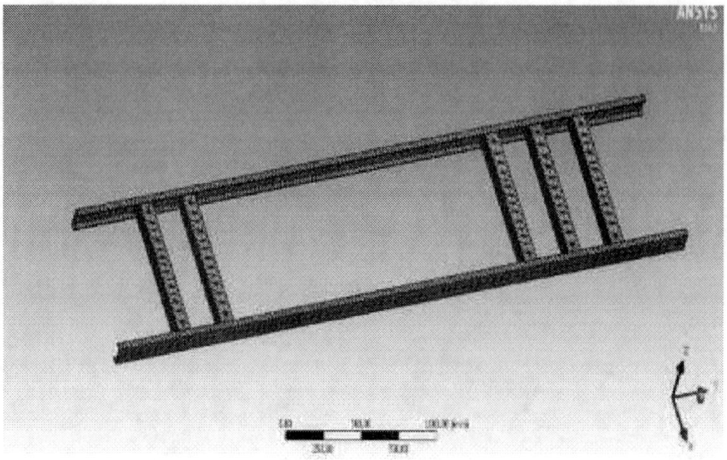

Figure 13. Mesh view of chassis using Aluminum Propylene

9.3 Load Applications and Boundary Conditions

Load in Tata Ace chassis is applied in two different chassis design in the form of pressure. Pressure of magnitude 0.017834 MPa is applied in Original C-Cross section chassis whereas 0.0105853MPa is applied in modified rectangular Cross section chassis. There are two boundary conditions which includes fixing the front and the rear axle.

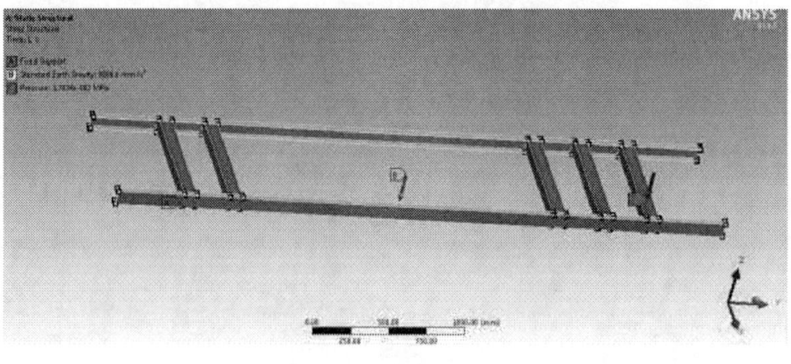

Figure 14. Loading and boundary Condition of C-section Tata-Ace chassis

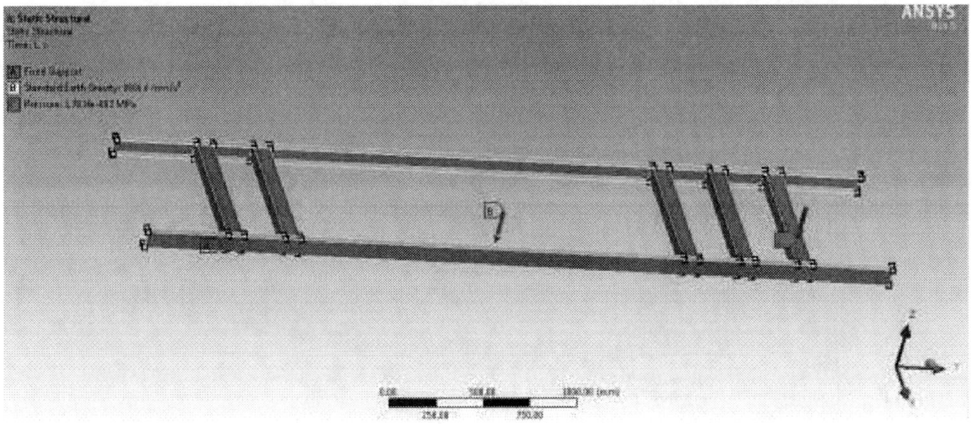

Figure 15. Loading and boundary Conditions of rectangular Cross-Section Chassis

10. ANALYSIS

Analysis is conducted by testing different materials on two different chassis and Equivalent stress (Von-misses stress), maximum principal stress, maximum shear stress and total deformation are found out. The results including above mentioned parameters are shown as follows.

10.1 Equivalent Stress (Von-misses) stress

- **Structural steel**

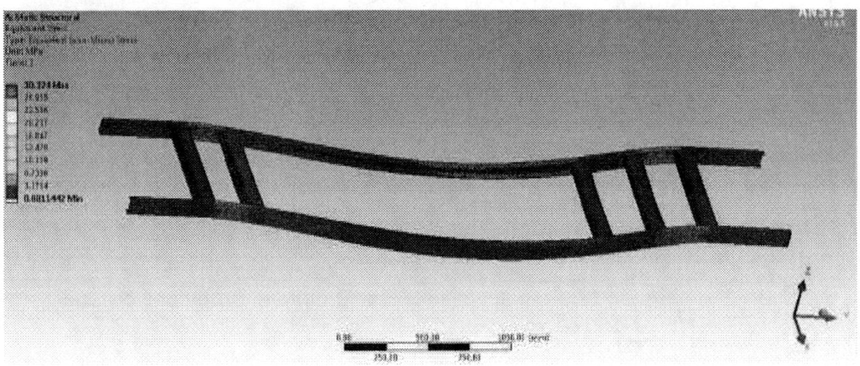

Figure 16. Equivalent stress in Structural steel C-section chassis

- **Aluminum epoxy**

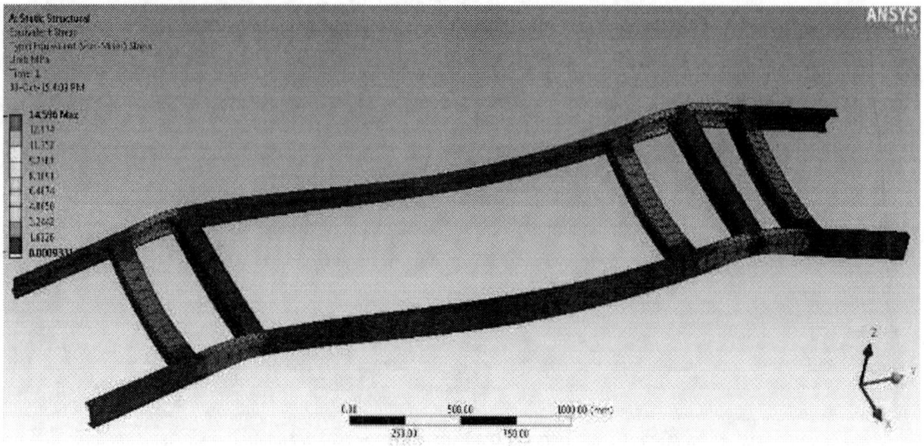

Figure 17. Equivalent stress in Aluminum Epoxy chassis

- **Aluminum propylene**

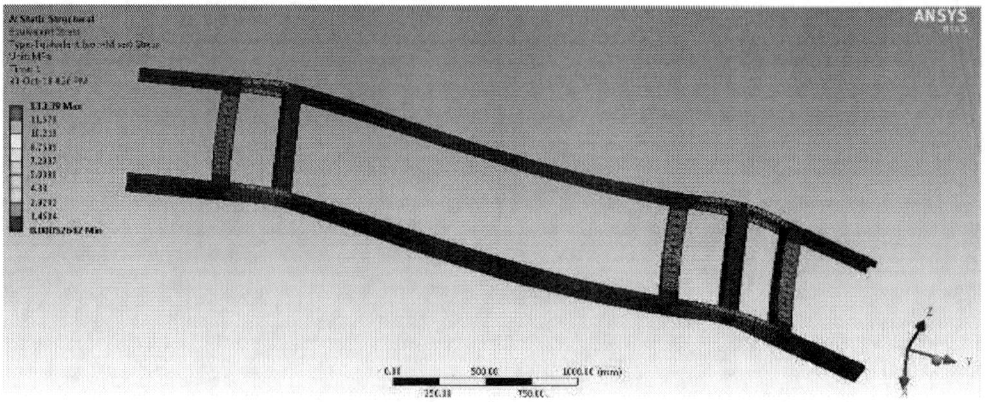

Figure 18. Equivalent stress in Aluminum propylene chassis

IOP Conf. Series: Materials Science and Engineering 1123 (2021) 012021 doi:10.1088/1757-899X/1123/1/012021

10.2 Maximum Principal Stress

- **Structural steel**

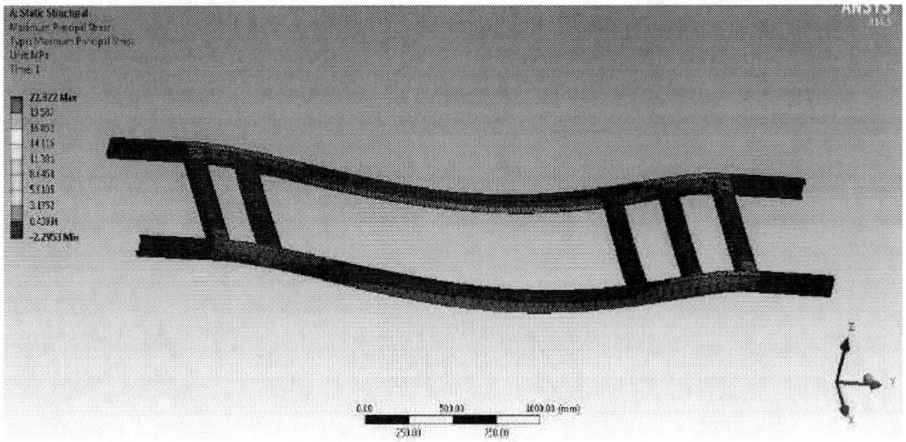

Figure 19. Maximum Principal Stress in Structural steel C-section chassis

- **Aluminum epoxy**

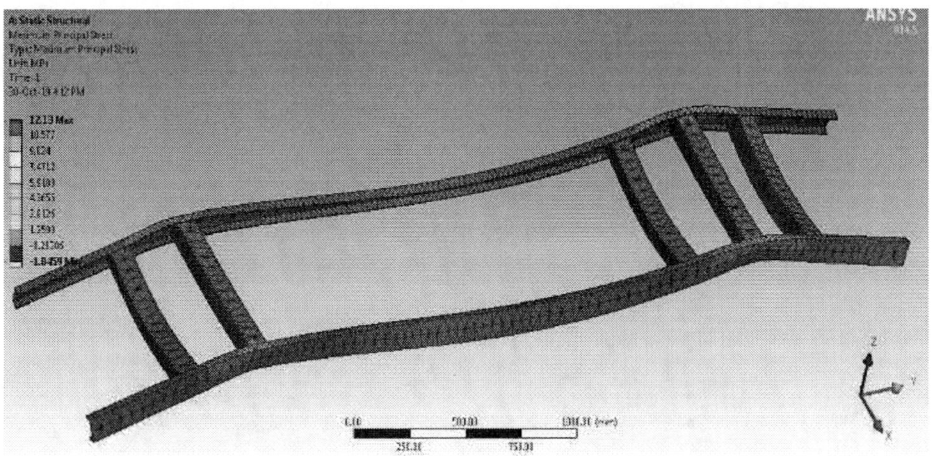

Figure 20. Maximum Principal Stress in Aluminum Epoxy chassis

➢ **Aluminum propylene**

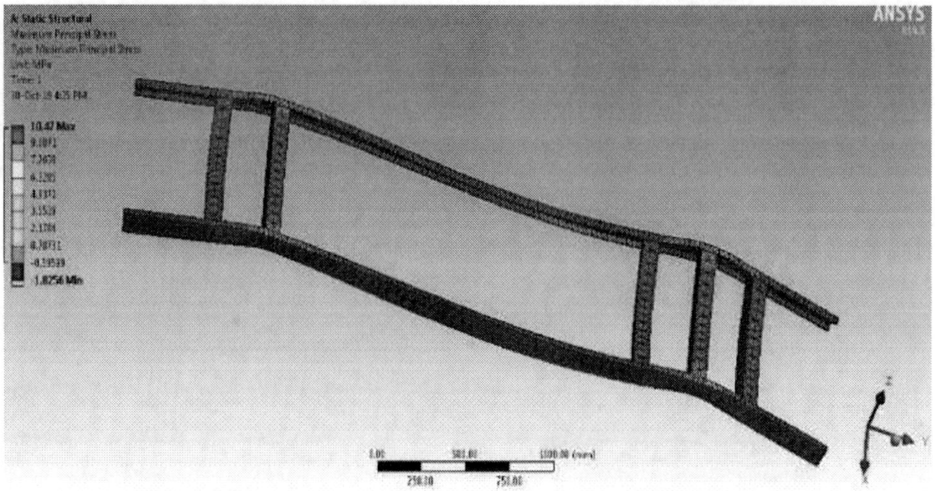

Figure 21. Maximum Principal Stress in Aluminum propylene chassis

10.3 Maximum Shear Stress

➢ **Structural steel**

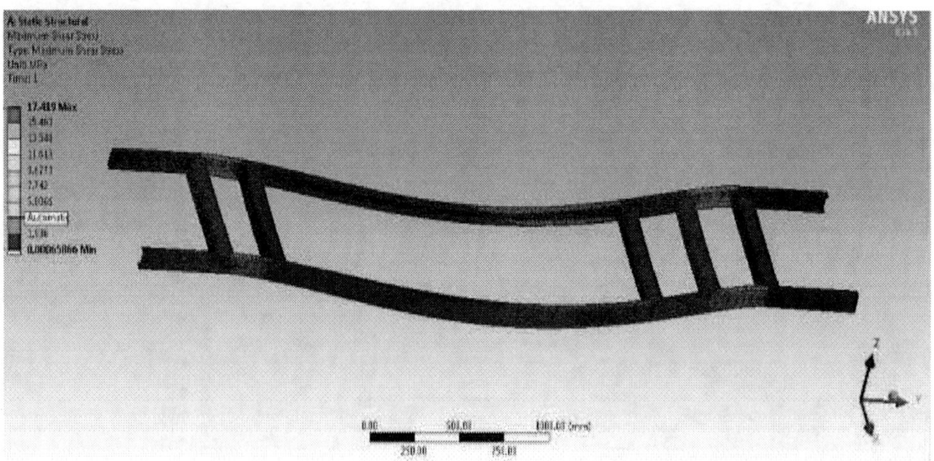

Figure 22. Maximum Shear stress in Structural steel C-section chassis

- **Aluminum epoxy**

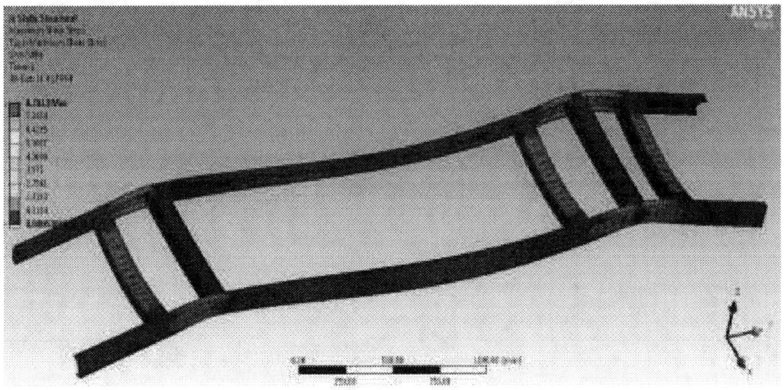

Figure 23. Maximum Shear stress in Aluminum Epoxy chassis

- **Aluminum propylene**

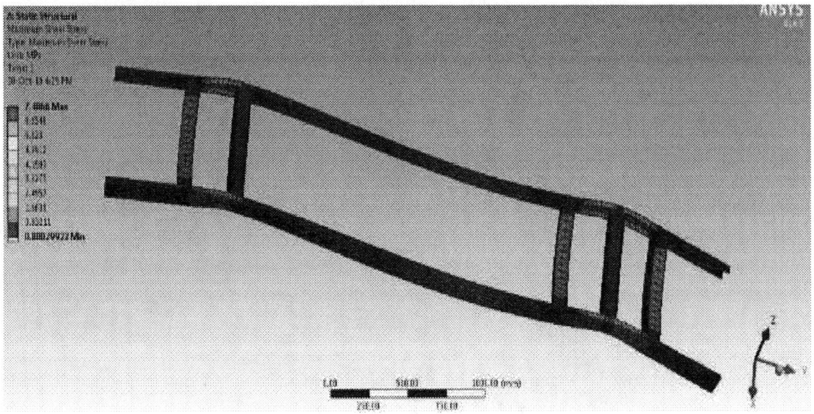

Figure 24. Maximum Shear Stress in Aluminum propylene chassis

10.4 Total Deformation

➢**Structural steel**

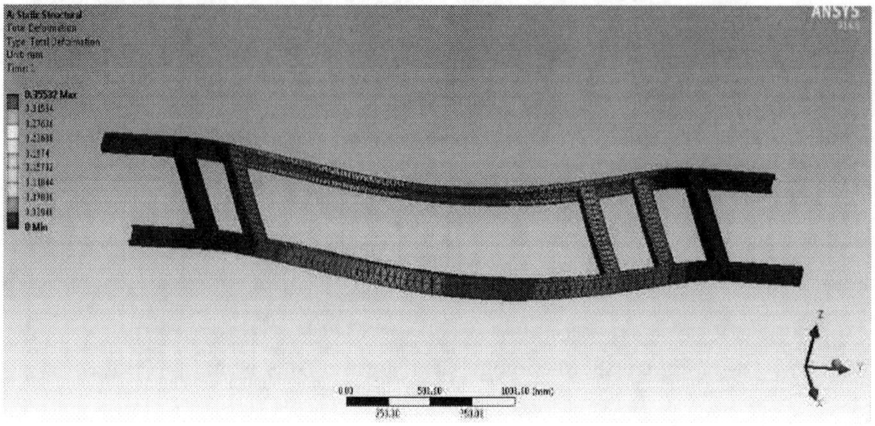

Figure 25. Total Deformation in C-section steel chassis

➢**Aluminum epoxy**

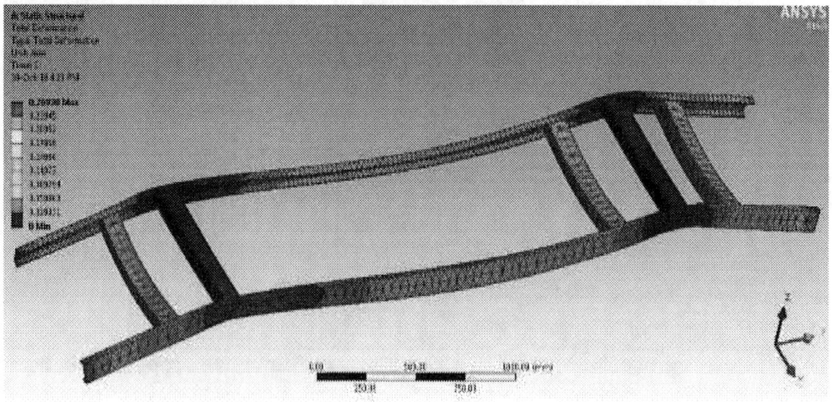

Figure 26. Total deformation in Aluminum Epoxy chassis

➢ **Aluminum propylene**

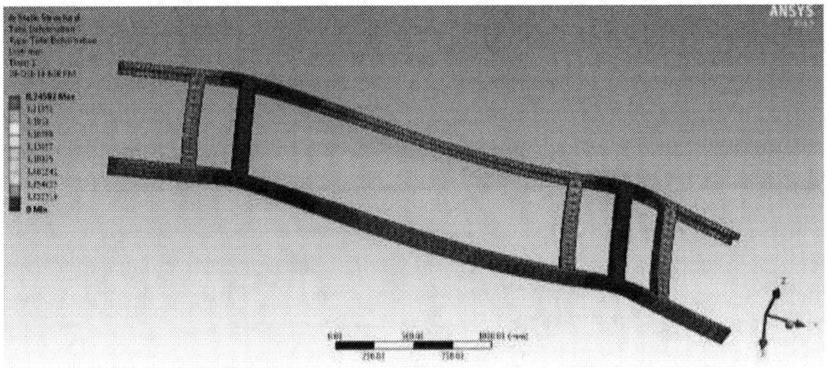

Figure 27. Total deformation in Aluminum propylene chassis

Table 3. Result comparison

Material	Equivalent stress (MPa)	Max principle stress (MPa)	Max shear stress (MPa)	Total deformation (mm)
Structural steel	30.248	22.32	17.419	0.35532
Aluminum Epoxy	14.596	12.13	8.2613	0.26938
Aluminum propylene	13.139	10.47	7.4806	0.24502

11. Conclusion

The design and Tata Ace chassis using C-section and rectangular cross section is done in SolidWorks and imported to ANSYS for finite element analysis. The analysis is conducted in conventional steel, Aluminum Epoxy and Aluminum Polypropylene respectively. After analysis a comparison is made between existing conventional steel chassis and composite materials in terms of deformation and stresses, to select the best one. From the calculations, it is found out that Aluminum Propylene has the lowest weight compared to steel and aluminum epoxy. Also, Aluminum Polypropylene has the lowest equivalent stress, principal stress, shear stress and lowest deformation induced. Based on the results it was inferred that Aluminum Polypropylene composites with rectangular box section has superior strength to withstand high load and induced low deformation and stress distribution when compared to steel.

References

[1] Mr. Jurgen Hirsch , Aluminum in Innovative Light-Weight Car , Hydro Aluminum Deutschland GmbH, R&D(2011)

[2] A.Hari Kumar V Deepanjali, Design & analysis of automobile chassis (2016)

[3] ShaikNeelophar Begum S.P. BhanuMurthy, Modelling and structural analysis of vehicle chassis frame made of polymeric composite material (2016).
[4] Andure M.W. Jirapure S.C. Dhamande L.P, Advance Automobile material for lightweight future. (2012).
[5] Priyanka Ranu, Ranjeev Singh Chauhan, Dr. V.N Bartaria , Mathematical Modelling and analysis of automotive chassis with composite materials using FEM (2017).
[6] Md.Azhar Kazmi ,Mr.V.Sunil, Mr.S.Udayabhaskar , Modelling and Analysis of Automobile Chassis using Composite materials (2017)
[7] Abhinn Bajaj, Shahnawaz Alam, Akshansh Uniyal, Static and Modal analysis of truck chassis (2016).
[8] Reija Suihkonen ,Katja Nevalainen, Olli Orell, Mari HonkanenLongcheng Tang, Hui Zhang, Zhong Zhang, Jyrki Vuorinen, Performance of epoxy filled with nano- and micro-sized Magnesium Hydroxide (2012)
[9] KwangSeon Shin, KeeJoo Kim, Suk-Woo Choi and MeungHoRhee, Mechanical Properties of Aluminum/Polypropylene/Aluminum Sandwich Sheets (1999).
[10] Manpreet Singh Bajwa, Yatin Raturi, Amit Joshi, Static load analysis of Tata Super Ace chassis and it's verification using solid Mechanics.

LIFE CYCLE ASSESSMENT OF NICKEL CADMIUM BATTERY

Nitin Kotkunde [1], K V Amrith Ashwin[2], Abhishek Gosavi [3], Mayank Gaur [4]

[1] Mechanical Engineering, BITS Pilani, Hyderabad, India, nitink@hyderabad.bits-pilani.ac.in

[2] Mechanical Engineering, BITS Pilani, Hyderabad, India, h20181060160@hyderabad.bits-pilani.ac.in

[3] Mechanical Engineering, BITS Pilani, Hyderabad, India, h20181060158@hyderabad.bits-pilani.ac.in

[4] Mechanical Engineering, BITS Pilani, Hyderabad, India, h20181060161@hyderabad.bits-pilani.ac.in.

Abstract. The life cycle assessment of the battery will help one to identify the energy and cost associated with the product from the cradle to gate processes. The ecological footprint of a company can reduce the power of a brand when consumers avoid unsustainable practices. The performance of the product can be compared with its competitors. The potential areas of improvement can be identified, which in turn helps to improve the performance of the product. The energy associated with the product during processing, manufacturing, usage and transportation stages has been identified in this paper. The energy pay back for the battery has been calculated, which can be used to compare batteries among each other. On finding the energy payback of the battery, a framework has been developed which can be used in future to compare different batteries. And the impact assessment helped to interpret the important areas of improvement and the sector producing maximum impact. And some methods which can be implemented to decrease the impact and burdens produced during the product development stages are discussed. Keywords: Energy payback, Impact potential, Specific energy consumption, Environmental impact

1. Introduction

In most of the countries, energy policy has mainly three objectives, they are, to prevent climatic changes, reduce the consumption of fossil fuels, providing required energy at affordable rate [1]. Technologies to store electricity plays a major role in providing secure, safe and sustainable energy [2]. It helps to increase the market share of renewable energy [1]. In low carbon energy generation, electrical storage mechanism is the deciding factor because of the intermittent or variable nature of power generation [3]. When batteries are used for partially or fully electrified vehicles, batteries always help to reduce the power requirement and carbon dioxide emissions. There are no more conventionally-fueled cars in cities, 40% use of sustainable low carbon fuels in aviation; at least 40% cut in shipping emissions, 50% shift of medium distance intercity passenger and freight journeys from road to rail and waterborne transport. All of which will contribute to a 60% cut in transport emissions by the middle of the century [4]. Currently different types of electric batteries are available in the market for meeting the needs and necessities of the customers. By improving the functionality of the batteries will increase the demand for the product and thereby helps to increase the market share of the product energy storage systems are improved for different applications such as in electric cars, trucks and hybrid vehicles and for naval applications. Analysis of the performance of electric cars and competing powertrain options in terms of greenhouse gas emissions over their life cycle. It also discusses key challenges in the transition to electric mobility and feasible solutions possible which includes vehicle and battery cost developments; supply and value chain sustainability of battery materials; implications of electric mobility for power systems; government revenue from taxation; and the interplay between electric, shared and automated mobility options [5].

Among different electric batteries available in the market, in this paper, life cycle assessment of Nickel cadmium battery is presented. Nickel cadmium pocket plate battery is a secondary battery. It is used for different applications such as fire alarm systems, signaling and telecommunication, Solar Photovoltaic, Instrumentation & Process control, Switchgear protection, Emergency lighting, Switchyard, UPS, Genset starting, Diesel Locomotive cranking, Electromagnetic lifters (EOT crane). It's used for different applications due to its ability to withstand high temperature, high reliability, long operating life, faster charging compared to other electric batteries and low maintenance [5]. But there are some problems faced, while using nickel cadmium battery, they are, Nickel and Cadmium are costly metals and are highly toxic, they get heated during its operation and can get into thermal runaway mode.

Earlier products developed and consumed in the word followed linear economy, where the raw materials required for the development of the product was extracted according to requirement which is later processed and manufactured using some desired technique, sold to customers who use it during its lifecycle and disposed to the environment after its operating period [6]. In order to minimize the wastage of natural resources, materials from the products which completed its lifecycle need to be recollected and recycled, so that it can reduce the extensive extraction of natural resources to meet the demand. By including the process, the product's economy can be changed from linear to circular. A product following circular economy helps in the development of the new product from the 'lifecycle completed products' [2]. Thus the requirement of freshly extracted materials to meet the demand can be reduced drastically. Understanding the positive outcomes of changing the product's economy from linear to circular, different governments around the world have passed different policies and laws for the conversion of the product's economy into a circular one [5]. Example, Dutch government announced a government-wide programme, which was developed for the circular economy, involving different departments to convert the Dutch economy to circular by 2050. It helps to provide improved economic benefits associated with the development of the product, increased growth of the economy with increase in spending due to lower prices of the product, appreciable saving of the material, the change in input and output leads to higher valuation of labor, which leads to increased employment opportunities and incentives for the innovative developments. Data to demonstrate how the global picture in research and innovation is changing and how Europe is positioned within the international context, underpinning the need for a more strategic approach to international cooperation in research and innovation in Europe [1].

Life cycle assessment of the Nickel Cadmium battery will help to analyses the energy input required by the product and the environmental impacts that the product tend to produce during its lifecycle. Cradle-to-gate life-cycle inventory analysis of lead-acid, nickel-cadmium, nickel-metal hydride, sodium-sulfur, and lithium-ion battery technologies has been reviewed and evaluated [6]. An algorithm was developed that will assess the ability of battery storage in capturing AC and DC power generated from photovoltaic cells, combined heat and power and wind turbines.

The assessment includes the impact that the storage element has on the import and export of energy from the electrical grid. It showed when producing on site electricity through microgeneration, suitably sized storage can reduce export substantially (by over 90% in some cases) and store this energy at a typical round trip efficiency of 70-72%. The developed model accounted typical losses in a battery storage system, including losses associated with inverters, efficiency of charge/discharge cycles and power electronics [7]. The currently available data is reviewed and the impact of the production of several types of battery in terms of energy, raw materials and greenhouse gases is calculated and highlighted. The impact of the production of batteries is examined and presented. It is shown that the most significant impact in many environmental areas in terms of production is caused by lithium based batteries. [8].

Life cycle assessment(LCA) is a tool which is used to identify the environmental impacts produced by the product during its 'cradle-to-gate' processes. This helps to quantify the environmental impact and environmental burden produced by the product, as shown in Figure 1. It helps to identify the sector, such as extraction, manufacturing, transportation, usage, etc., that is causing maximum impact which helps to identify the potential point of improvement. The energy payback is a measure of effectiveness of the product, lower the energy payback, better the product. By interpreting the gate-to-cradle process of the product, the different ways of improving the product's lifecycle can be understood. The proper marketing technique implemented with the help of the results obtained from the life cycle analysis will help to improve the economic performance of the product keeping the process performance same. An investigation through a comparison of families of solar assisted cooling systems (with solar thermal or PV) for studies concerning the research of effective and environmentally friendly systems that exploit solar radiation for cooling and heating purposes. It compares five solar assisted H/C systems with LCA methodology [9]. Assessing life cycle GHG emissions from Plug-in hybrid electric vehicles (PHEVs) and find that they reduce GHG emissions by 32% compared to conventional vehicles, but have small reductions compared to traditional hybrids. Also it shows GHGs associated with lithium-ion battery materials and production account for 2–5% of life cycle emissions from PHEVs [10]. A detailed lifecycle inventory of an electric battery and a rough LCA of Battery-powered electric cars (BEV) based on mobility was compiled. The study shows that the environmental burdens of mobility were dominated by the operation phase regardless of whether a gasoline-fueled ICEV or European electricity fueled BEV is used [11].

During the life cycle of the battery the battery will consume a finite amount energy and cause the evolution of certain gases that are toxic to the environment and produce harmful effects. Life cycle assessment helps to quantify those impacts and burdens produced by the product under consideration and helps to compare the same with the other electric batteries. It helps in identifying the product which produces less harm to the environment. This paper makes two contributions: 1) identifying the energy input required by the product and the environmental impact produced by the same; 2) it provides energy payback of the battery which can be used to measure the effectiveness of the battery; 3) it provides a set of energy and environmental outcomes and identifies the potential area of improvement.

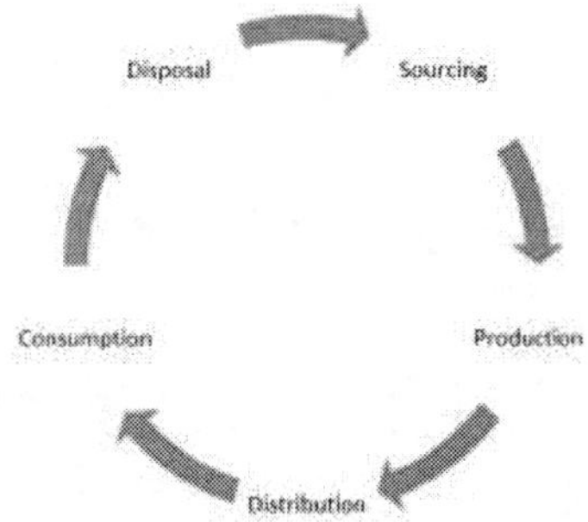

Figure 1. Different stages of a product during its life cycle which are considered for life cycle assessment

During the life cycle of the battery the battery will consume a finite amount energy and cause the evolution of certain gases that are toxic to the environment and produce harmful effects. Life cycle assessment helps to quantify those impacts and burdens produced by the product under consideration and helps to compare the same with the other electric batteries. It helps in identifying the product which produces less harm to the environment. This paper makes two contributions: 1) identifying the energy input required by the product and the environmental impact produced by the same; 2) it provides energy payback of the battery which can be used to measure the effectiveness of the battery; 3) it provides a set of energy and environmental outcomes and identifies the potential area of improvement.

2. Life Cycle Assessment

The material used in the battery are obtained according to the design of the battery. The energy consumption per kg of the battery is calculated and converted in terms of a battery. LCA mainly consist of four steps, firstly, goal and scope definition, where the goal and intended use of LCA is defined. Assessment is scoped in terms of a) boundaries of the product system under consideration b) temporal and technological scope of the process in the product system c) Assessment parameters to be considered in the assessment

2.1 Goal and Scope Definition

Here the system considered is a nickel cadmium battery of 3.2Kg Nickel Cadmium Pocket Plate Batteries, 140 Ah at 1.2 V. The percentage weights of different components in a Nickel Cadmium battery is shown in Figure 2.

Here the focus is on the battery and not on the application to which it is subjected. The material used in the battery are obtained according to the design of the battery. The energy consumption per kg of the battery is calculated and converted in terms of a battery.

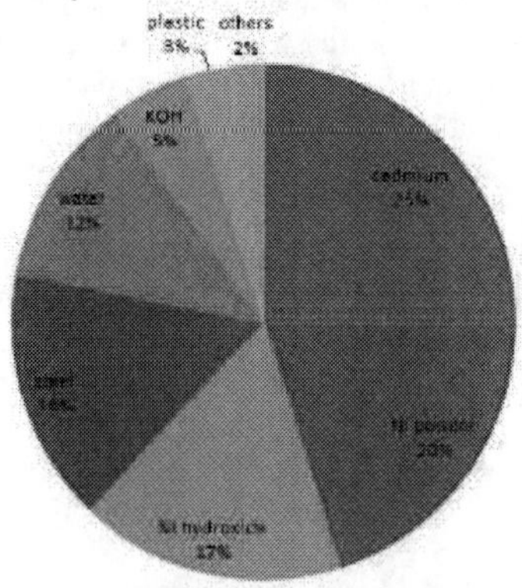

Figure 2. Frequency distribution of Percentage weights of different components in a Nickel Cadmium battery

Energy consumption is calculated in terms of the amount of electricity used, which is later converted in terms of cost, that is rupees, incurred for the development of the product. And Energy payback is calculated, in terms of energy input(E), (Wh), density of discharge(d), (%/100), mass of the battery(m), (kg), Specific Energy Density(w), (Wh/kg). Total energy input is the summation of energy consumption related to production, manufacturing, transport, use, miscellaneous operations. The energy output from the battery during its life time depends on the Specific energy density and the depth of discharge of the battery.

$$\text{Energy payback}=\frac{E}{w*m*d} \qquad \text{................} \ (1)$$

Secondly, life cycle inventory, the resource input and emission evolved from a system is considered and quantified, the Figure 3 shows the way of considering the inventory analysis of the system. It helps to identify the different raw materials and the corresponding quantity associated with the product. That is inventory analysis collects information on input and output, that is, environmental exchangers, of all the processes within the boundaries of the product system. The compilation of inventory data for each individual process quantifies he input and output associated with the reference flow of the products as derived from the functional unit. For example, the data is typically presented as the total amount of emission of substance X evolved during the product development process. Processes considered in the LCA are:

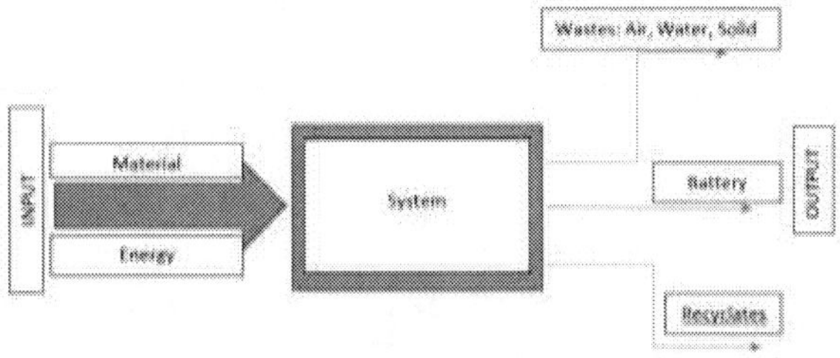

Figure 3. Boundaries assumed for the cradle-to-gate study evaluation

The function specificity is a fundamental characteristics of life cycle inventory and the ensuing impact assessment, and consistent with the purpose of LCA to evaluate the environmental impacts.

2.2 Inventory Analysis

Some of the steps used to calculate the manufacturing cost and energy consumption are explained as,

1.KOH(electrolyte): 268 kg of KOH is used to prepare 1kl of electrolyte, which can be used to produce 21428 number of batteries which requires 2.19 MJ/Kg.

2. Nickel: 0.8kg of $Ni(OH)_2$ is used for 140 Ah. Cost of $Ni(OH)_2$ cost Rs 900/Kg. electricity cost is taken as Rs 6.7/units. Converting 'Kwhr' into MJ/Kg, by multiplying with '3.6'. Equating energy consumed during product development process with the cost of electricity, gives 120.375 MJ/Kg for the battery considered.

3. Distilled water: 1.85 l of distilled water is used per battery. A litre of distilled water is assumed to cost Rs 20. It gives total cost associated with distilled water in the battery as Rs 3.03.

4.Use: The battery is assumed to operate for 2000 cycles. And assuming domestic cost of electricity/unit as Rs 5 therefore total cost of use/battery as Rs 1680.

5. Transportation: Assuming the product developed in Hyderabad to be sent to United States of America, where cost of transferring 2 kg is $ 51. Therefore for a battery weighing 3.2kg, cost about Rs 4896, assuming the conversion rate as Rs 60 for $1

6. Cadmium: About 0.7567kg of cadmium, costing 275RS/kg, is required for 1 battery considered. And the product consuming 34.875 MJ/Kg.

And there are some miscellaneous operations in accordance to the product development stages of the product in the industry, which sum up a total energy consumption in accordance to the development of the battery considered as an average of 2341.2064 MJ/Kg. and total life cycle cost of the battery a 10973.275Rs on an average. The Table1 shows the production and manufacturing energy associated with the battery considered.

Table 1. Production Energy and Manufacturing Energy associated with different materials used in the production of battery

Material	Production Energy(MJ/Kg)	Manufacturing Energy(MJ/Kg)
Cadmium	193	34.875
Ni sulphate-->Ni Hydroxide	90.6	120.375
Recycled Ni	37	96.48
KOH	38.2	2.19
Steel	37.2	1.507
Polypropylene	80	0.2

The production energy associated with the battery is obtained from the literature survey. Manufacturing energy is obtained from an industry leading the sector, in order to get a realistic point of view in accordance to the manufacturing sector, where the manufacturing energy associated with polypropylene is assumed as 0.2 MJ/Kg, after consulting with an engineer in the same industry that we considered to calculate manufacturing cost.

The production energy associated with the battery is obtained from the literature survey. And manufacturing energy is obtained from an industry leading the sector, in order to get a realistic point of view in accordance to the manufacturing sector, where the manufacturing energy associated with polypropylene is assumed as 0.2 MJ/Kg, after consulting with an engineer in the same industry that we considered to calculate manufacturing cost.

2.3 *IMPACT ASSESSMENT*

Environmental impact assessment is used to identify possible environmental impacts that the product will cause on to the environment. It helps in identifying the toxic gases evolved, solid and liquid wastes produced during it's lifecycle. It helps to identify the potential area of improvement and the environment implications that the changes will create on the environment. It helps in the decision making process of the management in deciding the sector where the highest amount of investment is required. The step also helps to design and re-design the product in terms of social, cultural and aesthetic perspectives. Thereby the input energy and cost requirements of the product can be reduced drastically. The step need to be implemented in the planning stage itself to improve the efficiency and to reduce the cost incurred in developing the product.

It addresses impacts produced due to evolution of toxic gases and other pollutants, such as global warming. Eutrophication of water bodies due to solid and liquid wastes, loss of natural resources including renewable and non-renewable, impacts on human health due to noise and radiation when exposed during operation. Environmental exchanges of all different various processes through which the product moves during it's life cycle is collected, such as inputs required, waste produced, recyclables, output product expected in each processes. The steps to correlate the emissions evolved in each process into the environmental impacts produced. Such as gases like carbon dioxide, methane has ability to absorb infrared radiation. Here the components such as volatile organic compound(VOC), carbon monoxide(CO), Nitrogen oxide(NO_2), Nitrous oxide(N_XO), particulate matter(PM), sulphur oxide, methane, carbon dioxide(CO_2), heavy metal such as cadmium, cobalt, nickel will be released during their entire life of the product. The quantity of the above said components are obtained from the literature survey, which is used to find the impact potential. The impact potential helps to identify the impact of these components creating on the environment. Table 2 shows the abbreviation of different gases evolved during the life cycle of the battery.

IOP Conf. Series: Materials Science and Engineering 1123 (2021) 012022 doi:10.1088/1757-899X/1123/1/012022

Table 2. Abbreviations of different gases evolved during the life cycle of Nickel Cadmium Battery

CO	Carbon Monoxide
CO2	Carbon Dioxide
CH4	Methane
PM	Particulate Matter
N2O	Nitrous oxide
NOx	Nitrogen oxides
SOx	Sulphur oxides
Cd	Cadmium
Ni	Nickel
Co	Cobalt

Table 3. Different impact potentials calculated in accordance to the battery

	G/KG	**Resource Depletion Potential**	**Global Warming Potential**	**Acidification Potential**	**Eutrophication Potential**	**Photochemical Potential**	**Human Toxicity**
Volatile Organic Compound (VOC)	0.9	-	9.9	-	-	0.0063	
CO	3.8	-	-	-	-	-	0.0456
NOX	11.1	-	-	7.77	1.443	-	8.658
PM	18.3	-	-			-	-
SOX	19.8	-	-	19.8	-	-	23.76
CH4	14.9	-	417.2	-		0.1043	-
N2O	0.1	-	26.5	-	-	-	-
CO2	9900	-	9900	-	-	-	-
Cd, Co, Ni	0.1	1.6072E-09	-	-	-	-	0.0057
Total		1.0672E-09	10353.6	27.57	1.443	0.1106	32.469

Where, CO means Carbon Monoxide, CO_2 means Carbon Dioxide, CH_4 means Methane, PM means Particulate Matter, N_2O means Nitrous oxide, NO_x means Nitrogen oxides, SO_x means Sulphur oxides, Cd means Cadmium, Ni means Nickel, Co means cobalt.

Different types of impact potentials are used to evaluate the product in terms of design for environment. The Table 3 shows the different potentials calculated for the battery considered during it is lifecycle using the equation 2,3,4,5,6,7. Some of them are discussed below:

a) Global warming potential (GWP):

The amount of CO2, N2O, CH4 and VOCs evolved during the gate-to-cradle processes are observed and its multiplied with their respective GWP potentials to obtain corresponding GWP. It's a factor measuring the amount of heat a greenhouse gas traps in the atmosphere up to a specific time horizon, relative to carbon dioxide.
where, ec2,k is the GWP factor of 'k' gas, and Bk is the quantity of gas in 'Kg'.

b) Human toxicity potential (HTP):
It is a quantitative toxic equivalency potential (TEP) that has been introduced to express the potential harm of a unit of chemical released into the environment. HTP includes both inherent toxicity and generic source-to-dose relationships for pollutant emissions. HTP is calculated by summing up the toxic releases in all different media, such as air, water and soil.

where *ec7,k*A, *ec7,k*W, and *ec7,k*S are human toxicological classification factors for the effects of the toxic emission to air, water and soil, respectively. *Bk*A, *Bk*W and *Bk*S represent the respective emissions of different toxic substances into the three media.

c)Photochemical Oxidants Creation Potential(POCP):
Photochemical oxidants creation potential (POCP), is usually expressed relative to the POCP classification factors of ethylene and is calculated as:
B(k) are the emission of VOCs. *ec6,k* is their photochemical oxidation factor.

d)Eutrophication Potential
Eutrophication is characterized by dense algal and plant growth owing to increased concentration of chemical nutrients needed for photosynthesis. Eutrophication potential (EP) is defined as the potential to cause over-fertilisation of water and soil, which can result in increased growth of biomass. It is calculated as:
where *Bk* is the emission of NO*x*, N, PO4 and COD and *ec5,k* are their respective eutrophication potentials. EP is expressed relative to PO_4^{3-}.

e)Acidification Potential
Acidification potential (AP) is based on the contributions of SO2, NO*x* , HCl, NH3. to the potential acid deposition, *i.e.* on their potential to form H+ ions. AP is calculated
where *ec4,k* represents the acidification potential of gas *k* expressed relative to the AP of SO2, and *Bk* is its emission in kg *per* functional unit.

f) Resource Depletion potential
Abiotic resource depletion includes depletion of nonrenewable resources, *i.e.* fossil fuels, metals and minerals. Abiotic resources are defined as natural sources (including energy sources), such as iron ore and crude oil, which are regarded as "nonliving". Abiotic depletion is one of the impact categories to be taken into account in Life Cycle Assessment. It is also one of the most frequently discussed impact categories. The total impact is calculated as:
where *Bk* is the quantity of a resource used *per* functional unit and *ec1,k* represents total estimated world reserves of that resource.
Table 4 shows the abbreviation of different impact potentials calculated for the battery during the life cycle of the battery.

Table 4. Abbreviations of different impact potentials considered in this paper

RDP	Resource Depletion potential
GWP	Global warming potential
AP	Acidification Potential

EP	Eutrophication Potential
POCP	Photochemical Oxidants Creation Potential

RDP $$E1 = \sum_{K=1}^{K} B(k)/ec(1,k) \quad (2)$$

GWP $$E(2) = \sum_{K=1}^{K} ec(2,K).B(K) \quad (3)$$

AP $$E4 = \sum_{K=1}^{K} ec(4,k)B(k) \quad (4)$$

EP $$E5 = \sum_{K=1}^{K} ec(5,k)Bk \quad (5)$$

POCP $$E6=\sum_{K=1}^{K} ec(6,k)B(k) \quad (6)$$

HTP $$E7 = \sum_{K=1}^{K} ec(7,ka)B(ka)+\sum_{K=1}^{K} ec(7,kw)B(kw)+\sum_{K=1}^{K} ec(7,ks)B(ks) \quad (7)$$

Energy payback= $\frac{1637802}{50*3.2*0.8}$ = 12795.33

where,

Energy input, E= 1637802 (Wh), Specific Energy Density(w), (Wh/kg)= 50 (Wh/kg), mass, m=3.2 (kg), density of discharge, d=80 %. The energy payback is calculated based on equation 1.

The waste, recyclables from each processes will produce an impact on the environment in their own ways. Characterization of these impacts will be based on the corresponding impact potential score of each gases.

2.4 *INTERPRETATION*

Interpretation, is the fourth and final stage in the LCA of the battery. The interpretation depends mainly on the different impact potentials evaluated. It helps to identify the processes to be modified inorder to develop the product more effectively with respect to design for environment. Calculating the energy payback of the battery, helps to understand the effectiveness of the battery. Maximum about of carbon emissions are due to transportation, a more efficient method of transferring product from one place to another helps to decrease the carbon dioxide emissions drastically. Salvage value is not considered during the LCA study, considering salvage value can increase energy payback due to increase in energy consumption and cost incurred during the product development stages. Most of the industries are conducting energy audition periodically, usually between 6 months, through an external energy auditing consultancy. Having an internal audit team can help to change the process layout of the product development process in such a way that the energy audit process can be less cumbersome and more effective, showing realistic value. Energy auditing a plant with a process layout designed for environment helps to identify the potential area of impact accurately.

Condition monitoring helps to increase the reliability of the machines, by predicting the fault in the machine and its location of the fault which helps to identify the severity of the the fault in the machine element and thereby planning the maintenance accordingly to reduce the non-production time or idle time and improving the production process and rate of production, and the process as a whole which helps to decrease the manufacturing and related overhead costs, thereby helps to decrease the total cost incurred on the product and which in turn helps to increase the market competitiveness and market share of the product drastically. Marketing the product developed through a process, techniques, and materials which produce a less environment impact and burden compared to its counterparts helps to improve its acceptance among the consumers. It helps to compare the same with different kinds of batteries to observe the relative advantage and disadvantage. Environmental burdens and impacts produced during the cradle to gate processes of the battery. Understanding the marketing strategy, that need to be applied, inorder to get a larger market share. Considering a mid-size car with a mileage of 7.5 litres per 100 kilometres. Then impact produced by the battery weighing 3.2Kg equals impact of running the car for 123.0469 km.

3. Conclusion

The work is mainly focused on life cycle assessment of nickel cadmium battery for various applications. The various important data are collected based on real time industrial consumptions. The following important conclusions are drawn from the present study:
The total energy input required for the development of nickel cadmium battery is 1637802 (Wh). The different types of impact potentials are calculated and quantified in detail. The considerable impact of the emissions during the product development stages, such as global warming potential as 10353.6, human toxicity potential summed up to 32.469, photochemical Oxidants Creation Potential equals 0.1106, total Eutrophication Potential produced by the product is 1.443, acidification Potential equals 27.57, and total resource Depletion potential as $1.6*10^{-9}$ due to heavy metal extraction and emission during different product development stages. Considering a mid-size car with a mileage of 7.5 liters per 100 kilometers. Then impact produced by the battery weighing 3.2Kg equals impact of running the car for 123.0469 km.

Future work involves comparison of LCA analysis of different grade of batteries. Identification of potential area of improvement in the process and product development stages and infrastructure required to meet the needs and requirements according to the guidelines of design for environment.

Acknowledgement

Authors are thankful to HBL Power Systems Limited for providing key inputs during the LCA study.

References

[1]. EuropeanCommission, 2012a. Commission Staff WorkingDocument Accompanying the Document Council and the European Parliament, the Council, the European Economic and Social Committee and the Committee of the Regions e Renewable Energy: A Major Player in the European Energy Market SWD (2012) 164 Final. Brussels, 6.6.2012.

[2]. Naish, C., McCubbin, I., Edberg, O., Harfoot, M.,Outlook of Energy Storage Technologies. European Parliament, Policy Department Economic and Scientific Policy, 2008.

[3]. Beccali, M., Brunone, S., Cellura, M., Franzitta, V., Energy, economic and environmental analysis on RET-hydrogen systems in residential buildings. Renew. Energy 33, 2008, 366e382.
[4]. EC, Transport 2050: Commission outlines ambitious plan to increase mobility and reduce emissions, European Commission, Brussels, Press Release; 2011.
[5]. IEA, Global EV Outlook. Understanding the electric vehicle landscape to 2020. International Energy Agency, Paris, France; 2013.
[6]. Argonne national laboratory, A review of battery life-cycle analysis: State of knowledge and critical needs, 2010.
[7]. Jenkins, D.P.,Fletcher, J. & Kane, J. (2008), Model for evaluating impact of battery storage on microgeneration systems in dwellings. Energy Conversion and Management 49 (2008) 2413–2424
[8]. Royters. UPDATE 1-Sanyo to boost output of auto Lithium batteries-Nikkei, [Online] Available from: http://www.reuters.com/article/2010/12/23/sanyo-idUSSGE6BM0BF20101223 [accessed 23.08.11].
[9]. Beccali, M., Cellura, M., Longo, S., Nocke, B., Finocchiaro, P., LCA of a solar heating and cooling system equipped with a small watereammonia absorption chiller. Solar Energy 86 (5), 2012, 1491e1503.
[10]. Samaras, C.; Meisterling, K. Life Cycle Assessment of Greenhouse Gas Emissions from Plug-in Hybrid Vehicles: Implications for Policy. Environ. Sci. Technol. 2008, 42, 3170–3176.
[11]. Notter, D. A.; Gauch, M.; Widmer, R.;Wager, P.; Stamp, A.; Zah, R.; Althaus, H.-J. Contribution of Li-ion batteries to the environmental impact of electric vehicles. Environ. Sci. Technol. 2010, 44, 6550–6.

MECHANICAL PROPERTIES OF GRAPHENE REINFORCED ABACA FIBER COMPOSITES

Shreyash Singh[1], Harsh Khatri[2], Shobhit Deshlahra[3], T. Narendiranath Babu[4]

School of Mechanical Engineering, VIT University, Vellore, India

Abstract. The main aim of this research work is to study the mechanical properties of abaca natural fiber composite reinforced with graphene. Natural fiber composites seek a numerous application nowadays due to its high strength is to weight ratio, high impact resistance, thermal stability, and recyclability. The percentage compositions of graphene reinforcement in the matrix are varied. Alkali treatment of abaca fiber is done to enhance bonding at the fiber-matrix interface. Conventional hand layup method is used for fabrication of the composite material. Moulds were made according to ASTM standard to test its tensile properties. The result showed that incorporation of glass fiber increases the performance of abaca fiber composites. Natural fiber composites have many potential applications in the automotive, aerospace and construction industries.

1. INTRODUCTION

Natural fibers are substances which can be directly obtained from living sources such as plants or animals. These materials are then greatly elongated to be used into making of threads, filaments and ropes. Natural fibers are a sustainable resource and have numerous benefits such as

High specific strength which comes from low specific weight

A renewable resource and does not require high energy for its processing

A desirable fiber aspect ratio

Biodegradable and from an economical view, have a low cost per unit volume.

Natural fibers are abundantly available and they have a broad spectrum, some very well-known natural fibers are cotton, jute, hemp, jute, wool, silk and bamboo. There is a growing interest in natural fiber reinforced with composites as we aim towards sustainable development[1]. The sustainable materials are becoming increasingly popular in materials technology both in production phase and product end life as we need a method that does not affect the ecology. The increased use of natural fiber has reduced the dependency on synthetic fibers which has drastically reduced the greenhouse emissions[2]

Natural fibers are starting to find a place for them in the world of composites. Natural fibre composites have been recognized as a low-impact opportunity to other synthetic composites. The replacement of glass fibres with herbal fibres for reinforcement in polymer composites appears to be a contemporary phenomenon. NFCs have huge ability as construction substances and as a substitute of other structural substances (eg, GRP, steel, concrete) in many programs .Their numerous applications extend to architecture, automotive as well as aerospace. The use of natural fibers have been noticeably effective in reducing the greenhouse emissions [3]. High Density Polythene (HDPE) can be replaced by natural fibers, various reasons include natural fibers being low density, low cost and high specific properties characteristics. HDPE which is a petroleum-based synthetic fiber has caused irreversible damage to our environment due to them being non-biodegradable. Natural fibers are evolving rapidly and soon may replace metals and ceramics.

There have been numerous research conducted on natural fibers but very few have been found for abaca fiber. Abaca fiber is obtained from the leaf sheath from the trunk of Musa textilis. Abaca is fine and light beige in appearance. It has a high lignin content, it can be as high as 16%. It is best known for its high strength and long fiber lengths (up to 3m). Abaca has a huge potential as it can substitute glass fiber in various automotive parts and it is currently known better for producing paper products, twines, nets and fishing nets. Some recent studies also show the potential of abaca fiber to replace the man made fibers and many other natural fibers in various other applications. It has been known that the tensile strength increases as we increase the percentage of fiber reinforcement[4].The waxy residuals are removed[5] after the alkali treatment (5% NaOH)which enhances the surface wettability, resulting in enhancement of interfacial adhesion. The fibers are able to rearrange themselves along the direction of tensile deformation due to the alkali treatment performed which removes noncellulosic impurities and materials from the interfibriller region[6]. The property of the composite is highly dependent on interfacial adhesion. The fiber is made more hydrophobic by removal of hydroxyl coat and hence it absorbs less moisture.

For the tensile test we try a combination of fiber-epoxy-graphene reinforced resin as we have been coming across the numerous features of graphene such as a young's modulus of 1.0 +- 0.1 TPa for a thickness of 0.355mm[7]. A monolayer of graphene has a inartistic strength of 130GPa. It has a fracture toughness which is measured as a critical stress intensity factor of 4.0 +- 0.6 MPa. We have used the process of sonification to make sure that the graphene is dispersed completely into the epoxy. Sonification uses sound waves at a predetermined frequency and for our findings we used a frequency of 35Khz and our mixture was immersed in a water bath for two hours for the sonification process to take place. We have made four different percentages of graphene and epoxy mixture. 0%, 1%, 2%, 3% and 4%.

In this paper, the tensile behavior of fiber-epoxy reinforcement and the behavior of fiber-epoxy-graphene reinforced resin for different percentages of graphene are analyzed. The comparisons are done and varying results are concluded.

2. MATERIALS

Viscous epoxy resin and Polyamine hardener is used for making the matrix system. Abaca fibre (also known as Manila hemp) is used. Research grade Graphene (Purity>99%, diameter: 0.8-2 nm) is used for experimental purpose. All the materials used in this research work is shown in the figure.

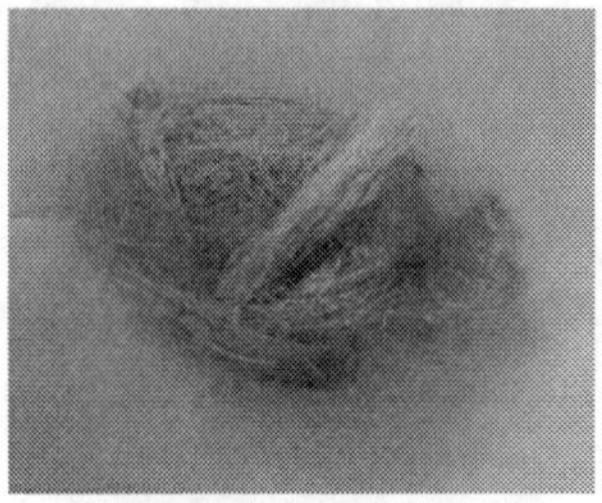

Figure 2.1 Abaca Fibre

Figure 2.2 Graphene

Figure 2.3 LY556 Epoxy Resin

Figure 2.4 Resin Hardener

3. Methods

a. Alkali Treatment of Fibre

About 300 gm of abaca fibre was treated in 1L 0.5% NaOH solution. The fibers were kept in alkaline solution for about 15 minutes at ambient temperature. The alkali treated fibers were then washed with distilled water. Then fibers are cured in room temperature for about 72 hours as shown in figure 2(b). Alkali treated fibers composites have better mechanical properties compared to composites containing untreated fibers.[8]

$$\text{Fiber} - \text{OH} + \text{NaOH} \rightarrow \text{Fiber -O- Na} + \text{H2O}$$

b. Preparation of Mould

Moulds were made according to ASTM standards D638 and D790 using wood as shown in the figure

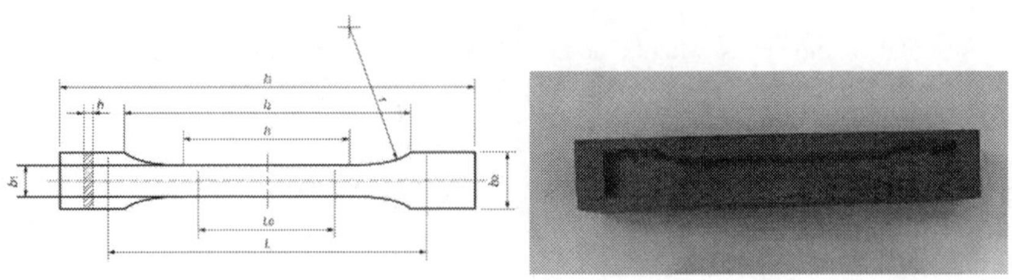

Figure 3.2. ASTM D638 standard mould

c. Ultrasonification

Ultrasonifaction of Graphene powder in epoxy resin is done prior to the fabrication of composites. An ultrasonic frequency was used to agitate Graphene particles in the epoxy solution. The research grade graphene with a purity of 99 wt.% was used for this research work. Concentration of 0-04% in the incremental steps of 1% by the weight % of the epoxy matrix was used. The Resultant mixture and hardener were mixed in the ratio 10:1. Dispersion of graphene in resin was done using sonication method as shown in the Figure 3.3. For further dispersion the mixture was mixed mechanically till it becomes homogeneous.

Figure 3.3: Ultrasonification Process.

d. Preperation of Composites

Steps involved in preparing the Graphene reinforced Abaca fibre composites are as follows
The alkali treated fibers are randomly arranged such that they form uniform strings.
Fibers are cut according to the dimension of ASTM D638 mould.
Epoxy-Hardener is mixed in the ratio of 10:1(%Vol) for minimizing the fiber pullout, this also increases the fiber matrix interface.
Fiber composites are prepared using a conventional hand lay-up method, followed by a compression molding technique. Composition of graphene is varied as shown in the table. The composites are varied as shown in the Figure 4.

Table 3. Sample Composition

Sample	Sample Composition	Alkali Treatment
S1	Epoxy + Abaca Fibre + 0% Graphene	5% NaOH Treated Abaca fiber
S2	Epoxy + Abaca Fibre + 1% Graphene	5% NaOH Treated Abaca fiber
S3	Epoxy + Abaca Fibre + 2% Graphene	5% NaOH Treated Abaca fiber
S4	Epoxy + Abaca Fibre + 3% Graphene	5% NaOH Treated Abaca fiber
S5	Epoxy + Abaca Fibre + 4% Graphene	5% NaOH Treated Abaca fiber

4. Experimental Procedure

The Specimens were made according to ASTM D638 standard and the tensile test was conducted on Digital tensile testing machine INSTRON 8801, as shown in Figure 5.2

The specimens were securely clamped by the top and bottom grips attached to the tensile testing machine as shown in the Figure 5.2.

During the tension test grips are moved apart at constant rate of 10mm/min to stretch the specimen.

The gage length of the tensile specimen was 80mm.

The force acting on the specimen and its displacement was analyzed and graph was plotted until the specimen fails.

After a certain load, the specimen breaks into two pieces as shown in the Figure 5.3.

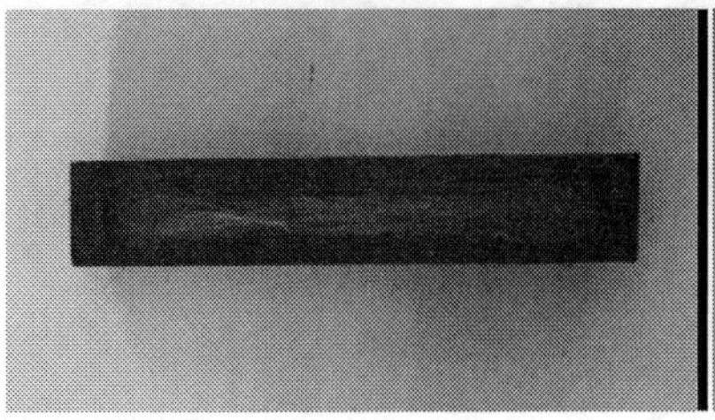

Figure 4.1 0% Graphene S1

Figure 4.2 1% Graphene S2

Figure 4.3 2% Graphene S3

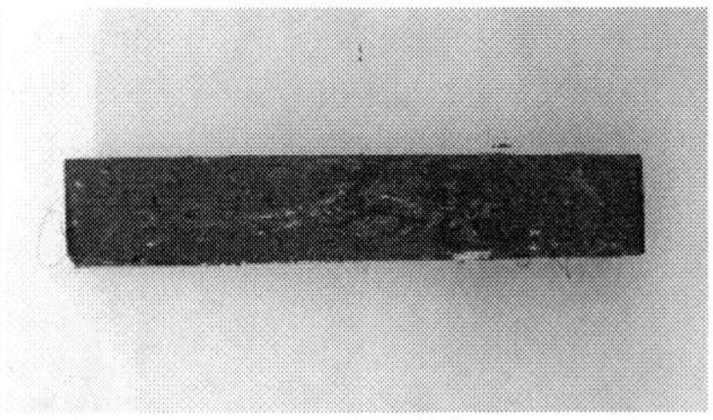

Figure 4.4 3% Graphene S4

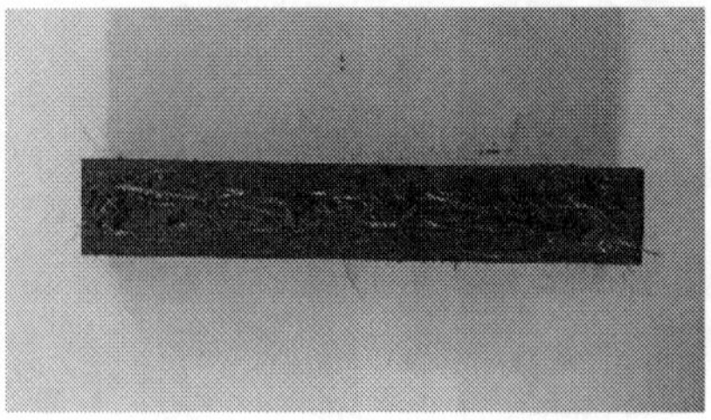

Figure 4.5 4% Graphene S5

Table 4. Specimen Characteristics for Tensile Testing

Characteristics	Particulars
Size	165mm x 20mm x 5mm
Epoxy Resin used	LY556
Resin Hardener	Polyamine Hardener(HY951)
Natural Fibre	Alkali Treated Abaca Fibre
Graphene %	0%, 1%, 2%, 3%, 4%

5. FABRICATION & TESTING OF COMPOSITES

Composites were made using convention hand layup Technique

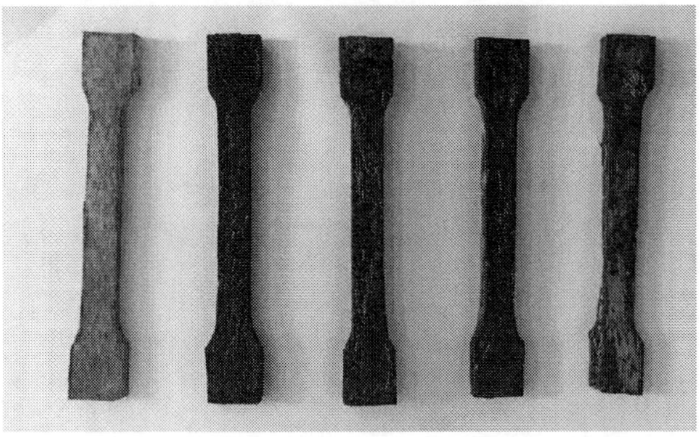

Figure 5.1 Test Specimen containing 0%,1%,2%,3%,4% Graphene Concentration respectively.

Figure 5.2 Tensile Testing was carried out in digital tensile machine INSTRON 8801.

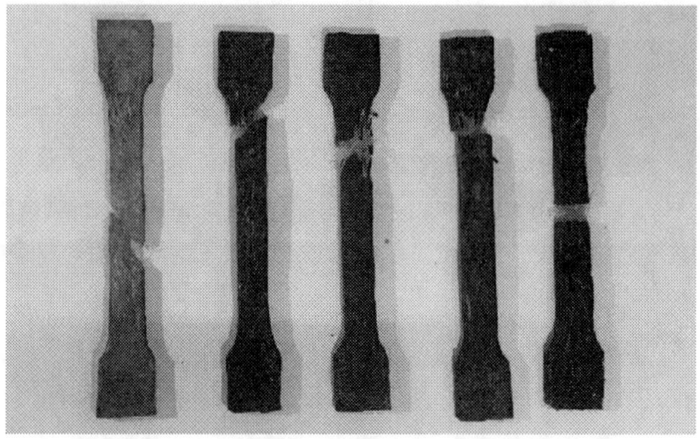

Figure 5.3 Tensile Test specimen after the test was carried out.

6. Result and Discussion

The load-displacement curves for the Abaca Fibres, composites are obtained from digital tensile testing machine and are shown in the following figures. The utmost stress is in accord with its maximal load. The concept of tensile strength used in this study, is defined as the maximum stress on the stress-displacement curve, being directly propositional to the maximum load.

It was observed that as the percentage of Graphine was increased by 1% in the ASTMD638 made of Abaca Fiber, Epoxy and Graphene, the strength of the composite increased.

The tensile strength of composite with 0% of Graphine was 402 N which is 10 % less than the Graphene with 1%. The maximum value of stress in case of Abaca Fiber-Epoxy-Reinforce Composite (Uniaxial Fiber Arrangement) , containing 4% Graphene, was 1045 N.

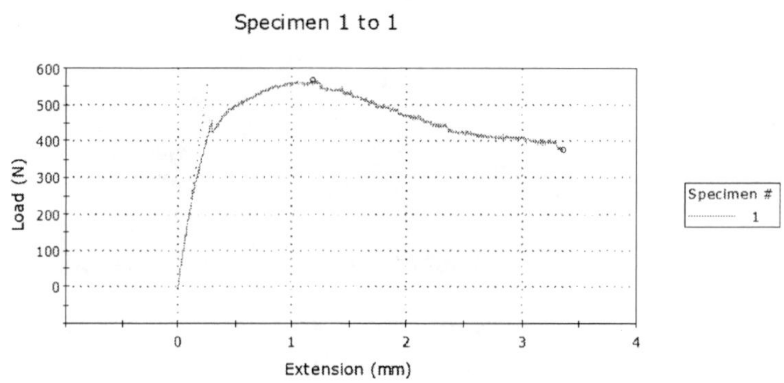

Figure 6.1 Load v/s Extension graph for 0% Graphene reinforced abaca fiber composite.

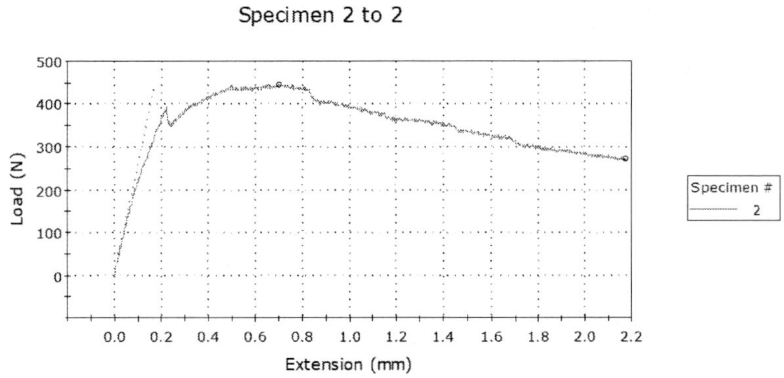

Figure 6.2 Load v/s Extension graph for 1% Graphene reinforced abaca fiber composite.

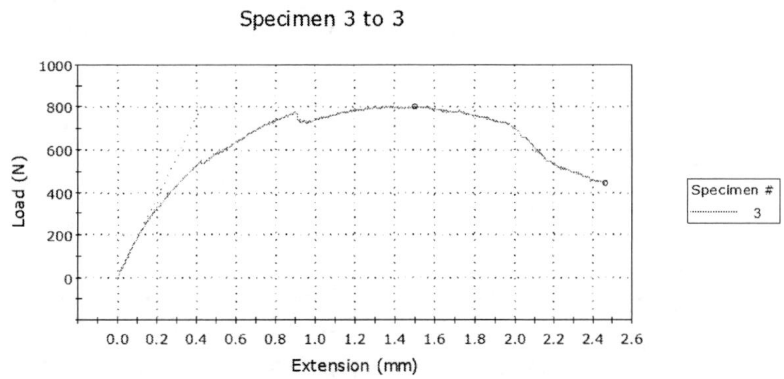

Figure 6.3 Load v/s Extension graph for 2% Graphene reinforced abaca fiber composite.

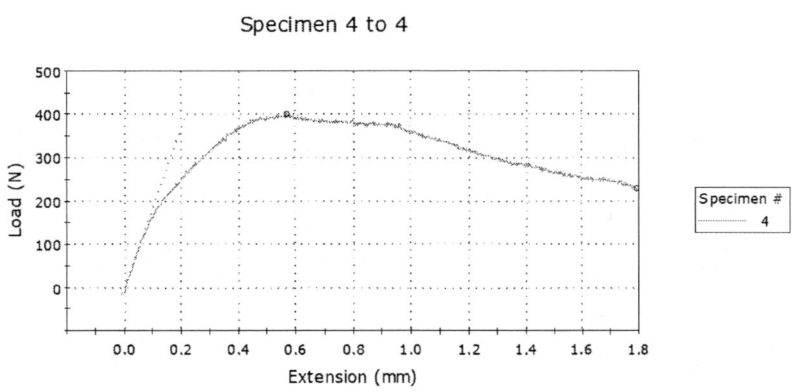

Figure 6.4 Load v/s Extension graph for 3% Graphene reinforced abaca fiber composite.

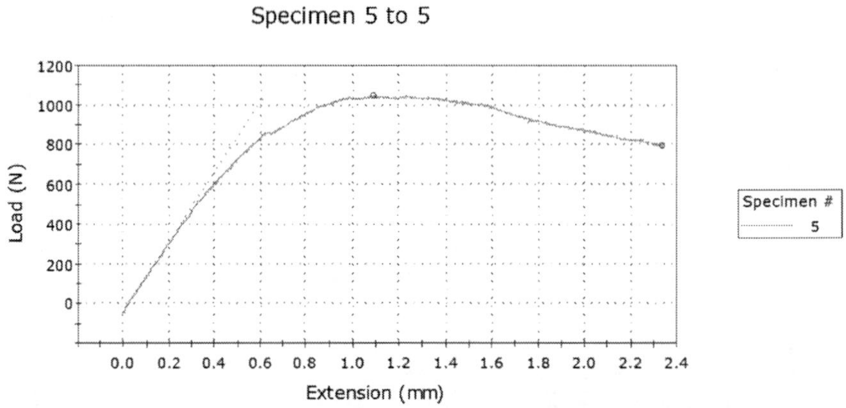

Figure 6.4 Load v/s Extension graph for 4% Graphene reinforced abaca fiber composite.

Table 6. Summary of experimental results of tensile test.

	Maximum Load (N)	UTS (GPa)	Modulus (Automatic Young's) (MPa)	Tensile strain at Break (Standard) (%)
Sample 1	402.65322	0.005	2293.250	1.792
Sample 2	447.75009	0.005	3135.018	2.174
Sample 3	570.11843	0.007	2525.061	3.359
Sample 4	805.49717	0.010	2284.226	2.461
Sample 5	1045.6919	0.013	2116.578	2.339
Maximum	1045.6919	0.013	3135.018	3.359
Mean	654.34217	0.008	2470.827	2.425
Minimum	402.65322	0.005	2116.578	1.792

7. Conclusion

The ASTM D638 specimen was prepared using, abaca fiber reinforced with epoxy and Graphene and its tensile strength was examined. It was evident that fibers treated with alkali sodium hydroxide (NaOH) indicate better mechanical properties as it enhance bonding at the fiber-matrix interface. Out of all the combinations used in the experiment with graphene, the composite with 4% Graphene had the maximum tensile strength. From this result, it can

be inferred that, Graphene increases the tensile strength of the composite, approximately by 15% (if increased by 1%each time). According to the application theory, the arrangement of the Uniaxial, provides the maximum tensile strength. An average of 20% increase was indicated in the experiment, after adding 1% graphene in each composite.
The composite created in the study shows its utilization in almost all the areas, apart from the area of medical sciences. It can be used as a substitute in the automobile Industry.

8. References

[1] S. T. Dadami, S. Matteppanavar, and I. Shivaraja, "Dielectric studies of Graphene and Glass Fiber reinforced composites Dielectric studies of Graphene and Glass Fiber reinforced composites," 2018.

[2] N. B. Thamba, T. Kucloury a, M. K. Jain, and R. V Mangalaraja, "Flexural properties of areca nut , sunn hemp and e-glass fibers reinforced with epoxy composites Flexural properties of areca nut , sunn hemp and e-glass fi bers reinforced with epoxy composites," 2019.

[3] T. N. Babu, T. Kuclourya, M. K. Jain, and R. V Mangalaraja, "Mechanical Behaviour of Areca Nut , Sunn Hemp Natural Fibers and E-Glass Fibers Reinforced With Epoxy Composites," no. 6, pp. 1092–1101, 2019.

[4] K. Korniejenko, M. Masri, I. Prabowo, J. N. Pratama, and M. Chalid, "Graphene Reinforced Natural Fiber Nanocomposites for Structural Applications Graphene Reinforced Natural Fiber Nanocomposites for Structural Applications," 2018.

[5] A. K. Sinha, H. K. Narang, and S. Bhattacharya, "ScienceDirect Effect of Alkali Treatment on Surface Morphology of Abaca Fibre," *Mater. Today Proc.*, vol. 4, no. 8, pp. 8993–8996, 2017.

[6] F. Sarker, P. Potluri, S. Afroj, V. Koncherry, K. S. Novoselov, and N. Karim, "Ultra-High Performance of Nano-Engineered Graphene-Based Natural Jute Fiber Composites."

[7] K. B. Reddy, A. Sai, K. Reddy, S. Bharath, S. Chand, and P. B. Kumar, "MECHANICAL PROPERTIES OF BAMBOO FIBRE WITH GRAPHENE AS A FILLER MATERIAL IN," vol. 119, no. 7, pp. 989–1000, 2018.

[8] Chikkol, Srinivasa. (2014). Abaca Fiber Reinforced Epoxy Composites: Evaluation of Impact Strength. International Journal of Sciences: Basic and Applied Research (IJSBAR). 18. 305-317.

[9] Stankovich, Sasha & Dikin, D. & Dommett, Geoffrey & Kohlhaas, Kevin & Zimney, Eric & Stach, Eric & Piner, Richard & Nguyen, Sonbinh & Ruoff, Rodney. (2006). Graphene-Based Composite Materials. Nature. 442. 282-6. 10.1038/nature04969.

[10] Bhardwaj, Sandeep & Kumar, Rajender & Sharma, Rupali. (2014). Natural Fibre Composites and Its Potential. Popular Kheti. 2. 230-232.

[11] Westman, Matthew & Fifield, Leonard & Simmons, Kevin & Laddha, Sachin &

Kafentzis, Tyler. (2010). Natural Fiber Composites: A Review. 10.2172/989448.

[12] Zini, Elisa & Scandola, Mariastella. (2011). Green Composites: An Overview. Polymer Composites. 32. 10.1002/pc.21224.

MODELLING AND SIMULATION OF AN UPPER HAND PROSTHETIC

Gaurav Mathur [1], Aman Chaure [2], Devyani Biswas[3], Dr. Raghuraman DRS[4]

[1] School of Mech Engg, Vellore Institute of Technology, Vellore, India, gauravmathgm@gmail.com

[2] School of Mech Engg, Vellore Institute of Technology, Vellore, India, amanchaure@gmail.com

[3] School of Mech Engg, Vellore Institute of Technology, Vellore, India, biswas.devyani24@gmail.com

[4] School of Mech Engg, Vellore Institute of Technology, Vellore, India, drsraghuraman@vit.ac.in

Abstract. This paper deals with the mechanical aspects that are taken into consideration while constructing a prosthetic arm. In the beginning, the design of the arm has been discussed at length. This section includes both -the part wise and the overall design made on modeling software. This is followed by explaining the movement of the arm. The parameters have been discussed in the section of the forward kinematics of the arm. The reverse engineering used in the design process is discussed. The material used has been covered towards the end. Finally, all these concepts have been put to use in a prototype developed by this team whose description is provided at the end of the paper.

1.INTRODUCTION

When an arm is lost, prosthetics play an important role in rehabilitation. An artificial arm improves the mobility of several amputees and aid in implementing daily activities. Prosthetics have evolved significantly from being fashioned from iron to wood and currently to a strong and durable lightweight material such as carbon fibre or plastics. In this paper, we have discussed about the material used in making of the prosthetic fore arm keeping in mind the importance of it being light in weight and durability for ease of use. The movements of the arm have been included in the robotics section of the paper, which comprises of forward and reverse kinematics. The dynamics section describes the effect of the subsequent movements and the possible failures the arm could encounter in real life. The final prototype is made to undergo a series of destructive and nondestructive tests to ensure durability.

2. DESIGN

The First design was based on the reverse engineering of a real hand with a digital Vernier robotics planning as well as the finite element analysis of the entire assembly and subassemblies. The bending in the x & y directions was analysed with the multiple simulations and the dimensions were iterated to get a sufficient factor of safety at the locations were the stress was concentrated.

2.1 Design of palm

Surface:

By the careful observation of the real palm contour, a replicated design was created to for the same in order to get a sufficient gripping action by the natural design curvature of the palm. This was done by merely taking the imprint of the palm over a clay and then taking the dimensions of the same using Vernier Calliper very carefully. The Figure 1 & Figure 2 shows the design of the palm contour and the wireframe design for the same. The fingers were designed such that they have a self-locking capability which constraints their motion after a certain limit in the backwards direction.

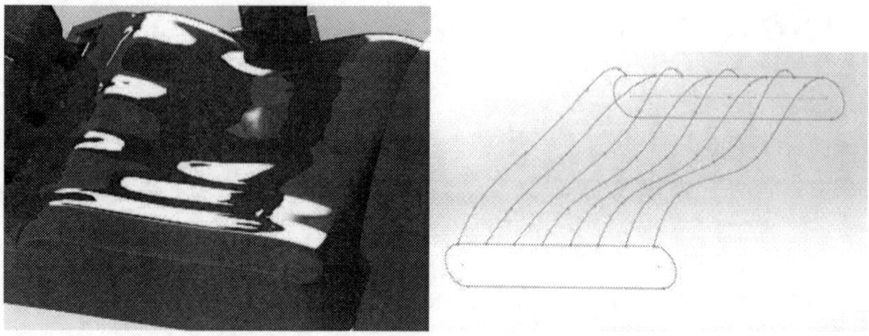

Figure 1. The Palm Contour & Wire Frame Structure

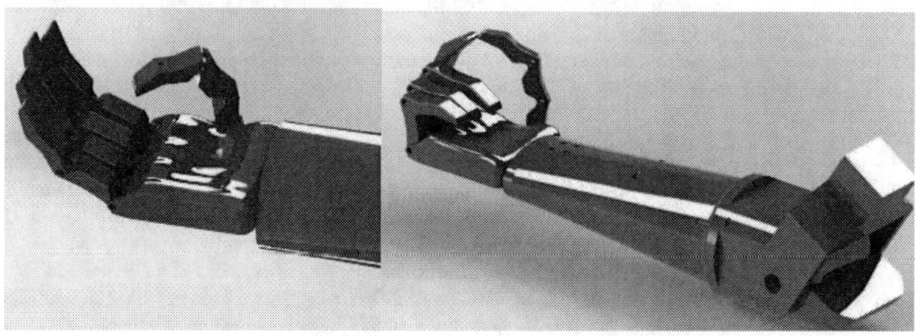

Figure 2. Overall Design of the hand

2.2 The Design of Thumb:

IOP Conf. Series: Materials Science and Engineering 1123 (2021) 012024 doi:10.1088/1757-899X/1123/1/012024

The Thumb is designed to have a maximum functionality with 2 degrees of freedom, which is then controlled by two motors, one having control over the bending of the upper two parts of the thumb and the other having control over the lower part for bending it transversely for reaching maximum points and thus having a bigger work volume. The figure shows the design of the thumb connected to the palm.

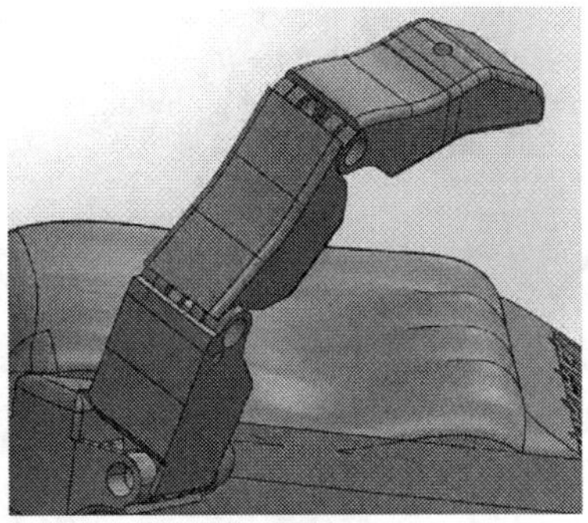

Figure 3. Thumb Design.

2.3 The design of Spring for reverse actuation:

The internal holes are designed to actuate the fingers with thread in the closing direction. The counter action of opening the fingers is achieved with the help of the loaded torsional springs. The simulation for determining the spring coefficient was determined with the help of SOLIDWORKS motion simulation. The Fingers were fixed under gravity along with a torsional spring torque acting on the joint to properly suspend their load.

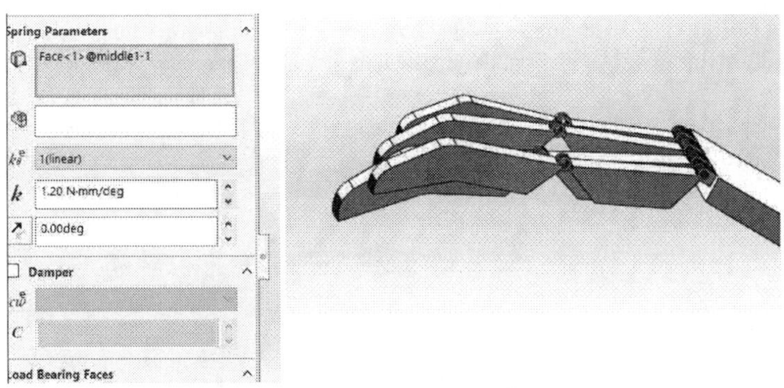

Figure 4. Spring Parameters Calculation

2.4 The design of Internal holes for Actuation:

The curved holes are designed such that they act maximum normal force can be generated by the tension in the thread created by the motors without hindering the design strength properties and proper control over the fingers. These curved holes are designed with the considerations of their additive manufacturability, the importance of this was that such a mechanism can only be 3D printed and cannot be manufactred by any other means. The threads pass through these holes and when sufficient tension is generated by the motors, the fingers close by cancelling the reverse torque of the torsional strings.

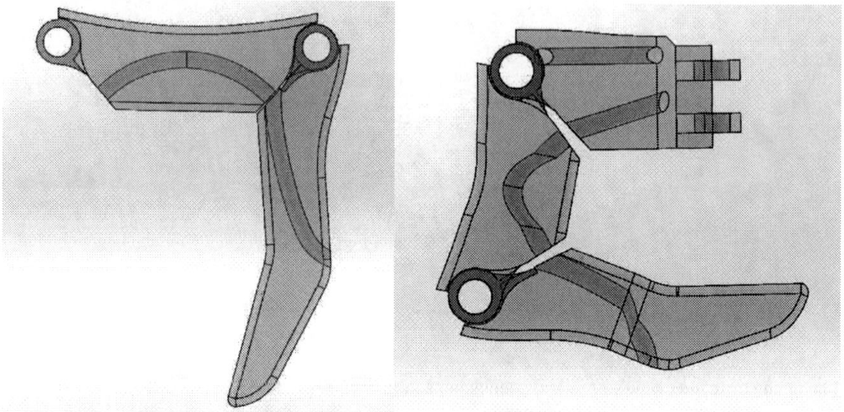

Figure 5. Finger Design(Index &Thumb)

2.5 The Design of palm:

The palm is having the most complex design for this product. The best possible design for accomodating five rotatory motors was developed along with its strength bearing capabilities and least possible weight with sufficient factor of safety. Palm also have several internal holes for the passage of 6 threads for 5 different fingers, one for each finger and two for thumb. All these holes are not intersecting and their one end opens at the finger joint and the other at the motor head. Some curvature was given for accomodating them in the lesser space and without interference with each other.

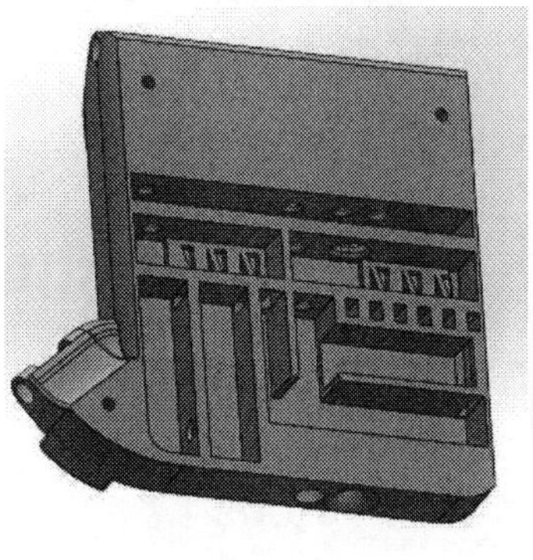

Figure 6. Finger Design(Index &Thumb)

Each hand finger consists of Two links and 1 degree of freedom. The four primary fingers are programmed to move in a way given in image below Figure 1. The actuation of the fingers is designed in such a way that their motion is tabulated in Table 2

Table 1. Link Lengths of the Fingers

Sr. No.	Name of the Finger	Link 1(mm)	Link 2(mm)	Link 3(mm)
1.	Thumb	36	36	50
2.	Inde	70	50	
3.	Middl	76	55	
4.	Rin	67	46	
5.	Littl	57	43	

Table 2. Rotation of Joints

Sr. No.	Simultaneous States	Joint1(Link 1 and Link2)	Joint 2(Link 2 and Main body) (Degrees)
1.	State	0	0
2.	State	30	0
3.	State	70	40
4.	State	70	70
5.	State	90	90

The Figure below (Figure 2) depicts the work volume of the Thumb designed. The thumb has two degrees of freedom and is designed to grip the objects from the opposing side.

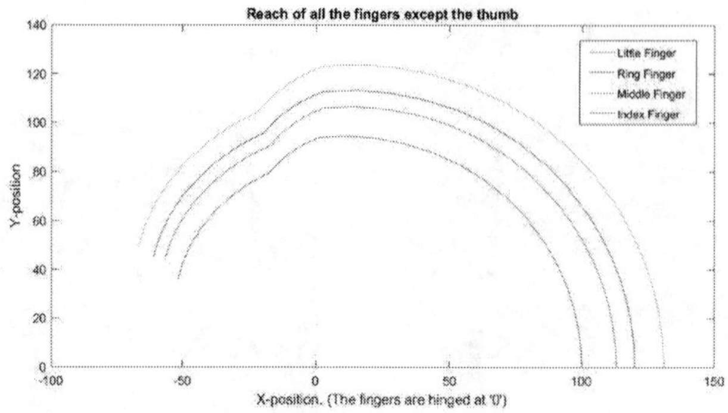

Figure 7. Image Depicting Forward Kinematics of all Fingers

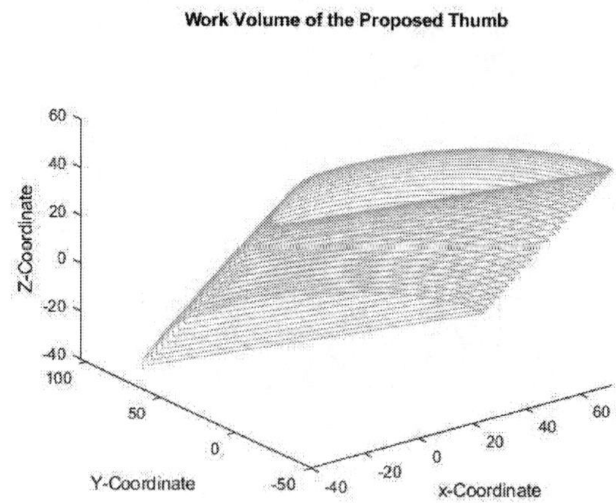

Figure 8. Work Volume of the Thumb

2.6 Finite Element Analysis:

Multiple Finite element Static Analysis were conducted for Multiple Combinations of the Fingers and Palm after which it was found that there was a sufficient factor of safety for the design for its stability and sustainability under the loads and a log life of fatigue.

Table 3. Mesh Information:

Sno	Mesh type	Solid Mesh
1.	Mesher Used:	Standard mesh
2.	Automatic Transition:	Off
3.	Include Mesh Auto Loops:	Off
4.	Jacobian points	4 Points
5.	Element Size	4.86577 mm
6.	Tolerance	0.243289 mm
7.	Mesh Quality Plot	High

IOP Conf. Series: Materials Science and Engineering 1123 (2021) 012024 doi:10.1088/1757-899X/1123/1/012024

8.	Total Nodes	40971
9.	Total Elements	24207
10.	Maximum Aspect Ratio	33.083
11.	% of elements with Aspect Ratio < 3	83.7
12.	% of elements with Aspect Ratio > 10	0.306
13.	% of distorted elements(Jacobian)	0

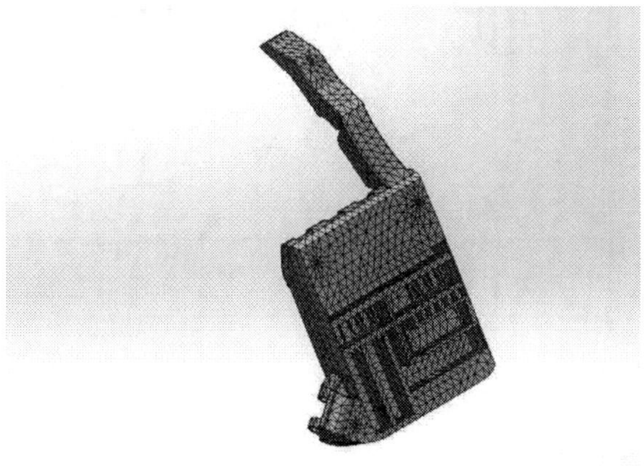

Figure 9. Meshed Image

Table 4. Material Properties

Model Reference	Properties
	Name: **Nylon 6/10** Model type: **Linear Elastic Isotropic** Default failure criterion: **Max von Mises Stress** Yield strength: **1.39043e+08 N/m^2** Tensile strength: **1.42559e+08 N/m^2** Elastic modulus: **8.3e+09 N/m^2** Poisson's ratio: **0.28** Mass density: **1,400 kg/m^3**

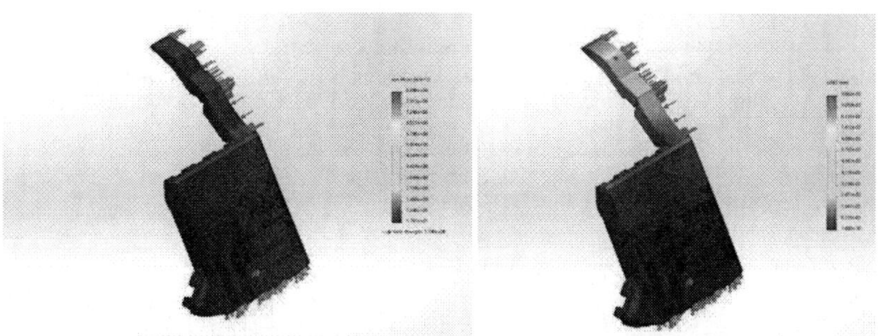

Figure 10. Images Depicting Various loads and Deflection due to them.

Table 5. Parameters of the Simulation:

Sim No:	TYPE	COMPONENTS	FORCE (N)
1	BENDING X	RING12 + PALM	10
2	BENDING X	INDEX12+PALM	10
3	BENDING X	MIDDLE12+PALM	10
4	BENDING X	LITTLE12+PALM	10
5	BENDING (+Y)	RING12 + PALM	10
6	BENDING (+Y)	INDEX12+PALM	10
7	BENDING (+Y)	MIDDLE12+PAL	10
8	BENDING (+Y)	LITTLE12+PALM	10
9	BENDING (-Y)	RING12 + PALM	10
10	BENDING (-Y)	INDEX12+PALM	10
11	BENDING (-Y)	MIDDLE12+PALM	10
12	BENDING (-Y)	LITTLE12+PALM	10

REFERENCES

Journal paper

[1]. H. B. Wang, Kinematics of a Five Degree-of-Freedom Prosthetic Arm, Mechanism and Machine Theory, 1 October 1998, Volume 33, Issue 7, Pages 895-908

Conference papers

[2]. Lokesh Saharan, Robotic Hand with Locking Mechanism Using TCP Muscles for Applications in Prosthetic Hand and Humanoids, Las Vegas, Nevada, United States Proceedings Volume 9797, Bioinspiration, Biomimetics, and Bioreplication 2016; 97970V (2016)

[3]. George P. Kontoudis, Open-Source, Anthropomorphic, Under actuated Robot Hands With a Selectively Lockable Differential Mechanism: Towards Affordable Prostheses, 2015 IEEE/RSJ International Conference on Intelligent Robots and Systems (IROS), Oct 2, 2015

Websites

[4]. Charles A. Harper (editor), 1999, Polyamide (PA, Nylon) 6/10 (610) [Online] Available from: https://www.makeitfrom.com/material-properties/Polyamide-PA-Nylon-6-10-610

[5]. Mechanics of Solids, ANSYS Workbench Guide, [Online] Available from: http://web.aeromech.usyd.edu.au/AMME5981/Course_Documents/files/Tutorial%20Week%203c%20-%20MECH3361%20Workbench%20Guide.pdf

[6]. ANSYS Training Manual, Static Structural Analysis[Online] Available from: http://mguler.etu.edu.tr/WB-Mech_120_Ch04_Static.pdf

IOP Conf. Series: Materials Science and Engineering 1123 (2021) 012025 doi:10.1088/1757-899X/1123/1/012025

OPTIMIZATION OF PROCESS PARAMETERS FOR PEAK TEMPERATURE DURING FRICTION STIR WELDING OF ALUMINUM ALLOY

Gurunath Shinde [1, 2], Ganesh Jadhav [2], Vahid Jamadar[2], Sarafaraj Mulani[2], Rachayya Arakerimath[1]

[1] Mechanical Engineering Department, G.H.Raisoni College of Engg. and Mgmt,Pune,India, Email:gurunathshinde@yahoo.com
[2] Mechanical Engineering Department, Dr.Daulatrao Aher College of Engg.,Karad, India

Abstract. Friction stir welding is a solid state joining method used to join materials which cannot be joined by fusion welding. In this joining method, heat produced due to the friction between tool and workpiece is used to weld workpiece material. Heat input depends upon amount of friction caused by process parameters such as rotational speed, weld speed, tool geometry, axial force etc. In the present study, effect of friction stir welding process parameters such as shoulder to pin diameter ratio, rotational speed and weld speed on peak temperature obtained during friction stir welding of aluminum alloy were investigated using taguchi L9 orthogonal array. ANOVA result shows that shoulder to pin diameter ratio (D/d) is the most significant factor affecting on peak temperature obtained during the FSW. **Keywords:** Friction Stir Welding, Peak Temperature, Taguchi L9 array.

1.INTRODUCTION

Friction stir welding is a solid state joining process used for wide applications with different aluminium alloys. Many researchers have successfully studied joining of different aluminium alloys for various applications such as automotive, aerospace, defence, structural, marine, transportation etc. The main principle of friction stir welding is the joining of materials using heat generation though friction, stirring and solidification after cooling. The amount of heat required should not reach melting point of the joining materials but it should be enough to reach plasticity zone of the joining the materials.Fig.1 explains working principle of friction stir welding process which shows FSW tool with shoulder and small pin plunged over the weld line of joining materials rigidly fixed. Rotational speed causes materials stirring action while weld speed gives joining direction and friction time. Amount of friction which causes peak temperature formation depends upon various process parameters such as rotational speed, weld speed, axial load, tool geometry etc. Shinde et al[8] explained friction stir welding can be performed on low cost simple machine tools like universal milling machine. Shinde et al [7] studied effect of various tool geometries and tool materials on friction stir welding and tool failures. Shinde et al [5,6] described effect of process parameters on heat input required for friction welding.Shinde et al[5] discussed reliability based design approach for development of FSW fixture to overcome issues of thermal instability, defect formation, better weld strength etc.

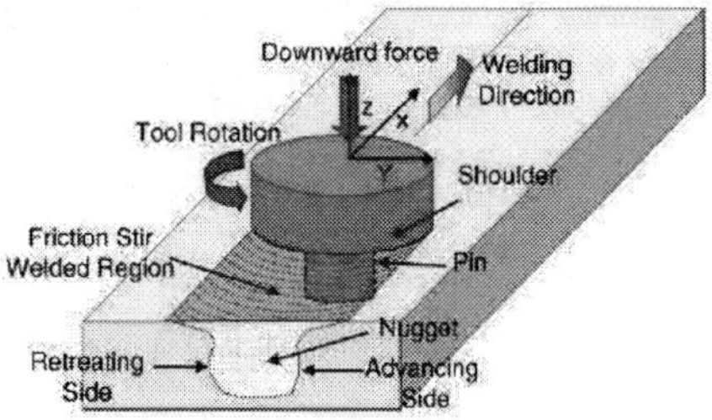

Figure 1. Working Principle of Friction Stir Welding

2. EXPERIMENTAL SET UP

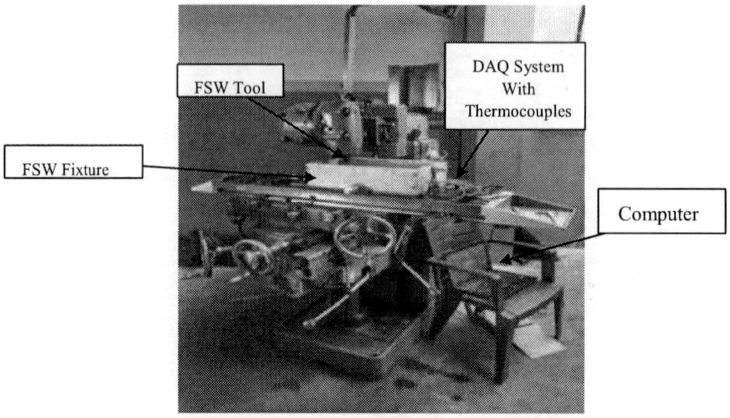

Figure 2. Experimental set up of Friction Stir Welding

Fig.2 shows experimental set up of friction stir welding developed on G Defour 221 N universal milling machine. It consists of FSW fixture to rigidly fix joining materials to avoid misalignment during the joining.FSW Tool was fixed in the tool adapter attached to the spindle head. Computer with Data Acquisition System was used along with K type thermocouples to measure temperature evolved during the joining. The machine has provision to vary process parameters such as rotational speed, weld speed, tilt angle, plunge depth etc. In this study we have investigated effect of three significant process parameters such as D/d ratio, rotational speed and weld speed on the peak temperature formation and weld quality. Aluminium AA3103 H12 alloy with 150mmX75mmX2mm was used for the study.FSW tools with straight cylindrical pin made up of H13 tool steel with 56 HRC hardness with D/d ratio of 4.38,5 and 5.62 were used during the study. Taguchi l9 orthogonal array was used as optimization method to study significance of process parameters on the FSW weld. Table 1 shows factors and their level for the taguchi L9 orthogonal array.

3. RESULTS AND DISCUSSION

The effect of process parameters such as D/d ratio, RPM and Weld Speed were studied from main effect plots of S/N Ratio and Means. S/N ratio and Means were calculated using nominal the better criteria to find objective of most suitable peak temperature required to produce better weld.Fig.3 shows FSW trials of Al-Al carried out as per the taguchi L9 orthogonal array. Maximum peak temperature of 365.5^{0}C at Al Advance Side and 364.75 at Al Retreating Side were obtained with trial no.9 and trial no.7 respectively.

Minimum Peak temperature of 244^{0}C at Al advance side and 231^{0}C at Al retreating side were observed with trial no.3.The objective of this study is to find out most suitable value of peak temperature during friction stir welding.

Higher temperature obtained during the friction stir welding may lead to melting of materials to be joined while lower temperature will lead to insufficient material stirring resulting into large number of defects. Hence we have used nominal the better criteria for carrying out the study. Some defects like flash, surface line, voids were also observed due to the thermal instability and improper material flow.

Table 1. .Factors and their Level

Factors	**Levels**		
	1	**2**	**3**
D/d	3.75	4.38	5
RPM	710	900	1180
Weld Speed	16	21	28 mm/min

The S/N ratio for Nominal is best is:

$$S/N = 10 \log_{10}\left(\frac{y^2}{s^2}\right)$$

Table 2. Taguchi L9 Orthogonal Array

Factors	**D/d**	**RPM**	**WS Mm/min**	**TPeak at Adv**	**TPeak at Retr**	**S/N Ratio**	**Means**
1	3.75	710	16	308	303.25	39.18	305.63
2	3.75	900	21	334.5	321.75	31.22	328.13
3	3.75	1180	28	231	244	28.24	237.50
4	4.38	710	21	314.25	290.25	25.01	302.25
5	4.38	900	28	266.75	254.5	29.57	260.63
6	4.38	1180	16	324	297.25	24.31	310.63
7	5	710	28	364.75	342.75	27.14	353.75
8	5	900	16	337.75	330	35.70	333.88
9	5	1180	21	363.25	365.5	47.20	364.38

Fig.2. Block Diagram of Temperature Measurement

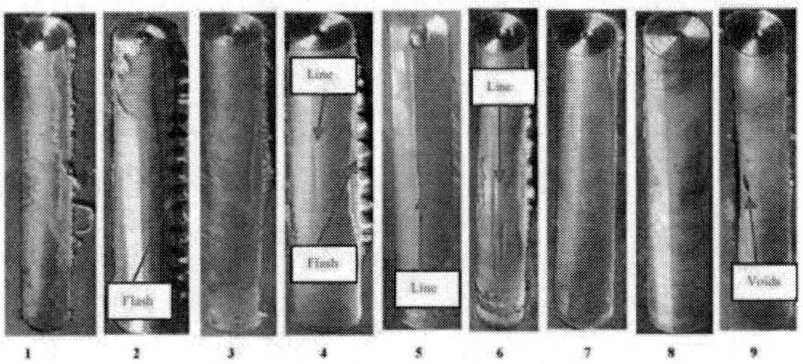

Figure 3. FSW of Al-Al as per Taguchi L9 Method

Rao et al [1] explained optimisation of process parameters for temperature using taguchi L9 orthogonal array method using Minitab. Kalvala et al [2] observed that the microstructural degradation in P91 welds can be controlled by low temperature friction stir welding technique.Gadakh et al[3]explained heat input modelling during friction stir welding using tapered pin tool profile.Biswas et al [4] studied effect of tool geometries on thermal histories during friction stir welding of AA1100. Tool geometry with concave shoulder and conical pin was found to be preferable for FSW of AA 1100 Analysis of Mean and S/N ratio gives optimum level of process parameters for most suitable peak temperature during friction stir welding.The Mean and S/N ratio calculations for each process parameter at all levels were performed using Minitab 18 software. Fig.4.a and Fig.4.b shows main effect plots for S/N ratio and Means. It was observed that D/d ratio is the most significant process parameter affecting peak temperature on both advancing and retreating side. Equation no.1 and 2 represent regression equation for peak temperature during friction stir welding on advancing side and retreating side respectively.

IOP Conf. Series: Materials Science and Engineering 1123 (2021) 012025 doi:10.1088/1757-899X/1123/1/012025

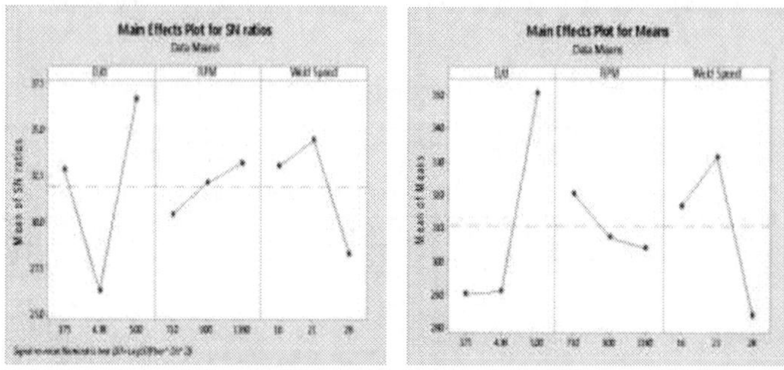

(a) (b)

Figure 4. Main Effect Plots for (a) SN Ratio (b) Means

Table 3. Response Table

S/N Ratio				Means			
Level	**D/d**	**RPM**	**Weld Speed**	**Level**	**D/d**	**RPM**	**Weld Speed**
1	**32.88**	30.44	**33.06**	**1**	290.4	320.5	**316.7**
2	26.30	**32.16**	34.48	**2**	**291.2**	**307.5**	331.6
3	36.68	33.25	28.32	**3**	350.7	304.2	284.0
Delta	10.38	2.81	6.16	**Delta**	60.3	16.4	47.6
Rank	1	3	2	**Rank**	1	3	2

Table 4. ANOVA Test Results for Advancing and Retreating Side

Advancing Side						Retreating Side					
Source	DF	Adj SS	Adj MS	F-Value	P-Value	**Source**	DF	Adj SS	Adj MS	F-Value	P-Value
D/d	2	7088.1	3544.0	2.13	0.320	**D/d**	2	7543.2	3771.59	4.69	0.176
RPM	2	829.0	414.5	0.25	0.801	**RPM**	2	196.7	98.36	0.12	0.891
Weld Speed	2	3959.8	1979.9	1.19	0.457	**Weld Speed**	2	3193.2	1596.59	1.99	0.335
Error	2	3329.2	1664.6			**Error**	2	1607.7	803.86		
Total	8	15206.1				**Total**	8	12540.8			

TPeak at Adv Side ^0C = 316.0 - 24.9 D/d_3.75 - 14.4 D/d_4.38 + 39.2 D/d_5.00 + 13.0 RPM_710- 3.0 RPM_900 - 9.9 RPM_1180+ 7.2 Weld Speed_16 + 21.3 Weld Speed_21- 28.5 Weld Speed_28..(1)

TPeak at Retr Side ^0C =305.47 - 15.8 D/d_3.75 - 24.8 D/d_4.38 + 40.6 D/d_5.00 + 6.6 RPM_710- 3.4 RPM_900 - 3.2 RPM_1180+ 4.7 Weld Speed_16 + 20.4 Weld Speed_21- 25.1 Weld Speed_28..(2)

Table 5. Model Summary

Advancing Side				Retreating Side			
S	R-sq	R-sq(adj)	R-sq(pred)	S	R-sq	R-sq(adj)	R-sq(pred)
40.7994	78.11%	12.42%	0.00%	28.3524	87.18%	48.72%	0.00%

4. CONCLUSIONS

In this study most suitable peak temperature were investigated for better weld quality during friction stir welding. Higher values of peak temperature will lead to melting or defects like line or flash in the weld while lower values of peak temperature may lead to the insufficient stirring of material which produces defects like voids, tunnel, crack etc. The process parameters D/d, RPM and Weld Speed were used for carrying out the study. The S/N ratio and Means were calculated by nominal the better criteria. Maximum peak temperature of 365.5^0C at Al Advance Side and 364.75 at Al Retreating Side were obtained with trial no.9 and trial no.7 respecively. Minimum Peak temperature of 244^0C at Al advance side and 231^0C at Al retreating side were observed with trial no.3.From response table of S/N ratio and Means, D/d ratio was found to be the most significant process parameter affecting peak temperature values on both advancing and retreating side

REFERENCES

[1]. R.Venkateswar Rao, Dr.M. Senthil Kumar,E. Shirisha "Optimize The Process Parameters Of Friction Stir Welded AA2014 & AA6061 Dissimilar Aluminium Alloys By Minitab Software " IOSR Journal of Mechanical and Civil Engineering (2016) PP. 66-71

[2]. Prasad Rao Kalvala,Javed Akram, Mano Misra, Damodaram amachandran, Janaki Ram Gabbita, "Low temperature friction stir welding of P91 steel" Defence Technology 12 (2016) 285–289.

[3]. Vijay Gadakh, Kumar Adepu, "Heat generation model for taper cylindrical pin profile in FSW" J mater res tech no l.(2013);2(4):370–375.

[4]. P. Biswas, N. R. Mandal, "Effect of Tool Geometries on Thermal History of FSW of AA1100" The Welding Journal,(2011) PP.129-S-135-S

[5]. Gurunath Shinde, Sarafaraj Mulani, Pradip Gunavant, Abhijeet Suryawanshi, (2018) "Experimental Investigation of Friction Welding on Aluminium AA5083 Alloy" International Conference on Energy and Environment.NIT Calicut Kerala.

[6]. Gurunath Shinde, Prakash Dabeer, (2017) "Review of Experimental Investigations in Friction Welding Technique" International Conference on Science and Engineering for Sustainable Development Bankok, Thailand,373-384.doi:10.21013/jte. ICSESD201736.

[7]. Gurunath Shinde, Prakash Dabeer,(2017)"Perspective of friction stir welding tools" ICMMM-2017, VIT, Vellore in Materials Today: Proceedings 5:13166–13176.

[8]. Gurunath Shinde, Prakash Dabeer, (2017) "Low-Cost Friction Stir Welding: A Review" ICAAMM-2016, Materials Today: Proceedings 4: 8901–8910.

[9]. Shinde G.V., Arakerimath R. (2020), "Reliability-Based Design Approach for Development of Friction Stir Welding Fixture. In: Gunjan V., Singh S., Duc-Tan T., Rincon Aponte G., Kumar A. (eds) ICRRM 2019 – System Reliability, Quality Control, Safety, Maintenance, and Management. ICRRM 2019. Springer, Singapore.

OPTIMIZING A PIEZOELECTRIC ENERGY HARVESTING SYSTEM

Anant R. Voleti [1], Shivam K. Singh [1], Raghuraman D. R. S [1, c*], Niranjan Vijayakumar [1], Tanushree D [2] and Balan A.S.S [1]

[1] School of Mechanical Engineering, Vellore Institute of Technology, Vellore, India

[2] School of Electrical Engineering, Vellore Institute of Technology, Vellore, India

[c] drsraghuraman@vit.ac.in

Abstract. The piezoelectric effect, through which some materials can convert mechanical stress to electrical energy and vice versa, sees a wide array of applications due to its ease of implementation and effectiveness at smaller scales. Piezoelectric materials have emerged in applications such as structural health monitoring and piezo-based smart structures. One such application is energy harvesting; piezoelectric materials can be used to provide power for micro scale devices, in part due to their high efficiency. However, the significantly lower power output of these transducers limits their applicability. This study focuses on comparing different configurations of a piezoelectric energy harvesting system for a motor. By modelling several configurations, this study aims to identify the configuration with the highest sustainable power output. **Keywords:** Energy Harvesting, Piezoelectric materials, Power generation, Vibrations

1. INTRODUCTION

Energy harvesting is the process of harnessing energy from ambient sources and storing the energy for future use. This is an old concept and has been achieved using windmills and water turbines. Energy harvesting did not gain enough attention because of the high-power requirements and the low energy conversion efficiencies of the harvesters; however, it has now gained significant attention. This is mainly due to the global movement towards green energy from ambient sources. There are three major mechanisms for energy harvesting, electromagnetic, piezoelectric and electrostatic harvesting [1]. Of these piezoelectric mechanisms show minimum material requirements with better energy conversion efficiencies.

Piezoelectric materials are a type of material that develop an electric field when they are stressed or deformed because of vibrations. This is known as the direct piezoelectric effect. Conversely, when an electric voltage is applied across the material, it deforms and vibrates [2]. The direct piezoelectric effect is of prime importance to the process of energy harvesting using piezoelectric materials. It is this mechanism that allows a piezoelectric material to absorb mechanical energy from a vibrating structure in contact and generate electrical energy from it. This is a class of energy harvesting called vibration – based energy harvesting. The harvester consists of a cantilever beam made up of layers of the piezoelectric material and one non – piezo material (shim) that provides strength and stiffness. This is connected to the vibrating host structure and an alternating voltage is induced in the electrodes surrounding the piezo layers due to the dynamic strain from the host structure. A large range of piezoelectric materials can be used to create the harvester.

1.1 Modelling Techniques

Erturk et. al. [3] presented a modelling method of the harvester beam based on the analytical modal analysis method, which is an accurate distributed parameter model. Initially, early models used a lumped parameter single degree of freedom structure, however it was later shown that these models provided very imprecise results for distributed parametric systems [4]. A more precise model for the distributed parameter system has been made using the Rayleigh Ritz formulation and then the Hamiltonian Principle to get the mass and stiffness matrices in the transformed space. [5]. A great improvement was made in the analytical modelling of the harvester by Erturk et. al. [6] by applying analytical modal analysis to the Euler – Bernoulli model of a unimorph with uniform section without a tip mass as shown in figure 1. The piezoelectric relations were added to precisely account for the coupling. The same technique is also applied to a bimorph. These methods are, however, approximate and validated over only a small frequency range. These parameter models are also based on a simple resistive impedance.

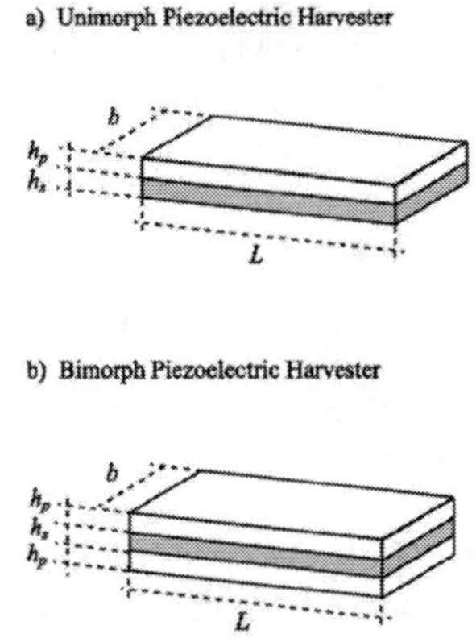

Figure 1. a) A unimorph and b) bimorph piezoelectric harvester [7].

1.2 Geometric Configurations

Various configurations of the energy harvester have been tested to find out the maximum energy harvesting capability. Different parameters are tested such as the tuning the resonant frequency of the piezoelectric harvester or changing the number of piezoelectric material layers in the harvester. Maximum power is generated when the resonant frequency of the harvester matches that of the input vibrations. There is considerable loss in power generated when the two frequencies vary [8]. The energy transformative ability of the piezoelectric material also depends on the coupling mode that the harvester uses. Two practical modes exist, which are the –31 mode and the –33 mode [9]. These are shown in fig. 2. The –31 mode is less efficient in coupling than the –33 mode, however the former is more efficient for energy harvesting in low intensity vibrations since it provides more strain in the harvester [10]. The –33 mode generates greater power output for high intensity vibrations condition.

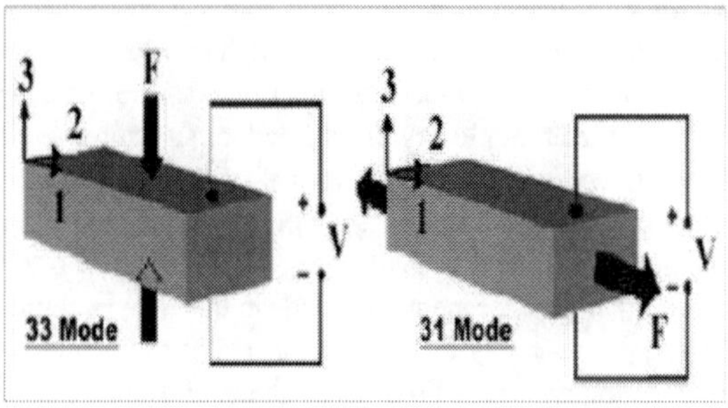

Figure 2. -31 and -33 piezoelectric coupling modes.

The shape of the harvester can also affect the power generation. The rectangular cantilever shape is most widely used by researchers of piezoelectric energy harvesters. However, using a trapezoidal shaped cantilever beam gives a continuous distribution of the strain throughout the structure unlike the rectangular shape which would contain a non-uniform strain distribution. The former configuration generated more power than the latter for the same volume of PZT material used [12].

2. METHODOLOGY

Piezoelectric elements generally consist of one or more layers of a piezoelectric material bonded to a shim. For the purposes of our study, we will consider a piezoelectric bimorph which consists of a stainless-steel shim, with PZT bonded to both sides of the shim. We aim to compare two separate configurations of the piezoelectric energy harvesting system.

The first configuration will consist of a single piezoelectric bimorph bonded to the top of the motor surface such that a sideways deflection of the motor (a deflection in the x-axis) will correspond to the 33 mode of the bimorph. The second configuration will consist of two bimorphs mounted on opposite sides of the motor, such that a vertical deflection of the motor (a deflection in the y-axis) will correspond to the 33 mode of the bimorph.

The bimorphs are the same thickness and width, but the length chosen is the optimal length for each configuration (70mm for a single bimorph and 60 mm for two bimorphs). These lengths were determined by directly optimising for maximum deformation at the base of the bimorph.

A preliminary modal analysis of the motor is conducted to obtain the mode shapes associated with the motor and its constraints. The motor is modelled as a steel stator and copper rotor spinning at 1500 rpm.

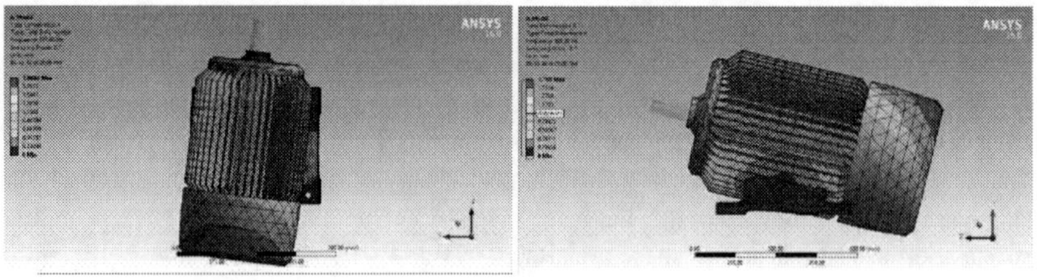

Figures 3 (a) (b). Different mode shapes of the motor.

IOP Conf. Series: Materials Science and Engineering 1123 (2021) 012026 doi:10.1088/1757-899X/1123/1/012026

Modal analyses of the two separate energy harvester configurations were then conducted to obtain the modes of vibration for each configuration, which can then be used to obtain the deformation frequency response to a specified vibration.

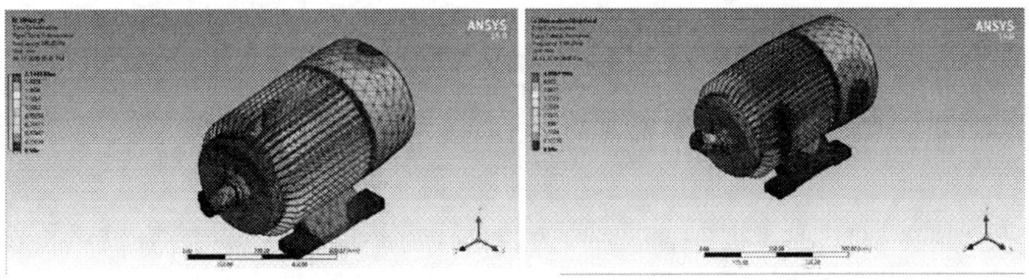

Figures 4 (a) (b). Comparison of mode shapes for a) single bimorph and b) double bimorph configurations.

We obtain the deformation frequency response of the base of the bimorph for a specified input vibration. For the same input vibration, the maximum base deformation of the bimorphs was 21.0 mm for the single bimorph configuration and 13mm for each bimorph in the two-bimorph configuration.

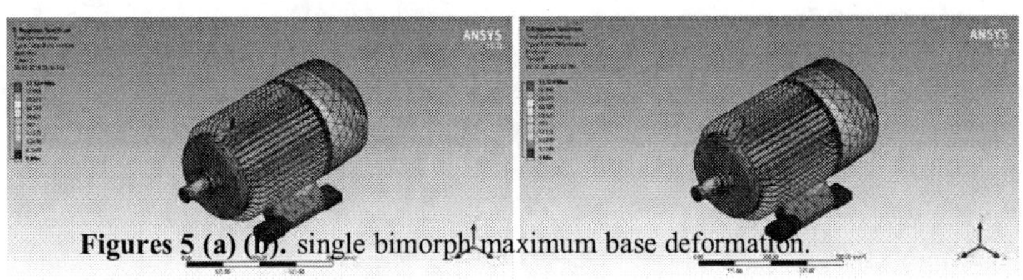

Figures 5 (a) (b). single bimorph maximum base deformation.

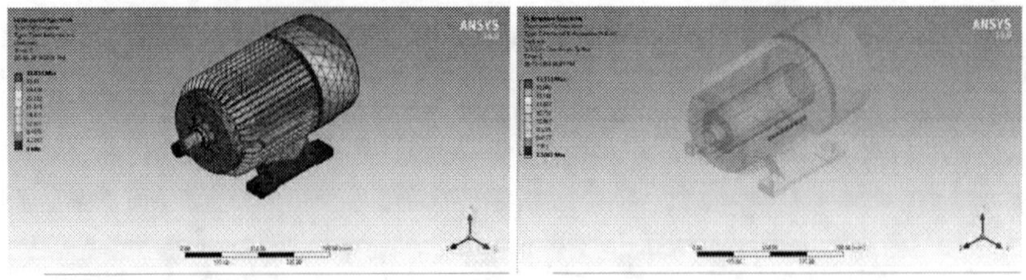

Figures 6 (a) (b). Double bimorph maximum base deformation.

The values of base deformation can be input into the transfer function to obtain the voltage generated by each configuration.

3. RESULTS

Given the same input vibration, the two-bimorph configuration marginally outperforms the single bimorph configuration, producing peak voltages of 41.5 V and 32.2V respectively. However, the model doesn't account for the damping effect of the piezoelectric elements and it can be assumed that the difference in voltage generated between the two configurations will be very small, making the first configuration more effective.

4. CONCLUSION AND FUTURE SCOPE

It can be concluded from the simulation study that there is no considerable voltage difference between that generated from the single bimorph configuration and the double bimorph configuration when accounting for damping due to electrical impedance. Therefore, accounting for vibration and the amount of material used, the single bimorph configuration is preferred.

The simulation can be further improved by considering the backward mechanical coupling. The study can be subsequently validated through experimentation. Considering the cost of lead zirconate titanate (PZT), a low-cost solution can be developed.

REFERENCES

[1] Williams, C. B., & Yates, R. B. (1996). Analysis of a micro-electric generator for microsystems. Sensors and Actuators, A: Physical, 52(1–3), 8–11.

[2] Anton, S. R., & Sodano, H. A. (2007). A review of power harvesting using piezoelectric materials (2003–2006). Smart Materials and Structures, 16(3), R1–R21.

[3] Erturk, A., & Inman, D. J. (2008). Issues in mathematical modeling of piezoelectric energy harvesters. Smart Materials & Structures, 17(6).

[4]Bonello, P., & Rafique, S. (2011). Modeling and analysis of piezoelectric energy harvesting beams using the dynamic stiffness and analytical modal analysis methods. Journal of Vibration and Acoustics, 133(1), 011009.

[5]. Lu, F. (2004). Modeling and analysis of micro piezoelectric power generators for micro-electromechanical-systems applications. Smart Materials and Structures, 13(1), 57–63.

[6] Erturk, A., & Inman, D. J. (2008). Distributed parameter electromechanical model for cantilevered piezoelectric energy harvesters. Journal of Vibration and Acoustics, 130(4), 041002–041002.

[7] Ruiz, R. O., & Meruane, V. (2017). *Uncertainties propagation and global sensitivity analysis of the frequency response function of piezoelectric energy harvesters. Smart Materials and Structures, 26(6), 065003.*

[8] Zhu, D., Tudor, M. J., & Beeby, S. P. (2010). Strategies for increasing the operating frequency range of vibration energy harvesters: A review. Measurement science & technology, 21(2).

[9] Zhu, D., Tudor, M. J., & Beeby, S. P. (2010). Strategies for increasing the operating frequency range of vibration energy harvesters: A review. Measurement science & technology, 21(2).
[10] Baker, J., Roundy, S., & Wright, P. (2005). Alternative geometries for increasing power density in vibration energy scavenging for wireless sensor networks. In Collection of Technical Papers—3rd International Energy Conversion Engineering Conference.
[11] Saadon S., Sidek O. B., & Hamad O. S. (2010). Vibration-based MEMS piezoelectric energy harvesters using cantilever beams. Journal of Optoelectronics and Advanced Materials, 4(8): 1219-1224.
[12] Roundy, S., Leland, E. S., Baker, J., Carleton, E., Reilly, E., Lai, E., et al. (2005). Improving power output for vibration-based energy scavengers. IEEE Pervasive Computing, 4(1), 28–36.
[13] MEMS and Nanotechnology Exchange. Material: Lead Zirconate Titanate (PZT), 2001. [Online] Available from: https://www.memsnet.org/material/ledzirconatetitanatepzt/ [Accessed 23 Nov 2019].

OPTIMIZING THE PROCESS PARAMETERS IN FRICTION STIR SPOT WELDING OF DISSIMILAR ALUMINUM ALLOYS USING GENETIC ALGORITHM

Suryanarayanan.R [1], V.G.Sridhar [2]

[1] School of Mechanical and Building Sciences, VIT University, Chennai, India, suryanaraynanr.2017@vitstudent.ac.in
[2] School of Mechanical and Building Sciences, VIT University, Chennai, India, vgsridhar@vit.ac.in

Abstract. Welding of dissimilar aluminum alloys by Resistance spot welding and Self-piercing rivets requires high energy consumption, increased cycle time, and costs. Friction stir spot welding (FSSW) is one of the best alternative welding techniques to join dissimilar aluminum alloys with the help of frictional heat and plastic deformation. However, the drawback of this process is the keyhole left behind by the tool. Hence this study focusses on avoiding the keyhole post welding process by the use of pinless tools to join thin sheets of Al 6061 and Al 5754. The influence of process parameters such as tool rotational speed, plunging speed, dwell time, plunge depth, and shoulder diameter was studied on the Tensile shear force of the joint. The optimal welding condition was predicted with the help of the Genetic algorithm. The confirmation tests showed that the predicted optimal conditions were in closer agreement to the experimental value.

1. INTRODUCTION

Joining of dissimilar aluminum alloys by conventional welding methods like Resistance spot welding (RSW) and Self-piercing rivets (SPR) requires high energy consumption, increased cycle time and costs. Hence an alternative welding technique like Friction stir spot welding (FSSW) is employed by the automakers to overcome the disadvantages of RSW and SPR [1]. The FSSW is a solid-state welding technique that joins similar and dissimilar materials with the help of frictional heat and plastic deformation. However, the keyhole left behind by the tool acts as a stress concentration factor and also acts as a corroding agent. To avoid this issue, Refill Friction stir spot welding (RFSSW) was developed to join weldments without producing the keyhole. However, this technique involves complex tool design, increased welding time and is suitable for joining thick plates [2]. For joining sheets with thickness <2 mm, pinless or flat tools are preferred. The use of pinless tools with and without features were reported to produce joints with higher load-bearing capacity when compared to that of joints welded with pin tools. Moreover, with the use of pinless tools also reduces the formation of the hook at the interface, which tends to act as the crack nucleation site and reduces the joint strength [3-4]. In pinless FSSW, the shoulder diameter and the shoulder plunge play a vital role in obtaining the joint with superior mechanical and metallurgical characteristics. The shoulder diameter and plunge depth help in inducing the plastic deformation and the flow of plasticized material flow to form stronger joints [5-6]. Various optimization techniques like Taguchi, Response Surface Methodology (RSM), Artificial neural network (ANN) and Fuzzy logic have been reported to find the optimal welding conditions in joining dissimilar aluminum alloys by FSSW [7-13]. Since there is scarce literature available in pinless FSSW of Al 6061 and Al 5754, in this paper, the optimal welding conditions are identified using the Genetic Algorithm (GA).

2. Experimental Methods

Al 6061 and Al 5754 alloys of thickness 1 mm, were placed onto the 10 mm thick mild steel backing plate and rigidly clamped onto the worktable. The FSSW machine of 30 kN axial force, 11 kW capacity, was utilized to join the dissimilar Al sheets with the help of pinless tools. Three pinless tools of H-13 steel with 52 HRC and different shoulder diameter were considered for this study. The chemical composition and mechanical properties of the alloys are listed in Table.1 and Table.2, respectively.

Table 1. Chemical composition of Al 6061 and Al 5754

Al 6061	**Element**	**Si**	**Fe**	**Mg**	**Cu**	**Mn**	**Cr**	**Zn**	**Al**
	Wt %	0.42	0.21	1.0	0.19	0.15	0.27	0.25	Bal
Al 5754	**Element**	**Si**	**Fe**	**Mg**	**Cu**	**Mn**	**Cr**	**Zn**	**Al**
	Wt %	0.25	0.31	3.2	0.02	0.38	0.23	0.18	Bal

Table. 2 Mechanical properties of Al 6061 alloys and Al 5754

Alloy	Yield Strength (MPa)	Ultimate Tensile Strength (MPa)
Al 6061	276	312
Al 5754	130	222

The experimentation sequence, Table.3, was done based on 5 factors, 3 levels Box-Behnken design, which is generated using Design Expert 11 software. Box-Behnken Design helps the experimenter to determine the second-order models instead of predicting the response values at the extreme points. The process parameters like tool rotational speed, plunging speed, dwell time, plunging depth, and shoulder diameter was considered for this study, as they were identified to affect the mechanical and metallurgical properties of the weldments from extensive experiments and literature survey. The mechanical testing of the weldments was done at a cross-head speed of 1 mm/min in a 50 kN capacity UTM machine. The Model-based calibration tool (MBC) in Matlab 2015 is employed for finding the optimal welding conditions by the Genetic algorithm, GA. The MBC toolbox utilizes the experimental sequence generated based on the Design of experiments to determine the optimal processing conditions.

Table 3. Experimental sequence with results

Ti	Tool Rotational Speed, A, RPM	Plunging Speed, B, mm/s	Dwell Time, C, s	Plunge depth, D, mm	Shoulder diameter, E, mm	Tensile shear Force, Ts, kN
1	1300	0.04	8	0.5	25	5.174
2	1000	0.04	4	0.75	20	3.979
3	1000	0.04	8	0.75	15	2.915
4	1000	0.02	8	0.5	25	3.484
5	1000	0.02	12	0.5	20	1.239
6	1000	0.04	4	0.25	20	2.525
7	700	0.04	8	0.75	20	4.589
8	1300	0.02	8	0.5	20	3.116
9	1300	0.04	4	0.5	20	4.460
10	1000	0.06	4	0.5	20	3.330
11	1000	0.04	4	0.5	25	5.704
12	1300	0.04	12	0.5	20	3.332
13	1000	0.04	12	0.75	20	4.898

14	1000	0.04	12	0.25	20	2.020
15	1000	0.04	8	0.25	25	2.330
16	1000	0.06	8	0.5	25	5.720
17	1000	0.04	8	0.5	20	2.617
18	1300	0.04	8	0.25	20	2.090
19	700	0.02	8	0.5	20	3.460
20	1000	0.04	8	0.5	20	4.840
21	1000	0.06	8	0.5	15	3.220
22	700	0.04	8	0.5	15	3.848
23	1000	0.04	8	0.5	20	3.193
24	1300	0.06	8	0.5	20	3.558
25	1000	0.02	8	0.5	15	3.190
26	700	0.04	4	0.5	20	3.850
27	1000	0.02	8	0.25	20	1.060
28	1000	0.04	8	0.5	20	2.613
29	1300	0.04	8	0.75	20	4.541
30	1000	0.04	8	0.75	25	6.262
31	1000	0.04	12	0.5	15	3.229
32	1000	0.04	8	0.5	20	2.650
33	1000	0.04	8	0.25	15	3.070
34	1000	0.06	8	0.25	20	3.300
35	1000	0.02	8	0.75	20	4.467
36	1000	0.02	4	0.5	20	1.974
37	1000	0.06	12	0.5	20	3.450
38	1000	0.04	8	0.5	20	4.140
39	1000	0.04	4	0.5	15	3.860
40	1300	0.04	8	0.5	15	2.260
41	700	0.04	12	0.5	20	3.560
42	700	0.04	8	0.25	20	3.460
43	1000	0.06	8	0.75	20	4.951
44	700	0.06	8	0.5	20	3.637
45	1000	0.04	12	0.5	25	5.696
46	700	0.04	8	0.5	25	6.323

From the above table, it is observed that maximum Ts of 6.323 kN was obtained for sample no. 46 at processing conditions of 700 RPM rotational speed, 0.04 mm/s plunging speed, 8 second dwell time, 0.5 plunge depth and 25 mm shoulder diameter. The sufficient heat input during and mechanical and metallurgical interlocking at the interface should have been achieved at these processing conditions to obtain a joint with the superior mechanical property.

3. Results and Discussion

The genetic algorithm (GA) is a metaheuristic optimization technique employed to solve constrained and unconstrained problems by biological evolution methods. The algorithm continuously regenerates itself from the population to obtain the optimal solution concerning its previous iterations. Between each iteration, the algorithm randomly chooses the solution from the population space, as the parents start to reproduce for the next iteration [14]. The steps in GA are listed in Fig.1.

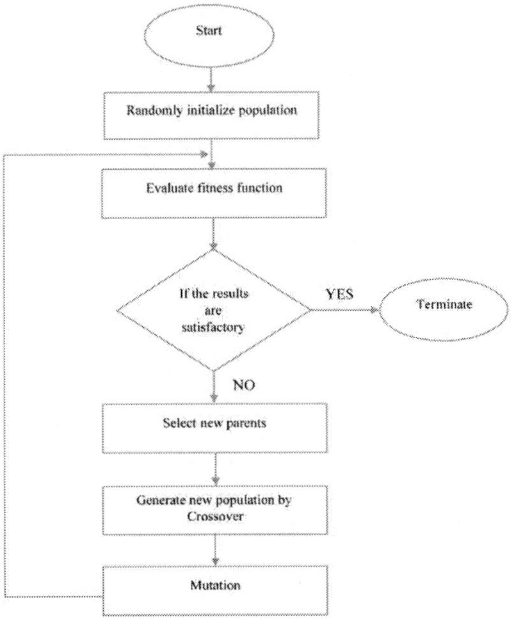

Figure 1. Steps in Genetic Algorithm

The experimental sequence generated from the Design expert software is stored in an excel file (.xlsx) format and then fed into the MBC toolbox in Matlab 2015. The MBC toolbox is a multipurpose toolbox designed to optimize the engine outputs based on the experiments conducted through the design of experiments. The objective of the genetic algorithm is to find the best set of process parameters that obtains the maximum Ts. The GA parameters are listed below.

Crossover function: scattered

Crossover fraction: 0.8

Mutation function: uniform

Selection function: tournament

Population size: 100

Generations: 500

The maximum Ts of 5.625 kN was predicted at the welding conditions of 999.415 RPM, 0.039 mm/s, 7.997 s, 0.75 mm and 24.845 mm. The confirmation test at the predicted processing conditions showed a Ts of 5.682 kN. The percentage error was computed to be 1.02 %, which shows that the GA's prediction is highly accurate.

3.1 Influence of Process parameters

The objective plot, Fig.2 explains the effects of the process parameters on the Ts of the joint. The tool rotational speed and plunging speed parameters play a vital role in heat generation during FSSW. The heat generated during FSSW is the combination of frictional heat from tool rotational speed and heat generated due to plastic deformation from the plunging speed. The Ts is observed to increase with the increase in the magnitude of plunging speed, whereas tool rotational speed has minimal effect on the load-bearing capacity of the joint, with the maximum contribution in heat generation is obtained from plastic deformation [15].

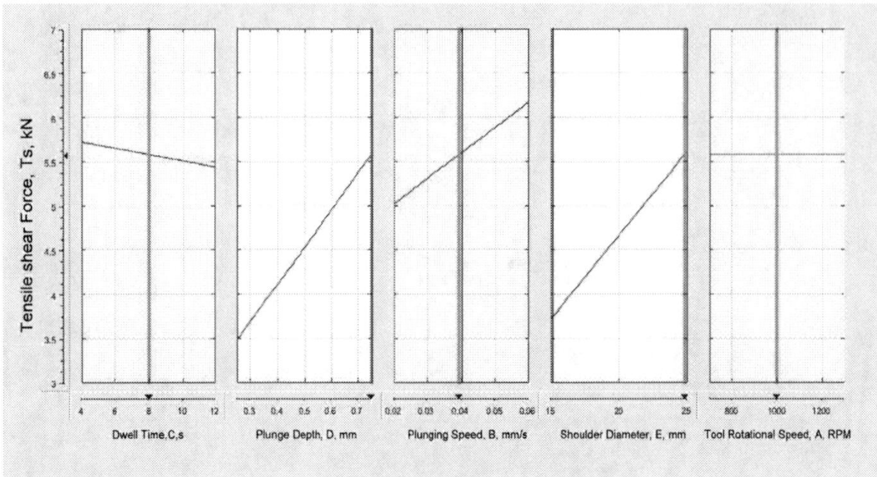

Figure 2. Objective Plots

The shoulder diameter has a significant role in frictional heat generation by establishing contact between the tool and the workpiece. In pinless FSSW, the frictional heat during welding is generated by shoulder diameter and tool rotational speed. Hence with the increase in shoulder diameter, there is sufficient heat achieved to form a stronger interfacial bonding between the two sheets [6]. The dwell time helps in the conduction of heat generated during welding, however, with the increase in dwell time results in increased thermal exposure that results in coarse grain microstructure and dents the load-bearing capacity of the weld joint. The plunge depth is vital in pinless FSSW as it helps to form a stronger interfacial bonding between the sheets [13]. With the absence of the pin in this process, the material flow between the upper and the lower sheet is entirely dependent on the plunge depth provided by the tool onto the top sheet. Hence a sufficient plunge depth is necessary during pinless FSSW to obtain a stronger weld joint [4]. However, an increase in these two parameters results in reduced joint strength due to and an increase in plunge depth resulting in the formation of a hook at the interface that acts as a crack nucleation agent [16]. Moreover, the increase in joint strength is also due to the formation of fully developed nugget with increased nugget width, which is credited to increased shoulder diameter and increased plunge depth.

Conclusions

Al 6061 and Al 5754 of 1 mm sheet thickness were joined by pinless FSSW. The genetic algorithm is employed to identify the optimal welding conditions. The maximum Ts of 5.625 kN was obtained at tool rotational speed of 999.415 RPM, plunging speed of 0.039 mm/s, Dwell time of 7.997 s, plunge depth of 0.75 mm and shoulder diameter of 24.845 mm. The confirmation test conducted at the predicted welding conditions showed that the prediction capability of the Genetic algorithm was highly accurate. The objective plots, shows that shoulder diameter, plunge depth, plunging speed and dwell time has a significant effect on the Ts, whereas tool rotational speed was observed to have minimal effect on the Ts.

References

[1] Kim D, Badarinarayan H, Ryu I, Kim J. H, Kim C, Okamoto K, and Chung K, Numerical simulation of friction stir spot welding process for aluminum alloys. Metals and Materials International, 2010, 16(2): 323-332.

[2] Farmanbar N, Mousavizade S.M, and Ezatpour H.R. Achieving special mechanical properties with considering dwell time of AA5052 sheets welded by a simple novel friction stir spot welding. Marine Structures, 2019, 65: 197-214.

[3] Tozaki Y, Uematsu Y, and Tokaji K. A newly developed tool without probe for friction stir spot welding and its performance. Journal of Materials Processing Technology, 2010, 210: 844-851.

[4] Bakavos. D, Chen. Y, Babout. L, and Prangnell. P, Material interactions in a novel pinless tool approach to friction stir spot welding thin aluminum sheet. Metallurgical and Materials Transactions A, 2011, 42(5). 1266-1282.

[5] Zhang G. F, Wei S. U, Zhang J, Wei Z. X, and Zhang J. X. Effects of shoulder on interfacial bonding during friction stir lap welding of aluminum thin sheets using tool without pin. Transactions of Nonferrous Metals Society China, 2010, 20. 2223-2228.

[6] Paidar M, Khodabandeh A., Sarab M. L, and Taheri M. Effect of welding parameters (plunge depths of shoulder, pin geometry, and tool rotational speed) on the failure mode and stir zone characteristics of friction stir spot welded aluminum 2024-T3 sheets. Journal of Mechanical Science and Technology, 2015, 29(11): 4639-4644.

[7] Bozkurt Y and Bilici M.K, Application of Taguchi approach to optimize of FSSW parameters on joint properties of dissimilar AA2024-T3 and AA5754-H22 aluminum alloys. Materials and Design, 2013, 51. 513-521.

[8] Karthikeyan R and Balasubramanian V, Predictions of the optimized friction stir spot welding process parameters for joining AA2024 aluminum alloy using RSM. International Journal of Advanced Manufacturing and Technology, 2010, 51. 173-183.

[9] Klobčar D, Tušek J, Smolej A, and Simončič S, Predictions of the optimized friction stir spot welding process parameters for joining AA2024 aluminum alloy using RSM. Welding in the World, 2014, 59. 269-281.

[10] de Castro C. C, Plaine A. de Alcântara H. N.G, and dos Santos J. F, Taguchi approach for the optimization of refill friction stir spot welding parameters for AA2198-T8 aluminum alloy. International Journal of Advanced Manufacturing and Technology, 2018, 99. 1927-1936.

[11] Pattanaik A. K, Pradhan. S, Panda. S. N, Bagal D.K., Pal. K, and Patnaik. D, Effect of Process Parameters on Friction Stir Spot Welding Using Grey Based Taguchi Methodology. Materials Today Proceedings, 2018, 5. 12098-12102.

[12] Zhang. A, Chen. X, Pan. K, and Wang. J, Multi-Objective Optimization of Friction Stir Spot-Welded Parameters on Aluminum Alloy Sheets Based on Automotive Joint Loads Metals (Basel). 2019, 520. 1-17.

[13] Chu Q., Li W.Y, Yang X. W, Shen J. J, Vairis. A, Feng W.Y, and Wang W. B, Microstructure and mechanical optimization of probeless friction stir spot welded joint of an Al-Li alloy. Journal of Materials Science and Technology, 2018, 34. 1739-1746.

[14] Babu K. K, Panneerselvam K, Sathiya P, Haq A. N., Sundarrajan S, Mastanaiah P, and Murthy C. S. Parameter optimization of friction stir welding of cryorolled AA2219 alloy using artificial neural network modeling with genetic algorithm. International Journal of Advanced Manufacturing and Technology, 2018, 94 (9-12): 3117-3129.

[15] Khosa S. U, Weinberger. T, and Enzinger. N, Thermo-mechanical investigations during Friction Stir Spot Welding (FSSW) of AA6082-T6. Welding in the World, 2010, 54. 134-146.

[16] Badarinarayan H, Yang Q, Zhu S. Effect of tool geometry on static strength of friction stir spot-welded aluminum alloy, International Journal of Machine Tools and Manufacture, 2009, 49 (2): 142-148.

REFRESHABLE BRAILLE MODULE USING CAM ACTUATED MECHANISM

Aryan [1], Rajkumar.E[2], Shivam Doshi [3]

[1] SMEC, Vellore Institute of Technology (VIT), Vellore, India,
aryan.vivarsh@gmail.com
[2]SMEC, Vellore Institute of Technology (VIT), Vellore, India,
rajkumar.e@vit.ac.in
[3] SMEC, Vellore Institute of Technology (VIT), Vellore, India,
shivam.doshi6588@gmail.com

Abstract. Less than 3% of 145 million blind people in developing countries at present are literate [1]. At present, in most of the cases, speech output has been the medium for blind people to access materials. For content which require deep understanding such as technical texts etc., it is not an appropriate modality, moreover blind people are more accustomed to braille transcripts and adaptability issues to all audio resources is difficult. And thus, arises the need for a cheap, easily adaptable, less complex solution to the problem. Braille points are varied arrangements of raised dots representing characters which are identified by touch by visually impaired people. Here we discuss in detail, development of a dynamic/refresh-able Braille display which presents Braille points by the up-and-down movement of pins using a completely new, first of its kind, Cam actuated mechanism with just two actuation points instead of a standard 6 used by every single refresh-able braille board in the world at present. Two nested shafts having 8 arrays of precisely calculated and placed cam, each were responsible for keeping the pins in up and down state. The shafts were actuated using two micro-servo motors placed next to each other. X degree rotation of servo after transmission through gear and subsequently shaft resulted in a particular configuration. Henceforth calibration was done so that all braille alphabet could be produced with specific servo rotations.Keywords

1. INTRODUCTION

According to the stats of World Health Organization about 285 million people are visually impaired worldwide. Further adding 7 lakhs to 9 lakhs to the figure are those are deaf and dumb. In this modern era of communication where the normal people can interact with their long distant relatives and friends through calls or messages, the disabled section of the society are not able to and are deprived of these basic opportunities. This way the talented ones miss the platform because of visual, speech or hearing inabilities. Braille is permutation and combination system of raised dots which correspond to different letter that allows blind people to read and write because of its versatility and elegance. But despite of that blind people don't have access to learn this sign code. Absence of knowledge in American Sign Language as a means of communication for deaf and dumb isolates them from the rest of the society. Our research and prototype brings to us not only the solution to cover up this flaw in our current system but also to make these victims aware of such sign languages and train them.

2. NOVELTY

We have planned to replace the heavy bulky models of braille board with our pocket prototype
Also no current technology has in cooperated a mechanism to aid both blind as well as deaf and dumb person at the same time.

IOP Conf. Series: Materials Science and Engineering 1123 (2021) 012028 doi:10.1088/1757-899X/1123/1/012028

Even if the person is not aware of the braille language or the American Sign Language, he/she can easily learn using the app

3. LITERATURE REVIEW

Before going into the death of this project we went through some research papers to bring out the uniqueness in the current system and our proposed system

Sr. No.	Paper Name	Head 1
1.	Designing of English Text to Braille Conversion System	In the current era, the world around us is going to be electronic. Everything is at present available at digital and virtual world and the whole world is taking the advantages of that but the problem is arising when the visually impaired person will be concerned about the electronic and digitized world. Approximately 84 million people in this world are not able to see and those blind persons could not be able to take the advantage of the electronic world like reading of digital data from the electronic thing. They use the Braille language to read the data with the sense of touching to it but the problem is arisen when the reading has to be done from the electronic content as they cannot sense it by touching to it.
2.	Aiding the Visually Impaired: Developing an efficient Braille Printer	With the large number of partially or completely visually impaired persons in society, their integration as productive, educated and capable members of society is hampered heavily by a pervasively high level of braille illiteracy. This problem is further compounded by the fact that braille printers are prohibitively expensive – generally starting from two thousand US dollars, beyond the reach of the common man. Over the period of a year, the authors have tried to develop a Braille printer which attempts to overcome the problems inherent in commercial printers. The purpose of this paper, therefore, is to introduce two prototypes – the first with an emphasis of cost effectiveness, and the second prototype, which is more experimental and aims to eliminate several demerits of Braille printing. The first prototype has been constructed at a cost significantly less than the existing commercial Braille printers.
3.	Braille Teaching Electronic Prototype	The developed project aims to introduce an electronic device, in a blind people community, an electronic device, capable of generating an autonomous learning of Braille reading system. In addition, it was decided to develop this project, taking as reference the current rate of 0.3% of visually impaired people in Ecuador, the government's willingness to support projects that involve social inclusion; and the University commitment, to improve the lifestyle of people with disabilities.The electronic board Arduino Mega, is responsible for receiving the input signals and activate the peripherals, which generate the learning of the Braille code.
4.	Braille Display by Rotating Multi-	Braille display is a device from which the visually challenged people can actually read the text output from the computer in the form of Braille character But the commercially available Braille display price is in the range of □ 100000 to □ 500000 so it can't be easily avail for poor

	Octagonal Segment	visually challenged people. So, this paper discusses about the making of new type of Braille display, by rotating multioctagonal segments, with low cost. Here the text is converted into Braille character by rotating each octagonal segment, which contains all possible combination of raised dots of one column of Braille cell on its faces, to a particular degree. The required Braille character appears on the top face of the octagonal segments.
5.	Braille Grade 1 Learning And Monitoring System	Braille is a standard code for people with visual disabilities that consists of an array of high points. This paper explains the implementation of a low-cost prototype for reading, writing, and audio-assisted evaluation. This didactic and user-friendly system aims to improve the learning of disabled people by providing teachers new and modern alternatives to lead and monitor the student's activities. The Braille Grade 1 system consists of the following four stages: (a) writing, by the use of 10 buttons placed in matrix form, (b) reading, through 10 push-pull solenoids, (c) memorization, and (d) evaluation by using a touch display where the generated letters are displayed and assisted by audio messages
6.	Blind Aid: A Self-Learning Braille System for Visually Impaired	Braille is vital to all visually impaired individuals and it's the only system through which visually impaired children can learn to read and write, yet the rate of Braille literacy among visually impaired people belonging to developing countries including Pakistan is alarming low. Thus, Braille literacy is the key to employment and full participation in society. This research paper presents the design of a low-cost, low-power, portable, self- learning, and user friendly Braille system. The designed system serves as Braille writing and reading tutor, so visually impaired people can enhance their Braille writing and reading skills without the assistance of a Braille teacher. The designed system takes the input through Braille keyboard and produces the speech output and it also has the capability to read documents. It is believed that by implementing the designed Braille system in schools and homes, Braille literacy rate can be increased and visually impaired people can be employed and can fully participate in society.
7.	Research Investigating Implications of Adopting the Unified English Braille Code	The purpose of this report is to present the outcome of a project supported by the Braille Authority of North America (BANA) to conduct research on the Unified English Code (UEBC). We will begin by presenting a listing of the research questions given to us by BANA that we felt we were able to address within the limits and resources available for these studies. Following this list is a brief description of the structure of the research that was carried out. There are three distinct phases to this project which include focus groups, experimental reading studies, and text analysis. Each phase is reported separately. Finally, we will end with implications of our findings and concerns for needed further research on issues related to the adoption of UEBC.
8.	Multi	India is now a home to the world's largest number of blind people.

	Character Identification for Visually Impaired Using Braille System	Technologies are developed day by day principally in communication mobile phones which plays a crucial role. But visually impaired people can't able to use message application in mobile phones. The proposed system is to help the blind to know the letters through voice communication. This system designed so as to make thing easy for the blind people. In this system we use 6 dots Braille system which mostly used in blind school. Normally speech recognition software, audio interface is mostly used features to use computers by the blind. Here the same features are introducing in the mobile with Braille system. In this system, we receive messages by using modem. All that messages are recognized by audio interface and sending acknowledgment for that particular message by using speech recognition with help of Braille system. So with this system blind peoples can able to use all the applications in mobile phone as a normal people.in future this system will be helpful to blind people to communicate whole the world with simple interface.

4. INTEGRATED IMPLEMENTATIONOF MODULES

The entire mechanism has been divided into three modules. Each module has its own contribution in bringing the universality and elegance of the system.

3.1) The Mechanical Design
3.2) The Android App
3.3) Actuation of the servo motors.

3.1 THE MECHANICAL PROTOTYPE

The basic concept of cam shafts are used to design the dynamic braille board, we are using two cam shafts driven individually by one servo motor each. One shaft is used to control 3 pins out of the six pins in the braille board .Cam shaft id a cylindrical rod with different cam geometries placed around it according to the application. A cam is a curved eccentric shape placed around a shaft to move the follower (in this case pins) as per the requirement. Cam shafts are widely used in many mechanical actuation units such as in IC engines to timely open and close the inlet and outlet valve. The complete design of the mechanism is designed on Solidworks

Figure 1. Side by side view of the Solidworks model and the actual manufactured module

Each shaft has three portions of cam divided lengthwise for controlling each pin as shown in Fig 1. In order to obtain all 2^3=8(since each pin can be up or down at a time) combinations for the three pins controlled by one shaft the cams are placed at 360/8=45 degrees angle around the shaft, each cam has maximum extension of 3mm for the pin to be raised and 2mm width for ensuring that the required pin is raised even when the angle of rotation is not 100% accurate .The length of ne cam is 4 mm to ensure free axial moment of the shaft so that the motor can rotate the shaft without taming much load as to when the shaft is not free to move and may get stuck in between to put stall on the motor .The shape of the cam and the positioning of the cam are designed so that minimum angle of rotation and minimum torque is required to move pin in between orientation for different braille alphabets .
Since the design of the cam shaft and the whole assemble is no conventional , small and complex , it is not a feasible option to manufacture it by conventional manufacturing and machining processes therefore this model is made by additive manufacturing (Fusion deposition modelling 3d printing) technology . This model is manufactured using PLA(polylactic acid) plastic filament since it is the most widely used and available used material in the market but ABS filament can also be used for better strength and durability of the model
.for printing this model the desktop printer Creality Ender3 was used , for printing the parts all the Solidworks design files are saved in STL file format and then sliced in the software Cura to obtain the g- code for running the printer , the part orientation , layer thickness and infill percentage are set for each part according to the requirement and after printing all the parts are assembled together with the servo motor to obtain the braille board .

3.2 THE ANDROID APPLICATION

The implementation of the voice assistance app was a tedious task as the app menu had all icons which were controlled through gesture.

a)Contacts
b)Personal Assistance
c)Speech to Text
d)Book
e)Keyboard
f)Text to speech
g)Camera

Gestures like right swipe allowed the user to select a particular icon and left swipe to return back to the previous menu. There is a counter which detects every right sweep icon as it is updated after every consecutive swipe. Recent contacts has also been in cooperated which allows the user to call as shownMethods to control the gesture like up down right left swipe were merged. Up swipe and down swipe.The KeyBoard part as shown allows the user to text and at the same time learn the ASL if he/she sis unaware about it. Book reading is another key feature We have firebase real time storage, we have certain number of books uploaded, as soon as the user double swipes right, we convert the book into text. To sync the letters with braille we have in built Google speech to text after that app is connected to real time firebase. in braille child the value is updated then the data in python is also simultaneously updated.

3.3 ACTUATION OF SERVO MOTORS

IOP Conf. Series: Materials Science and Engineering 1123 (2021) 012028 doi:10.1088/1757-899X/1123/1/012028

In order to rotate the two cam shafts two servo motors are used (in this case micro servo sg90) to get a closed lopped system for accurate rotation of shaft by closed loop system . The servo motor used offers the rotation range of 0-180 degree but in order to obtain all orientations of pins the shaft should be rotated complete 360 degree. In order to obtain this complete 360 degree rotation of shaft using 180 degree rotation a set of gears are used with reduction ratio 2. The module of gear used in this model is 0.5, the bigger gear which is attached to the servo motor has 24 teeth and teeth smaller one which is attached to the cam shaft has 12 teeth, both the gears have 4 mm face width .The gears are meshed together and thus the partial rotation of servo is converted to complete rotation of the cam shaft. The step of angle required to change the orientation of pins to the next alphabets is 45/2=22.5 degree. With each rotation of 22.5 degree of the servo motor next alphabetic orientation is obtained.

4 GENERAL ARCHITECTURE

4.1 Ways of interpersonal communication from a disabled to a normal person

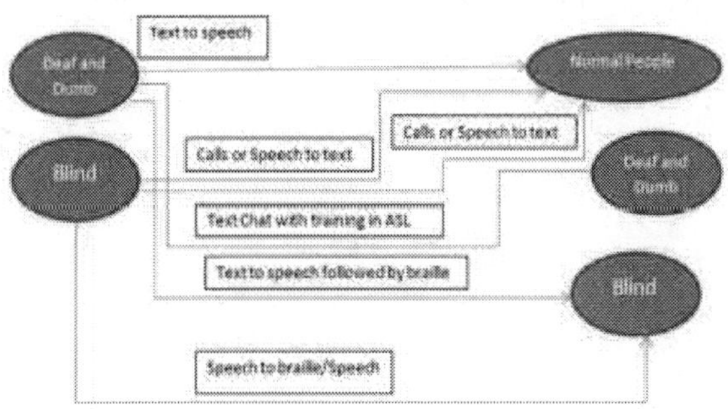

Figure 2. Ways of interpersonal communication from a disabled to a normal person

4.2 Ways of interpersonal communication from a normal to a disabled person

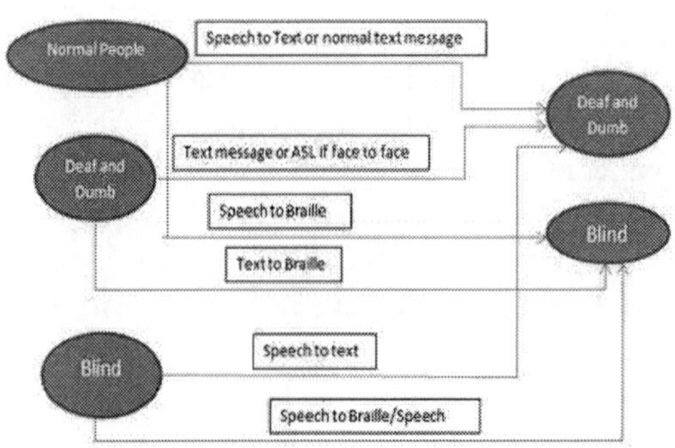

Figure 3.Ways of interpersonal communication from a normal to disabled person

5. METHODOLOGY

We have several cases to be considered.

-Case1: Firstly a blind person interacting with a normal
-Case2: A blind person interacting with a blind
-Case3: A visually impaired person interacting with a deaf and dumb.
-Case4: A deaf and dumb interacting with a normal person.
-Case5: A deaf and dumb interacting with someone having the same problem.
-Case6: Deaf and dumb interacting with a blind

For the above mentioned cases we planned to develop a universal system in which we have integrated several modules into one. Now case by case we will solve the mentioned cases. Taking Case1 into consideration whenever a blind person wants to talk, he/she can simply make a phone call through the app. The app has a gesture control feature with sweep down and sweep up feature which allows the user to choose the various icons available. And right sweep allows selecting the icon. Now coming to case 2 where a blind person one side speaks words which are converted letter by letter into corresponding braille segments so that the blind person on the other side can decrypt the message. Now case3 is a condition where the person on the other side is deaf and dumb. In this case google speech to text has been in cooperated in the app which allows the person to see the message on app. Now we will be seeing into the cases where deaf and dumb will be interacting. Focusing on Case 4 where the deaf and dumb can simply text his or her message through the keyboard feature on the app. Now Case 5 can be solved through simple text to text conversation as the person on both side are having same disabilities. Last case is when the person on the other side is blind. In such case we expect the deaf and dumb will be texting message will be letter by letter converted into braille characters.
The cases have been depicted in the figure. Fig 3. shows an interpersonal communication between a deaf and dumb with a normal person, a deaf and dumb and a blind person.

6. Conclusion

Using innovation in multiple areas and using first of a kind actuation mechanisms we were able came up with a low cost high performance model which integrates the solution of all the disabilities at a time. This would overall serve as a great aid to people facing issue with daily interpersonal communications, learning and other areas which people could implement this system in.

REFERENCES

[1] http://ieeexplore.ieee.org/abstract/document/7193267/
[2] http://ieeexplore.ieee.org/document/7830573/
[3] https://ieeexplore.ieee.org/document/8244896/
[4] http://ijergs.org/files/documents/BLIND-79.pdf

ROBUST FRACTIONAL ORDER CONTROLLER FOR AN EXPENDABLE LAUNCH VEHICLE

Elizabeth Varghese [1], Dr S Dasgupta [2], Dr J S Savier[3]

[1] Department of Electrical & Electronics Engineering, Mar Baselios College of Engineering & Technology, Thiruvananthapuram, India, eeliza.v@gmail.com
[2] Former Deputy Director, VSSC, Thiruvananthapuram, India, santanudsgpt@yahoo.co.in
[3] Department of Electrical & Electronics Engineering, College of Engineering Trivandrum, Thiruvananthapuram, India, savierjs@cet.ac.in

Abstract. Fractional calculus is an interdisciplinary area with multifarious applications especially in control systems due to more flexible parameters to adjust the dynamic behaviour of all the physical systems. This paper presents a fractional order controller structure for an expendable launch vehicle during ascent phase. The launch vehicle has highly flexible and unstable dynamic model, and in addition to that unpredictable control issues are caused by the sloshing of liquid propellants and the inertia problems of the engines. Under these circumstances controller with large robustness margins are required to meet deviations in model parameters which are unknown before the actual flight. The rigid body dynamics of the vehicle is considered for designing fractional order controller to overcome all these control issues. The robustness of fractional order controller is compared with existing classical controller for 20% variations in the aerodynamic coefficients. The stability margins of the open loop transfer function with all plant dynamics shows the supremacy of the proposed controller over the classical controller. The control inputs to launch vehicles, in the form of attitude commands, are affected by the wind. Hence the designed fractional controller simulated under the wind disturbance input and the result shows disturbance rejection capability of controller.
Keywords. Attitude Control, Fractional Order Controllers, Launch Vehicle

1. INTRODUCTION

Launch vehicles are used to transport and put satellites or spacecraft into space. When vehicle flies through the atmosphere it experiences aerodynamic forces and moments. Vehicle is aerodynamically stable if it can maintain its orientation under wind disturbances. Vehicle is aerodynamically unstable if its orientation diverges due to angle of attack by wind disturbances. In general, launch vehicles are aerodynamically unstable[1],[2]. The launch vehicle guidance system provides the inputs to the autopilot necessary to fly the vehicle in a predefined direction. These control inputs are generally in the form of attitude commands.The attitude control problem is to develop a controller with good tracking performance, adequate stability, rapid and well damped response to input command with moderate insensitivity to external disturbances [3]. Due to the aerodynamically unstable model of launch vehicle, the control design becomes more complicated. The attitude control problem of the launch vehicles requires an effective controller to track the recommended trajectory. The prime significance lies in building up an autopilot arrangement that accomplishes sufficient short period stability and performance. This makes utilization of the alleged time slice approach, in which the time varying mass and inertial properties are considered as constant over a short period of time. [1].

IOP Conf. Series: Materials Science and Engineering 1123 (2021) 012029 doi:10.1088/1757-899X/1123/1/012029

A conventional PID controller is easy to design and implement to the system. The stability of the system can be assured by properly designing the proportional controller. But it introduces a constant steady state error. The steady state error can be eliminated by the use of integral control, but it deteriorates stability of system. The derivative controller helps in predicting the future error depending on the current error rate. Thus, a suitable combination of the three basic modes can be improve all aspects of the system performance. [3]Though the researches are going on fractional Calculus over fast few centuries, the recent advancements in computational tools attracted more work on this area. The classical PID controllers are mostly used in industry due to its simplicity in design and more systematic procedures available in design of controllers. The design of Fractional PID controller can be done using frequency domain approach or time domain approach. The combination of time domain and frequency domain specifications also can be used as constraint for choosing control parameters. The most commonly used method is to design FOPID controller parameters using tuning methods such as genetic algorithm or particle swarm optimization.[9] The experimental validation of FOPID controllers performed in some industrial applications and experimental platforms such as Magnetic Levitation Systems and Inverted Pendulum, reinforces the capability of fractional order controllers.[4] Podlubny has proposed a generalisation of the PID controller as controller. He also demonstrated that the fractional-order PID ($PI^{\lambda}D^{\mu}$) controller has better controller action than classical PID controller [7]. In addition to the fractional controller design the concept of fractional calculus can be applied to system modelling also. Conventionally all the systems are model by ordinary differential equation of integer order. But the developments in the area of fractional calculus, helps to model the system as non-integer model [5],[6].Since solution methods for fractional differential equations were not developed, the conventional integer models where used to represent the complex systems. The different definitions of non-integer order differentiation and integration has developed in past few decades. This can be used to apply the fractional calculus in to wide area of control applications [6].

2. LAUNCH VEHICLEMODELLING

Launch vehicle attitude dynamics is varying as the time changes. The effects of aerodynamics, actuator dynamics, engine inertia and vehicle flexibility causes the variation in attitude dynamics of launch vehicle. Referring to [1], for preliminary design of autopilot, the launch vehicle is assumed to be rigid. The simplified dynamic model of the launch vehicle in the pitch plane obtained by the equilibrium moment equations and force equations. Also assumed that the parameters of the model are time invariant for short time period under consideration

.

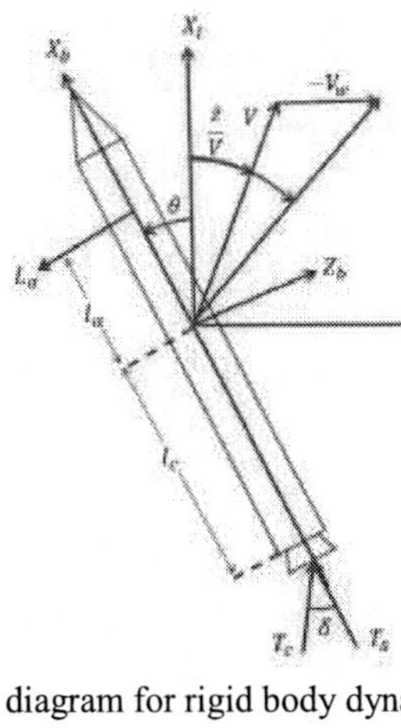

Figure 1. Geometrical diagram for rigid body dynamic model[2]

The linearized equations of motions of the launch vehicle in the pitch plane can be derived by considering the force equation and moment equations

The force equation,

$$\ddot{z} = \frac{(T_T - D)}{m}\theta - \frac{L_\alpha}{m}\alpha + \frac{T_c}{m}\delta \quad (1)$$

T_c, control thrust, L_α ,the normal aerodynamic force /angle of attack, θ the actual altitude, T_T, total thrust, D the drag, α, angle of attack and δ rocket engine deflection angle.

The moment equations

$$\ddot{\theta} = \mu_\alpha \alpha + \mu_c \delta \quad (2)$$

$$\alpha = \theta + \frac{\dot{z}}{U_0} + \alpha_w \quad (3)$$

The rigid body equation is given by

$$G(s) = \frac{\theta(s)}{\delta(s)} = \frac{\left[\mu_c s + \frac{Z_\alpha \mu_c + Z_c \mu_\alpha}{U_0}\right]}{\left[s^3 + \frac{Z_\alpha}{U_0}s^2 - \mu_\alpha s + \frac{\mu_\alpha g cos\theta_0}{U_0}\right]} \quad (4)$$

U_0 , longitudinal velocity, g , the gravitational acceleration. The other parameter and their values during the time period is given in table1.

At the initial stage of the flight, the aero dynamic pressure is negligible and the equation (4) reduces to

$$G(s) = \frac{\theta(s)}{\delta(s)} = \frac{1}{s^2} \quad (5)$$

During ascent phase of the flight, the aero dynamic pressure is considerable, the forward velocity is U_0 is very high and the rigid body equation can be represented as

$$G(s) = \frac{\theta(s)}{\delta(s)} = \frac{\mu_c}{s^2 - \mu_\alpha} \quad (6)$$

Though the launch vehicle is a complex flexible body structure, for preliminary analysis it is assumed that vehicle is completely rigid [1].The sloshing and engine inertia effects are also removed and reduced mass m0 of the vehicle is considered. The rocket deflection angle δ is given by an electrohydraulic action having first order lag as transfer function. The actuator is first order lag transfer function with bandwidth KA , given as

$$G_a(s) = \frac{K_A}{s + K_A} \quad (7)$$

Table 1. Rigid Body Parameters For a Typical Launch Vehicle

Sr. No.	Parameter	Symbol	Units	Time period (72-152) sec
1.	Reduced mass of the model	m0	slugs	5058
2.	control thrust	T_c	lb	342000
3.	the normal aerodynamic force /angle of attack	μ_α L_α	lb/rad	198000
4.	the aerodynamic moment /angle of attack		sec^{-2}	2.8
5.	control moment coefficient /angle	μ_c	sec^{-2}	4.56

3. CONTROL SYSTEM DESIGN

3.1 SIMPLIFIED AUTOPILOT DESIGN

The vehicle dynamics represented by equation (6) has one poles in the right hand side and hence the system is unstable. Classical controller design method based on pole placement is used to design the autopilot for the launch vehicle. The block diagram of the simplified rigid body along with actuator and the feedback is shown in the figure 2.

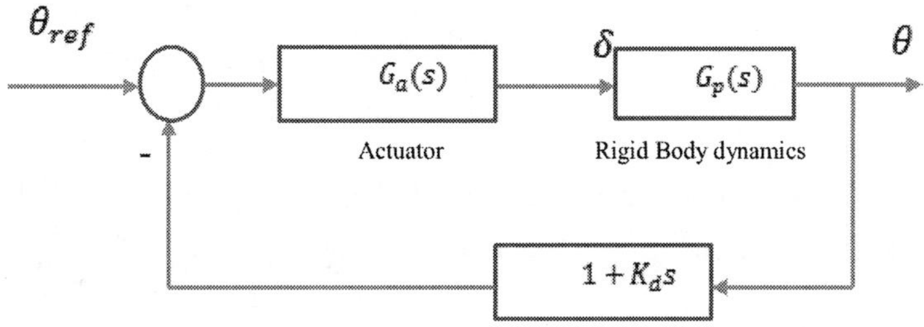

Figure 2. Autopilot for Simplified Rigid Body [1]

The attitude angle and rate of angle is sensed and feedback so as to make the system closed loop with PD controller. The closed loop transfer function of the system.
The closed loop characteristic equation of the systems is given as

$$1 + \left(\frac{K_A}{s + K_A}\right)\left(\frac{\mu_c}{s^2 - \mu_\alpha}\right)(1 + K_d s) = 0 \quad (8)$$

The values of K_A and K_d are designed such that the dominant poles of the closed loop system has damping ratio ξ =0.7 natural frequency, $\omega_n = 3.8 rad/sec$. The step response of the systems is shown in figure 3.

The step response shows that the systems gives a steady state error of 30% and settling time is also high 2 sec.The response of the autopilot needs to show good tracking performance and faster response to commands. In order to improve the tracking performance with reduced steady state error and faster response, PID controller is designed

3.2 PID CONTROLLER DESIGN

The PID controller produces control signal $u(t)$ consisting of three terms-one proportional to input signal, $e(t)$ another one proportional to integral of input signal, $e(t)$ and the third one proportional to derivative of input signal, $e(t)$ In PID controller.

The transfer function of PID- controller,

$$u(t) = K_p e(t) + K_i \int e(t)dt + K_d \frac{d}{dt} e(t) \tag{9}$$

$$G_c(s) = \frac{U(s)}{E(s)} = K_p + \frac{K_i}{s} + K_d s \tag{10}$$

The PID controller with the system is shown in Figure 4

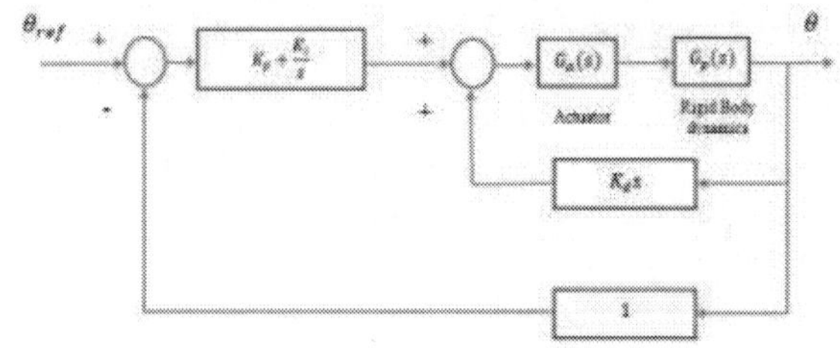

Figure 3. Block diagram of rigid body with P

In this design a PI controller is introduced in cascade with rigid body and actuator so that the pair of dominant closed loop poles satisfies specified time domain specifications ζ and ω_n. Here, the desired system is a second order system with a damping ratio, $\zeta = 0.7$ and natural frequency, $\omega_n = 3\ rad/sec$ and steady state error, $e_{ss} \le 1\%$. Designed values for $K_i = 23.7354$, $K_p = 14.1143$ and $K_d = 0.35$.

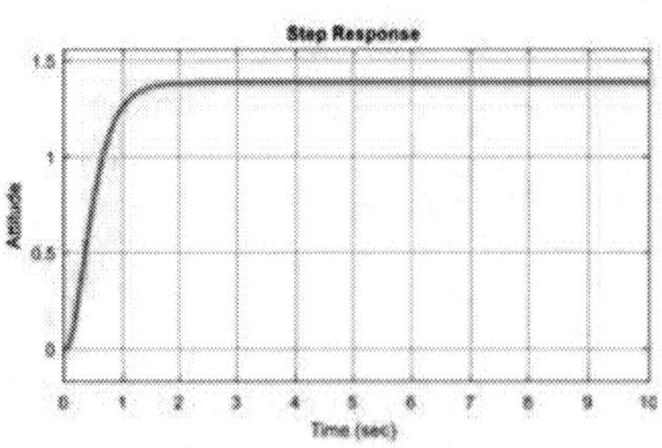

Figure 4. Step Response of the autopilot for the simplified rigid body

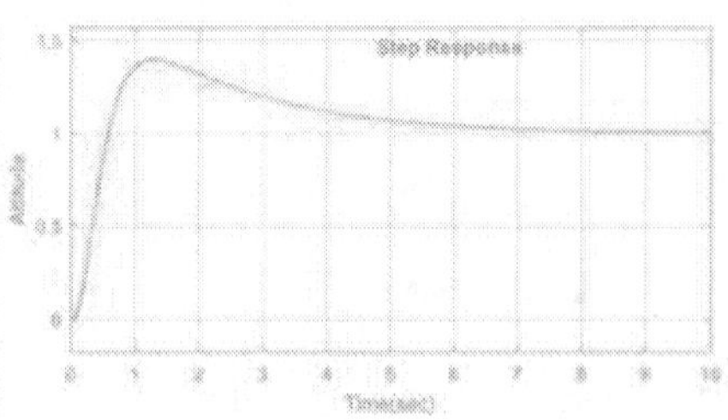

Figure 5: Step Response of rigid body with PID Control

3.3 FRACTIONAL PID DESIGN

Fractional order PID (FOPID) controller is a generalized form of classical PID controller [4]. In this type of controller instead of taking integral or differential of error signal, non-integer order integral or differential of the error is input is calculated for finding out the control input to the plant[5].As demonstrated in many literatures [6],[7], FOPID outperforms conventional PID due to the presence of additional design parameters. While designing FOPID controller, the order of

IOP Conf. Series: Materials Science and Engineering 1123 (2021) 012029 doi:10.1088/1757-899X/1123/1/012029

integration and differentiation can also be chosen as control parameters, in addition to the proportional, integral and derivative constants in the conventional PID.

The time domain equation of FOPID controller input is given as equation (11).

$$u_{FOPID}(t) = K_p e(t) + \frac{d^{-\lambda}}{dt^{-\lambda}} K_i + K_d \frac{d^{\mu}}{dt^{\mu}} e(t) \quad (11)$$

$$G_{FOPID}(s) = \frac{U(s)}{E(s)} = K_p + \frac{K_i}{s^{\lambda}} + K_d s^{\mu} \quad (12)$$

where $K_p, K_i, K_d, \lambda, \mu$ are the proportional gain, integral gain, and derivative gain, order of integration and order of differentiation respectively. The equation (13) represents the closed loop characteristic equation of the system .It is a nonlinear equation hence the initial solution of controller parameters are obtained by fsolve () function in MATLAB. Further the accuracy of the controller is obtained by optimization of the system using genetic algorithm to minimize ITAE (Integral Time Absolute Error) as objective function[10].

$$1 + \left(\frac{K_A}{s + K_A}\right)\left(\frac{\mu_c}{s^2 - \mu_\alpha}\right)\left(K_p + \frac{K_i}{s^{\lambda}} + K_d s^{\mu}\right) = 0 \quad (13)$$

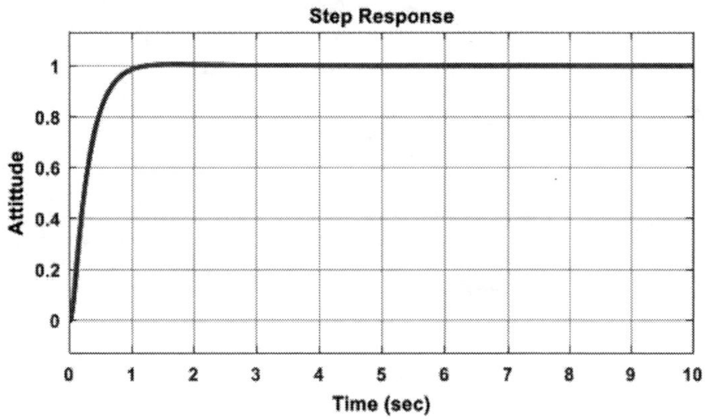

Figure 6. Step Response of rigid body with Fractional PID Control

PERFORMANCE EVALUATION AND ROBUSTNESS ANALYSIS

In this section the performance evaluation of three controller with rigid body dynamics is discussed. The main requirement of the attitude control system is to track the guidance command with less steady state error. While maintaining accurate tracking the actuator angle constraint also to be satisfactory. The actuator angle constraint is 5^0[1].

4.1 TRACKING PERFORMANCE

The unit step response of Rigid Body with nominal parameter for three types of controllers is shown in the figure 7a.The system with PD control has large steady state error. Though the introduction of PID controller reduced the steady state error, it has overshoot in the response. The response with FOPID controller settles fast with zero steady state error. The figure 7b shows the actuator angle response with three controllers. The angle for FOPID is higher while comparing other two controller. But the actuator angle is within the permissible limit.

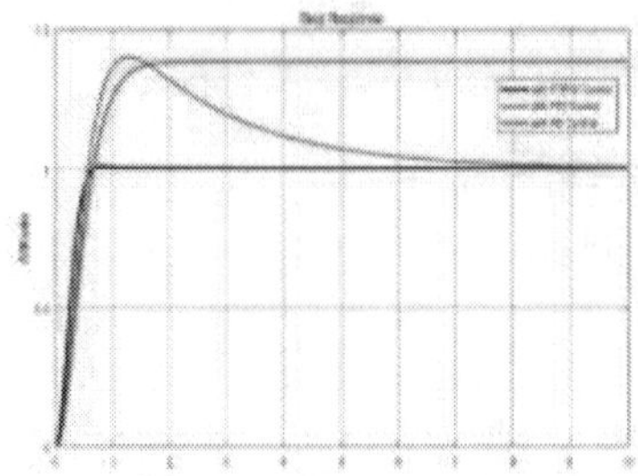

Figure 7a: Step Response of Rigid Body with nominal for different controllers

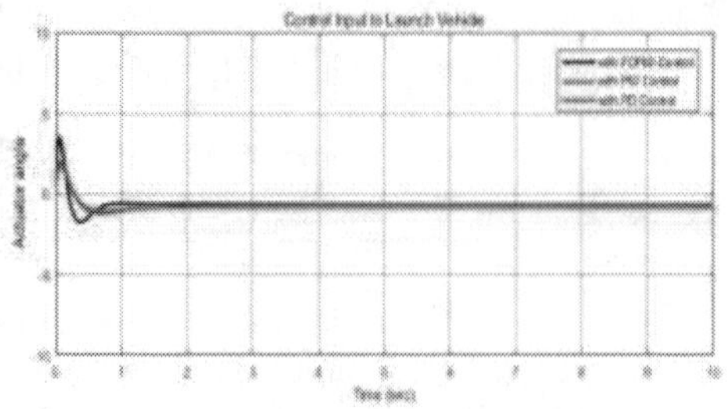

Figure 7b: Actuator Angle Response of Rigid Body with nominal for different controllers

4.2 ROBUSTNESS TO PARAMETER VARIATIONS

Robustness to parameters variations is also important in the attitude control system of launch vehicle since the vehicle parameters are not known exactly. Hence the response of the system with off nominal values also analyzed without changing the controller parameters. The aerodynamic moment (μ_a) is increased 35% and the control moment coefficient (μ_δ) is decreased by 25%.The simulation results shows the better response of system with FOPID control, without any drastic change in the response from the nominal response.

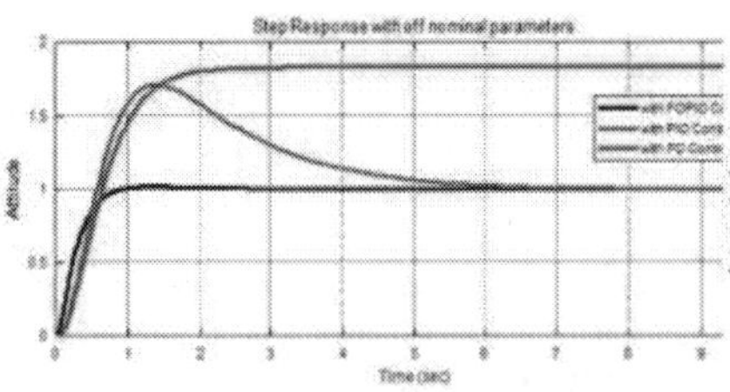

Figure 8 (a). Step Response of Rigid Body with off nominal for different controllers

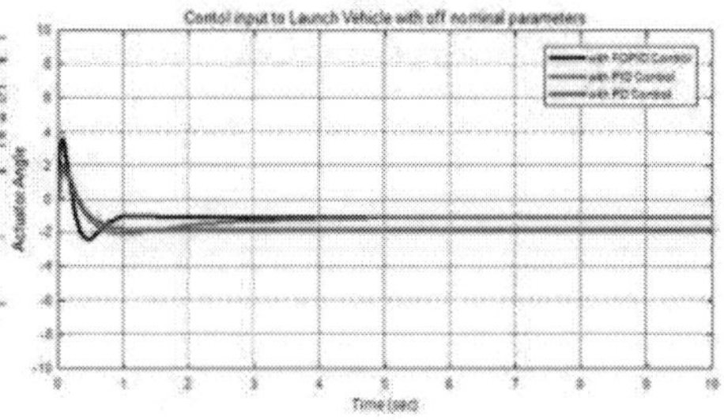

Figure 8 (b). Actuator Angle Response of Rigid Body with off nominal for different controllers

4.3 DISTURBANCE REJECTION

The disturbance rejection capability of the controller is analyzed by adding wind disturbance input at the input side of launch vehicle along with controller input. The response is analyzed with nominal and off nominal plant parameters. During atmospheric phase of attitude control wind velocity effects the longitudinal velocity. In all cases the tracking performance of the FOPID controller is better than the other two controllers. Though the magnitude of actuator angle is high for FOPID controller, it lies within the permissible limit.

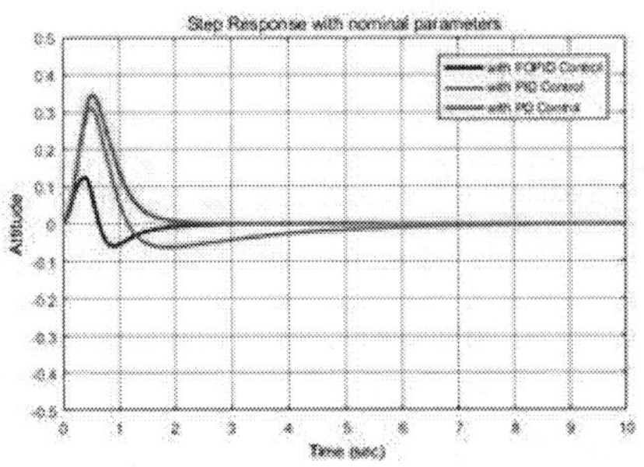

Figure 9 (a). Step response from wind disturbance (nominal values) for different controllers

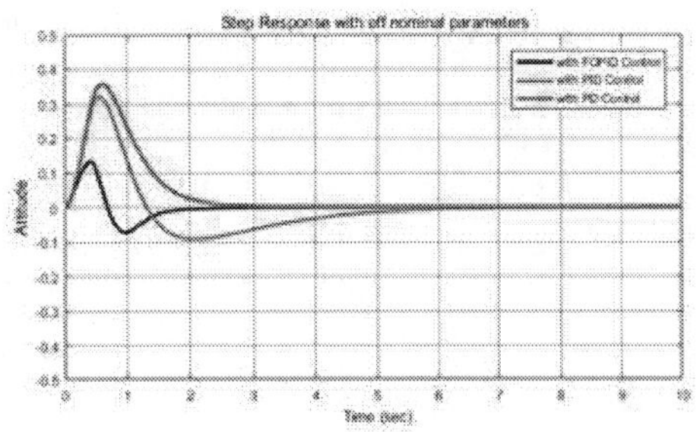

Figure 9 (b). Actuator Angle Response for wind disturbance (nominal values) for different controllers

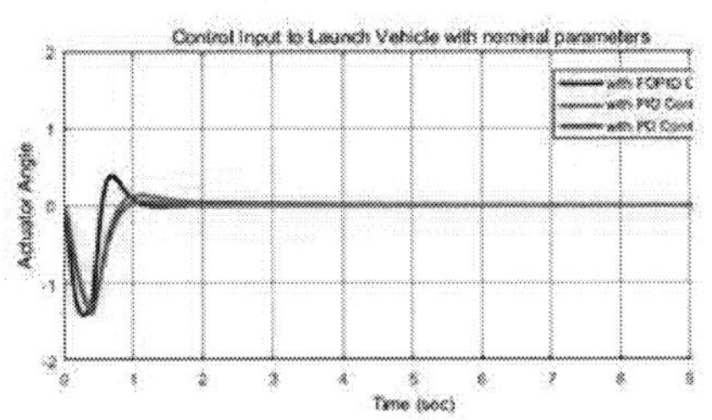

Figure 10 (a). Step response from wind disturbance (off nominal values) for different controllers

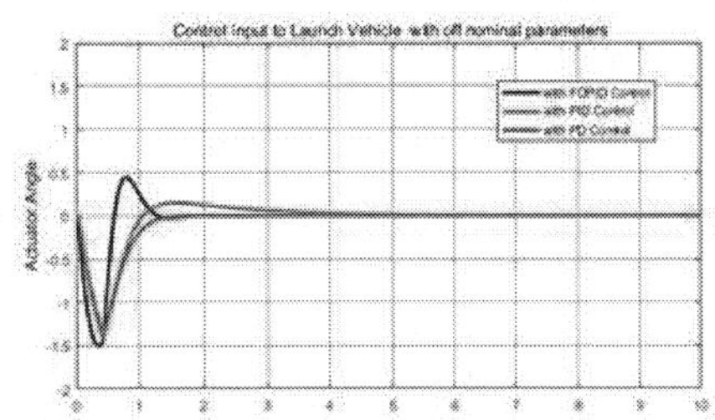

Figure 10(b). Actuator Angle Response for wind disturbance (off nominal values) for different controllers

CONCLUSION

The attitude controller for the launch vehicle, during the highly unstable atmospheric phase is designed. The unstable system is stabilized using PD controller in the feedback path. The tracking controller for the stabilized system is designed using PID controller and FOPID controller. In both cases the controllers are designed to meet the time domain specifications and the controller parameters are optimized using genetic algorithm for minimizing the error indices. The tracking performance and disturbance rejection capability of these controllers where analyzed for nominal and off nominal cases. The simulation results demonstrates that the FOPID controller has better robust behaviour than the conventional PID controller. The performance of the FOPID controller can be further modified by using optimizing weighed fitness function of both error signal and control input to launch vehicle.

REFERENCES

[1]. Arthur L Greensite, Analysis and Design of Space vehicle Flight Control Systems, Vol-VII, Attitude control during launch, Spartan Books, US, July 1967.

[2]. Bong Wie,Space Vehicle Dynamics and Control, the American Institute of Aeronautics and Astronautics, Inc. Second Edition, 2008.

[3]. Kadam V.N.,Practical Design of Flight Control Systems for Launch Vehicles and Missiles, Allied Publishers ,India,2009

[4]. Silviu Folea, Cristina I. Muresan, Robin De Keyser,and Clara M. Ionescu, Theoretical Analysis and Experimental Validation of a Simplified Fractional Order Controller for a Magnetic Levitation System, IEEE transactions on control systems technology, May 21, 2015.

[5]. Yang Quan Chen, Ivo Petras and Dingyu Xue ,Fractional Order Control - A Tutorial,American Control Conference Hyatt Regency Riverfront, St. Louis, MO, USA June 10-12, 2009,pp 1397-1411.

[6]. Yeroglu N. Tan, Note on fractional-order proportional–integral–differential controller design, IET Control Theory Application, vol. 5, 2011, pp. 1978–1989.

[7]. Podlubny, "Fractional-Order Systems and $PI^{\lambda}D^{\mu}$ Controllers," IEEE Transactions on Automatic Control, vol. 44,1999, pp. 208-214.

[8]. Dingyu Xue, YangQuan Chen and Derek P. Atherton, "Linear feedback control-Analysis and Design with MATLAB,the Society for Industrial and Applied Mathematics, 2007

[9]. Saptarshi Das, Suman Saha, Shantanu Das and Amitava Guptaa, On the selection of tuning methodology of FOPID controllers for the control of higher order processes., ISA Transactions, Vol. 50, Issue 3, pp. 376-388, 2011.

[10].Anagha P. Antony and E. Varghese, "Comparison of performance indices of PID controller with different tuning methods," 2016 International Conference on Circuit, Power and Computing Technologies (ICCPCT), Nagercoil, 2016, pp. 1-6.

SIMULATION OF BALLISTIC IMPACT ON DIFFERENT COMPOSITE SAMPLES OF BULLET PROOF VEST

[1]Narendiranath Babu T, [2]Anhad Preet Singh, [3]Nagireddy Sudershan Reddy, [4]Dhairya Murpani

[1]Professor, Department of Mechanical Engineering, Vellore Institute of Technology, Vellore, Tamilnadu, India.

[2,3,4] UG Students, Department of Mechanical Engineering, Vellore Institute of Technology, Vellore, Tamilnadu, India.
narendiranathbabu.t@vit.ac.in

Abstract. This paper presents the simulation results of ballistic impact tests on sample of composite material. The materials used are Kevlar 29, Carbon Fiber, E -Glass Fiber and Steel 1006. In simulation test, we used 9mm parabellum bullet having velocity of 400 m/s. The cad and analysis were done in Ansys Workbench Explicit Dynamics. The result was compared within different combinations of composite material and the most efficient and cost - effective composite material was chosen.
Keywords. Ballistic, Kevlar 29, Carbon Fiber, E- Glass Fiber, parabellum bullet, Explicit Dynamics, composite

1. Introduction

The main use of bullet proof vests is preventing entry of bullets into the protected area. The bullet proof vest to be utilized ought to be sturdy and simple to create, use and repair. The use of thick steel plates is the traditional method of armouring. Later, multilayer bullet proof vest was created which improved the penetration resistance/weight ratio relative to steel.

In multi-layered armours, the outer ceramic layer deflects the bullets and the fibres hold the bullet particles and the metal part stops the bullet. These kinds of bullet proof vests generally weigh less than single layer steel armours providing same protection.

So, to enhance the ballistic resistance of the bullet proof vest, different combination of composite materials together was used. Composite multi layered bullet proof vest is generally light weight and durable. The composite materials used are- carbon fiber, Kevlar, E-glass and ballistic steel 1006. The composite materials used here are not that expensive and are easily available. Software's like Solidworks, Catia, Ansys workbench were used for modelling and analysing.

In this study, multi layered bullet proof vest was designed with different combinations of composites where composites were stacked up uni-directionally and multi-directionally and simulations on software were performed and the results are presented.

2. Design of Armour

The armour design is usually done in two phases specially for those that may have multiple stacks for maximum efficiency. First phase of armor design is the selection of optimum combination of materials to be used and the second phase includes simulation and testing of the structure for the ballistic resistance of the material being used in the designing of the bullet proof vest and to check the total efficiency of the armor. The bullet proof Armor or vest was designed using several reinforced prepeg fibers like carbon fiber, Kevlar, E-glass and ballistic steel 1006. Different combinations of these composite materials were designed in two ways- unidirectional and other is multidirectional. The first combination of composite material for the armor is Kevlar-Ballistic Steel composite. In this combination, 2 steel plates of 0.5 mm thickness and 8 kevlar29 fiber of 0.25mm thickness is used for the design. The second combination used for the design is Carbon Fiber Ballistic Steel Composite. In this combination, 2 steel plates of 0.5 mm thickness and 8 carbon fiber of 0.25mm thickness is used for the design. The third combination used for the design is carbon fibre-glass fiber-steel composite. In this combination, 2 steel plates of 0.5 mm thickness and 4 carbon fibers and 4 glass fibers of 0.25mm thickness is used for the design. The fourth combination used for the design is Kevlar-carbon steel composite. In this combination, 2 steel plates of 0.5 mm thickness and 4 carbon fibers and 4 Kevlar fibers of 0.25mm thickness is used for the design. The fifth combination used for the design is Kevlar/glass steel composite. In this combination, 2 steel plates of 0.5 mm thickness and 4 glass fibers and 4 Kevlar fibers of 0.25mm thickness is used for the design. All these combinations were stacked up in uni-directional layup and then in multi directional layup and the simulation was done on the samples to find the minimum deformation in the sample, hence to provide maximum protection.

3. Modelling and impact simulation

3.1 Modelling methodology

The bullet-proof vest has been modelled with Ansys Workbench 19. First, the geometry of both the bullet and the vest has been modelled with Ansys Design Modeler, and then simulation of the projectile against each sample stacked up in unidirectional layup has been done. Which is followed by simulation of the projectile against samples stacked up in multi- directional layup is done.

Ansys Explicit Dynamics has been used to solve the model and generate modelling results. The materials of each part have been modelled into Ansys engineering data, and then the model properties like mesh and contact conditions between parts have been setup. Finally, the solver properties like initial conditions, system statics, and dynamic properties, and desired output have been defined.

3.2. Projectile

The bullet is made according to the dimensions of 9mm parabellum bullet. The model is made in Ansys software and finite meshing is done. A material lead alloy is assigned to bullet model and is ready for the analysis part. The bullet which has to hit the plate is made of lead alloy having a speed of 400 m/s.

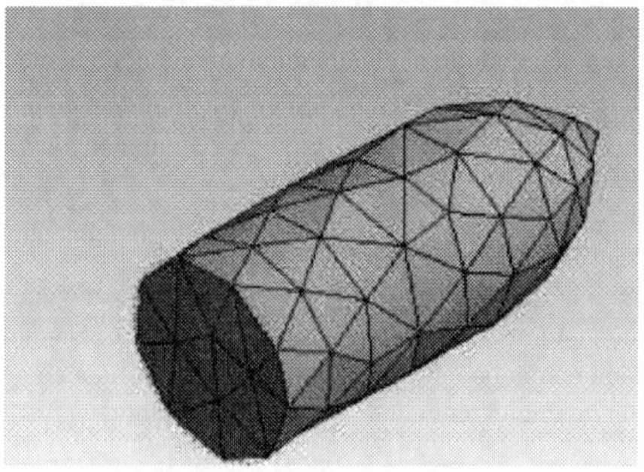

Figure 1. Bullet meshing:

3.3. Ballistic Armor

The Armor is made with several combinations of materials like carbon fiber, Kevlar, E-glass and ballistic steel 1006. Each material had to be modelled separately into Ansys material engineering data. Five samples of vest are made using different combinations of materials. All these materials are stacked up first in unidirectional layup and then multidirectional layup.

Since a small part of material also exhibits the same properties as original, we took a 10*10 cm sample plate of the selected material for simulation instead of whole bullet proof vest. And that 10*10 cm sample is made by composites of different material mentioned above.

Each sample is made by taking different materials combining together by layup. Each sample is of 3 mm thickness.

3.4. Kevlar-29 200gsm

Kevlar has a unique combination of high strength, high modulus, toughness and thermal stability. It was developed for demanding industrial and advanced-technology applications. Currently, many types of Kevlar are produced to meet a broad range of end uses. Kevlar mainly use for two reasons, and both are about performance: It's lightweight and easy to integrate. A thin blanket can serve as structural reinforcement or ballistic protection, everywhere from seismic shear walls to bank counters. Sprinkle the fibers into carbon composites to cut weight and boost strength. The grades Kevlar 49 and 149 are the lightest and most robust; Kevlar 29 is comparable impotency to glass fibre, but weighs less. Kevlar 29 mechanical properties are as shown as below-

3.5. Carbon fiber UD prepreg 230 gsm

Carbon fibers or carbon fibres are fibers about 5–10 micrometres in diameter and composed mostly of carbon atoms. Carbon fibers have several advantages including high stiffness, high tensile strength, low weight, high chemical resistance, high temperature tolerance and low thermal expansion. These properties have made carbon fiber very popular in aerospace, civil engineering, military, and motorsports, along with other competition sports. However, they are relatively expensive. When compared with similar fibers, such as glass fibers or plastic fibers.

3.6. Glass Fiber

The glass fiber composites strength/weight ratios are higher than those of most other materials and their impact resistance is phenomenal. Further they possess good electrical properties, resistance to moisture and outdoor weathering and resistance to heat and chemicals. These properties are coupled with ease of fabrication. Properties: controlled flexibility, high mechanical strength, lightness of weight, easy formability, durability, molded in colors excellent surface and resistant to corrosion and wear.

3.7. Steel1006

In the simulation steel was modelled as an explicit material with grade code SAE 1006 from Ansys Explicit material data sources. This grade of steel is mostly used for ballistic impacts because of its properties.

3.8 Simulation

The entire layup is done using Acp analysis in Ansys software. The model has been solved in Ansys Explicit dynamics in Ansys workbench 19. The analysis is given necessary boundary conditions like cycle time of 7e-03 sec , speed of bullet of 400m/s, fixed support to plate on all edges. The initial condition of the test has been simulated by assigning an initial component velocity to the bullet parts equals to 400 m/s in the Z direction and fixed support boundary condition at X and Y faces of the stack. The simulation is first done for unidirectional layup and then for multi directional layup and difference between the layup is found using the result.

And the simulation is performed

Each time different sample is analyzed and different deformations are observed. Out of all we considered the best when compared to quality and cost and we selected five samples of certain layup pattern.

Result and Discussion

After performing the simulation for each sample which are stacked in unidirectional way the results are shown below. In addition, the simulation model could predict the total and directional deformations.
Sample 1:
Every sample layup is symmetrical (such that i and n-i materials are same)
First sample layup consists of two different types of materials one is Kevlar and the other is ballistic steel. These are arranged in a pattern in a stackup such that the top most two layers are Kevlar prepeg

fibers followed by steel followed by two layers of Kevlar. Two layers of this stackup is used to make the sample of bullet proof vest. Here we used Kevlar and steel as materials.

Kevlar sample pattern after the impact has max deformation of 6.8077 mm. Since the sample is hit by a bullet at 400 m/s we can see a the deformation in the figure below. The region with red colour has maximum deformation and light blue region has min deformation.

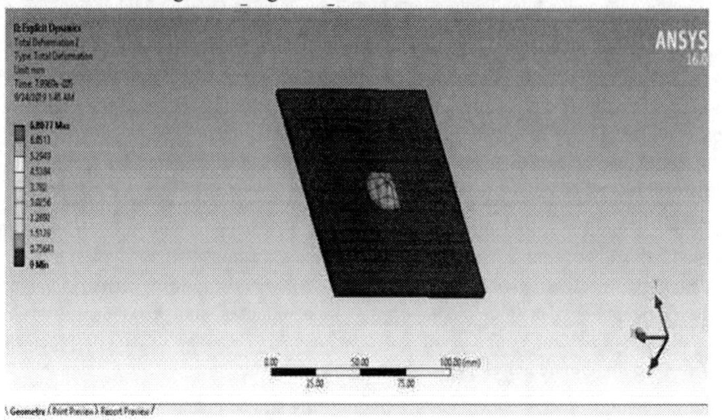

Figure 2. Sample 2

Sample two layup consists of three different types of materials one is Kevlar, one is carbon fiber and the other is ballistic steel. These are arranged in a pattern in a stackup such that the top most two layers are Kevlar prepeg fibers followed by one layer of steel followed by two layers of carbon fiber. Two layers of this stackup arranged in opposite orientations such that outer surface exposed to environment is Kevlar and is used to make the sample of bullet proof vest. Here we used Kevlar, carbon fiber and steel as layup materials. And the reason

why we used Kevlar fiber exposed to bullet is it has good piercing resistance when compared to carbon fiber.Kevlar Carbon sample after the impact has max deformation of 5.3268 mm. Since the sample is hit by a bullet at 400 m/s we can see a the deformation in the figure below. The region with red colour has maximum deformation and light blue region has min deformation. We have done the analysis for carbon fiber as top layer it showed more deformation than this Kevlar carbon sample in which Kevlar fibers are as top most layers.

Figure 3. Sample 3

Sample Three consists of three different types of materials one is Kevlar, one is Glass fiber and the other is Ballistic steel. These are arranged in a pattern in a stackup such that the top most two layers are Kevlar prepeg fibers followed by one layer of steel followed by two layers of Glass fiber. Two layers of this stackup arranged in opposite orientations such that outer surface exposed to environment is Kevlar and is used to make the sample of bullet proof vest. Here we used Kevlar, Glass fiber and Steel as layup materials. And the reason why we used Kevlar fiber exposed to bullet is it has very good properties when compared to Glass fiber.
Kevlar Glass sample after the impact has max deformation of 5.8439 mm. Since the sample is hit by a bullet at 400 m/s we can see a the deformation in the figure below. The region with red colour has maximum deformation and light blue region has min deformation.

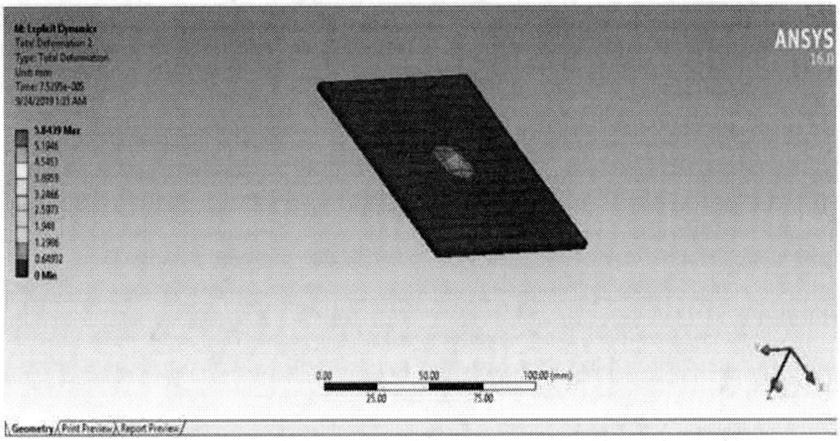

Figure 4. Sample 4

Sample Three consists of three different types of materials one is Carbon fiber, one is Glass fiber and the other is Ballistic steel. These are arranged in a pattern in a stackup such that the top most two layers are Glass prepeg fibers followed by one layer of steel followed by two layers of Carbon fiber. Two layers of this stackup arranged in opposite orientations such that outer surface exposed to environment is Glass and is used to make the sample of bullet proof vest. Here we used Carbon fiber, Glass fiber and Steel as layup materials.

Glass Carbon sample after the impact has max deformation of 14.836 mm. Since the sample is hit by a bullet at 400 m/s we can see a the deformation in the figure below. The region with red colour has maximum deformation and light blue region has min deformation. The reason why this sample experienced such deformation is glass fiber is used as top layer which does not has that good impact resistance and even Carbon fiber has less impact strength than Kevlar.

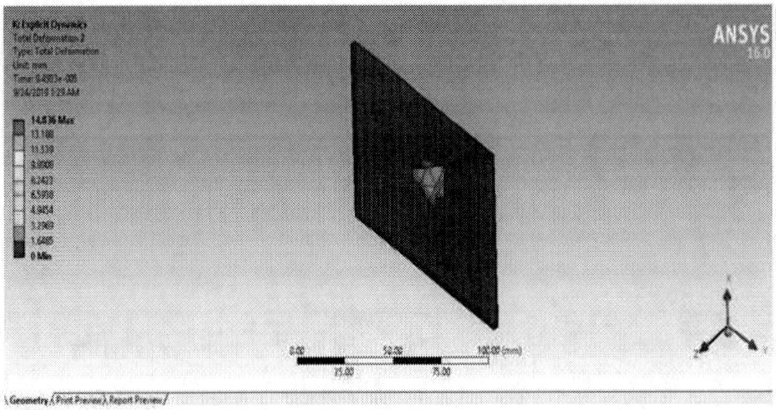

Figure 5. Sample 5

Fifth sample consists of two different types of materials one is Carbon fiber and the other is Ballistic steel. These are arranged in a pattern in a stackup such that the top most two layers are Carbon prepeg fibers followed by Steel followed by two layers of Carbon fibers. Two layers of this stackup is used to make the sample of bullet proof vest. Here we used Carbon fiber and steel as materials.

Carbon sample pattern after the impact has max deformation of 14.067 mm. Since the sample is hit by a bullet at 400 m/s we can see a the deformation in the figure below. The region with red colour has maximum deformation and light blue region has min deformation. Since the carbon fiber has less piercing resistance than Kevlar it showed much deformation than carbon sample.

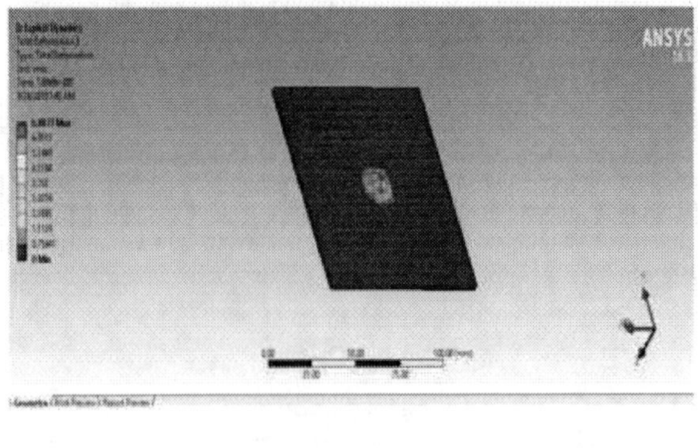

Figure 6. Deformation analysis

The strength can be further improved by increasing the thickness.Based on the above simulations we observed that kevlar carbon fiber is most effective.

The explicit analysis is done again by following multi directional layup. Since multi directional layup has some effect on it, we used it in the analysis.

The deformations after analysis are:

Sample 1:

Kevlar sample pattern after the impact has max deformation of 6.8077 mm. It has no effect for only one type of material fiber (Kevlar) used in the layup process.

Figure 7. Deformation analysis sample 1

Sample 2

Kevlar carbon sample deformation is 4.9383 mm. This sample had shown less deformation than the one done in unidirectional layup process. Since the thickness is very less the deformation is only slightly reduced. By increasing the thickness we can reduce the deformation further.

Figure 8. Deformation analysis sample 2

Sample 3:

Kevlar glass sample deformation is 5.486 mm. Even for this sample the multidirectional layup has shown some effect.

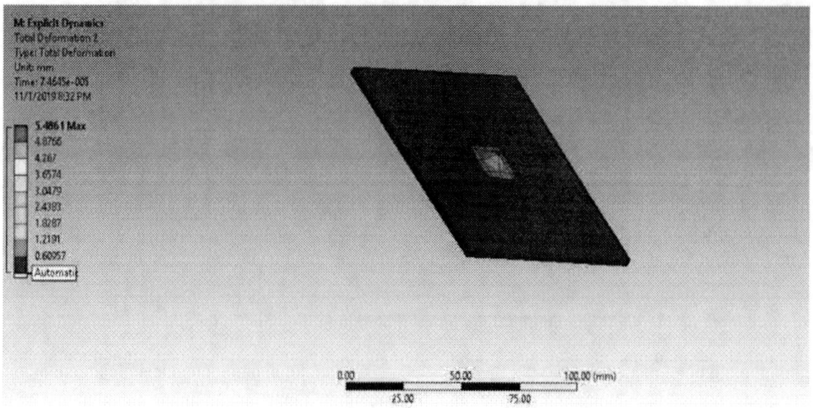

Figure 9. Deformation analysis sample 3

Sample 4:

Carbon glass sample deformation is 14.583 mm. It is same for this sample too. The deformation is reduced for multi directional layup.

Figure 10. Deformation analysis sample 4

Sample 5:

Carbon sample has a deformation of 14.107 mm. As it is discussed above same material fibers sample has shown no effect on the deformation. This sample deformation is same as the unidirectional sample.

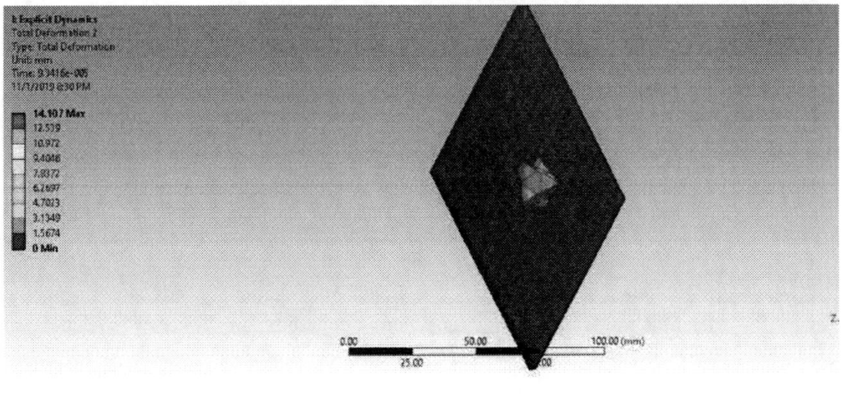

Figure 11. Deformation analysis sample 5

As we can see there is not much deformation for same material (kevlar sample and carbon sample) multi directional layup. The deformation slightly decreased for other type of composite samples since its thickness is less there is not much change. From this we know that multi directional layup can be helpful for larger thickness and useless for same material composites.

Conclusion

Different materials and geometry designs were developed using composite materials for ballistic structural armors. In this research, a 100 cm^2 sample of a laminate plate consists of different materials. Composite laminate structures with like carbon fiber, Kevlar, E-glass and steel layers were modeled with Ansys simulation program to check the technical feasibility of the armor design. The strength can be further improved by increasing the thickness of the plates. Based on the above simulations we observed that Kevlar carbon fiber is most effective.

Acknowledgments

The authors thank Mr. Narendiranath Babu T from Vellore Institute of Technology for his efforts and help for the completion of this project.

References:

[1] Simulation and Experimental Tests of Ballistic Impact on Composite Laminate Armor

Ali Murat Soydan,[1] Bahadır Tunaboylu,[2] Ahmed Galal Elsabagh,[2] Abdul Kadir Sarı,[3] and Recep Akdeniz[3]

[2] Michael Cohen, Laminated Armor, U.S. Patent no. 6,497,966, 2001.
M. A. G. Silva, C. Cismaşiu, and C. G. Chiorean, "Numerical simulation of ballistic impact on composite laminates," International Journal of Impact Engineering, vol. 31, no. 3, pp. 289–306, 2005.

[3] A. Banerjee, S. Dhar, S. Acharyya, D. Datta, and N. Nayak, "Numerical simulation of ballistic impact of armour steel plate by typical armour piercing projectile," Procedia Engineering, vol. 173, pp. 347–354, 2017.

[4] PP. M. Cunniff, "An analysis of the system effects in woven fabrics under ballistic impact," Textile Research Journal, vol. 62, no. 9, pp. 495–509, 1992.

[5] S. L. V. Shashi Kant, "A review on analysis and design of bullet resistant jacket-ballistic analysis," International Advanced Research Journal in Science, Engineering and Technology, vol. 4, pp. 1–10, 2017.

[6] C. Y. Tham, V. B. C. Tan, and H. P. Lee, "Ballistic impact of a KEVLAR® helmet: experiment
and simulations," International Journal of Impact Engineering, vol. 35, no. 5, pp. 304–318, 2008.

[7] M. Saleh, L. Edwards, and I. G. Crouch, "Numerical modelling and computer simulations," in The Science of Armour Materials, pp. 483–579, Elsevier, New York, NY, USA, 1st edition, 2017.

[8] T. Belytschko, R. Gracie, and G. Ventura, "A review of extended/generalized finite element methods for material modeling," Modelling and Simulation in Materials Science and Engineering, vol. 17, no. 4, Article ID 043001, 2009.

9)A. Tabiei and G. Nilakantan, "Ballistic impact of dry woven fabric composites: a review," Applied Mechanics Reviews, vol. 61, no. 1, Article ID 010801, 2008.

10)A. D. Resnyansky and G. Katselis, Ballistic and Material Testing Procedures and Test Results for Composite Samples for the TIGER Helicopter Vulnerability Project, DSTO Systems Sciences Laboratory, Edinburgh, SA, Australia, 2004.

11)W. J. Tomlinson and I. Collier, "The mechanical properties and microstructures of copper and brass joints soldered with eutectic tin-bismuth solder," Journal of Materials Science, vol. 22, no. 5, pp. 1835–1839, 1987.

12)H. Ma and J. C. Suhling, "A review of mechanical properties of lead-free solders for electronic packaging," Journal of Materials Science, vol. 44, no. 5, pp. 1141–1158, 2009.

SOFT COMPUTING APPROACH TO DESIGN ENERGY EFFICIENT PROTOCOL IN WIRELESS SENSOR NETWORK

Santosh Anand [1], Somnath sinha [2]

Department of Computer Science Amrita School of Arts and Sciences, Mysore Amrita Vishwa Vidyapeetham, India.
[1]santoshanand.jha @gmail.com, [2]ssin.mca @gmail.com

Abstract. The major limitation of Wireless Sensor Network (WSN) is the efficient utilization of energy resources. Maximum energy is lost during communication in terms of transmission power. The minimum and maximum value of transmission power depends on the node density for any particular network. In this research work the transmission power which is generally considered to be fixed for any node is subjected to be changed by taking random values between minimum and maximum transmission power as directed by the proposed neural network model. The idea cab be used in any simulation technique to enhance the lifetime of WSN. It proves that the network can save the energy considerably.
Keywords. Wireless sensor network (WSN), Routing Protocol, Energy Efficient, Neural Network (NN), AODV

1. INTRODUCTION

WSN is a network in which nodes are distributed randomly to monitor the environmental parameters. In this nodes are capable to sense for the targeted parameters, process it and send to the sink through wirelessly via neighbor nodes. Sensor nodes in WSN having several major constraints in the terms of memory, computing and energy. Because each and every sensor nodes works on the limited power supply given by battery so nodes should be properly and effectively use for the communication. Else it will be degrade the lifetime of WSN and it will come to one point that WSN will be not do any communications. So energy is the major research challenges. Every nodes in WSN is tiny in nature so memory and processing capability is minimal. But WSN having the redundant nodes which can help to give the alternate path for communication. Nodes receives and transmit the information to the neighbor nodes according to the transmission range. In general if the neighbor nodes in high distance nodes require more transmission power, if less than its requires less transmission power. But according to the survey and existing work done, each and every nodes having fix transmission power for minimum and maximum distance of neighbor nodes. Which is not appropriate in the terms of saving energy and enhancing the lifetime of WSN.
So there is a need of system which make nodes to set the transmission range dynamically or randomly according to the distance of neighbor nodes. So if nodes in less distance than it will help to achieve the energy target. This can be done with the help of routing protocols, in the below section survey is done on various algorithm, mechanism and routing protocols. In all, neural network is the best suitable for the dynamic or random transmission range according to the neighbor nodes.

2. Literature Survey

In this work [1], authors proposed distributed algorithms to reduce energy consumption during the communication in WSNs by minimizing the total transmission power of sensors without affecting the connectivity of the network. DTRNG algorithm is used based on the Relative Neighborhood Graph (RNG) in graph theory. With the help of this algorithm a node can dynamically adjust its transmission power if it is involved in a topological change. DTCYC algorithm is used in this work to determine the Transmission power by removing the largest edge in Cycles which minimize total transmission power. Therefore, the DTCYC algorithm is efficient in saving energy and can thus increase's the lifetime of the network. This work having various advantages such as

1. Distributed algorithms which will reduce communication energy consumption in WSNs.
2. Cycles will increase the lifespan of the network.
3. Smaller range will consume more energy

Limitation of work: there is no clearly analyses the algorithm and choosing the network range need to be done properly, large range consume more energy.

In this work [2], authors have used the smart antenna system to decrease the energy consumption. In this energy consumption of the home or office is compared using smart antenna system (SAS) and omnidirectional systems (OAS) statistically. The wireless sensor network allows multiple access which increases the overall transmission capacity providing better service quality. Due to the problems in wireless systems and reducing the deployment costs of wireless access, technologies such as ZigBee, MTM, HSPA etc. will become a remote access Trojan. So smart antenna system can increase the capacity of wireless network, as the SAS allows energy to be transmitted or received in one direction rather than disseminated in all directions. There are several features of this work

1. Potentialities of integration of smart antenna system in wireless sensor networks.
2. Performance is traced by comparison of WSN using SAS and OAS.
3. SAS is more energy efficient.

In this work [3], authors extended the MTUS (modified triple umpiring system), which repair the broken route if the broken node is near to the destination node. Otherwise, if the broken node is far away from the destination node, a Route Error (RRER) message is sent back to the source node, and the source node rebroadcasts RREQ to find a new route. Self-recovery MTUS can repair break route without considering the distance between the broken node and the destination node. Author propose MTUS with optimal Handoff-based self-recover feature, in this MTUS link breakage recovery mechanism is implemented. Source nodes broadcast RREQ message to find new route to destination when there is a link break. MTUS takes intermediate node to detects link breaks which help to repair the break route. But in optimal handoff based self-recovery maintain the route before the complete failure of the link with the help of node's neighbor power list.

3. Characteristic, Challenges and Design Objectives of WSN

In area of WSN, applications vary and depend on a wide range of technologies, characteristics, design objectives and challenges of WSN. Therefore application requirements on WSNs differ from one application to another. A few explorations that should be taken in consideration is mentioned and discussed below.

3.1 CHARACTERISTICS

WSN is the most effective way of the communication in any industries such as telecommunication. Because it is highly reliable, compatible, feasible and less cost. It is the best suitable for ant task oriented telecommunication or to monitoring purpose in the affordable price such as forest monitoring, sea monitoring communication and border monitoring in defense department . Based on the review and exiting work, WSN characteristics are defined as the follows:

1. **Dense self-deployment:** Dense having the major role while deploying the nodes in WSN. Because nodes are deployed randomly in any particular arear with redundant nature. So nodes can configure automatically and communicate with neighbor nodes.
2. **Topological inconstancy:** Nodes are deployed in the place where human cant reach unstructured WSN, so deployment will be done through any machinery devices. Which will be hard, and as it is unstructured topology will be change according to the routing mechanism schedule the path.
3. **Limited transmission range:** Every node can transmit according to the battery power capacity and majorly it is static. So this is the major research challenges to make it dynamic according to the capacity of battery available.

4. **Application centric**: WSN is always designed and deployed for some specific tasks so it is always hard to change or modify
5. **Less amount of storage and processing:** Nodes having the less memory and processing, due to tiny size.
6. **Limited energy resources:** Sensor nodes in unstructured WSN works on the battery which having the limited power and there is no chance of replacing and charging the battery.
7. **Data redundancy:** Same data can be received from the redundant nodes in WSN.
8. **Broadcast communication:** Flooding technique is used generally in all the existing WSN to send the information to sink or to create the paths.

3.2 Objectives of WSN in the Terms of Design

Any proposed WSN will gives the area where to deploy and specify the target application. So according to the requirements design objective of WSN is decided.
There are the most common objectives to design any WSN as follows:

1. **Deployment area, available resources and allocated budget:** Deployment area of WSN is mainly depends upon the area geographically. So area is directly proportional to the cost and number of sensor nodes. As number of nodes is directly proportional to the cost of WSN. So all three are linked tightly with each other.
2. **Topology used in WSN**: The size of the network and the area of interest determine the network topology. WSN topology is a dynamic topology that may be simple with single or few numbers in direct communication hops between the nodes or complex with multi-hop complex topological architecture.
3. **Power consumption**: Sensors mainly depend on batteries as power suppliers. It's not possible to change the battery or recharge.
4. **Coverage range:** In order to preserve the network consumed energy and to increase its productivity and reliability, network coverage range should be selectively determined. Small transmission range between nodes will decrease the amount of needed power for transmission between directly connected nodes. The huge coverage areas usually cased an eavesdropping.
5. **Quality of service:** The frequent changes in the sensed data are highly effected with the time factor. Reliability and usability usually depend on QoS.
6. **Simplicity**: The heterogeneous and autonomous nature of sensors in WSN as well as the complex topological nature requires simple and convenient communication, processing and power consumption models in order to ease and increase the efficient utilization of the network.
7. **Mobility:** Each sensor has the flexibility to change its location based on some environmental factors that strongly affect nodes movements are varies based on the application area. The extent of dynamic in WSN as well as the speed of mobility frequently influences the size, design and protocols of the network.
8. **Fault tolerance**: The ability to preserve the network performance and functionality even after individual node failure or congestion in some of parts of the network. The adaptability of WSN can be achieved by using efficient routing protocols, power management approaches and communication establishments.

3.3 Challenges
Deployment of WSN face a number of great challenges that urge researches to focus on it in order to reach an optimal performance of WSN, few challenges are listed below [9, 10, 11].
1. Hardware constraint: Since WSNs depend on battery based power devices. The less energy consumption devices in WSN are the most efficient and lasting WSN. Because computational capabilities and storage capacity will also affect life time of WSN. Therefore size, processing, cost and the number of the sensors in the applied environment should be considered.
2. Flexibility: Flexibility are needed for different network deployment schemes and topologies, routing protocols, power management methods and so on.
3. Reliability: The functionality and performance to WSNs should not be affected negatively in the time of any node failure. So there is a need of fault tolerance techniques to ensure reliability.
4. Connectivity: Maintain connectivity among all sensor nodes through the network life time is a very challenging issue. Because every node data is important. Some sleep modes can be practiced by some nodes in order to reduce the rate of harvested energy.

5. Power consumption: Sensor works on limited energy so the researchers attentions to power conservation and power management approaches which prolong the WSN lifetime.
6. Deployment: The WSN deployment can be densely by a huge number of sensors in applied area of application or sparse network with a few and limited number of sensors. Communication in WSNs is achieved by single or multi number of hops between sensors. So there is a need of system which should control the deployment cost without affecting the application functionality.
7. Scalability: WSNs should be able to support variety of routing protocols, huge nodes number and wide area of application as well as the frequent increases of network expansion.
8. Lifetime: The longevity and coverage of the WSN should be guaranteed. The main emphasis is to prolong the network lifetime. Some adapting mechanisms such as power management techniques and adaptive routing protocols are used to overcome the limited resources efficiently and to ensure the maximum network lifetime.

4. WIRELESS ROUTING PROTOCOLS

In the WSN energy resources play the vital role because each node in the WSN works having one fixed amount of energy which is provided by the battery. There is no chances of replace the battery and charge it. If energy is properly and effectively utilized then it prolongs the lifetime of WSN.

4.1 ROUTINGTECHNIQUE

In general, there are several routing protocols are used in wireless sensor network to improve the quality and make more efficient network. These protocols are designed based on the energy, packet delivery ratio, packet drop ratio, communication links and paths, scheduling and so on. Such protocols are Directed Diffusion routing protocol, sensor protocols for information via negotiation routing protocol, gradient based routing protocol, low energy adaptive clustering hierarchy routing protocol, self-organizing, routing protocol, threshold sensitive energy efficient routing protocol, virtual grid architecture routing protocol and so on. But when energy is concerned than routing protocols are designed on the basis of the operation as well as network structure. It classified in the following:

1. Structure base network routing: In this WSN routing is based on the nodes set up and location or position of the nodes. Routing protocols are classified as flat-based routing, location based routing and hierarchical based routing.
2. Operation base network routing: In this protocol is designed for specific targeted operation. These routing protocols are classified as QoS based routing protocol, Negotiation based routing protocol, Query based routing protocol, Coherent based routing protocol and Multipath based routing protocol.

4.2 SOFT COMPUTING PROTOTYPES IN WSN ROUTING:

In the recent work observation, researchers are using the soft computing parameters to optimize the routing of WSN, while considering the enhancement of lifetime, major challenges for design and nodes deployment.
There are several soft computing techniques which may be deployed in WSN to enhance the lifetime such as
Neural networks, fuzzy logic, swarm intelligence, reinforcement learning.

1. Reinforcement Learning (RL) Approach: It is an area of machine learning. This approach can be deployed in any software and hardware to find the best possible path for a particular situation. In this there no training is compulsory and it is differ from the supervised learning. If training data is not available than RL can take the decision from experience means past data. If this approach can apply in WSN than it can help to reduce the latency as well as best for traffic load balancing. Its performance can be better than traditional Ad-Hoc Network. It can only help in the scenario of one source to many sink or one sink to many nodes. But not suitable for all the scenario of WSN. According to the result from the existing work, it shows that RL can increase the lifetime of WSN by 50% without learning technique.
2. Swarm Intelligence (SI) Approach: This approach is based on the Artificial Intelligence system. It can be applied as optimal routing technique in WSN. There are several approaches based on the SI. Such as Ant based routing protocol, sensor driven and cost aware routing protocol, flooded forward ant routing protocol, energy efficient ant based routing approach. SI routing provide better flexibility in routing but it generates heavy traffic in Network. Which makes the

loss of energy and packet loss. So this routing protocol can not be able to deploy if considering the packet delivery ratio and energy because of backward and forward movements in the network.
3. Evolutionary Algorithm for routing in WSN: It is also concept of artificial intelligence to do the optimization for the sensor nodes in the terms of path scheduling. This is similar to VBS(Virtual backbone scheduling) and DAODV(Dijkstra AODV) for the path creation and path selection. But in this optimized path based on the energy of the link nodes which is shortest has to be selected for the communication. Shortest path is a good selection when the node having enough amount of power source but not for fixed power supply.
4. Fuzzy Logic (FL) routing approach: This approach is completely based on mathematical and conceptual model. It can deploy in the WSN where environment of network is undefined. It minimize the energy consumption of node and effectively utilizing the nodes energy in WSN.
5. Neural Network Approach for routing in WSN: This approach is used to enhance the lifetime of WSN using clustering technique. It is based on arithmetic and solve the complex relation in the terms of energy calculation and optimization. It is the best approach for any WSN to solve the energy research challenges. This approach is based on the input provided by the sensor, it can apply to sink to take the decision. Sink will have the nodes details such as power, transmission range, signal strength and data transmission rates, sink will give these details to neural network and final neural network has to provide the optimal path for transmission.

There are several features of Artificial Neural Network model if it can deploy in WSN:

1. Minimal degradation of overall performance of the network
2. Learning of the network is unsupervised continuous process.
3. Network connections (communication paths) can be able to modify according to the experience.
4. Node in WSN can parallel distributed to receive and transmit the information.
5. Sink will be highly and strongly connected with all other nodes in WSN.

Due to the above features of Neural Network which can be deploy in the WSN to enhance the lifetime and make strong and flexible communication in the terms of shifting path from one to another. So below section is proposed method using Neural Network, explained in below section.

5. PROPOSED METHODOLOGY:

Generally in WSN, paths are created by CDS (connected dominant set) and scheduling is done according to protocol, it may be VBS, AODV, DOADV,LEACH etc. But every node having there on fixed radio transmission range which is major drawback in the terms of energy. Because transmission is directly proportional to the power, if node sends the information to more distance it will take more energy and if small distance it will take less energy. So there is a need of dynamic transmission range selection.

In this work, Neural Network methodology is proposed to predict the optimal path according to the distance of the sensor nodes. Below figure1 shows the neural network deployment in WSN.

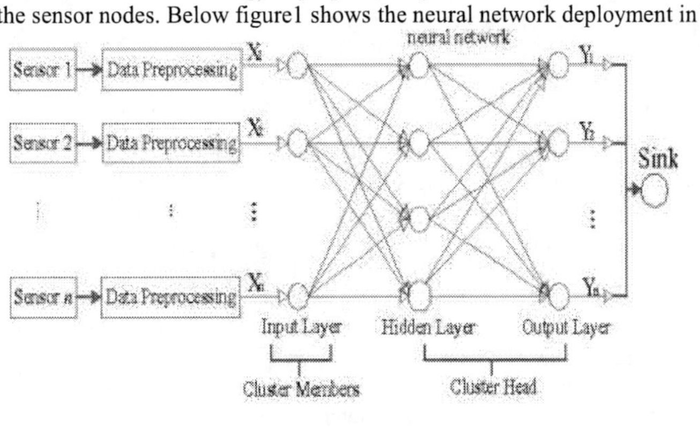

Figure 1. Neural network for wireless sensor network

5.1 Proposed algorithm

1. Set up the n number of nodes in WSN with redundant nodes.
2. Sink will keeps the details of each nodes.
3. Nodes parameters such as transmission range, energy of node, distance of neighbor nodes and details of all neighbor nodes.
4. Now sink will may have the details of each node, these parameters will give to the neural network model.
5. Neural network model will decide the path according to the energy level of nodes, range, distance and number of the nodes in the path.
6. Optimal path will be decided by neural network model.
7. This path will have the selection of dynamic transmission range.
8. This path will be enhanced the lifetime of WSN

6. CONCLUSION

This work is comprehensive study of all the intelligence techniques which is used in WSN to improve the lifetime. This paper is more specific study of neural network, genetic algorithm, artificial intelligence and fuzzy logic. This work brings the neural network in WSN to predict the optimal path for the communication on the basis of transmission range, density, distance and energy level.

REFERENCES

[1]. Xiao Chen and Neil C. Rowe, "Saving Energy by Adjusting Transmission Power in Wireless Sensor Networks", IEEE Globecom 2011.

[2]. Mohamed Hanaoui, Rachid Aouami, Mounir Rifi,,"Smart antenna system for wireless sensor network to improve energy efficiency, IJEAT,ISSN:2249-8958, Volume:5, Issue-3, Feb 2016.

[3]. S. Ganesh and R. Amutha, "Efficient and Secure Routing Protocol for Wireless Sensor Networks through Optimal Power Control and Optimal Handoff-Based Recovery Mechanism", Hindwai publishing corporation, journal of computer networks and communications, volume 2012, Article ID 971685, DOI 10.1155/2012/971685

[4]. I F Akyildiz, W Su, Y Sankarasubramaniam and E Cayirci (2002) "Wireless sensor networks: a survey", Computer Networks Vol. 38, No. 4, pp. 393 – 422.

[5]. Averkin, A. N., Belenki, A. G., &Zubkov, G. (2007) "Soft Computing in Wireless Sensors Networks", In Conference of European Society for Fuzzy Logic and Technology (EUSFLAT), No.1, pp. 387-390.

[6]. Zheng, J., &Jamalipour, A. (2009)"Wireless sensor networks: a networking perspective", published by A John & Sons, Inc., Wiley. Com

[7]. Romer, K., &Mattern, F, (2004)"The design space of wireless sensor networks", Wireless Communications, IEEE, Vol. 11, No. 6, pp. 54–61.

[8]. Yoneki, E., & Bacon, J. (2005)"A survey of Wireless Sensor Network technologies: research trends and middleware's role", University of Cambridge TR, 646.

[9]. Seah, W. K., Eu, Z. A., & Tan, H. P. (2009)"Wireless sensor networks powered by ambient energy harvesting (WSN-HEAP)-Survey and challenges", Proceedings of the 1st International Conference on Wireless Communications, Vehicular Technology, Information Theory and Aerospace &Electronic Systems Technology (Wireless VITAE), pp.1-5.

[10]. Kulkarni, R. V., Forster, A., &Venayagamoorthy, G. K. (2011)"Computational intelligence in wireless sensor networks: A survey", Communications Surveys & Tutorials, IEEE, Vol. 13, No. 1, pp. 68 – 96.

[11]. A.Alkhatib and G.Baicher, (2012)"An Overview of Wireless Sensor Networks", 2012 International Conference on Computer Networks and Communication Systems (CNCS 2012) IPCSIT, Vol. 35.

SMART SOLAR POWERED TENT

Rahul Sridharan, Maruti Sriram, Rajkumar.E
[1,2] Under Graduate Students, SMEC, VIT Vellore , India
[3]Associate Professor, Department of Design and Automation ,SMEC, VIT Vellore, India Email :- rahul.sridharan2017@vitstudent.ac.in , sriram.ayyagari2017@vitstudent.ac.in, rajkumar.e@vit.ac.in

Abstract: Many Engineering Technologies and day to day applications are now moving towards a eco friendly and environment free solutions for the routine processes. This paper investigates on how the environment friendly products which doesn't hinder the day to day activities done by the humans without causing any harm to the nature by involving unique engineering technologies which can help the product to be sold in the market and helping the humans to utilize the nature's resource and fulfilling the daily activities. This main objective of the regular tent existing in the market are based on different needs and specifications but it isn't able to fulfill the purpose either it isn't long lasting or not portable to carry for long distance travelling or on tough mountains for hiking. Therefore, to put an end to this everlasting issue this product is developed to cater all the needs of a traveler irrespective of the place he goes. The unique part of this tent is that it is completely powered by solar energy as it will provide in and around electricity in the tent as it has solar panel and LED lights are stuck around it to provide light. Researchers have compared with regular tents with the solar powered tents and notice that apart from lighting the place it can kill the bacteria by ultraviolet rays and keeping the place clean and can be used for necessary chores done during travelling. The Optimized form of product is that it's a multifunctional and can be used in any sort of place and any sort of travelling.

Keywords:- Solar Panel , Batteries ,LED lights , Energy storage element , miniaturized

1. INTRODUCTION

The idea is to introduce perovskite and flexible solar panels to produce maximum solar energy in out tents, to power our smart and safety devices. We foresee to produce enough power to provide users with lighting, heating, Wi-Fi (if permissible in that area), cooking, an IOT device and many other such safety features too. Our goal is to produce green power and ultimately be able to produce such large tent at cost effective rates, allowing it to be used in rural areas which have scope for solar energy.

2. LITERATURE REVIEW

Kevin bulls [1] proposed a material that could make solar power dirt cheap Perovskites have been known for over a century, but no one thought to try them in solar cells until relatively recently. That particular material is very good at absorbing light. While conventional silicon solar panels use materials that are about 180 micrometers thick, the new solar cells use less than one micrometer of material to capture the same amount of sunlight. Satoshi Uchida [2] paper on material challenges for cells in twenty -first century progress in materials and production processes has played an important part in this development. Yet, there are many challenges before photovoltaics could provide clean, abundant, and cheap energy. We review this research direction, Next PV working on promising solar cell technologies. The cooperation was focused on efficient photovoltaic devices, such as multifunction, ultrathin, intermediate band, and hot-carrier solar cells, and on printable solar cell materials such as colloidal quantum dots. Istiak Hussain [3] paper on functional materials ,device architecture and flexibility of perovskites solar cells Photovoltaic solar cells work by converting the energy of incident photon's directly to electricity the development of different photovoltaic technologies can be roughly classified into three discrete generations advanced techniques and research trends related to this emerging photo novel PSC architectures, methods that increase overall cell efficiency, and substrates that allow for enhanced device flexibility.

3. WORKING PRINCIPLE

The perovskite will be embedded with nylon to make the main part of the tent, we will use medium carbon steel for the frames on the tent, above the tent will be placed three flexible solar panels. The electricity produced by these components will power all our smart features on the tent. The perovskite used gives us scope to create much more electricity required to run our smart components without any problems.

4. METHODOLOGY

Steps are shown in figure 1.

Step1:- Manufacture tent frame

Step2:- Cover the frame with the perovskite fabric

Step3:- The flexible solar panel is placed over the perovskite material

Step4:- The panels and the perovskite are connected to the battery

Step5:- All appliances are finally connected to the battery.

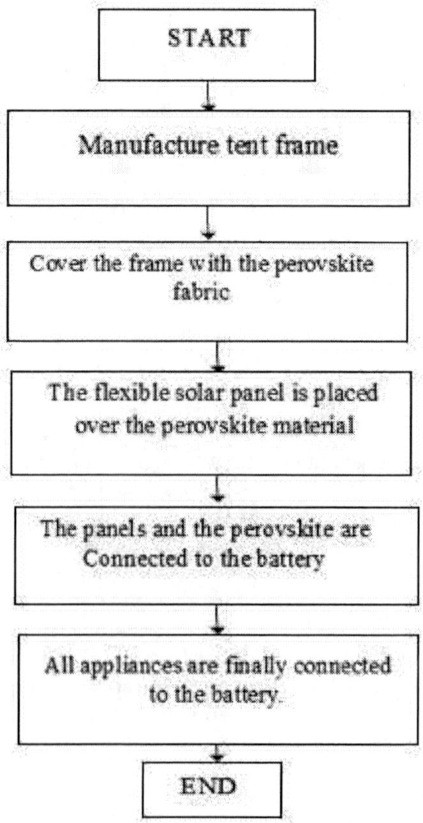

Figure 1: Methodology

Figure 2. CAD Model 1

Figure 3. CAD view 2

4. CALCULATIONS AND POWER CONSUMPTION

Flexible cells or PV devices are translucent they use little brittle materials and they require low light and can generate electricity. A lithium ion battery with a high capacity and a low power rating would deliver a low amount of electricity which is enough to run a few crucial apparatus. They can degrade in high heat and humidity. It give an efficiency of 22% Lead and Tin advance the technology and it runs on DC Current It can produce up to 300-400 watts in a good sunlight for a size of 46 square meters

Electricity Generation Calculation:

1) Upper Bound – 18,400 Watts

2) Lower Bound - = 46 x 30 = 13,800 Watts

Total Electricity can be used for 8-10 hrs. as it is completely proportional to the weather conditions as it can generate electricity in low light conditions as well. As it runs on DC Current now the perovskites cells can produce from 13,800 to 18,000 watts of electricity. From the electricity generated the here are the basic products which are installed in the tent This is an estimated calculation of each device / Product of how much electricity is consumed per product.

1) IOT Device :- 2 – 4 Watts

2) LED Lights – 60 Watts

3) Charging Ports – 2.5 Watts each

4) Induction Stove - 300 Watts

5) Wifi (Router)- 2 to 60 Watts

Solar cells give an attractive option for directly photo-charging lithium-ion batteries, the use of perovskite solar cell packs with four single CH3NH3PbI3 based solar cells connected in series for directly photo-charging lithium-ion batteries assembled with a LiFePO4 cathode and a Li4Ti5O12 anode. Our device shows a very high photo-electric conversion and storage efficiency of 7.80% and very good cycling stability, which d can do better than other reported lithium-ion batteries, lithium–air batteries, flow batteries and super-capacitors integrated with a photo-charging component. The newly developed self-chargeable units based on integrated perovskite solar cells and lithium-ion batteries hold promise for various potential application.

5. PRODUCT ARCHIITECTURE

Flexible Solar Panels
Specification Flexible solar panel:
1. This contains back contact solar panels for both on and off grid
2. 19%-23% high efficiency.
3. Flexible and light weight

Features Flexible solar panel:
An efficiency of upto 23% has been held by this solar PV module, which has also been recorded as the world's highest efficiency for a cell. While compared with the traditional models, the above has a higher average efficiency of 25-30%.

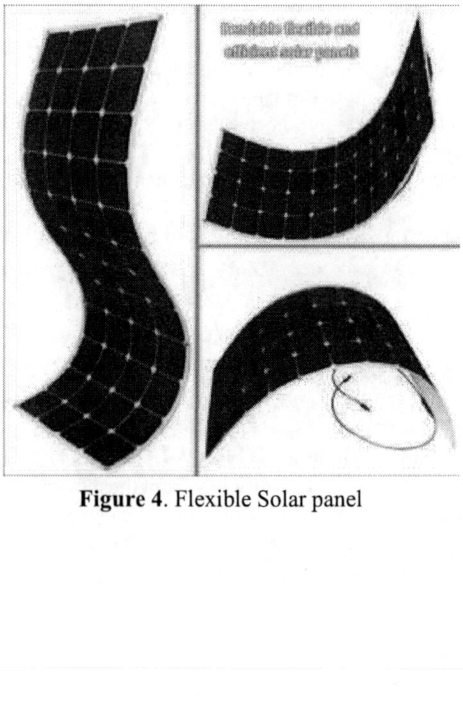

Figure 4. Flexible Solar panel

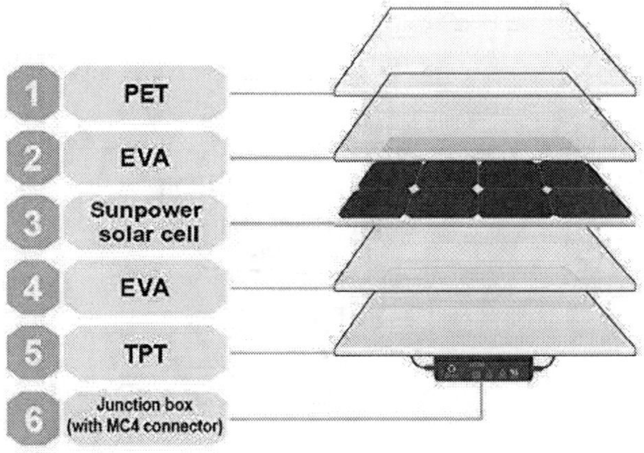

Figure 5. Layers in the solar panel

Place of Origin: Guangdong, China Brand
Name: SHINE SOLAR OR OEM Model
Number: SN- H110W
Type: Standard Solar
Panel Size: 1060*540*3mm

Certificate: TUV/IEC61215/IEC61730/CE/ISO
Color: Black and white/ black
Connector: MC4
Weight (Kg):1.6
Efficiency: 19.9%
Solar cell: SunPower from USA Voltage: 12V/24V
Application: Solar Home System
Cell type: Monocrystalline
Max Power: 110W
Price: $100

Perovskite Fabric

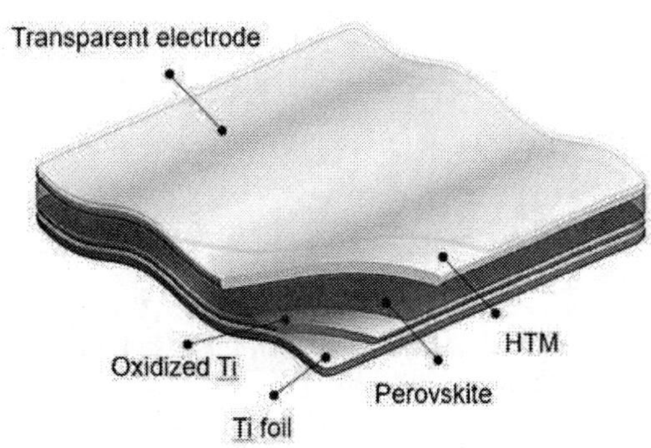

Figure 6. Perovskite fabric deconstruction\

The perovskite fabric is embedded into the nylon of the tent which provides instant electricity. The efficiency of the tent is much higher, over 25%. It also provides instant lighting. This material is yet to be integrated and mass produced so it's exact specifications yet to be determined. Nevertheless, it is indeed promising.

IOP Conf. Series: Materials Science and Engineering 1123 (2021) 012032 doi:10.1088/1757-899X/1123/1/012032

Battery

Figure 7. Battery

Can store up to 1800 watt of power and provide electricity for up to 24hrs on 100 watt load.

GOOGLE HOME MINI
The IOT device used here is the google home mini. It is a wireless smart speaker with inbuilt google assistant with a highly advanced voice UI. Price: Rs 2799

Figure 8. Google home mini

Model: Home Mini Smart Speaker
Model Name Home Mini Color White
Power Requirement 5V
OS Compatibility :Android and iOS Touch **Pad** No
Speaker included Yes
Speaker size 1.4
Display included No
Network connectivity Bluetooth Connectivity
Alexa Calling No
LED Lights
Solimo Smart LED Light, 12Watt, B22 Holder, It is also IOT enabled for customizability 16 million shades of RGB light available
Warranty: 1 year
Wattage: 12W; Lumens: 1080lm; Bulb base: B22
The Smart life App for this product is available for both android and IOS , Compatible with Alexa, can be used to give voice commands to adjust brightness, set timers, change colours and much more Connects with 2.4 GHz WiFi; no need for additional gateways or hubs Price: Rs649

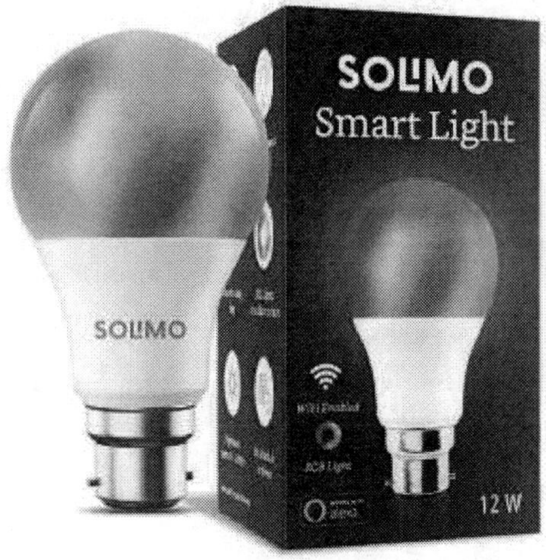

Figure 9. IOT light bulb

Heater

Figure 10. IOT Light bulb

[MOVING HEAD DESIGN]: With a moving function, the heater can shake the head left and right, letting the warm hot air blow to every corner

[MULTIPLE SAFETY PROTECTION]: Safety first, our heater body is made of high quality materials with no bad chemical odor.Space heaters for indoor use worrying about scorched floors is avoided thanks to the built in tip-over protection switch. Even when it is tipped over, the switch automatically turns it off.

[CUSTOMIZE YOUR COMFORT WITH 2 MODES]: This space heater air cooler electric has cool Fan and hot mode to choose, so it is available for four seasons. It has an indicator light which depicts the temperature mode it is on. Price Rs 2699

6. MAIN FEATURES

The solar-powered tent contains photovoltaic fabric to generate solar energy, heater, wireless hub and magnetic induction charging pouch. Whenever you will active the feature, the tent dome will glow gently, pointing you back to your tent. In case of any emergency the tent will shoot out a flare into the sky to notify people around. There is a basic lock mechanism in the lock for safety of the things inside the tent. If signal catching is possible in the area Wi-Fi will work No need of extra cooking equipment an electric pan can be used. Lighting in and around the tent is taken care of.

7. CONCLUSION

We believe we have made a smart, eco-friendly and sustainable tent and with further research and resources we would be able to develop this to much higher levels with the ultimate goal of implementing this technology at large scale in rural areas to the less fortunate as homes.

REFERENCES

[1] A Material That Could Make Solar Power "Dirt Cheap". Energy News by Kelvin Bullis on August 8, 2013.

[2] UCLA researchers improve process for manufacturing highly efficient solar cells. New technique holds promise for commercial production of perovskite photovoltaics. Bill Kisliuk | February 03, 2014.

[3] A. Mei, X. Li, L. Liu, Z. Ku, T. Liu, Y. Rong, M. Xu, M. Hu, J. Chen, Y. Yang, M. Gratzel, H. Han. A hole-conductor-free, fully printable mesoscopic perovskite solar cell with high stability. Science, 2014; 345 (6194): 295 DOI: 10.1126/science.1254763.

[4] EVANSTON, Ill., May 12, 2014 — A new solar cell promises enhanced efficiency and a smaller environmental footprint voltaic technology, with a focus on the diversity of functional materials used for the various layers of PSC devices.

SUITABILITY OF STANDARD FRACTURE TEST SPECIMENS FOR LOW CONSTRAINT CONDITIONS

Nagaraj Ekabote[1], Krishnaraja G. Kodancha [2], S.K. Kudari[3]

[1, 2]School of Mechanical Engineering, KLE Technological University, Hubballi, India.

[3] Department of Mechanical Engineering, CVR College of Engineering, Hyderabad, India ekabotenagaraj@gmail.com
krishnaraja@kletech.ac.in,
s.kudari@rediffmail.com

Abstract. Test specimen selection plays a vital role in successful transformation of laboratory fracture parameter findings to the real fracture behavior of any component. The ASTM and BSI Standard test methods for measurement of fracture toughness are dominated by high constraint specimens like Compact Tension (C(T)) and Single Edge Notch Bend (SEN(B)). However, most cracks in real time structure are shallow and surrounded by low constraint. Hence, the standard specimen like Middle Crack Tension (M(T)) in ASTM and Clamped Single Edge Notch Tension (Clamped SEN(T)) specimen in BSI, yield fracture toughness value to be test specimen dependent. The Clamped SEN(T) specimen suitable for various real pipeline (low constraint conditions) application has been widely reported in literature, that however in not the case for M(T) specimen which has scarcely reported. The 3D numerical analysis of SEN(T) and M(T) specimens with an identical flaw size for fracture toughness and constraint parameters (T-stress) under linear-elastic conditions are dealt in this article. The standard specimen that emulates low constraint fracture behaviour forms the basis for recommending suitable specimen for defined real time conditions. The conservatism associated with usage of standard specimen to find fracture parameters for an application has been quantified in this work for field implementation by practitioners in fracture mechanics.**Keywords**: Clamped SEN(T), Fracture toughness, Low constraint condition, M(T).

1. INTRODUCTION

The structures such as pipelines and reactor pressure vessel (RPV) that are crack imbibed due to manufacturing processes and service conditions get susceptible to fracture failure [1, 2]. A regular inspection of these components becomes mandatory to ensure that crack shape and size do not reach critical proportions that damage the part/structure during operation. Growth of the crack ensues if the operating conditions are enough to overcome the material fracture toughness. However, study on replicating real time crack behaviour through laboratory test specimen has its own challenges on account of difficulty to depict true behaviour of crack under these loads. American Society for Testing and Materials (ASTM) and the British Standards Institution (BSI) have recommended standards for different crack behaviour of components based on material and loading conditions [3]. These standards quantify plane strain fracture toughness by stress intensity factor (KIC) in linear elastic conditions and J integral (JIC), Crack opening displacement (COD), Crack-tip- opening angle (CTOA) in elastic-plastic conditions for Mode-I loading. Standard specimens used to replicate true nature of crack behaviour result in a conservative fracture toughness value. A single parameter from these standards to quantify crack behaviour is not adequate as it provides too little insight into the true nature of crack.

The researchers have proposed new parameters apart from fracture toughness to quantify true nature of crack under various loading conditions. The new parameters provide scope to define plasticity around crack tip in terms of stress, strain, and displacement collectively designated as constraint parameters. In case of experimental methods constraint parameters are difficult to be defined, but analytical and numerical methods for standard specimens are easy to be comprehended. The Tij stresses, Q stress, A, and A2constraint parameters are well defined in literature for measures of in-plane and out-of-plane constraints near crack tip/front [4]. Each constraint parameter has its own limitations whose discussion lies beyond the scope in this article. The defined constraint parameters along with fracture toughness quantify crack tip/front growth, under the nomenclature of two-parameter and three-parameter fracture assessment [5].

The standard specimen selection to replicate component crack-behaviour plays a vital role, apart from the chosen type of fracture assessment. Thus selection of standard specimens and their limitations to experimentally measure actual fracture toughness, on basis of prevailing constraint at real crack has led to evolution of constraint-corrected fracture assessment of structure. The numerical approach to assess fracture through constraint correction needs high skill and precise computational ability, although data related to standard specimens and values concerning crack shape and geometry are found in literature [6]. The transferability of constraint corrected assessment of standard test specimen to actual structure is extensively researched over last two decades [3, 7].

ASTM and BSI recommend few standard specimens for high and low constraint conditions. High constraint specimens in these standards are Compact Tension (C(T)), Single Edge Notch Bend (SEN(B)), Disc Shaped compact (DC(T)), and Arc Shaped Tension (A(T)). These standard specimens ensure minimum fracture toughness for various geometry of crack under linear elastic fracture mechanics conditions (LEFM).The real practical components susceptible to fracture still use this fracture toughness value leading to design conservatism. Most practical components are under low constraint conditions and possess high toughness for shallow cracks. To overcome the conservatism associated with test specimen, BS 8571:2018 suggests usage of low constraint Single Edge Notch Tension (SENT) specimen [8]. BS 8571:2018 ensures single fracture toughness value for metallic materials under LEFM (in-terms of COD) condition and a resistance curve (R- curve) for ductile materials under Elastic-Plastic Fracture Mechanics (EPFM) condition.

Similarly, ASTM E2472 – 12(2018) recommends Middle Crack Tension (M(T))/ Center-Cracked Tension (CCT) specimen to measure fracture toughness (in-terms of CTOA) for metallic materials under EPFM condition [9]. Both standards are defined (or revised) recently, to acquire fracture toughness data of low constraint to replicate real practical situations. The ASTM standard for metallic materials on low constraint condition under LEFM regime has not been developed with a standard specimen for single value of fracture toughness. The non-usage of SENT specimen in ASTM standards has compelled researchers choose M(T) as low constraint specimen but only under EPFM conditions.

S. Cicero et al. [10] reviewed the fracture assessment of low constraint specimen adopted by industry. The available fracture assessment methods considered ensured structure safety under given load conditions at the expense of excessive conservatism. Matthias Verstraete et al. [11] have suggested a framework to select efficient laboratory test specimen for large scale structure with fracture toughness data adopted to compare different fracture toughness trajectories of laboratory test specimen. The authors recommended suitable test specimen based on preparation cost, time, material needed, and required test capabilities. Zhu X K et al. [12] have reviewed that many researchers claimed suitability of clamped SEN(T) through experimental, numerical and analytical methods for fracture assessment in low constraint applications. However, lack of SEN(T) based ASTM standards necessitates thoughtful relook and comparison of low constraint specimen like M(T) and SEN(T).

Yan Di et al. [13] have suggested use of modified C(T) specimen to measure CTOA for X80 pipeline steel with recommendation for successful conduction of ASTM E1820 test. J.M. Warrenet al. [14] have used Finite Element Analysis (FEA) to predict failure of C(T) and M(T) specimens on basis of CTOA fracture criteria. Lu, L. and Wang, S [15] have conducted tests on C(T) and M(T) specimens for various thicknesses (plane stress and plane strain) to collect K-R curves by ASTM E1820. A novel method was proposed to find CTOA from K- R experimental data. Similarly, K. Lu and T. Meshi [16] have investigated T-stress variation in Center-Cracked Tension (CCT) specimens for various values of the ratios of crack length to width (a/W) and thickness to width (B/W) of the specimen as per ASTM E2472. The T- stress components were negative for all a/W, and B/W considered in the numerical study. The extensive analysis of M(T) specimen is reported in literature, unlike the constraint comparison of M(T) with SENT (low constraint) specimen that finds scarce reported work. The research gap for selection of appropriate standard and test specimens based on application can be helpful if constraint comparison between similar standard test specimens were available.

The selection of standard specimen for low constraint application on SENT and M(T) becomes crucial owing to limited literature available, necessitating user to draw better interpretations with respect to their suitability for an application. A thorough 3D numerical analysis was performed on identical flaw size, fracture toughness and constraint parameters under LEFM conditions for both specimens. A comparison of the constraint parameters through Tij stresses at the crack front were conducted for various a/W ratio. The equations of both specimens for fracture toughness and Tij stress preferred 3D analysis to be adopted. The low constraint specimen was recommended based on constraint parameter trajectory and fracture toughness for various a/W for both standard specimen.

2. STANDARD SPECIMENS AND DIMENSIONS

The 3-D models of Clamped SEN(T) and M(T) specimens are shown in Figure 1, with both specimens under remote tension that ensures Mode-I loading. In clamped SEN(T) specimen, **H** refers to specimen height (daylight of the specimen), **H*** is the clamped distance referred to as gripped region held by the machine, **a** is the initial crack length before loading, **W** is the specimen width, and **B** is specimen thickness. **H/W**=10 and **H*/W** = 4 were considered as per BS 8571:2018 also adopted by C. Bassindale et al. [17]. The square cross- section (**B**=**W**) was considered in this work. Similarly, for M(T) specimen, according to ASTM E2472 –12(2018), **2a** was initial crack length before loading, **2W** was specimen width, **2H** was specimen height and **B** was specimen thickness. In both specimens, **P** or **σ** was applied remote load. Though in ASTM E2472 – 12(2018) clamped M(T) specimen was recommended, direct loading without clamp also resulted in identical values with negligible variation (<0.1%) compared to clamped M(T). The further analysis had M(T) specimen without clamp for easy and refined mesh purposes.

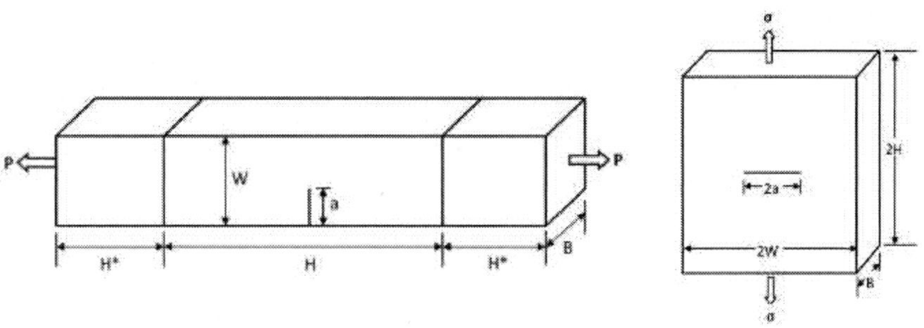

Figure 1. Standard Specimens (a) Clamped SEN(T) (b) M(T)

Analytical solution for clamped SEN(T), suggested by Zhu's [18] closed-form polynomial was used to compare the obtained results. Zhu proposed a sixth-order polynomial to obtain geometrical factor **F(a/W)** that further resulted in KI value based on equation (1). Similarly, for M(T) specimen, closed-form solution mentioned in equation (2) was used [19] for comparison with obtained results.

$$K_I = \frac{P}{B\sqrt{W}} F(\frac{a}{W}) \quad (1)$$

$$K_I = \sigma\sqrt{\pi a}\, F(\frac{a}{W}) \quad (2)$$

The geometry factor **F(a/W)** depends on standard specimen type, crack length, and specimen width.

3. FINITE ELEMENT MODEL AND ANALYSIS METHODOLOGY

3.1 Model Parameters and Computational Procedure

The comparative study of standard specimen fracture toughness in terms of constraint level at crack front with real structure was adopted by virtue of taking both clamped SEN(T) and M(T) specimens with identical initial crack length and thickness. Since real structure (pipe) has a definite thickness with variable crack length as defects, thickness of both specimens were kept constant, and only a/W was varied. Crack length to width ratio (a/W) was varied between 0.25 to 0.7 in increments of 0.05 keeping the specimen thickness as 25.4mm. Other geometrical parameters were derived as per standard relationship. Structural steel with Young's modulus of 207 GPa and Poisson's ratio 0.3 was used during the study [17].

Abaqus 6.14-1commercial FE package was used for all numerical calculations and computational analysis [20]. Fatigue pre-crack was modeled considering seam crack to ensure a sharp crack structure. To maintain crack front singularity, collapsed elements with 0.25 mid-node shift towards crack was used [21]. The counter integral method was used to evaluate both stress intensity factor and Tij-stresses. T11-stress represents stress parallel to crack faces used as a constraint parameter. Similarly, T33-stress was evaluated by relationship in equation (3). In equation (3) E represents Young's modulus, ε33 strain in the z-direction, and v poisson ratio. Both T11 and T33 stresses confirm measurement of in-plane and out of plane constraints at crack tip/front in LEFM[22].

$$T33=E\,\varepsilon 33 + vT11$$

The Tensile load (P) on clamped SEN(T) and equivalent remote stress (σ) for M(T) Specimen were applied . The load considered in the analysis confirmed the LEFM condition at the crack front [23]. To Simulate boundary conditions for clamped regions only load direction movement was allowed and other degrees of freedom was restricted.

3.2 Finite Element Model & Mesh

In case of clamped SEN(T), half symmetry model and in M(T),a quarter symmetry model was used. A typical half symmetry model with FE mesh for clamped C(T) specimen for a/W= 0.5 is shown in Figure 2. Similarly, quarter symmetry FE meshed model of M(T) specimen for a/W=0.5 is shown in Figure 3. Twenty-node quadratic brick elements with reduced integration (C3D20R) were used in the analysis [20]. Near crack front, these nodes were collapsed at one end (near the crack tip side) for efficient replication of singularity. The dense mesh was used near crack front surface and structured mesh maintained in the entire FE model for both specimens. The number of elements was between 8,000 and 20,000 for all a/W considered in both specimens. More details about the mesh generation in cracked models can be obtained from Abaqus Manual [24]

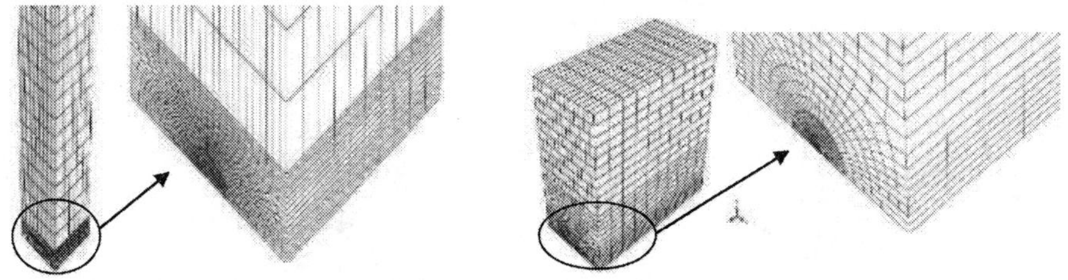

Figure 2. FE meshed model of Clamped SEN(T) **Figure 3.** FE meshed model of M(T)

Standard mesh convergence was performed for both specimens. The stress intensity factor (SIF) values were unaffected beyond certain significant fine mesh refinement at the crack front. The SIF value deviation for mesh refinement was examined and maintained within 0.1% for the entire investigation.

4. RESULTS AND DISCUSSION

The values of KI for both clamped SEN(T), and M(T) specimens were calculated using equations (1) and (2) for various a/W considered. But these values suited for 2-D specimen considered. The 3-D model indicated maximum KI along crack front (present analysis) and KI at crack tip (from equations (1) and (2)) for both specimens with reference to a/W considered as shown in Figure 4. These equations do not consider the effect of constraint originated due to specimen thickness. The present KI (max KI) results on 3-D model have higher values in comparison to Zhu's solutions for all a/W considered for clamped SEN(T).Similarly, for M(T) specimen, all a/W resulted in higher KI values as compared to analytical results. Higher KI in 3-D analysis was attributed to constraint present near the crack front and can be further quantified by constraint parameters.

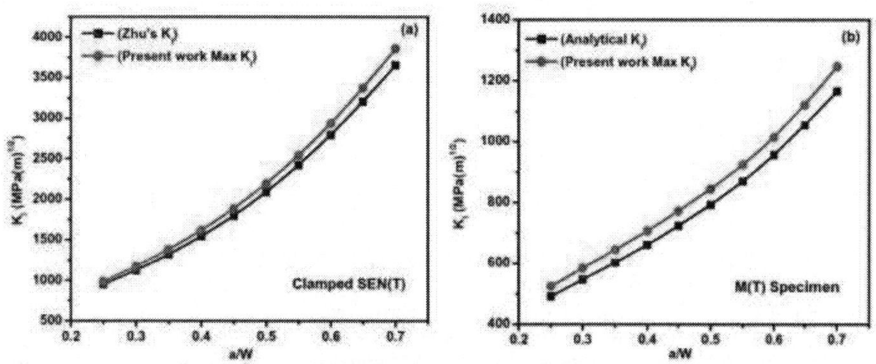

Figure 4. Variation of K_I with a/W (a) Clamped SEN(T) (b) M(T)

Similarly, maximum value of T11 obtained from present analysis and T33 (using equation (3)) along the crack front for both specimens is shown in Figure 5. In both specimens, the T11 and T33 values were negative indicating the low constraint specimen. In clamped SEN(T), T11 increased with increase in a/W and a reverse trend was observed for out-of- plane constraint, T33. Similar kind of variation was observed in Z. Liu et al. [25]

IOP Conf. Series: Materials Science and Engineering 1123 (2021) 012033 doi:10.1088/1757-899X/1123/1/012033

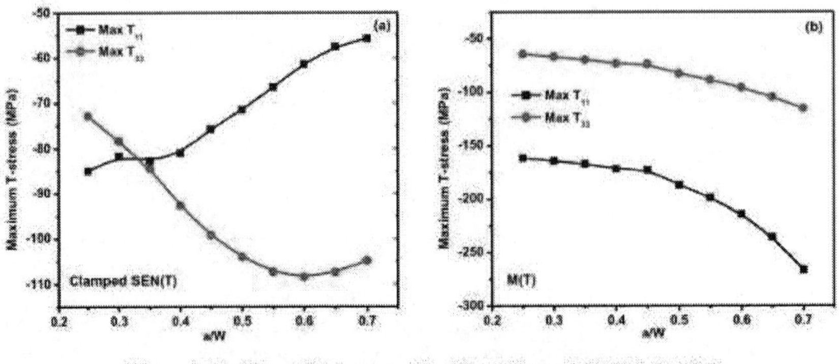

Figure 5. Variation of T-stresses with a/W (a) Clamped SEN(T) (b) M(T)

However, for M(T) specimen, T11 and T33 both decreased with increase in a/W and are in agreement with Henry and Luxmoore [26]. Clamped SEN(T) specimen's T11 value increasing nature with a/W, making it vulnerable in-plane constraint specimen compared to that of M(T).

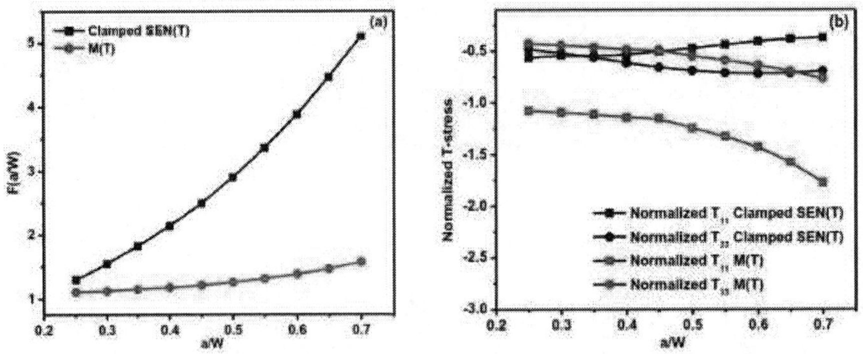

Figure 6. Comparison of Clamped SEN(T) and M(T) specimen (a) F(a/W) (b) Normalized T-stresses

For a given load and a flaw size, selection of laboratory standard specimen influence on fracture toughness values and their constraint at the crack front, can be compared using normalized values of KI, T11 and T33. The Figure 6 shows variation of normalized KI (in terms of F(a/W)) and normalized T-stress with a/W for both specimens. The trajectory of F(a/W) was higher in Clamped SEN(T) compared to M(T) specimen, also the trajectory of normalized T11 was also higher in clamped SEN(T). However, normalized T33 values of M(T) were higher and trajectory attained a similar magnitude for higher a/W with clamped SEN(T). The out-of-plane constraint effect was almost same in both specimens because of slight difference between T33 trajectories. Overall, M(T) specimen constraint values/trajectory exhibited low constraint values as compared to clamped SEN(T) for same crack length and specimen thickness. Based on numerical analysis, the M(T) specimen was a better representation of low constraint compared to clamped SEN(T).

For Clamped SEN(T)

$$F\left(\frac{a}{W}\right) = -9.6572(a/W)^4 + 24.899(a/W)^3 - 11.768(a/W)^2 + 6.5834(a/W) + 0.0396 \quad (3a)$$

$$Maximum\ T_{11} = -2589.2(a/W)^4 + 4362(a/W)^3 - 2499(a/W)^2 + 622.83(a/W) - 141.94 \quad (3b)$$

$$Maximum\ T_{33} = -2084(a/W)^4 + 4598(a/W)^3 - 3369(a/W)^2 + 878.54(a/W) - 145.62 \quad (3c)$$

For M(T)

$$F\left(\frac{a}{W}\right) = 3.381(a/W)^4 - 3.201(a/W)^3 + 1.857(a/W)^2 - 0.173(a/W) + 1.070 \quad (4a)$$

$$Maximum\ T_{11} = -85.65(a/W)^4 - 968.2(a/W)^3 + 841.5(a/W)^2 - 281.6(a/W) - 128.4 \quad (4b)$$

$$Maximum\ T_{33} = 967.9(a/W)^4 - 2012(a/W)^3 + 1290(a/W)^2 - 382.8(a/W) - 21.82 \quad (4c)$$

To avoid numerical calculations and complex analysis for both specimens, it was proposed to develop equations (3) and (4) for F(a/W), T11, and T33 based on curve fitting. The error analysis on the proposed equations and numerical findings indicated proposed equations to have a maximum of 1.3% and a minimum of 0.002% error. The results strongly indicated the suitability of proposed equations as an alternative to complex simulations and numerical analyses.

5. CONCLUSIONS

The conclusions based on the computational studies to compare two different standard specimens used for low constraint conditions in terms of constraint trajectory are presented. The 3-D analysis of both the specimens for the same load and flaw size was considered to estimate the KI and T- stresses parameters. In most cases, maximum KI and T-stress were at center of crack front line owing to crack-tunneling effect at the center. Based on numerical analysis, the M(T) specimen was a better representation of low constraint compared to clamped SEN(T). But preparation of M(T) specimen compared to Clamped SEN(T) was challenging due to fatigue pre- crack initiation and experimental measurement of KIC.

The numerical analysis requires fine mesh and complex simulation to analyze crack geometry and extract the required parameters. The proposed SEN(T) and M(T) mathematical equations based on computational studies avoids complex simulations to offer an easier investigation to practitioners in the field of fracture mechanics. The margins of error associated with these equations are within acceptable industrial margins.

6. ACKNOWLEDGMENTS

KLE Technological University has partially supported this work under "Capacity Building Project" grants.Authors thank KLE Society and KLE Technological University, Hubballi, for the funds and support.

7. REFERENCES

[1]. Khattak MA, Mukhtar A, Khan KA. Common root causes of pressure vessel failures: a review Advanced Research in Applied Mechanics, 2016, 21(1), pp.22-37.

[2]. Tyson, W.R., Newman Jr, J.C. and Xu, S, Characterization of stable ductile crack propagation by CTOA: Review of theory and applications, Fatigue & Fracture of Engineering Materials & Structures, 2018, 41(12), pp.2421-2437.

[3]. Zhu, X.K., Review of fracture toughness test methods for ductile materials in low-constraint conditions, International Journal of Pressure Vessels and Piping, 2016, 139, pp.173-183.

[4]. Matvienko, Y.,Comparison of the constraint parameters in elastic-plastic fracture mechanics, FratturaedIntegritàStrutturale, 2019, 13(49), pp.36-43.

[5]. Malikova, L, Multi-parameter fracture criteria for the estimation of crack propagation direction applied to a mixed-mode geometry, Engineering fracture mechanics, 2015, 143, pp.32-46.

[6]. Gupta, M., Alderliesten, R.C. and Benedictus, R, A review of T-stress and its effects in fracture mechanics. Engineering Fracture Mechanics, 2015, 134, pp.218-241.
[7]. Qian, G., Cao, Y., Niffenegger, M., Chao, Y.J. and Wu, W., Comparison of constraint analyses with global and local approaches under uniaxial and biaxial loadings. European Journal of Mechanics- A/Solids, 2018, 69, pp.135-146.
[8]. BS 8571:2018, Method of test for determination of fracture toughness in metallic materials using single edge notched tension (SENT) specimens, British Standard Institution.
[9]. ASTM E2472–12(2018), Standard Test Method for Determination of Resistance to Stable Crack Extension under Low-Constraint Conditions, ASTM International, West Conshohocken, PA, 2018, www.astm.org.
[10]. Cicero, S., Ainsworth, R.A. and Gutiérrez-Solana, F., Engineering approaches for the assessment of low constraint fracture conditions: A critical review, Engineering Fracture Mechanics, 2010, 77(8),pp.1360-1374.
[11]. Verstraete, M., Hertelé, S., Van Minnebruggen, K., Denys, R. and De Waele, W., Considerations in selecting laboratory scale test specimens for evaluation of fracture toughness, In Sustainable Construction and Design (SCAD-2013).
[12]. Zhu XK, McGaughy T. A review of fracture toughness testing and evaluation using SENT specimens, 10th International Pipeline Conference 2014 Sep 29. American Society of Mechanical Engineers Digital Collection.
[13]. Di, Y., Shuai, J., Wang, J. and Tu, S., A new specimen for high-grade pipeline steels CTOA test, Engineering Fracture Mechanics, 2015, 148, pp.203-212.
[14]. Warren, J.M., Lacy, T. and Newman Jr, J.C., Validation of the Two-Parameter Fracture Criterion using 3D finite-element analyses with the critical CTOA fracture criterion, *Engineering Fracture Mechanics*, 2016, 151, pp.130-137.
[15]. Lu, L. and Wang, S., Relationship between crack growth resistance curves and critical CTOA, *Engineering Fracture Mechanics*, 2017, 173, pp.146-156
[16]. Lu, K. and Meshii, T., A systematic investigation of T-stresses for a variety of center-cracked tension specimens, *Theoretical and Applied Fracture Mechanics*, 2015, 77, pp.74-81.
[17]. Bassindale, C., Wang, X., Tyson, W.R. and Xu, S., Numerical verification of stress intensity factor solution for clamped single edge notched tension (SENT) specimens, *Fatigue & Fracture of Engineering Materials & Structures*, 2018, 41(2), pp.494-499.
[18]. Zhu, X.K., Corrected stress intensity factor solution for a British standard single edge notched tension (SENT) specimen, *Fatigue & Fracture of Engineering Materials & Structures*, 2016, 39(1), pp.120-131.
[19]. Prashant Kumar, SIF of more complex cases, *Elements of fracture mechanics*, Mcgraw Hill Education (India), 2017.
[20]. Abaqus version 6.14-2, DassaultSystemes, Providence, 2016.
[21]. Kudari, S. K. and Kodancha, K. G. 3D Stress intensity factor and T-stresses (T11 and T33) formulations for a Compact Tension specimen, *FratturaedIntegritàStrutturale*, 2016, 11(39), pp. Pages 216-225.
[22]. Meshii, T. and Tanaka, T., Experimental T33-stress formulation of test specimen thickness effect on fracture toughness in the transition temperature region, *Engineering Fracture Mechanics*, 2010, 77(5), pp.867-877.
[23]. Larsson, S.G. and Carlsson, A., Influence of non-singular stress terms and specimen geometry on small-scale yielding at crack tips in elastic-plastic materials. *Journal of the Mechanics and Physics of Solids*, 1973, 21(4), pp.263-277.
[24]. Abaqus/CAE user's guide, DassaultSystemes, Providence, 2014.
[25]. Liu, Z., Wang, X., Tang, J., Deng, C., Zhao, H. and Chen, X.,. The effects of in-plane and out-of-plane constraints on JR curves for X80 steel: A study using clamped SENT specimens. *Engineering Fracture Mechanics*, 2019, 206, pp.342-358.
[26]. Henry, B.S. and Luxmoore, A.R., Three-dimensional evaluation of the T-stress in centre cracked plates. *International journal of fracture*, 1994, 70(1), pp.35-50

Tensile behavior of lotus natural fiber and e-glass fibers reinforced with epoxy composites

A V Abhijith[1], Abhishith Dharmarajan Thayyil[1] and T Narendiranath Babu[1]

[1] Department of Mechanical Engineering, Vellore Institute of Technology, Vellore, Tamil Nadu, 632014, India
Email: narendiranathbabu.t@vit.ac.in

Abstract. This paper is concerned with the study of tensile properties of lotus fiber reinforced with epoxy composites. Specimens of epoxy composites were made from natural fibers by arranging them in uniaxial, biaxial as well as in criss-cross manner with and without E-glass fiber. To enhance the mechanical properties, fibers were undergone alkaline treatment (NaOH) and later they were fabricated using hand lay-up technique. It is understood from the results that better tensile properties were shown by uniaxial and minimal by biaxial arrangement.

Keywords: Lotus, Composite, Tensile Strength, E-glass Fiber, Criss – cross, Uniaxial, Biaxial, Specimen

1. INTRODUCTION

Today, non-biodegradable materials became the major constituent in almost all products in the day-to-day human life. Increasing environmental awareness is forcing humans to seek more eco- friendly materials as a substitute such as natural fibers and other organic substances. The natural fibers include hemp, coir, jute, kapok, kenaf, flax, nettle, bamboo, sisal, ramie, etc. They show many advantages as compared to synthetic fibers such as biodegradability, recyclability and low density. They have similar specific tensile properties, and are also non-abrasive to the equipment causing minimal health risk [1]. Since they are cheaper and lighter, they have low mechanical properties as that of glass fiber. But this can be eased by hybridization of the natural fibers with glass fiber composites. Composites contain reinforcement (strong load carrying material) which is incorporated in matrix (weaker material). While reinforcement imparts strength, matrix sustains the orientation and position of reinforcement [2].

Academic and industrial areas have an increased interest in natural fibers as reinforcement in composites. They have many advantages over synthetic fibers. Even though moisture sensitivity stands out to be their major disadvantage, acceptable mechanical properties makes it consider them as a good choice in various domains. They have much wide range of applications in packing, building construction, cosmetics. Nowadays even the interior portions of Automotive and aircrafts are designed using natural fibers [3]. Car manufacturers across the globe have started accepting natural fiber composites for dashboards, door panels, headliners and interior parts [4].

Lotus (Nelumbo nucifera) is a flowering plant with high amount of fibrous content in stem and root. They come under the category of perennial aquatic herb which is mainly found in India and China [5]. The fibers used in this paper were extracted from the stems which were soaked in water for 2 nights and were washed with a piece of linen until all thorns were removed [6]. These stems contain large amount of cellulose which can be used in paper, textile and medical industries. Moreover lotus fibers are becoming popular in ayurvastra and also in making luxury garments, since they are eco-friendly and comfortable to use [7, 8]. Even though they contain 17 kinds of amino acids, they are insusceptible to skin and are healthy for human body [9]. Since these fibers have low density (1.18 gcm^{-3}) they are suitable for making lightweight composites [10].

Glass fiber reinforcement is known to improve the mechanical properties of natural fiber composites. There are many types of glass fibers but in this paper e - glass fibers are used. Some of the merits of glass fibers are:

- Excellent chemical resistivity towards acids and solvents

- Low dielectric constant, low moisture absorption and high strength to weight ratio.
- They are cheap and easy to trim even after curing.

Effect of glass fiber inclusion was measured by making few composites using e – glass fibers. Lotus fiber orientation was also examined by arranging them in 3 different ways such as uniaxial, criss – cross and biaxial.

2. MATERIALS

Glass fibers used were of two types- uniaxial and biaxial. These glass fibers were of 5 – 10 μm diameter and had a surface density of 360 gram per square meter. Moderate adhesive epoxy resin (LY556) and polyamine hardener (HY951) which was cured under room temperature was used as the matrix. Natural fiber used here was lotus which is shown in figure 1.

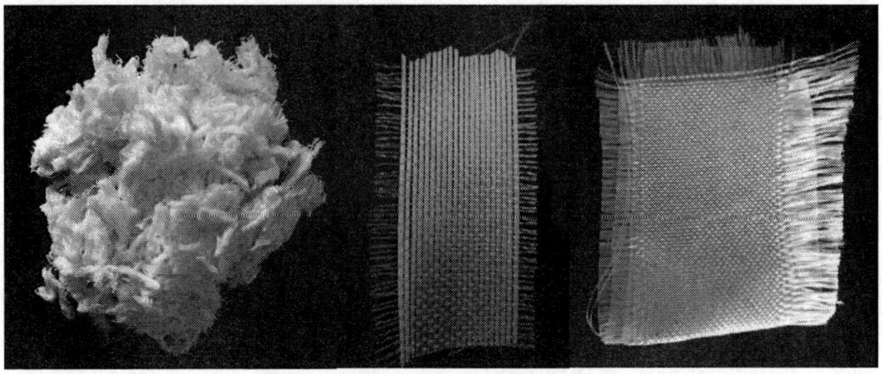

Figure 1. (a) Lotus fiber (b) Uniaxial glass fibers (c) Biaxial glass fibers.

3. METHODOLOGY

3.1 Alkali treatment

About 250 grams of lotus fiber was soaked in 1 L of 0.5 % NaOH solution for 30 minutes. At 65 % humidity and at normal room temperature, fibers were allowed to dry for 3 days. Through alkalization, increase in quality of the composite was expected due to the improved adherence of fiber with the matrix. So the fiber will show better mechanical properties because of the removal of non-cellulosic content such as lignin and hemicellulose from the natural fiber which affects their tensile properties [11].

3.2 Preparation of mold

For making the composite specimens of desired shape, molding boxes were prepared which was made of wood. ASTM standards - D638 was followed to get the dimensions of the cavity (figure 2).

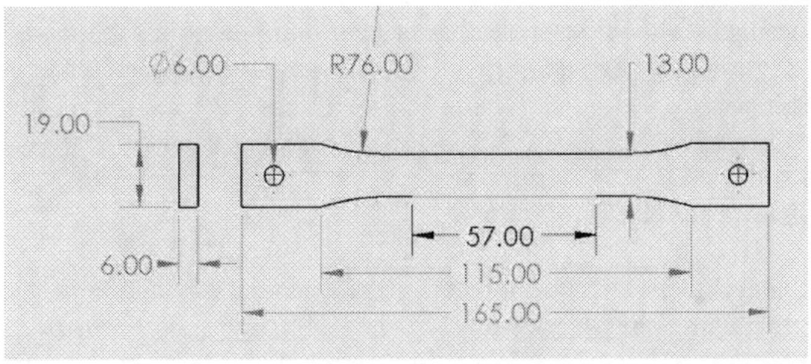

Figure 2. Tensile test mold dimensions (A S T M standard D 638).

3.3 Preparation of composites

Since alkali treated fibers were in disordered manner, uniform threads were formed by twisting them. They were then cut into pieces of different length as per the 3 different arrangements (uniaxial, biaxial and criss-cross). Epoxy and hardener were mixed in 10:1 (v/v) ratio to make a solution for minimizing the fiber's detachment. Three alternative layers of epoxy-hardener solution and lotus fiber were placed in the mold and this was done by hand lay-up technique followed by compression molding. Figure 3 shows the prepared specimens. Addition of epoxy increases adhesion effect which further leads to satisfactory reinforcement. This makes desirable fiber distribution and effective bridging of gaps in between those natural fibers [12].

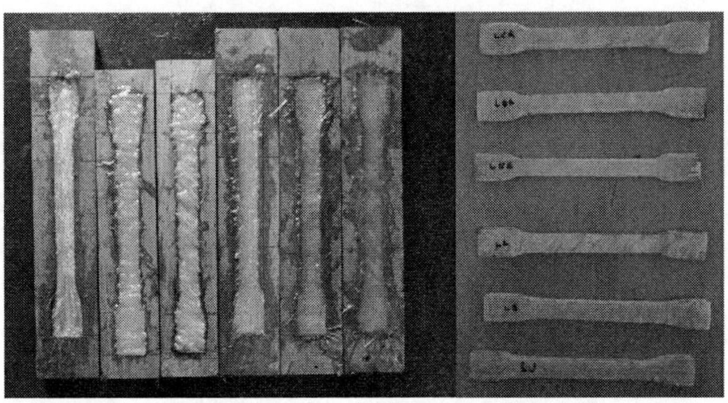

Figure 3. (a) Lotus composites; (b) Specimens prepared for testing.

3.3.1 Lotus fiber composite (no glass fiber).

- Base of cavity in mold was coated with epoxy-hardener solution. Alternative layer by layer arrangement finally formed three layers of fiber and four layers epoxy-hardener in the cavity.
- It lead to an overall fiber to epoxy-hardener volume by volume ratio of 60:40 which is shown in table 1.
- 5 threads of lotus fiber were placed along the length in each layer to form uniaxial arrangement of specimen.
- In biaxial arrangement, twisted threads were cut according to the width of the cavity, which were then placed along the length of the mold. Approximately 60-70 pieces were cut and placed accordingly.

- For criss-cross arrangement of fiber, threads were cut into pieces in such a way that it can fit across the width at an angle of 45 degrees. Around 65-75 pieces of these cut fibers were placed in tilted manner along the length.

3.3.2 Lotus fiber composite containing glass fiber.

- Epoxy-hardener, lotus fibers, epoxy-hardener, glass fibers, epoxy-hardener layers were arranged in the cavity by starting with epoxy-hardener.
- So in total 2, 5, 2 layers of lotus fiber, epoxy-hardener, glass fiber were obtained respectively which lead to a volume by volume ratio of 45:45 for lotus and epoxy-hardener in mold cavity.
- In uniaxially arranged specimen, 165 mm long lotus fibers were measured and cut to arrange along the length. Each layer carried 5 threads and uniaxial e - glass fibers were attached accordingly.
- For biaxial arrangement, fibers were cut and arranged perpendicular to cavity's length. Biaxial e - glass fibers were inserted and almost 60-70 pieces of fiber were required.
- Fibers were cut and arranged in the cavity at an angle of 45 degrees to the length. Criss-cross arrangement was done by using 65-75 pieces of lotus fiber and biaxial E-glass fiber.

Table 1. Compositions of composites.

Sl. No.	*Composite description*	*Lotus fiber (% v/v)*	*Epoxy-hardener (% volume / volume)*	*E - glass fibers (% volume / volume)*
1	Lotus fiber composites with epoxy(no glass fibers)	60	40	0
2	Lotus fiber composites with epoxy containing glass fibers)	45	45	10

They were allowed to dry at atmospheric temperature for 13 days and were taken out from the mold to get the specimen for performing the mechanical test. Involved steps in the preparation of specimens are abbreviated in flow chart (figure 4).

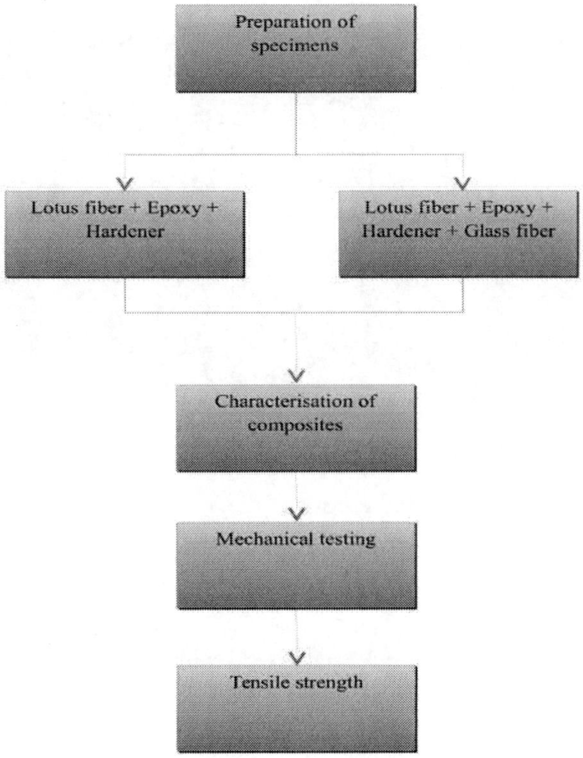

Figure 4. Methodology.

4. EXPERIMENTAL PROCEDURE

1. Digital tensile testing machine in laboratory was used to perform tensile test of ASTM D638 standard specimens. Details of machine are provided in table 2.
2. Grips in the machine securely clamp 2 ends of the specimen as shown in figure 5.
3. The gauge length of the specimen was 115 mm.
4. Grips of the machine pull the specimen at a constant rate (1mm min^{-1}).
5. Force applied and it's elongation of the specimen were continuously recorded and a graph was drawn.
6. When the force applied in the specimen exceeded critical load, failure occurred and it got broke.

Table 2. Specifications of machine used.

Equipment	Digital Tensile Testing Machine
Model number	8801
Load cell	Dynacell load cell highlights compensation for inertial loads brought about by overwhelming fixtures and grips
Maximum working pressure	207 bar
Capacity of dynamic load	Axial force up to ± 100 kN
Computer software used	Bluehill : Static test & Wavematrix : Dynamic test
Supplier	Instron

IOP Conf. Series: Materials Science and Engineering 1123 (2021) 012034 doi:10.1088/1757-899X/1123/1/012034

Figure 5. (a)Specimen clamped in machine (b) Tensile testing machine (c) Specimen after failure (Courtesy: Materials testing lab, VIT Vellore).

5. RESULTS

Load-extension curve of lotus fiber composite, with and without E-glass fibers obtained from the digital tensile testing machine are given in figures 6-8.

Largest value of tensile stress that a substance could withstand before failure occurs is known as tensile strength. Stress is directly proportional to the load applied. Hence maximum load corresponds to maximum tensile strength. It gives the tendency to elongate the material since force was acting along the length in opposite direction. Usually natural fiber composites show somewhat lesser mechanical properties due to decrease in stiffness and brittleness [1]. But incorporation of E-glass fiber showed greater improvement in it.

Plotted graphs showed a linear relationship until the breaking point where it showed a sudden depression. To get more precise result, load was applied at a very small rate of 1mm per minute. Graphs obtained clearly implied that the largest value of tensile strength (18.91 MPa) was shown by the composite in uniaxial natural fiber arrangement (in presence of glass fibers). On the other hand, criss-cross and biaxial arrangement showed a lesser value of 18.38 MPa and 11.92 MPa respectively. In case of Lotus - epoxy composites, uniaxial, criss-cross and biaxial structure show values of 17.15, 13.73 and 9.14 MPa.

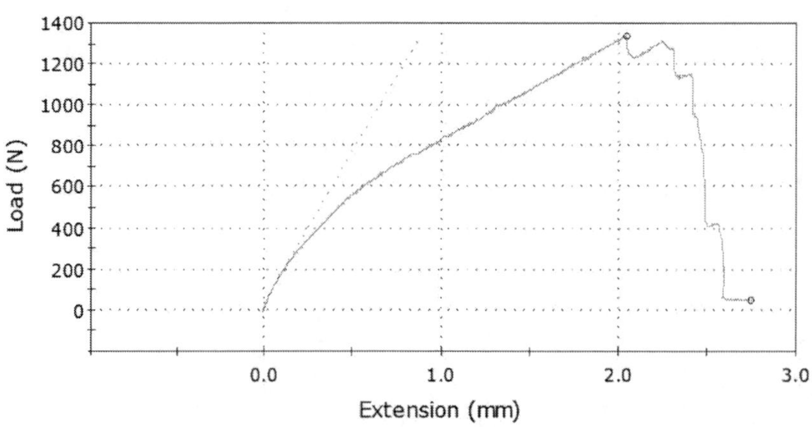

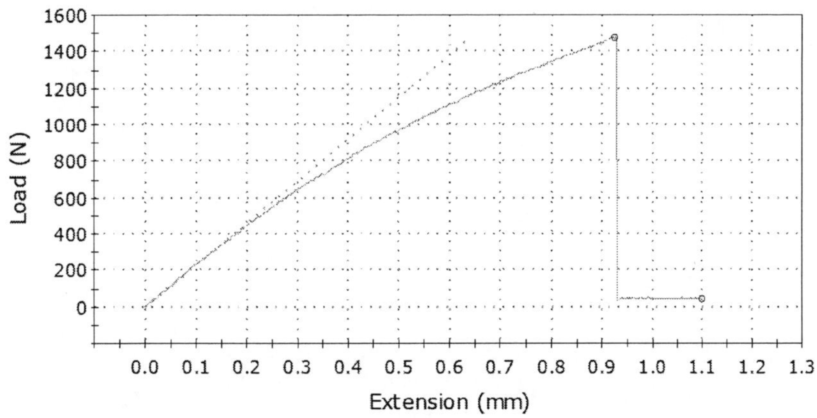

Figure 6. (a) Lotus and epoxy composite (Lotus fiber arrangement in uniaxial manner); (b) Lotus and epoxy composites with glass fiber (Lotus fiber arrangement in uniaxial manner)

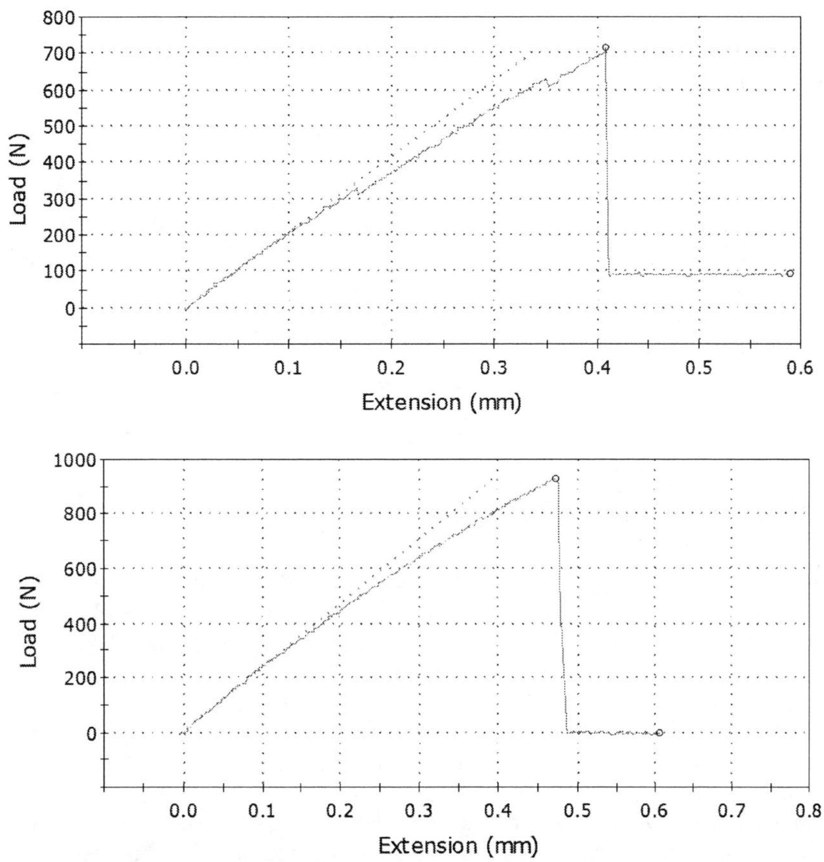

Figure 7. (a) Lotus - epoxy reinforced fiber composite (Lotus fiber arrangement in biaxial manner); (b) Lotus-epoxy-glass fiber reinforced composite (Lotus fiber arrangement in biaxial manner)

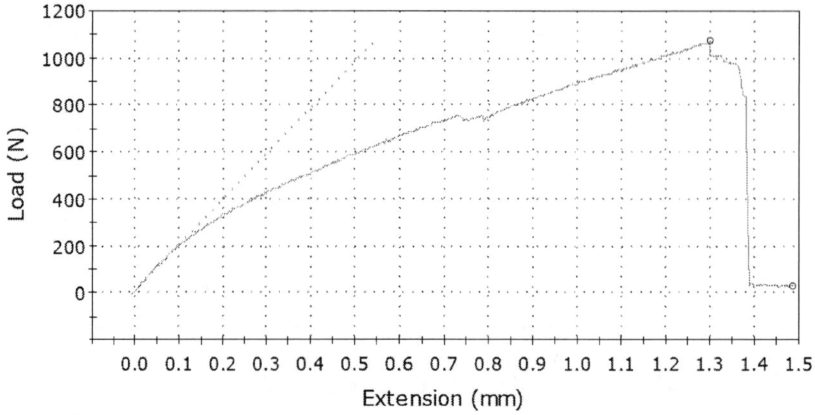

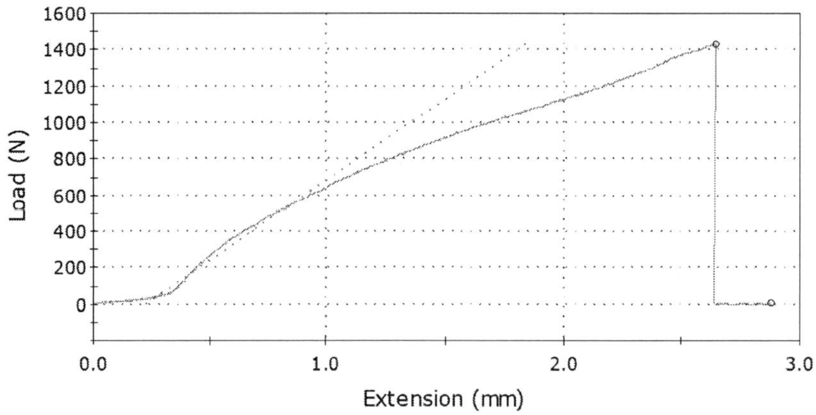

Figure 8. (a) Lotus-epoxy reinforced fiber composite (Lotus fiber arrangement in criss - cross manner); (b) Lotus-epoxy-glass fiber reinforced composite (Lotus fiber arrangement in criss - cross manner)

Orientation of fibers and glass fiber inclusion were the key factors for the variations seen in the graphs obtained.

5.1 Comparison based on presence of glass fiber in composites

Hybridization with glass fiber showed significant increase in composite mechanical properties. 6 Specimens were made with a fixed volume percentage of 60 and 40 of lotus and epoxy-hardener respectively. Similarly another 6 specimens containing lotus, epoxy-hardener, E-glass fiber in 45, 45, 10 volume % were prepared respectively. Addition of glass fibers in the composite always played a vital role in slightly increasing the tensile strength of the natural fiber [13].

It was evident from the graph that uniaxial arrangement without glass fiber can withstand a tensile strength of 17.15 MPa as compared to a specimen having the same arrangement incorporating glass fiber whose strength was 18.91 MPa. When criss-cross arrangement showed a tensile strength of 13.73 MPa, presence of glass fiber increased it to 18.38 MPa, whereas in case of biaxial arrangement, it went up to 11.92 MPa from 9.14 MPa. It concludes a drop of 9.3%, 25.3% and 23.32% breaking stress in uniaxial, criss-cross and biaxial arrangement respectively. These values are summarized in Figure 9.

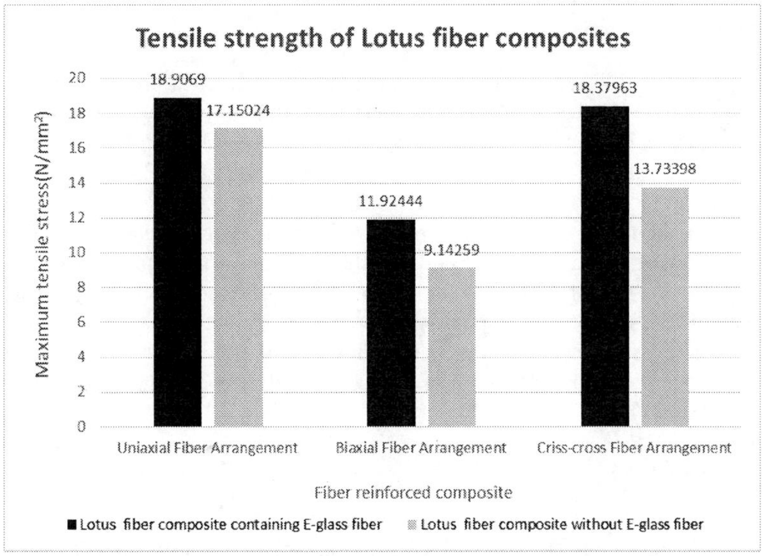

Figure 9. Maximum tensile stresses of different specimens.

Among all these specimens, it was well-defined that specimens with E-glass fiber had higher tensile strength.

5.2 Comparison based on orientation of fibers

For all composites, with and without E-glass fiber, highest tensile strength was shown for uniaxial, followed by criss-cross and least was for biaxial arrangement. When the fibers were in uniaxial orientation, fiber matrix interface were bonded with appreciable compatibility as compared to biaxial and criss-cross arrangement. Experimental values of breaking load and stress are listed in the table 3 for all specimens.

Table 3. Experimental results of tensile test.

Sl No.	*Specimen Composite components (fiber orientation)*	*Breaking load (N)*	*Breaking tensile stress (MPa)*
1	Lotus and epoxy (uniaxial)	1337.72	17.15024
2	Lotus and epoxy (biaxial)	713.12	9.14259
3	Lotus and epoxy (criss-cross)	1071.25	13.73398
4	Lotus, epoxy and glass fiber (uniaxial)	1474.74	18.9069
5	Lotus, epoxy and glass fiber (biaxial)	930.11	11.92444
6	Lotus, epoxy and glass fiber (criss-cross)	1433.61	18.37963
Mean		1160.09	14.87

Uniaxially oriented lotus-epoxy-glass fiber reinforced composite failed at a tensile stress of 18.91 MPa. Criss-cross and biaxial specimens having E-glass fiber, showed 2.79% and 36.93% drop in breaking strength compared to this. Whereas uniaxially oriented lotus –epoxy reinforced composite showed 17.15 MPa and in case of criss-cross and biaxial orientation it decreased to 19.92% and 46.69% respectively.

6. CONCLUSION

ASTM D638 standard was followed to make specimens of lotus fiber in uniaxial, criss-cross and biaxial orientation with and without E-glass fiber. Lotus fibers were treated with NaOH solution before the mold was prepared to enhance bonding at the fiber-matrix interface. When glass fiber was added 45:45 volume/volume ratio of lotus fiber and epoxy-hardener was followed. They were arranged in 60:40 volume / volume ratio in the absence of E-glass fiber. The fibers which were cut according to the required dimensions were arranged in layer by layer manner to fill the cavity. Results obtained from the tensile test performed in digital tensile testing machine showed that uniaxial orientation of fibers withstood greater tensile stress. Moreover it was enhanced by the addition of E-glass fiber. Tensile strength of the specimens in uniaxial, criss-cross and biaxial were increased by 10.24 %, 33.82% and 30.43 % by the addition of glass fiber. These composites have a huge number of applications such as in medical field and in textile industry. Since the load it can withstand was limited, it can be used for the production of soft carry bags.

7. ACKNOWLEDGEMENT

The authors are extremely thankful for the support from the management of VIT Vellore in helping for accomplishing this research. They are also thank Vibration lab, VIT Vellore for providing required epoxy, hardener and E-glass fibers. The authors are grateful for the assistance in providing the tensile testing machine in advanced materials testing lab, VIT Vellore. All other cooperation from lab technicians and faculties are sincerely acknowledged.

8. REFERENCES

[1] Sanjay, M.R., Arpitha, G.R. and Yogesha, B., 2015. Study on mechanical properties of natural-glass fibre reinforced polymer hybrid composites: A review. *Materials today: proceedings*, ***2*(4-5)**, pp.2959-67.

[2] Taj, S., Munawar, M.A. and Khan, S., 2007. Natural fiber-reinforced polymer composites. *Proceedings-Pakistan Academy of Sciences*, ***44*(2)**, p.129.

[3] Sanjay, M.R., Arpitha, G.R., Naik, L.L., Gopalakrishna, K. and Yogesha, B., 2016. Applications of natural fibers and its composites: An overview. *Natural Resources*, ***7*(03)**, p.108.

[4] Holbery, J. and Houston, D., 2006. Natural-fiber-reinforced polymer composites in automotive applications. *Jom*, ***58*(11)**, pp.80-86.

[5] Cheng, C., Guo, R., Lan, J. and Jiang, S., 2017. Extraction of lotus fibres from lotus stems under microwave irradiation. *Royal Society open science*, ***4*(9)**, p.170747.

[6] Hlaing, C.S., 2016. Lotus Robe in Kyaing Khan Village Innlay Lake, Shan State (South): An Anthropological Perspective. *Dagon University Research Journal*, ***7*(1)**, p.102.

[7] Wu, M., Shuai, H., Cheng, Q. and Jiang, L., 2014. Bioinspired green composite lotus fibers. *Angewandte Chemie International Edition*, ***53*(13)**, pp.3358-61.

[8] Gong, Y., Han, G.T., Zhang, Y.M., Zhang, J.F., Jiang, W. and Pan, Y., 2015. Research on the degradation performance of the lotus nanofibers-alginate porous materials. *Polymer degradation and stability*, ***118***, pp.104-10.

[9] Chen, D.S., Gan, Y.J. and Yuan, X.H., 2012. Research on Structure and Properties of Lotus Fibers. In *Advanced Materials Research* (Vol. **476**, pp. 1948-1954). Trans Tech Publications.

[10] Wu, M., Shuai, H., Cheng, Q. and Jiang, L., 2014. Bioinspired green composite lotus fibers. *Angewandte Chemie International Edition*, ***53*(13)**, pp.3358-61.

[11] Bledzki, A.K. and Gassan, J., 1999. Composites reinforced with cellulose based fibres. *Progress in polymer science*, ***24*(2)**, pp.221-74.

[12] Babu, N.T., Kuclourya, T., Kumar, M.J. and Mangalaraja, R.V., 2019. Mechanical Behaviour of Areca Nut, Sunn Hemp Natural Fibers and E-Glass Fibers Reinforced With Epoxy Composites. *Engineering and advanced tehnology*, **8(6)**, pp.2249-8958.

[13] Jarukumjorn, K. and Suppakarn, N., 2009. Effect of glass fiber hybridization on properties of sisal fiber–polypropylene composites. *Composites Part B: Engineering*, ***40*(7)**, pp.623-27.

TOPOLOGY OPTIMIZATION OF THE BELL CRANK & BRAKE PEDAL

Tanmay Nandanwar, Keyour Waghela, Eshaan Gupta, Narendiranath Babu T

Department of Mechanical Engineering, Vellore Institute of technology, Vellore, India, tanmayvit802620@gmail.com
Department of Mechanical Engineering, Vellore Institute of technology, Vellore, India
Department of Mechanical Engineering, Vellore Institute of technology, Vellore, India
Associate professor, Department of Mechanical Engineering, Vellore Institute of Technology, Vellore, India, narendiranathbabu.t@vit.ac.in

Abstract. Topology Optimization has been an effective tool for weight reduction and performance design and has essential applications in automobile industry. The main objective of this paper is to design a lighter weight bell crank and brake pedal than the existing design which is generally used in the automobile parts. The newly optimized design of bell crank and brake pedal is compared with the existing design. The CAD model of the bell crank and brake pedal is modeled using SOLIDWORKS software and the analysis of the newly optimized model is carried out using the Altair INSPIRE 9.5 . Using the above methodology, a light weightbell crank and brake pedal model is designed and the FEA analysis is used for obtaining the maximum and minimum value of displacement, % of yield, tension/compression, shear stress and von mises stress on the optimized design. The results of the design reflect that the weight of the bell crank has been reduced to 140 grams which is about 24% less weight then the existing design and the optimized weight of the brake pedal is reduced from 363g to 211g.

Keywords: Altair Inspire, Bell crank, brake pedal, FEA and Topology Optimization

1. INTRODUCTION

Topology Optimization is a computational methodology for the optimization of the geometries of various components. It is a mathematical procedure that optimizes the component layout within a given space constraints for a given set of loading and boundary conditions with a aim of maximizing the performance of the component. It optimizes the shape of the component in such a way that its weight is reduced and it performs in much similar way to the existing design without negotiating with the strength. Finite element analysis is used for analysis to implement the generated topologies [1]. Topology optimization has started given much importance since the last decade and hence consumers are looking for best of the products. Hence topology optimization plays a crucial role in the automotive industry. TO of various component is to make the less time to produce the product that is stronger, lighter and with less cost. The design and weight of that component has an essential influence on the performance of the cars. In order to reduce the cost of the manufacturing component the TO of the particular component has to take care of the manufacturing constraints. By reducing the weight of the vehicle using TO we reduce the fuel consumption and hence enhance the performance. When the mass of the vehicle is reduced the inertial forces required to be overcome by the engine is reduced and hence the energy required for moving the vehicle is reduced. Generally, the fuel consumption reduced by 5% by reducing the weight of the vehicle by 7-10% [2]. By reducing the weight of the single component may not give significant impact on the performance and fuel consumption efficiency of the vehicle, hence there is a requirement of optimization of various components as good as possible. Cavazzuti used the

method of TO to design the high-performance automotive chassis [3]. J. Bauer used the topology optimization method on structural trusses by taking descrete design variable [4]. Dongkaixu, Junchen and Yuchengtang, Jiancao have designed the stamping dies for automotive vehicles by using topology optimization so as to reduce the die weight and also its cost in order to improve fuel efficiency and vehicle safety.[5]. Kaya N. re-designed the failed clutch fork using topology and shape optimization by the response surface method [6]. Swapnil Thigale reduced the weight of Brake Disc using Topology Optimization.[7]. Lindan Yi used ANSYS software, takes topology optimization design of automobile engine bracket, and makes the strength analysis, fatigue analysis and model analysis of the new optimized bracket. The new bracket is having less stress concentration and its fatigue analysis are not based on real road load spectrum.[8]. Zheng used topology optimization in engine support structure of construction machinery [9]. The older TO formulation use the Finite Element Analysis to evaluate the design related performance of the component. TO has a wide range of applications in various sectors. This paper presents TO of the bell crank used in the suspension system of automobiles. Bell crank is a lever or rocker having two arms meeting at a pivot at a right angle, and is used for transmission of motion between two parts meeting at an angle. It is a type of crank that can change motion through an angle. The degree of angle can vary from 0 to 360 degrees but 90 and 180 degrees are most common. It converts the translation motion of one component into the translation motion of other component operating at different angles. However, the wider the angle extended by the crank ,the more non-linear the direction of motion which builds up high stresses at the angles overtime [10].A good example of bell crank is a bicycle brake where the force coming from the handlebar lever is rotated at an angle of 90 degrees to push the brake block against the wheel rim. It is being used in the suspension system in automobile, where it actuates the spring-damper unit with the help of either pushrod or pull rod. It is one of the most important components of the automotive suspension system.[10]. The results show that the newly optimized model of bell crank has similar performance characteristics as compared to the existing design without negotiating the required strength. The brake pedal finds its use in automotive applications for brake mechanism. The mechanism of braking system involves conversion of kinetic energy in motion into brake energy commonly known as heat [11]. Mohd Nizam performed topology optimization of brake pedal with changing the material and found out that the new optimized brake pedal was 22% lighter as compared to the existing model [12]. K K Dhande, N I Jamadar and Sandeep Ghatge performed the topology optimization on brake pedal so as to reduce the weight as compared to the stock model without compromising on the performance efficiency of the brake pedal [13].
Currently used brake pedal contains excess material than the performance requirement leading to an increase in the weight of the vehicle. The brake pedals are largely made from metals but composite clutches and accelerator pedals have started using effectively in automotive sectors. This study is mainly concentrated on variable-material for the conceptual design, brake pedal profile, which is achieved by 78% lighter in weight as compared to the existing metallic pedals and 64% lighter weight as compared to the aluminum according to the General Motors specification. In Design, with the application of optimization, the aims of this paper is to reduce the weight of the existing brake pedal design of a car without the substitution of material. By reviewing the design constraints, load and boundary condition the optimization was run and the analysis was carried out using Altair Inspire. However, the result of the optimization process needs further refinement as it has a manufacturability deficiency. Thus, knowledge of design engineers to interpret and refine the proposed design is vital to ensure it is possible for production. In conclusion, the application of optimization with the integration of engineering knowledge of design engineer will be able to produce an optimal brake pedal design in a short time.

2. TOPOLOGY OPTIMIZATION OF THE BELL CRANK/BRAKE PEDAAL

Process flow for topology in Altair inspire

1) First step is developed the CAD model of the required part and generate a STEP format file.
2) Open Altair inspire and import the STEP file in it.
3) Now the software differentiates the model into 2 parts the design space and non-design space.
4) Design space is the selected region where topology optimization needs to occur, and the non-design space are the region which remain unchanged like mounting points, junctions etc.

IOP Conf. Series: Materials Science and Engineering 1123 (2021) 012035 doi:10.1088/1757-899X/1123/1/012035

5) The next step is to apply constraints for simulation, like support, force, pressure or moment etc. The material properties are also selected for the part.
6) Once all the above steps are completed icon for topology optimization is selected. Now there are 2 purposes minimize weight or maximize stiffness (for selected percentage of weight) as shown in the image below.

Figure 1: Selection of stiffness

7) For the bell crank/brake pedal we have selected for maximum stiffness for given percentage of weight.
8) Once the simulation is done, we are presented with topology optimized rough surfaces.
9) Now in this software there is a feature called polynurbs which smoothens the surface to improve its mechanical property and it make manufacturable. After finishing this step, the part is completed and for 3D printing.

3. RESULTS AND DISCUSSION

We have done our FEA analysis on ALTAIR Inspire. The CAD model and the simulations are shown as follows:

Analysis of bell crank

Figure 2. Stock model of bell crank

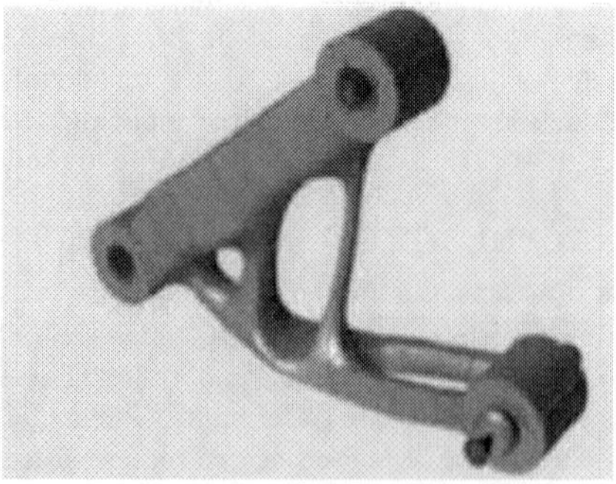

Figure 2 a. Optimized design of bell crank

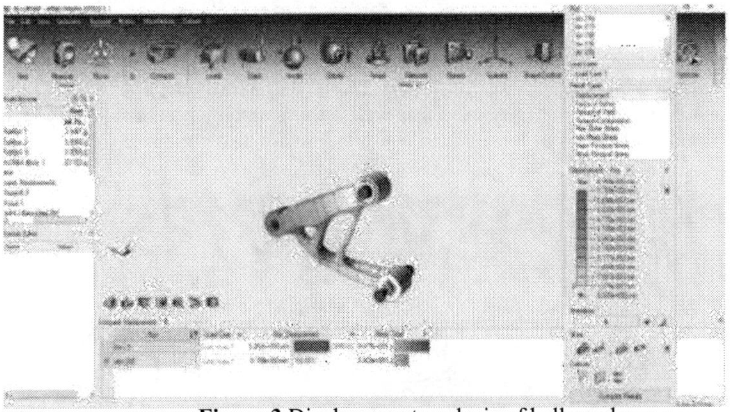

Figure 3. Displacement analysis of bell crank

The displacement analysis(fig3) shows a maximum displacement of **6.75e^-2mm** and minimum of **1.27e^-3 mm.**These displacement values won't have much effect on the components working thus proving a good stiffness of the design.

Figure 4. Yield analysis of bell crank

In the above figure we can see a maximum yield of 28.62% and minimum 0.06%.

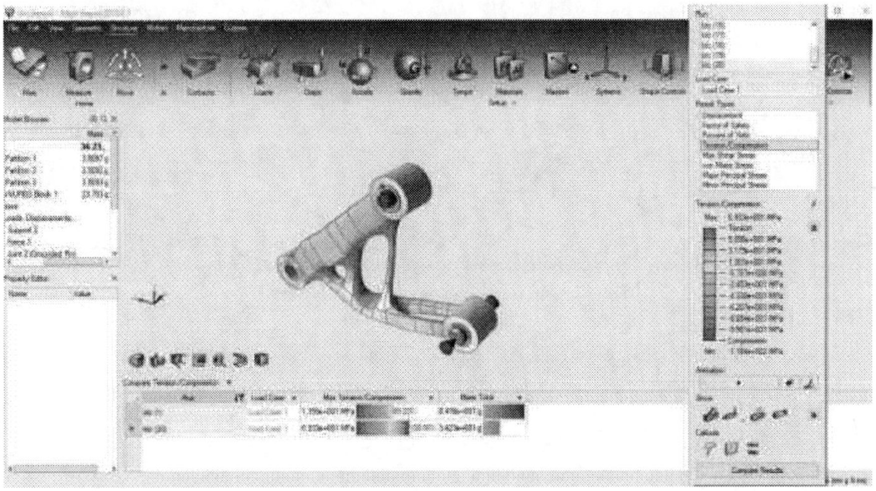

Figure 5. Tension/Compression analysis of bell crank

In figure 5 the tension compression regions are highlighted this tells about the load pattern in bellcrank which also helps in failure analysis as well.a comparison with the stock model is also shown in the figure and values of tension and compression are also mentioned.

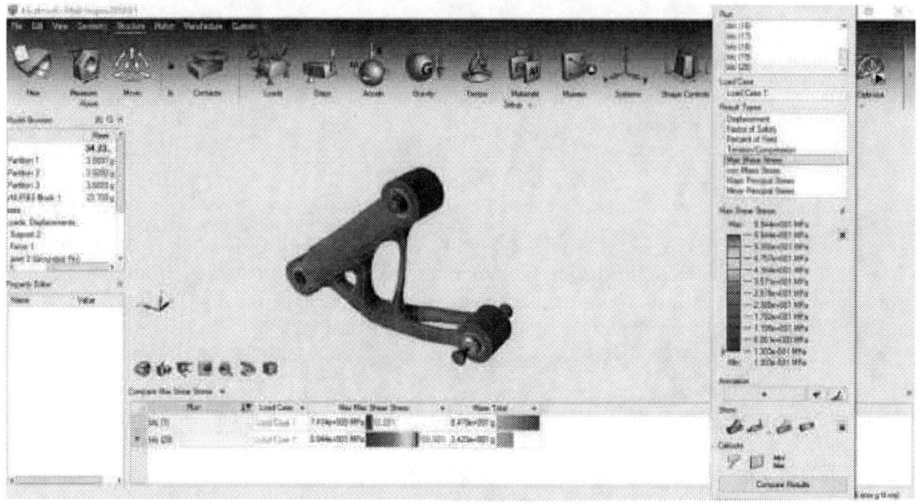

Figure 6. Shear stress analysis of bell crank

Max shear stress analysis is shown in the above figure the maximum shear stress observed was **5.94e^1MPa**and minimum **1.303e^-1MPa**

Figure 7. Von mises analysis of bell crank
(Min: 2.42e^-1MPa & Max:1.18e^2MPa)

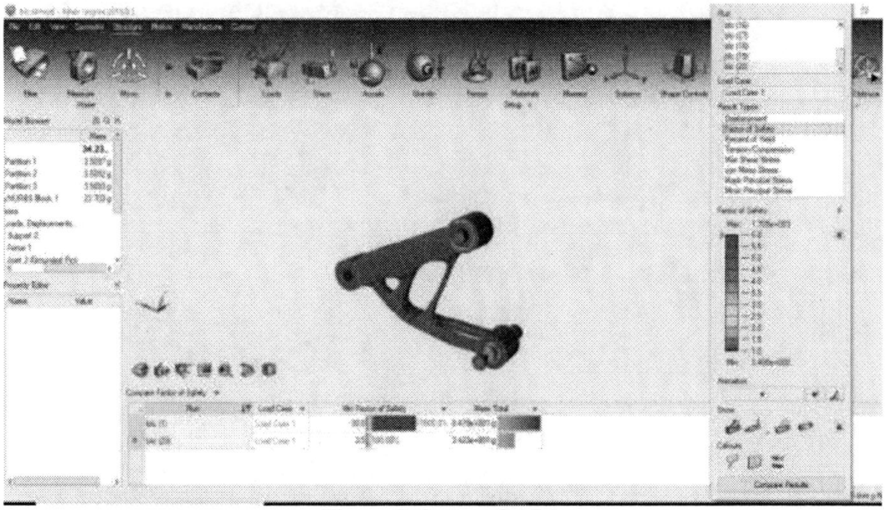

Figure 8: factor of safety

From the above results we observe significant weight reduction upto **59%**,factor of safety of 3.5, displacement all are within permissible limits.

Analysis of brake pedal:

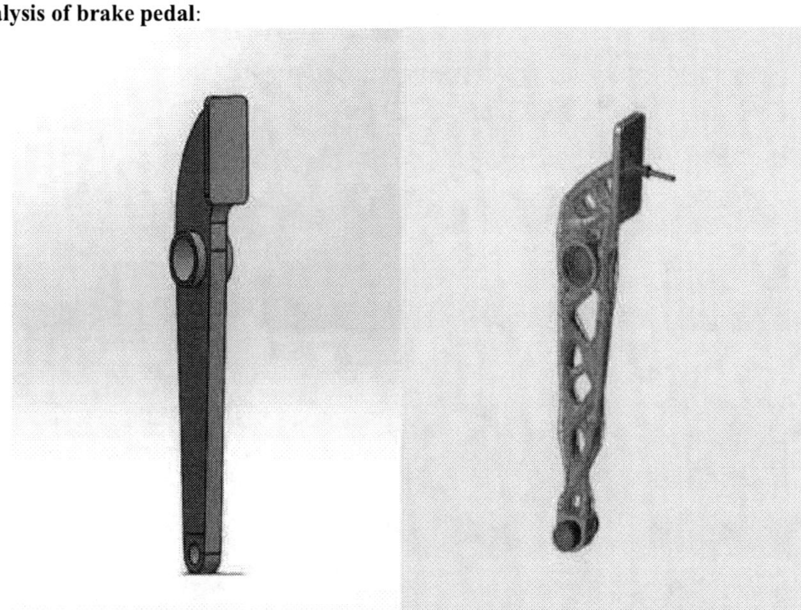

Figure 9. Existing and optimized model of the brake pedal

In figure 9 the left figure shows stock model of brake pedal and right model is topology optimized, as we can see a truss patterns is stiffest iteration for this part within the given parameter of having a min factor of safety of 1.3.

IOP Conf. Series: Materials Science and Engineering 1123 (2021) 012035 doi:10.1088/1757-899X/1123/1/012035

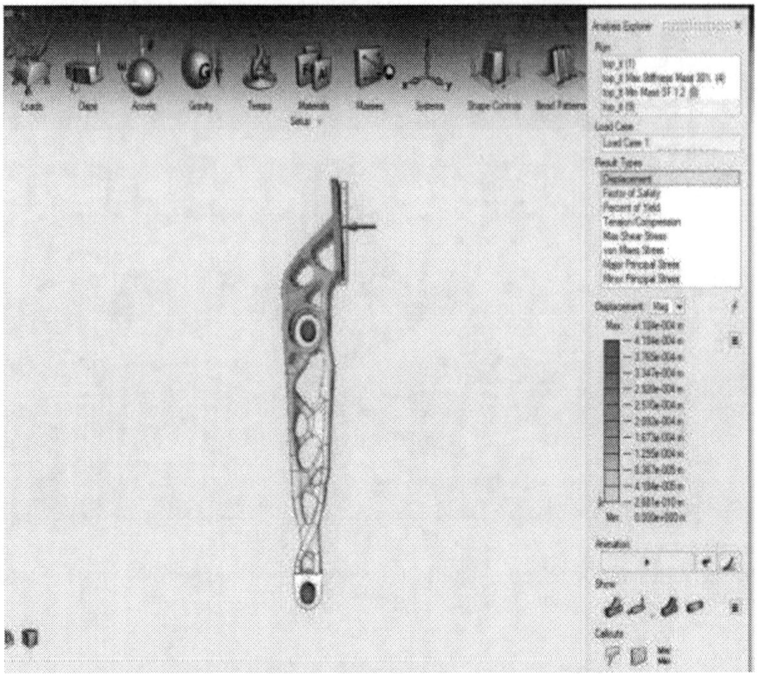

Figure 9. Displacement analysis of the optimized brake pedal

Maximum and minimum displacement were observed to be 4.184e-004m and 0m respectively, these values are negligible and don't affect the functioning of brake pedal.

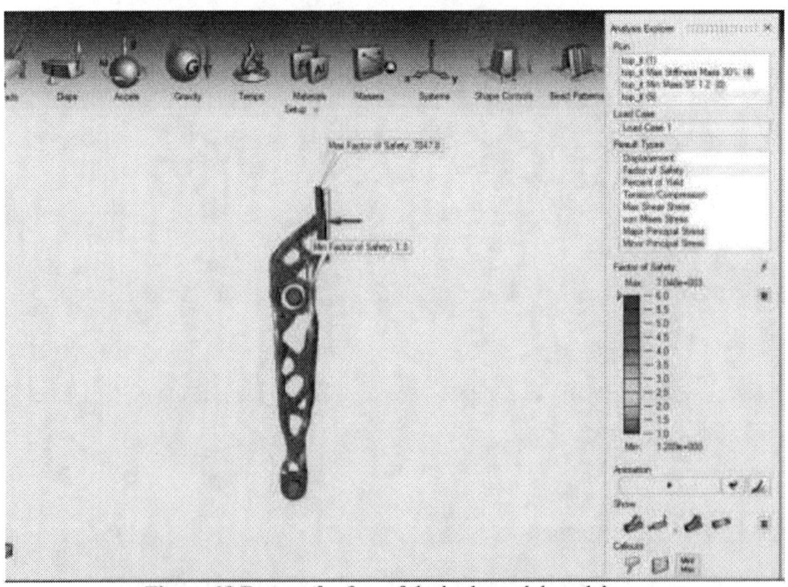

Figure 10. Factor of safety of the brake pedal model

As discussed previously during topology optimization we enter a minimum factor of safety value and the optimization is bassed around two goals i.e either to achieve max stiffness or min weight , since the above pedal was simulated for max stiffness and a min factor of safety was entered 1.3 we received the above result.

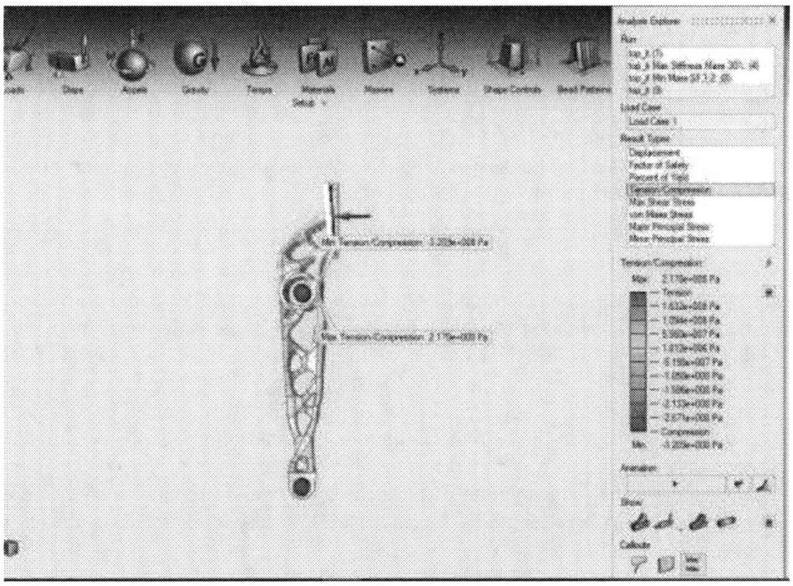

Figure 11. Tension/Compression analysis

figure 11 shows tension and compression distribution around the pedal which helps in failure analysis as well.

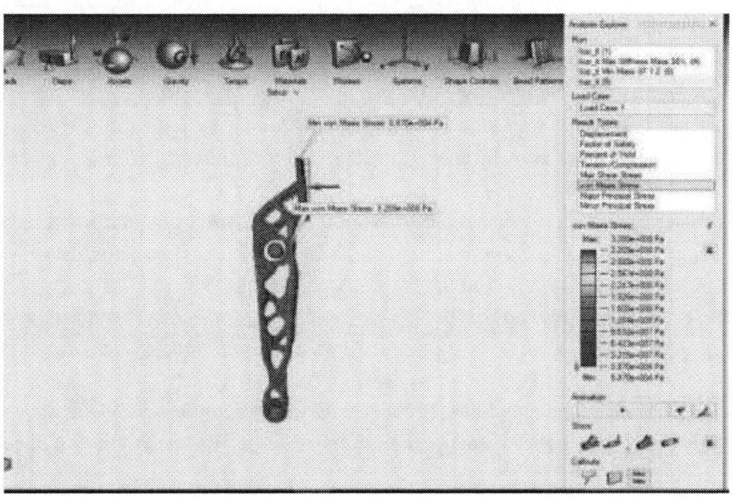

Figure 12.Von mises analysis of the brake pedal

As we can see their is significant difference between the two design of the brake pedal with help of topology optimization techniques the final weight has been reduced by 42%, the min factor of safety(see figure10) is 1.3(approx) and max displacement are within the safety range,the tension and compression simulation shows the load distribution at different regions of the pedal which important for failure analysis.

Our study compares the two different models for the bell crank lever under the same working load conditions using the Altair Inspire. The results drawn are compared for the deformation, over all von mises stress, Compression/tension. It's found that the second model is optimized keeping the stress and deformation in the safe limits.So if we apply same techniques for various components of a car such as uprights,brake caliper,seat mounts etc, we can achieve huge amount of weight and improve efficiency and performance.

4. CONCLUSION

The topology optimization was performed on the bell crank and brake pedal with two different goals, the bellcrank was simulated for minimum weight and brake pedal was simulated for max stiffness, this was visible in the final design of these components .The weight of the newly optimized bell crank is 140g less than that of the existing design and hence there is reduction in the weight of the component by 59%(approx).The weight of the newly optimized brake pedal reduced from 363g to 211 g. Also, with the help to TO there is an enhancement in the performance of the newly optimized design. In this way if perform topology optimization on various components of a car we can reduce the fuel consumption of the vehicle by weight reduction of the components without negotiating with the strength of the material, and also improve the vehicles dynamic performance.

5. REFERENCES

[1] Bendsøe, M.P. and Sigmund, O. (2003) *Topology optimization - theory, methods and applications.* Springer Verlag, Berlin Heidelberg".

[2] Ashley, S. *Shedding pounds on a magnesium diet (2010).* Automotive Engineering International, pp. 34- 6"

[3] Cavazzuti, M., Baldini, A., Bertocchi, E., Costi, D., Torricelli, E., & Moruzzi, P. (2011a) *High performance automotive chassis design: a topology optimization-based approach. Structural and Multidisciplinary Optimization*, Vol.44 (1), pp45- 56.B".

[4] J. Bauer, W. Gutkowski, Z. Iwanow, (1981) *A discrete method for lattice structures optimization, Eng. Optim.* 5 (2) 121–127.

[5] Dongkaixu, Junchen, Yuchengtang, Jiancao,(2012) *Topology optimization of die weight reduction for high-strength sheet metal stamping* International Journal of Mechanical Sciences, Volume 59, Issue 1, (P 73-82).

[6] Kaya, N., Karen, İ., & Öztürk, F. (2010). *Re-design of a failed clutch fork using topology and shape optimization by the response surface method. Materials & Design*, 31(6), 3008-3014".

[7] Swapnil Thigale, Chinmay Shah, (2016) *Weight Reduction in Brake Disc using Topology Optimization,* IJRET: International Journal of Research in Engineering and Technology.

[8] Lin, D.Y. (2013) *Automotive engine bracket topology optimization and finite element analysis.* Zhejiang University of Technology, Hangzhou

[9] Zheng, S.P., Lv, C. and Liu, H.G. (2013) *Topology optimization of engine support structure. construction machinery*, 12, 18-21.

[10] D&T (1997*) Mechanisms: Bell Crank.* [online] available from [22 March 2015]
[11] *The Physics of Braking Systems* by James Walker, (2005), Stoptech LLC".
[12] Mohd Nizam Sudin, Musthafah Mohd Tahir, Faiz Redza Ramli, Shamsul Anuar Shamsuddin, Topology Optimization in Automotive Brake Pedal Redesign International Journal of Engineering and Technology (IJET)
[13] K K Dhande, N I Jamadar and Sandeep Ghatge, *"Design and analysis of composite brake pedal : AN ERGONOMIC APPROACH",* International Journal of Mechanical Engineering and Robotic Research, Volume 3, No. 3, July 2014, ISSN 2278 –0149".

Trust Based Fake Node Identification in Social Networking Sites

Spoorthy A S[1], **Somnath sinha**[2]

[1] Department of Computer Science, Amrita School of Arts and Sciences, Mysore, India, spoorthydechu@gmail.com
[2] Department of Computer Science, Amrita School of Arts and Sciences, Mysore, India, ssin.mca@gmail.com

Abstract. In recent years Social Networking is very commonly used for communications. It is the platform where users are free to join and share their data between groups of people. Due to the openness of this network any one can use multiple fake identifications and fraud the people into its group. These types of fake profiles are required to be identified and wipe out from the network. Some of the Social Networking Sites (SNS) that are commonly used are Twitter, Facebook, Instagram, WhatsApp, WeChat etc. These SNSs are already have some mechanism to identify fake profiles and prevent them from such type of malicious activities but are not sufficiently efficient. This paper represents comprehensive literature review for identification mechanism of fake account on Facebook and proposes a new mechanism to stop this type of harmful activities by using graph based activity diagram.

***Keywords*:** Face book, Fake node, Online Social Networking Site (OSNs), Social networking site (SNS)

1. Introduction

SNS are the applications that are commonly used by the individual mainly for the communication purpose. The most commonly used ones are Twitter, Facebook, Instagram, WeChat, WhatsApp and so on. These are the networks where we can create multiple accounts and communicate with the users. In SNS a user has the ability to create multiple accounts, either fake or genuine. Fake profiles are used to fraud the users whereas the genuine profiles are used by the individuals in their day to day life. The SNS has many strength as well as shortcomings, strength like we can contact a person who is very far away from us where as the shortcomings is that people are getting fraud by the fake accounts through sharing their information to the unknown accounts. The statistic for the social networking site users according to the recent studies is shown below (Table 1.1)

Table 1.1: Statics of the Social Networking Sites as per 2019.

Rank	Network	Number of Daily Users	Monthly Visits
1	Facebook	1.6 billion	2.4 billion
2	Google+	2 billion	395 million
3	Twitter	134 million	330 million
4	Linked in	5.3 million	3.3 million
5	Pinterest	1 billion	291 million

This particular paper concentrates on one such SNS that is Facebook. Facebook was first introduced on February 4, 2004 by Mark Zuckerberg. It was mainly invented to help the college students to connect with each other. But recently it is used by every individual across the world to communicate with each other and to perform more activities like uploading photos, videos, tagging their friends and many more. Since it is open to every individual, recently there are many attacks that are taking place in this Facebook by getting the details of genuine user by a non-legitimate user.

There are many researches taking place in this area. Some of the common papers are by [1] Mohammadreza Mohammadrezaei proposed some of the methods like graph analysis and the classification algorithm to improve the efficiency of detecting the Fake profiles in the SNS, [2] Priyanka Kumari proposed method that identifies the fake profiles present in the users list using a "Social Media" application [3] Hani M. Ibrahim the proposed method uses a Novel Detection Technique that is Fake profile recognizer to verify the identity of the fake profiles. These papers give some of the ideas of eradicating the fake profile from the SNS. The main objective of the particular paper is to find these non-legitimate or fake nodes and completely eradicate its presence from the Facebook Network.

2. LITERATURE REVIEW

Facebook is one of the best SNS. Its benefits are many, may be in Business fields, entertainment field, in case of Journalism field, communication field and so on. Till date there is no proper procedure or a method that is found to eradicate the fake nodes from the Facebook. Some of the Existing systems are listed below to identify the Fake node.

The security challenges that occur in the social networking sites and about the fake nodes are discussed in by the authors Mohammadrezaei, M., Shiri, M. E., & Rahmani, A. M in the paper [1]. Some of the similarity measures such as common friends, cosine, and Jaccard, L1-measure and weight similarity are calculated. The proposed model was evaluated using Twitter dataset. Finally they have used Medium Gaussian SVM algorithm to predict the fake accounts. The paper covers some of the very important concepts like Graph Analysis and Similarity types, Introducing of Resampling, Principle Component Analysis and Machine Learning. The efficiency of the paper was expertise by Machine Learning method. The ratio of Fake accounts were identified in twitter data is 1:100.the main weakness f the proposed method is that fake account had to work in a network and from there they were able to find a legitimate or a fake account by analyzing their friends account. The future work of the paper was to present a method to identify the fake or legitimate account before performing any activities by the user account or to find the fake account during the time of the registration process.

Fake accounts are identified based on the activities and the account information is proposed by the authors Kumari, P., & Rathore, N. C in the paper [2]. To show the efficiency of the model, they developed an application that is similar to Facebook called"SocialMedia" to eliminate the Fake accounts that are found in the users friend list. The effectiveness of the proposed concept was given by this app.Online evaluation was made to identify the usage of such app in the network and the accuracy was found to be 87.5%. In the proposed system the parameters used was number of mutual friends, is family, age group, average like and average comment. After getting the parameters the trust weight calculation was done between the user's friends. If the friends are having low trust weight then there is a high possibility of the account being a fake one. The future work was to collect other attributes of SNS users and to find the fake account with its help.

A Novel detection technique called Fake Profile Recognizer (FPR) which verify the profiles identity, and detect the fake profiles in Online Social Networking Sites is discussed by the author Meligy, A. M in the paper [3]. The detection of fake profile is based on utilizing Regular Expression (RE) and Deterministic Finite Automaton (DFA) methods. The proposed technique was evaluated on three datasets of Online Social Networking Sites: Facebook, Google+, and Twitter. The results reconnoitered high Precision, Recall, accuracy, and low False Positive Rates (FPR) of detecting Fake Profiles in the three datasets. In twitter dataset there was a problem where Support Vector machine dint work correctly and which was later replaced by Social Honey pot.

The OSNs which are tremendously increasing the way in which people communicate and divide their personal, professional and political information is discussed by the author Conti, M., Poovendran, R., & Secchiero, M. in the paper [4] . As per the cyberspace the OSNs are attracting the Interest are trying to utilize this weakness. Recently the security and privacy issues are increasing day by day, the researchers are trying to find these and erase. The paper created an analysis of social network graphs[5] from background of privacy threats. In the future work the plan will be extended to characterize the online interaction with the user and strengthen the link and the quality of the mechanism.

The potential attacker can present the automated crawling and identity thefts that occur in SNS to get access to personal information of the user is discussed by the authors Leyla Bilge, Thorsten Strufe, Davide Balzarotti, Engin Kirda Eurecom in the paper [6]. The attack that was presented as identity theft of existing user profiles automatically and sending of friend requests to the contacts of the duplicated victim. The acceptance of friend request is the hope from the attackers. By beginning a relationship as a friend with the contacts of a victim, the attacker is able to access all the personal information provided by them.

The other method[7] was to present a valuable and feasible automated launch of cross-site profile cloning attack. So that while attacking they are able to create a fake profile in a network where the victim has not registered but contact the victim's friend who registered on both the network.

The threats that occur in SNS and they have gone through some of the targets about how the attackers does the attack and how they perform it is discussed by the author Weimin Luo, Jingbo Liu, Jing Liu in the paper[8]. The paper is divided into 2 parts: User and Social Networking Sites. And there are some discussions on countermeasures against the attacks in Social Network and also the paper proposes a security framework towards the end. The disadvantage of the paper is towards the end they have designed a framework where the user must have some ideas on the security of their account without this the calculation of this stage becomes difficult.

The result of a research towards data security factors and wakefulness of some of the risk of SNS, and the most notable of identity theft is discussed by the author Lang, M., Devitt, J., Kelly, S., Kinneen, A., O'Malley, J., & Prunty, D in the paper[9]. The research was conducted for the people of 18-24 age groups residing in Ireland. Some of the findings were that many people had a informal outlook towards data backup and password protection, some were not aware about the attacks that was taking place through the virus transmitted by the portable device. The main disadvantage of this paper is the researcher have taken just 120 accounts to find the statistical correlation which not an effective one. Future study to prepare the questionnaire for larger number of people worldwide and to extend the study.

The SNS, fake account creation are becoming more harm than a cyber crime. Some of the algorithms and methods are being proposed for detection of the fake profile is discussed by the author Ramalingam, D., & Chinnaiah, V in the paper [10]. The paper presents some of the existing techniques to detect the fake profile. The disadvantages of the paper is that the prevention of fake account should be done rather than detecting them this method is not proposed in the paper

An attack so called automated Social engineering which explains how the SNS can be used for social engineering is discussed by the authors Huber, M., Kowalski, S., Nohlbe;rg, M., & Tjoa, S in the paper [11]. The particular research takes engineering to one step advance by automating tasks that are time intensive. The implementation was based on prototypical implementation (ASE bot), which had two experiments that was conducted. The first experiment [12]which have examined the gathering information capabilities of the bot. In the second experiment prototype performs a Turing test. Making Facebook with restrictive security and policy may become destructive for the account this is the main disadvantage. The future work of the paper is to implement the original ASE bot to get a valuable insight.

Some of the security threats and some of the defense mechanisms regarding popular OSNs is discussed by the authors Gao, H., Hu, J., Huang, T., Wang, J., & Chen, Y. in the paper [13]. It tells about wide range of attacks and some of the defense mechanism for those attacks. The author categorize these attacks into four categories - privacy breaches, viral marketing, network structural attacks and malware attacks and focus mainly on privacy concerns. The future work is to create a SNS that has which lots of entertainment services and to secure the data of the user without any issues.

After going through these entire articles there is no perfect mechanism till date to eradicate the fake accounts from Facebook. And this becomes the main advantage of the particular article to eradicate the fake account from the Facebook using a trust based method.

3. FACEBOOK

The most popular SNS is the Facebook. And it has almost 2.41 billion active users as per the survey made in the year 2019. Facebook is not only used for entertainment purpose it has many activities in it. And it consists about 28 activities that are used by the user. Some of the activities that are performed by the Facebook are: Timeline, post view, post update, like, notification, login, log out, gaming, reading an article, Movies links, videos, profile update, comment, share posts, sending friend request, message, memories, history, tag photo, markets and many more.

Out of these activities some are more frequently used by the user are mentioned below:

a. **Post count:** Number of post uploaded by the user in his/her timeline. The Fake accounts are predictable to have lesser count of posts in his/her account. The count can be collected from the timeline of the users account.
b. **Comment count:** The average number of comments received for each post. The fake accounts are expected to have unwanted post with lesser number of comments. The counts of the comments are collected from the user timeline.
c. **Friendsversary:** The post that celebrates the day when they became friends. The Fake account does not care about celebrating anniversary. The counts of Friendversary are collected from the user timeline.
d. **Followers:** The count of the follower following the user. The Fake accounts will have a less followers count. The counts of the followers are taken from the timeline.
e. **Events:** The Number of events the user have attended. The Fake accounts do not update any events that they have attended. The number of event attended is collected from the user timeline.
f. **Friend request:** The number of friend requests accepted will be more. But in rear cases the request may be less. The Friend Request count is taken from the user's timeline.

CHALLENGES OF FACEBOOK:

Some of the most popular challenges that the Facebook is facing is listed below:

1. **Battling Election Interference and Misreport:**
   Limits the spread of misreport and removes millions of the fake accounts daily. Uses some of the technology and some findings by the experts to remove bad actors. Some of the government partners, other companies, NGO's helps in increasing the security.
2. **Strengthening the privacy and protection the information of the people:**
   It is mandatory to secure the data that is being given to the Facebook by the account holders. More than 400 apps were suspended who violated the policy. Rolled out some of the

GDRP style all over the world and tried making some of the privacy tools much easier to use. Established some of the Data use team for the maintenance purpose.

3. Prioritize the users safety and well being:

Over 30,000 people are working for the privacy control. While working for the privacy its been detected that for about 99% of the terrorist related content is taking place in Facebook. And for about 97% of violence and the graphical contents were found. For this reason Facebook have launched an appeal to remove all these contents.

4. Giving more information to the user about the ads they see:

It's common that the number of ads increases in the Facebook and many among this are fake and many are the originals. Recently a new technique was introduced where it allows the user to watch those ads and report if they are fake ones. And there was a new creation of the political ads which was very transparent and it was started at US. In some of the countries the ads where posted by paying the Facebook.

5. Seeking the effective regulation:

Facebook is working with lawmakers to make an effective and smart legislation to strike the balance between the competing tensions. It is mainly working with government for the security purpose of the platform.

4. Proposed Methodology

The main objective of the research is to identify the suspicious account in Facebook. Here two stage mechanisms are used for identifying malicious account or node. The stages are described in the subsequent sections.

4.1 Stage 1: Measuring the authenticity of the account.

In Facebook Most of the people are used to like the post that is posted by individual where the total number of people "likes" for the post goes to about 80%, and about 54% of people are watching "videos" for entrainment as well as knowledge purpose and some are interested to "read the article" where the count for this particular category lies for about 47% and in order to build social connections with beloved one's, people do "messages" and the survey gives data about 47% and about 46% of people use Facebook to "news updates". The graph of the activities is given in the Figure 4.1

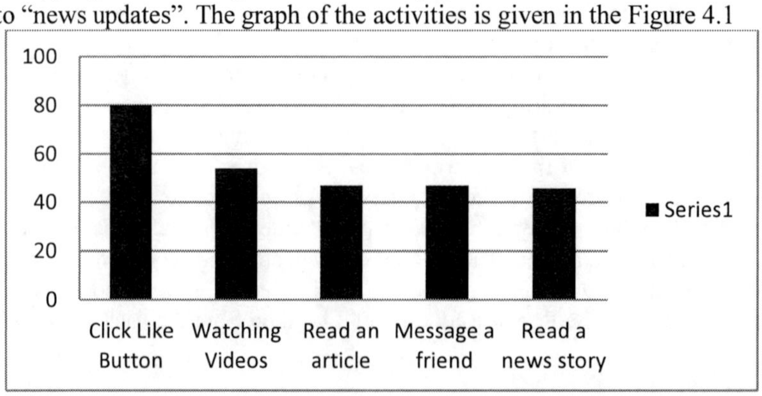

Figure 4.1: Popular activities in Facebook.

Out of all these five categories we are just considering "Like" and "Videos". Only these 2 parameters are considered because these both are most commonly used by every individual and the usage accuracy is also above 50%. Rest of the three that is reading an article, messages, news updates are below 50% hence these parameters cannot be considered.

Considering the Conditional probability model the probability of the of the like and videos are calculated where the probability of watching video is divided as regular and suspicious nodes where the regular node is further classified as watching videos and not watching videos. Similarly the suspicious node is also divided as watching videos and not watching videos. Here the average count of the users liking the post is .8% and those who don't like the post are .2%. And the average count of watching video is .54% and the users who don't watch video is about .46%. And the probability calculation is shown below (Figure 4.2)

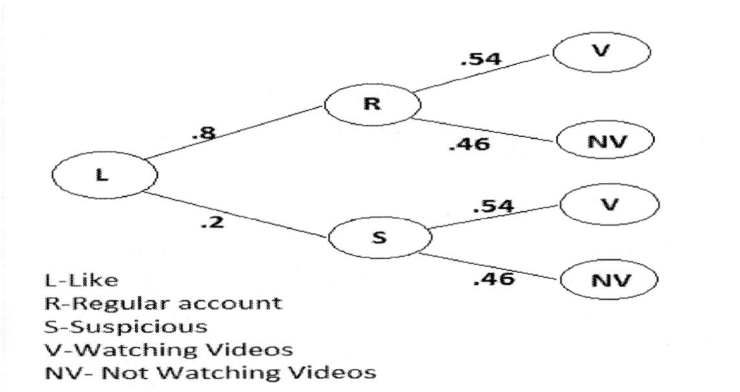

Figure 4.2: Measuring the Probability of the account.

The probability is calculated as:
$P(S \cap NV) = 0.2 \times 0.46 = 0.092$
$P(R \cap NV) = 0.8 \times 0.46 = 0.368$
$P(S \cap V) = 0.2 \times 0.54 = 0.108$
$P(R \cap V) = 0.8 \times 0.54 = 0.432$

From the above observation it is clearly explained that the probability $P(S \cap NV)$ is a highly suspicious accounts, $P(R \cap NV)$ and $P(S \cap V)$ are moderately suspicious accounts and $P(R \cap V)$ are the genuine accounts. All the Genuine accounts are removed from this stage and the highly suspicious and moderately suspicious accounts are carried forwarded to the Stage 2. Here we are just categorizing some of the accounts as the highly suspicious and moderately suspicious ones. Complete identification of the fake nodes is done in the stage 2.

4.2 Stage 2: Detection of the fake nodes using direct graph.

The detection of fake node is done using the direct graph where the moderately suspicious and highly suspicious accounts are added to the direct graph where the connectivity of nodes is shown below (Figure 4.3).

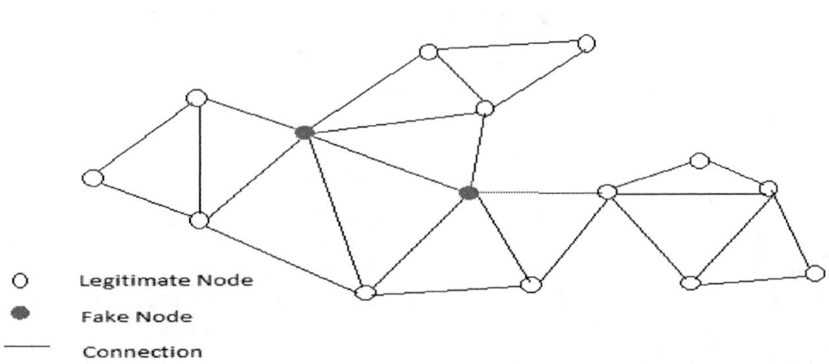

Figure 4.3: Direct graph for Fake account detection

The fake nodes are detected by calculating the affinity[12]. Consider the 2 nodes as i and j between which the affinity value is calculated as

$$Aij = (Tij - 2Lij)\frac{Tij+Lij}{N} \tag{1}$$

Where,

T_{ij} = Number of similarity between event i & j.
L_{ij} = Number of dissimilarity between event i & j.
N = Total number of events

After calculating the affinity (A_{ij}) using equation(1) if the value is ≤ 0 then it is a suspicious account and if the value is >0 then it is a genuine account. And if the value occurs in terms of Positive result then it has high affinity and if the value is in terms of negative then it has lesser affinity or no similarity between the nodes. If the more number of connections with other node is present and they are similar then the node is detected as a fake one and it is eliminated from the network else it is considered as a genuine account.

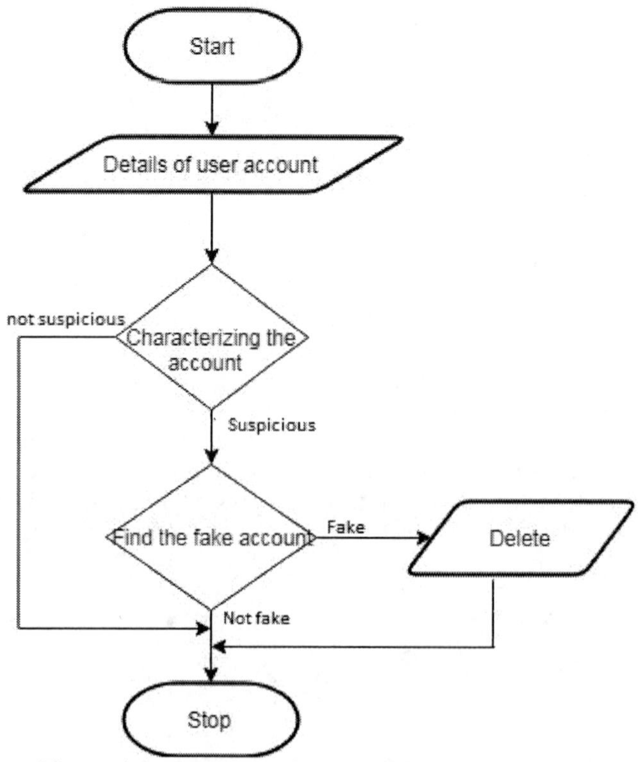

Figure 4.4: Flowchart of the identifying fake accounts

Algorithm:

Step 1: Gets the details of the user fake account.

Step 2: Identifies the suspicious accounts using the conditional probability model and categorize the account as suspicious or highly suspicious.

Step 3: The suspicious and highly suspicious accounts are checked for similarity testing by using the graphical relation.

Step 4: The accounts having more connectivity with the other account are taken as malicious account.

5. CONCLUSION AND FUTURE WORK

Throughout the paper the detection of fake nodes that are present in the Facebook is done. The literature review suggests some of the existing methods and their disadvantages in detecting the fake nodes. The paper has two stage mechanisms where the fake node can be detected. In the first stage some of the attributes of the accounts chosen and probability is calculated for those activities and in the second stage the detected fake nodes are done through direct graph by calculating the affinity. If more number of clones is present in multiple accounts then it is considered as a fake node. The paper detects the fake nodes where these fake nodes can be eliminated easily from the SNS. The Future work for the paper is to increase the accuracy by adding more general activities, and when it is detected further prevention technique can be added using some of the soft computing approach.

REFERENCES

[1] M. Mohammadrezaei, M. E. Shiri, and A. M. Rahmani, "*Identifying Fake Accounts on Social Networks Based on Graph Analysis and Classification Algorithms*", Hindawi Security and Communication Networks, 2018.

[2] P. Kumari and N. C. Rathore, "*Fake Pro fi le Identification on Facebook Through Social Media App,* " Information and Communication Technology for Sustainable Development,pp. 83–90, 2018.

[3] A. M. Meligy, "*Identity Verification Mechanism for Detecting Fake Profiles in Online Social Networks,*" no. January 2017, 2018.

[4] M. Secchiero, "*FakeBook : Detecting Fake Profiles in On-line Social Networks,*" International Conference on Advances in Social Networks Analysis and Mining 2012.

[5] K. Sreelakshmi, S. Anand, and S. Sinha, "*Black Hole Attack in Mobile Ad Hoc Network – Analysis and Detection,*" International Journal of Recent Technology and Engineering (IJRTE) no. February, 2019.

[6] L. Bilge, T. Strufe, D. Balzarotti, E. Kirda, and S. Antipolis, "*All Your Contacts Are Belong to Us : Automated Identity Theft Attacks on Social Networks,*" International World Wide Web Conference Committee (IW3C2) 2009.

[7] S. Sinha, A. Paul, S. Pal, and C. Engineering, "*THE SYBIL ATTACK IN MOBILE ADHOC NETWORK : ANALYSIS AND.*" "*International Conference on Recent Trends in Communication and Computer Networks - ComNet 2013*"organized by ACEEE,

[8] W. Luo, J. Liu, J. Liu, and C. Fan, "*An Analysis of Security in Social Networks,*" IEEE International Conference on Dependable, Autonomic and Secure Computing pp. 648–651, 2009.

[9] M. Lang, J. Devitt, S. Kelly, A. Kinneen, J. O. Malley, and D. Prunty, "*Social Networking and Personal Data Security : A Study of Attitudes and Public Awareness in Ireland,*" International Conference on Management of e-Commerce and e-Governmentno. January 2009, 2014.

[10] D. Ramalingam and V. Chinnaiah, "*Fake profile detection techniques in large-scale online social networks: A comprehensive review,*" Comput. Electr. Eng., vol. 65, pp. 165–177, Jan. 2018.

[11] M. Huber, S. Kowalski, M. Nohlberg, and S. Tjoa, "*Towards Automating Social Engineering Using Social Networking Sites,* " pp. 117–124, 2009.

[12] R. Kulsum, S. Anand, and S. Sinha, "*A Soft Computing Approach to Analyse Aodv Routing Protocol,* " International Journal of Innovative Technology and Exploring Engineering (IJITEE) no. 8, pp. 1443–1446, 2019.

[13] H. Gao, J. Hu, T. Huang, J. Wang, Y. Chen, and A. Osns, "*Security Issues in Online Social Networks,*" Published by the IEEE Computer Society 2011.

[14] Facebook dataset| kaggle, 2018 [online]. Available from: https://www.kaggle.com/sheenabatra/facebook-data [Accessed on: Jul 15,2018]

[15] Dustin scout , 2019 [online]. Available from :https://dustinstout.com/social-media-statistics/#google-plus-stats [Accessed on: July 2019]
[16] Facebook, 2019 [online]. Available from: https://newsroom.fb.com/news/2019/02/addressing-challenges/ [Accessed on: Feb 4, 2019]
[17] Statista Research Department. Statista, 2016 [Online]. Available from: https://www.statista.com/statistics/420714/top-facebook-activities-worldwide/ [Accessed on: Apr 16, 2016]

USE OF CORRUGATED PIPE HEAT EXCHANGERS IN WASTE HEAT RECOVERY STEAM GENERATORS

Sukanta Nayak[1], Satyasreet Jena[2], Abu Sharique Shamshad Khan[3], Mani Kant Paswan[4] and Vinod Kumar Sharma[5]

[1,3,4] National Institute of Technology Jamshedpur, India
[2,5] Vellore Institute of Technology, Vellore, India

Abstract. Heat exchangers are used for a variety of purposes in various domestic, commercial& industrial applications. Some common examples are sensible heating & sensible cooling in chemical processes, condensation and steam generation in power plants, and recovery of waste heat etc. Heat exchangers performance enhancement can lead to more economical design of heat exchanger which increases energy, material & cost savings related to heat exchange process. Making corrugation on pipe surface is one such method. Numerous works has been carried out in this area, where parameters such as inside diameter of pipe, corrugation pitch, corrugation depth, temperature and phase of liquid has been studied. Corrugation of tubes can increase both the inner and outer surface convective heat transfer coefficient without any considerable rise in pressure drop. It also results in increase in mixing of fluid, flow turbulence, unsteadiness by limiting the fluid boundary layer growth adjacent to the tube surface.

Keywords: *Heat exchanger, Corrugation pitch, Corrugation depth, Pressure drop etc.*

1. INTRODUCTION

With population growth, the demand for electricity has increased rapidly. This has motivated the researchers to look for more efficient and sustainable technologies for power generation thereby reducing the cost of electric power generation and reducing emissions containing harmful pollutants. Some of these technologies are power plants with waste heat recovery systems consisting of a gas cycle and steam cycle, or a steam power plant linked by a Heat Recovery Steam Generator (HRSG). The HRSG produces steam at a suitable temperature and pressure by recovering heat from the flue gases coming out of the gas turbine.

1.1. Origin of the project

Heat exchangers are used in different purposes ranging from conversions, utilization & recovery of thermal energy. Some common examples are sensible heating & sensible cooling in chemical processes, condensation and steam generation in power plants, and recovery of waste heat etc. The techniques used for increasing the heat exchanger performance are also referred as eat transfer augmentation techniques or Heat transfer intensification techniques. Heat transfer through convection is improved by decreasing the thermal resistance in the heat exchanger. Heat transfer augmentation techniques lead to increase in heat transfer coefficient but at the same time leads to increase in pressure drop. Thus, the heat transfer rate and pressure drop are the main parameters which are to be considered during design of heat exchanger using any of these techniques. There are three main categories into which Existing enhancement techniques can be classified: Passive Method, Active Method, and Compound Method.

1.2. Objectives

The Objective of the project includes the following:

- Comparative study of corrugated pipe and plain pipe heat exchangers in HRSG.

1.3. Literature review

Use of corrugated tube can produce superior heat exchanger performance. Corrugation of tubes can increase both the inner and outer surface convective heat transfer coefficient without any considerable rise in pressure drop. Increase in mixing of fluid, flow turbulence, unsteadiness by limiting the fluid boundary layer growth adjacent to the tube surface. To reduce the size of industrial shell and tube heat exchangers corrugated tubes are chosen. The use of corrugated tubes is still a new method for enhancing the performance of heat exchangers and a lot of work is going on in this field.

Rainieri and Pagliarini conducted experiments to study the heat transfer characteristics of tubes with corrugation. Transverse and helical corrugations were made on the smooth surface of the pipe with different pitches with ethylene glycol as the fluid used. Among the smooth tube, helical corrugated tubes and transverse corrugated tubes, results showed that helical transverse corrugated tubes were having the highest heat transfer enhancement.

Laohalertdecha and Wongwises investigated the behavior of heat transfer coefficient and pressure drop due to change in pitch in condensation and evaporation process in corrugated tubes with R-134a as working fluid. For all experiments performed, corrugated tube is better than the smooth pipe in terms of heat transfer coefficient and pressure drop.

Targanski and Cieslinski studied evaporation inside a smooth copper pipe, smooth stainless steel pipe and two enhanced pipes for R-407C/oil mixtures and pure R-407C as the working fluids. Results obtained were in favour of corrugated tubes.

1.4. Methodology

The design of HRSG involves following steps:

- Determination of design input parameters.
- Working Fluid selection.
- Selection of working pressure.
- Determination of temperature profile.
- Determination of Convection Heat Transfer Coefficient.
- Heat exchanger Area calculation.
- Determination of number of pipes.
- Calculation and Comparison of Heat Exchanger Performance.

2. Mathematical Modeling

Representation of the processes by a mathematical model helps in simulating the system

2.1. *Assumptions*

In the development of mathematical model certain assumptions were made. These are as follows:

- Water and flue gas are in steady flow within the the system.
- Unfired HRSG.
- No loss of thermal energy from the heat exchangers.
- To prevent condesation, the temperature of flue gases in the chimney is assumed to be more than 100°C.
- Staggered arrangement of tubes is selected for better heat transfer.

Table 1. Parameters

Sr. No.	Design Input Parameters	Value
1.	Flow rate of water at the tube inlet, m_w	1 kg/s
2.	Flow rate of flue gas at the shell inlet, m_g	1.573 kg/s
3.	Water temperature at the tube inlet, T_{c1}	242°C
4.	Flue Gas temperature at the shell inlet, T_{h1}	852°C
5.	Flue gas temperature at the shell outlet, T_{h2}	420°C
6.	Tube inside Dia, D_i	25.4 mm
7.	Tube outside dia, D_o	31 mm
8.	Shell dia	838 mm
9.	Pressure inside tube	160 bar
10.	Pressure outside tube	1 bar

3. Temperature Profile

The determination of temperature profile is the first step in the design of HRSG. The inlet and outlet temperatures of the flue gases and working fluid are calculated by energy balance method. The calculation should be done by using a suitable pinch point (Difference between the gas temperature at the evaporator outlet and the temperature of the fluid at the evaporator inlet). Equating the heat lost to the heat gained

$$Q_{cold} = Q_{hot}$$
$$mw(h_{cold,o} - h_{cold,in}) = mg(h_{hot,in} - h_{hot,o})$$
$$mwcw(T_{c,2} - T_{c,1}) = mgcg(T_{h,1} - T_{h,2}) \tag{1}$$

Substituting the values from Tables 2.1, 2.3 in the above equation the water temperature at the tube outlet is found to be 347.3°C.

3.1. *Overall Heat Transfer Coefficients*

Calculation of the overall heat transfer coefficient (U) can be done as shown below:

$$\frac{1}{UA} = \frac{1}{h_o A_o} + \frac{\ln(\frac{D_o}{D_i})}{2\pi k l} + \frac{1}{h_i A_i} \tag{2}$$

Thin steel tubes having high thermal conductivity were used in the heat exchanger. Therefore, the second term in the above equation can be neglected. Therefore the equation for U gets simplified to the form:

$$\frac{1}{U} = \frac{1}{h_o} + \frac{1}{h_i} \tag{3}$$

To calculate the inner heat transfer coefficient for steady state incompressible flows inside a tube with uniform cross-sectional area, the flow intensity can be calculated using the Reynolds number:

$$Re_D = \frac{4m_w}{\pi D_i \mu} \tag{4}$$

The present calculation shows that the Reynold number is 9.13×10^3, thus the flow is turbulent. Now, for smooth circular pipe with turbulent flow the Nusselt number (Nu_D) is given by DittusBoelter Equation

$$Nu_D = 0.023 Re_D^{0.8} Pr^n \tag{5}$$

For cooling, n = 0.3, and for heating, n = 0.4. The value of Nu_D was found to be 33.34. The heat transfer coefficient for convective heat transfer inside the pipe is calculated as:

$$h_i = Nu_D \frac{K}{D_i} \tag{6}$$

The value of internal heat transfer coefficient was found to be 721.3 W/mK.

Next, the heat transfer coefficient for the convective heat transfer outside the bank tubes is calculated. The heat transfer in flow over tube bundles depends largely on the flow pattern and the degree of turbulence, which in turn are functions of the velocity of the fluid and the size and arrangement of the tubes. The tube bundles can be in-line or staggered arrangements. The bundle geometry is characterized by the Transverse pitch S_T and the longitudinal pitch S_L between the tube centers. The diagonal pitch S_D is sometimes used for the staggered arrangement. To define the Reynolds number for flow through the tube bank, the flow velocity is based on the minimum flow area. The Reynolds number is

$$Re_{D,max} = \frac{V_{max} D}{\nu} \tag{7}$$

In the case of staggered configuration, the maximum velocity could be at either the diagonal plane A_2 or the transverse plane A_1 of Figure. V_{max} will be at A_2 if the spacing between the rows is such that such that

$$2(S_D - D) < (S_T - D)$$

$$S_D = \sqrt{(S_L^2 + (\frac{S_T}{2})^2)} < \frac{S_T + D}{2} \tag{8}$$

In our case S_T=S_L=3D=93 mm substituting these values in equation 8 it was found that V_{max} occurs at A_1 and is given by

$$V_{max} = \frac{S_T}{(S_T - D)}V \tag{9}$$

The V_{max} was found to be 12.08 m/s. Putting this value in equation 7, the value of $Re_{D,max}$ is 3780. Zukauskas developed a correlation for average Nusselt number for flow over a bank of tube

$$\overline{Nu_D} = C_1 Re_{d,max}^m Pr^{0.36}(\frac{Pr}{Pr_w})^{0.25} \tag{10}$$

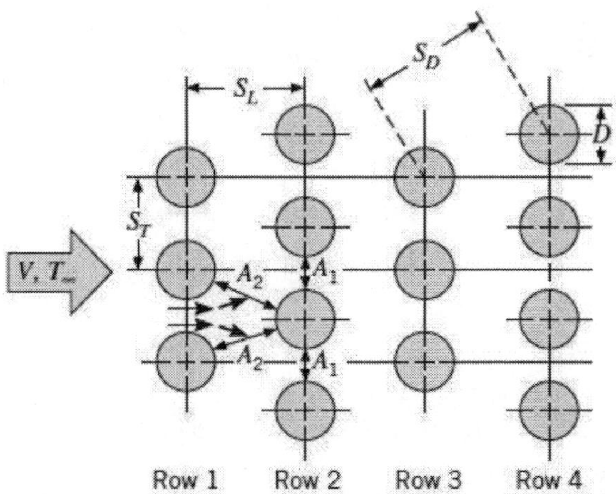

Figure 1. Staggered tube arrangement inside the shell

$n_R \geq 20$,
$0.7 \leq Pr \leq 500$,
$10 \leq ReD,max \leq 2 \times 106$

where n_R is the number of tube rows, all properties are at the bulk fluid temperature (average of the fluid outlet T_o and inlet T_i temperatures) except Pr_w ($\frac{Pr}{Pr_w}$=0.94) which is evaluated at the tube wall temperature, and the constants m and C_1 are listed in Table below.

Table 2. Values of Constants

Configuration	*ReD,max*	C_1	m
Staggered(S_T/S_L <2)	$10^3 - 2 \times 10^5$	0.35(*ST* /*SL*)0.2	0.60

If the number of rows of tubes, n_R,is less than 20 there is a decrease in the average heat transfer coefficient, thus a correction factor is incorporated such that

$$\overline{NuD},(n_R<20) = C_2\overline{NuD},(n_R \geq 20) \tag{11}$$

where C_2 is given in Table below.

n_R	1	2	3	4	5	10	13	16
Aligned	0.70	0.80	0.86	0.90	0.92	0.95	0.97	0.98	0.99
Staggered	0.64	0.76	0.84	0.89	0.92	0.95	0.97	0.98	0.99

Nu_D value was found to be 40.58. The convection heat transfer coefficient h_o can then be calculated using

$$h_o = Nu_D \frac{K}{D_o} \tag{12}$$

h_o was found to be 100.69 W/m^2K.

Table 3. Fluid Properties

Properties	Water	Flue gas
Density, ρ (kg/m^3)	641	0.384
Dynamic Viscosity, μ (Ns/m^2)	57×10^{-6}	39.26×10^{-6}
Specific heat capacity, c (KJ/KgK)	8	1.226
Thermal conductivity, k (W/mk)	25.4×10^{-3}	78.4×10^{-3}
Prandtl number, Pr	0.96	0.615

From equation 3, 6, 12 the overall heat transfer coefficient U is calculated to be 88.35 W/m^2K.

3.2 Area Calculation
The total heat transfer is given by

$$Q = FUoAo(\Delta T_{LM})counterflow \tag{13}$$

where

$$(\Delta T_{LM})_{counterflow} = \frac{(T_{h_2} - T_{c_1}) - (T_{h_1} - T_{c_2})}{\ln \frac{T_{h_2} - T_{c_1}}{T_{h_1} - T_{c_2}}}$$

The value of correction factor F was determined using the plot of F for single pass cross-flow heat exchanger (one fluid mixed and other unmixed). Using equation 13 the outer area required for the heat transfer is calculated. The values of Q, U,$(\Delta T_{LM})_{counterflow}$ are 842.4 KW, 88.35 W/m^2K and 313.5K respectively. Then, the length of tubes inside the shell was to be 2.71 m as the number of tubes and standard tube diameters have already been assumed (57 and 31 mm respectively).

4. Design of Corrugated Pipe

In mathematical modeling the analysis was done by assuming the cross- section of tubes inside the shell to be plain. By using the same length and inlet conditions for a corrugated pipe the heat transfer can be enhanced, which increases the outlet temperature of water.In design view, mainly 3parameters has to be specified for the cross- section of corrugated pipe. They are inner diameter (d_i), pitch(p), height(e).

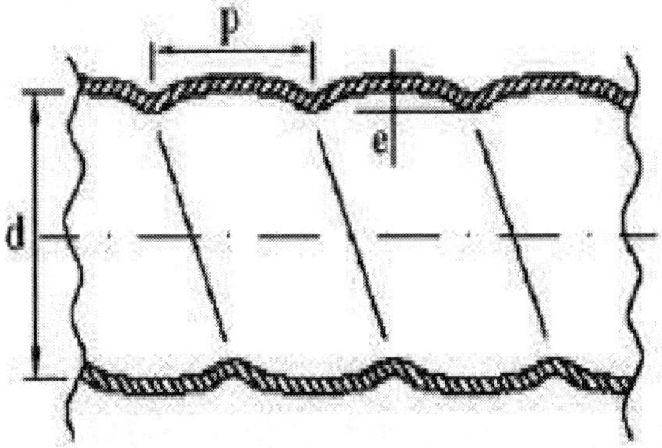

Figure 2. Schematic diagram of a corrugated pipe

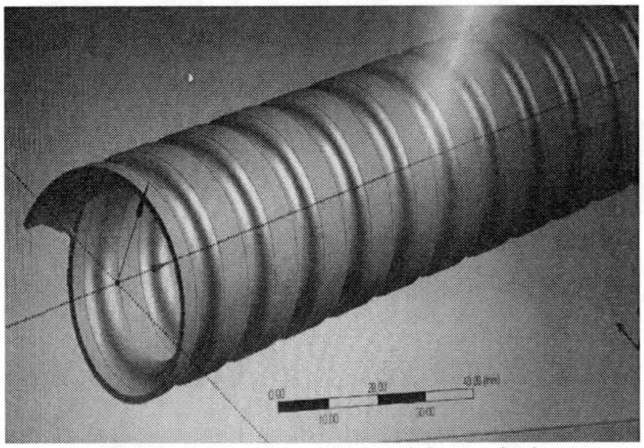

Figure 3. Design considered for analysis

The Fig.3 shows the design, which we have considered by taking the parameters as d_i= 25.4 mm, p= 20 mm, e= 2.085 mm.

5. Results and Discussion

IOP Conf. Series: Materials Science and Engineering 1123 (2021) 012037 doi:10.1088/1757-899X/1123/1/012037

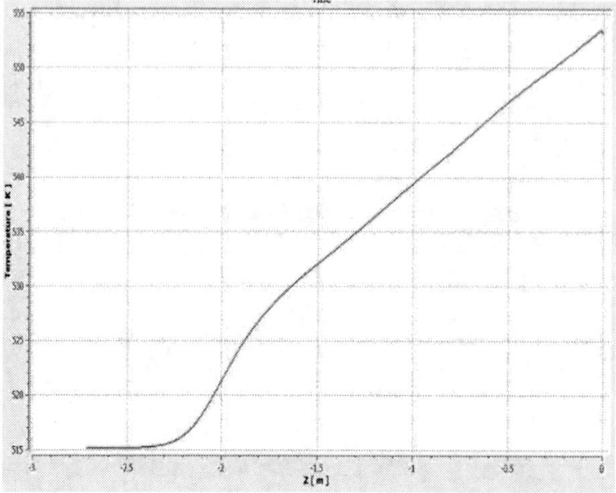

Figure 4. Temperature profile along the axis of plain pipe

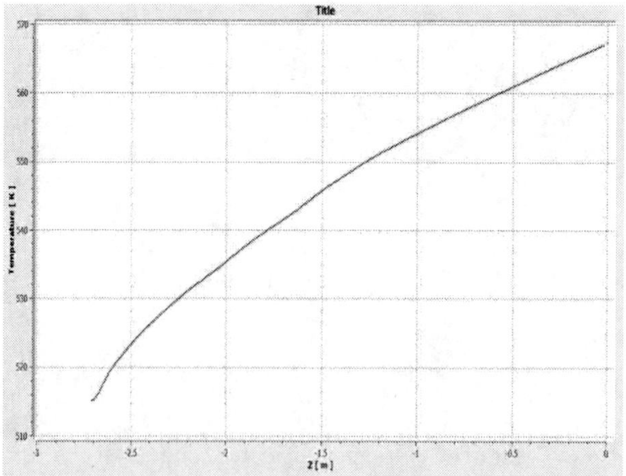

Figure 5. Temperature profile along the axis of corrugated pipe

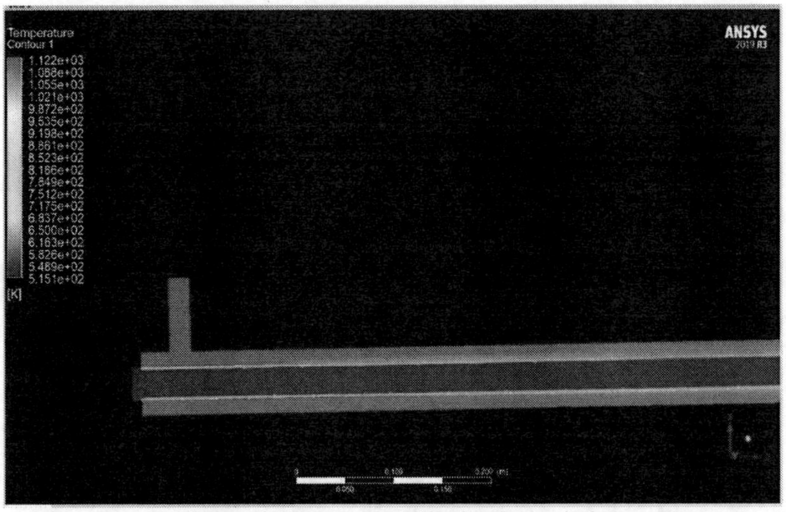

Figure 6. Temperature distribution at intlet

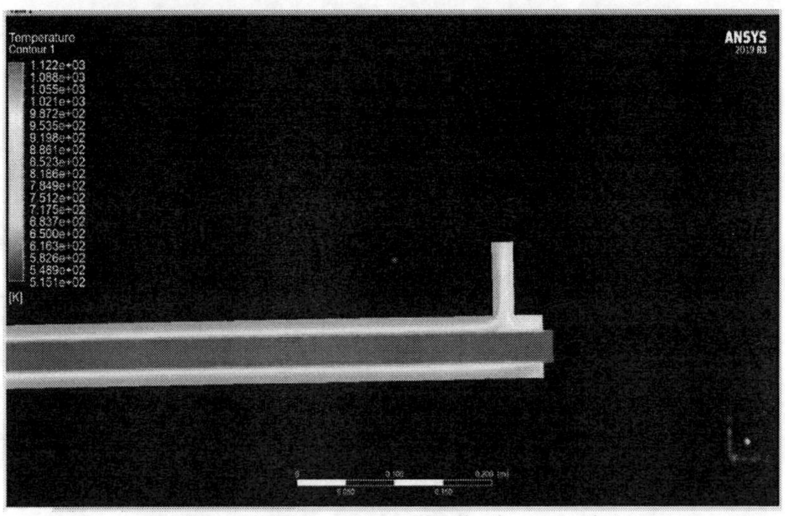

Figure 7. Temperature distribution at outlet

IOP Conf. Series: Materials Science and Engineering 1123 (2021) 012037 doi:10.1088/1757-899X/1123/1/012037

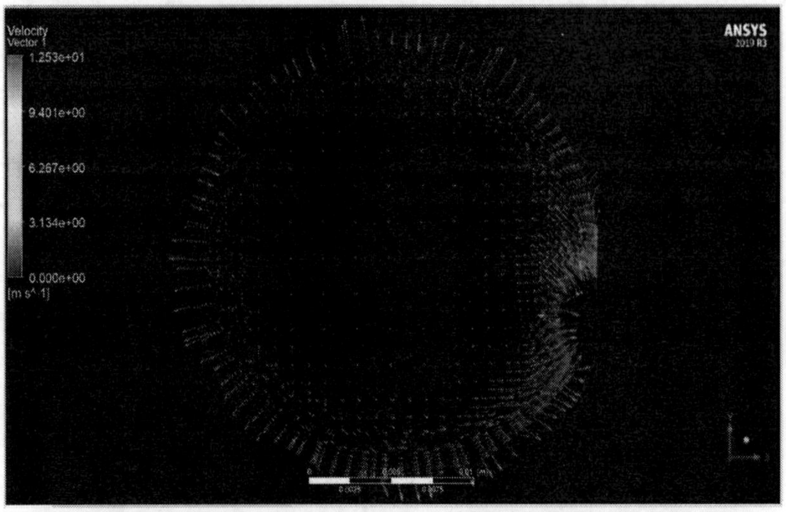

Figure 8. Effect on velocity vector of fluid due to corrugation

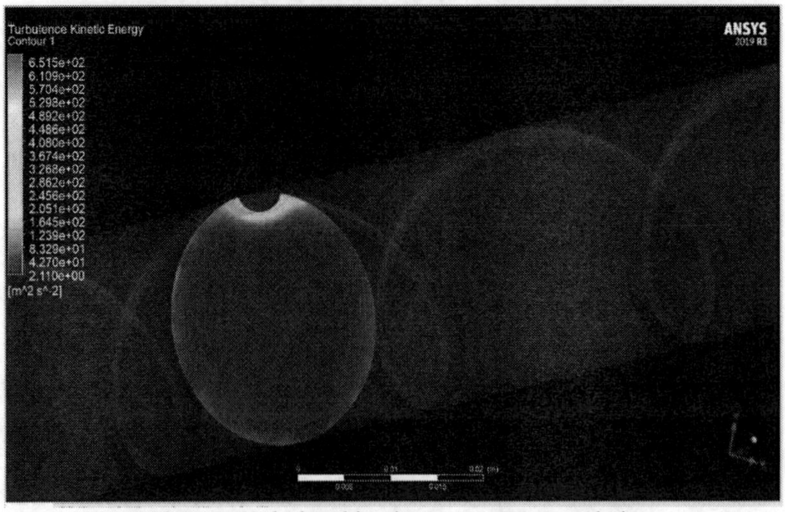

Figure 9. Turbulent kinetic energy corrugated pipe

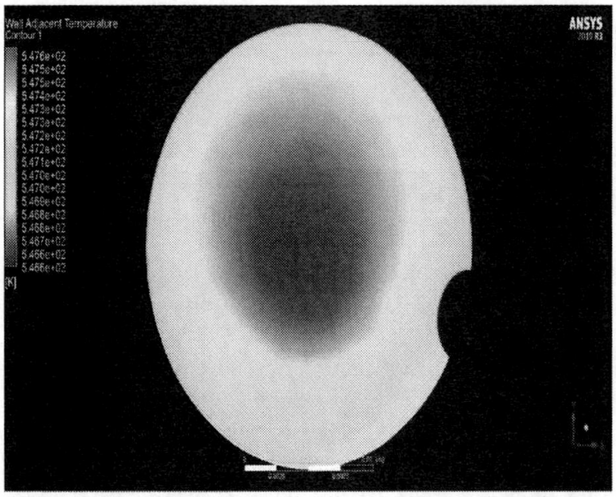

Figure 10. Temperature distribution in corrugated pipe

Fig.4- Fig.10 provided above are the results obtained from the analysis done with given inlet conditions. The values obtained for different parameters after the analysis is mentioned in the table below:

Table 4. Resulting Outlet Conditions

Parameter	Plain Tube	Corrugated Tube
Tube Outlet Temperature	285°C	294.3°C
Shell Outlet Temperature	759°C	740°C
Pressure drop in the tube	4413 Pa	192678 Pa

6. Conclusions

The effect of corrugation on the heat transfer and the pressure drop in the pipe was investigated numerically. It has been seen that the temperature rise is more in the case of corrugated pipe as compared to plain pipe due to enhanced heat transfer but the pressure drop is also increased. Hence, care must be taken that the heat transfer must be economical and not at the cost of significant increase in pumping power.

7. References

[1] Barba A., Rainieri S., and Spiga M., "*Heat transfer enhancement in a corrugated tube*", International Communication on Heat Mass Transfer, 2002, 29, 313-322.

[2] Bhoi, Shailesh. M, Patel, Devendra., Jejurkar Avdhoot,"Review on Heat Transfer Phenomena using corrugated tubes in Heat Exchangers", International Journal of Fluid and Thermal Engineering, 2016, 05, 023-029.

[3] Dong Y., Huixiong L., and Tingkuan, C., "*Pressure drop, heat transfer and performance of single phase turbulent flow in spirally corrugated tubes*", Journal of Experimental Thermodynamic Fluid Science, 2001, 24, 131-138.

[4] Laohalertdecha S., and Wongwises S., "*The effects of corrugation pitch on the condensation heat transfer coefficient and pressure drop of R-134a inside horizontal corrugated tube*", International Journal of Heat Mass Transfer, 2010, 53, 2924-2931.

[5] Rainieri S., Pagliarini G., "*Convective heat transfer to temperature dependent property fluids in the entry region of corrugated tubes*", International Journal of Heat Mass Transfer, 2002, 45, 4525-4536.

[6] Sheikholeslami M., Bandpy M.G., and Domiriganji D., "*Review of heat transfer enhancement methods: Focus on passive methods using swirl flow devices*", Renewable and Sustainable Energy Reviews, 2015, 49, 444-469.

[7] S.W. Hong, A.E. Bergles, Augmentation of laminar flow heat transfer in tubes by means of twisted-tape inserts, J. Heat Transfer 98 (1976) 251–256.

[8] S. Garimella, V. Chandrachood, R.N. Christensen, D.E. Richards, Investigation of heat transfer and pressure drop augmentation for turbulent flow in spirally enhanced tubes, ASHRAE Trans. 94(2) (1988).

[9] Targanski T., and Cieslinski J.T., "Evaporation of R407C/oil mixtures inside corrugated and micro-fin tubes", Applied Thermodynamics Engineering, 2007, 27, 2226-2232.

[10] Vicente P.G., Garcia A., and Viedma A., "*Experimental investigation on heat transfer and friction factor characteristics spirally corrugated tubes in turbulent flow at different Prandtl number*", International Journal of Heat Mass Transfer, 2004, 47, 671-681.

[11] W. Nakayama, K. Takahashi and T. Daikoku, Spiral ribbing to enhance single-phase heat transfer inside tubes, Proceedings of the ASME–JSME Thermal Engineering Joint Conference, Honolulu, 1983.

[12] Zimparov V., "*Enhancement of heat transfer by combination of three start spirally corrugated tubes with a twisted tape*", International Journal of Heat Mass Transfer, 2001, 44, 551-574.

Vision Based Self Adaptive Algorithm for 6 Axis ABB Industrial Welding Robot

Parameshwaran.R [1], Tony Thomas.A [2], Madhu Vishal.V[3]

[1] Department of Mechatronics Engineering, Kongu Engineering College, Perundurai, TamilNadu-638060, India, parameshwaran76@gmail.com
[2] Department of Mechatronics Engineering, Kongu Engineering College, Perundurai, TamilNadu-638060, India, dec84tony@gmail.com
[3] Department of Mechatronics Engineering, Kongu Engineering College, Perundurai, TamilNadu-638060, India, madhuvela12@gmail.com

Abstract. The ultimate aim of this work "Vision Based Self Adaptive Algorithm for 6 Axis ABB Industrial Welding Robot" is to develop a self-adaptive RAPID algorithm based on the captured image of the work piece. The image of the work piece is captured using NI Guppy pro F031C camera which has 300 DPI with 120 FPS resolution. The captured image is transferred to the LabVIEW software for developing the self-adaptive algorithm through RS232 serial communication protocol. The LabVIEW vision assistant module is used to develop an algorithm based on the geometry of the captured image. The various image processing operations like Thresholding, Morphological operation (Thinning), and edge detection are carried out. The caliper tool is used to measure the distance between the coordinate points (welding distance). With the aid of Visual Basic, the measured values are converted into coordinates. The coordinates are used to develop a RAPID program with the aid of ABB Robot Studio library functions. The virtual server is established between the ABB Robot Studio and IRC5 Controller with the use of MOD-BUS protocol with VISA. After receiving the data from the MOD-BUS, the IRC5 controller moves the ABB Industrial Robot End effectors along with the welding gun for the welding purpose. The work piece is located on the designed jigs and fixtures. The different types of welding with different design can be incorporated with the Vision Assistant module for further development. To illustrate the developed algorithm, robot assisted MIG welding process is carried out. The welded work-pieces are tested for its strength quality and the results are verified. Optimal welding parameters for the good quality of weldment identified by using Taguchi method of optimization.
***Keywords**: NI Guppy pro; Thresholding; Edge Detection; Thinning; MOD-BUS; IRC5 Controller; Taguchi Optimization*

1. Introduction
The Robotic system considered in this study consists of the robotic arm namely the ABB IRB 1410 with 6 axes. Robot studio software is used for effective control of the robot. The IRC5 controller is the ABB's fifth generation controller. Its motion control technology, True-Move and Quick-Move, are key to the robot's performance in terms of accuracy, speed, cycle-time, programmability and synchronization with external devices. All ABB robot systems are programmed with RAPID (ABB's flexible high-level programming language). It also incorporates powerful support for the most

common robot process applications such as welding and assembly. The ABB IRB 1410 is a 6 axis industrial welding robot with a payload of 5kg. It is being used in the industries for the continuous welding operations. It has a high accurate control and easy movement, the speed and response of the robot manipulator is also pretty impressive while compared to other models in that category.

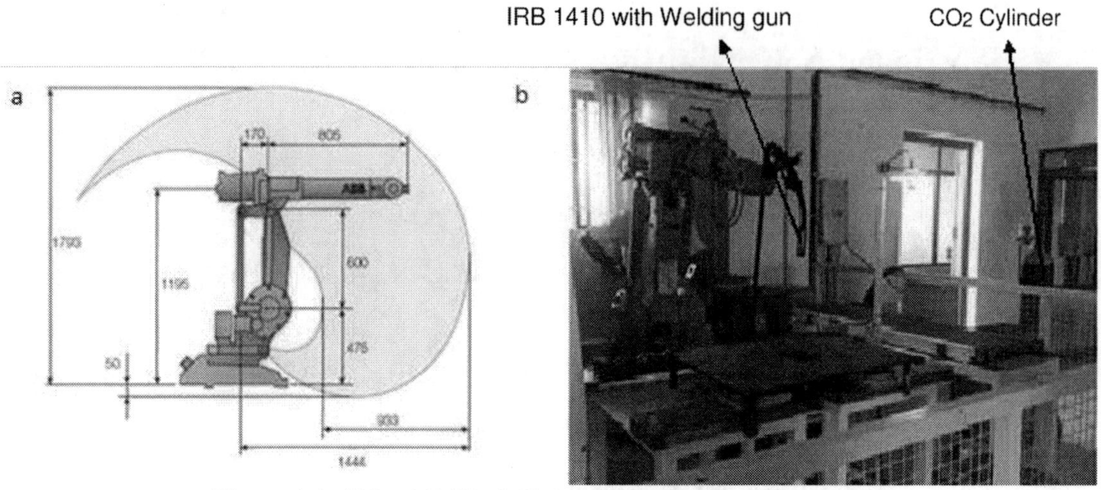

Figure 1.(a)IRB 1410 Work Volume; (b) Welding Robot.

ABB IRB 1410 work range is shown in the figure 1 (a). It gives the idea about the maximum reach of the end effectors and the work volume that the robot manipulator can reach in order to make the work done. IRB 1410 is the 6 axis robot used for the purpose of welding. The robot is proven in MIG welding applications and provides outstanding performance. The IRB 1410 has integrated wire feed cabling and mounting holes for optimized assembly of process equipment on the arm. MIG welding functions are included as standard in the IRC5 robot controller. They are made available via the operation interface unit nothing but the flex pendent. The real time integration of the welding gun with the robot end-effectors is shown in the figure1 (b). Enhanced welding automation, safety from dangerous fumes, higher quality welds and more efficient processes are just some of the advantages of it. lee et al (2010) proposed an intelligent welding robot in shipbuilding, the RRXC which consist of the industrial robot and auxiliary transport device in this process the six axis ABB industrial robot is being used. Paes et al (2014) on his work with trajectory control of the ABB industrial robot, describes about the path planning and development of dynamic robot module. Trajectory optimization technique is done by using the dynamic parameter of the robot module developed. Muzan et al (2012) on his work on using robots for painting application, discuss about precise movement of the end effectors for specific target letter which are pre-programmed into the controller system. Later the end effector and manipulator is imported into the robot studio for RAPID programming. Antonelli et al (2013) the algorithm developed which filters inaccurate and noisy data, re-orders points and computes a trajectory suitable for a welding robot, which is done by using the Bluetooth for transfer and manipulation of the data. Antonelli et al (2016) proposed a system on the qualification of a collaborative human-robot welding cell and the welding process is more reliable. Luo et al (2014) on their work presents a weld pool edge detection technique based on an off axial green illumination laser and a coaxial image capturing system that consists of a CMOS camera and optic filters. By studying the reflection of the green laser illumination the appropriate width of the keyhole generated by the welding process is identified. Tsai et al (2011) on his work presents the path planning of golf club head welding robot by using machine vision system, the robotic welding path is obtained by rotations and translation along the axes of the task frame according to the requirement of the welding attitude. The respective mathematical frame is designed for the transformation of the vision data into the path planning coordinate value. Luo et al (2015) proposed a system on the Vision–based weld pool boundary extraction and width measurement during keyhole fiber laser welding in which the green laser illumination the appropriate width of the keyhole generated by the welding process is identified. Dinham et al (2013) presented a paper work on the autonomous weld seam identification and localisation using eye-in-hand streo vision for robotic arc welding that enables the robust identification of narrow weld seams for ferrous materials combined with reliable image matching and triangulation through the use of 2D homograph. Zhang et al (2013) presented on the real-time seam tracking control technology during welding robot GTAW process based on passive vision sensor that enables the analyzing the features of welding images, a new improved Canny algorithm has been proposed to detect the edges of seam and pool, and extract the characteristic parameters of welding images. Special PID controller is used. Bragun et al (2015) proposed asystem on the Stereo vision based measuring system for welding path inspection which measures the arc position in 3D space based on the stereo vision principle. The simulated irregularities are visible as anomalies in the welding path. Xu et al (2012) presents a technology about real-time seam tracking, a set of vision sensor system has been designed for the welding robot. By analyzing the features of welding images, a new improved Canny algorithm has been proposed to detect the edges of seam and pool, and extract the characteristic parameters of welding images. A specifically designed PID controller is being used for the proper functioning of the algorithm. Zhang et al (2014) on there work uses the common CCD camera for through image processing and calibration, the shapes and sizes of both the keyhole and the weld pool were measured under different welding conditions. The observation results lay solid foundation for optimizing the welding process parameters and implementing process control of plasma arc welding. Bracun et al (2015) on their work on weld path inspection uses the stereo vision based system, which measures the arc position in 3D space based on the stereo vision principle. The simulated irregularities are visible as anomalies in the welding path.

2. PROPOSED SOLUTION

In this system, the robot and the vision camera are incorporated together into a single system for the effective control of the robot. Figure 2 shows the block diagram of the incorporated system with the data flow in it. Initially, the with the help of the NI camera the image is captured automatically by the triggering Lab VIEW's virtual tool kit. The captured image is then passed through series of process under the vision assistant in which thresholding, morphology, edge detection and caliper process are carried out. Then the system would refer the captured image with the predetermined sample images for the final change in the coordinate system. These coordinates are then transferred to the IRC5 controller through the Robot Studio software.

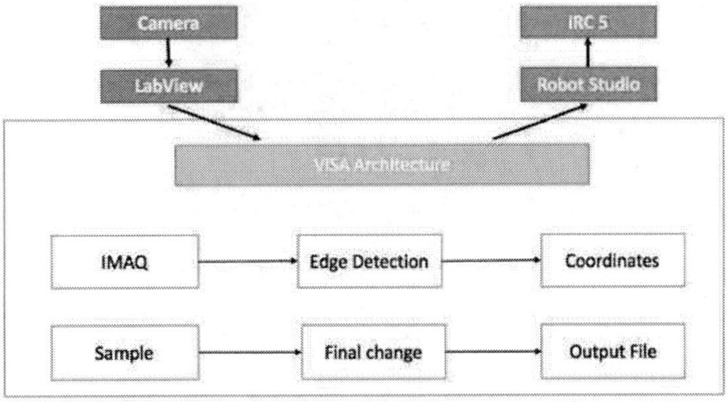

Figure 2. System Architecture and work flow.

3. SYSTEM DEVELOPMENT

3.1 Vision Development Module

The initial step of the entire process is to capture the image in the allotted perimeter for the further processing. LabVIEW is the tool used for the image capturing and processing. In LabVIEW, the vision assistant is responsible for accessing the hardware unit i.e. the camera by the software for capturing the image. The Vision development module uses the vision assistant for the development of the various sub level process in the image processing section. The captured image is regulated by the threshold value so that the proper monochromatic image with singularity image detection is done. Then in the next process of morphology the noise in the image is reduced by the thinning process of the edges and is represented in the figure 3.

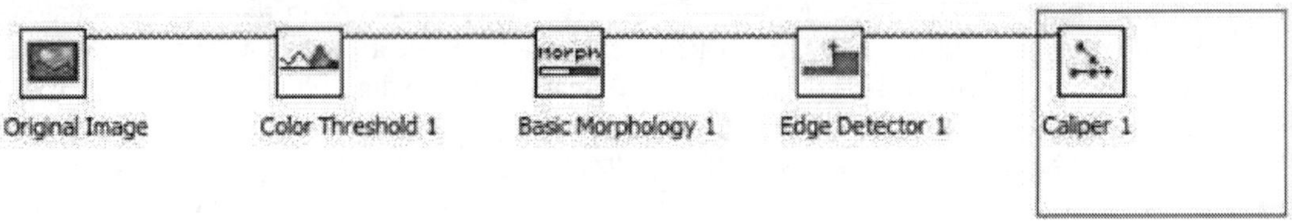

Figure 3. Image Development Process

Then the two edges as described in the sampling process are detected automatically by the edge detector unit of the vision assistant. The final image after all the process flow is shown in the figure 4.

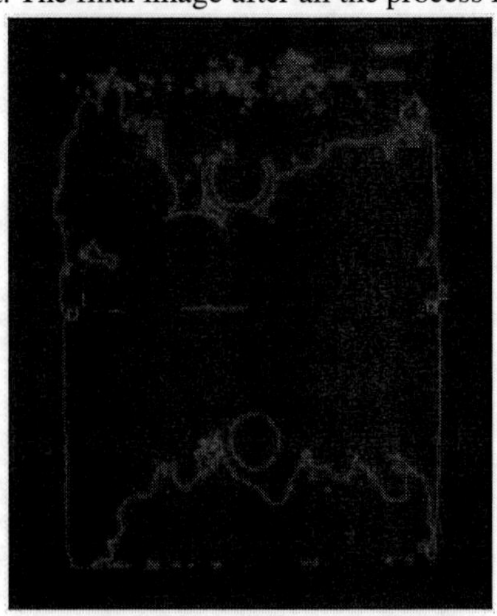

Figure 4. Processed Images

The image mentioned above is the final processed image in which the points marked with green boxes as number 1 and 2 represents the joints of the welding process. The point 1 indicates the starting point and the point 2 indicates the destination point. Then the Caliper tool is being used for the identification of distance between these points for the purpose of coordinate development.

3.2 Pixel Conversion

The output value from the vision assistant is the distance between the two points but the system provides the value in the form of pixels. But the robotic module does not understand the values in pixels. So these pixels are to be converted into mm (millimeter) unit for the transformation of the

IOP Conf. Series: Materials Science and Engineering 1123 (2021) 012038 doi:10.1088/1757-899X/1123/1/012038

values to the robotic unit. For the conversion of pixels to mm, some of the parameters of the camera are required such as the DPI (Dots per Inch). Dpi is nothing but the representation of quantity or number of dots or pixels present in the square of side one inch. The Dpi of the NI Guppy pro is 300 from that the formula is derived to converting the pixel to mm. For this conversion, separate mathematical functions are performed under coordinate transfer algorithm.

$$1\,mm = (Pixels * 25.4), \quad (1)$$

3.3 *Algorithm Development*

The algorithm is developed to get the data from processed image and to transfer it into programming coordinates. Value of X, Y and Z form and replace the values in the RAPID program. The main part of the system is the coordinate development algorithm which is shown in the figure 5. The image is captured and evaluated in the vision development module and the final image is converted into necessary data values as per the coordinate system of the robot controller for the easy communication of the system. Initially the vision acquisition block is used for the capturing of the image by accessing the camera; the captured image is displayed in the display unit with the help of the display block. Once the image is captured it is sent into the Vision assistant where the image processing technique is done and as an output from the block it gives the measurement value in pixel unit. Later on the pixel unit is converted into millimeter by means of the formula derived with the help of dpi of the camera. Then the value is sent into the bundle to binary unit block. In that block, the measurement values are changed to coordinates with respect to the two dimensional image captured.

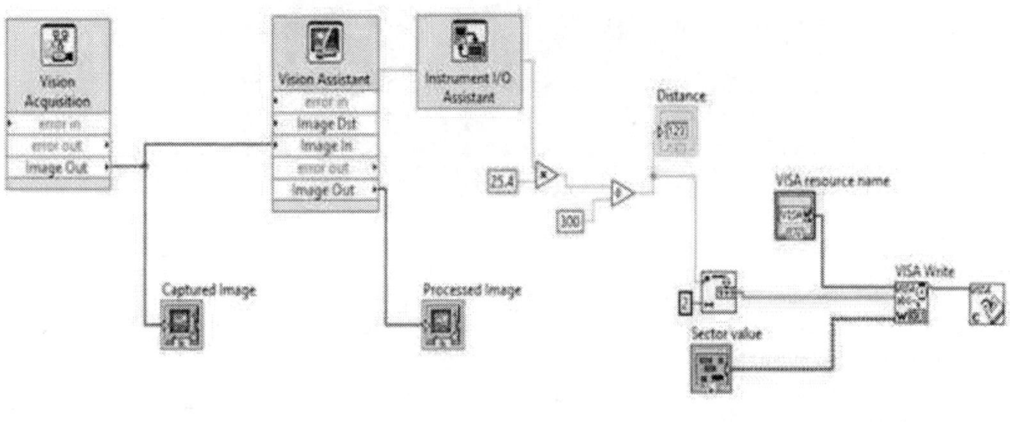

Figure 5. Co-ordinate Development Algorithms

3.4 *Communication Module Development*

The communication between these two software is done with help of the virtual server named as MODBUS. In this case, LabVIEW acts as server and the robot controller acts as the slave. With the help of the developed virtual environment, the robot controller is controlled by the LabVIEW. Figure 6 shows the designed Communication Protocol.

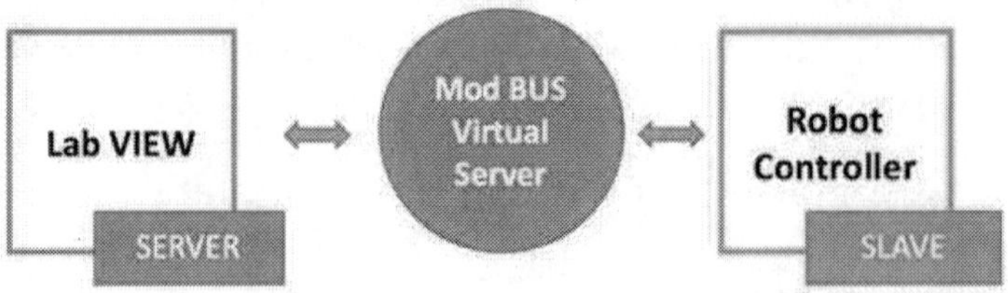

Figure 6. Communication Protocol

3.5 Development of end effector

The welding gun must be attached to the robot manipulator; a separate end effector must be designed. The end effector is one of the important parts of the robot manipulator. In robotics, an end effector is the device at the end of a robotic arm, designed to interact with the environment. The end effector is manufactured with the CNC machining process. Since the robot manipulator is highly precisely manufactured and operated in all condition, good quality of manufacturing is necessary for the perfect match of the end effector with the manipulator. Figure 7 shows the view of the manipulator where the welding gun is placed horizontally.

Figure 7. End Effector

3.6 Program Execution Module

The coordinates calculated from the LabVIEW are uploaded into the Robot Studio by means of three coordinate transform variables. For three separate x, y and z coordinate values separate variables are assigned. The Sub-program variables for the data transfer are:

- Visionpick.trans.x
- Visionpick.trans.y
- Visionpick.trans.z

These three variables carry the data from the LabVIEW script to the robot studio to generate the RAPID program for the robot controller to operate. Then the robot manipulator will move along the X, Y and Z axis to reach the starting point and move to the ending point. This movement will make the welding process to complete. The real time integration of the welding gun with the robot end-effector is done. Due to this integration, the welding automation, safety from dangerous fumes and weld quality are enhanced.

4. EXPERIMENTATION

Once the welding process is ready to operate, the optimal welding parameter for the effective and better welding is to be identified. The parameter which affects the welding process is identified initially. In case of the MIG welding process using robot, three parameters are considered:

- Gas flow rate
- Current rating
- Feed rate

In the automated MIG welding process, the speed of the welding gun is constant since it is controlled by the robot. Based on these values, the design of experiments is used to identify the number of experiments using Minitab optimization software. Table 1 shows the values for the design of experiments.

Table 1. Taguchi Experimental design.

Ex.No	Gas Flow (kg/cm^2)	Current Rating (A)	Feed Rate (mm/min)
1	1	1	1
2	1	1	2
3	2	2	1
4	2	2	2
5	1	2	1
6	1	2	2

4.1 Material Processing and Testing

For the purpose of welding, the mild steel material is chosen and the dimensions of the work piece are same for all the eight experiments designed. For the effective penetration of the welding, the chamfering process is done at the welding joints for all the work pieces. The values for the welding parameters are identified and tabulated in the table 2. These values are changed in the real time system and welding process is carried out.

Table 2. Welding Parameters

Gas Flow (kg/cm^2)	Current Rating (A)	Feed Rate (mm/min)
4	40	1.8
4	40	3.7
6	60	3.7
4	60	3.7
6	40	1.8
6	40	3.7

Based on these parameters, the eight experiments are done one by one with respective change in the welding parameters. The eight welded work-piece are shown in the figure 8.

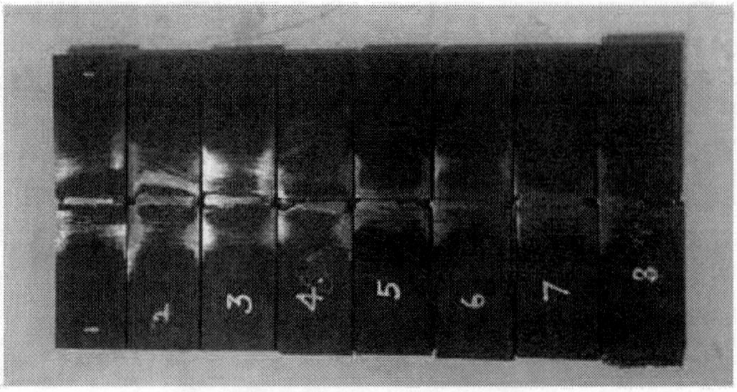

Figure 8. Welded Work piece

After the completion of the welding process, the eight work-pieces are made to undergo hardness test and tensile strength test. This is done in order to identify the work-piece with highest strength and hardness, so that the particular parameter of welding could be used for the effective welding purpose and to obtain a good quality of welding.

5. Selected Results

After the completion of the test over the work-piece, the tensile strength and hardness values are enumerated in the table 3.

Table 3. Test Values

Experiment	Tensile Strength (N/mm²)	Hardness
1	2.012	83.334
2	0.321	80.667
3	0.724	77.334
4	6.197	85.000
5	10.060	77.000
6	2.012	82.000
7	1.529	64.000

The test values enumerated from the table are uploaded into the Minitab optimization software in order to apply the Taguchi method of optimization to the system. So the three parameter and two level processes is selected and the respective values are uploaded into the processing area. Then the Taguchi method of optimization is applied over the values for the identification of the optimistic parameter.

Table 4. S/N ratio for tensile strength

Experiment	Tensile Strength (N/mm²)	S/N Ratio
1	2.012	2.434
2	0.321	2.112
3	0.724	4.728
4	6.197	4.406
5	10.060	5.090

The signal to noise ratio for the parameter tensile strength is calculated and are tabulated in the table 4.

Table 5. S/N ratio for Hardness

Experiment	Hardness	S/N Ratio
1	83.334	75.834
2	80.667	82.416
3	77.334	75.000
4	85.000	81.583
5	77.000	79.083

Similarly, for the hardness the S/N ratio is calculated and is represented in the table 5.

6. Discussion

The Signal to noise ratio is considered for the optimization process because if the signal to noise is higher than the output parameter will also be higher. The noise in the system must be lower than the signal in the system in that approach the signal to noise ratio is considered. So from the Signal to Noise ratio obtained from the hardness and tensile strength as an influencing factor the graphical representations are made. They are shown in the figure 9 and 10. In both the cases the better result is considered as higher. Because higher the tensile strength higher the quality of welding. Similarly higher the hardness value delivers the good quality of welding.

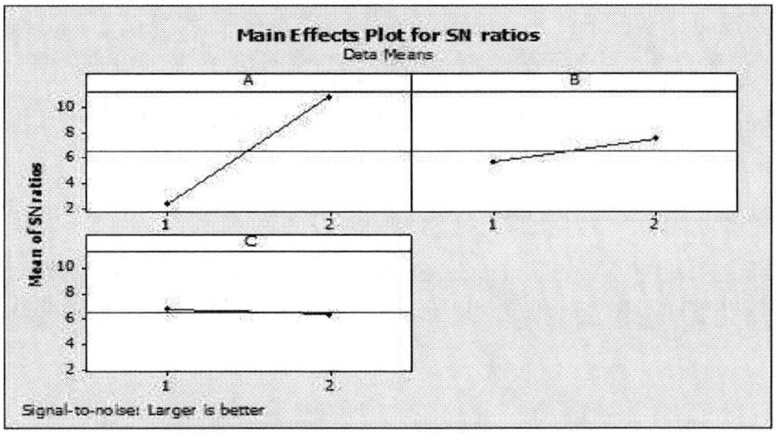

Figure 9. S/N ratios for Tensile Strength

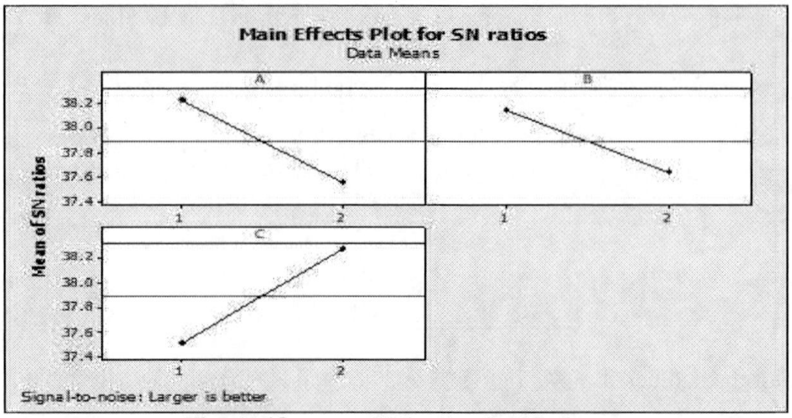

Figure 10. S/N ratios for Hardness

7. Discussion

From both the graphical and numerical representation, the optimal parameters can be obtained. In case of the tensile strength the experiment number five with tensile strength of 10.06 N/mm^2 provides better result. In case of the hardness value the experiment number six with hardness value of 82 provides better result.

8. References

[1] Lewis A. Muth, Robert B. Peck, Robust separation of background and target signals in radar cross section measurements, IEEE transactions instrumentation and measurement, 2005, 54 (6), p.2462.

[2] S. Soliman and C. Wheatley, Frequency coordination between CDMA and non-CDMA systems. In Proceedings of the MTT-S Symposium on technologies for wireless applications digest, San Diego,CA, USA, 20-22 Feb. 1995, pp.123-130.

[3] Robert B. Peck, Foundation engineering, 2nd edition, New York: McGraw-Hill, 1972, pp. 230–292.

[4] M. Farkas, Implementing network security in a local bank, In: R. Grande (ed.), Enterprise resource planning system: a theoretical perspective, 3rd ed. Oxford: Oxford University Press, 2003, pp.23-42.

[5] S. Hawking. The Beginning of Time. A public lecture. Professor Stephen Hawking's website, 2000. [Online] Available from: http://www.hawking.org.uk/home/hindex.html [Accessed 20 Nov 2006

IMPROVED SENTIMENT ANALYSIS FOR DRAVIDIAN LANGUAGE-KANNADA USING DICISION TREE ALGORITHM WITH EFFICIENT DATA DICTIONARY

Ranjitha P[1] and Bhanu K N[2]

Department of computer science, Amrita Vishwa Vidyapeetham, Mysuru, India
[1] ranjitha_p@asas.mysore.amrita.edu
[2] knbhanu@asas.mysore.amrita .edu

Abstract. The trend of exploring the micro blogging services is being hit door to door. The creation and sharing of short text messages is at its top gear. Reviews, being one of the fastest growing online networking services, which receives around millions of reviews per day, has emerged as an invaluable source of the sentiment analysis. Sentiment analysis refers to recognize the sentiment, polarity, opinion or emotion of a text. Websites like news Kannada, prajavani, wedunia.com, and oneindianews have reviews written in Kannada which has a higher preference than the common languages. In the current system, research are carried in sentiment analysis for regional language such as Hindi, Bengali, Kannada etc., are very few compared to English language. Here we propose a system which is computationally identifying and categorizing opinions expressed in a piece of text written in Kannada language, especially in order to determine whether the writer's attitude towards a particular topic, product, etc. is positive, negative, or neutral. Keywords: Natural language processing, Sentiment analysis, data dictionary, Feature extraction, Tokenizing, decision tree.

1 Introduction

In the current system, research works done in sentiment analysis for regional language such as Hindi, Bengali etc., are not too many when compared to English language. These languages were analyzed by equating lexicon words using part of speech tagging technique. This represents emotions as adjectives pattern suffixed word. These semantic words are rated through machine learning approach. Machine learning includes algorithms such as rule based algorithm, maximum entropy etc.

India being multi-linguistic nation sentiment analysis in English forms a disadvantages. For analyzing the documents or piece of text that are of particular regional language such as Kannada language, it required to have sentiment analysis by using datasets in Kannada. Websites like news Kannada, prajavani, wedunia.com, and oneindianews have reviews in Kannada which has a higher landing than the ordinary languages. Even though Sentiment analysis in Kannada with English translation produces similar results with generic set of sentiment words but does not consider the domain of the text which is the feature of an entity called subjective in sentiment analysis. So our proposed model uses the Kannada words without translating to English and trains the model using efficient data dictionary by decision tree classifier. Here Data dictionary is the one where we store set of words which is labelled as positive/negative words and each word is assigned with certain score (polarity).

2. Related Work

Sentiment analysis is a current trending topic for the research. There are many unsupervised algorithm which is used to classify a review as recommended or not recommended. As an initial step, the text data is POS tagged in order to identify the adjectives, noun and adverbs and then the semantic orientation of these words are carried out. Here PMI-IR algorithm is used for calculation of semantic orientation between words. [2]

A novel approach for machine translation of simple sentences from Kannada to English language is introduced. In the first step method execute lexical analysis to process input transliterated Kannada text to recognize and to classify tokens using newly constructed source language lexicon KannadaWordNet. Then, newly designed recursive descent parser for the domain validates the transliterated Kannada text with the syntactic constructs/grammar rules. [3]

Tools like Paraphrase generator are very essential component of any NLP application. Paraphrase generator tool for south Indian language of Kannada language using dictionary lookup approach. By using the morphological analyzer, dictionary look up and morphological generator approach, the Paraphrase are generated for Kannada language. [5]

Polysemy word are the word which has multiple meanings. The same word can be treated differently at different instance. Consider a text from Kannada paragraph, a person with no prior knowledge of Kannada language cannot distinguish the difference in the meaning of a polysemy word. It can be concluded that in a huge document with multiple polysemy words cannot be summarized very easily. [6]

In Hindi language sentiment lexicon called Hindi-Senti Word Net (H-SWN) was introduced using English Senti Word Net (SWN) and English- Hindi Word Net linking. The sentiment analysis of documents was analyzed using sentiment annotated corpora, Machine translation and resource based sentiment analysis methods. [8]

Sentiment Analysis in Indian Languages is a trending topic for research area. Word Net, POS tagger, Senti Word Net and many other resources are available for very few languages. A Senti Word Net development for Indian Languages like Bengali, Hindi and Telugu is tried using Word Net, dictionary based and Corpus based methods in a Hindi sentiment lexicon called Hindi-SentiWord-Net (H-SWN) was developed using English SentiWordNet (SWN) and English- Hindi WordNet linking. The sentiment analysis of documents was performed using sentiment annotated corpora, Machine translation and resource based sentiment analysis methods. [9]

3 Methodology

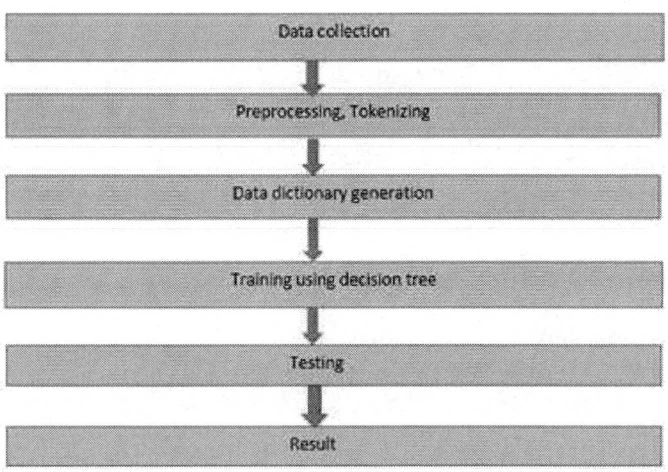

Fig. 1. Workflow of the proposed work

Sentiment analysis follows the steps as shown above in Fig 1. Data is collected from various websites like prajavani, wedunia.com, and oneindianews is considered as the dataset for training. Collected dataset is first subjected to the pre-processing.Pre-processed data is taken for data dictionary generation fig shows the data dictionary of the proposed system. Once the data dictionary is ready is it is trained by data mining algorithm-decision tree. Classification requires foregoing learned labeled data in a particular form Known as Trained data, which are attributes and its values along with target function called Class Label. Text is tokenized during pre-processing of data set and attaches tags to each tokenized words using part of speech taggers [1]. These tagged words are identified as features or opinion words according to role in their text. Decision tree algorithm for Kannada sentiment analysis is used to train and test the data and classify into different labels. Decision tree algorithm for Kannada sentiment analysis is explained in the algorithm 1.

Algorithm1: Decision tree algorithm for Kannada sentiment analysis

---

Input: Randomly chosen Kannada review over internet.
Output: classifies the review into positive / negative review.

Let R be the review
S be the set of words.
S= {w1, w2, w3--------wn}
Pc=Positive counter, Nc=Negative counter, Pv=Polarity values, Pdd=Positive word
In the data dictionary, Ndd=Negative word in data dictionary, Vp=Polarity value of
Words, Pr=Positive, Review, Nr=Negative Review.

1: Tokenize R
  Set Pc to zero and Nc to zero.
2: For each W in S
3: If W is found in Pdd
    Increment Pc to value of Vp
4: Else W is found in Ndd
    Increment Nc to value of Vp
    End If
    End For
5: If Pc>Nc
    Return Pr
  Else reurn Nr
    End If

IOP Conf. Series: Materials Science and Engineering 1123 (2021) 012039 doi:10.1088/1757-899X/1123/1/012039

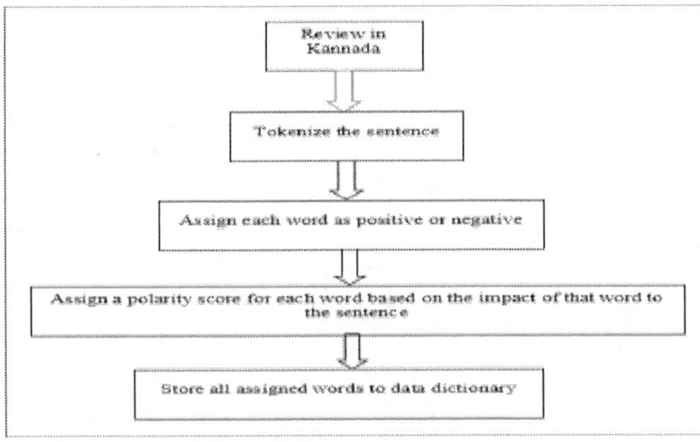

Figure. 2 Steps to create data dictionary.

Fig 2 shows the steps to create the data dictionary for Kannada sentiment analysis. In the first step reviews are randomly chosen from the internet. Chosen review subjected to the tokenization. Later each word is labelled as positive or negative and assigned polarity values for each word and stored it the data dictionary. Table 1 shows the sample data dictionary where each word as word type and polarity score.

Table 1. Sample data dictionary words

Kannada words	Word Type	Polarity score
ಭೇಧ	Negative	2
ಸೊಗಸು	Positive	3
ದ್ರೋಹಿ	Negative	3
ಒಲವು	Positive	2
ಸುಂದರಃ	Positive	4

IOP Conf. Series: Materials Science and Engineering 1123 (2021) 012039 doi:10.1088/1757-899X/1123/1/012039

4 Results and discussion

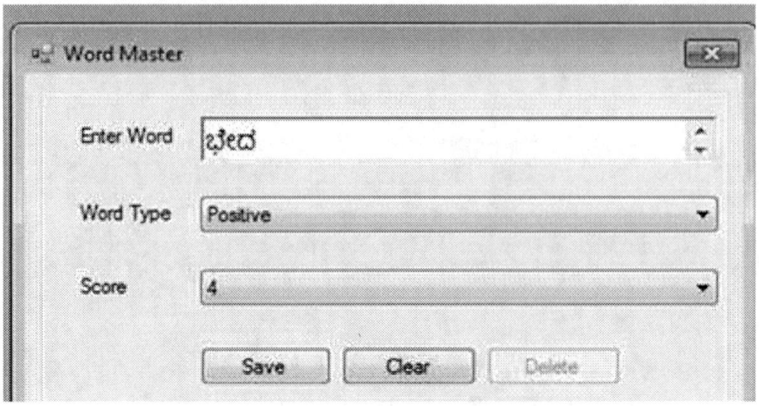

Figure. 3 Adding the word to data dictionary

Fig 3 shows the creation of data dictionary manually. Here we have created data dictionary which consists of more than 500 words for each positive and negative category .Each word will be given one score.

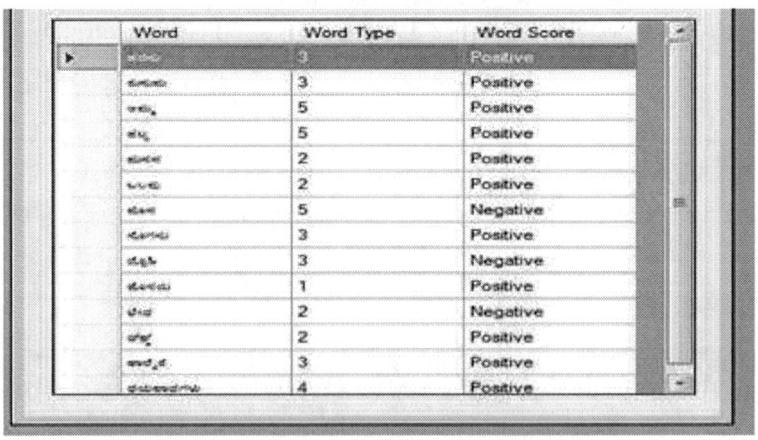

Word	Word Type	Word Score
[illegible]	3	Positive
[illegible]	3	Positive
[illegible]	5	Positive
[illegible]	5	Positive
[illegible]	2	Positive
[illegible]	2	Positive
[illegible]	5	Negative
[illegible]	3	Positive
[illegible]	3	Negative
[illegible]	1	Positive
[illegible]	2	Negative
[illegible]	2	Positive
[illegible]	3	Positive
[illegible]	4	Positive

Figure. 4 shows the list of words in dictionary

IOP Conf. Series: Materials Science and Engineering 1123 (2021) 012039 doi:10.1088/1757-899X/1123/1/012039

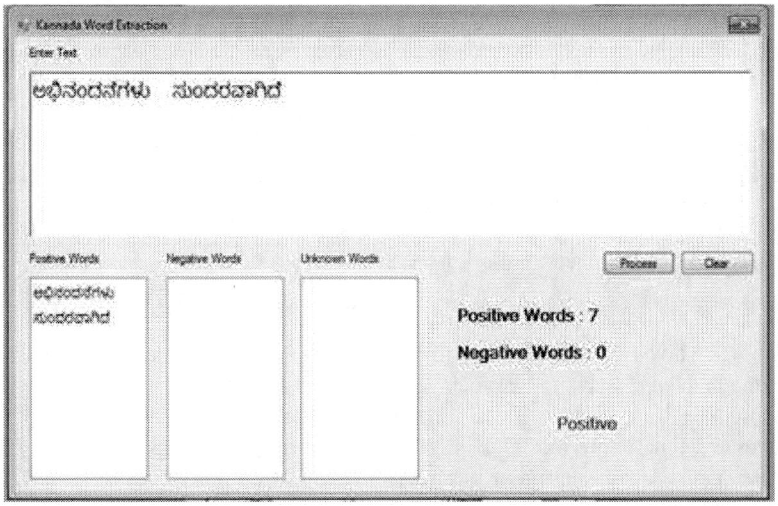

Figure. 5 showing the test data is classified to positive

Fig 5 shows the working of the proposed system. When the test data is given the trained system or model. First it will list number of positive word and negative word in the given text data and later it will classify the text into positive or negative. The trained model is providing up to 85% of accuracy.

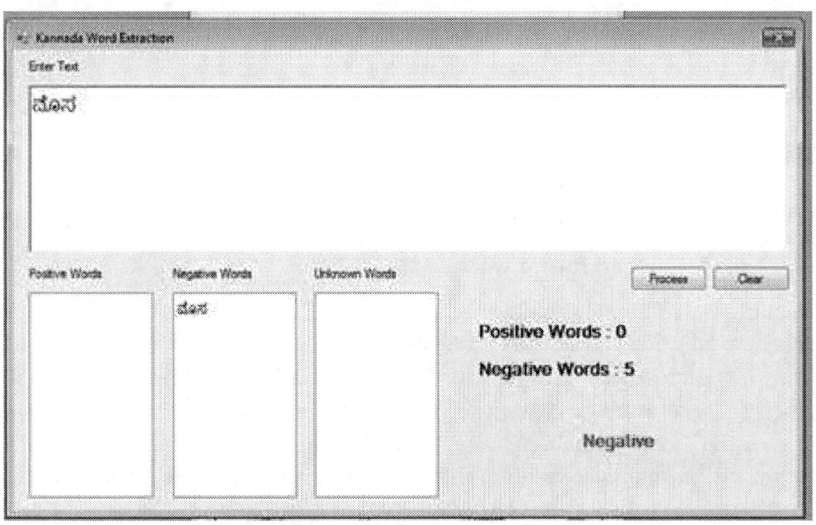

Figure. 6 showing the test data is classified to positive

Table 2. Comparision on different algorithms.

Methods	**Accuracy**
Naïve bayes	65%
Negator Algorithm	53%
Decision tree using data dictionary	85%

Table 2 shows the result by comparing different algorithm. By translating kannada to english using Naïve bayes they have achieved 65% of accuracy[11]. Using Negator algorithm for kannda sentiment analysis [11]resulted in 53% of accuracy. Finally our model resulted in the accuracy of 85%.By comparing this three methods for sentiment analysis decision tree algorithm is giving more accuate result.

Table. 3 Comparison on Precision and Recall

Method	**Precision**	**Recall**
Decision tree	0.78[1]	0.79[1]
Decision tree with efficient data dictionary	0.85	0.84

Table 3 shows the comparision on precision and recall between two methods. Using decision tree results in precision of 0.78 and recall of 0.79[1]. By manipulating decision tree with efficient data dictionary we can achieve 0.85 precision and 0.84 recall.

5 Conclusion and future work

Features based opinion mining is a challenging task in Kannada language and there are only few tools available such as Training data set specific to a domain, part of speech Tagger. Certain words in Kannada like "ಕಥೆಗೊಂದ ", "ಎನ ನವುದಕೊಂ", "ಮನೆಯುಂದರಲ್ಲದ","ಚಚೆರಂಯಾಗ್ ತತವೆ" on machine translation produced ambiguous text, which forms one of the major drawbacks and result in improper results. Thus sentiment analysis in regional language gives better accurate results compared to Machine Translated English Language. Here by creating the effiecient data dictionary Accurancy of 85% can be achieved.Only few analyses is done in Indian regional languages, hence this work can be extended to other regional languages especially in south Indian languages for the enhanced version of sentimental analyzer.

References

[1] Rohini V, Merin Thomas and Dr. Latha C A 2016 Domain based Sentiment Analysis in Regional Language-Kannad *Int. J. of Engg. Research & Tech.* (IJERT) Vol 4 Issue 22

[2] .Hemalatha, Dr. Saradhi Varma G P and Dr. Govardhan A Sentiement analysis tool using machine learning algorithm *Int. J. of emerging Trends and Tech. in computer science* vol 2 Issue 2

[3] Peter Turney D 2002 Thumbs Up and Thumbs Down? Semantic orientation applied to Unsupervised Classification of Reviews",Institute for Information Technology, National Research Council of Canada Ottawa,Proceeding of $oth *Annual Meeting of the Association for computional Linguistics* Philadelphia

[4] Solamki Yogesh Ganeshbhai,Bhumika and Shah K 2015 Feature based opinion mining:A Survey Surat India

[5] Aditya Joshi, Balamurali A R and Pushpak Bhattacharyya, 2010 A Fallback Strategy for Sentiment Analysis in hindi: a Case Study",IIT Bombay.Proceedings of ICON 2010: 8th *Int. Conf. on Natural Langugae Processing*

[6] Anil Kumar K M, Rajasimha N 2015 Analysis of user's sentiment for kannada web documents",Sciencedirect, Procedia computer science *Eleventh Int. Multi-Conf. on Information Processing* Vol 54 pp 247- 256

[7] Ankit singh and Md.Enayat Ullah Aspect based Seniment Analysis ,Indian Institute, Kanpur CS365A http://goo.gl/qdTRhf.

[8] Jeevanandam Jotheswaran and Dr.Y.S.Kumaraswamy Opinion mining using decision tree based feature selection through manhattan hierarchical cluster *measure J. of theoretical and applied science* Vol 58

[9] Tien-thanh Vu,Huyen-Trang Pham, Cong-To Luu 2011A Feature-Based Opinion Mining Model on Product Reviews in Vietnamese *Vietname National University, Hanoi(VNU), College of technology,Semantic Methods SCI*381 pp.23-33,springerlink.com

[10] Prabhu Palanisamy, Vineet Yadav and Harsha Elchuri, Serendio:Simple and Practical Lexicon Based approach to Sentiment analysis Serendio Software Pvt Lid Guindy Chennai

[11] Remya Govindan1 and Rosna Haroon P 2017 A Survey on Sentiment and Emotion Classification in Indo-Dravidian Languages Imperial Journal of Interdisciplinary Research (IJIR) Vol 3 Issue .

ANALYSIS OF LOW-VELOCITY IMPACT RESPONSE IN AA5083 PLATE

V. Sairam [1], S. Kishor Kanna [2], P.S. Samuel Ratna Kumar [3]

[1] Department of Mechanical Engineering, Kumaraguru College of Technology, Coimbatore, India, sairam.17me@kct.ac.in

[2] Department of Mechanical Engineering, Kumaraguru College of Technology, Coimbatore, India, kishore.17me@kct.ac.in

[3] Department of Mechanical Engineering, Kumaraguru College of Technology, Coimbatore, India, Samuelratnakumar.ps.mec@kct.ac.in

\

ABSTRACT

This paper investigates the low velocity impact performance of AA5083 using ABAQUS software. The impact resistance was measured in terms of depth of indentation created over AA5083. Low-velocity impact results indicates, that the angle of inclination of the impact resisting body have a great influence on the impact resistance properties. The analysis study was carried out for different orientations of the alloy plate with respect to the horizontal datum namely 30° ,60° ,90° respectively. A Finite Element Method (FEM) model has been developed using ABAQUS software tool to analyse and study the low- velocity impact test on the AA5083 plate. Results from the FEM model reveals that the depth of indentation over AA5083 is minimum when the angle is inclined at 90°.

***Keywords**: Low-velocity impact, AA5083, Angle of inclination, ABAQUS.*

1. INTRODUCTION

Alloys are the combination of one or more metal with different atomic radius and electronegativity elements. Alloys have occupied various industrial applications such as aerospace, automobiles, aircrafts, defense, sports equipment and manufacturing of machinery component. Due to its less weight, high strength, and good material properties like isotropic, crystalline structure and ease to machine [1]. In Boeing 787, most of the components are made up of alloy materials like aluminium and titanium which leads to reduction in fuel consumption by 20% [2]. AA5083 aluminium alloy have magnesium as a second major element and traces of manganese and chromium. Aluminium alloys are used in different applications and one of the most important applications are aircraft structure. The properties of AA5083 alloy are high strength, high stiffness, easy to machine with high dimensional stability [1-3].

The main risks of aircraft structure are structural failure and damage. The reasons for the failure of aircraft structure are formation of voids in the microstructure of the materials error in manufacturing process [4]. The impact loading can be categorized into low-velocity impact, intermediate velocity impact, high-velocity impact and hyper-velocity impact. The range of low-velocity impact is below 10 m/s. The intermediate velocity is occurred between the 10 m/s and 50m/s, high velocity impact range is between 50 m/s and 1000 m/s and hyper velocity impact range is between 2 km/s to 5 km/s. In this paper low-velocity impact analysis were discussed. The low-velocity impact causes unseen impairment with a large area of dynamic structural response, like impact by a bird during takeoff and landing, tool drop [6-8]. By finite element analysis method deformation produced by low-velocity impact test can be predicted. This method reduces time consumption and reduce cost to conduct experimental study [9]. In this paper ABAQUS software is used to analyse the low-velocity impact test to predict the deformation produced.

2. ANALYSIS PROCEDURE

The 3D finite element modelling is a better way for simulating impact damage on an alloy, especially with regards to the delamination of layers because it is the main failure mechanism and which causes more impact energy absorption [9-10].

For high strain rate deformation in many finite element programs, the Johnson-Cook plasticity model were incorporated. Being an empirical or purely phenomenological model, it is heavily dependent on experimental data [12-13]. Despite this, its parameters are more easily obtainable than those in physical-based constitutive models. The material law was defined using Johnson-Cook formulation. This is given in equation 1.

$$\sigma = (a + b\varepsilon^{n}p)(1+c\ln \dot{\varepsilon}/\dot{\varepsilon}0)(1-T^{*m}) \quad (1)$$

σ = true stress level, a = yield stress, εp = true plastic strain, b= Hardening modulus, n = Hardening Exponent, c= Strain rate coefficient, $\dot{\varepsilon}$ = Strain rate, $\dot{\varepsilon}0$ = Reference strain rate .

In the finite element design, the low-velocity impact on Aluminium alloy was analysed using ABAQUS software. The bullet was used as a three dimensional(3D) rigid body while the plate was used as a 3D deformable part. The solid model of an impactor(bullet) and plate is created by ABAQUS software. The diameter and length of the bullet was defined as 5mm and 20mm. The diameter of the plate was given as 100×100×10mm. The velocity at which the impactor strikes the plate is 3 m/s.

The material property for both the impactor(bullet) and plate was assigned using PROPERTY module in the menu. The material properties were assigned as per the table 1.

Table 1. Summary of material parameter for AA5083

Category	Parameter name	Value	Source
Physical properties	Density	2660 kg/m3	United states of America Department of Defense 1998
Thermal properties	Thermal	120 W/ (mK)	
	Coefficient of thermal expansion	2.34x10-7 /K	
	Inelastic heat	0.9	
	Specific heat	900 J/ (kg. K)	
Isotropic Elastic Model	Young's modulus	71.7 GPa	
	Poisson's ratio	0.33	
Johnson Cook plasticity model	A	137.89 MPa	Calculated from experimentation at the University of Manitoba
	B	216.73 MPa	
	N	0.4845	
	C	0.002	
	M	1.225	
	Tm	933 K	United states of America Department of Defense 1998
	Tref	293 K	
Johnson Cook Dynamic Failure Model	D1	0.178	Failure parameters for AA5083-H116(GRYTTEN , et al.2007)
	D2	0.389	
	D3	-2.246	
	D4	0	
	D5	0	

The solid models for the simulation were assembled in the required orientation using ASSEMBLY option in the module dropdown box. In first case, the plate was inclined to the axis at 30⁰ and the orientation of the bullet was 90⁰ to the axis of the plate. It is shown in figure 1. In second case, the plate inclination was changed to 60⁰ to its axis and the impactor angle remains same, as shown in figure 2. In third case, the inclination of the plate was further changed to 90⁰ to its axis and the impactor remains same, as shown in figure 3

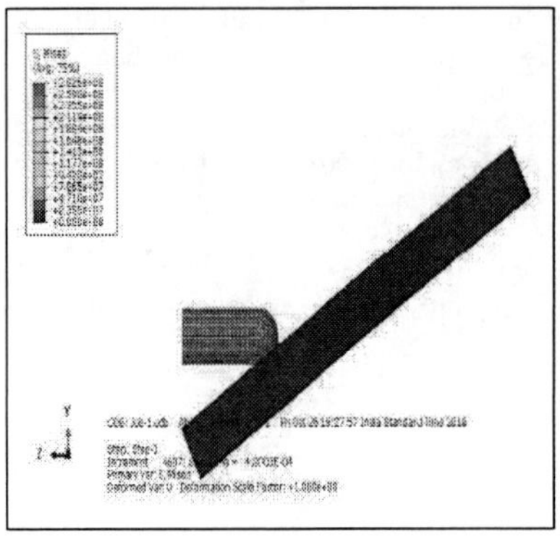

Figure 1. Plate at 30⁰ inclination

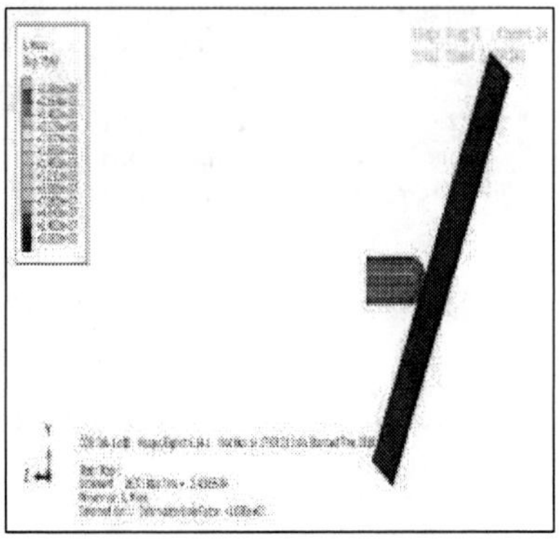

Figure 2. Plate at 60⁰ inclination

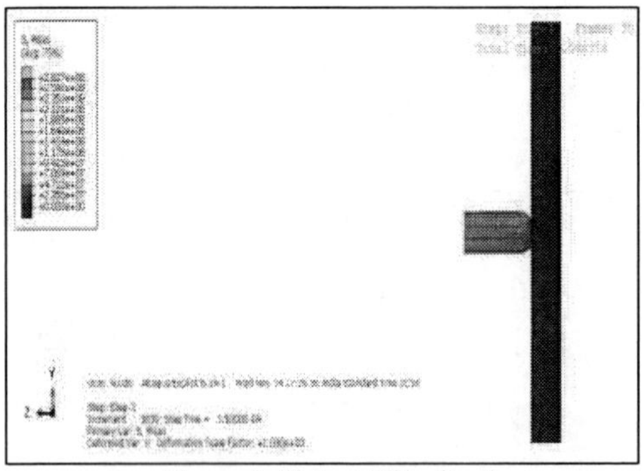

Figure 3. Plate at 90⁰ inclination

The STEP section of the module was used to create each frame in which the dynamic explicit models are created. In the interaction section of the module the type of the contact between the two models were assigned. In load section at module the boundary condition of plate was assigned to arrest all degree of freedom.

In the MESH section define the type of mesh, element type and the dynamic model to achieve accurate results. The plate was designed using 46, 080 C3D8R elements with 8-node 3D linear brick. The interference between the impactor and an alloy was analysed using COH3D8 cohesive elements

IOP Conf. Series: Materials Science and Engineering 1123 (2021) 012040 doi:10.1088/1757-899X/1123/1/012040

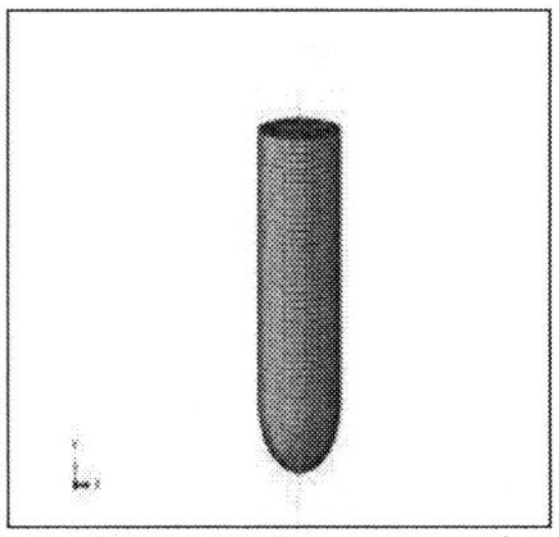

Figure 4. Impactor mesh

Figure 5. Plate mesh

The job section was used to create analysis. After assigning all the required parameters for analysis, the analysis part was done. Based on the number of elements the size and the time requirement are increased for analysis.

3. RESULT AND DISCUSSION

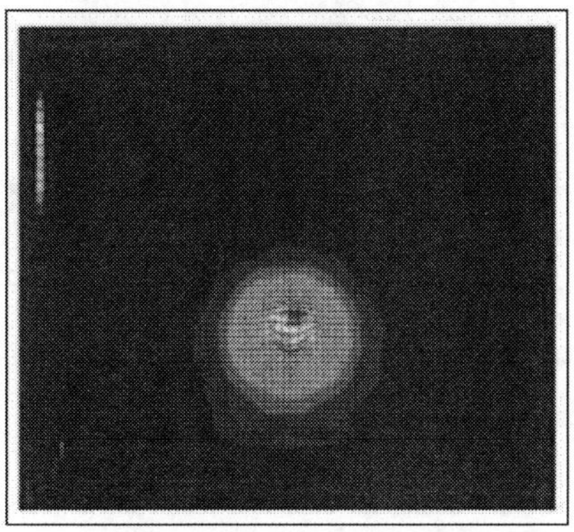

Figure 6. Deformation of the plate at 30°

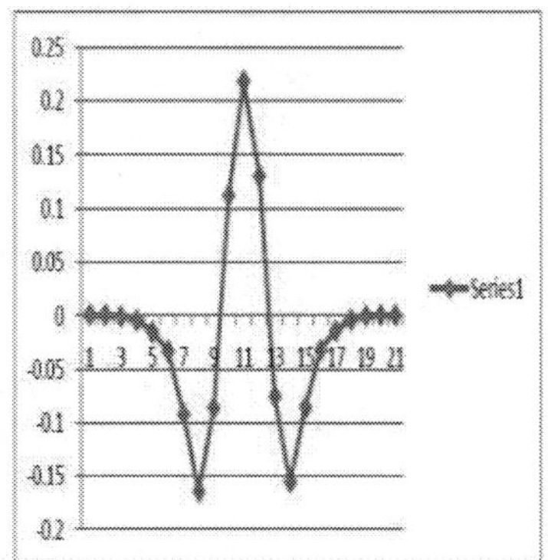

Figure 7. Depth of indentation at 30°

As per the result obtained from the simulation, the deformation produced in the plate at 30° inclination is high as is shown in Figure.6. The shear stress produced in 30° is 2.826*10e8 and it is high compared to the other degrees of inclination such as 60° and 90°. From the graph shown in Figure.7, it is observed that the depth of indentation is 2.694*10e-1. It is much higher depth of indentation than the other two degrees of inclination.

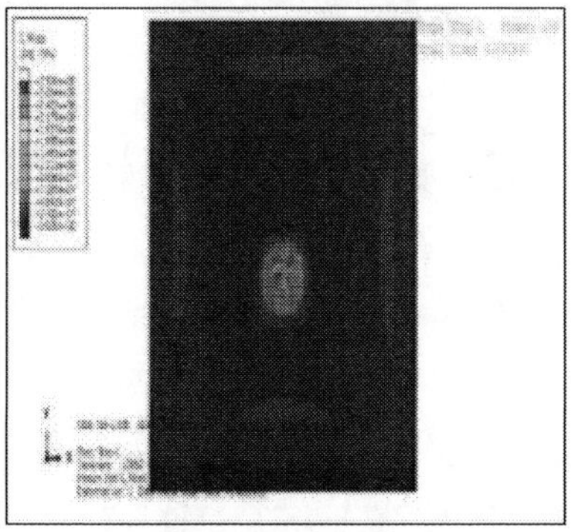

Figure 8. Deformation of the plate at 60°

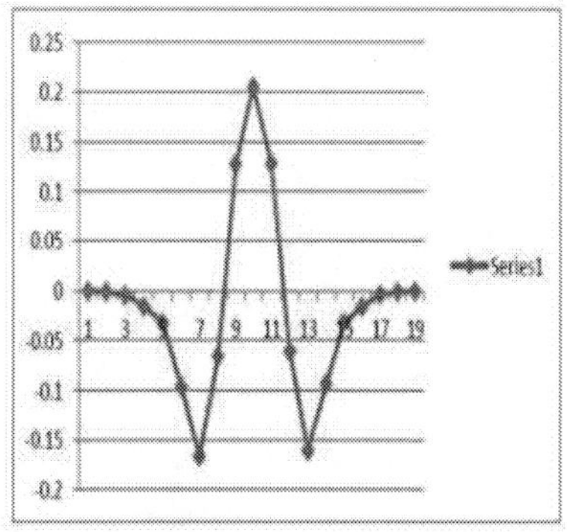

Figure 9. Depth of the indentation at 60°

From the simulation results, it was found that the deformation produced in the plate at 60° inclination is higher than 90° inclination. It is shown in Figure.8 The shear stress produced in 60° is 2.728*10e8. It is more when compared to the 90° inclination and less than 30°. From the graph, shown in Figure.9, it is observed that the depth of inclination is 2.277*10e-1. It is higher depth of indentation than 90° inclination but lower than the 60° inclination.

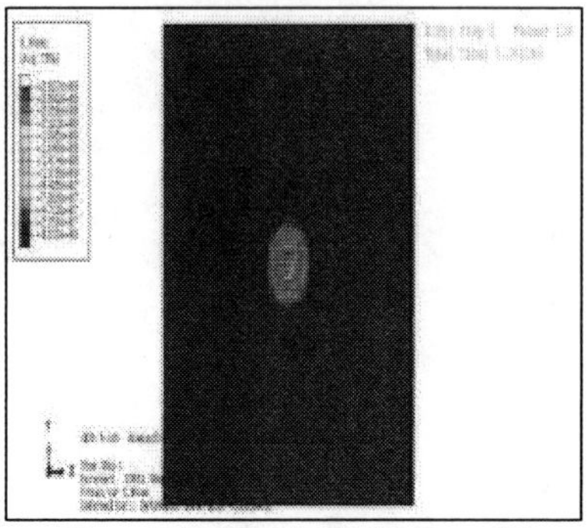

Figure 10. Deformation of the plate at 90°

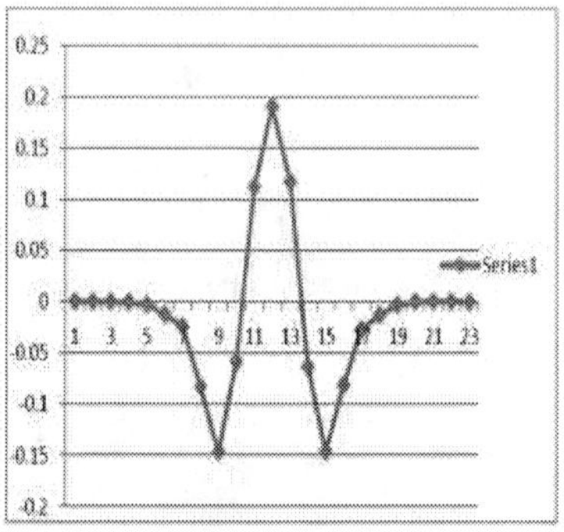

Figure 11. Depth of indentation at 90°

As per the result obtained from the simulation, the deformation produced in the plate at 90° inclination is low. It is shown in Figure.10. The shear stress produced in 90° is 2.442*10e8. It is less when compared to the other degrees of inclination say 60° and 90°. From the graph, shown in Figure.11, it is observed that the depth of inclination is 2.08810e-1. It is considerably less depth of indentation than the other two degrees of inclination.

Table 2. ABAQUS analysis result

S.NO	**Angle (degree)**	**Depth of indentation (mm)**	**Stress value (N/mm^2)**
1.	30	2.694*10e-1	2.826*10e8
2.	60	2.277*10e-1	2.728*10e8
3.	90	2.088*10e-1	2.442*10e8

From the table 2, it is clear that the depth of indentation is minimum when the impactor strikes the plate which is inclined at an angle of 90° as well as the stress value is also low, when compared to other two inclination plates say 60° and 90°.

4. Conclusion

Complete design was done by ABAQUS software. After the designing, the analysis of plate with three different inclination (30°,60°,90°) was done to find the depth of indentation as well as the amount of stress produced. The depth of indentation is 2.694*10e-1 at 30°, 2.277*10e-1 at 60°, 2.088*10e-1 at 90° respectively. The stress value is 2.826*10e8 at 30°, 2.728*10e8 at 60°, 2.442*10e8 at 90° respectively. It was found that the depth of indentation produced in 90° inclination plate is minimum when compared to other two degrees of inclination.

At the peak, when the plate is 90° perpendicular to the impactor, the depth of indentation and stress is minimum. Futher in future, the composite materials are used instead of alloy material for the analysis purpose, it will give more progress in the material properties and its application.

5. ACKNOWLEDGMENT

The authors thank the management of Kumaraguru College of Technology (KCT), Coimbatore, India, for the promotion of research and KCT CAD laboratory for allowing us to utilize the software facilities throughout the research.

REFERENCES

[1].Y Zhao T, S L Zhang, G Chen, X N Cheng, C Q Wang. "In situ (Al2O3+Al3Zr) NP /Al nano composites synthesized by magneto chemical melt reaction", Composite Science Technology, 68, 2008, 1463-1470.

[2].Al-Bahrani, M., & Cree, A. (2019). A simple criterion to evaluate the degree of damage in composite materials after sudden impact loads by exploiting the MWCNTs piezoresistive property. Carbon, 150, 505–517.doi: 10.1016/j.carbon.2019.05.053.

[3].Mazahery Ali, Shabani Mohsen Ostad. "Characterization of cast A356 alloy reinforced with nano SiCp composites", Transactions Nonferrous Metallurgical Society, China 22, 2012, 275-280.

[4].Findlay, S. J. and Harrison, N. D. "Why aircraft fail", s.l.: Elsevier Science Limited, November 2002, Materials Today, Vol 5, pp. 18-25.

[5].Hirai, Yasunobu, Hamada, Hiroyuki and Kim, Jang-Kyo. Impact Response of Woven Glass Fabric Composites - 1 Effect of Fibre Surface Treatment. 1998, Composites Science and Technology, pp. 91-104.

[6].Vaidya, Uday K. Impact Response of Laminated and Sandwich Composites. [book auth.] Serge Abrate. Impact Engineering of Composite Structures. Brimingham: Springer Wien New York, 2011, pp. 97-191.

[7].S.N.A. Safri, M.T.H. Sultan, N. Yidris, and F. Mustapha. Low Velocity and High Velocity Impact Test on Composite Materials - A review.

[8].Naik, N. K. and Shrirao, P. 2004.Composite structures under ballistic impact, Composite Structures, pp. 579-590.

[9].Furqan Ahmada, Fethi Abbassia B, Myung Kyun Parkc, Jae-Wook Jungd and JungWuk Hong. Finite element analysis for the evaluation of the low-velocity impact response of a composite plate.

[10].Cheng W, Hallquist J. Implementation of three-dimensional composite failure model into DYNA3D. Livermore Software Technol Corporation. 2004.

[11].Y. Shi, C. Pinna, C. Soutis (2016). "Low-velocity impact of composite laminates damage evolution".

[12].A. Dorogoy, D. Rittel. Determination of the Johnson–Cook material parameters using the SCS specimen, Exp. Mech. 49 (2009) 881-885.

[13].Murugesan M, Jung DW. Johnson cook material and failure model parameters estimation of AISI-1045 medium carbon steel for metal forming applications.

[14].Er. Karanjeet Singh, Er. Ramandeep Singh Dhaliwal, Er. Didar Singh, Er. Satvirpal Singh Grewal. Design, Analysis and Optimization of Non-Standard Weld Neck Body Flange in Small pressure Vessel.

[15].Wiegand J, Hornig A, Gerlach R, et al. An experimental method for dynamic delamination analysis of composite materials by impact bending. Mechanics Adv Mater Struct. 2015. DOI:10.1080/15376494.2012.73606.

[16].D. Düring, L. Weiß, D. Stefaniak, N. Jordan , C. Hühne (2015) title of the paper is "Low-velocity impact response of composite laminates with steel and elastomer protective layer".

IOP Conf. Series: Materials Science and Engineering 1123 (2021) 012041 doi:10.1088/1757-899X/1123/1/012041

Investigation of aluminium LM6 metal matrix composites reinforced with graphene flakes using stir casting process

Narendiranath babu.T [1],Yadhukrishnan [2],Ishaan Rahul Saxena[3],Drumit B Modi[4]

[1]SMEC, VIT UNIVERSITY, VELLORE, INDIA, narendiranathbabu.t@vit.ac.in
[2]SMEC, VIT UNIVERSITY, VELLORE, INDIA, yadhukrishnan.2017@vitstudent.ac.in
[3]SMEC, VIT UNIVERSITY, VELLORE, INDIA, ishaan.saxena2017@vitstudent.ac.in
[4]SMEC, VIT UNIVERSITY, VELLORE, INDIA, bharatkumar.modi2017@vitstudent.ac.in

Abstract. Through the past decades Aluminium has been used in several manufacturing sectors because of its unique properties like light weight, highly malleable and ductile, high strength etc. Metal Matrix Composite (MMC) is being in trend due to its ability to inhibit good properties of its constituents . The purpose of this paper is to manufacture Aluminium LM6 MMC with infused Graphene Flakes(in concentration of 0.1%,0.25%,0.4% by weight) by Stir Casting Process and to study the changes in the mechanical characteristics and compare with the former. Secondly as Graphene is highly expensive material we will be using Exfoliation method to extract Graphene Flakes from the graphite rods.

1. Introduction

Today in the modern industries Aluminium is the most important engineering metal and is used in various engineering fields such as the automotive , aerospace industries etc because of their outstanding properties namely, high specific strength, high wear resistances , high stiffness, high dimensional stability , high corrosion resistance and light weight product.[1-2]. Since late eighty this has been becomes the most researched field topic to increase the strength of the Aluminium metal without compromising its ductility and its always been a challenge to the scientist. Aluminium metal matrix composite (hybrid aluminium matrix) became an important research work conducted by many people. In Aluminium matrix composite aluminium of serves as the parent material and an reinforcement is added with it in specific measurements to boost up its mechanical properties, there are many researchers trying out different materials as reinforcement to get the desired output, as each reinforcement produces a different set of material properties.[4-5].Commonly carbides of transition metals or non-metals like silicon carbide(SiC), Tungsten carbide(WC), Titanium Carbide(TiC) or simply the non-metals that tend to increase the bonding strength are used as reinforcement. Generally hybrid aluminium composites are manufactured by various process like powder metallurgy ,melting and squeeze casting stir casting, powder metallurgy and heat sintering etc[7]. Generally powder metallurgy and cryomiling techniques are used to achieve uniform mixing of the reinforcement with the base material followed by sintering to strengthen the bonding and to reduce the void the lattice structure. Graphene an allotrope of carbon with single layered one atom thick lattice structure with sp2 hybridisation[11], it has been an intensive research material due to its high electron mobility of 15000 $cm^2/(V.s)$ and its approximately 10 times higher than the normal semiconductor wafer, being a single layered structure it has a low density of 2.2 g/cm^3,extremely high thermal conductivity 3000 W/(m.K), high electrical conductivity, and although being a non-metallic compound it has high tensile strength reaching up to 1060 GPa and its one of the hardest material besides diamond[10].These splendiferous properties of graphene makes it the most expedient reinforcement with several base material alloys. Muhammet et al. [12] investigated on the hot extruded pure aluminium matrix composite with reinforcement as single dimensional multi-wall carbon nanotube and two-dimensional graphene nanoplatelets(GNP),according to his research theory there was an substantial increase in the mechanical properties of the pure aluminium when reinforced with the graphene and also inferred that there was no formation Al_4C_3 compound during the x-ray diffraction analysis. Sreehari Peddavarapu et al. [13] they investigated the tribological properties of the aluminium 6082 with graphene nanoflakes as reinforcement using the stir casting machine, they tested the samples using a Pin-on Disc tribometer to determine the wear properties at different velocity and load conditions. According to them there was a quite improvement in the wear loss properties and heat generation compared to non-reinforced aluminium alloy and it was due to the lubricating properties of the graphene that the wear was reduced to a significant amount. Ajay Kumar P et al.[14] they reinforced graphene nanoflakes with the aluminium A356 using friction stir casting machine and investigated the mechanical properties like tensile strength, elongation, hardness test, and performed Transmission Electron Microscopy(TEM), fractography analysis and XRD analysis. After the experimentation they inferred that the after the friction stir casting the TEM analysis showed that the graphene was encapsulated in aluminium matrix and there were reduction the voids and defects between the graphene and aluminium, then there was 12% increase in the ductility of the reinforced aluminium matrix, the grain size was reduced and work hardening rate was increased. Mohammad Khoshghadam-Pireyousefan et al.[15] investigated on the advanced method of producing aluminium graphene reinforced matrix incorporating high energy ball mill and molecular level mixing and followed by spark plasma sintering, they inferred the manufacturing of uniform distribution of Cu nanoparticle with graphene oxide nanosheets and there was no traces of Al_4C_3 during the XRD analysis, secondly the Al_2Cu particles were precipitated homogenously after the spark powder sintering and there was around 78 % increase in the yield strength of the composite thus formed. Yu Zhang Ng et al.[16] they performed an investigatory experiment on the process of electrochemical exfoliation of graphene with an stabilizer like diethanolamine(DEA) and the structural properties were analysed by performing FESEM , Raman spectrometer. The Liquid-phase exfoliation enforced with electrochemical process and constant current setup produced uniform graphene sheet by alleviating the heat absorption by the graphene nanoflakes, it was also inferred that with the addition of the stabilizer there was agglomeration of graphene flakes and thus resulting in non-uniform sheets compared to the one formed without the use of stabilizer. From the literature so far done it is observed that most of the experimental analysis using the graphene nanoflakes as reinforcement are done on the wrought aluminium alloys with low silicon content. For our experimental and investigation work the Stir Casting machine was used for preparing the Aluminium LM6 samples with 0.1, 0.25 and 0.4 wt% of graphene as reinforcement. The main idea behind using the casting alloy is to mainly observe the effect of graphene reinforcement in the presence of silicon as the second major constituent of the alloy. The Graphene flakes are generally produced by various industrial methods for instance electrochemical exfoliation of the graphite sheets or rods, or by the reduction of graphene oxide in colloidal solution via chemical reaction called Hummers Reaction. We adopted the method of electrochemical exfoliation, in this process high voltage is passed across the graphite films with zinc sulphate as electrolyte slow dissociation of the carbon atoms form the graphite rod produces 3-5 layered stacked carbon atoms which on further processing through ultrasonicator yields graphene nanoflakes. These graphene flakes are majorly used to fabricate graphene layered composite , hybrid Aluminium composites and metal matrix composites[6]. The crucial task is to carefully diffuse the graphene flakes with the Aluminium metal so to provide required mechanical strength without inhibiting the brittle behaviour to the metal composite thus produced. The mechanical properties are determined by the uniformity and diffusion of the graphene flakes in the Aluminium matrix. This research investigation mainly focused on the effect of mechanical and materialistic properties of casting Aluminium LM6 casting alloys with graphene flake reinforcement as it contains 10-13% silicon, secondly there was a major challenge to dominate the mechanical strength over the brittle behaviour and to observe the microstructure of the Aluminium composite using field emission scanning electron microscope(FESEM) with electronic discharge spectroscopy(EDS).

2. Materials And Methods

2.1. Properties of Materials

For the investigation purpose we decided to use casting aluminium alloys. Cast Aluminium alloys mostly constitutes of the Al-Si matrix, they are used in every manufacturing industry for the casting purpose because of their high mechanical strength, ease in casting process, better weldability, corrosion

IOP Conf. Series: Materials Science and Engineering 1123 (2021) 012041 doi:10.1088/1757-899X/1123/1/012041

resistant. Among the various casting alloys we choose the LM6 alloy as its commercially used for casting and its economic. LM6 alloys are generally used in automotive industries, aerospace industries and marine applications.

Table 1. Material Properties of the LM6 alloy[18]

Material Properties	Properties
Density	2650 Kg/m^3
Thermal Conductivity	0.34 $cal/cm^2/°C$
Tensile Strength	170-180 N/mm^2
Elongation(in %)	7
Brinell Hardness	55-60
Elastic Modulus	71
Impact Strength	9.0 N-m

Table 2. Percentage of constituents present in the Aluminium LM6 alloy

ALUMINIUM LM6 SPECTROSCOPY REPORT	
CONSTITUENT	PERCENTAGE
Silicon	12.92%
Copper	0.66%
Iron	0.23%
Zinc	0.10%
Magnesium	1.71%
Tin	0.11%
Titanium	0.04%
Vanadium	0.02%
Polonium	0.09%
Nickel	0.32%
Manganese	0.20%
Aluminium	83.53%

2.2. Stir Casting Process

Stir casting is the most economical and widely used commercial method for the manufacturing of Aluminium Matrix Composite or hybrid Aluminium matrix composites. It's also classified as the liquid state method of composite material fabrications. Generally Aluminium metal is placed in a ceramic crucible and is heated to temperature of 800°C - 900°C i.e. just above the melting point of the Aluminium. Then the required reinforcement is added to the molten metal and with the help of the mechanical stirrer set at the required rpm is used to properly mix the reinforcement with the liquid metal. This technique is also referred to as Vortex technique that creates a small vortex so that the reinforcement can uniformly dispersed into the molten metal. The stir casting process is limited to certain amount of dispersion and there is not complete homogeneity. There are chances that the reinforcement particles might settle at the bottom or form clusters due to difference in the densities and gravity. Although there are some drawbacks of stir casting machine it's still used for mass production of AMC and MMC. In our setup, we have placed the LM6 ingot inside the furnace, as per its dimensions. Then the machine is switched on and the temperature is raised to 897°C. During melting, the molten aluminium will react with the atmosphere to form a layer of aluminium oxide which will shield the surface from further reaction with the atmosphere. The metal mould is placed in an oven to be heated to about 450°C, so that when we pour the molten metal into the metal, the possibility of bubble formation is reduced. The graphene flakes are then placed inside a granite cup and heated to 200°C. After the matrix metal have completely melted, graphene flakes is slowly dispersed into the molten matrix metal. Instantly, the stirrer is brought down via a remote and placed in position inside the furnace. The stirrer is turned on with a RPM of 334 which will enable uniform dispersion of reinforcement. Degassifier powder was dispersed into the molten metal, while the stirrer is still active. Wetting agents like borax and magnesium are added so that the formation of aluminium oxide does not take place, as Al2O3 will prevent the uniform mixture of graphene particles and the matrix metal. After 4 minutes, the metal is poured into the mould(which is to be removed when the metal is ready to be poured).

Figure 1. Stir casting machine

2.3. Graphene Preparation

There are various industrial methods to produce graphene nanoflakes like chemical vapor deposition(CVD), mechanical and thermal exfoliation process, Hummer's method that basically involves the catalytic reduction of Graphene Oxide. Prashant Tripathi et al.[17] investigated on the process of electrochemical exfoliation of the graphene using graphite rods in an alkaline electrolytic solution and analyzed the graphene sample thus produced using TEM images and HRTEM micrographs, according to them by applying the DC bias voltage across the platinum and graphite electrode the kinetics of the reaction were enhanced and produced 1-4 layered graphene sheets. In depth study of the electrochemical exfoliation we decided to proceed with this method. In our experimental setup graphite rods are connected to positive and negative terminal and then a high DC voltage is applied. The setup is placed in an acidic electrolytic medium (0.5M Na_2SO_4 solution). The graphite rod in the presence of sulphate ions reduced to lesser stacked graphene sheets losing electrons. The graphene produced through the exfoliation process is further processed under ultrasonic sonicator to break the graphene stacks to 1-4 layered structures. The process continued for two days and the electrolyte was periodically changed so to stabilize the kinetics of electrochemical electrolysis. The solution is filtered out using filter paper and the graphene sample is cleansed and sterilized with the conc. HCl to ensure there is no impurities left out. Graphene sample is mixed with acetone solution and is placed in a ultrasound machine for couple of time and is operated at frequency of 40KHz. Then solution is again filtered to accrue the graphene sample and dried in a furnace at 120°C.

Figure 2. Filtration of graphene sludge

Figure 3. Graphene after ultrasound

2.4 Aluminium Sample Preparation

This task was done with the help of Stir Casting machine as mentioned earlier because of its simplicity and less time consumption its best suited for Aluminium Matrix composite fabrications. The Aluminium we used was casting Aluminium (LM6 alloy) generally used for casting purpose. The alloy was first cut into billets of 1 kg each and then preheated in crucible in the Stir Casting machine for three hours so that the metal gets completely liquified. We made three Aluminium samples with varying concentration of graphene powder(0.1%, 0.25% and 0.4% wt.).The graphene powder was also preheated and then mixed with the molten metal, degasifier agent was also added to the molten metal so that the trapped gases can escaped out. The uniform mixing was achieved with the help of mechanical stirrer that was rotated at 400 rpm for 15 minutes. Then the molten metal with reinforcement was poured in the heated metal die to get a cylindrical cast and was left aside to get solidified. The cast was then machined out to get a test specimen for Tensile Test, Charpy Test, Brinell Hardness Test and for observing the microstructure of the metal composite.

3. Results And Discussion

The mechanical properties of the Aluminium LM6 alloy with varying concentration of graphene nanoflakes as reinforcement(0.1,0.25,0.4 wt%) are analysed by performing Tensile strength, Brinell Hardness Test , Impact Test. The morphological of the matrix formed was determined from the results obtained by FESEM (Field Emission Scanning Electron Microscope) and the spectroscopy report from the XRD. The data has been presented with brief discussion in the following section.

3.1. Tensile Test

The mechanical properties of cast samples such as tensile strength was determined using the UTM(Universal Testing Machine)[8].The tensile tests were conducted on tension test specimens of dumbbell shape with the gauge thickness of 10mm and length 45mm and the total length of sample 75mm and outer diameter 20mm. During tensile test, a sample is subjected to controlled tension until it attains failure. we can find the ultimate tensile strength, maximum elongation, reduction area and various other properties from this test. For an isotropic material, uniaxial tensile testing machine is the most used. This test is conducted to find a suitable material which will perform under normal and extreme forces. A tensile specimen is generally a standard piece, with 2 shoulders and a gage, where the shoulders are used so that they can be readily gripped, and the gage section is where the deformation occurs.

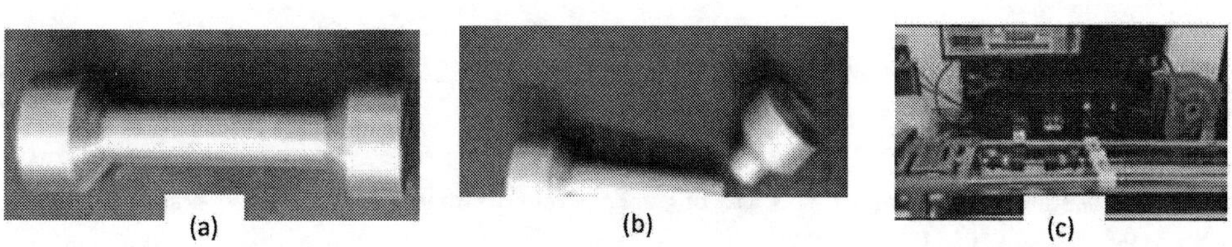

Figure 4. Tensile test sample for with graphene reinforcement before (a), after the testing(b) and the experimental setup(c)

IOP Conf. Series: Materials Science and Engineering 1123 (2021) 012041 doi:10.1088/1757-899X/1123/1/012041

Table 3. presents the data obtained by the Tensile Test of the Aluminium graphene metal matrix sample with 0.1,0.25 and 0.4% wt reinforcement of graphene and the base aluminium LM6 alloy.

Table 3. Tensile Test Results

Sample Prepared	Ultimate tensile strength
LM6 alloy without graphene reinforcement	175MPA
0.4% graphene	190MPA
0.25% graphene	217MPA
0.1% graphene	246MPA

Analysing the results obtained it can be inferred that there are a increase in the tensile strength of the LM6 metal matrix infused with 0.1%wt graphene, subsequently with further increase in the percentage of Graphene reinforcement there was a gradual decrease in the strength of the metal matrix. This decrease in the tensile strength with increase in the graphene reinforcement is due to the dominance of the brittle nature of the Graphene over the Al-Si dominating matrix properties. It can also be inferred from the results that around 0.1% there was a maximum increase in the tensile strength so the reinforcement ratio should be maintained in the range of 99.9% base metal and 0.1% approx. Graphene reinforcement. Apart from the Tensile strength the brittle nature of the matrix alloy was confirmed by the fracture pattern, surface finish of the crack and necking process, it was observed that the sample broke at an instant without undergoing a gradual necking process. It can be stated based on the spectroscopy report of the compound that there might the possibility of formation of the Si-C compound due to excess amount of Silicon present in the alloy mixture this in turn adds to the brittle nature of the matrix thus formed. The figure 11 shows the graph obtained after the tensile strength testing of the samples.

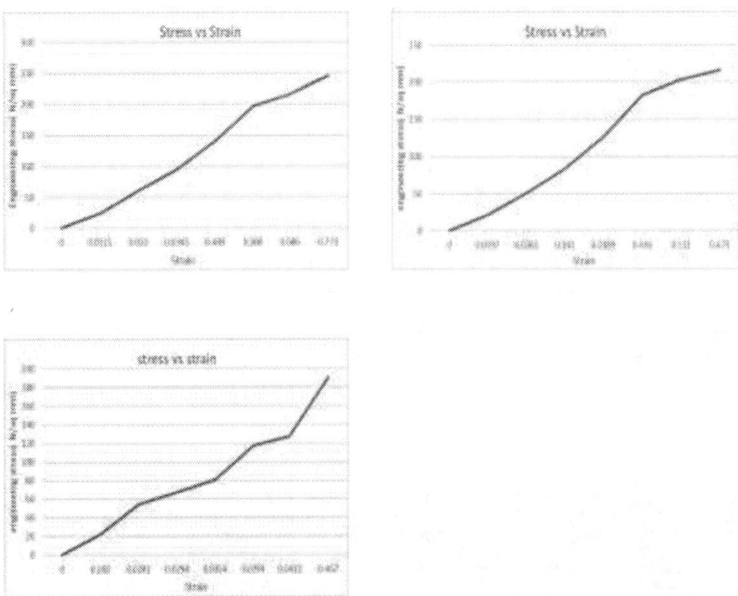

Figure 5. Graph of Engineering Stress vs Strain (a) 0.1 wt% of graphene reinforcements. (b) 0.25 wt% of graphene reinforcements. (c) 0.4 wt% graphene reinforcements

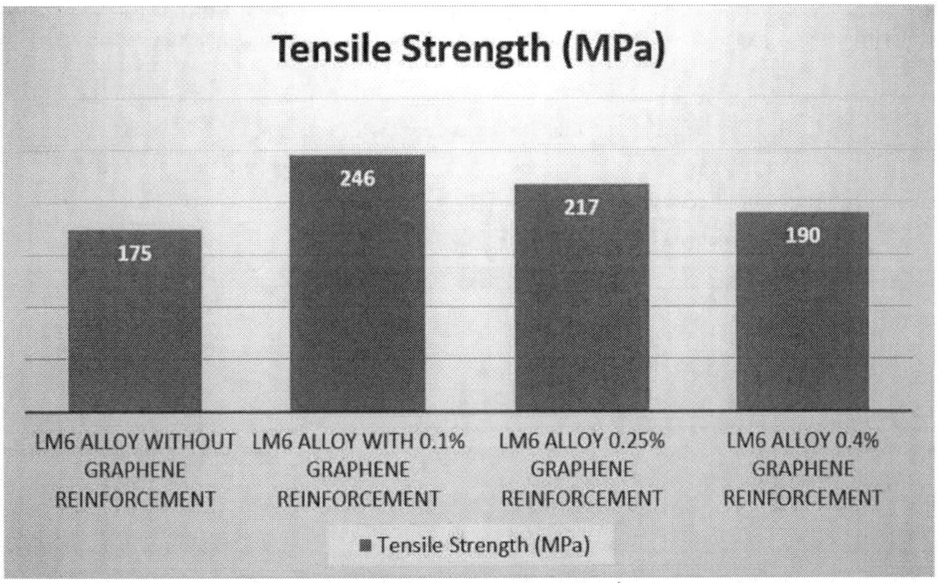

Figure6: Graph comparing the tensile strength for (a) LM6 alloy without reinforcement (b) 0.1 wt% of graphene reinforcement (c) 0.25% wt% of graphene reinforcement (d) 0.4 wt% of graphene reinforcement

The graph above shows the tensile strength of the composite materials manufactured. As per this graph it can be implied that the strength of the material increased when doped with 0.1 percent of graphene and while increasing the concentration , the strength is shown to be decreasing.

3.2. Impact Test

Impact testing were done using the Charpy Impact Testing Apparatus and amount of energy absorbed by the material during a fracture (impact) and the nature of fracture were obtained. The test samples after the testing are shown below with the crack pattern. Samples were prepared as per ASTM: D256 Standard. An arm of the machine is held at the specified height (with constant potential energy) and released so it hits the sample, the sample absorbs certain amount of energy which is then noted. Impact test was the only possible test to prove that with the reinforcement of graphene and silicon in the Aluminium base metal introduced the brittle nature and after certain concentration limit the brittle nature surpassed the ductile behaviour of the metal and made it to lose its ductility. The figure 7 shows the Impact strength samples after the experiment and from the crack pattern the brittle nature of the matrix can be inferred.

Figure 7. Charpy test sample for 0.4 wt.% graphene reinforcement

Figure 8. Charpy test sample for 0.25 wt.% graphene reinforcement

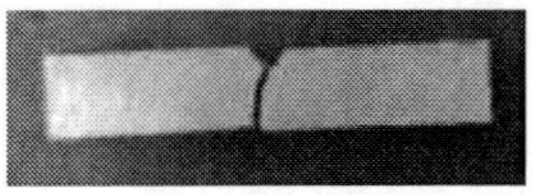

Figure 9. Charpy test sample for 0.1 wt.% graphene reinforcement

The Table 4. Below shows the Impact force sustained by the Aluminium graphene metal matrix sample with 0.1,0.25 and 0.4% wt reinforcement of graphene and the base aluminium LM6 alloy. The impact force is in KGM unit(1 KGM = 9.8066 NM).

Table 4. Impact Test Results

Sample Prepared	Impact force sustained (KGM)
LM6 alloy without reinforcement	0.6
0.4% graphene	0.3
0.25% graphene	0.7
0.1% graphene	0.9

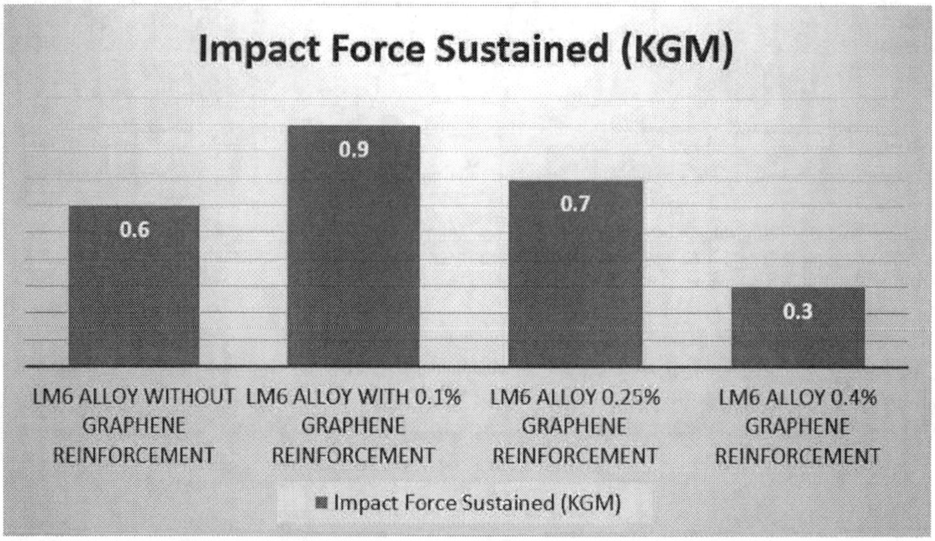

Figure10: Graph comparing the test results for Impact Force Sustained(KGM) for (a) LM6 alloy without reinforcement (b) 0.1 wt% of graphene reinforcement (c) 0.25% wt% of graphene reinforcement (d) 0.4 wt% of graphene reinforcement

The graph above shows the impact strength of the composite materials manufactured. As per this graph it can be implied that the strength of the material increased when doped with 0.1 percent of graphene and while increasing the concentration , the strength is shown to decrease.

3.3. Hardness Test

Hardness is an attribute of a material, not a key physical property. Hardness is defined as the resistance to indentation, and it is calculated by measuring the permanent depth of the indentation. Hardness is the internal resistance to localized plastic deformation produced by mechanical indentation or abrasion [science direct reference]. Hardness of a material depends on its ductility, elastic stiffness, viscosity etc. Hardness test values is extremely useful for selecting materials, because the values indicate how well the material can be machined and how well it will wear. Hardness test was performed by using a Rockwell hardness testing machine using a 1.5875 mm diameter steel ball indenter and 100Kg load is applied, for each sample two set of reading were taken and then the average is considered as hardness of the particular sample[8,9]. The figure 4 shows the Brinell Hardness test samples with the indentations marked.

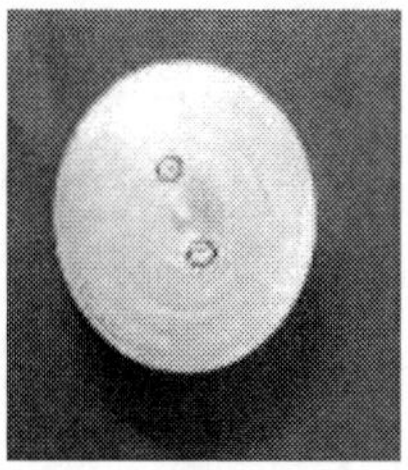

Figure 11. Brinell hardness test sample for 0.1 wt.% graphene reinforcement

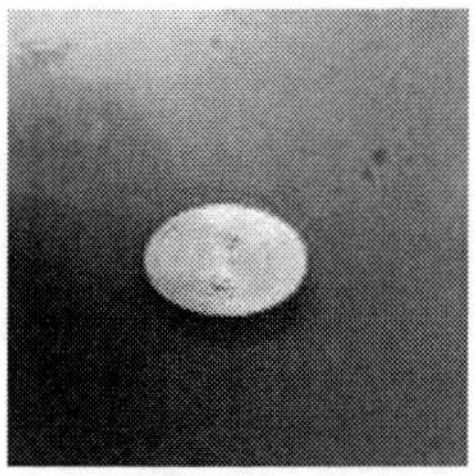

Figure 12. Brinell hardness test sample for 0.25 wt.% graphene reinforcement

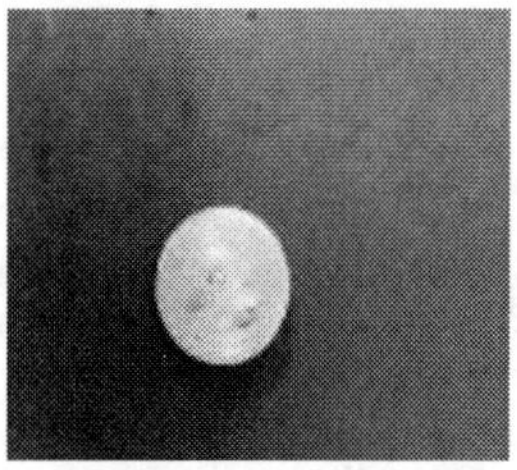

Figure 13. Brinell hardness test sample for 0.4 wt.% graphene reinforcement

The Table 3. Below shows the Hardness Value of the Aluminium graphene metal matrix sample with 0.1,0.25 and 0.4% wt reinforcement of graphene and the base aluminium LM6 alloy obtained at 100N load.

Table 5. Hardness Test Results

Sample Prepared	Hardness value (HRB)
LM6 alloy without reinforcement	56
0.4% graphene	55
0.25% graphene	54.5
0.1% graphene	56.7

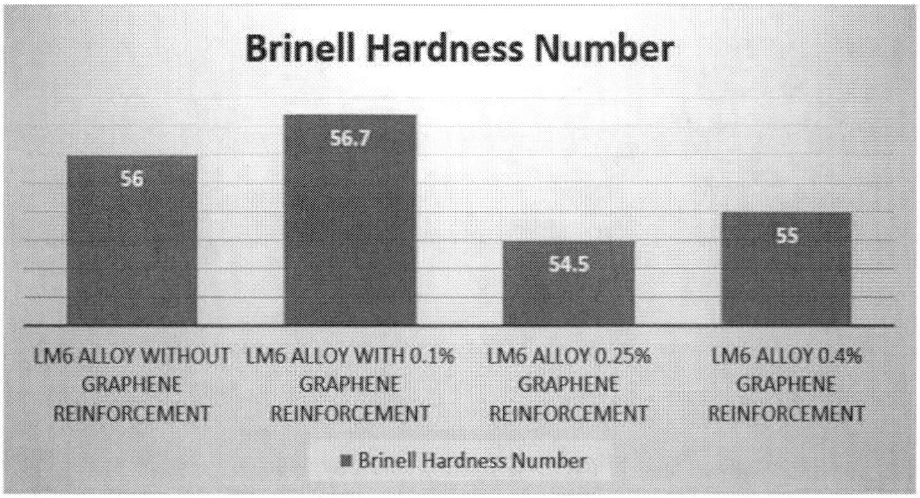

Figure14: Graph comparing the test results for Brinell Hardness Number for (a) LM6 alloy without reinforcement (b) 0.1 wt% of graphene reinforcement (c) 0.25% wt% of graphene reinforcement (d) 0.4 wt% of graphene reinforcement

The results thus obtained showed that there was a minimal increase in the hardness of the metal matrix with 0.1% reinforcement of Graphene and with further increases in the weight percentage of the reinforcement there was a decreasing pattern followed up by the metal matrix. Although the graphene sheets in in the metal matrix adds hardness up to a certain level increase in the percentage also increases the chance of inheriting the brittle nature to the base metal thus reducing the hardness of the final matrix alloy.

After comparing the results of hardness , impact and tensile strength it was confirmed that with 0.1% weight graphene reinforcement there was betterment in the mechanical properties of the matrix but as the concentration of the graphene increased the brittle nature starts dominating over the other mechanical properties. The images shown below fortifies the brittle nature of the matrix because of the irregular crack pattern. The Aluminium matrix is not so capable to be used for impact-based purposes due to low impact absorption strength. With 0.4% Graphene reinforcement there was a large reduction in the impact strength.

3.4. Microstructural Analysis

The microstructure analysis was performed using scanning electron microscope equipped with high resolution digital camera. The surface of the sample was first polished with 200,500,1000 grains size emery paper and etched with Keller solution. The microstructure study was carried out using Zeiss metallurgical microscope. Figure 12 (a), (b) & (c) shows the microstructure of LM6 alloy reinforced with 0.1,0.25,0.4 percent graphene. The base metal is aluminium and the dendritic structure in yellow is the silicone. Observing graphene in this is very difficult. Some dark spots were found which could indicated the presence of graphene particles in a clustered form.

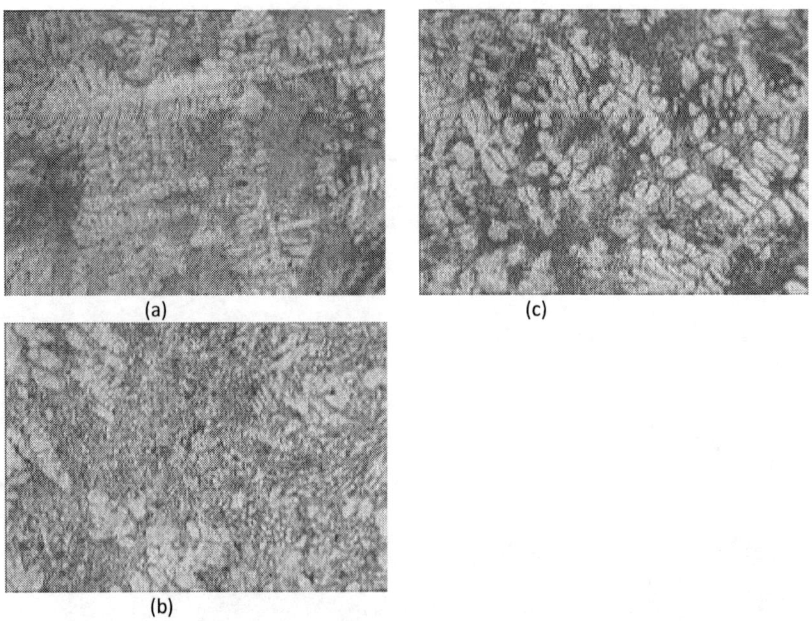

Figure 15. (a) Microstructure of LM6 alloy with 0.1% graphene reinforcement , (b) Microstructure of LM6 alloy with 0.25% graphene reinforcement , (c) Microstructure of LM6 alloy with 0.4% graphene reinforcement

To confirm the report, FESEM and XRD analysis was conducted on each sample. XRD report, presence of carbon confirms the graphene is present and FESEM analysis is conducted to see the structure of graphene fused with the base metal in matrix. FESEM was used to obtain high resolution images of each sample. In the microstructure. It is seen that the graphene flakes are distributed randomly. Since the method of casting was stir casting and the quantity if graphene used was only 0.1,0,25,0.4 percent, detecting graphene flakes bonded with Matrix metals was difficult.. After the casting was completed, the cast was machined so that we can perform the necessary test. The presence of graphene in every layer of the cast is highly unlikely. The graphene appears to be clustered together in some places which can be seen from the FESEM images. The bond between graphene and matrix metal is important as the carbon fillers helps in load transfer between reinforcements and matrix. Interfaces and strong interfacial between LM6 alloy and the reinforcement plays a crucial role in finding the tensile strength of the composite. Reinforcement are uniformly mixed in the matrix metal due to the action of stirring in the stir casting machine. The stirring restricts the particles to float on the matrix, as the stirrer will create a vortex, which will result in the particles getting sucked in to the centre and getting mixed uniformly. The reinforcements are fully wet due to the molten matrix metal which lead to a better bonding of the particles with the metal.

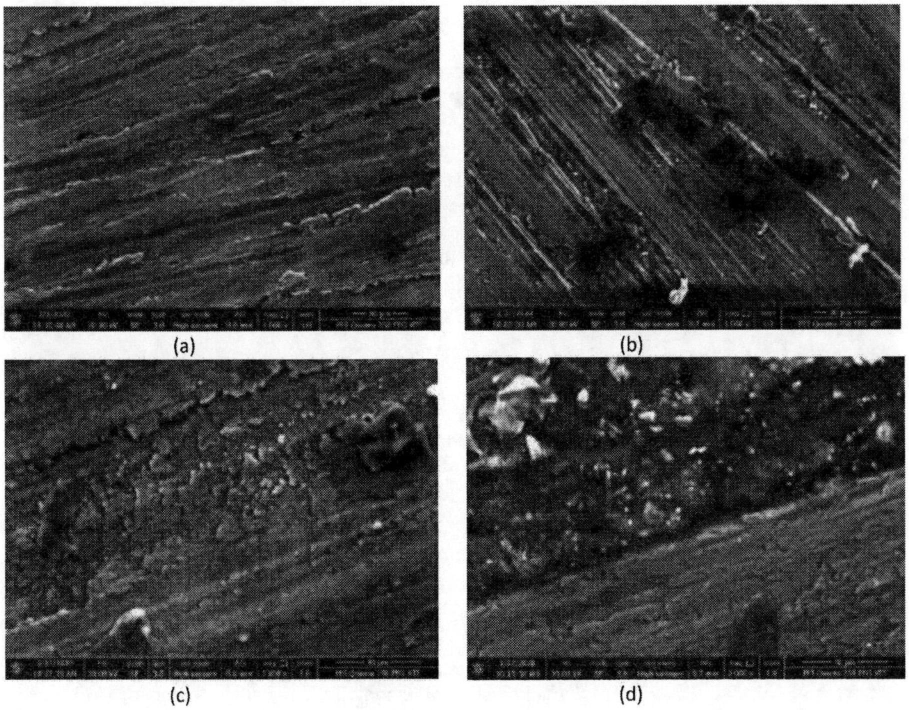

Figure 16. (a) FESEM of LM6 alloy base metal(at 1000x), (b) FESEM of LM6 alloy with 0.1% graphene reinforcement(at 1000x), (c) FESEM of LM6 alloy with 0.25% graphene reinforcement(at 2600x), (d) FESEM of LM6 alloy with 0.4% graphene reinforcement(at 2000x)

Figure 13(a-d) presents the FESEM images of the Aluminium LM6 base alloy, and the metal matrix composite formed reinforcing Graphene in the LM6 alloy with the help of Stir Casting machine. The dendritic structure in the 13(a) is due to the presence of Silicon in the alloy. The graphene produced is in the cluster form instead of the layered structure. The uniform diffusion of graphene particles in the LM6 alloy matrix is achieved with the automated stir using the rotor at a uniform speed of 340 rpm. Figure 13(b) shows the clustered graphene particles on the surface of the LM6 alloy. From the SEM reports it is also inferred that with help of the stir casting and solidification using the gravity die casting reduces the number of casting defects compared to the conventional casting.

3.5. XRD Examination

X-ray Diffractograms were prepared using the X'Pert^3 MRD equipment using Copper K-alpha wavelength(1.54068 Å). Xray diffractograms are used to determine the crystalline structure of the specimen and the unit cell dimensions. The peaks in the diffractograms determine whether the material is amorphous of crystalline. If there are continuous peaks in the diffractogram then it's a crystalline structure and each peak denotes the lattice parameter of the crystalline structure which can be calculated by the Bragg's Equation. The theta obtained is the angle between the incident ray and the diffracted ray from the specimen so while calculating the interplanar distance and based on that the lattice parameter(h,k,l).The machine directs X-ray towards the specimen and measures the intensity of the diffracted ray, and provides a characteristics diffraction pattern. The machine collect data from range of

2Θ(ranging from 10-90°). The X-ray diffractograms for the matrix formed by reinforcing graphene in the Aluminium LM6 base metal are presented.

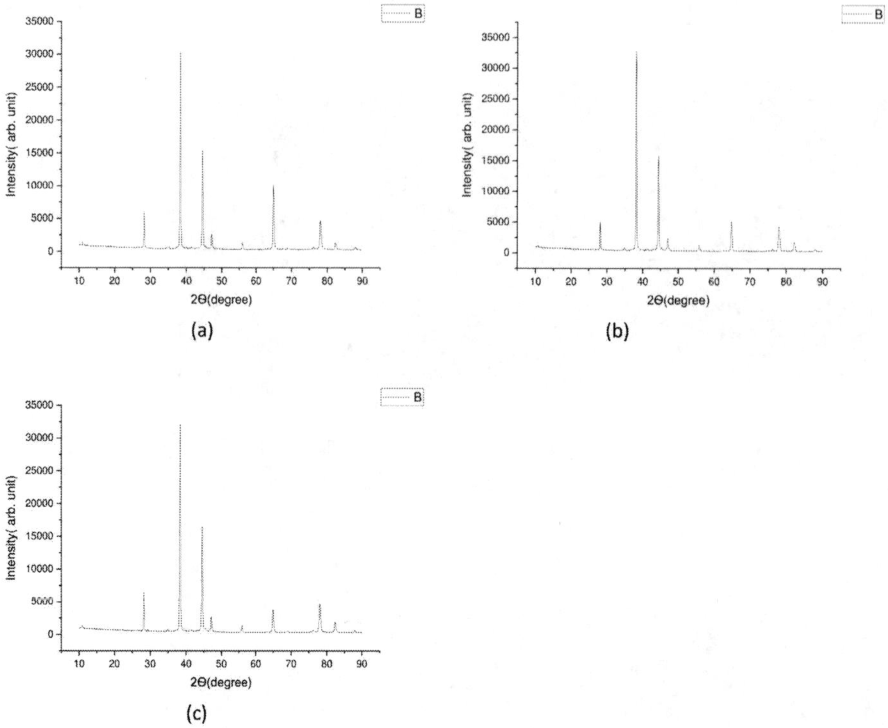

Figure 17. (a) X-ray diffractogram of LM6 alloy with 0.1% graphene reinforcement , (b) X-ray diffractogram of LM6 alloy with 0.25% graphene reinforcement , (c) X-ray diffractogram of LM6 alloy with 0.4% graphene reinforcement

From the diffractogram presented above it can be inferred that the matrix composite formed by reinforcing graphene with the Aluminium LM6 base alloy has crystalline structure. The peak intensities for the matrix were obtained at 2Θ equivalent to 38.3 °, 38.15 °, 38.4 ° for 0.1, 0.25 and, 0.4% graphene weight reinforcement with the base alloy corresponding to the lattice parameter (111) [18], apart from the peak values the other corresponding peak values for 2Θ~28.27 °,44.56 °,47 °,55.9 °,64.97 °,78.06 °,82.26 °. It can also be inferred from the diffractograms that after reinforcing with the graphene particles the metal matrix didn't lose its crystalline structure but there were some changes in the lattice orientation and the diffractograms were almost similar for the metal matrix.

4. Conclusion

The aluminium-graphene metal matrix composite was successfully created by electrochemical exfoliation graphene extraction process and stir casting method. Increase in hardness and tensile strength values indicate that the graphene can be used as an effective reinforcement for aluminium matrix. The tensile strength of the specimen with 0.1%wt of graphene reinforcement increased up to 30% as compared its parent alloy, for 0.25%wt graphene reinforcement specimen it is increased up to 25% and for the 0.4%wt reinforcement it increased up to 8% and can be stated that with increase in the reinforcement percentage of graphene in the parent alloy with the mechanical properties starts to deteriorate due to dominance of the brittle non-metallic nature of graphene as it's one of the allotropic form of carbon. The hardness for the casted specimen with reinforcement also showed some plausible results with increase in their hardness but for the 0.4%wt it got reduced due to increase in the graphene concentration, likewise their was increase in the impact strength of the specimens, but reinforcing with the graphene particles it also imparted some brittle nature that was observed by the crack pattern on the specimen while performing the Charpy test. The microstructure of the sample denotes that defects during the casting and solidification were reduced with the help of stir casting and there was a uniform diffusion of the graphene in the aluminium LM6 alloy, and the graphene produced was in a cluster formed form instead of the planar orientation which was inferred from the FESEM images. The X-ray diffractograms showed that the metal matrix composite thus formed using the graphene reinforcement had crystalline structure with the maximum peak near around 38° having aluminium lattice parameter of (111).

References

[1] A. Saboori, C. Novara, M. Pavese, et al. *An investigation on the sinterability and the compaction behavior of aluminum/graphene nanoplatelets (GNPs) prepared by powder metallurgy, Journal of Materials Engineering and Performance*, Vol. **26**, No. 3, 993-999, 2017.

[2] S.Gopalakrishnan, *Production and wear characterization of AA 6061 matrix titanium carbide particulate reinforced composite by enhanced stir casting method.* Composites: Part B 43 -2012 302- 308

[3] S. E. Shin and D. H. Bae,*Deformation behavior of Aluminium alloy matrix composites reinforced with few-layer graphene*,Composites Part A-Applied Sci. and Man., vol. **78**, pp. 42-47, 2015

[4] Iman S. El-Mahallawi , Ahmed Yehia Shash and Amer Eid Amer,*Nanoreinforced Cast Al-Si Alloys with Al2O3, TiO2 and ZrO2 Nanoparticles-* Metals,2015, 5, 802-821; doi:10.3390/met5020802.

[5] El Mahallawi, I.S.; Egenfeld, K.; Kouta, F.H.; Hussein, A.; Mahmoud, T.S.; Rashad, R.M.; Shash, A.Y.; Abou-AL-Hassan, W. *Synthesis and Characterization of New Cast A356/(Al2O3) p Metal Matrix Nano- Composites.* In Proceedings of the 2nd Multifunctional Nanocomposites & Nanomaterials: International Conference& Exhibition MN2008, Cairo, Egypt, 11–13 January 2008. Prabu, S.B. Influence of stirring speed and stirring time on distribution of particles in cast MMC. J. Mater. Process. Technol. 2006, 171, 268–273.

[6] El-Mahallawi, I.; Shash, Y.; Eigenfeld, K.; Mahmoud, T.; Rashad, R.; Shash, A.; El Saeed, M. *Influence of Nano-dispersions on Strength Ductility Properties of Semi-solid Cast A356 Al Alloy.* Mater. Sci. Technol. 2010, 26, 1226–1231. M. Farkas, Implementing network security in a local bank, In: R. Grande (ed.), Enterprise resource planning system: a theoretical perspective, 3rd ed. Oxford: Oxford University Press, 2003, pp.23-42.

[7] El-Mahallawi, I.; Abdelkader, H.; Yousef, L.; Amer, A.; Mayer, J.; Schwedt, A. *Influence of Al2O3 nanodispersions on microstructure features and mechanical properties of cast and T6 heat-treated Al Si hypoeutectic alloys.* Mater. Sci. Eng. A 2012, 556, 76– 87.M. Rashad, F. Pan, A. Tang et al. Effect of graphene nanoplatelets addition on mechanical properties of pure aluminium using a semi-powder method, Progress in Natural Science-Materials International, Vol. 24, 101–108, 2014

[8] Sajjad Amirkhanlou, *High-strength and highly-uniform composites produced by compocasting and cold rolling processes*. Materials and Design 32 (2011) 2085-2090

[9] S.A Sajjadi, *Microstructure and mechanical properties of Al-Al2O3 micro and nano composites fabricated by stir casting.* Materials science and engineering A-528 (2011)8765-8771.

[10] Liang-Xu DONG, *Properties, synthesis, and characterization of graphene*, Qiang CHEN Laboratory of Plasma Physics and Materials, Beijing Institute of Graphic Communication, Beijing 102600, China.

[11] S.J. Yan, S.L. Dai, X.Y. Zhang, C. Yang, Q.H. Hong, J.Z. Chen, Z.M. Lin, *Investigating aluminum alloy reinforced by graphene nanoflakes*, Mater. Sci. Eng. A 612 (2014) 440e444.

[12] *Investigation of mechanical properties of carbonaceous (MWCNT, GNPs and C60) reinforced hot-extruded aluminum matrix composites*,Muhammet Emre Turan Karabuk University, Metallurgical and Materials Engineering Department, Karabuk, Turkey

[13] Sreehari Peddavarapu, Muthu Veerappan A., Raghuraman S. School of Mechanical Engineering, SASTRA Deemed University, Thanjavur 613401, India,*Tribological study of Graphene reinforced AA6082 surface composite processed through Friction Stir Processing.*

[14] Ajay Kumar P , Madhu H C, Abhishek Pariyar , Chandra S. Perugu , Satish V. Kailas , Uma Garg , Pradeep Rohatgi, *Friction stir processing of squeeze cast A356 with surface compacted graphene nanoplatelets (GNPs) for the synthesis of metal matrix composites*.

[15] Mohammad Khoshghadam-Pireyousefana , Roohollah Rahmanifarda,1, Lubomir Orovcikb , Peter Švecc , Volker Klemm, *Application of a novel method for fabrication of graphene reinforced aluminum matrix nanocomposites: synthesis, microstructure, and mechanical properties*,S0921-5093(19)31605-3

[16] Yu Zhang Ng1, a), Khi Poay Beh1 , Faris Hidayat Ahmad Suhaimi1 , Chee Kiat Tan1 , Fong Kwong Yam1 , Hwee San Lim1 and Mohd. Zubir Mat Jafri1 1 School of Physics, Universiti Sains Malaysia, Penang, Malaysia, *Investigation of Electrochemical-Based Exfoliation of Graphene with The Aid of Stabilizer.*

[17] Prashant Tripathi, Ch. Ravi Prakash Patel, M. A. Shaz*and O. N. Srivastava* Nanoscience Centre, Department of Physics (Centre of Advanced Studies), Banaras Hindu University, Varanasi-220005,India,*Synthesis of High-Quality Graphene through Electrochemical Exfoliation of Graphite in Alkaline Electrolyte.*

[18] S. Venkatesan, M. Anthony Xavior, School of Mechanical Engineering, VIT University, Vellore 632 014, TN, India, *Tensile behavior of aluminum alloy (AA7050) metal matrix composite reinforced with graphene fabricated by stir and squeeze cast processes.*

IOP Conf. Series: Materials Science and Engineering 1123 (2021) 012042 doi:10.1088/1757-899X/1123/1/012042

Design of battery pack & controller for combined auto driven mini-harvester

Rahul S Warghane [1], Rajkumar E [2]

[1] Ph.D. Scholar, Department of Mechanical Engineering , Vellore Institute of Technology-, VIT University Vellore,Tamil-Nadu,632014 India warghanerahul1@gmail.com
[2] Associate Professor , Faculty of Mechanical Engineering , Vellore Institute of Technology- VIT University , Vellore,Tamil-Nadu,632014, India ,rajkumar.e@vit.ac.in

Abstract: The mechanized farming can promisingly increase the profitability and prevent the dependency of farmers on foreign labours. The design and implementation of Battery pack and controller for specially designed combined auto-driven mini harvester is presented in this paper. The mechanical load estimation and its mapping with electrical drives are formulated. The interfacing of electrical drives with remote operation is achieved by Electronic Arduino system. The relay control system is designed with individual drive system. The sketches are mapped with the functionality of drives. The drive selection is a dependant function of mechanical load requirement. The mechanical load estimation is done with actual mechanical failure consideration and design approach. The pin-out connections for all the ports are done on the basis of logical sketch command. The open source Arduino software is used to run the sketches. The sketch written for all the drives is compiled and demo prototype machine is tested for functioning. The run trail is successful and can conclude that the Arduino mega gives the sound interface with 8 channel relay. The computation and measurement results of power requirements for the controller and drive are within the limiting conditions.

***Keywords**: Arduino Sketch development, Battery-pack design, Electro-mechanical load mapping, Relay interface, Remote operation*

1. Introduction

The explosion of population causes upmost requirement of food & agricultural goods all over the world. The shrinking fertile land size is making farmers incapable for adoption of bigger farm machineries. Though the advancement and development in the field of agriculture taking place promisingly still the adoption and implementation for the small countries like India, Indonesia, Brazil is rare or mostly unseen [1]. The combined harvester available is limited for large farm fields and rich farmers. Though attempts have been made to address the issue of small farmers still the problem remain unsolved. [1] The post harvesting operation is most labour intensive work [2]. The post harvesting has to be performed in the hot dry whether to trace the maximum yield. The favourable condition leads to the operating temp about 40 ^{0} C [3]. The manual operation at such an operating temperature leads to severe health issue or may cause death due to heat stroke. The detailed literature review and market survey gives the traces for unavailability of combined mini harvester. The available combined harvester with their capacity and capability & constraints are deeply studied and presented by present researcher in open source journal. The findings lead to the conclusion that no combined mini harvester is readily available to fulfil the requirement. [4]

To take up the issue an attempt is made to design the combined auto driven harvester. The combined auto-driven harvester is very novel machine specially designed for the small famers. The machine performs the combined operation of cutting, threshing, grain separation, husk trapper and husk compacting at the same time in the farm field itself. The mechanical design and mechanism simulation trial is over and work is presented in conference journal by present researcher [5]

The said machine is auto-driven with all the drives run through remote controlled Bluetooth interface Arduino controller. The electrical load requirements for the different mechanisms are mapped with the mechanical load. The electrical drives serve the mechanical power requirements. The reference of auto-driven robot cars and remote controlled mechanism is taken to formulate and initiate the design [6]. The concept of using Arduino for developing microcontroller based projects has been extended in this research.

For technical development of the model the design constrains are given in the form of geometrical configuration. Taking into consideration about its compact design the harvesting blade span is limited for covering three crops in single trail. The mechanical and thereby electrical load requirements are thus formulated for the strength and some physical parameters for the said configuration.

1.1. Estimation of load requirement

The configuration of combined harvester with its major components is shown in the figure below. The machine is designed with a view to develop the combined mini harvester for small farmers. The load requirement is considered on the basis of design calculations formulated for the mini harvester. The reference design calculation provides a basis for load estimation in present work [5]. The given mechanical load requirements are fulfilled through the electrical drives. The load estimation for individual drives is formulated with the concept of system design. In system design the whole machine is subdivided in small machine subassembly and mechanisms. Adopting the same concept for electric load calculation the major mechanical load requirements are identified as

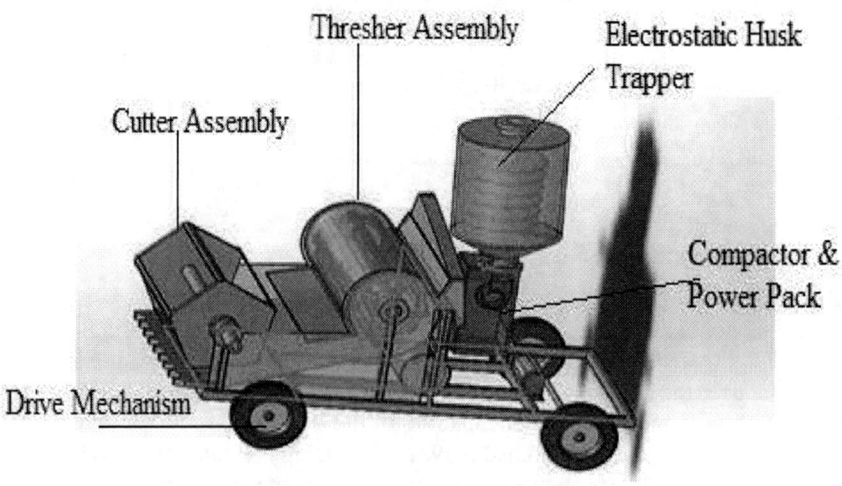

Figure 1. Combined Harvester

1. Cutter drive
2. Thresher drive
3. Electrostatic husk trapper
4. Dive for Dry compactor
5. The Drive mechanism

1.2. Load calculation for thresher drive

The thresher mechanism is a drum assembly driven by electric motor to impart the contact friction and impact shear on the crop bulk [7]. The electric load requirement is formulated with the reference of mechanical load calculation. The mechanical & electrical load requirement are formulated as under

$$Total\ load\ = Static\ load + Dynamic\ load \quad (1)$$

$$Static\ load = Self\ weight + Inertia\ force \quad (2)$$

$$Self\ weight = m\ x\ g \quad (3)$$

$$Mass\ (m)\ =\ Density\ (\varrho)\ x\ Volume\ (V) \quad (4)$$

With the reference of design calculation Thresher drum is made up of cast steel. So, ϱ is 7800 kg/m^3. The mass of Thresher assembly is 49.68 kg. With consideration of maximum 10 % inertia force on the self load the static load is 303.88 N.

The dynamic force for performing the threshing is 130.9 N.

Total force requirement is 434.9 N. Thresher drum speed is 748 rpm. With the torque radius of 175 mm gives the total power requirement of 7.5 Kw at 750 rpm standard electric motor. With speed reduction gearbox having velocity ratio of 6 the desired motor required is 1.25 Kw having rpm 6000.

Figure 2. Thresher Drive

1.3. Electrostatic husk trapper mechanism

With the reference of design formulation & calculation for static cling the power requirement is given as

$$P\ =\ Current\ (I)\ *\ Volt\ (V) \quad (5)$$

.

For the current of 16.485 amps and 48 V gives the power requirement as 810 W.

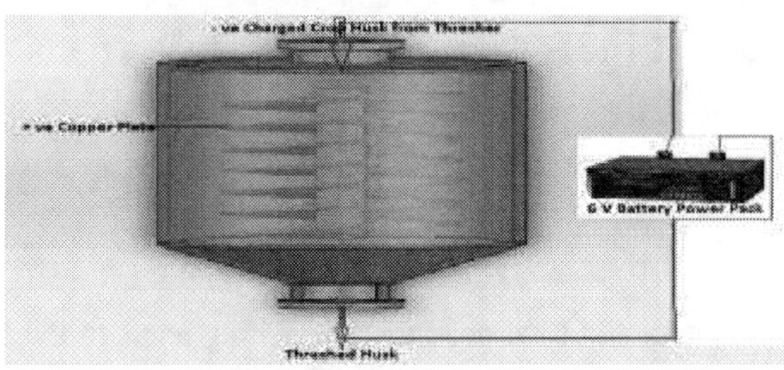

Figure 3. Husk Trapper

1.4. Drive for dry compactor

The test trail results for making the husk bricks gives the dead load requirement of 8 KN force acting on brick area of 400 mm^2. The given load requirement leads to the selection of hydraulic power pack of 3.2 bar capacity. The power requirement for power pack motor is 750 W.

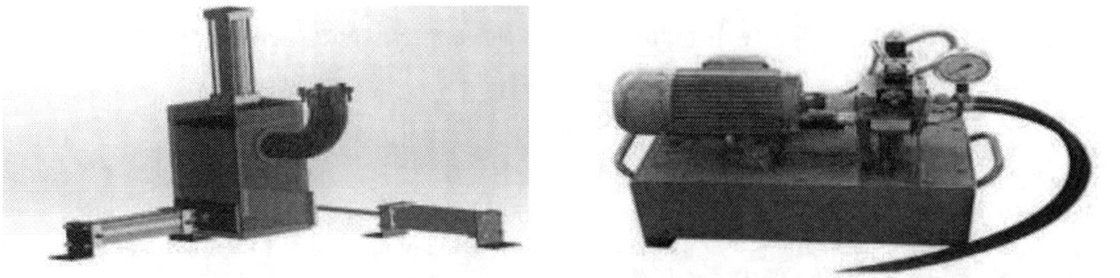

Figure 4. Compactor Power Pack

1.5. Drive for cutter

The cutter drive is a blade assembly impart dynamic shear on the crop bulk. The concept mechanism diagram shown in Figure 05 . The dynamic shear is applied by drum mechanism driven by an electric motor coupled through gearbox.

With the reference of geometrical configuration the weight of drum is 54.55 kg . With the consideration of 10 % inertia load gives the static load on cutter drum is 633 N.

The dynamic force for cutting is calculated with the reference of crop shear strength and core stem area. Considering the strongest crop stem for design formulation i.e. soybean with shear strength of 2.67Mpa [8] and maximum core area of 20 mm gives the cutting force requirement of 2515.5 N for cutting three corps in a row.

Total force acting on drum assembly is 3148 N. To provide the cutting thrust the torque applied on 450 mm cutter drum is 633.44 Nm. For the designed peripheral speed of 6 m/sec with 450 mm drum the speed of drum is 254.77 rpm.

The power requirement for cutter assembly is 16.20Kw at 254 rpm. The speed reduction gearbox leads to the power requirement. Using gearbox with velocity ratio of 10 reduces the power requirement to 1 Kw double speeds motor.

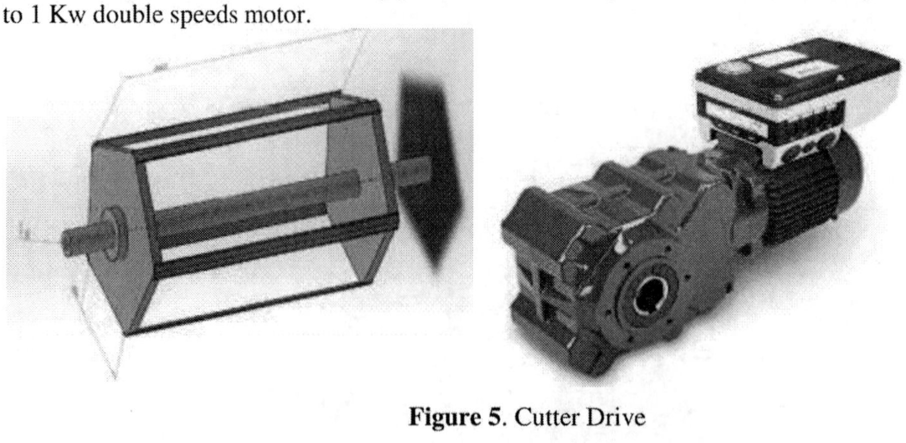

Figure 5. Cutter Drive

1.6. Drive mechanism

With 4 x 4 transmission system and total dead load of 300 kg leads to the static load on machine as 2.943 KN. With the reference of tyre rim assembly selected for the load and application leads to the torque requirement of 441Nm. The propelling speed of machine is 1.4 f/sec gives the power requirement of motor with 2 Kw capacity and 6 speed reduction gearbox. The all wheel drive approach gives the power requirement for each drive motor as 500 Watt.

IOP Conf. Series: Materials Science and Engineering 1123 (2021) 012042 doi:10.1088/1757-899X/1123/1/012042

Figure 6. Drive Motors

2. Selection of battery pack

The analysis of mechanical failure and estimation of actual electrical load gives an account of true electrical load requirement. The true design power load requirement for given machine considering the different drives leads to the total power load requirement as 7.5 Kw battery pack system. With the consideration of general power factor of 0.86 [9] for single Phase motor the actual power requirement is 8.5 Kw. The line diagram for single main power distribution is as shown below.

Figure 7. Battery Pack

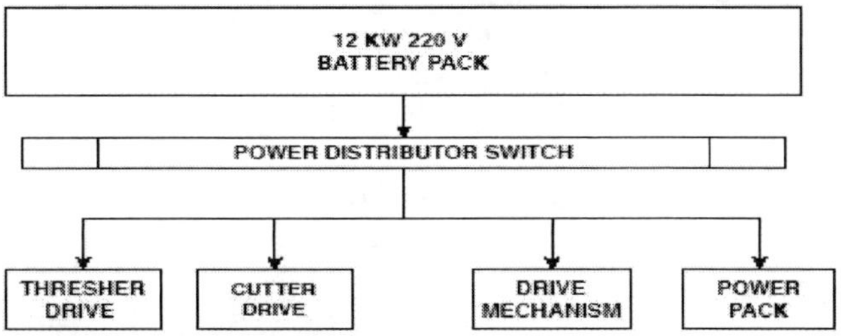

3. Simulation of drives

To inline the total drives in machine with standalone power source leads to the design of monotonous controlling system. Also, the machine operates in the dry hot weather condition with minimum operating temperature of above 40 degree [3]. The stated constraint leads to the requirement of remote operation. To fulfil the demand the system is integrated with the remote controlled drives through Arduino interface. The Arduino controlled power management system operates the function of high voltage drives via low voltage electric module i.e. relays [10]. The serial data feed into the Arduino through the program designed for individual drive. The microprocessors fetch the command and perform the logical operation. The logical pin-out connection fetches the data input for individual port and processes for operating the relays [10].

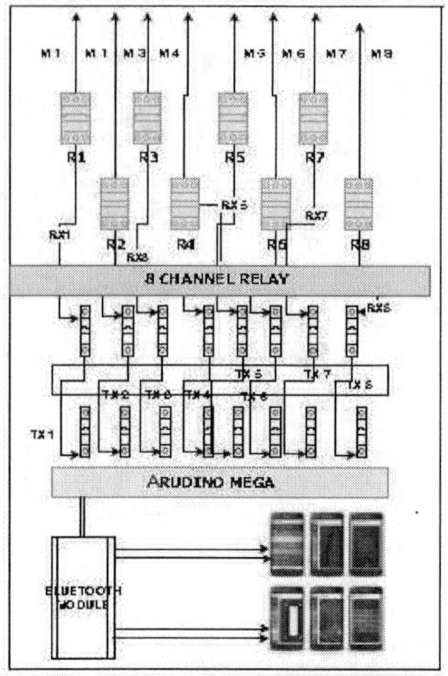

Figure 8. Drives Logical Configuration

4. DESIGN OF CONTROLLER

The logical relationship between drives and relay shown in fig.08 gives the address note of high data transmission between the ports. The 8 relay operation for the controlling of machine with minimum lag demands strong data processor. The high end processor available with Arduino Mega serves the purpose with an added benefit of 18 serial (TX) transmission ports [I].

The operation of individual drives is analysed and algorithm is plotted for all the drives. The sketch is developed for all the drives taking into consideration of functionality and logical screening of all the ports. The digital data transmission is addressed in the sketch for minimum lag [I]. The logical pin-out connection for all the ports is as shown

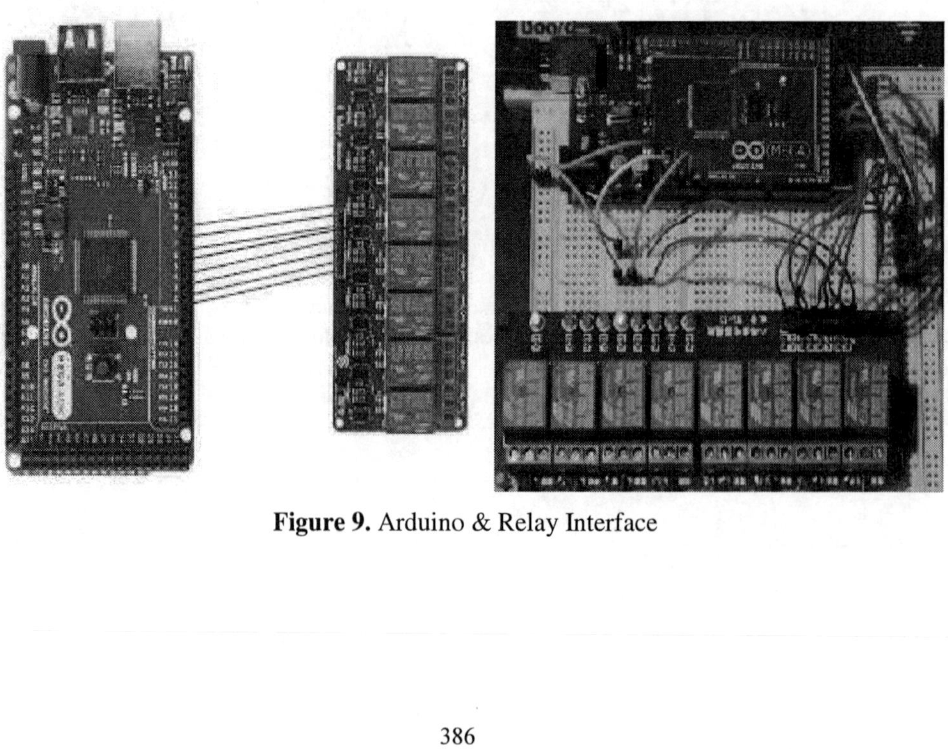

Figure 9. Arduino & Relay Interface

5. Selection of relay channel kit

With the reference configuration diagram and electrical load requirement leads to the motor requirement as follows

- Drive motor – 500 W 3000 rpm
- Thresher motor – 1.25 KW at 6000 rpm
- Hydraulic power pack motor – 750 W at 1000 rpm.
- Cutter motor – 1 KW at 6000 rpm.
- Electrostatic husk trapper – 810 W

The high power management leads to the use of relay circuit. The readily available 8 channel relay circuit serves the purpose.

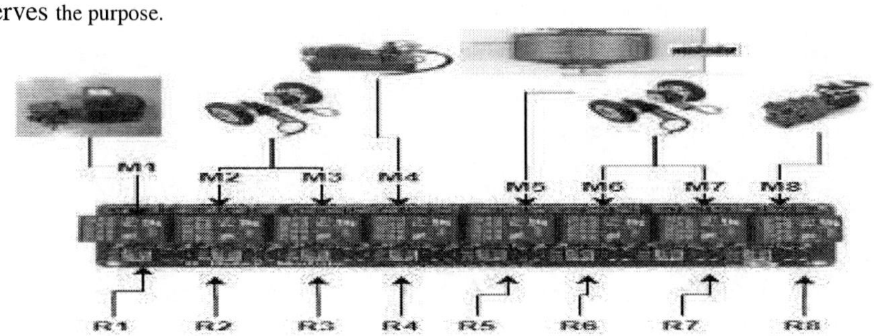

Figure 10. Relay & Drives Interface

The 8 channel relay kit provides an interface for connection of high power equipment with low power controlling equipment [II]. In present work all the ports of 8 channels relay module is used thus provided with 8 pin-out connection including the ground and VCC pin-out connection [12]. The logical connection pin-out junction is stated as below

- Logic GND: Connected to GND on Arduino.
- Input 1 (IN 1): Connected to digital pin D1 on Arduino.
- Input 2 (IN 2): Connected to digital pin D2 on Arduino.
- Input 3 (IN 3): Connected to digital pin D3 on Arduino.
- Input 1 (IN 4): Connected to digital pin D4 on Arduino.
- Input 2 (IN 5): Connected to digital pin D5 on Arduino.
- Input 3 (IN 6): Connected to digital pin D6 on Arduino.
- Input 2 (IN 7): Connected to digital pin D7 on Arduino.
- Input 3 (IN 8): Connected to digital pin D8 on Arduino.

```
#define RELAY1 7 //Defining the pin 7 of the Arduino for the 8 relay module
 #define RELAY2 6 //Defining the pin 6 of the Arduino for the 8 relay module
 #define RELAY3 5 //Defining the pin 5 of the Arduino for the 8 relay module
 #define RELAY4 4 //Defining the pin 4 of the Arduino for the 8 relay module
void setup()
 {
 pinMode(RELAY1, OUTPUT); //Defining the pin 1 of the Arduino as output
 pinMode(RELAY2, OUTPUT); //Defining the pin 2 of the Arduino as output
 pinMode(RELAY3, OUTPUT); //Defining the pin 3 of the Arduino as output
 pinMode(RELAY4, OUTPUT); //Defining the pin 4 of the Arduino as output
 }
```

6. Arduino coding

The combined harvester consists of 8 machine drive components [5]. The 4 machine drive components listed as R2 ,R3 ,R6 ,R7 used to drive the machine (machine propelling).Whereas the remaining machine drive listed as R1 , R4 ,R5 ,R8 used to run the cutter, power pack and husk trapper & the thresher ref. Fig 1 .

The machine drive motors needs to be programmed separately with the concept of robot cars.

The logical approach is as below

- Drive motors - R2, R3, R6, R7.
- R 2,R 3 -Left Drive motors
- R6 , R7 - Right Drive motors
- Forward motion - R2, R3 , R6 , R7 High
- Left turn - R6 , R7 High
- Right turn - R 2, R 3 High.
- Reverse Motion - H Bridge activate.

```
char t;
void setup() {
pinMode(2,OUTPUT);    //left motors forward
pinMode(3,OUTPUT);    //left motors reverse
pinMode(6,OUTPUT);    //right motors forward
pinMode(7,OUTPUT);    //right motors reverse
pinMode(9,OUTPUT);    //Hbridge
Serial.begin(9600);
```

The machine drives R1, R4, R5, and R8 are constant drive motor runs on the constant speed input, pinned connection as D1 , D4 , D5 , D8. The relay acts as a switch and controls the power for motor. The Arduino digital pin used to turn the relay on and off and is given the name 'motorpin' in the sketch. In order to maintain the constant speed of motor a potentiometer is connected in-line with the drive motor to ensure the constant torque on mechanises

```
int motorPin = 1,4,5,8; //Relay assign
void setup()
pinMode(motorPin, OUTPUT);
Serial.begin(9600); //Set constant moor speed
while (! Serial);
Serial.println("Speed 0 to 255");//Potientiometer set speed 255
void loop()
if (Serial.available())
int speed = Serial.parseInt();
if (speed >= 0 && speed <= 255)
analogWrite(motorPin, speed);
```

7. Bluetooth interface

The Bluetooth module HC 05 is used to provide the wireless interface between the remote mobile control and Arduino setup. HC 05 receive the serial communication command through the android app is designed for sending serial data to the Bluetooth module. The data received at Bluetooth module send to Arduino through the TX pin of Bluetooth module (RX pin of Arduino). The Code received at Arduino check the received data and processes the command.

```
data = Serial.read();        //Read the  incoming  data and store it into
    variable data
    Serial.print(data);      //Print Value inside data in Serial monitor
    Serial.print("\n");      //New line
    if(data == '1')          // Checks whether value of data is equal to 1
    digitalWrite(13, HIGH); //If value is 1 then LED turns ON
    else if(data == '0')     //  Checks  whether value of data is equal to 0
```

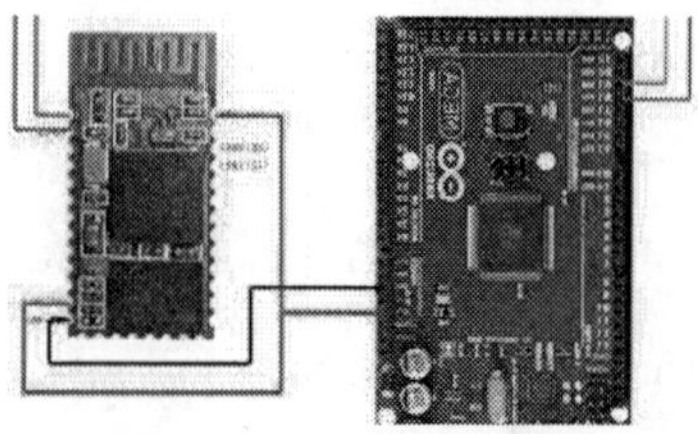

Figure 11. Bluetooth Interface

8. Results & discussion

The 8 channel relay kit is wired as per the drive design and the logical pin-out connection is done. The Arduino sketch is compiled and run through the interfacing Bluetooth module. The true operating parameters in the form of required current and voltage required is computed. For each drive the operating parameters are calculated with the power factor of 0.86 and efficiency of drive as 90 %. The power requirement on drives and controller relay module are as under.

Table 1. Relay & Drive Control parameters

Relay port	Drive name	Electrical load (KW)	Drive parameter		Controller parameter	
			Current (Amp)	Voltage (V)	Current (Amp)	Voltage (V)
1	Thresher Drive	1.25	41.35	37	6.46	4
2	Left Drive Motor 1	0.5	55.64	11	3.22	2
3	Left Drive Motor 2	0.5	55.64	11	3.22	2
4	Power Pack	0.75	54	17	4.84	4
5	Husk Trapper	810 W	20.65	48	5.23	3
6	Right Drive Motor 3	0.5	47.08	13	3.48	2
7	Right Drive Motor 4	0.5	47.08	13	3.48	2
8	Cutter Drive	1	58.29	21	6.45	4

The power requirement for each drive is calculated and the desired controller requirement is mapped. The calculation and measurement result shows that, the power requirements for relays are under the limiting conditions of relays.

9. Conclusion

The method of mechanical design load estimation is stated in this work. The mechanical and electrical load mapping is done and stated the drive system requirement. The power load calculation and selection of power-pack is done with suitable power factor for actual drive system.
The formulation of method for sketch development is stated and sketch is run on Arduino free source software. The sketch developed for drive interface is run and trail is successful .Arduino mega provides successful interface with 8 channel relay board. The lag time for drive control is almost negligible.

Acknowledgments
This work was supported by the Research Program of the School of Mechanical Engineering, VIT University, Vellore, Tamil-Nadu -India

References

[1] Department of agriculture and Farmers Welfare 2019 List of Combine Harvesters Tested at the Farm Machinery Training & Testing Institutes *Mechanization & Technology Division, Department of Agriculture & Cooperation, New Delhi* **June 2019**
[2] Hegazy, Rashad. 2013 . Post-harvest Situation and Losses in India. *Technical Report · June 2013 10.6084/m9.figshare.3206851*.**v1**.
[3] Omafra staff. 2014 Soybeans: Harvest and Storage. *Agronomy Library Harvest Timing* http://www.omafra.gov.on.ca/ Additional Sources: **August 2014**

[4] Rahul S. Warghane et.al. 2018, Review on Design and Optimization of Thresher In Crop Cutting Machine *International Journal of Mechanical Engineering and Technology (IJMET)* Volume 8, Issue 11, November 2017, pp. **1020–1028.**

[5] Rahul Warghane ,Rajkumar E 2018 'Design and Synthesis of Crop Cutting Machine' *International Journal of Advanced and Innovation Research* ,Volume 03 Issue 06 Pg **92-94** ISSN NO :2278-7844

[6] Bagloee, S.A., Tavana, M., Asadi, M. 2016 Autonomous vehicles: challenges, opportunities, and future implications for transportation policies. *J. Mod. Transport. 24,* ***284–303*** *(2016).* https://doi.org/10.1007/s40534-016-0117-**3**

[7] Brandini, Adhemar, 1969 . Corn kernel forces during impact shelling , *Retrospective Theses and Dissertations.* **14482**. https://lib.dr.iastate.edu/rtd/14482

[8] Piotr kuźniar 2016, Physical and chemical properties of soybean seeds determine their susceptibility to mechanical damage. *Zemdirbyste-Agriculture ISSN 1392-3196 Vol. 103, No.* **2**

[9] Khan, Muhammad. 201). Speed Control and Power Factor Improvement of Single Phase AC Motor. *Universal Journal of Electrical and Electronic Engineering.* 2. **328-336**. 10.13189/ujeee.2014.020804.

[10] Oussama ben belgith, lasaad sbita, 2014 Remote GSM module monitoring and Photovoltaic System control , *First International Conference on Green Energy (ICGE)* ***2014***

Digital handbook:

I. Digital Acoustics Corporation Text Speak Design Group, *Westport, CT* 06880 U.S.A.

II. Handson Technology "8 Channel 5V Optical Isolated Relay Module" *http://www.handsontec.com/*

DESIGN AND ANALYSIS OF MODIFIED FOUR STROKE FOUR REVOLUTION ENGINE

Prashant S. Jadhav[1,2] and Chinmaya P. Mohanty[1]

[1]School of Mechanical Engineering, VIT University, Vellore-632014, India
[2]Department of Mechanical Engineering, Rajarambapu Institute of Technology, Rajaramnagar, Shivaji University, Kolhapur, Maharashtra,415 414, India

Abstract. In today's world, one of the important tasks is to handle the energy available and to minimize the heat losses. In automotive sector, engine is the primary energy source and there are many losses associated with an engine out of which engine cooling and exhaust losses takes major share. In four-stroke SI/CI engines, four strokes of reciprocating cylinder involve two rotations of the crankshaft. These strokes are suction, compression, power and exhaust respectively. A large amount of fuel energy is wasted when fuel is not compressed completely and also at the exhaust of internal combustion engines. This paper focusses on minimizing losses of internal combustion engine by secondary expansion of exhaust gases through 180° of crank revolution. For this purpose, four stroke four revolution mechanism is proposed and this is achieved by modifying current slider crank mechanism with double connecting rod arrangement. This modified engine shows an increase in the air standard cycle efficiency by 4.10% where as fuel air cycle efficiency rise is observed to be 5.25%. There is also significant rise observed in the modified engine over conventional engine in terms of gross indicated thermal efficiency and brake thermal efficiency as well.

Introduction

Transportation is very important activity in human transformation. The modes of transportation of humans and goods have changed gradually over the years. Earlier travelling was done by walking using legs. Then there was invention of wheel which was revolutionary. With the help of wheels, forming cart and it was powered by muscular energy of animals like bulls and horse. Then James Watt introduced the first Engine which totally evolved the automotive sector. Heat engines perform their work by chemical combustion of fuel and converting it to heat energy. Heat engines are of mainly two type's external combustion engines and internal combustion engines. In an external combustion engine, combustion reaction takes place outside the engine while in an internal combustion engines combustion reaction takes place inside the engine. Both engines have various applications in many fields.

Now a day for transportation, vehicles are used. These vehicles are powered by internal combustion engines. Internal combustion engines, works on fossil fuels like petrol, diesel etc. The fuel is burned inside the engine and with help of I.C engine components, rotary motion is achieved which drives the wheels of automobiles. There are total four strokes in an engine suction, compression, expansion or power and exhaust. In suction the air fuel mixture is taken inside the combustion chamber as piston inside cylinder moves from top dead center (TDC) to bottom dead center (BDC). In compression stroke, the piston moves from BDC to TDC and fuel air mixture gets compressed and burned. In Power stroke the burned fuel emits energy and moves piston towards BDC. Power stroke is only the work producing stroke. In exhaust stroke, piston moves from BDC to TDC thereby releasing the gases through exit port. And one cycle completes with these strokes and this process is repeated again. Piston transfers force to connecting rod which is connected to crankshaft. The mechanism converts this reciprocating motion into rotary motion. The engine which works on spark ignition is called as S.I. engine while which works on compression ignition is called as C.I. Engine. There are two types of engines viz; two stroke engine and four stroke engine. In two stroke engine, suction and compression performed in one stroke while power and exhaust in next stroke. In this way two strokes are completed in one revolution of crankshaft. While in four stroke engines four strokes are completed in two revolutions of crankshaft. I.C. Engines works with above processes.

It is estimated that within next 50 years the fossil fuels are going to exhaust from earth. Engine works on fossil fuels such as petrol and diesel. The efficiency of SI engine ranges between 26-30% while that of C I Engine ranges between 30-35 %. It means that if 100 kJ of energy is produced then only 30 to 35 kJ energy is only utilized for desired work. All other energy wastes in the form of different losses such as frictional loss, cooling loss, blow by loss. Along with this, the number of vehicles in the world is increasing every day which also increases fuel demands. To overcome this issue, either non-conventional energy sources are required to be used more or else the efficiency of existing engine needs to be increased.

A considerable amount of literature has been published on the influences of basic engine parameters on efficiency, performance and emission of multi-cylinder engines and larger displacements. V. Hariram et al. [1] presented that the varying the compression ratios (CR) have substantial effect on the performance, exhaust and ignition parameters. In this work, single cylinder direct CI engine is observed with fluctuating CR viz; 18, 17 and 16 at changing loads. Here, the ignition and variable performance on changing the CR were studied particularly. Decrease in brake thermal efficiency and rise in exhaust gas temperatures were measured when CR was lowered from 18 to 16. The brake specific fuel consumption was seen to be increased on decreasing the compression ratio. Decrease in peak cylinder pressure and increase in ignition delay period was seen on lessening of CR. The peak heat release rate was closer to TDC on changing CR from 16 to 18. The rate of pressure growth was also studied and maximum of 5.38 bar/CA and minimum of 0.78 bar/CA was observed on above CR.

J.S Jadhao et al. [2] presented the inexorably global issue on fast growing economy, a general deficiency of energy, the inner ignition motor smoke dissipate heat and natural contamination has been progressively stressed vigorously as of late. Amount of complete heat provided to the vehicles, on an average, 32 to 45% is transformed into an useful mechanical work; the remaining heat is expelled to the atmosphere through exhaust gases and vehicle cooling frameworks, which leads to entropy rise and natural pollution, hence, it is required to re-use waste heat as an useful work. The recovery and usage of waste heat saves fuel yet, in addition, decreases the measure of waste heat and ozone-depleting ingredients. The study validates the availability and credibility of waste heat from the inward burning motor additionally portray the loss of smoke gas vitality of an interior ignition motor. Possible techniques to regain the waste heat from burning motor and implementation and emanations of the inside ignition motor.

Joshi invented [3] "A mechanism for converting single or each reciprocating motion to two times rotation of the crank by providing a double connecting rod" which can be applicable for all types of IC engines to increase their efficiency. This mechanism comprises of a Piston reciprocating within a cylinder. A double connecting rod which is pivoted to the piston. A cam is mounted to the second connecting rod. The second connecting rod is provided to a crank wheel, which in turn mounted to the shaft end. As piston start moving from TDC to BDC, 180° rotation of crank is achieved and when it returns to TDC again, a complete rotation of crank is over.

Sun J and Wang [4] highlighted the amelioration with globalization and economic policies that have transformed the trend and momentum of world economy in the last couple of decades. The main focus is on product quality, rapid design and development and finally an economical product. To survive in this challenging market, industries have to upgrade with technology, skills to offer effective products. Design and analysis of intricate components has now become convenient with the application of Finite Element Method. Automobile Industry has transformed drastically with the evolution of next generation of CAD/CAM technologies, from engine performance to aesthetics. The lead time to manufacture automobiles has dropped marginally over the years; because of betterment with Geometric modelling and Computer Integrated Manufacturing (CIM). Design of Internal Combustion (I.C) Engine parts play a vital role in betterment of the functioning of vehicles. Design and modifications of all the important components are done in view to increase the performance and efficiency.

Siliveri naresh et al. [5] presented, Piston is a part of To-fro engines, siphons, Air compressors and pneumatic cylinders among other comparable mechanisms of engines, its aim is to exchange constrain from extending gas in the cylinders to the crankshaft by means of a cylinder pole or associating pole. Here the cylinder is structured, broke down and the assembling procedure has been examined. Cylinder temperature has impressive impact on proficiency, emanation, execution of the motor. Motivation behind the examination is estimation of cylinder transient temperature at a few points on the cylinder, from virus begin to unfaltering condition and correlation with the consequences of limited component analysis.in this undertaking the cylinder is demonstrated and collected with the assistance of CATIA programming and segment is coincided and investigation is done in ANSYS programming and the warm and static conduct is contemplated and the outcomes are organized.

Jovan Ž. DORIĆ et al. [6] displayed the reproduction of the working procedure in another IC motor idea. The essential element of this new motor idea is the acknowledgment of variable development of the cylinder-piston. The capricious cylinder development, it is anything but difficult to give a variable pressure proportion, adjustable uprooting, and burning amid consistent volume. These preferences over standard cylinder component are accomplished through an amalgamation of the two sets of non-roundabout apparatuses. The introducer system is intended to get a particular movement law which gives better fuel utilization of motors. In this study, Ricardo/WAVE programming was utilized, which gives a completely coordinated treatment of time-subordinate liquid elements and thermodynamics by methods for the one-dimensional detailing. The outcomes acquired in this incorporate the proficiency normal for this new warmth motor idea. The outcomes demonstrate that ignition amid steady volume, variable pressure proportion, and variable removals significantly affect the improvement of fuel utilization.

Vikas Kumar et al. [7] reviewed at the point when fuel is scorched in an engine, heat is delivered. Extra heat is likewise created by grinding between the moving parts. Just roughly 30% of the energy discharged is changed over into valuable work while staying 70% must be expelled from the engine to keep the parts from dissolving. In air-cooled I.C Engines, expanded surfaces called fin are given at the circumference of engine chamber to build heat exchange rate. That is the reason the investigation of fins is vital to enhance heat dissipation rate. The fundamental of the point of this work is to think about different looks into done in past to improve heat exchange rate of cooling fins by changing cylinder geometry and material.

Kolchin et al. [8] clarified the crankshaft is a standout amongst the most fundamentally stacked segments as it encounters cyclic loads through twisting and torsion amid its life. Its failure will make genuine harm the engine so it's vital at the season of configuration to confirm exhaustion quality and torsional vibration and physical approval on proving ground just as on field vehicles. More difficulties in crankshaft plan because of expanding vehicle payloads, lower weight prerequisite, higher effectiveness and longer solidness life. This paper will give a general rule for the understudy and industry engineers for planning the crankshaft which can serve the more extended toughness existence with no failing.

Methodology

This paper has adopted reverse engineering strategy. Reverse engineering is the procedure of extricating learning or structure data from anything man-made and re-creating anything dependent on the extracted data. For this project the Piaggio Ape Passenger Vehicle engine is selected. The selection is made by considering stroke length, stroke to bore ratio.

Selection

The primary requirement for this project is under square engine. The problem of clashing of the connecting with the cylinder wall is not present in the under square engine. The diesel has higher knock ratings; hence it gives wide range compression ratio selection. Complexity in designing is eliminated as it is naturally aspirated and naturally cooled engine. Single cylinder engine eliminates complex design procedure like multi-cylinder engine.

Ape Piaggio Passenger Vehicle Engine features:
- Diesel engine
- Naturally cooled engine
- Single cylinder naturally aspirated engine.
- Stroke to bore ratio is 0.87 which is under square engine.

Modified engine with four strokes and four revolution engine is compared with the theoretical efficiencies of the existing engine. The assumptions in Modified Engine are:
- Same amount of fuel injected.
- Cut-off ratio =5%

Comparison of diesel and modified diesel engine

Compared values of existing diesel and modified diesel engine are shown in Table 1

Table 1. Parameter values comparison

Sr. No	Parameter	Existing engine (%)	Modified engine (%)
1	Air Standard cycle efficiency	60.65	64.5
2	Losses due to variation of specific heat	13	13
3	Loss due to progressive combustion	5.5	5.5
4	Loss due to incomplete combustion	3.5	3.5*
5	Direct heat loss	4.5	4.5*
6	Exhaust heat loss	1.5	1.5*
7	Pumping Loss	0.5	0.5
8	Rubbing friction loss	3.5	3.5*
9	Fuel Air cycle efficiency	47.65	51.5
10	Gross indicated thermal efficiency	32.65	36.5
11	Brake thermal efficiency	28.65	32.5

IOP Conf. Series: Materials Science and Engineering 1123 (2021) 012043 doi:10.1088/1757-899X/1123/1/012043

Design of modified engine components

Variable parameters

Large number of varying parameters are to be selected for functional design i.e. estimation of swept volume capacity which the volume displaced by piston in the slider crank mechanism selected to carry out thermodynamic cycle. A piston reciprocates in the cylinder to take the air in and compress it. Thereafter combustion releases the heat and forces the piston to deliver mechanical work to output shaft through a connecting rod and crankshaft. The design process is complete after establishing the dimensions of cylinder, piston, connecting rod, crankshaft and supporting structure to maintain the kinematic relation between these components.

Mean effective pressure**:** Mean Effective pressure is the normal pressure inside the cylinders of an I.C Engine dependent on the determined or estimated power yield. It rises as manifold pressure increases. For any engine, working at given speed and power yield will be explicit shown mean effective pressure, IMEP and relating brake mean effective pressure, BMEP.

Cylinder bore (d): The inside diameter of the working cylinder is called as cylinder bore.

Stroke length (L): The distance through which a piston moves between two consecutive reversals of its direction of motion is called stroke length

Piston area (A): The area of a circle of diameter equal to the cylinder bore is called the piston area.

Design of gear components:

The main running gear components of reciprocating IC engine such as slider crank, four bar chain configurations are to be designed to bear the load to which they are subjected primary elements are

- Cylinder
- Cylinder Head
- Piston Assembly
- Connecting Rod Assembly
- Crankshaft

Design of cylinder:

Fine grained cast iron or cast steel is the most suitable material for cylinder liner. It offers better resistance to wear due to sliding of piston in it and corrosion when corrosive residual gases are left during non-operation. The material should have ultimate tensile strength of 300Mpa. The liner has to withstand maximum gas pressure during the cycle. It will be designed for static strength under maximum gas pressure with factor of safety so that fatigue will be covered.

t = 5 mm (Thickness of Cylinder)

Design of cylinder head:

Initially the head is to be treated as a flat circular plate secured to one end of the cylinder, by cylinder bolts at pitch circle diameter and subjected to uniform maximum gas pressure. Actual cylinder head has circular holes to accommodate inlet and exhaust valves and the fuel injector. These will weaken the head. However, reinforcement of the port passage will compensate for loss of strength.

Design of cylinder bolts:

Bolts withstand the entire gas pressure force during the operation which varies between 0.1 MPa (atmospheric to 8MPa) during each cycle of operation.
Using the Soderberg Equation

For a smaller engine number of bolts selected is generally four.
Minimum no. of studs = 0.01D + 4 = 4.86.
Maximum no. of studs = 0.02D + 4 = 5.72.
Therefore the no. of studs are = 5
$Core\ Diameter\ dc = 8.6 \times 10{-}3 = 9\ m\ (approximately)$

Piston design:

The cylinder-piston needs to transmit the gas power to the interfacing bar through the pivot association known as cylinder stick or little end of interfacing bar. The gas drive follows up on the best surface of the cylinder known as a crown and is transmitted by means of the supervisors holding the cylinder stick and the bearing and after that on to the little end of the associating pole. The rest of the barrel some portion of the cylinder far from a crown, past focus of stick goes about as a guide for straight movement and takes the side push because of obliquity of the associating pole to transmit it to the chamber liner.

$$Crown\ thickness = dc = 8.42\ mm$$

Barrel thickness:

Taking the mean breadth of that area as 85% of the cylinder-piston distance across, the thickness of this segment can be determined. 10% of that area is decreased give to a radial allowance in this area.

$$\text{Thickness } (tw) = 4.21 \text{ mm}$$

Unless radial width of piston ring which to be accommodated in grove is decided, barrel thickness cannot be finalised.

Piston ring design:

General practice is to choose 2 to 4 gas rings and one oil ring. Oil ring is obliged in the skirt some portion of the cylinder.

$$Barrel\ thickness = t = 3.64\ mm$$

Hence, barrel thickness = 4.21+3.64=7.85 mm (approx.8 mm)
In this calculation, w does not play any role and is adopted equal to t = 4 mm.

Hence barrel thickness = 6+4 = 10 mm.

Frees size of ring:

The ring needs to apply the required pressure force when fitted in section. This is acquired by initial pressure of ring in the cylinder bore diameter. The gap between free ends before assembly and after assembly is given as-

Total Gap (Ig) = 11.50×0.5 = 12 mm

IOP Conf. Series: Materials Science and Engineering 1123 (2021) 012043 doi:10.1088/1757-899X/1123/1/012043

Piston pin design:

The overall length of the pin is about 90% of cylinder width, to accommodate circlips and to prevent the pin from fouling with cylinder liner.

$\therefore$ $Length\ of\ pin$ =77 mm

Inner Diameter of piston pin = di=36 mm

Outer Diameter of piston pin = do=35 mm

Design of connecting rod:

The connecting rod is designed for a varying load from 0 to 53kN for every working cycle. The connecting rod is subjected to whipping force in the plane of rotation due to the mass inertia of rod itself, which further introduces the bending stresses in the rod. Section selected is I section.

Area =1.394×10−4 $m2$

This area can be provided by selecting any cross-section- circular, elliptical, rectangular or I section. Thickness of I section (t) = 3.5 mm

Buckling stress in XX plane (σxx) =259.11 MPa

Buckling stress in YY plane (σyy) =185.84 MPa

Checking for whip stress:

Connecting rod is the only component in the engine which is subjected to very complex kinematics and dynamics.

$BMmax$=11.24 Nm

σtc=9.41 $MPa.$

$\therefore$Angle α, s=9.46o

$Compressive\ Force$ =14.133 KN

$Compressive\ Stress$=73.61 M Pa

$Total\ force$ =73.61+9.41=83.02 MPa. (This is safe.)

Design of small end:

Length and diameter of bearing in connecting rod has been already established.

Volume of piston material =309.62 CC

$Mass$=839.06 gm

$Mass\ of\ piston\ pin$ =158.85 gm

$Mass\ of\ piston\ assembly$=997.55 gm

$Inertia\ Force\ (F)$ = 6.64 KN

$Tensile\ stress\ in\ eye$=58.3 MPa

This value is very small compared with strength of forgeable steel Sy=750 MPa, Se=150 MPa

Design of big end:

Design consideration for big end of the connecting rod is based on bearing requirement. The bearing load continuously varies during the cycle.

Gas pressure force (pb) = 30.735 $N/mm2$

Inner diameter of crank pin eye, where thickness of bushing is approx. 2 mm

Is =42+2×2 =46 $mm.$

The thickness is maximum at centre where the B.M is maximum, but can be reduced to about 7 mm at the end of diameter. The cap is subjected to a fluctuating stress 0 to 100 MPa.
By Soderberg equation, Factor of safety = f.s =2.45

Design of bolts:

Two bolts are used to hold the cap rigidly, with tightening force twice that of the inertia force required to be transmitted to the piston to retard it.

$$\therefore Initial\ Tightening\ Force = 2\times 6.61 = 13.322\ KN$$

$$Fmax = 19.983\ KN$$

$$Fvary = 3.3305\ KN$$

Soderberg equation gives:

Core diameter (dc) = 7.518 mm

∴ Approximate diameter = 8 mm.

Previously assumed diameter to estimate centre distance was 10 mm, to calculate the bending movement of cap is on higher side. Hence the design is safe.

Design of crankshaft:

It has been already stated during the bearing design process that rigidity of crankshaft is the most significant parameter of the crankshaft design. This leads to under stressing the crankshaft. However, one has to check the bearings of shaft support. For making the design compact, l/d ratio for the main bearing is also much less than 0.75.

Therefore, core diameter of crankshaft (d)=38.125 mm

Design of telescopic rod:
Assuming 70% of pressure falls down during first expansion

$Fmax$=10455.9 N and $Fmin$=0 N

By Soderberg Equation, assuming the cross-sectional area is to be circular,

$$(\pi/4)d^2 = 8.3648\times 10{-}5$$

$$d = 10.32\ mm$$

The fig 1 shows front view and detailed 3 D view of modified engine.

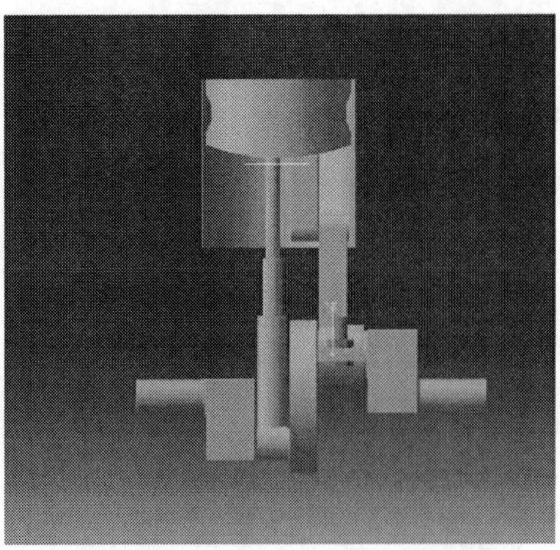

Fig. 1 Modified Engine view (frontal)

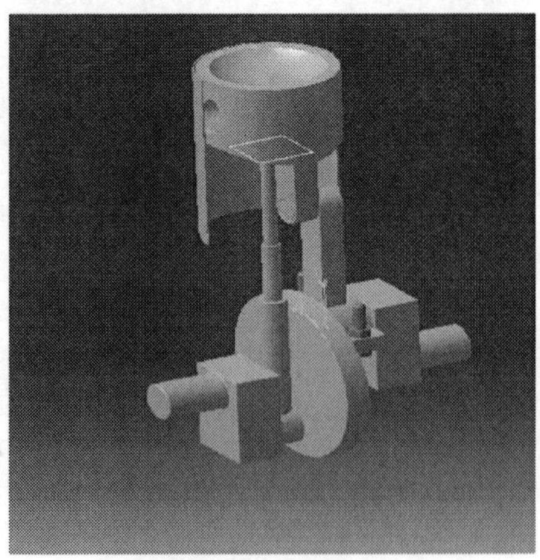

Detailed 3D view of Engine

Conclusion

The four stroke four revolution engine mechanism is mainly developed for getting more work output than traditional engine. Theoretically it can be justified that efficiency increases by 4% and can give improved work output. However, there are some practical limitations in working of model viz. precision of telescopic rod, plunger pin. Also, the engine is bulkier. If these limitations are overcome by alternatives in the future then this engine will stand tall as the milestone in the field of IC engine.

REFERENCES:

[1] Hariram V and Shangar RV 2015 *Alexandria Engineering Journal* **54** **pp** 807-814.

[2] Jadhao J and Thombare D 2013 *International Journal of Engineering and Innovative Technology* **p2**

[3] Joshi J, Patel D and Lcit B 2017 *International Journal of Engineering Science* **pp** 43-48

[4] Sun J and Wang J 2004 *Internal Combustion Engines* **p** 18

[5] Siliveri naresh Mr. V. Srinivasa Rao, *Design and Thermal Analysis of I.C Engine Piston* **p 23**

[6] Dorić J and Klinar I 2014 *Thermal Science* **18** **pp** 113-127.

[7] Kumar V Jain D and Lomash D 2016 *International Journal of Recent Development in Engineering and Technology* **p 5**

[8] Kolchin, A.I. and Demidov, V.P., 1984. *Design of automotive engines.* **vol 3** (MIR publishers) pp 131-171

Improvement of hot corrosion resistance of dissimilar weldments by current pulsing in a simulated power plant environment

S M Muthu[1], Arivarasu M[2*], Saurav Mohanty[3], and N Arivazhagan[1]

[1] School of Mechanical Engineering (SMEC),
Vellore Institute of Technology (VIT), Vellore,
India – 632014.

[2] Center for Innovative Manufacturing and Research,
Vellore Institute of Technology, Vellore
India- 632 014,
*Email: arivarasu.m@vit.ac.in

[3] University of Texas
Austin, USA

Abstract. Hot corrosion studies are of utmost importance for welded components in the power plant industry as well as for industries which deal with exposure to the high-temperature corrosive environment. The failure of welded components at high temperatures has often led to the reduced service life of a component and catastrophic disasters. This has been a significant engineering challenge in the design of welded components. It is generally seen that the failure of the welded sections mostly occurs because of the deposition of molten salt over the weldments which accelerate the hot corrosion reactions. In this present research work, two dissimilar base materials, AISI 4340 and AISI 304L have been welded with two different welding processes namely continuous current gas tungsten arc welding (CCGTAW) and pulsed current gas tungsten arc welding (PCGTAW) while making use of ERNiCr-3 as filler material. A comparative study of the dissimilar welded plates was carried out by simulating a molten salt environment at 600°C involving hot corrosion behavior. By analyzing the microstructure and thermo-gravimetric data, it is concluded that the pulsed current gas tungsten arc welding process has superior corrosion resistance properties while using ERNiCr-3 as the filler material as compared to the continuous current gas tungsten arc welding process.

Keywords: Dissimilar weldments; Weldment corrosion; Hot corrosion; Thermogravimetric analysis

1. Introduction

AISI 304L austenitic stainless steel has electric industrial applications in power plants, aerospace and chemical industries owing to its excellent weldability, excellent corrosion resistance, and high-temperature strength while on the other hand AISI 4340 finds its application in the design of aerospace components such as landing gear. Due to their high impact strength and superior fatigue properties, these steels are also preferred to manufacture components pertaining to aerospace, nuclear and defense applications. Due to design constraints and industrial applications yielding to design flexibility, dissimilar welding processes are employed which make the joining of two dissimilar weldments possible. Several industrial applications need welding of dissimilar joints because of its economic benefits [1-2]. Joining materials with different elemental compositions pose a challenging task, since they have a huge variation in the co-efficient of thermal expansion and thermal conductivity. Even if successful joint is made between them, the metallurgical, mechanical and corrosion properties will be inferior and a detailed study of each dissimilar joint in its environmental condition is of paramount importance to analyse the service life of the joints. One of the important ways to improve these properties is the selection of proper joining processes and welding filler rod. However, dissimilar weldments tend to fail at high temperatures because of the aggressive corrosive nature of the industrial environments. The hot corrosion of these dissimilar joints exposed to chlorine is unfavorable for the service life of power plant components, where the fuel comprises of coal which contains chlorides [3-4].

Various researchers have reported the failure of components in these service conditions [5]. Failure is generally observed in the weld zone since most of the times it is inferior to the base material. The morphology of the weldment undergoes a transformation on exposure to high-temperature conditions. Recent advances in research on high-temperature corrosion primarily focus on the co-relation of the service life failure of operational welded joints to the degradation of microstructural properties during welding [5-6]. It has been stated that the highest operational temperature of welded joints of the existing materials employed in the power plants are below 600°C [7]. The hot corrosion effects are more aggressive at 600°C because of the high sulphidation tendency which is caused due to generation of low-melting alkali sulphate of eutectic mixtures [8-9]. By studying the past research works, it can be said that the dissimilar joints of AISI 304L and AISI 4340 are used in wide ranging corrosive environments. Therefore, it is of utmost importance to study the hot corrosion properties of these joints at elevated temperatures. Also, various researchers have reported that the Pulsed Current Gas Tungsten Arc Welding (PCGTAW) methods deliver higher cooling rates and shorter solidification times which support enhanced weld mechanical properties than the continuous current gas tungsten (CCGTA) welding [10-11]. However, reports related to the systematic investigation of dissimilar weldment corrosion resistance employing these welding techniques using different filler wire materials are not available at present.

In this research work, CCGTA and PCGTA welding processes were employed using ERNiCr-3 as the filler material. The hot corrosion studies were carried out in a molten salt environment of K_2SO_4 + 60% NaCl at 600°C to simulate a power plant environment as pointed out by Neilsen. The objective of this work is to identify and compare the two welding processes (CCGTA and PCGTA) and conclude which one of them is better suitable for the filler material being used (ERNiCr-3) for the welding processes. The corrosive behaviour on the weldments was analysed extensively using SEM/EDS and XRD to provide a brief report about the extent of the corrosive reaction on the weldment.

2. Experimental procedure

2.1 Specimen preparation

For proceeding with the experiment, two dissimilar base metal plates of AISI 304L stainless steel and AISI 4340 aeronautical steel were taken. The two dissimilar plates were then joined by GTA welding process using continuous and pulsed current mode by using ERNiCr-3 as the filler material. The elemental composition of the parent materials used for this experiment is listed in Table 1. The macrographs of the weldments were taken, and the root dimensions were measured which is shown in Figure 1. Successful dissimilar joints of these materials were obtained by continuous current gas tungsten arc welding (CCGTA) process and pulsed current gas tungsten arc welding (PCGTA) process.

Table 1 Elemental composition of the parent material

	Elemental composition (wt.%)						
Base metal	C	Si	Mn	Cr	Ni	Mo	Fe
AISI 4340	0.31	0.23	0.64	0.98	1.34	0.23	Bal.
AISI 304L	0.037	0.46	0.95	19.14	8.35	0.17	Bal.

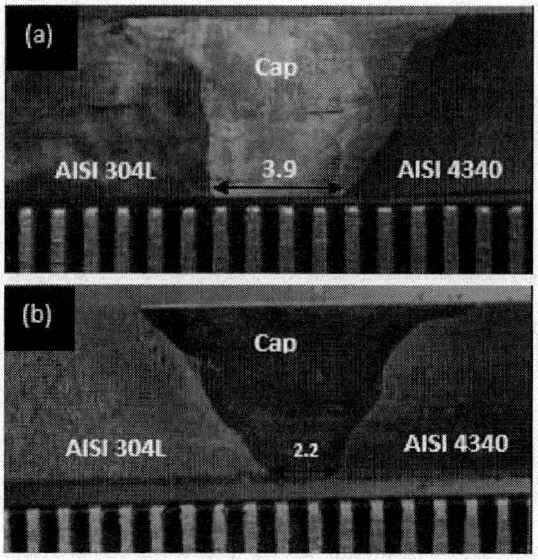

Figure 1. Macrographs of dissimilar weldments of AISI 304L and AISI 4340 welded using (a) CCGTA-ERNiCr-3; (b) PCGTA-ERNiCr-3

2.2 Hot corrosion

Cyclic hot corrosion studies were carried out on samples (30 × 15 × 5 mm) which were cut out from the dissimilar weldments of AISI 304L and AISI 4340 using a wire cut electrical discharge machine (EDM). The samples were polished using emery paper of 400, 600 and 1000 grit size. A eutectic solution of K_2SO_4 and 60% NaCl salt was applied over the entire surface of the hot corrosion study sample to simulate a corrosive environment. The surface density of the salt layer was measured to be around 3-5 mg cm^{-2}. The samples were preheated to 200°C to remove any residual moisture before subjecting them to the cyclic hot corrosion tests. For each cycle, the sample was put in a hot tubular furnace and subjected at 600°C for 1 hour continuously. It was immediately followed by cooling the sample at room temperature for 20 mins. The change in weight of the sample was then recorded using an electronic analytical weighing balance to carry out the thermo-gravimetric analysis further and find out the corrosion rate. This whole process was repeated for 50 cycles (Figure 5 (a) and 5 (b)) to get an extensive data set for investigation of the hot corrosion behavior. Metallurgical characterization of the hot corroded samples after the 50 cycles was executed using a scanning electron microscope (SEM), energy dispersive spectroscopy (EDS) and analyzing the spalled off corrosion products using X-Ray Diffraction (XRD) as well.

3. Results and discussions

3.1 Microstructure

In the CCGTA welding process, the microstructures were observed to exhibit d-ferrite clusters with a well-defined weld interface at the heat affected zone in the AISI 304L region (Figure 2-a). The weld-zone can be perceived to be austenitic. Cellular structures can be observed in the weld-zone along-with the presence of martensitic grain boundaries adjacent to the weld interface. While on the other hand, the AISI 4340 region (Figure 2-c) exhibited a prominent martensitic phase along-with dispersed carbides. In the PCGTA welding process, the microstructures showed similar d-ferrite clusters as that in the CCGTA welding process around the AISI 304L region. However, the weld zone in the PCGTA welding process showed the presence of very fine equiaxed grain growth (Figure 3-b) which can be cited as a reason for the improvement of corrosion resistance properties [2, 10]. The AISI 4340 region showed similar martensitic phase as that in the CCGTA welding process along-with dark dispersed carbides [2].

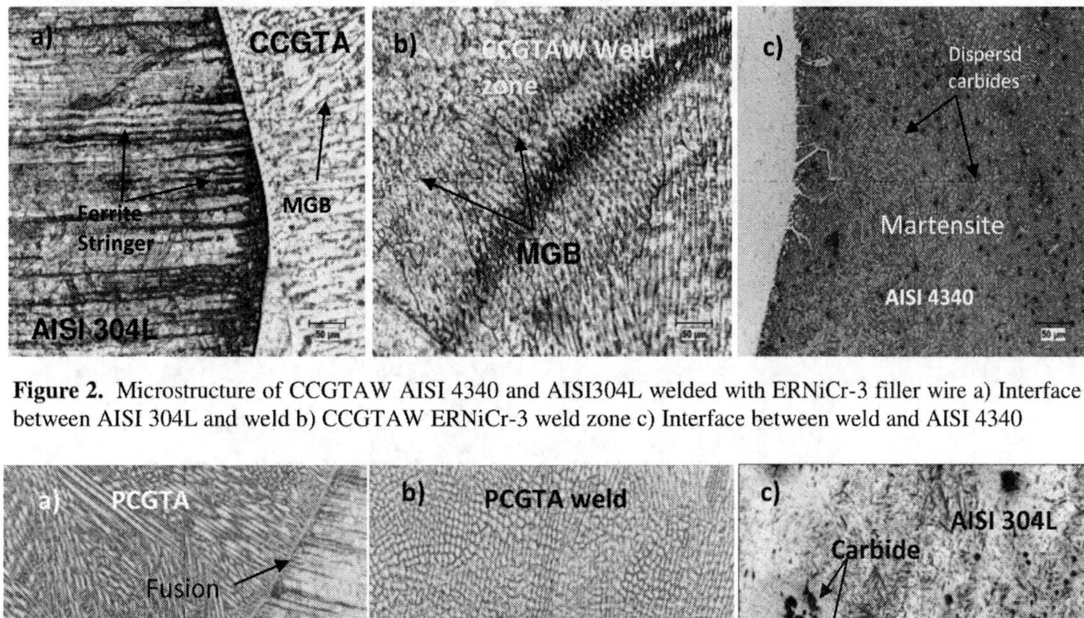

Figure 2. Microstructure of CCGTAW AISI 4340 and AISI304L welded with ERNiCr-3 filler wire a) Interface between AISI 304L and weld b) CCGTAW ERNiCr-3 weld zone c) Interface between weld and AISI 4340

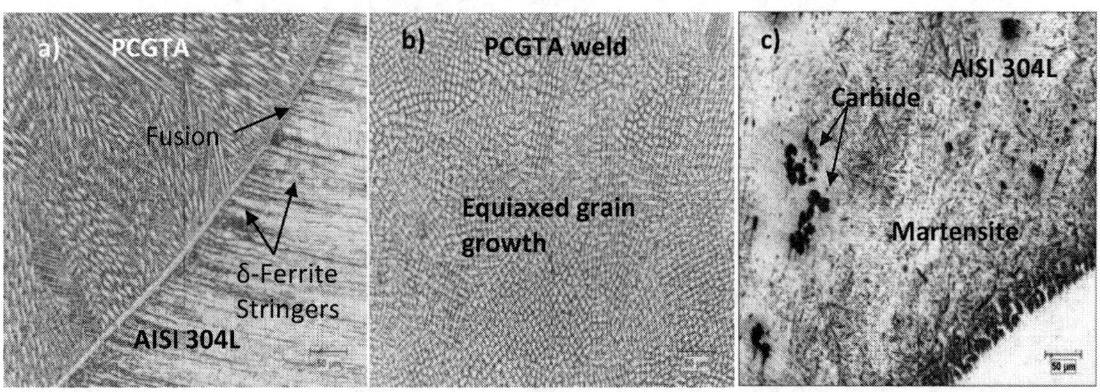

Figure 3. Microstructure of PCGTAW AISI 4340 and AISI 304L welded with ERNiCr-3 filler wire a) Interface between AISI 304L and weld b) PCGTAW ERNiCr-3 weld zone c) Interface between weld and AISI 4340

3.2 Hot corrosion

It can be visibly noticed to some extent that the weld-zone with the ERNiCr-3 filler material by the PCGTA welding process (Figure 4) has relatively minor corrosion aggravation than the weld zone by the CCGTA welding process (Figure 4) around the 30th cycle. In the 50th cycle, it can be distinctly noticed that the weld-zone in the PGGTA welding process has better corrosion resistance than CCGTA welding process.

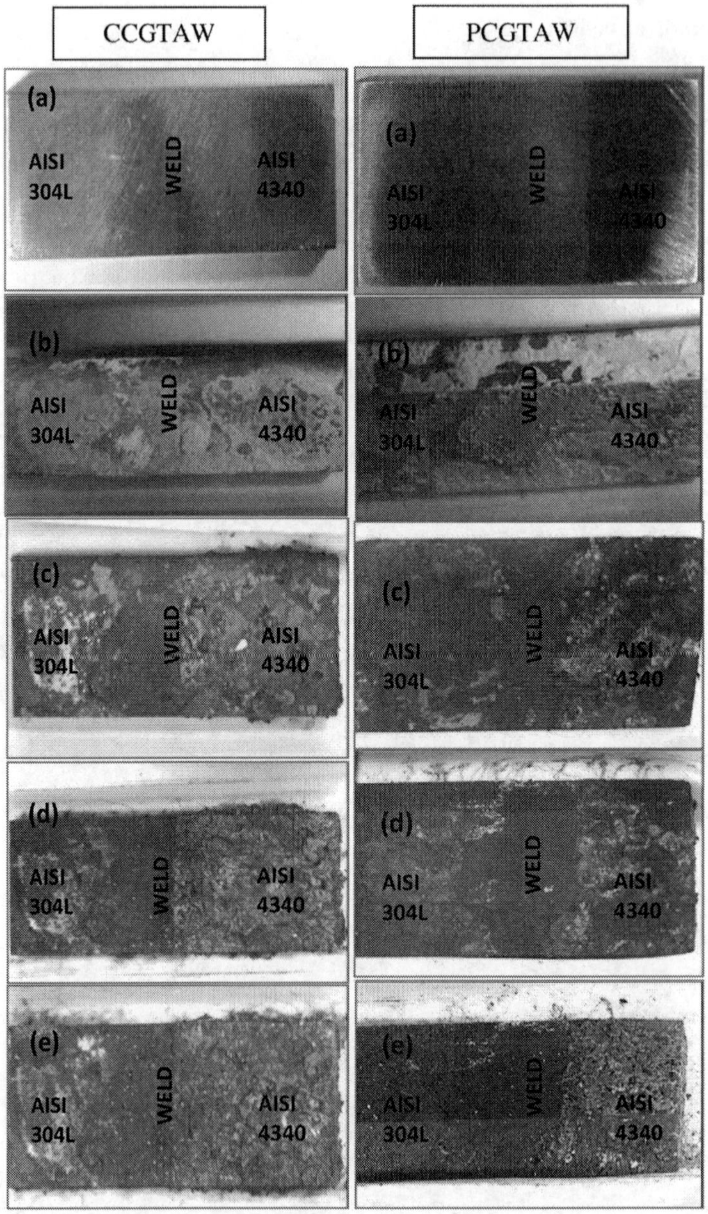

Figure 4. Hot corroded dissimilar joints welded with CCGTA and PCGTA ERNiCr-3 filler (a) after polishing (b) after salt coating and at the end of (c) 10th cycle (d) 30th cycle (e) 50th cycle.

3.3 Thermo-gravimetric analysis

The corrosion kinetics for ERNiCr-3 filler wire is shown in Figure 5 (a) and (b). It is evident from the plot that it follows the parabolic rate law. A plot calculated the Kp (parabolic rate constant) for weight gain square vs. time linear least-square algorithm to a function in the form of $(\Delta W/A)^2 = K_p \times t$. The corrosion rate ($K_p$) for the corresponding bimetallic joints employing CCGTA and PCGTA welding process was calculated to be 63.73×10^{-2} mg^2 cm^{-4} s^{-1} and 17.14×10^{-2} mg^2 cm^{-4} s^{-1}.

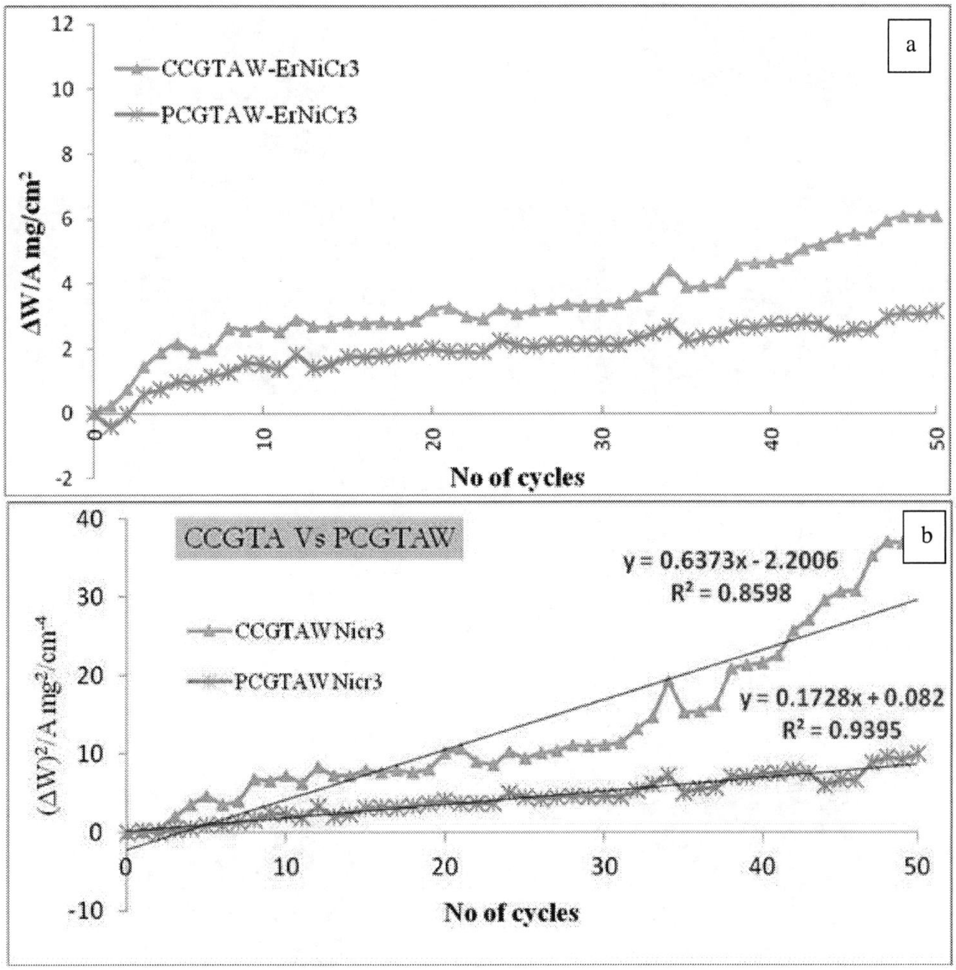

Figure 5. Plot of (a) cumulative weight gain as a function of time and (b) weight gain square Vs time for dissimilar joint between AISI 304L and AISI 4340 using ERNiCr-3

3.4 SEM/EDS

The SEM morphology of CCGTA sample has been shown in Figure 6. While observing the heat affected zone of the AISI 304L region, it was seen that the region had undergone significant corrosion and the corrosion products were discernible in the region. By analyzing Figure 6 (a) and (b), it can be said that the corrosion was uniform in both the Z1 and Z2 regions. However, it can be distinctively seen that the corrosion was more severe in the AISI 304L weld interface as compared to the weld interface of AISI 4340. From the EDS elemental composition, it was seen that AISI 4340 region had comparatively very less Cr content while it had the presence of a more significant amount of iron oxides as compared to that of the AISI 304L region.

The SEM morphology of PCGTA sample has been shown in Figure 7. Evidently, the corrosion products have reduced as compared to that of the CCGTA process. It is observed that the oxide grains over the weld-zone are fine-grained which was also observed in the microstructure image (Figure 3). High concentrations of Fe_2O_3 was found in the scales over the weld zone while the scales over the HAZ of the AISI 304L region chiefly consisted Cr_2O_3 [2].

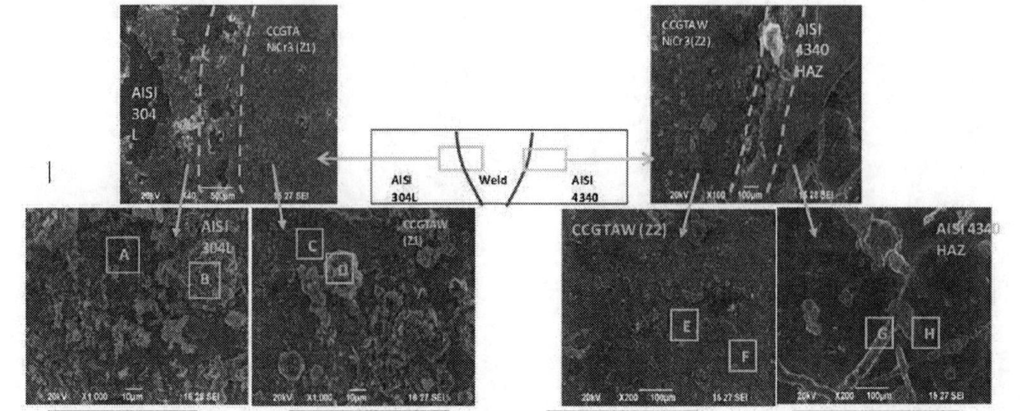

	SOI A	SOI B
Element	Weight %	
O K	7.93	6.7
Na K	0.32	3.83
Si K	0.13	0.04
S K	0.06	0
Cl K	0.29	0.08
Cr K	1.22	3.41
Mn K	0.88	1.24
Fe K	88.67	84.62
Ni K	0.1	0
Nb L	0.4	0.08

	SOI C	SOI D
Element	Weight %	
O K	26.82	35.68
Na K	12.2	12.76
Si K	0.22	0
S K	0.15	0.34
Cl K	0.57	9.76
K K	1.34	4.2
Cr K	3.79	2.52
Mn K	7.92	5.18
Fe K	38.89	27.4
Ni K	7.01	1.4
Nb L	1.1	0.76

	SOI E	SOI F
Element	Wt %	
O K	36.74	37.23
Na K	9.94	16.95
Si K	0.17	0.28
S K	0.08	1.37
Cl K	0.45	18.86
K K	0.58	7.24
Cr K	3.54	1.55
Mn K	5.07	2.34
Fe K	31.26	10.65
Ni K	11.63	2.71
Nb L	0.54	0.82

	SOI G	SOI H
Element	Wt %	
O K	31.65	14.93
Na K	0.85	4.98
Si K	0.04	0.35
S K	0.06	0.45
Cl K	0	7.19
K K	0.05	3.22
Cr K	0	0.36
Mn K	1.12	1.17
Fe K	66.07	66.87
Ni K	NA	0.57
Nb L	0.16	0.21

Figure 6. SEM/EDS results of CCGTA ERNiCr-3 at interface between a) AISI 304L and weld zone; b) Weld zone and AISI 4340.

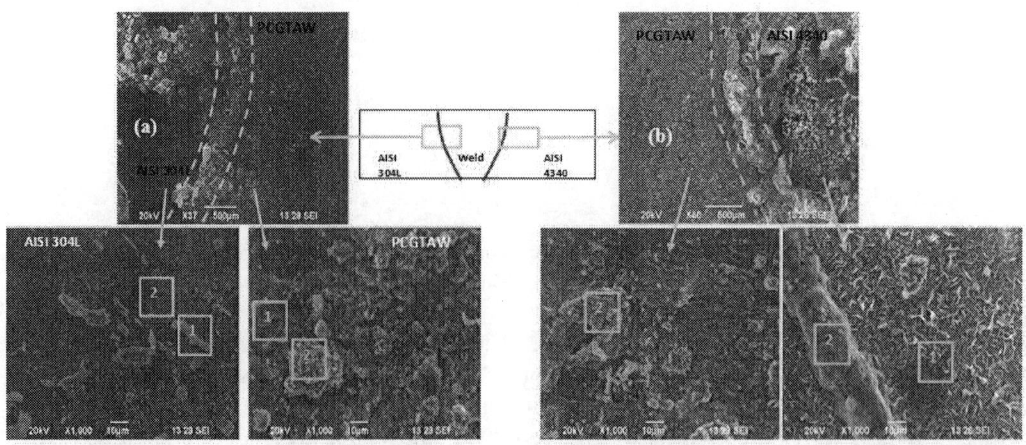

	SOI 1	SOI 2
Element	Wt %	Wt %
O K	36.79	12.83
Na K	NA	0.94
Si K	0.55	0.15
S K	0.94	0.21
Cl K	9.8	0.8
K K	1.16	0.31
Cr K	12.17	18.08
Mn K	1.89	2.42
Fe K	31.25	57.76
Ni K	5.08	6.33
Nb L	0.37	0.17

	SOI 2	SOI 1
Element	Wt %	Wt %
O K	23.44	41.62
Na K	8.47	13.63
Si K	0.23	0.15
S K	0.43	1.15
Cl K	5.36	6.8
K K	3.44	8.43
Cr K	4.52	6.05
Mn K	15	7.58
Fe K	22.75	6.69
Ni K	14.47	6.32
Nb L	1.88	1.59

	SOI 2	SOI 1
Element	Wt %	Wt %
O K	30.27	32.43
Na K	6.7	3.75
Si K	0.18	0.23
S K	1.73	0.17
Cl K	3.77	1.36
K K	6.36	0.78
Cr K	3.8	13.72
Mn K	8.33	5
Fe K	31.5	38.88
Ni K	6.2	3.06
Nb L	1.15	0.62

	SOI 2	SOI 1
Element	Wt %	Wt %
O K	37.06	6.75
Na K	1.05	20.32
Si K	0.05	NA
S K	0.1	0.63
Cl K	0.71	51.93
K K	0.2	15.95
Cr K	0.08	0.19
Mn K	1.15	0.23
Fe K	59.57	3
Ni K	0.02	1.01

Figure 7. SEM/EDS results of PCGTAW ENiCr-3 at interface between a) AISI 304L and weld zone; b) Weld zone and AISI 4340.

3.5 XRD

The XRD pattern of the spalled off surface products from the weldments produced by both CCGTA and PCGTA welding processes on exposure to the molten salt environment have been shown in Figure 8. On analysis of the corrosion products that spall off after the hot corrosion studies, the following metal oxides were found to exist: Fe_2O_3, Cr_2O_3, $FeCr_2O_4$, $NiFe_2O_4$, and Cr_2NiO_4. The oxides of Si and Mn were also found to co-exist in the corrosion products, which are known to degrade the corrosion resistance properties. XRD revealed the presence of metallic chlorides such as $FeCl_2$, $CrCl_2$ and $NiCl_2$ on all the weldments [10].

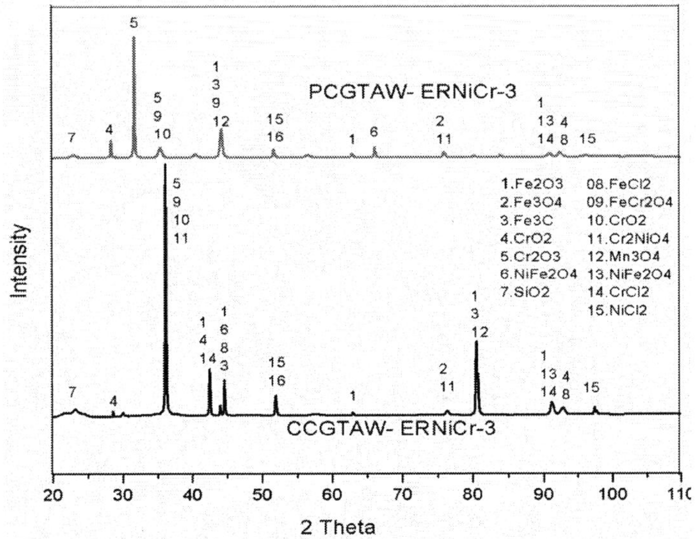

Figure 8. XRD analysis of peaks for the dissimilar AISI 4340 and AISI 304L hot corroded joint welded with ERNiCr-3 filler.

4. Conclusions

After the comparative hot corrosion behavioral studies of the dissimilar weldments of AISI 304L and AISI 4340 welded by PCGTA and CCGTA processes while making use of ERNiCr-3 as the filler material, the following conclusions can be derived:

- A sound weld joint beween AISI 4340 and AISI 304L by employing ERNiCr-3 filler was fabricated.
- The K_p (corrosion rate) for the PGTA welding process (17.14×10^{-2} mg^2 cm^{-4} s^{-1}) was found to be less than the conventional current process (63.73×10^{-2} mg^2 cm^{-4} s^{-1}). This shows that corrosion resistance of the joints fabricated by pulsed current has better hot corrosion behavior.
- Oxides formed on the AISI-4340 was permeable and were spalling leading to severe corrosion and loss of material.
- A strong protective oxide layers of Cr_2O_3 and NiO were observed on the base material AISI-304L and the fusion zone.
- In a molten salt high-temperature environment, the corrosion resistance of pulsed current ERNiCr-3 weldment is more prominent than continuous current ERNiCr-3 weldment. The occurrence of fine equiaxed grains in the PCGTA ERNiCr-3 can be cited as a reason for superior corrosion resistance.
- The corrosion resistance and aggregate mechanical properties can be attributed to the use of ERNiCr-3 filler wire which is suitable for industrial applications making use of AISI 4340 and 304L in pulsed current mode.

References

1. M Arivarasu, K Gokulkumar, KD Ramkumar, N Arivazhagan 2016 *High Temp. Corros.* **39**.
2. Arivarasu, M., Venkatesh Kannan, M., Devendranath Ramkumar, K. and Arivazhagan, N 2017 *Corros. Eng., Sci. Technol.* ***52*(2)** 114-123.
3. Eliaz, N., Shemesh, G. and Latanision, R.M. 2002 *Eng. Fail. Anal.* **9(1)** 31-43.
4. Hara M 1991 *J. Jpn. Inst. Met. Mater.* **55** 1207–1215.
5. Wang, H., Wang, H., Kan, P., Zhao, Q., Zhang, X. and Ding, Y., 2018 *Oxid. Met.* **90(5-6)** 571-583.
6. Zhou, W., Zhou, K., Deng, C., Zeng, K. and Li, Y. 2017 *Ceram. Int.* **43(12)** 9390-9400.
7. Michelsen HP, Frandsen F, Dam-Johansen K, 1998 *Fuel Process. Technol.* **54** 95–108.
8. Montgomery, M., Karlsson, A. and Larsen, O.H. 2002. *Mater. Corros*, ***53*(2)** 121-131.
9. Das, D., Balasubramaniam, R. and Mungole, M.N. 2002 *Mater. Sci. Eng.* **A 338(1-2)** 24-32.
10. Arivazhagan, N., Singh, S., Prakash, S. and Reddy, G.M. 2011 *Mater. Des.* ***32*(5)** 3036-3050.
11. Muthu, S.M. and Arivarasu, M. 2019. *Mater. Res. Express*, ***6*(11),** p.116577.

Numerical investigation on air-cooling enhancement of a motor cycle engine by varying fins geometry

Bibhuti B Sahoo[1,4], **Chandrakanta Nayak**[2] **and Mukul Shrivastava**[3]

[1]School of Mechanical Engineering, Vellore Institute of Technology, Vellore – 632014, India, bbs.iitg@gmail.com
[2]Department of Mechanical Engineering, Gandhi Institute of Technological Advancement, Bhubaneswar – 752054, India. chandra_kec@rediffmail.com
[3]Department of Mechanical Engineering, Vardhaman College of Engineering, Hyderabad – 501218, India, mukul@vardhaman.org
[4]Corresponding author

Abstract. Two-wheeler vehicles are equipped with an air-cooled cooling system. Internal combustion engines generate heat to obtain mechanical energy but non-convertible heat energy is dissipated from engine heat and walls. Air-cooled fins provide extra material with an extended surface area to enhance heat transfer rate. The performance of fins can be increased by changing different parameters like material, geometry, number of fins etc. Present work combines the changes in the fin geometry, shape, and pitch to maximize the heat transfer rate as compared to earlier available works. Modelling of 100 cc engine with cooling fins is recreated using CAD software SOLIDWORKS. A heat thermal analysis of original and modified models was done with Mechanical ANSYS and Fluent. In the thermal analysis, temperature variation along the length of the fins was analyzed. The modified model was able to achieve an 8.5 % increase in heat transfer rate with air-motion as compared to the existing original 100 cc engine cylinder.

***Keywords**: Two wheeler, Cooling fins, Heat transfer, Modelling, Air-cooling*

1. Introduction and background

The reciprocating internal combustion (IC) engines produce high temperature and pressure during the combustion of fuel. This force imparts on the piston in the expansion stroke which produces mechanical energy. The melting point of the engine cylinder material is generally lower than this high IC engine combustion temperature. Hence, practically, the cylinder temperature is reduced by some cooling methods to avoid a complete seizure of the piston, bearing and other important parts. Also, higher engine temperatures decrease engine performance by lowering the volumetric efficiency, promoting pre-ignition and detonation. Hence, IC engines are deliberately designed for heat losses of approximately 70% through cooling systems to prevent the thermal damage of cylinder material. Only about 30% of engine heat is converted for useful mechanical power [1, 2]. The air-cooling method is mainly adopted in small capacity IC motor bike engines effectively due to space constraints and sufficient air flow availability. In this method, the extended surfaces from the combustion chamber called 'fins' are used to cool the engine. Air flowing over and around the cylinder carried away engine heat through convective heat transfer. Fins are cast on the cylinder block and head to increase the heat transfer rate by increasing the convective surface area. The amount of engine heat to be dissipated through fins depends on some factors such as geometry, shape, size, material, wind speed, etc. Some numerical and experimental studies were done by the researchers to optimize these fin factors to enhance the engine heat transfer rate [3-8].

In experimental work, the elliptical fin was found to be more efficient in transferring engine heat over a circular fin [9]. The gap between two consecutive fins was concluded as the main parameter for engine heat dissipation rather than fin quantity in experimental work by Yosidha *et al.* [10]. When the ambient temperature is reduced to a very low value, the small engines are usually overcooled and the fin performance increases with an increase in ambient temperature [1, 11]. In another experimental work, when fin geometry was changed from regular rectangular to porous fins, it produced a 5.6% higher heat transfer rate [12]. The fins surface with holes or perforation fins reduces the materials use and also, increases the heat transfer rate from the engine. It was confirmed in a motor bike modelling work where the researcher found 5% to 6% increase in heat transfer rate for holed fins design [2, 13-14]. Due to the decrease in Nusselt number, it was found in numerical analysis that the permeable fins produced significant heat rate enhancement over the solid fins [15]. Aluminum fins produce a high heat transfer rate as compared to copper fins. But, the copper material is more stable at the lowest temperature [16].

Fin configuration is one of the important parameters which decides the heat transfer rate [17-19]. Using software simulation of an air-cooled engine, a higher heat transfer rate was achieved for perpendicular aligned circular shaped fins instead of rectangular shaped fins and fin materials [20]. In another simulation work using ANSYS software, the trapezium shape fins were produced 4% to 13% more heat transfer rate than the triangle, rectangle and circular fins [21]. Researchers have also found that convex and conical shaped fins are more efficient in heat transfer from the slam air-cooled engine [22-23].

The previous research works on air-cooled fin for the IC engine have demonstrated the enhanced heat transfer rate through varying on the fin parameters such as material, geometry and shape, surface area, perforation, etc. individually. Hence, to enhance the heat transfer rate to a maximum level, a combination of fin parameter variation can be adopted on the air-cooled fins. To achieve this objective in this work, we have combined the changes in the fin geometry, shape, and pitch of a 100 cc motor bike engine through 3D modelling SOLIDWORKS. Subsequently, its heat transfer performance was analysed and compared through ANSYS mechanical and FLUENT software.

2. Methodology

2.1. CAD model

The two-wheeler engine taken here for the study was a 100 cc capacity Honda Make motor cycle. The complete dimensions of air-cooling arrangement of this engine considered for CAD modelling is presented in Table 1. The original design of air-cooling was modified by changing different parameters of fins, such as shape, surface and pitch. The modified fin arrangement with the engine cylinder is also modelled as per the dimensions given in Table 1. The CAD modelled designs are presented in Figures 1 and 2.

2.2. ANSYS simulation

The CAD models were imported to ANSYS workbench for the heat transfer analysis with help of APDL 16.0. The ANSYS software is provided with simple meshing methods where node sizes can be varied according to need in the same part with a varying shape with internal temperature 300°C and convection coefficient of air. Theoretical methods were used in the ANSYS simulation of models without air flow. The surface area of the original and modified model was calculated as 0.13565436 m^2 and 0.135877.5 m^2 respectively. Whereas, the total heat flux over the area of the original and modified model was calculated as 6328.2 W/m^2 and 6579.4 W/m^2 respectively.

IOP Conf. Series: Materials Science and Engineering 1123 (2021) 012045 doi:10.1088/1757-899X/1123/1/012045

Table 1. Dimensions Details of CAD Models

Fin Parameter	Original Design	Modified Design
Number of fins	30	45
Pitch	7.5	3.5
Surface area	1350.60 cm^2	1821.18 cm^2
Volume	330.98 cm^3	373.05 cm^3
Number of holes	0	12/ fin
Fin end shape	Rectangular	Triangular
Extended surface	No	Trapezoidal
Area/ volume	4.1	4.9

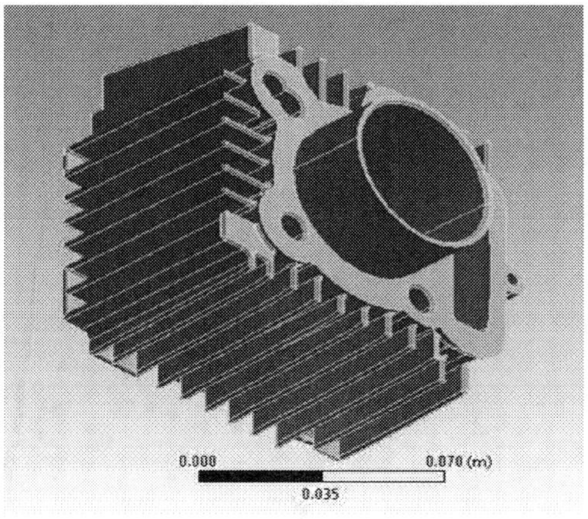

Figure 1. Original model of engine

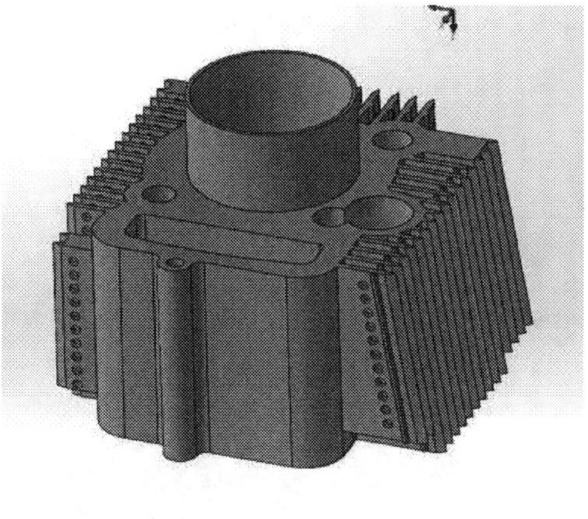

Figure 2. Modified model of engine

Method 1: Theoretical method for temperature distribution data obtained from ANSYS simulation.

Heat transfer from fins, $$Q = \sqrt{(hpKA} \times \theta \times \frac{[\tanh(ml) + (h/mK)]}{1 + ((h/mk) \times \tanh(ml))}$$

where, $\theta = (T_b - T_s)$

p = perimeter (m),

l = length (m),

A= surface area (m^2),

$m = \sqrt{(hp/KA)}$

h = convective heat transfer coefficient (W/m^2K), and

K = thermal conductivity (W/mK).

The temperature gradient for both models using the above mathematical relations is presented in Figures 3 and 4.

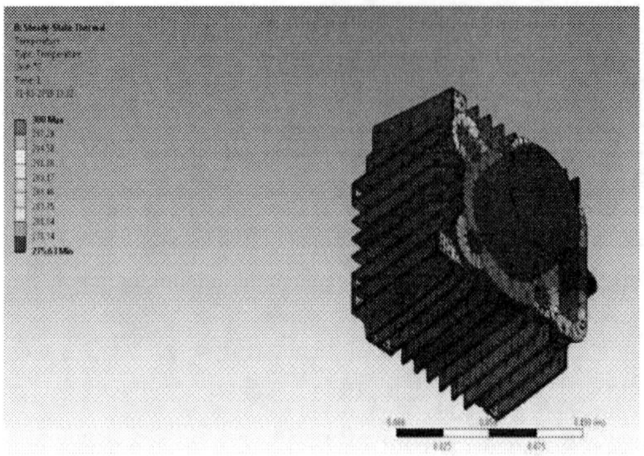

Figure 3. Temperature gradient of original model

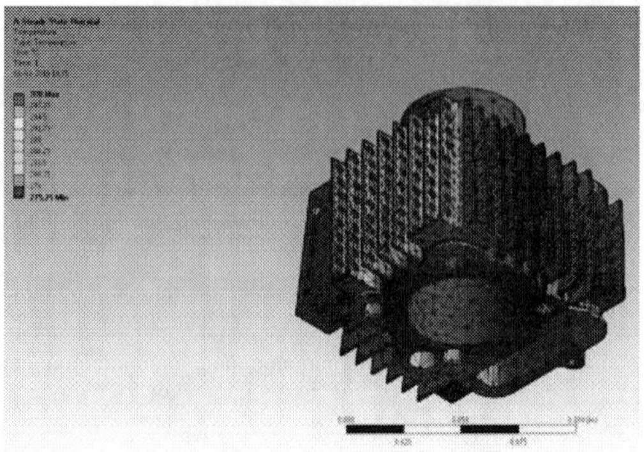

Figure 4. Temperature gradient of modified model

Method 2: Calculating total heat flux over the outer surface area of the fin model.

$$\text{Flux of F through S} = \iint_S (F \cdot n)dS$$

Where, F= heat flux (W/m^2), S= surface area (m^2), n = normal direction vector.

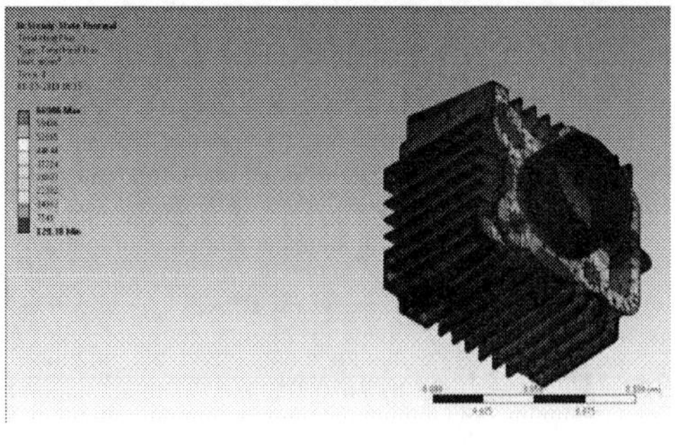

Figure 5. Heat flux gradient of the original model

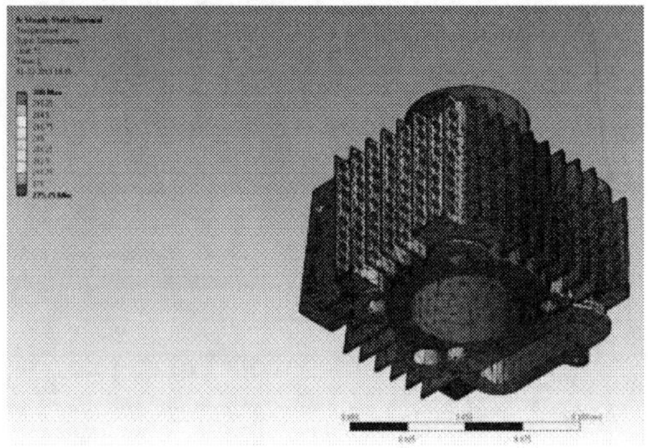

Figure 6. Heat flux gradient of the modified model

2.3. ANSYS Fluent analysis

The CAD model was imported into ANSYS Fluent as shown in Figure 7. Smart meshing is done on the model can be seen in Figure 8. Boundary conditions were applied to the model as shown in Figure 9. The inner engine surface temperature was assumed for this study to be constant at 300 °C. Air velocity was taken as 45 km/hr with 22 °C ambient temperature. Total heat flux was calculated over the outside surface area as shown in Figure 10.

Surface integration methods were used along the surface, unlike normal integration methods. Surface integrals were already present independent of each other but scattered over the given surface area. Heat integrals are vector quantities that can be surface integrated over a given surface area to calculate total heat flux over the required amount of area of the body of any shape. *The* air Mass flow rate was calculated near the fin surface area. The final FLUENT ANSYS reports both original and modified models were generated from the analysis and presented in Figures 11 and 12 respectively.

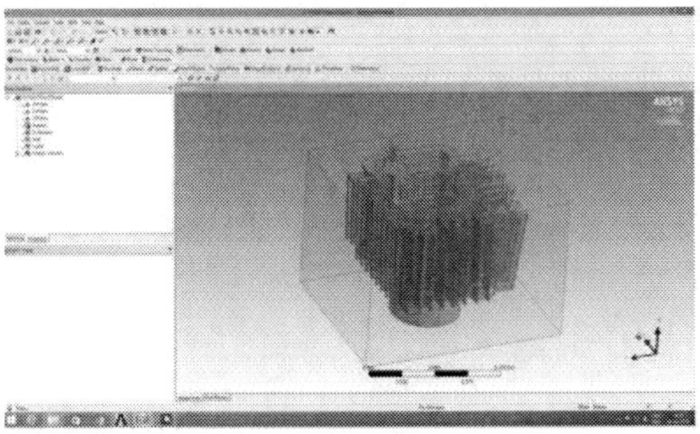

Figure 7. Enclosure in Fluent

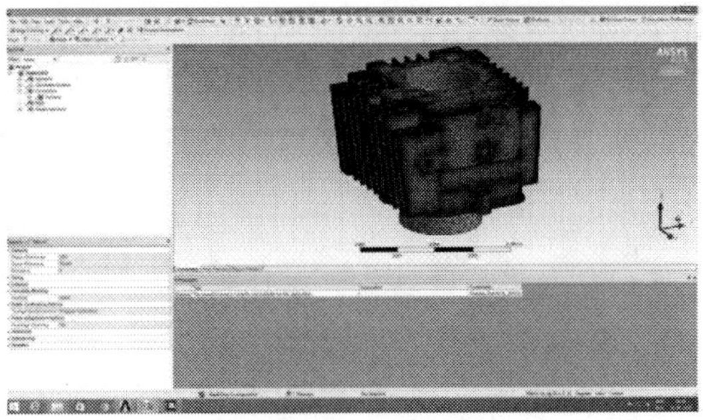

Figure 8. Meshing in Fluent

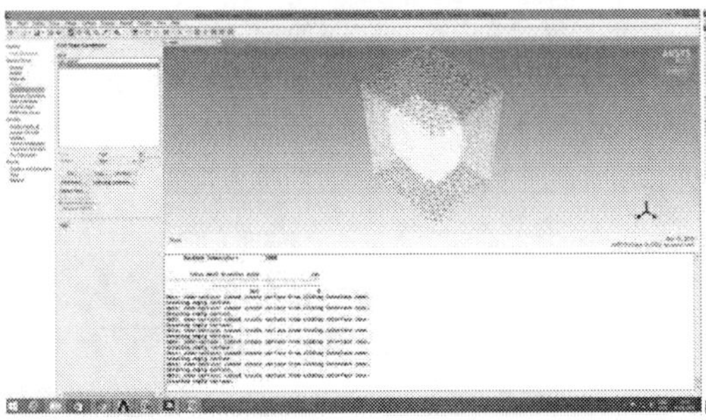

Figure 9. Boundary condition setup

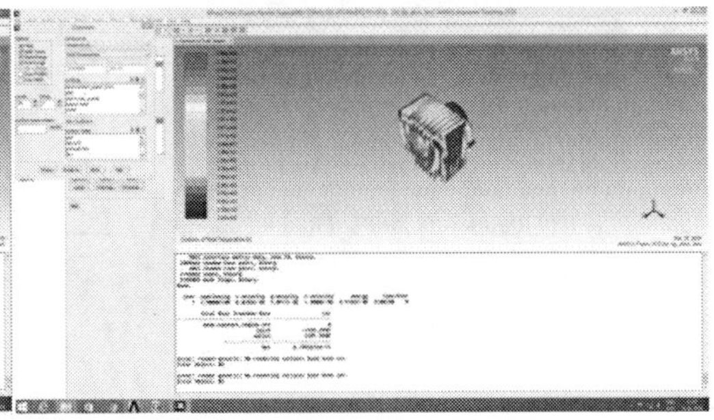

Figure 10. Heat flux model

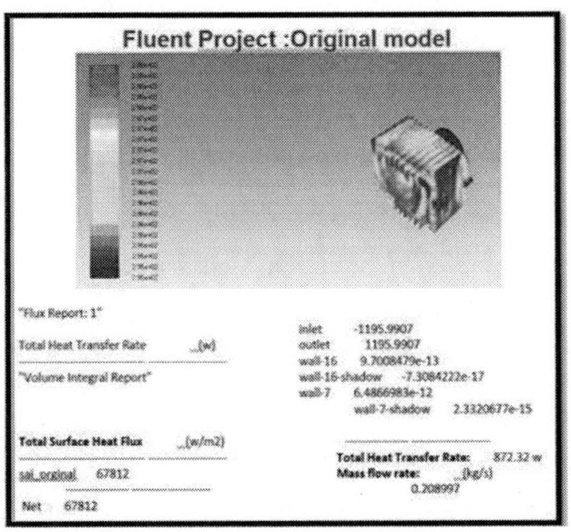

Figure 11. ANSYS report of original model

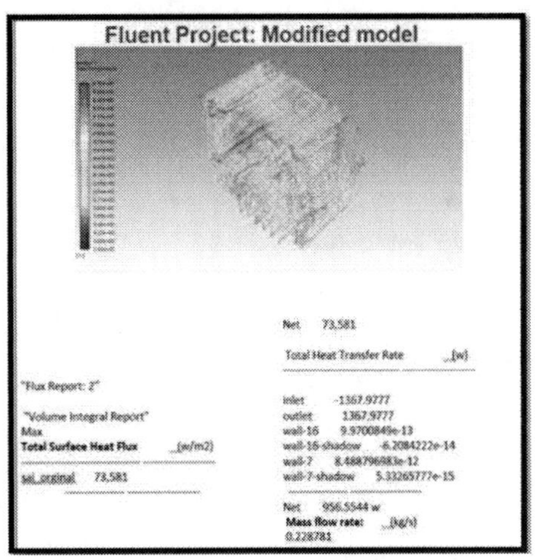

Figure 12. ANSYS report of modified model

3. Results and discussion

The complete results of the numerical analysis done on both original and modified engine models by using ANSYS Mechanical and Fluent software are presented in Table 2. Considering the surface area and total heat flux over the area of each model, the ANSYS Mechanical analysis showed that the amount of heat transfer from the original and modified engine model was 858.44 J and 894.32 J respectively. There is an increase of 4.18 % heat transfer from fins with the modified model when air flow over the fins was not taken into account.

When the simulation was completed using ANSYS Fluent it was possible to consider the engine in a moving state. Here, the air flow over the fins plays an important role in increasing the heat transfer rate. It was found that the heat transfer rate from the original model was 872.32 J increased to 956.54J with the modified model which is an increase of 8.51%. Using ANSYS Fluent analysis, it was found that the rise in air flow which causes a 4% rise in the coefficient of heat transfer is directly proportional to the heat transfer rate.

Table 2. Technical Results of various models

ANSYS Software	Model	Total heat flux (W/m^2)	Net heat transfer rate (J)	Mass air flow at fin wall (kg/s)
Mechanical	Original	67411	858.44	----
Mechanical	Modified	68121	894.32	----
Fluent	Original	67812	872.32	0.208997
Fluent	Modified	73581	956.54	0.228781

4. Conclusions

The following conclusion was made from the numerical investigation on heat transfer from air-cooled motor cycle engine through fins.

i. It is validated through the above research work that the use of more than one fin design modifications techniques helps to improve the comparative heat transfer rate than the original engine head.

ii. Net change in the heat transfer rate of 3% to 4% was observed in all previous research works when the individual modification method was carried out. But, after adopting combined fins modification techniques, we observed a net rise of 8.5% in heat transfer rate is observed which is significant.

iii. The reasons for the rise in heat transfer rate achieved in the current modified model due to the following parameters as given below.

- *Area:* Use of trapezoidal extensions on the fin body increases the surface area for the same available volume.
- *Shape:* Spike shapes always enhances the heat transfer rate. So, the end shape of the fin was designed as triangular shapes instead of blunt faces.
- *Pitch:* Reducing the pitch to 3.5 mm, we were able to include more number of fins for the same surface area which directly increased the heat transfer rate
- *Holes:* A rise in air flow along with an increase in surface area was achieved when holes were made in the fins.
- *Air velocity:* Velocity of air and convection coefficient rise and hence, higher heat transfer rate due to the turbulence in air flow was created due to less fin pitch and holes.

Acknowledgments

The authors wish to thank and acknowledge their B.Tech. and M.Tech students, Vellore Institute of Technology Vellore for their assistance towards the numerical analysis.

References

[1] Pulkit A, Mayur S and Srinivasan P *Proceedings of the World Congress on Engineering* 6-8 July London UK.

[2] Thornhill D and May A 1999 *SAE Paper 1999-01-3307.*

[3] Rathod HD, Modi AJ and Rathod PP *International Journal of Mechanical Engineering and Technology (IJMET)* **4(2)** 328-333.

[4] Raja Rao PN and Vishnu Vardhan T 2013 *International Journal of Engineering and Technology (IJERT)* **2(8)** 404-412.

[5] Rustum IM and Soliman HM 1988 *Journal of Heat Transfer* **110(2)** 366-372.

[6] Sparrow EM and Bahrami PA 2004 *International Journal of Heat and Mass Transfer* **23(11)** 1555–1560.

[7] Myhren JA and Holmberg S 2011*International Journal of Thermal Sciences* **50(2)** 115-123.

[8] Petroski J 2013 *ASME 2013 International Technical Conference and Exhibition on Packaging and Integration of Electronic and Photonic Microsystems* 16-18 July Burlingame California USA.

[9] Nagarani N and Mayilsamy K 2010 *International Journal of Engineering Science and Technology* **2(7)** 2839-2845.

[10] Yoshida M, Ishihara S, Murakami Y, Nakashima K and Yamamoto M 2006 *JSME International Journal Series B- Fluids and Thermal Engineering* **49(3)** 869-875.

[11] Kundu B and Bhanja D 2010 *International Journal of Heat and Mass Transfer* **53(1-3)** 254-267.

[12] Pise A and Umesh A 2010 *International Journal of Mechanical Engineering and Technology (IJMET)* **1** 238-247.

[13] Kumbhar DG, Sane DNK and Chavan ST 2009 *Proceedings of the International Conference on Advances in Mechanical Engineering* 3-5 August Gujarat India.

[14] Babu G and Lavakumar M 2013 *IOSR Journal of Mechanical and Civil Engineering (IOSR-JMCE)* **7(4)** 24-29.

[15] Abu Hijleh AKB 2003 *ASME Journal of Heat Transfer* **125(5)** 804-811.

[16] Matkar M and Ravanan PM 2011 *International Conference on Operations and Quantitative Management* 28-30 June Nasik India.

[17] Fernando Illan and Alarcon M 2010 *Journal of Thermal Science and Engineering Applications* **2(4)** 041002 (7 pages).

[18] Sood Pakdee D, Behnia M and Copeland DW 2001*The International Journal of Microcircuits and Electronic Packaging* **24(1)** 77-83.

[19] Wange S and Metkar R 2013 *International Journal of Engineering and Innovative Technology* **2(11)** 328-333.

[20] Gupta D and Wankhade SR 2015 *International Journal on Mechanical Engineering and Robotics (IJMER)* **3(2)** 1-4.

[21] Sonkar A, Singh Rajput I, Dhruw J and Sahu KK 2017 *International Journal of Engineering Research & Technology (IJERT)* **6(4)** 324-331.

[22] Sagar P, Teotia P, Sahlot A and Thakur HC 2017 *Materials Today: Proceedings* **4(8)** 8558-8564.

[23] Jagdale M, Nitnaware P and Bharati R 2018 *International Research Journal of Engineering and Technology (IRJET)* **5(5)** 2560-2562.

Design and value engineering of a high temperature high pressure multifunctional compression moulding unit to manufacture zero defect parts

K V Gopinath [1], K Padmanabhan [2]
SMEC, VIT, Vellore, India, Gopinath.kv@vit.ac.in
SMEC, VIT, Vellore, India, Padmanabhan.k@vit.ac.in

Abstract. Hot compression moulding is one of the better techniques employed to make zero defect products. This paper mainly discusses about the conversion of a conventional hydraulic press into a high pressure, high temperature compression moulding unit at a low cost using local state of the art methodology. A normal press used for blanking, drawing and various metal forming operations is modified here by addition of top and bottom platens with heating coils controlled by PID controllers to operate up to 500°c within which high Temperature Plastics, Aluminium, Tin, Lead, Magnesium etc. can be hot compacted with foil films and reinforcements or can be hot forged with matching dies. In this paper, hot compression moulding as a sound technique for manufacturing zero defect thermoplastic and advanced thermoplastic based composites is demonstrated and emphasized based on a defect evaluation of the moulded parts using an ultrasound scanning technique.

Keywords: *High Pressure Hot Compression Moulding, Thermoplastics, Composites, Zero defect, C Scan, Value Engineering.*

1. Introduction

The potential of high performance materials can be improved to maximum utility by choosing suitable manufacturing techniques based on their applications. Compression moulding is one of the most common manufacturing techniques used for products made of plastics, metals, ceramics and composites. The quality of the product depends on various parameters of moulding such as temperature, product size, pressure, material properties, cycle time and porosity levels. Compression moulding has been in use from making swords in ancient days till recently to make automobile and aircraft parts and in many areas where advanced materials or high performance parts are fabricated and used [1]. With the advent of material processing science and special moulding methods [2][3], enhanced mechanical properties can be achieved which in turn are necessary for novel high end applications.

A hydraulic press is a machine used to shape, mould, straighten and crush materials by applying compressive force through platens or a bed. It works on the basis of Pascal's law which states that force acting on confined fluid at any point is uniformly transmitted throughout the fluid. Thus pressure created by a small actuated effort in a confined area on a small piston is transferred through the hydraulic fluid to apply greater force on the larger piston [4]. This compressive force is primarily utilized in hydraulic pressing for metal forming operations, plastics moulding, forging, blanking, shaping, deep drawing and shearing.

Hence, the focus of the present study is about upgradation of an existing conventional hydraulic press into a high pressure high temperature compression moulding unit at an affordable cost. The importance of hot and high pressure compaction to make superior quality zero defect products are further established through non-destructive ultrasound testing of the fabricated laminates. The significance of this work is the cost effective and value engineering [5] based manner in which the parameters required to fabricate zero defect products are achieved. A brief statement of the investment costs for the design and development of such a new platen press and its modification is also provided at the end.

2. Experimental details of the platen press

A four column pillar hydraulic press in the lab consists of two cylinders, piston pumps, relief valves, a control unit, oil reservoir, ejector, pressure gauges and two flat normal dies. Since heating facility is not available in the existing dies and platens, it can only be used for shaping, powder compacting and other simple pressing processes for which only a compressive force is required between the platens with no rise in temperature.

Further, the existing press is only intended for use in fabrication processes like blanking and powder compaction by applying high speed compressive forces for a short period of time and achieve mass production through short cycle times. Table 1 and Figure 1 provide the details of the existing unmodified press.

Table 1. Existing Hydraulic Press Specifications

Type of Press	Hydraulic four pillar compacting press
Capacity	15 Tons
Stroke	300 mm
Day light	300 mm
Slide size	400 mm × 400 mm
Table size	400 mm × 400 mm
Power	5 HP
Supplier	Press Form Industries, Chennai, India.

IOP Conf. Series: Materials Science and Engineering 1123 (2021) 012046 doi:10.1088/1757-899X/1123/1/012046

Figure 1. Existing Press without heated platens and high pressure check valve

3. Upgradation of the existing press

High pressure hot compaction method is very effective for fabrication of certain advanced materials and parts with zero defect. Hot compaction of certain materials for instance polymers, plastics and composites requires precise temperature and pressure settings to be applied for forming. The objective of this work is to integrate high pressures with high temperatures and achieve a controlled synergic effect on the moulding material and improve its structural integrity and quality. The formability of polymers owing to temperature variations differ from one another. With the advent of high temperature plastics that require forming temperatures in excess of 400 ° C a precise control of temperature and pressure are priority. Hence a precise control of temperature over and pressure over a wide range is necessary. At the same time caution is to be taken to avoid very high temperature at which an otherwise normal press would suffer thermal loads and a rise in hydraulic fluid temperature higher than its normal working level. In the present investigation, 53° C which is the maximum permissible temperature for the typical hydraulic oil used, would affect the functionality due to evaporation and boiling in a confined tube or piston that would alter the partial pressures and affect the control over the pressure of forming.

Based on the above factors the platen materials and heater coils to be used in platens are selected. Hardened mild steel integrated with Kanthal heating coil is selected and installed as the heated platens. The design requirements and fabrication work were provided to an external small scale casting industry "Vellore Foundries" and `ready to fix heated platens' were procured from them.

In order to control the temperature precisely a PID – Proportional Integral Differential controller which is a closed loop mechanism that employs feedback of a continuously calculated difference between a desired set point and measured process variables is applied. Here, the input values and corrections based on Proportional, Integral and Differential terms maintain the required temperature precisely. Though Kanthal heating element can be used to heat up to 1000°C, the maximum temperature permitted here is restricted to 450°C beyond which hardened mild steel platens would not be suitable for forming due to hot deformation related issues.

The next challenge was to hold a high compression pressure of the heated platens for a required duration of time contrary to the existing design. Normally the ram up and ram down movements are hydraulically actuated through the oil pumped by the motor in the cylinder manifold block. With the existing design, for prolonged compression the ram down switch in the control unit needs to be operated manually keeping the motor running continuously as long as a high compression up to 120 Bars is needed. This is a tedious process when the operating pressures are higher as the motor coil may heat up and incur damages during sustained operations.

Hence it was proposed to hold the pressure created in the main cylinder by installing a piloted check valve which would arrest the hydraulic fluid flow from the main cylinder unless a ram up switch is pressed in the control unit. So, the ram down condition with high pressure acting on a platen can be maintained for a sustained duration and the part material being moulded by the platens will be under a precise compressive pressure at the desired temperature. A piloted check valve was installed in the hydraulic press. It was attached to the manifold block of the main cylinder. Figures 2 and 3 show the technical details of the manifold block and the installed check valve. The pressure lines shown in the Figure 4 allow the hydraulically actuated oil flow to the main cylinder and the relief valve releases the oil back to the reservoir tank. Both the pressure line and relief valve were connected through the piloted check valve. During the ram down operation the hydraulic oil from the reservoir is pumped into the main cylinder through the check valve by the motor and then it is turned off. At the same time the relief valve is closed thus maintaining the pressure as long as the ram up switch is ON. Once the compression moulding is completed the motor can be turned on again to allow a ram up and release the moulding pressure which would allow the oil to flow back to the reservoir through the piloted check valve. The maximum capacity of the press is 120 bars. The pressure can be increased gradually or through stages from the atmospheric value independent of the temperature settings. At the room temperature loading the pressure slightly drops from a high 120 bars by a few bars because the oil in the main cylinder is maintained at room temperature which would be low in partial pressure. This compels the use of the ram up switch manually to maintain the pressure constant. But at higher temperatures in the main cylinder this issue is not observed as the applied high pressure is stable due to the generation of partial pressure of the gas out of evaporation of oil. However, the loss of oil due to possible leaks and continuous evaporation are to be topped up periodically. The oil temperature gauge and the level indicators in the main tank assist in the monitoring and maintenance of these requirements. Thus, the existing press was upgraded for a high pressure high temperature compression moulding operation that facilitates multifaceted solutions in the manufacturing of zero defect parts.

Figure 2. The Manifold block In Blue

Figure 3. Check valve to maintain moulding pressure.

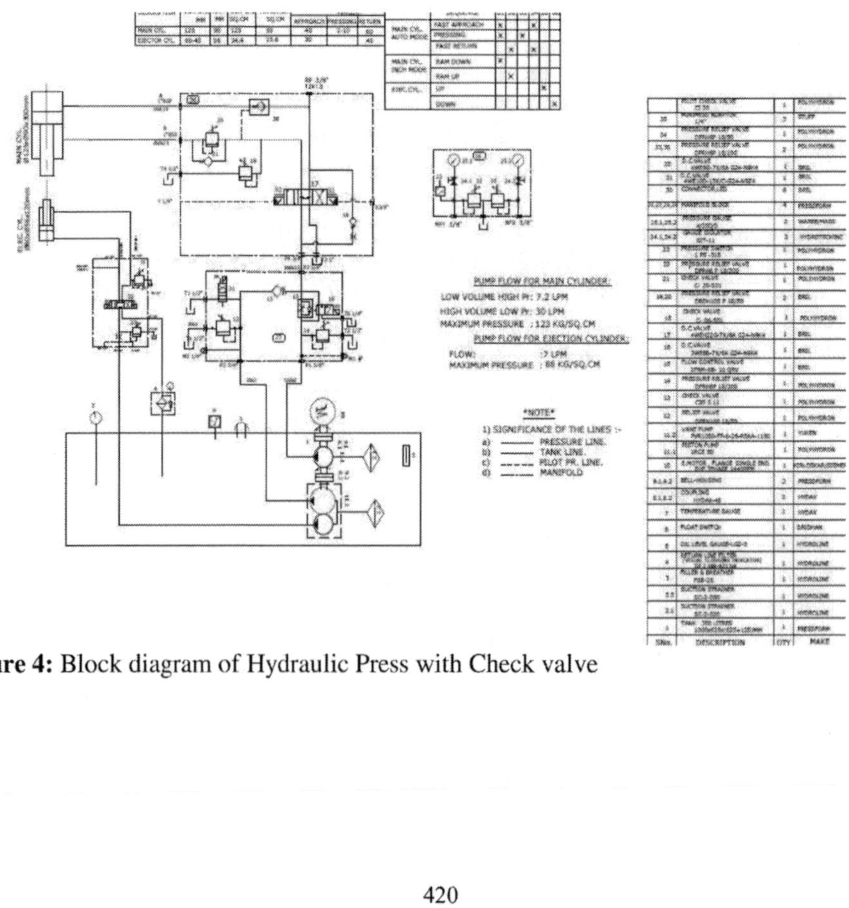

Figure 4: Block diagram of Hydraulic Press with Check valve

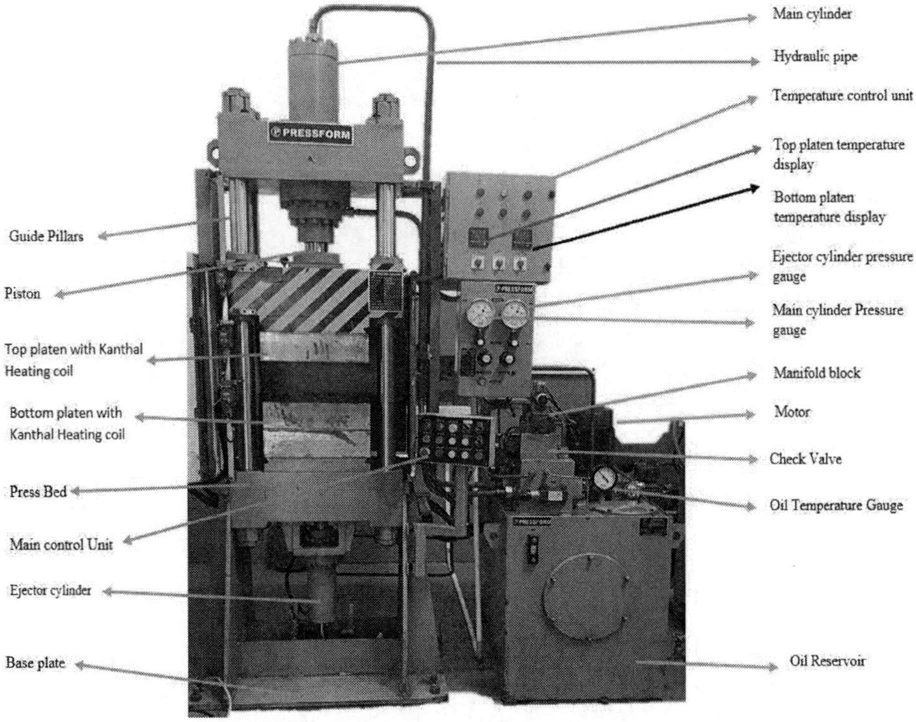

Figure 5. Modified Press with piloted check valve

4. Fabrication of zero defect composite laminates

After successful installation of heated platens and check valve, the press was operated and checked for any leakages with and without loading. Leakage if any was successfully stopped through detection and repair. It was decided to fabricate a flat laminated composite plate made of fibre reinforced thermoplastic material of thickness 3mm. Kevlar -29 bidirectional fabric and polyethylene films were chosen as the moulding materials. The layup sequence followed was that of two layers of polyethylene film and one layer of Kevlar fabric in a sequence till a 3 mm thickness was moulded [6]. Initially the heater was turned on for an hour so as to reach the desired temperature of hot compaction of polyethylene [7][7]. Then readily cut, oriented and stacked Kevlar fabric and polyethylene sheets were placed between the platens and release films. A pressure of 100 bars was applied for 30 minutes by turning on the ram down switch in the control box and maintaining the pressure through the check valve operation. Then the sample was left to cool down in the platen at a reduced pressure until it reached room temperature. The fabricated sample was trimmed at the edges and after visual inspection it was checked for defects through an ultrasound non-destructive scanning technique applied to laminated composites [8].

5. Results and discussion

The sample was inspected using Ultrasonic C scan. The scan parameters are given in Table 2 as follows;

Table 2. C-Scan parameters

Scan parameters	
Probe type	Pulse echo[9]
Pulse type	Bipolar
Sampling frequency(MHz)	25
Probe frequency(MHz)	2.3

The sound attenuation C- scan images obtained for the various laminate layers at specified depths are illustrated as follows;

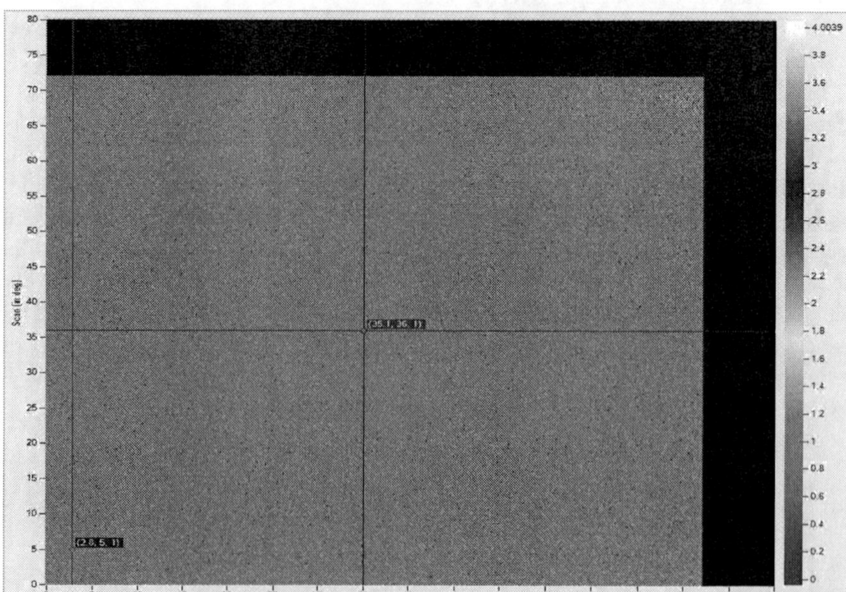

Figure 6. At a depth of 1.2mm for top.

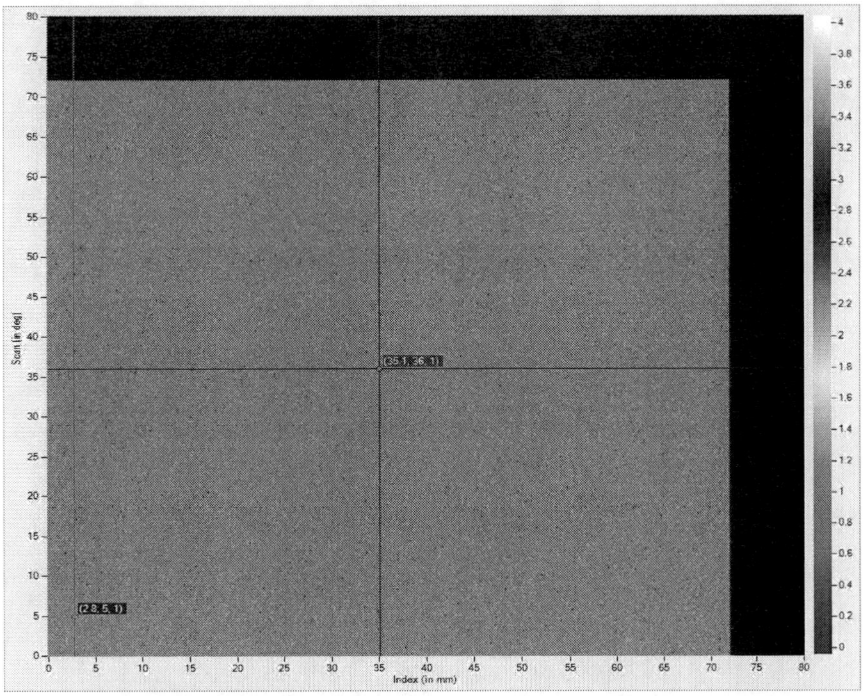

Figure 7. At a depth of 1.8mm from top

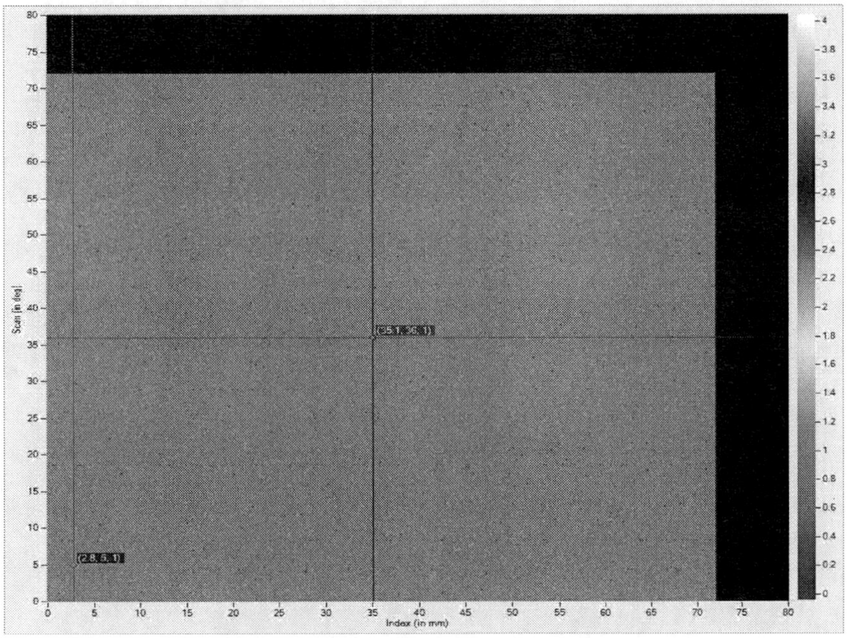

Figure 8. At a depth of 2.0mm from top

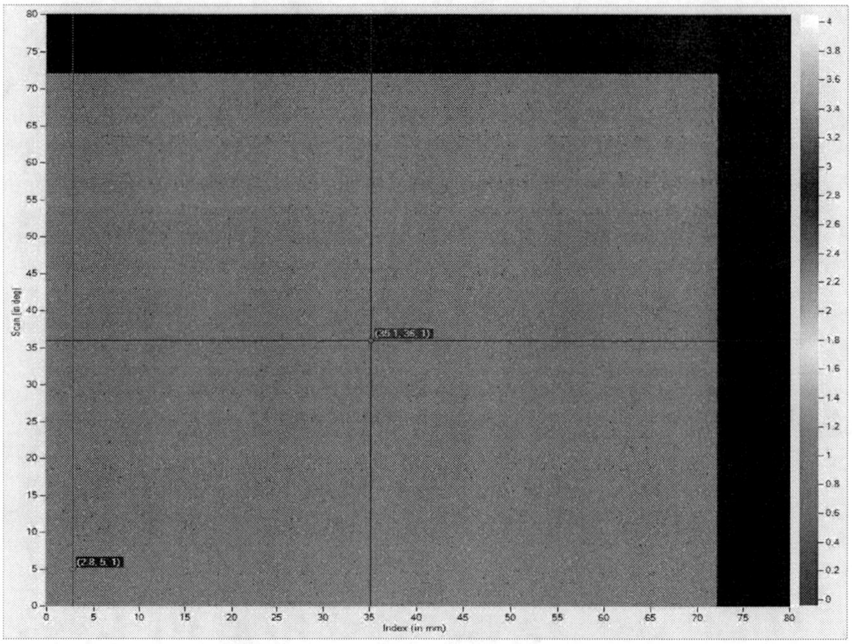
Figure 9. At a depth of 2.5mm from the top

Figure 6 to 9 show a depth profile of the sound attenuation plots at specified depths of the fabricated composite laminate. The scanned images are obtained by through the thickness scanning [10] of the specimen immersed in water which is used as a coupling fluid. The transducer runs perpendicular to the surface of the specimen. The ultrasonic sound reflected from the specimen layers is captured by the transducer. Based on the amplitude of the received sound energy the data is presented by an Automatic Data Acquisition (ADA) system integrated with the computer. The scan images at different layer depths show uniformity in the attenuation patterns. It is evident from the scanning results that the sample is virtually defect free through the layers and the attenuation patterns are acceptable and very minimal as evidenced through excellent bonding between the layers. No delaminations were observed that otherwise would have produced a characteristic attenuation pattern as seen in the blue-green and blue regions of the colour bar in the C Scan images. The colour bar on the right side of scan images is the scale for determining the defects in the specimen. Bright red colour at the bottom most part of the colour bar with zero mark is an indication of smooth transit of ultrasonic sound waves. In the bar, the colour changes from bright red to normal red, yellow, green, thick blue, light blue and finally to white depending on debonding , delaminations and the associated sound attenuation patterns. Thus, as seen from the C scan images this machine qualifies for use in the high pressure, high temperature moulding of ballistic grade parts and gadgets. As mentioned earlier, this is a multifunctional equipment which can also manufacture hot forged, stamped, blanked, diffusion bonded, moulded and hot compacted parts normally not manufactured through conventional compression moulding machines.

6. Capital investment and costing

The essence of this project is the cost savings at which zero defect products are achieved. A brief note on the capital investment costs will give clarity. The cost of existing unmodified press is about 15 lakhs and the cost of upgradation is about 3.5 lakhs. Hence, the total cost incurred for an upgraded high pressure high temperature compression moulding press is about 18.5 lakhs. The cost of a new custom made high pressure press with an inbuilt check valve and heated platen unit is about 30 lakhs for the temperature and pressure ranges specified here, in the market. Currently, the available compression moulding presses in the market can fabricate parts only up to 300° C and 25 bars of pressure in the 10 tons capacity range. Thus at least about 65 % cost savings were achieved in the present upgradation exercise, comparatively. As this equipment can be more than substitute for a composite fabrication autoclave with similar capabilities, a costing comparison reveals that such an autoclave would cost a minimum of Rs.70 Lakhs in the market. A technology transfer and commercial sale only would increase the costs of the equipment upgraded here which is not the intention of the present exercise. The margin of returns for a superior quality of products is hence very high and the goals of value engineering stand achieved through industry partnerships

.

7. High end applications

Fabrication of products by the hot compaction method using upgraded hydraulic press with high temperature and high pressure control is now a proven method for producing superior quality and zero defect composite products. A wide variety of materials including fiber-metal composites [11], metal-metal laminates[11], ballistic materials, thermoplastics [12], several thermosetting plastics and high temperature advanced composites can be processed using this heated platen press for novel applications. It is to be emphasised that the above said parts cannot be fabricated or manufactured even by advanced additive manufacturing methods using the existing state of the art knowhow in deposition as compaction and consolidation are still issues.

Acknowledgement

We wish to acknowledge "Vellore foundries and Thambanu Castings" -a small scale casting industry for installing the heated platens and PID temperature controllers and Press Form Industries, Chennai for installing the check valve for pressure control as per our design requirements.

References

[1] Sumaila M and Akii Ibhadode A O 2011 Design and Manufacture of a 30-ton Hydraulic Press *Au J.T* **14** 196–200.

[2] Murugan R, Ramesh R and Padmanabhan K 2014 Investigation on Static and Dynamic Mechanical Properties of Epoxy Based Woven Fabric Glass/Carbon Hybrid Composite Laminates *Procedia Engineering* **97** 459-468.

[3] Uday C, Varma C S T, Varma B N S K, Ramya M and Padmanabhan K 2014 The Influence of Rigid Foam Density on the Flexural Properties of Glass Fabric/Epoxy-Polyurethane Foam Sandwich Composites *Inlt. Jl. Chem. Tech Res* **6** Issue (6) 3314-17.

[4] Phelan RM 2019 Hydraulic press. *In: AccessScience.* McGraw-Hill Education

[5] Anil Kumar M 2014 *Value Engineering Mastermind* (Sage Publications, International).

[6] Murugan R., Ramesh R and Padmanabhan K 2020 Influence of Stacking Sequence on Mechanical Properties and Vibration Characteristics of Glass/Carbon Hybrid Plates with Different Fabric Areal Densities. *Structural Integrity Assessment. Lecture Notes in Mechanical Engineering.* Springer, Singapore.

[7] Hine P and Ward Iani. 2004 Hot compaction of woven poly (ethylene terephthalate) multifilaments. *Journal of Applied Polymer Science*. **91**. 2223

[8] Sharan Chandran M 2020 Preparation and Characterization of Self Reinforced Fibre Polymer Composites with Emphasis on the Fibre/Matrix Interface, PhD Thesis, VIT University, Vellore, India.

[9] Padmanabhan K, 1995 Failure Analysis of Glass, Carbon or Kevlar Fibre Reinforced Epoxy Based Composites in Static Loading Conditions, PhD Thesis, IISc, Bangalore, India.

[10] Padmanabhan K and Kishore 1995 Failure behaviour of carbon fibre/epoxy composites in pin-ended buckling and bending tests *Composites* **26** Issue (3) 201-206.

[11] Kumaran D, Rajadurai A and Padmanabhan K 2020 Processing and evaluation of thermal-sprayed high-performance polymer fiber reinforced Zn metal matrix composites. *J Comp Mater* 54 Issue (3), 3171-3180.

[12] Matheswaran, M., Padmanabhan K and Kishore 1995 Static and impact behaviour of thermoplastic modified glass fabric/epoxy composites. *J Mater Sci Lett* **14,** 951–954.

Speech Recognition based Industrial Cloud Robot for Service-Oriented Sustainable Manufacturing

Kalaiarassan G[1], Franklin Prashanth C[1], Prakash M[1] and Shreya Phirke[1]

[1]Vellore Institute of Technology, Katpadi, Vellore, Tamil Nadu, India

E-mail: kalaiarassan.g@vit.ac.in

Abstract. Industrial Cloud Robotics is an amalgamation of cloud computing and an industrial robot to establish a service-oriented manufacturing system. Teleoperation of an industrial robot – the ability to control a robot from a remote location, is facilitated through this system. In this paper, a framework is proposed by utilizing speech recognition and industrial cloud robots for the applications of sustainable manufacturing. We have used Google's speech recognition services to control the robot manipulator. We have employed Android Speech API in a custom android application that receives the speech signal, transcribes, and forwards it to a server. The application can be viewed or accessed by any host computer, which primarily serves as a user. The monitoring unit and the data is fetched by the Robot Speech Interface unit via the internet and web sockets. The interface triggers the required action of the robot through the relay board actuation of the digital input facility of the robot. Through this work, it is realized that, despite the disturbances and noise interferences, speed and reliability are not compromised.

1. Introduction

ISO 8372:2012 defines Industrial Robots as "an automatically controlled, reprogrammable, multipurpose manipulator, programmable in three or more axes, which can be either fixed in place or mobile for use in industrial automation applications" [13]. An archetypal robot manipulator is generally mounted on a fixed base and is connected to links that move in various directions about its axes, allowing the robot to work. The end-effector can be selected and customized based on the application of the robot. Ever since the discovery of the first industrial robot in 1954, industrial robots have found use in applications such as welding, painting, assembling, palletizing, pick and place, product testing, and more.

The demand and use of robots have dramatically burgeoned in the last few decades. Robots are being used for various applications that range from medical to manufacturing, to improve productivity and product quality. Due to the surge in demand, Industrial Robots are designed to make them user friendly in terms of the control features, decentralized accessibility aspects (remote access), and flexibility of operations. The rapid shift from manual labor to automation has given rise to a new category of robots, the Intelligent Robots. Intelligent Robots perform several complex robotic tasks that involve perceiving and reasoning the environmental factors by making use of sensor technology and control algorithms. Human-Machine Interaction (HMI) is a major feature that promotes effortless and safe interaction experience with robots. Machine Learning, HMI, and Cyber-Physical Systems are increasingly incorporated in industrial robots as they facilitate ease of functionality, flexibility, accessibility, and other control features.

In this paper, the domain of teleoperation is explored for convenient and sustainable manufacturing. Teleoperation is the ability to control a robot from a remote location. It is useful in addressing the issues related to inspection, programming, and servicing of the robot. The remote-control signals are sent through a wire via a local wireless system, over the internet. Teleoperated robots currently are extensively used in the field of medical treatment and surgeries, where the instructions given by the doctor are interpreted by the master and slave robots who treat the patients. Furthermore, these teleoperated robots are also used in industries that possess a threat to human health such as highly toxic areas and nuclear reactors.

Attributing to the convenience of operation that it offers, there have been expeditious advancements in the field of speech recognition and transcription by major Artificial Intelligence researchers, developers, and providers. Natural Language Recognitions of over 80 languages including its most common dialects is now possible. This inspires a new field of research and provides a broader scope for controlling various machines by implementing speech recognition technology. Moreover, the use of voice as input to control various devices and processes has been extensively developed as an effective tool for the human-machine interface (HMI). Siri, Cortana, Google Assistant are the pioneers of the field and have reached the very hands of the masses. Especially, Google's 'Google Home' and Amazon's 'Echo' have become part of many households. These systems offer AI integrated Natural Language Recognition features that understand the commands and respond using intelligent reasoning and algorithms. This paper attempts an integration of two features namely, IoT system and voice or speech control in the domain of robotics.

2. Overview of related works

L. Z. L. Zhizeng et.al [1] in 2004 successfully developed a speech recognition system to be applied for robots in their research. But the speech recognition was limited to only 50 words maximum. Commands which may be considered basic now, like the 'left', 'right', 'front' and 'back' were incorporated in the robot system and tested for its efficiency and reliability of the system. They implemented the DTW method of speech recognition. The time varying digital filter represented the shape of vocal cords and the Gain factor represented intensity of the voice. Endpoint-detect, signal-strengthen, windowing, framing of the signal were the signal pre-processing steps executed before feature extraction and DTW. The reciprocal spectrum coefficient of linear prediction co-efficient was used for efficient in the recognition of isolated word.

B. M. Chinnathurai et.al [10] have worked on building a robot for the sorting work, in a waste segregation process, in order to aid and amplify the process of recycling. The robot system had a vision system which was programmed using a Raspberry pi on Linux OS for object detection. Once the object was detected the robot collected the required objects.

S. N. A. Mohamad et.al [8] developed word semantic models that signified the relation between semantic databases that represented the meaning of a word with the registered word from the speech recognizer, using MATLAB. The noise signals were filtered using the MFCC algorithm by extracting the sequence of features for every short-time frame of the input signal.

3. Problem Definition

Increased manufacturing requirements and technological advancements have led to the expansion of the Industrial Robots industry. It is estimated that the market size Industrial Robot industry will be around 24500 Million USD by the end of 2025, with a CAGR of 10.2% between 2019 to 2025. Industrial robots are being deployed in automotive as well as several non-automotive sectors. While installing a robot may increase productivity by reducing the cycle time, it manifests several financial and logistical challenges like a large upfront investment, employee skillset training, implementation of safety measures, among others. Furthermore, the scant number of expert robot programmers or service engineers in this field can lead to delayed response time against service requests, high service charges, increased downtime, and accumulation of work that affects the quality of the product.

Moreover, every industrial robot is copyrighted by their respective manufacturers and provide little or no scope for customization. Additional financial investments have to be made to customize the robot as per the requirement. Different types of equipment have to be purchased from the Original Equipment Manufacturer (OEM) for every customization required. The functionality of the robot is restricted due to factors such as controller limitation: manufacturers have dedicated controllers for each model, and they are non-interchangeable. The compatibility of various features is pre-fixed within the controller. For instance, if an industrial robot does not facilitate speech recognition if the controller has no provisions for that feature. Additionally, if speech-recognition is required, a new robot has to be purchased.

Existing HMI techniques like touch and gesture require higher human effort. All other HMIs require an embedded system, usually wired to the robot or sometimes are wireless till a small range. However, provisions for distant remote control and command of the robot are not available

4. Methodology

The proposed methodology for the project is given below:

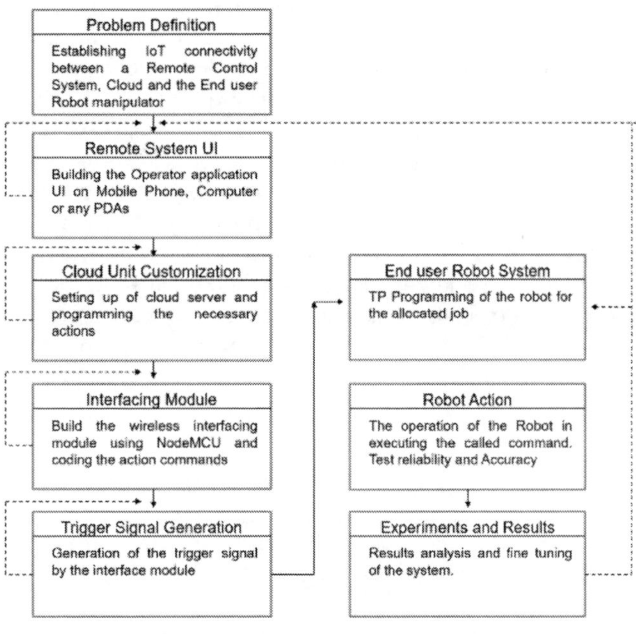

Figure 1. Methodology of proposed work

5. The System

5.1. System Layout

The basic system layout is as given in the below figure 2. Here, the major three components are the Input Unit, which in our project is an android mobile phone, the intermediate cloud and the robot system. The robot in return is continuously monitored by an IP camera system which again transfers the data to the cloud.

IOP Conf. Series: Materials Science and Engineering 1123 (2021) 012047 doi:10.1088/1757-899X/1123/1/012047

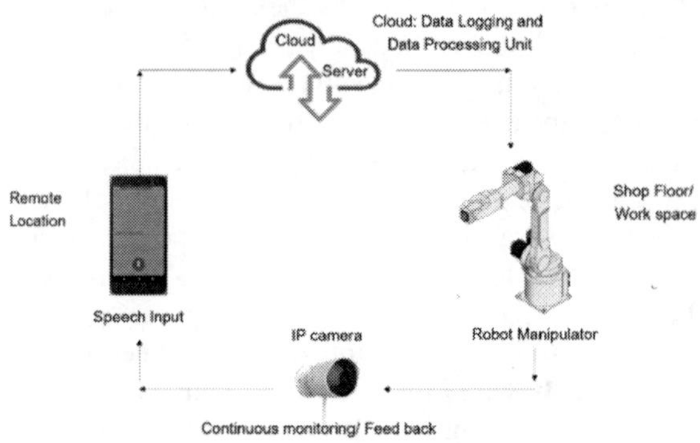

Figure 2. The System Layout

5.2. System Framework

The remote-control unit is the first subsystem of the framework, it is accessed and used by the programmer or operator of the robot service company. The primary element of the system is internet-enabled applications compatible with devices like mobile phones, computers, or any PDAs. For this particular project, we focus on the mobile device as our preferred mode of the user interface for the system. An android application is developed for the same, which is powered by the Android Speech recognition API. This application acts as our core engine for speech recognition since we focus on speech as our desired mode of input with other methods like text and UI commands available.

The cloud is a remote server that receives the data emitted from the mobile phone and processes it for keyword detection, which specifies the nature of the job. (like pick, take, write, move, etc). The cloud is programmed to identify the keywords and separate them from the objects in the command. The cloud passes the command or data to the robot via web sockets. The system is monitored by a camera that uploads the data to the cloud. This data viewed by the operator to verify the robot's performance according to his commands.

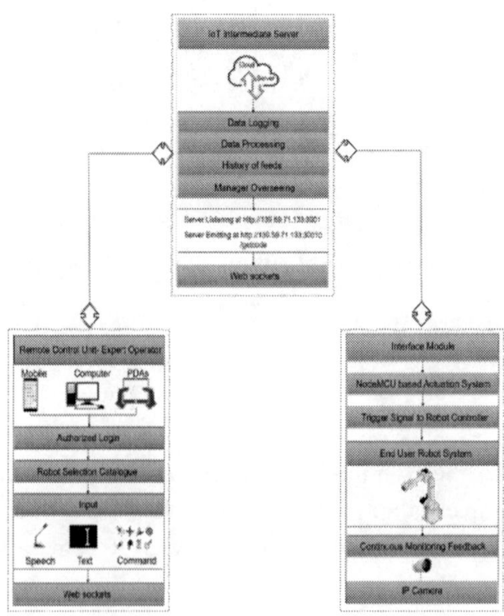

Figure 3. Proposed System Framework

6. The Interface Module

The interface module is constructed using NodeMCU as the base brain. The microcontroller is programmed to receive the data signals from the server. A relay board is actuated by the triggering signals from the microcontroller. Relay is used as the Robot takes input as only 24V through its digital input ports. Hence, the MCU triggers the relays to give the input to the robot. The interface setup is as shown in Figure. 4. This setup is fixed on the robot's peripheral mount head on J3.

Figure 4. The interface module

7. Experimentation and Result

The system was subjected to various tests below to check the performance and the reliability against different variables.

7.1. Environmental Conditions

The working environment is the first test, where the system was subjected to ideal quiet and noisy conditions. 10 samples were taken in a trial for each condition and the observation in as given in Table

Table 1. Reliability for speech recognition at ideal and noisy conditions

Conditions		Trial 1		Trial 2	
		Number of correct detections	%Correct detection	Number of correct detections	%Correct detection
Normal inbuilt phone mic	Ideal condition	18	90	19	95
	With noise disturbances	15	75	16	80
Dedicated mic (headsets)	Ideal condition	19	95	20	100
	With noise disturbances	17	85	18	90

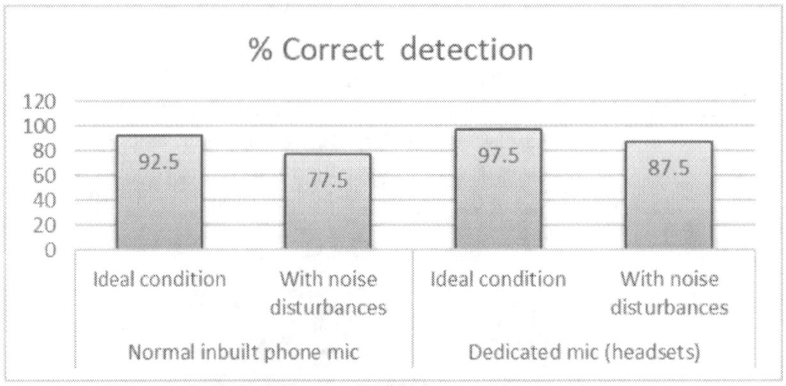

Figure 5. Reliability at different working environments

7.2. Internet Band Width
The maximum data transfer rate of a connection or network is described by Bandwidth. It measures the amount of data that can be sent over a particular network in given time period.

Table 2. Speech Recognition Reliability check against Internet bandwidth

Internet upload speed	Trial 1 Time in Sec	Trial 2 Time in Sec
3.5 MBPS (Jio 4G net)	3	4
6.8 MBPS (Idea 4G net)	3	3

As the data transfer required is very small (the data is only audio and text), the changes in internet band width does not affect the performance of the system, unless the bandwidth falls very low. When the upload speed is less than 0.8 MBPS, the speech recognition is slow.

8. Conclusion and Future Scope
The proposed framework using speech as the input has shown desirable performance under the various experimental conditions and thus is reliable. Resources spent on service engineers can be minimized by implementing this system. This system further allows easy energy optimization, resource sharing, and convenience of manufacturing by enabling remote robot access. It can save time, monetary resources, labour costs, and provide the client with the best possible service. Hence, this tele-commanding of the robot is a better alternative to the conventional system of programming debugging, inspection, and service.

The system is to be developed as a cyber-physical system, where the robot controller communicates with the cloud server directly. The aim is to eliminate the need to program the local robot. Expert programmers can program energy and memory-efficient algorithms capable of performing a predefined set of applications. The program is stored on the cloud server and the task of the controller is to select the program required and download it from the cloud.

The integration of Artificial Intelligence for the vision of the system, using ImageNet, will enable image recognition of various objects that are trained. Almost all real-time objects can be detected, and thus be acted upon.

References
[1] L. Z. L. Zhizeng and Z. J. Z. Jingbing 2004 Speech recognition and its application in voice based robot control system *2004 Int. Conf. Intell. Mechatronics Autom. 2004. Proceedings* pp 960-963
[2] Q Hong, C Zhang, X Chen and Y Chen 2007 Embedded speech recognition system for

intelligent robot *Proc. 14th Int. Conf. Mechatronics Mach. Vis. Pract. M2VIP2007* pp 35-38

[3] P Mayorga O, J M Olguín E, A M Hernández C and J I Flores M 2007 Gaussian components optimization for a robot controlled by speech commands in Mexican Spanish *Electron. Robot. Automot. Mech. Conf. CERMA 2007 - Proc.* pp 324–329

[4] Detlef RICHTER 1998 Robot Control by Digital Binocular Computer Vision and Digital Speech Recognition *Optimization of Electrical and Electronic Equipments – Bravov*. pp 617-622

[5] Rihab Chaâri, Fatma Ellouze, Anis Koubâa, Basit Qureshi, Nuno Pereira, Habib Youssef and Eduardo Tovar 2016 Cyber-physical systems clouds: A survey *Comput. Networks* **108** pp 260-278

[6] I V Blagouchine and E Moreau 2010 Control of a speech robot via an optimum neural-network-based internal model with constraints *IEEE Trans. Robot* **26** pp 142-159

[7] X Liu, S S Ge, R Jiang and C H Goh 2016 Intelligent speech control system for human-robot interaction *Chinese Control Conf. CCC* pp 6154-6159

[8] S N A Mohamad, A A Jamaludin and K Isa 2017 Speech semantic recognition: A control mechanism for assistive robot *USYS 2016 - 2016 IEEE 6th Int. Conf. Underw. Syst. Technol. Theory Appl* pp 139-144

[9] O Rogalla, M Ehrenmann, R Zollner, R Becher and R Dillmann 2002 Using Gesture and Speech Control for Commanding a Robot Assistant *Proceedings afthe 2002 IEEE Int. Workshop on Robot and Human Interactive Communication* vol. pp 454-459

[10] B M Chinnathurai, R Sivakumar, S Sadagopan and J M Conrad 2016 Design and implementation of a semi-autonomous waste segregation robot *Conf. Proc. - IEEE SOUTHEASTCON* pp 1-6

[11] R Baheti and H Gill 2011 Cyber-physical Systems *Impact Control Technol.* pp 161-166

[12] S N A Mohamad, A A Jamaludin and K Isa 2017 Speech semantic recognition: A control mecshanism for assistive robot *USYS 2016 - 2016 IEEE 6th Int. Conf. Underw. Syst. Technol. Theory Appl.* pp 139-144

[13] ISO 2012 Robots and Robotic Devices – Vocablary (Online: International Organization for Standardization)

Effect of fiber orientation on the tensile and flexural properties of vetiverfiber reinforced epoxy composites

Senthilnathan Natarajan[1], Narendiranath Babu Thamba[2], Sukalpan Nandi[3], Kavitha K.V.N[4], Srinath Ayyagari[5]

[1,2,3,4,5]Vellore Institute of Technology, Vellore, India
Email:senthil_nsn75@yahoo.co.in
Email:narendiranathbabu.t@vit.ac.in
Email:sukalpan97@gmail.com
Email:kvnkavitha@yahoo.co.in
Email:srinath.ayyagari2016@vitstud
ent.ac.in

Abstract - Natural fibre composites find application in some areas of aerospace, automotive, marine, electrical and sports equipments. They replace traditional materials due to the advantage of bio-degradability and environmental durability. However, better understanding of the composite behavior is required to ascertain its durability in the field of applications. In this work, vetiver fibre composites are prepared and tested for tensile and flexural strength. Vetiver fibers of length 15-16cm were used as reinforcements and epoxy resin with hardener is used as the matrix. Three types of composites were prepared using three different arrangements of the Vetiver fibers. The orientation of the fibres include longitudinal, transverse and a third arrangement combining both the directions. The flexural and tensile tests of these three types of composites are carried out and the results are compared with each other and useful conclusions were drawn from them.

Keywords: Vetiver fiber, composite, epoxy, hand layup technique, tensile test, flexural test.

1. Introduction

Natural fibre composites are viable alternatives to synthetic fibre reinforced materials owing to thebenefits of low weight, reduced cost, reasonable strength and modulus, environment friendly and easy to recycle. The fibres are generally used as reinforcements with the resin matrix. In some applications it is used in the form of fillers in plastic polymers [1]. The recent development in natural fibre processing and composite science provides good opportunity to explore new materials which can be recycled and are biodegradable. They are emerging as a good substitute for glass fibres mainly in the automotive industry where they are used in door panels, seat coverings and back rest, floor mats etc [2,3].Several research works based on the reinforcement of natural fibres has been carried out in the recent decade. The behaviour of the polymer composites reinforced with fibres such as jute, kenaf, hemp, flax, sisal, oil palm fibre, coir, banana, bamboo has been studied [4]. Some of the composites with natural fibre reinforcement are found to be commercially viable due to their low market price. But still the market share of these composites is low due to their low thermal stability, low impact strength and high water absorption [5]. The strength of these composites depend on the chemical treatment of the fibers, fiber weight or volume fraction, piling of the layers of fiber and bonding between matrix and fiber [6,7]. To enhance their mechanical properties, composites of hybrid type are also being explored by conjoining two kinds of natural fiber or a natural fiber with glass fiber [8,9].

Vetiver with its biological name chrysopogon zizanioides, is a grass variety belonging to the family of Poaceae [10]. Its origin is reported to be from India. Its grass and thick aromatic roots are highly valued and mainly used for medicinal purposes [11].The root extract is also used for textile finishing [12]. Vetiver fibers as strengthening agent in conventional concrete have been found to increase the compressive strength and flexural strength [13]. A study on vetiver-polypropylene composite revealed substantial increase in tensile strength and youngs modulus compared to neat polypropylene material

[14]. The vetiver was mixed with polypropylene in the form of grass and fiber in various percentages. The chemical treatment on vetiver grass was also found to enhance the mechanical properties of the composite. It was reported that the reinforcement of natural fibers with polypropylene has aided in higher rate of crystallization [15]. The characterization of alkaline treated vetiverfiber revealed modifications in its physical properties. It has resulted in higher decomposition temperature compared to untreated fiber. It has also resulted in enhancement in the crystalline structure of the fiber [16]. A hybrid composite of vetiver-jute-glass was developed and the analysis showed improvement in flexural and tensile behaviour of the composite in comparison with bifibered composites made of vetiver-glass or jute-glass [17].

The work on vetiver fiber as a natural fiber composite is limited. Nihal et al reported the tensile properties and impact strength of polymer plastic composite with varying weight percentage of vetiver fibre [18]. Although, many studies have been made into the mechanical behaviour of natural fibre composites with respect to their weight percentage, the variation in mechanical properties corresponding to the orientation of the fibre in the composite is not so well explored. V.G.Geethamma et al has used natural rubber with coir fibers to study the effects of fibre orientation [19].The current effort deals with the study of tensile and flexural behaviour of vetiver fiber in different orientations.

2. Materials and Methods

The materials used in the fabrication of the composites are vetiver natural fiber, epoxy resin of specification LY556, hardener of specification HY951[20] and wood for making the mould. Vetiver fibers used in the preparation of the composite were first cleaned throroughly with distilled water and then treated with 0.5M NaOH solution to remove dust, cellulose, lignin, hemicelluloses from the outer surface of the fibers. Chemical treatment of natural fibres using NaOH is known to raise the strength of natural fibre composite [17, 21]. After 2 hours of treatment, the fibers are cleaned in distilled water to eradicate surplus NaOH and dried for 24 hours at 60□ C by heating in a furnace. Hand layup procedure was used for preparing the natural composites [22, 23]. Epoxy resin and hardener were mixed in the ratio of 10:1 to prepare the matrix for the samples.

2.1 Preparation of Mould

The moulds required for tensile and flexural tests were fabricated conforming to ASTM D638 and ASTM D790 standards respectively using low density fiber board. The dimensional model was prepared as per the standards shown in figure 1 and figure 2 using solidworks software and CNC lathe machine was used to get the required shape of the mould from the wooden base structure. The mould was then cleaned thoroughly to remove chips and dusts over the surface. These moulds are used for preparing the composites using hand lay-up process. For breaking the mould upon completion of the hand layup process, a cutting machine was utilized to cut away the wooden mould. During the cutting process, extra care was taken so as not to damage the samples.

2.2 Preparation of Composites

Three types of composites were prepared using three different arrangements of the Vetiverfibers as shown in figure 3.
Composite Type Longitudinal: Long Vetiverfibers of length 15-16cm were placed longitudinally throughout the length of the sample.
Composite Type Transverse: Short Vetiverfibers of length 1-1.2cm were placed transversely throughout the length of the sample.
Composite Type Bi-directional: This sample was prepared using both short and long fibres. The short fibres were placed transversely and the long fibres were placed longitudinally.

The fibres with uniform shape and size were selected and cut to the required length. The epoxy-hardener solution is formulated in the ratio of 10:1 [24]. The composites were prepared using hand layup technique and then subjected to soft compression moulding method. For preparing the composite type longitudinal, the epoxy solution was laid first, followed by the layer with longitudinal fibres. This alternate arrangement of epoxy solution and vettiver fibres were carried out till 4 layers of epoxy and 3 layers of fibers were obtained. For the longitudinal arrangement, 5 threads of fibers were placed in each layer.

For preparing the composite type transverse, similar procedure was followed by placing the fibers in the transverse direction. The fibers were cut to fit in the width of the mould. Around 35 cut fibers were used in each layer of preparation. The composite type bi-directional was prepared by placing the fibres in both longitudinal and transverse directions alternatively. After the first epoxy layer, the fibres are placed in

IOP Conf. Series: Materials Science and Engineering 1123 (2021) 012048 doi:10.1088/1757-899X/1123/1/012048

longitudinal direction with 5 threads in the layer. This was followed by second epoxy layer and the fiber layer in transverse direction. This procedure was continued till two layers of longitudinal fiber and one layer of transverse fibers were obtained. The epoxy and fiber ratio were maintained at 40:60.

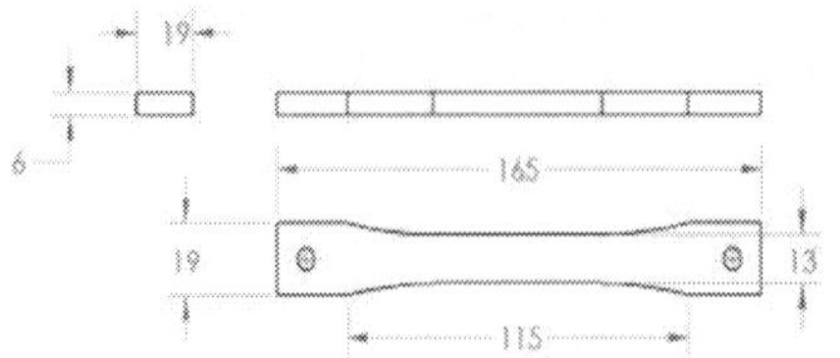

Figure. 1 Specifications of tensile test mould

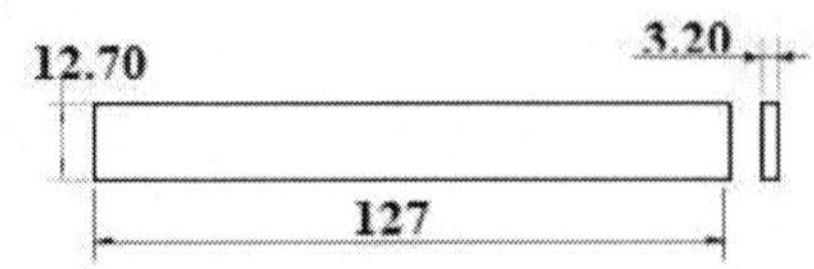

Figure. 2 Specifications of the flexural test mould

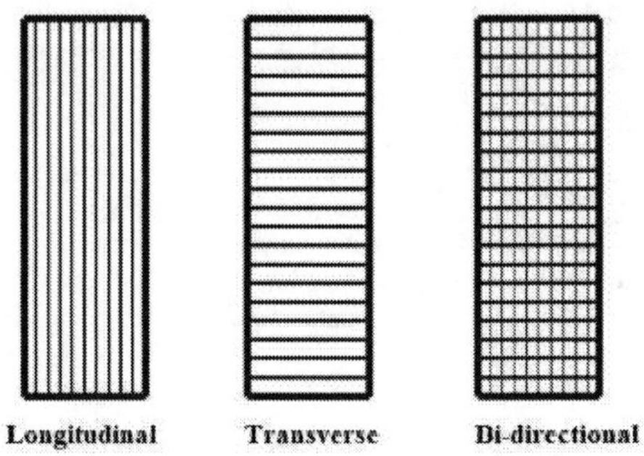

Figure. 3 Orientation of the vetiver fibres used for testing

3. Experimentation

To understand the tensile behaviour of the vetiver fibre composites, tensile test was done in Instron 8801 machine shown in figure 4. Samples were prepared according to the tensile test standard for natural composites. The tensile test was completed at a rate of 1mm/min at room temperature to find out the ultimate tensile strength of the specimen. According to the standard, the gauge length used for the specimen is 50mm. The three point bending test was also performed in the Instron 8801 machine. The samples were prepared according to the test standard for natural composites. The testing was done at a strain rate of 2mm/min by applying the load in the center of the span of the specimen.

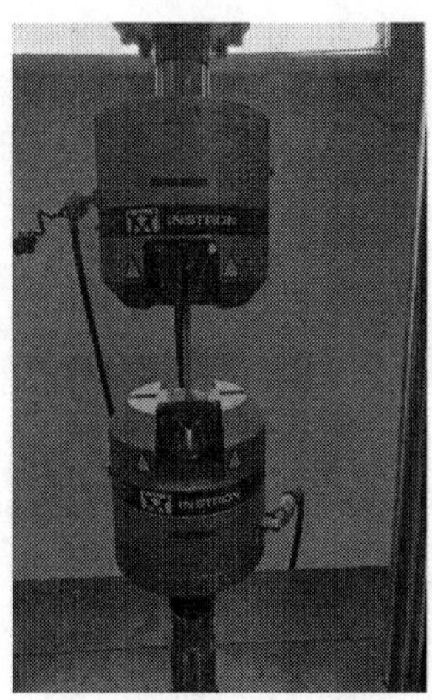

Figure. 4 Tensile test performed in Instron universal testing machine

4. Results and Discussion

4.1 Tensile test results

The load-displacement behaviour of the vetiver fibre composite in tensile testing with three different orientations are shown in figure 5. A portion of the testing samples are shown in figure 6.

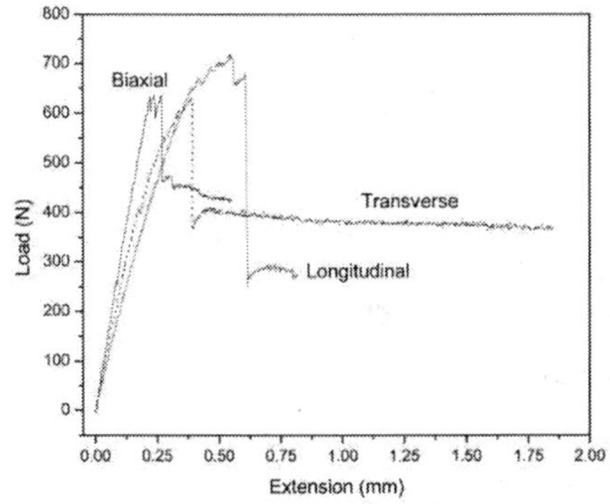

Figure. 5 Load-Extension variation for the three orientations of the tensile samples

The tensile test data are tabulated in table 1. The tensile strength of the fibre composite is found to be in the range of 7MPa to 9MPa. Nihar et al has reported a tensile strength of 25.4MPa to 26.9MPa with the use of polypropelene as the matrix with 35% of fibre. The investigations on different orientations with epoxy resin as matrix has not been carried out. The fibre orientation with longitudinal direction is found to give the highest tensile strength of 9.624MPa [18]. Saravanna et al performed the tensile strength of Vetiver fibre composite with matrix of epoxy polymer Lapox-L12 and Lapox K-6 and found to be 11.9MPa [23].

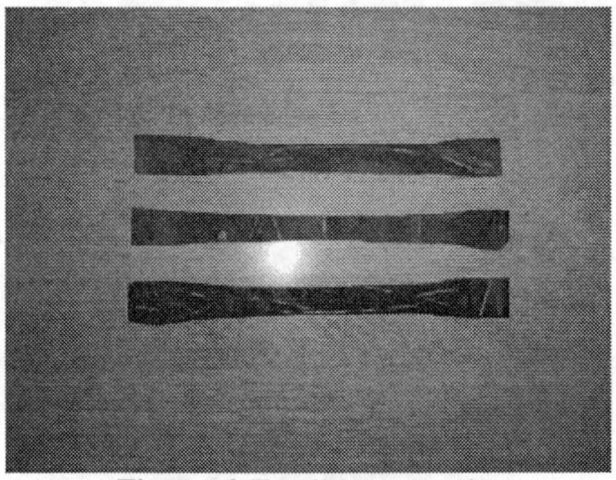

Figure. 6 Tensile test samples

Table 1. Tensile test data

Type of composite	Maximum Load (N)	Ultimate Tensile Strength (MPa)	Young's Modulus (MPa)	Tensile strain at break (%)
Composite Type Longitudinal	719.48767	9.624	1792.234	1.162
Composite Type Transverse	630.40257	8.932	2447.313	2.638
Composite Type Bi-directional	638.56840	7.867	2538.860	0.786

4.2 Bending test results

Flexural strength is an important aspect to gauge stiffness of a given composite and is determined using the slope acquired from the load–displacement curve. Such flexural behaviour is generally analysed by means of a three-point bending test. During this test, the layers are subjected to different stresses, the mid layer experiences shear, the top layer compression and the bottom layer to tension. The resulting failure of composite is thus attributed to both bending and shear. Various important material properties such as modulus of elasticity in bending, flexural stress and strain, stress–strain response of the specimen can be calculated from three-point bending test. In this work, the flexural test of three-point bending test was performed on a universal testing machine. The variations in flexural stress as a function of flexural strain of the vetiver fibre reinforced epoxy composites for the three specimens of different orientations are shown in figure 7. The tested samples are shown in figure 8.

The results in figure 7 shows that the flexural strength attains a maximum value of 45.74247 MPa for composite specimen with longitudinal configuration but it exhibits a minimum value of 17.43875 MPa in the transverse configuration. The sample with transverse lay of fibres show better flexural strength than the other two samples. The highest strain to failure, shows a ductile behaviour. However, the use of resin also plays a critical role in maintaining the fibers as straight structure and to prevent from bending.

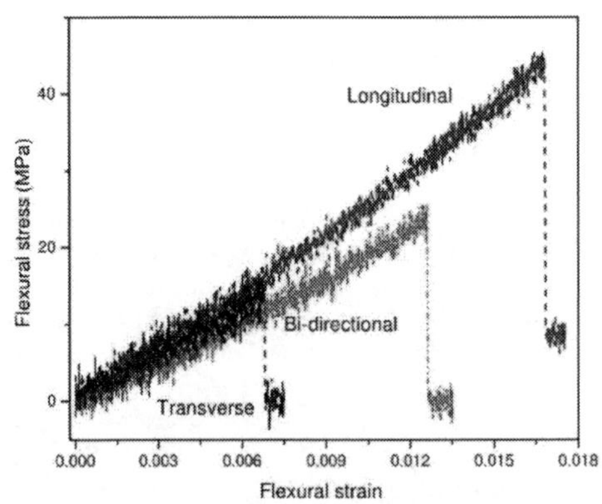

Figure. 7 Flexural stress and flexural strain diagram for the three orientations of bending samples

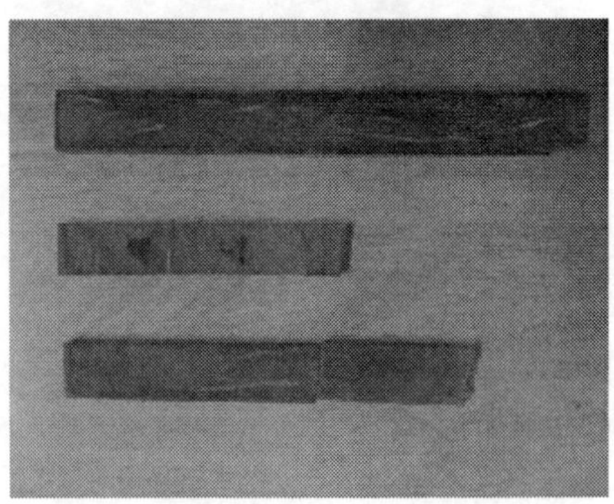

Figure. 8 Samples tested for flexural strength

The data acquired from the flexural test is tabulated in table 2. The flexural modulus is found to vary between 2.1-3.1GPa. Flexural modulus helps to determine the degree of deformation of a material under bending stress and provides a measure of ductility of the material. Saravana Bavan et al carried out flexural tests on vetiver fibre composite after chemically treating the fibre using alkali method and preparing the composite with epoxy resin and hardener. The flexural modulus was reported to be 1.1-1.2GPa [23].Composite type with longitudinal fibres has the highest value of flexural modulus of 3111.089 MPa and thus, has the highest stiffness of all the samples. While composite type with bi-directional fibres has the lowest flexural modulus value of all the samples of 2118.366 MPa.

Table 2. Flexural test data

Type of composite	Maximum Load (N)	Maximum Flexure extension (mm)	Modulus (Automatic) (MPa)	Maximum Flexure stress (MPa)	Flexure strain (mm/mm)
Composite Type Longitudinal	103.69	2.746	3111.08984	45.74247	0.01676
Composite Type Transverse	27.64	1.100	2318.03887	17.43895	0.00611
Composite Type Bi-directional	58.37	2.069	2118.36602	25.74904	0.01262

5. Conclusions

The tensile tests of the vetiver fiber epoxy composites showed a higher value of tensile strength for the composite with longitudinal fibers. The composites can be further tested for different proportions of fiber and epoxy. The flexural strength of the longitudinal fiber composite is found to be 62% more than the composite with transverse fibers and 44% more than the bi-directional fiber composites.

Acknowledgments

The authors express their gratitude and appreciation for the support extended by the management of Vellore Institute of Technology-VIT.

References:

[1] Jawaid M, Khalil A H P S and Bakar A A, 2011 Woven hybrid composites: Tensile and flexural properties of oil palm-woven jute fibres based epoxy composites *Materials Science and Engineering: A* **528** 5190-5195.

[2] Mohanty A K, Misra M and Drzal L T 2005 Natural fibers, biopolymers, and biocomposites *CRC press*, **1** 1-36.

[3] Norhidayah M H, Hambali A A, Bin Yaakob M Y, Zolkarnain M and Saifuddin H Y 2014 A Review of Current Development in Natural Fiber Composites in Automotive Applications *Applied Mechanics and Materials* **564** 3-7.

[4] Nunna S, Chandra P R, Shrivastava S and Jalan A K 2012 A review on mechanical behavior of natural fiber based hybrid composites *Journal of Reinforced Plastics and Composites* **31** 759-769.

[5] Demir H, Atikler U, Balköse D and Tıhmınlıoğlu F 2006 The effect of fiber surface treatments on the tensile and water sorption properties of polypropylene–luffafiber composites *Composites Part A: Applied Science and Manufacturing* **37** 447-456.

[6] Ajay D and Naidu L 2018 A study on different chemical treatments for natural fiber reinforced composites *International Journal of Mechanical and Production Engineering Research and Development* **8** 143-152.

[7] Sood M and Dwivedi G 2017 Effect of fiber treatment on flexural properties of natural fiber reinforced composites: A review *Egyptian journal of petroleum* **27** 775-783.

[8] Hafiz Z M, Akil H M, Bakar A A, Ishak Z A M and Cheng L W 2015 Effect of water absorption on pultruded jute/glass fiber-reinforced unsaturated polyester hybrid composites *Journal of composite materials* **46** 51-61.

[9] Dong C 2018 Review of natural fibre-reinforced hybrid composites *Journal of Reinforced Plastics and Composites* **37** 331-348.

[10] Chahal K K, Bhardwaj U, Kaushal S and Sandhu A.K 2015 Chemical composition and biological properties of Chrysopogonzizanioides (L.) Roberty syn. Vetiveriazizanioides (L.) Nash *Indian Journal of Natural Products and Resources* **6** 251-260.

[11] Gnansounou E, Alves C M and Raman J K 2017 Multiple applications of vetiver grass–a review *International Journal of Environmental Science* **2** 125-141.

[12] Naikwadi S, Sannapapamma K J and Venugopal C K 2017 Optimization of Vetiver Root Extract for Textile Finishing *Int. J. Curr. Micro biol. App. Sci.* **6** 2009-2022.

[13] Kumaraguru R, Santhoshkuma S, Kuralamuthan S and Anand M 2017 Experimental Investigation on Vetiver as Fibre in Conventional Concrete *International journal of engineering research & technology* **6.**

[14] Ruksakulpiwat Y, Suppakarn N, Sutapun W and Thomthong W 2007 Vetiver–polypropylene composites: physical and mechanical properties, Composites Part A: *Applied Science and Manufacturing* **38** 590-601.

[15] Somnuk U, Eder G, Phinyocheep P, Suppakarn N, Sutapun W and Ruksakulpiwat Y 2007 Quiescent crystallization of natural fibers–polypropylene composites *Journal of applied polymer science* **106** 2997-3006.

[16] Sutapun W, Suppakarn N and Ruksakulpiwat Y 2010 Study of characteristic of vetiver fiber before and after alkaline treatment *Advanced Materials Research* **123** 1191-1194.

[17] Vinayagamoorthy R and Rajeswari N 2014 Mechanical performance studies on Vetiveria zizanioides/jute/glass fiber-reinforced hybrid polymeric composites *Journal of Reinforced Plastics and Composites* **33** 81-92.

[18] Rastogi N, Kakkar A and Shakthivel S 2019 Characteristic investigation of 50-50% vetiver fiber reinforced polymer composites *International journal of Mechanical and Production Engineering Research and Development* 274-283

[19] Geethamma V G, Mathew K T, Lakshminarayanan R and Thomas S 1998 Composite of short coir fibres and natural rubber: effect of chemical modification, loading and orientation of fibre *Polymer* **39** 1483-1491.

[20] Banga H, Singh V K and Choudhary S K 2015 Fabrication and study of mechanical properties of bamboo fibre reinforced bio-composites *Innovative System Desing and Enginnering* **6** 84-98.

[21] Hossain S I, Hasan M, and Hassan A 2013 Effect of chemical treatment on physical, mechanical and thermal properties of ladies finger natural fiber *Advances in materials science and engineering*.

[22] Patil A and Chheparge C K 2018 Fabrication and Characterization of natural fiber reinforced polyester composites *IOSR Journal of Mechanical and Civil Engineering* 64-70.

[23] Bavan D S and Kumar G M 2014 Tensile and Thermal Degradation Properties of Vetiver Fiber Composites *Procedia Materials Science* **5** 605-611.

[24] Thamba N B, Kuclourya T, Jain M K and Mangalaraja R V 2019 Flexural properties of areca nut, sunn hemp and e-glass fibers reinforced with epoxy composites *Materials Research Express* **6** 105344.

IOP Conf. Series: Materials Science and Engineering 1123 (2021) 012049 doi:10.1088/1757-899X/1123/1/012049

Application of active vibration control on single stage spur gear

Manas Raman Menon, Tanay Hariprasad, Karun Thomas, Ramesh Babu Vemuluri, Siva Prasad Darla*, Edwin Sudhagar P and Ashok B
School of Mechanical Engineering, Vellore Institute of Technology, Vellore-632014, Tamil Nadu, India.

*Corresponding author email ID: sivaprasaddarla@vit.ac.in

Abstract. Vibrations in a mechanical system can cause power losses or vibrational fatigue which leads to failure in the material. In the current scenario automotive vehicles majorly use passive vibrational control techniques for vibration attenuation. Passive vibrational control tend to be heavy in nature and can only control frequencies in the range of 100 – 200 Hz. Conversely, active vibrational control is the potential alternative to passive vibrational control, they consist of strain sensor, piezoelectric actuator, a control system and a power amplifier to draw the control actuator. The application of active vibrational control is flourishing due to the demand in critical vibrational control in extreme conditions (underwater, polar, space). This paper provides an overview of the applications of active vibration control on single stage spur gear. The vibrations emitted in a single stage spur gear is optimised by active vibration control.

1. Introduction

Gears are one of the most widely used mechanical transmission systems due to their ability of transmitting large power and efficiency and have a longer service life than belt drive or chin drive. The spur gear drive transmits power between two parallel shafts which gives it the highest efficiency compared to all the other gears (bevel, helical, worm, cycloid, and hypoid) of about 94% to 98%. The power is lost is in the form of vibration and noise. Vibrations in the single stage spur gear system are measured by modal analysis and harmonic analysis of the system. The contact stress and bending stresses are induced in the gear due to the tangential force acting on the gears. If the contact stresses on the gear is more than the wear resisting power of material, then the gear failure might take place this process is called as wear or pitting failure of the gear. This is primarily discussed in Patil and Chappar [1].

Bryan and Mark [2] discussed the development of an active system for vibration control over the chassis and reduce noise in the passenger cabin and adaptive filter theory. Further, identified optimal transducer locations to help in reduction of vibrations. Bianchini [3] discussed the implementation of a cost effective active vehicle control system which was incorporated into the steering wheel column, which specifically eliminates abundant idle engine vibrations transmitted to the steering wheel. Further, presented the numerical analysis of the actuator placement and sizing. The optimal actuator location for each mode of vibration was the one that exhibits the highest localized strain at particular frequency of interest. Tzou *et. al.,* [4], discussed the various materials used in small and microscale transducers. It also provided an overview of the smart materials and strutonic systems available in the market. Sam *et. al.,* [5] did a comparative study of active and passive suspensions focusing on ride quality improvement by using Linear Quadratic Regulator (LQR) control for the actuator in active suspension. Active suspension increases tire to road contact, makes vehicle more stable. Timo *et. al.,* [6] developed a multi-channel adaptive feed forward controller based on LMS algorithm, controller uses commuted motor as an actuator. Experiments showed that a housing vibration reduction of 31 dB at second gear mesh frequency the controller reduces the structure borne sound by 18 dB and airborne sound by 15 dB. Arjon [7] presented an active engine mount system that reduces the transmission of engine vibration to the chassis and the noises in the passenger cabin. The adaptive feed forward control enabled the actuator to reduce the force transmitted to chassis. Simulation results showed significant reduction in transmission of vibrations and the response time of control signal by improving the inherent delayed control signal. Dogruer and Pirosoltan [8] showed the periodic time-varying mesh stiffness and elastic modes of spur gear drive by using a non-linear controller which had full access to the absolute angles of spur gear pair from a reference point. The non-linear controller cancels the nonlinear force that acts on gear and pinion teeth. The system becomes a linear one, which opened up a room for the application of linear control techniques. Hatch [9] explained state space equations of various models using ANSYS and MATLAB. Khot *et. al.,* [10] discussed the extraction of full and reduced models of a cantilever beam by using ANSYS which helped to reduce the computational time. Narayankar and Mangrulkar [11] showed the importance of considering contact stress and bending stress during the mating of gears as a parameter of gear design. Further, validated with analytical results which shown well agreement. Ali *et. al.,* [12] presented the appropriate methodology in order to conduct a modal analysis and hence compare the performance of a conventional spur gear with a modified involute spur gear. Chouhan and Lodwa [13] shown the importance of finding the natural frequencies and mode shapes of a two stage spur gear system as a design parameter by using mathematical formulation. Kowelczyk *et.al.,* [14] discussed the advancements in active vibration control systems used in automotive industries. Further, gave an overview of the developments and research behind this systems. Adel *et. al.,* [15] developed a dynamic model for a composite cantilever beam and used this to create a feedback active vibration control loop using a PID controller in a Simulink environment. Maslan *et. al.,* [16] formulated a transfer function for a piezo electric patch with appropriate sizing and location by using a system identification technique. Yanjun *et. al.,* [17] used finite element method (FEM) in order to obtain the natural frequencies of gear and rack power transmission system at varying working speeds in order to avoid resonance. Hassan *et. al.,* [18] evaluated the dynamic responses and natural frequencies of a section of spur gear drive using a two dimensional finite element model. Further, evaluated the optimum operating speed. Ramesh Babu *et. al.,* [19] developed a PID controller loop with a negative feedback mechanism for vibration control in composite beams which is being used extensively in many engineering fields. Gabbert and Ringwelski [20] discussed the use of a design approach for smart-lightweight structures that were made of sheets of metal or fibre which was reinforced with plastics and equipped with thin piezo-electric wafers that acted as actuators and sensors that aided in controlling vibrations and noise. Validation of the results was done by comparing the simulation output with experimental data. Experimental and numerical results showed good agreement. With a velocity feedback control attenuations up to 4dB in the vibration level were achieved at the resonance frequency regions of the most dominant modes.

In this paper, the source of vibration within transmission systems such as single stage spur gear is considered and its noise levels are measured. Further, simulate an open active vibration control system with feed forward strategy with the help of Simulink module in MATLAB with piezoelectric material as actuators that can help reduce transmission error in a single stage spur gear pair. Also, performs modal analysis of the spur gear system with ANSYS, results are validated with analytical results and there is good agreement is observed. Further, the reduction in vibration amplitudes is observed after usage of piezoelectric patches and actuators.

2. Modelling

2.1. Gear drive modelling

A spur gear drive with gear ratio 1:3 is designed using the following dimensions on SOLIDWORKS, material used for the gear drive is grey cast iron. The distance between both the centres of the gears is

40 mm. The complete gearbox model is shown is figure 1 and the dimensions of the drive are shown in table 1.

Table 1. Dimensions of the spur gear drive.

Parameter	Gear	Pinion
Module	1	1
Pressure angle	20 degrees	20 degrees
Pitch diameter	60 mm	20 mm
Number of tooth	60	20
Width	12 mm	12 mm
Radius of the inner hole	8 mm	4 mm

This gear system is placed inside a gear casing also made of grey cast iron with stub axial diameter for gear and pinion of 17 mm and 8 mm respectively. Four ball bearings are also attached on both either sides of the gear and pinion with thickness 9 mm and 4 mm respectively. The gears are placed in a closed casing but the figure shows the gear placed in only the bottom casing. Meshing is an important part of the analysis, for an accurate result fine mesh is necessary. The finer the mesh, higher the accuracy but with more analysis run time. The SOLIDWORKS model is imported into ANSYS workbench for the meshing and further analysis procedures. The meshing of the spur gears is shown in the figure 2.

Figure 1. Complete gearbox model.

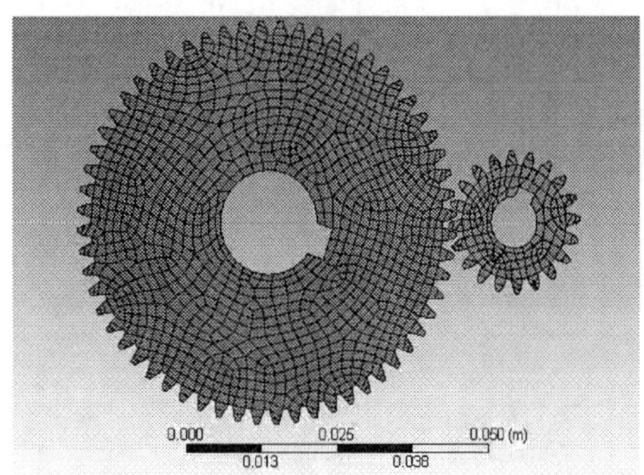

Figure 2. Gear mesh.

Any component kept in space has 6 Degrees of Freedom (DOF). Three DOF are translation along X, Y, Z axes and three rotation along the same axes. The pinion gear is given fixed support for analysis as shown in figure 3. In ANSYS, fixed support constraints all the DOF. Frictionless support is applied to the gear model as the force is applied in a tangential direction and the surface body is allowed to rotate and movie freely [1]. A moment is applied to the part where frictionless support is applied. Due to this moment or torque the gears start rotating.

2.2. Modal analysis

A modal analysis is performed in ANSYS workbench in order to determine the frequencies at each mode. The boundary conditions and constraints are set accordingly. Torque of 300 Nm is applied. The first three mode shapes are shown in figure 4, 5 and 6. The values of first three natural frequencies are presented in table 2.

Figure 3. Fixed support.

Figure 4. 1st mode.

Figure 5. 2nd mode shape of spur gear drive.

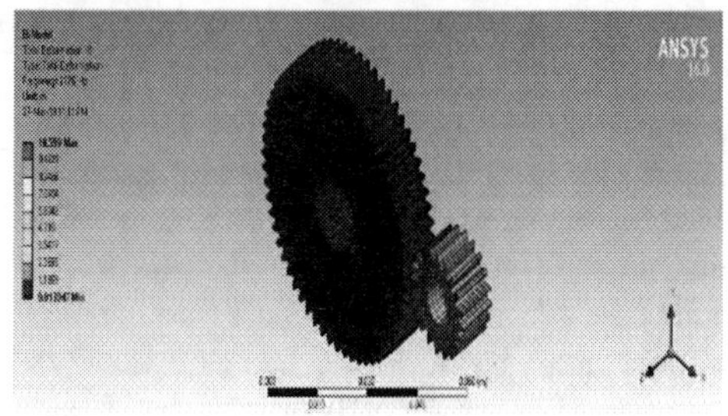

Figure 6. 3rd mode shape of spur gear drive.

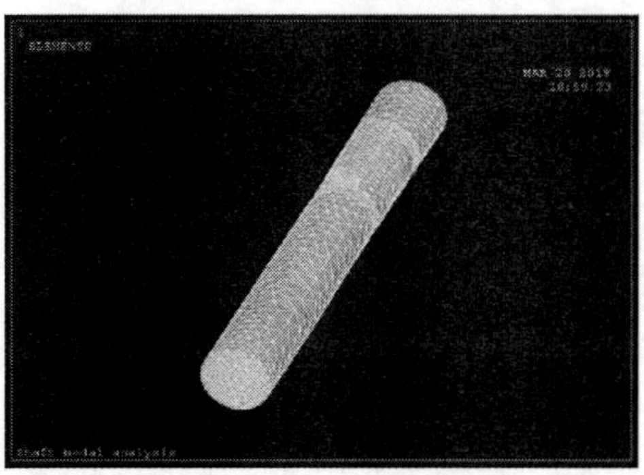

Figure 7. Meshing of the shaft.

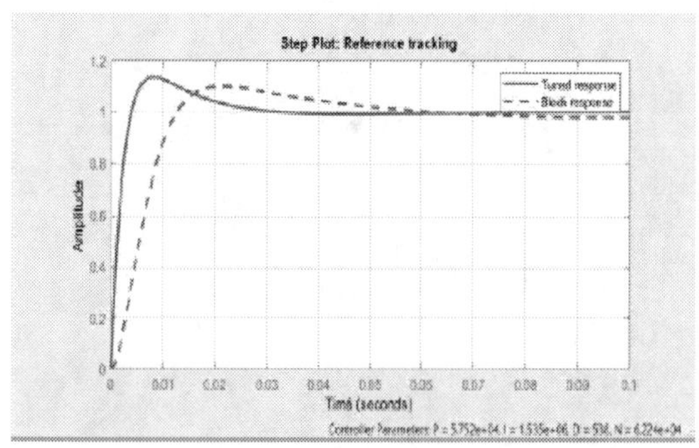

Figure 8. Simulink modal of piezoelectric patch actuator.

Table 2. Frequency at each mode shape of the gear.

Frequency number	Hertz	Rad/sec
1	82.2	516.4
2	686.1	4310.8
3	2176.0	13672.2

IOP Conf. Series: Materials Science and Engineering 1123 (2021) 012049 doi:10.1088/1757-899X/1123/1/012049

The modal analysis of the shaft also performed in ANSYS APDL environment, with the material properties of grey cast iron, and the modelled as 10 sections as shown in figure 7. After meshing the model the modal analysis is performed. The first three natural frequencies at particular modes are extracted using the Block Lanczos Solver method and presented in table 3. Further, a MATLAB code to extract the Eigen values and Eigen vectors to form the state space matrix.

Table 3. Frequency at each mode shape of the shaft.

Frequency number	Hertz	Radians/sec
1	2.3	14.5
2	69.5	436.4
3	545.7	3428.5

3. State space formulation

The results of the FEM analysis are then converted into Eigen values and saved, which are readable by MATLAB. The state space of the *model* is represented as follows.

$$\dot{X} = AX + BU \tag{1}$$

$$Y = CX + DU \tag{2}$$

where A is system Matrix, B is input Matrix, C is output Matrix and D is direct transmission Matrix. X represents the column vector that in turn represents the state of the given system, $X^{\&}$ represents the differentiation of column vector, Y is the output matrix and U is the input matrix. The equations of motion of a multi degree of freedom [10] system acted on by external loads is represented by

$$[m]x^{\&\&}+[c]x^{\&}+[k]x = F \tag{3}$$

The above equation leads to 'n' coupled second order differential equations. This equation can be uncoupled by solving the eigenvalue problem for (3). Eigen vectors x(1), x(2),... are obtained. This forms a modal matrix

$$[x]_n = \{x_1, x_2, x_3, \ldots\ldots x_n\} \tag{4}$$

The solution of equation (3) is expressed as a linear combination of the normal modes as

$$x(t) = [x]_n \, x_p(t) \tag{5}$$

Here, $x_p(t)$ is known as the displacement in principle coordinates.

4. Validation – Hertz contact stress formula

Hertzian contact stress is something that tackles studies of mating parts in mechanical engineering and tribology. This is considered to be a stress formed within mating parts. If not accounted for properly it can usually cause serious problems to machinery. Here, the contact stress between two spheres of different sizes is considered. The contact stress equation is as follows:

$$\sigma_c = \left[\frac{W_t}{\pi}\times F\cos\phi\times\frac{\left[\frac{1}{r_1}+\frac{1}{r_2}\right]}{\left[\frac{(1-\nu_1^2)}{E_1}\right]+\left[\frac{(1-\nu_2^2)}{E_2}\right]}\right]^{\frac{1}{2}} \tag{6}$$

where, $W_t = \frac{2T}{D_p}$, $r_1 = r_2 = \frac{D_p \sin\phi}{2}$, F = Face width = 12 mm, ϕ = Pressure angle = 20°, D_p = pitch diameter = 21.5 mm, ν = Poisson's ratio = 0.21, E =Young's modulus= 1.8 x 10^5 MPa, T = Torque= 250 N-m, W_t = load.

By substituting the above values in the equation (6), from the calculation it is found that the contact stress comes to about 2914.98 MPa.

Table 4. Validation of contact stress values.

Materials	ANSYS (MPa)	SOLIDWORKS (MPa)	Hertz Contact Stress	Percentage of deviation	
				ANSYS	SOLIDWORKS
Grey Cast Iron	2732	2934	2915	6%	1%

From the table 4, the error percentage is quite less when the results are compared and therefore the analytical formula validates the results obtained.

5. Results and discussion – Simulink control loop

It is a graphical programming domain where dynamic systems can be modelled and analysed. Simulink provides many features which help us to develop control loops which can be used in many real time practical situations. Simulink library provides most commonly used blocks in the control environment. PID controller expels the steady-state error and reduces framework settling times while keeping up a sensible transient reaction. Trial and error method is integrated for tuning of PID. The proportional, integral and derivative are used together to find the output of the controller. K_p, K_i and K_d represents proportional gain, integral gain and derivative gain respectively.

The transfer function should be modelled out of the parameters piezoelectric patch and is given below:

$$\text{Transfer function} = G(s) = \frac{Y(s)}{U(s)} \quad (7)$$

Here G represents the Laplace transform. U(s) is the input and Y(s) is the output. The transfer function of the Piezo electric patch is given below:

$$G(s) = \frac{0.03747s^2 - 0.07066s + 0.03319}{0.4882s^2 - 1.492s + 0.0039} \quad (8)$$

These functions i.e. equation (7) and (8) will be inputted in the Simulink environment for further study. The actuator (piezoelectric patch) to be perfectly bonded to the casing of the spur gear pair [3]. The control force acts on the nodes in the X-direction. It acts at the end of the actuator. The output displacement and the exciting force are at points of maximum stiffness, calculated as Eigen values at each mode shape, these acts in the Z-axis. Using Simulink, a feedback loop is created that will help cancel out the transmission vibration error due to meshing issues. A feedback loop is shown in figure 9. The PID controller used here is tuned according to the output signal of the spur gear pair. The PID controller accordingly corrects the error in the signal with the specified K_p, K_i and K_d values. Tuning the PID controller helps in providing the controlled vibrations by optimizing the parametric values through selecting targeted frequency ranges. The plots from Figure 8 helps to give a detailed analysis and what parameters are required to control it. The PID controller is once again tuned according to get an optimized values as shown in table 5.

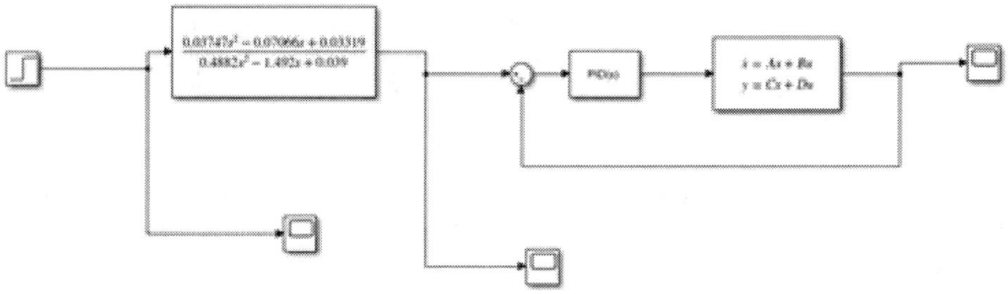

Figure 9. Simulink modal of DC motor actuator loop

Table 5. Piezoelectric patch PID.

Parameters	Values
K_p	5472
K_i	1535
K_d	538
N	6434

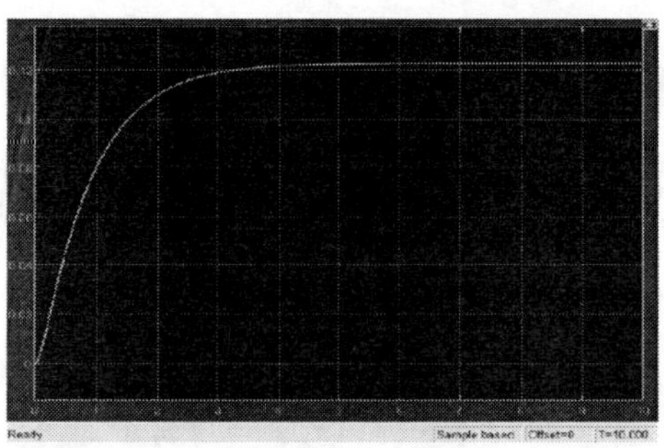

Figure 10. Amplitude versus time graph for Piezoelectric patch without PID.

Figure 11. Amplitude versus time graph for Piezoelectric patch with PID.

The plot in figure 10 and 11 shows a decent decrease in the curve of the amplitude versus time graph, showing how the PID controller helps in reducing the vibrations. This shows that there is a significant reduction in vibrations when a PID controller is used. Steady state is achieved much faster with lesser frequency changes. The input is given to the system as a step block which simulates the vibrations in the gear pair.

6. Conclusion

In this paper, a feedback control loop that can be used to reduce the vibrational noise produced by the transmission errors within a spur gear drive is created. Using the modal analysis of the spur gear system which is incorporated into the Simulink environment through state space form to obtain the desired results. From the graphs it can be inferred that using piezoelectric patch covers on a large surface area and therefore can cover more strain points. The piezoelectric patch provides a low overshoot than any other actuator making the patch more efficient to use. Use of a PID controller is necessary to create an output signal that can have lesser vibration patterns and therefore do a better job than using an actuator alone. The controller helps to achieve the steady state in a short time, therefore decreasing the settling time of the control system. This project helps in understand the importance of the behaviour of a system at various mode shapes and how it effects the system overall. This paper also realises the importance of appropriate tuning of PID controller to obtain a rectifying signal that helps attenuate the unwanted noise. This work can helps the manufacturers to make machines more reliable, efficient and effective by explaining the need for a tuneable controller that can help generate better results when combined with an equally good actuator. It also portrays the power of the Simulink environment and its ease of use. It also explains the need to study modal analysis of systems in order to understand the behaviour of systems and make modelling more effective and reliable.

7. References

[1] Patil S and Chappar S Static and frequency response analysis of spur gear for different materials 2017 *Int. Research J. of Engineering and Technology (IRJET)* **4** 1341- 48

[2] Bryan Riley and Mark Bodie An adaptive strategy for vehicle vibration and noise cancellation 1996 *Proc. of the IEEE National Aerospace and Electronics Conf. NAECON* 836-44

[3] Bianchini E Active vibration control of automotive steering wheels 2005 *SAE Technical Paper* 2005-01-2546 1-9

[4] Tzou H S, Le H J and Arnold S M Smart materials, precision sensors/actuators, smart structures and structonic systems 2004 *Mechanics of Advanced Materials and Structures* **11** 367- 95

[5] Sam Y M, Ghani M R H A and Ahmad N LQR controller for active car suspension 2002 *IEEE conf. TENCON Intelligent Systems and Technologies for the New Millennium* 441-46

[6] Timo B and Andreas M Multi-channel active gear pair vibration control with an electronically commutated motor as actuator 2014 *IEEE Int. Electric Vehicle Conf. (IEVC)* 978-96

[7] Arjon T Adaptive feed forward controller applied for active engine mount system 2015 *Int. Conf. on Intelligent Unmanned Systems* 1-5

[8] Dogruer C U and Pirosoltan A K Active vibration control of a single-stage spur gearbox 2017 *Mechanical Systems and Signal Processing* **85** 429-44

[9] Hatch M R *Vibration simulation using MATLAB and ANSYS* 2001 Chapman and Hall 567-630

[10] Khot S M, Yelve N P, Tomar R, Desai S and Vittal S Active vibration control of cantilever beam by using PID based output feedback controller 2012 *J. of Vibration and Control* **18** 366-72

[11] Narayankar N and Mangrulkar K Contact stress and bending stress analysis of spur gear by analytical method 2017 *Int. J. on Theoretical and Applied Research in Mechanical Engineering* **6** 1-3

[12] Ali N, Sarfraz A and Dolas D R Comparative modal analysis of conventional spur gear with modified involute spur gear 2015 *Int. Research J. of Engineering and Technology* **2** 1752-56

[13] Chouhan D and Lodwa A Finite element analysis of two stage gear pair to finding natural frequencies with effect of design parameters 2016 *Int. J. of Science and Research* 2016 **7** 1119-22

[14] Kowalczyk K, Svaricek F, Bohn C and Karkosch H An overview of recent automotive applications of active vibration control 2004 *RTO AVT Symposium RTO-MP-AVT-110* 241-6

[15] Adel M, Ibrahim K, Gad A R and Khalil A E M Development of dynamic model for vibration control using finite element analysis 2016 *J. of Engineering Sciences* 1-9

[16] Maslan M N, Mailah M and Darus I Z M Identification and control of a piezoelectric patch actuator 2012 *Latest Advances in Systems Science and Computational Intelligence* 204-9

[17] Yanjun X, Lihu H, Jiayu Z and Yanchun X Dynamic analysis of gear and rack transmission system 2014 *The Open Mechanical Engineering Journal* **8** 662-67

[18] Hassan A, Thanigaiyarasu G, and Ramamurti V Effects of natural frequency and rotational speed on dynamic stress in spur gear 2008 *International Journal of Mechanical and Mechatronics Engineering* **2** 1310-18

[19] Ramesh Babu V, Prasanth T, Chowdary Y and Reddy V Active vibration control of a composite beam 2018 *Int. J. of Mechanical Engineering and Technology* **9** 353-63.
[20] Gabbert U and Ringwelski S Active vibration and noise control of a car engine: modeling and experimental validation 2014 *Mechanics and Model-Based Control of Advanced Engineering Systems,* 123-35

IOP Conf. Series: Materials Science and Engineering 1123 (2021) 012050 doi:10.1088/1757-899X/1123/1/012050

Front, rear and side impact analysis of backbone chassis of a compact sports car

Umang Bharatkumar Ramaiya, Ramesh Babu Vemuluri*, SivaPrasad Darla, Edwin Sudhagar P, Ashok B
School of Mechanical Engineering, Vellore Institute of Technology, Vellore-632014, TamilNadu, India.

Corresponding Author email ID*: rameshbabu.v@vit.ac.in

Abstract. The automotive chassis is the main load carrier and energy absorbing component in all crash events namely, frontal impacts, rear impacts, side impacts and roll over. It is important to have a strong and light weight material for the chassis. The purpose of this paper is to observe how a backbone chassis behaves during collision. This work involves static, dynamic and modal analysis of a backbone chassis of Lotus Europa, a compact sports car. Deformation, stress and strain distribution are evaluated and factor of safety is calculated using static structural analysis. In modal analysis mode shapes for different natural frequencies of the system are presented. Further, impact analysis performed using explicit dynamic analysis compares the deformation and stress distribution for front, rear and side impact for different impact velocities. Two materials namely mild steel and aluminium are compared for impact analysis.

1. Introduction

An automotive chassis also known as a vehicle frame is the skeleton system of a vehicle which holds all the components of a vehicle intact ranging from powertrain to vehicle controls. It is the primary structural component where all automotive components are attached either by welding or by bolted joints. An automotive chassis not only acts as a carrying unit but also provides protection to passengers during impacts or collisions. The three major types of impacts experienced by an automobile are frontal impacts, rear impacts and the deadliest - side impacts. An ideal automotive chassis has the perfect characteristics and is most suited for an automobile. It is light weight and at the same time strong, easy and less expensive to manufacture, provides good ventilation and space to passengers. Most importantly it has the ability to withstand static, dynamic loads and provides protection from impacts. The automotive chassis is basically of two types conventional and non-conventional. The backbone chassis is a type of conventional chassis and has a tubular backbone which helps connect the front and rear suspensions attachment areas. It has a very compact design hence can be used for small sports cars. Ananda *et. al.,* [1] designed a front car bumper to withstand frontal impacts. Static analysis was performed to get maximum stress and deformation for different shell thicknesses, to check if design is safe based on strength criteria. Dynamic analysis was performed to check maximum amplitude and dynamic conditions for safe design. Tyagi [2] designed a chrome alloy steel space frame tubular chassis for a Formula 1 student car and analysed it for maximum stress and deformation. Boundary conditions for the analysis were loads acting on the chassis under stationary condition. Finally, the design was proven safe for manufacturing. Arshad [3] studied the dynamic response of a frame structure subjected to point loads by performing vibration analysis using finite element and numerical time integration methods. The effect of speed of moving load was investigated along with the spring stiffness effect attached to the frame. Chandan *et. al.,* [4] had made an approach to design a space frame formulae student car. Analysis was performed to evaluate stresses during front, rear and side impacts. Modal analysis was also performed to get displacement at different modes. Dasaradh *et. al.,* [5] had designed a heavy vehicle ladder chassis using SolidWorks and performed static, dynamic and modal analysis using ANSYS to get stress distributions and deformations for different materials and three different thickness of crossbar. Chinnamaddaiaah *et. al.,* [6] designed a go-kart chassis using Catia V5 and analysed it for front, rear and side impacts to find stress concentrations and weak points of the chassis. Someswara *et. al.,* [7] had designed a standard ladder chassis and analyzed it for maximum shear stress, Von-Mises stress and deformation by assigning different materials such as aluminum and steel alloys under static loading. The design was based on the least weight and highest strength. Veeraswamy and Sudheer [8] designed a side door composite beam to replace existing steel beam to minimize weight and improve resistance against side impact. They performed static analysis to compare the total energy absorptions of the beams. Carbon/epoxy was used as replacement for steel. Manpreet *et. al.,* [9] designed a ladder chassis and performed static structural analysis to find the maximum stress acting and the hotspots where design modifications would be required. Further, verified the results with theoretical results and found good agreement. Monika and Razik [10] reviewed various analysis techniques used to analyze a ladder chassis by referring the literature available. Ravi *et. al.*, [11] designed and performed structural analysis on heavy vehicle ladder chassis made of steel and different polymer composites having C-section, I-section and box channel cross-sections. Further, various studies such as deformation versus load and stress versus load graphs, a comparison of stiffness, strength and weight were presented. Finally, chassis design with least weight and highest strength and stiffness was chosen. Piyush and Akshay [12] designed a space frame chassis for a formula student car and then calculated stress and deformations based on front, rear, side impact, torsional and rollover using finite element analysis. Baskara *et. al.*, [13] made a 3-dimensional computer aided design (CAD) model of a space frame chassis and analyzed it for front and side impacts using ANSYS software. Further, torsional stiffness was also evaluated for the chassis. Prajwal *et. al.*, [14] first modelled a space frame tubular chassis and analysed it for stress and deformation using static analysis and then manufactured the chassis and performed actual testing on it for verification. Sairam and Gopinath [15] performed static and dynamic analysis on Tatra chassis which is of backbone type to calculate maximum stress acting. Limited modifications were carried out by adding stiffeners of various thicknesses to the C-channel to enhance the load carrying capacity. Chandrakant and Prashant [16] designed a door intrusion member for side impact protection of Hyundai Verna car and analyzed it for deformation and stress distribution during side impact loading. The results were verified by Universal Testing Machine (UTM) testing results of the intrusion member. By observing the above literature it is evident that static structural, modal analysis and impact analysis on the backbone chassis is not presented.

In the present paper, static structural analysis, modal analysis and impact analysis of the backbone chassis is presented. The main aim of this paper is to observe how a backbone chassis Lotus Europa behaves during collision. Deformation, stress and strain distribution are evaluated and factor of safety is calculated using static structural analysis. In modal analysis mode shapes for different natural frequencies of the system are presented. Further, impact analysis performed using explicit dynamic analysis compares the deformation and stress distribution for front, rear and side impact for different impact velocities. Two materials namely mild steel and aluminium are compared for impact analysis.

2. Modelling

A 3D model of a backbone chassis frame is designed using SolidWorks software as shown in figure 1. Chassis dimensions are from the Lotus Europa backbone chassis referred from a manufacturing manual of Lotus Europa chassis. The static structural analysis, modal analysis and impact analysis is performed by using ANSYS 16. For impact analysis, a separate impact wall is modelled and considered for front, side and rear impacts. Boundary conditions for front impacts with wall at front end of the chassis is shown in figure 2. Two different velocities of approaches such as 50 kilometer per hour (kmph) and 100 kmph of the chassis for front and rear impact are set in a single direction, impact wall approaches the chassis with same velocities for side impact.

IOP Conf. Series: Materials Science and Engineering 1123 (2021) 012050 doi:10.1088/1757-899X/1123/1/012050

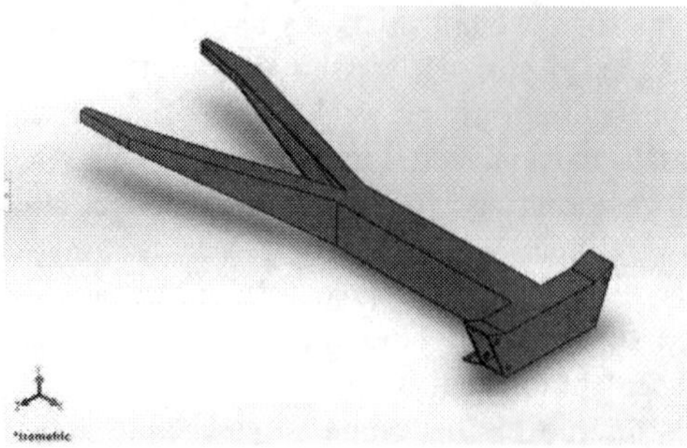

Figure 1. Backbone chassis CAD model

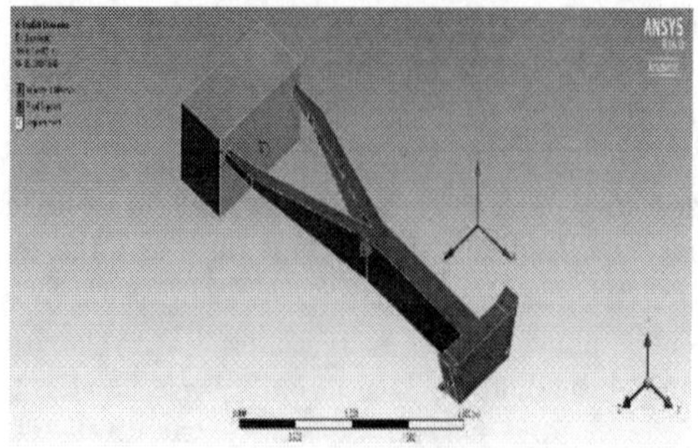

Figure 2. Boundary conditions – front impact

3. Results and Discussion

3.1. Static Structural Analysis

Static structural analysis is performed to get total deformation as shown in figure 3, stress as shown in figure 4 and strain. Factor of safety is calculated to check the design safety. The material used for analysis is mild steel.

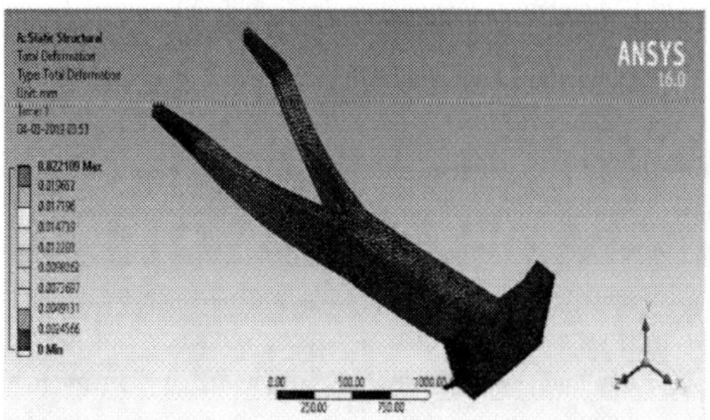

Figure 3. Total deformation – static structural

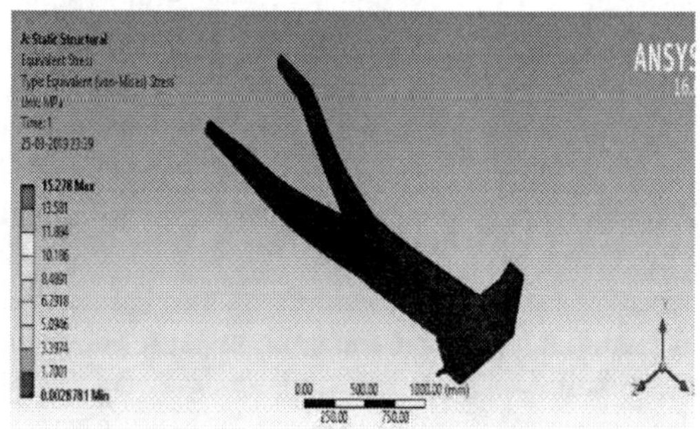

Figure 4. Von-Mises stress – static structural

Since only body forces are acting and no particular external forces are applied, hence the deformation is negligible so is the strain. The maximum induced stress is only 15.27 MPa as shown in table 1. Hence, the factor of safety which is the ratio yield strength and induced stress is 16.06. The design is thereby safe.

Table 1. Static structural analysis result.

Title	Deformation (mm)	Stress (MPa)	Strain (mm/mm)
Maximum	0.022	15.278	6.72×10^{-5}
Minimum	0	0.004	2.63×10^{-8}

4. Modal Analysis

Analysis is run and five different mode shapes at particular natural frequencies are obtained. Figure 5 shows the mode shape and Von-Mises at fifth natural frequency. Deformation and stress for these five mode shapes are noted. Mild steel is the material assigned. Induced stresses get amplified at specific natural frequencies for different mode shapes. The deformations at these frequencies are also at peak. Following table 2 describes the deformations and stress values of first five mode shapes.

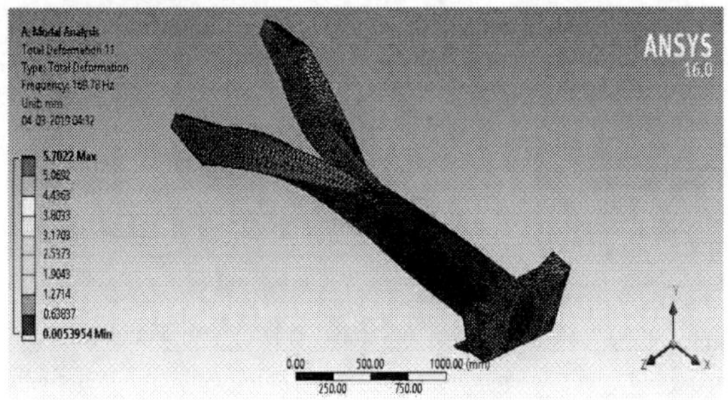

Figure 5. Mode Shape and Von-Mises Stress

Table 2. Modal Analysis Result.

Mode shape	Natural frequency (Hz)	Maximum Stress (MPa)	Maximum Deformation (mm)
1	46.21	311.33	6.60
2	81.48	318.04	5.92
3	110.87	258.09	7.29
4	141.79	368.34	7.50
5	169.78	898.68	5.70

5. Impact Analysis –Front End

It is observed that the deformations increase as the velocity of impact increases. Interestingly stress increases for mild steel and for aluminium stress decreases with increase of impact velocity. Figure 6-9 shows the total deformation and Von-Mises stress for mild steel material at 50 kmph and 100 kmph impacts. Similarly, figure 10-13 shows the results for aluminium material. Summarised results of front impact for mild steel and aluminium are presented in table 3 and table 4 respectively.

Figure 6. Total deformation for MS at 50 kmph

Figure 7. Von-Mises stress for MS at 50 kmph

Figure 8. Total deformation for MS at 100 kmph

Figure 9. Von-Mises stress for MS at 100 kmph

Figure 10. Total deformation for AL at 50 kmph

Figure 11. Von-Mises stress for AL at 50 kmph

Figure 12. Total deformation for AL at 100 kmph

Figure 13. Von-Mises stress for AL at 100 kmph

Table 3. Front Impact Analysis – Mild Steel.

Impact Velocity (kmph)	Maximum total deformation (mm)	Maximum stress (Pa)
50	239.66	2.782×10^{11}
100	284.07	2.903×10^{11}

Table 4. Front Impact Analysis – Aluminum.

Impact Velocity (kmph)	Maximum total deformation (mm)	Maximum stress (Pa)
50	319.15	2.906×10^{11}
100	378.15	2.804×10^{11}

Front impact leads to chassis failure as the induced stress is much higher than yield strength of both the materials. Aluminium chassis is more vulnerable to failure.

6. Impact Analysis –Rear End

For crash analysis at the rear end of the chassis, it is observed that the deformations as well as induced stresses increase with the increase of impact velocity from 50 kmph to 100 kmph. Figure 14 and 15 shows the total deformation and Von-Mises stress for mild steel at 50 kmph impact velocity from rear end respectively and similarly, figure 16 and figure 17 shows for aluminium. Summarised results of rear impact for mild steel and aluminium are presented in table 5 and table 6 respectively.

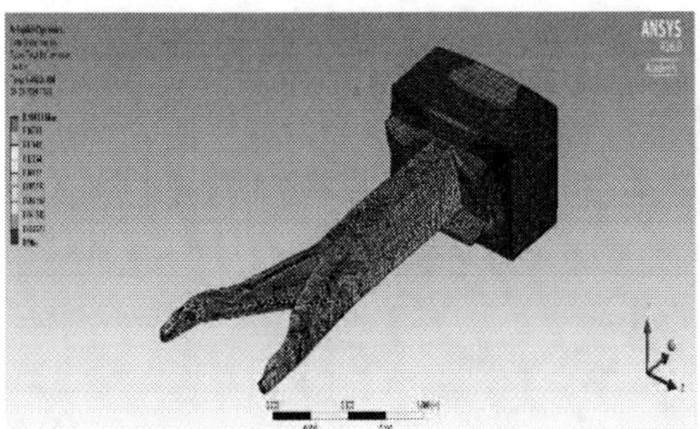

Figure 14. Total deformation of MS – rear end

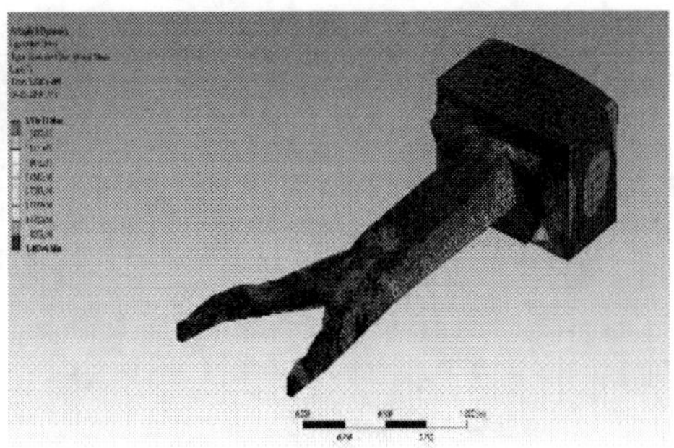

Figure 15. Von-Mises stress of MS – rear end

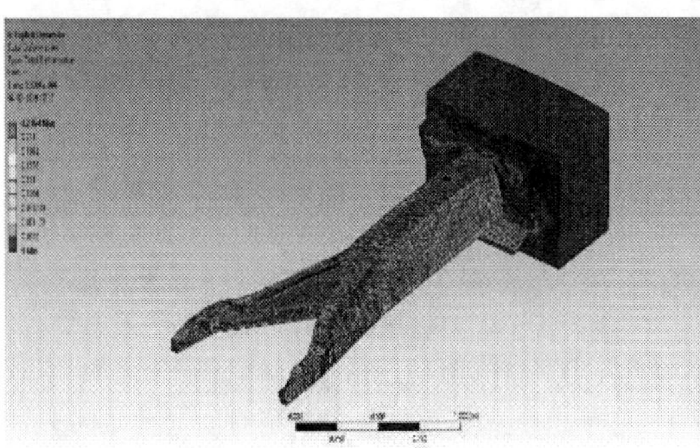

Figure 16. Total deformation of AL – rear end

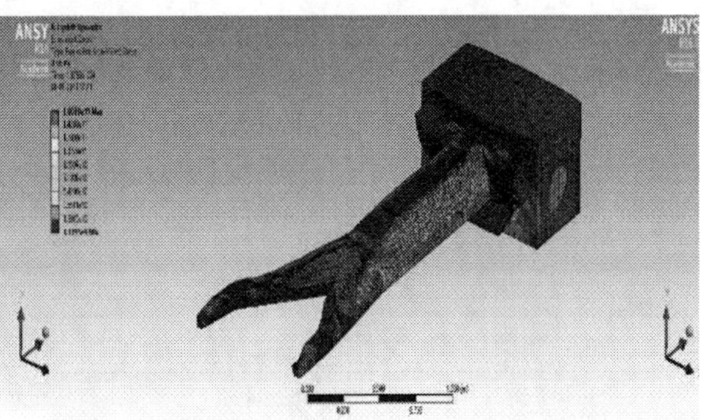

Figure 17. Von-Mises stress of AL – rear end

Table 5. Rear Impact Analysis – Mild Steel.

Impact Velocity (kmph)	Maximum total deformation (mm)	Maximum stress (Pa)
50	188.31	1.51×10^{11}
100	238.70	1.606×10^{11}

Table 6. Rear Impact Analysis – Aluminum.

Impact Velocity (kmph)	Maximum total deformation (mm)	Maximum stress (Pa)
50	239.40	1.603×10^{11}
100	305.72	1.757×10^{11}

Rear impact leads to chassis failure as the induced stress is much higher than yield strength of both the materials. Aluminium chassis is more vulnerable to failure. Moreover, deformations obtained during rear impact are lesser than front impact.

7. Impact Analysis – Side

It is observed that the deformations as well as induced stresses increase as the velocity of impact is increased. Figure 18 and 19 shows the total deformation and Von-Mises stress for mild steel at 50 kmph impact velocity from side of the chassis respectively and similarly, figure 20 and figure 21 shows for aluminium. Summarised results of side impact for mild steel and aluminium are presented in table 5 and table 6 respectively.

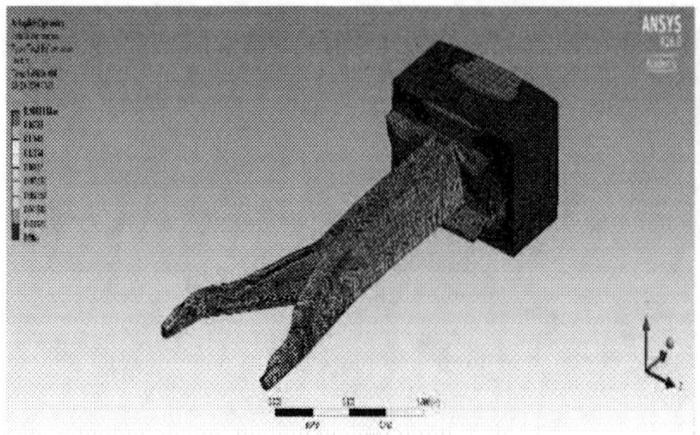

Figure 18. Total deformation of MS – Side

Figure 19. Von-Mises stress of MS-side

Figure 20. Total deformation of AL – Side

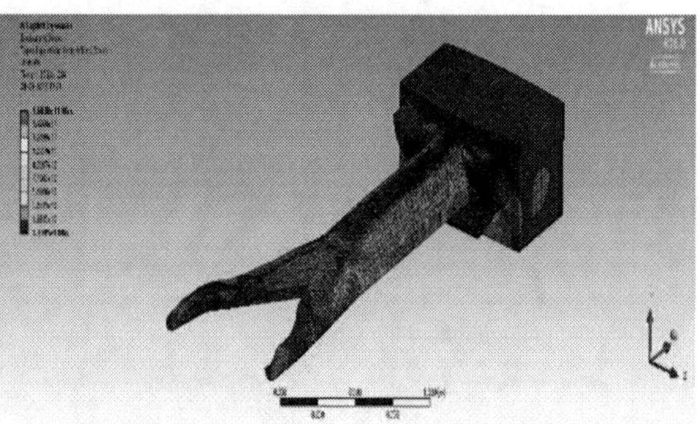

Figure 21. Von-Mises stress of AL – Side

Table 7. Side Impact Analysis – Mild Steel.

Impact Velocity (kmph)	Maximum total deformation (mm)	Maximum stress (Pa)
50	188.31	1.51×10^{11}
100	238.70	1.606×10^{11}

Table 8. Side Impact Analysis – Aluminum.

Impact Velocity (kmph)	Maximum total deformation (mm)	Maximum stress (Pa)
50	239.40	1.603×10^{11}
100	305.72	1.757×10^{11}

Rear impact leads to chassis failure as the induced stress is much higher than yield strength of both the materials. Aluminium chassis is more vulnerable to failure. Moreover, deformations obtained during rear impact are lesser than front impact. It is observed that the overall deformation for all impact velocities for side impact is less compared to front and rear impact. This could be due to bending resistance is higher for this chassis. Furthermore, there is not much difference observed in the results obtained for mild steel and aluminium materials.

8. Conclusion

In this paper, Backbone chassis is considered for the static, modal and impact analysis by varying materials such as mild steel and aluminium. Through static structural analysis it is found that the design is safe in terms of factor of safety. Mode shapes and their natural frequencies are found using modal analysis these are the instances where any load applied on the chassis gets amplified. Impact analysis using explicit dynamics for front, rear and side impact illustrates a comparison of stress and deformation of chassis under the action impacts at different velocities such as 50 and 100 kmph. By comparing the materials used in impact analysis it is concluded that mild steel is more preferable over aluminium since it is less vulnerable to failure. Finally, Front impact causes the highest damage to the backbone chassis and lowest damage in side impact. This article will help the researchers or industrialist who would like to work with backbone chassis.

9. Codes and Standards

The following standards have been used through the paper to conduct various analysis in ANSYS 16 Academic.

1. ASTM A36 (Mild Steel)
2. ASTM B209 – 14 (Aluminum)

10. References

[1] Ananda Babu, Vijay Praveen and Venkateswarao M Crash analysis of car chassis frame using finite element method 2012 *International journal of engineering research and technology*, **1** 1-8

[2] Apoorva Tyagi Design and analysis of a space frame tubular chassis for a formula student car 2016 *International journal of innovative research in science, engineering and technology*, **5** 16451-6

[3] Arshad Mehmood Using finite element method vibration analysis of frame structure subjected to moving loads 2015 *International journal of mechanical engineering and robotics research,* **4** 50-65

[4] Chandan S N, Sandeep G M and Vinayaka N Design, analysis and optimization of race car chassis for its structural performance 2016 *International journal of engineering research and technology* **5** 361-7

[5] Dasaradh Palagiri, Chavali Joy Davidson and Ramesh Babu Vemuluri Static, dynamic and harmonic analysis of heavy vehicle chassis system 2017 *International journal of research in mechanical engineering and technology* **70** 44-7

[6] Chinnamaddaiaah, K, Lakshmipathi Y and Ravikanth Raju P Modelling and structural analysis of a go-kart vehicle chassis frame 2017 *International journal of mechanical engineering and technology* **8** 305-11

[7] Someswara Rao K, Pradeep Kumar and Sai Kumar B Design and analysis of light weighted Chassis 2017 *International journal of mechanical engineering and technology*, **8** 96-103

[8] Veeraswamy K and Venkata Sudheer Design and analysis of a composite beam for side impact protection of a car door 2016 *International research journal of engineering and technology* **3** 464-9

[9] Manpreet Singh Bajwa, Yatin Raturi and Amit Joshi Static load analysis of tata super ace chassis and its verification using solid mechanics 2013 *International of mechanical and production engineering* **1** 55-8

[10] Monika S Agarwal and Md Razik A review on study of analysis of chassis 2013 *International journal of modern engineering research* **3** 1135-8

[11] Ravi Chandra M, Sreenivasulu S and Syed Altaf Hussain Modelling and structural analysis for heavy vehicle chassis made of polymeric composite material by three different cross sections 2012 *International journal of modern engineering research*, **2** 2594-600

[12] Piyush Ram Shahade and Akshay Kumar Kaware Structural performance analysis of formulae SAE car 2014 *International journal of pure and applied research in engineering and technology* **2** 46-61

[13] Baskara Sethupathi P, Chandradass J and Shubham Sharma Design and optimization of FSAE chassis using FEA 2018 *In proceedings of international conference on advances in mechanical engineering* **402** 012184

[14] Prajwal Kumar, Vivek Muralidharan and Madhusudhana G Design and analysis of tubular space frame chassis of high performance race car 2014 *International journal of research in engineering and technology* **3** 497-501

[15] Sairam Kotari and Gopinath V Static and dynamic analysis on tatra chassis 2015 *International journal of modern engineering research* **2** 86-94

[16] Sarang C and Prashantkumar Design and experimentation of side impact beam for Hyundai Verna 2017 *International research journal of engineering and technology* **4** 62-9

A review on Graphene Reinforced Composites for Ballistic applications

Tanay Kuclourya[1], Shubham Mudliar[2], Mohit Kumar Jain[3] and Narendiranath Babu Thamba[4]

[1,2,3,4]School of Mechanical Engineering, Vellore Institute of Technology, Vellore, India
Email:tanay.kuclourya@gmail.com
Email:shubham55086@gmail.com
Email:mohitkumarjain275@gmail.com
Email:narendiranathbabu.t@vit.ac.in

Abstract - The review paper emphasizes mainly on the hybridization of coir fiber and spider silk with carbon fiber and different percentage compositions of graphene, with an ultimate aim as to how the strength of these hybridized composites can contribute in the fields of material and manufacturing engineering. Firstly, there is an analysis of different properties of coir, carbon fibers, graphene and spider silk. Secondly, there is an analysis of the ballistic test performed on the specimens. In this review, some very basic yet important information is collected and discussed which will help the future researchers in the preparation of hybrid composites involving graphene. Being a natural fiber, the bristle coir fibers are pre-treated with alkali (NaOH) and then acidic treatment is done to remove the excess alkali. After the fibers become neutral, permanganate treatment is done. In case of spider silk, no pre-treatment is done. These fibers are reinforced with epoxy and different weight percentage of graphene (1wt%, 2wt%, 4wt%, 6wt %). Carbon fibre is incorporated in between using conventional hand layup technique. The epoxy and graphene mixture is sonicated to make sure that a uniform and good mixture is obtained. The review facilitates to perform ballistic test on the samples as the results from this test can substantiate to deliver a vital improvement in the field of material selection for body armors.

Keywords: hybridization, coir fiber, spider silk, carbon fiber, graphene, ballistic test, alkali treatment, acidic treatment, permanganate treatment, epoxy, hand layup technique, sonicated and body armors.

1. Introduction

In the coming years, a number of industrial sectors will be utilizing natural fibers, including automotive, building, packaging, construction and aerospace industries. Natural fiber composites can be made renewable with comparatively low prices and low material density [1]. Even High Density Polythene (HDPE) which is non-biodegradable can be replaced with the biodegradable natural fibers [2]. However, the challenge still remains to expand the usage of green fibers composites. This can be done only when the cost of production with the properties of these matrixes likes mechanical and heat are parallel to the traditionally used materials [3]. Rural areas have developed their economy by virtue of incorporating natural fibers in various different applications and industries [4]. Natural fibers exhibit an upper hand over man-made fibers as they are cheaper, lighter, smooth with the equipment, safe to the skin, have comparable mechanical properties, less energy consumption and biodegradable [5]. However, having said so, there is still a hybridization of manufactured or artificial fibers such as e-glass fibers and carbon fibers with natural fibers to enhance the properties of the latter in the area of strength. Graphene is often used as reinforcing material to impart strength. From studies, it has been found that interface to volume ratio along with filler size are the governing factors for enhancing the mechanical strength of the composites. Graphene has emerged as outstanding filler for matrix enhancement. The oxygen functional groups in the Graphene oxide impart strong binding to the matrix [6]. Many researches have even shown that the hybrid composites of Carbon fiber and epoxy have better mechanical properties than plain composites [7]. Whenever there is a requirement of a combination of low density and good mechanical property, Carbon fibers (CFs) have an edge over glass fibers due to their high stiffness [8]. From these references (6-8), it is clear that till now there is no such systematic mechanical testing done on fiber composites involving carbon fiber and graphene with any other natural fiber. Through this review, some very important characteristics of Coir fiber, carbon fiber, Graphene and spider silk are discussed, with a view that in coming future, researchers can think of forming hybrid composites consisting of carbon fiber, graphene and coir fiber (natural fiber).

1.1 Coir

When talked about natural fibers, one of the very potential fibers is the coir fiber. It is the hairy casing of the coconut shell. It possesses about 48% of lignin increasing strength and elasticity of fibre [9]. There is an annual production of around 250,000 tons of coir. One very interesting fact is out of the total production of the white coir in the entire world, 60% is produced from an Indian state- Kerala. Our neighboring country, Sri Lanka too has a significant contribution of 36% in the production of brown coir throughout the world. However, half of the total coir fiber produced in a year across the world is consumed by India alone [10].

1.1.1. Coir hybrid composites

Synthetic fibers if replaced by natural fibers, will contribute towards sustainable development and reduction in greenhouse emissions [11]. When a coir fiber is added into a plastic material its tensile strength, impact strength and hardness increases in comparison with pure plastic specimen because fibers provide strength to reinforced composites.

Figure. 1 Bristle coir fibers

As the fiber wt % increases in composite interface bonding between fiber and resin keeps on decreasing after a point [12]. An investigation deduced that there is a direct relation between the distance and amount of the fiber with water absorption rate and void content in coir fiber-epoxy composites [13-14]. Arrangement of fibers in the synthesis have a great effect on the composites related to mechanical behavior specially strength, it is observed that longitudinal direction of fibers displayed better mechanical properties than in the transverse direction [15]. Composite is considered good when we have proper interfacial bonding in composite with less dissipation of energy. [16]. Whenever there is an application of tensile load on the short fiber composites, there is a transfer of load to fiber-matrix surface. It takes place by virtue of shear force. The stresses are zero at the end of the composites, while it increases along the fiber length, hence the composite should have a minimum length which is called as the critical length (Lc) so that the tensile loading process is done precisely [17].

1.1.2. Alkali treatment of coir fibers

Raw unprocessed fibers are treated with chemicals in order to be used for any research purpose. The sole aim is to remove the excess lignin content from the fibers. Having said so, there are ill-effects of using excessive alkali. Abundance of oxidant concentration may lead to the loss of the fibrous amount of the fiber which ultimately reduces the crystallinity of the end product. Some researchers have deduced that, when raw fibers are treated with 5% NaOH, composites showed an increase by 34X in tensile strength whereas the load carrying capacity increases 9 folds when compared with composites made from the non-treated fibers. After the alkaline treatment, the strength can further be increased, if the fibers are treated with 3% Permanganate. This increases the load carrying capacity by 1.3X. It also reduces the deflection by 2.9 times with respect to the pure epoxy sample. Surface roughness by virtue of alkaline treatment is the basic reason for the easy bonding of the fibers [18]. As discussed earlier, alkaline treatment helps to remove excess lignin as well as wax and oils. These constituents otherwise contribute in the formation of a weak boundary layer as they reduce the bonding among the fibers with the composite [19]. There has been a substantial improvement observed in the flexural as well as tensile properties of sisal-carbon hybridized composites when alkali treated fibers were used in place of untreated fibers. The reason was again the increase in the surface roughness topography by the action of hemicellulose removal. This topography imparted better adhesion to the fibers which led to enhanced mechanical properties [20].

1.1.3. Coir fiber mass content

Lignin content is high in the case of coir fiber which is contrasting to other natural fibers, which makes it more durable than other fibers [21]. Coir fiber mass content is important in deciding the synthesis properties like mechanical. When coir fiber mass content was taken as 25%, optimum mechanical properties were obtained as the tensile strength increased by 54.5% and flexural strength by 45.7%, when compared to the pure resin [22]. From various researches, it was concluded that nano-graphite particles when used like strengthening substance, highly increased the synthesis behavioral properties in context to mechanical. [19].

1.2. Carbon fiber

From various studies, it is found that properties of any polymer matrix can be enhanced by hybridization in context to mechanical [23]. For advanced structural applications, plant and synthetic fibers are being hybridized in the nano filler modified polymeric matrix. This interesting research approach has succeeded in obtaining excellent viscoelastic, mechanical and thermal properties. The process of hybridization has reduced the degradation of environment as these have reduced the dependency on synthetic fibers. Carbon fiber has been shown in figure 2.

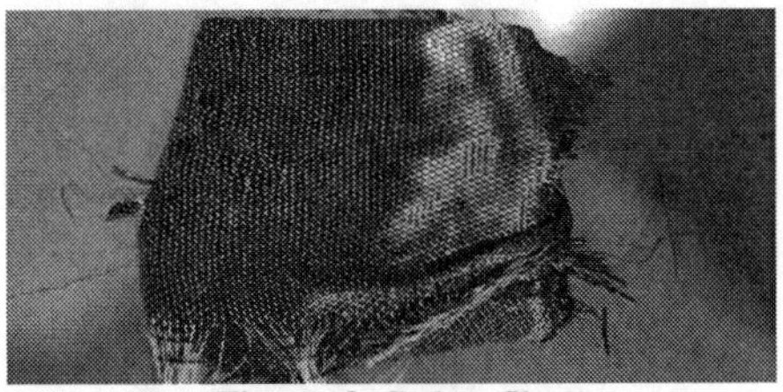

Figure. 2 Carbon fiber

From studies conducting dynamic and thermogravimetic mechanical analysis, it was observed, at elevated temperatures, Kevlar/epoxy polymeric structures can be efficiently replaced by the hybrid composites [24]. One of the most potential fibers of the hybridization process is carbon nano-fiber (CNF). These fibers, as nano fillers, have found an effective use in enhancing the laminated composite's shear strength between the laminar. From studies, it has been observed that on addition of organo-clay, properties such as fracture toughness and crack growth resistance underwent significant improvement related to carbon and different polymer matrixes. The mixing of GnPs into epoxy matrix can be done through ultrasonic sonication [25]. Use of natural plant fibers actually positively adds up to the carbon expense as many researches have shown confiscating carbon dioxide nature of these strands of natural fibres [26]. The epoxy resins which are reinforced with carbon fibers are used for many useful purposes such as bridge construction, fuel cells and even aerospace industry [27].

1.3. Graphene

Graphene, two dimensional single atom thick honeycomb material has been considered as a highly reliable substance because of its excellent electrical conductivity, unique thermal and mechanical properties. It has many potential applications in clean energy, sensors, electronic and photonic devices [28]. Graphene is the strongest materials of carbon nanotechnology. Graphene oxide (GO) powder can be shown in figure 3.

1.3.1. Graphene reinforced composites

The thickness of a monolayer of graphene 2D isotropic sheet is around 0.335mm, but still it has an exceptional Young's modulus of 1 TPa and strength of 125 GPa. With a calculated value of 0.25-0.30, graphene has a very high failure strain [29]. Tensile strength reaches maximum with a small amount of graphene fillers added, beyond that strength decreases [30].

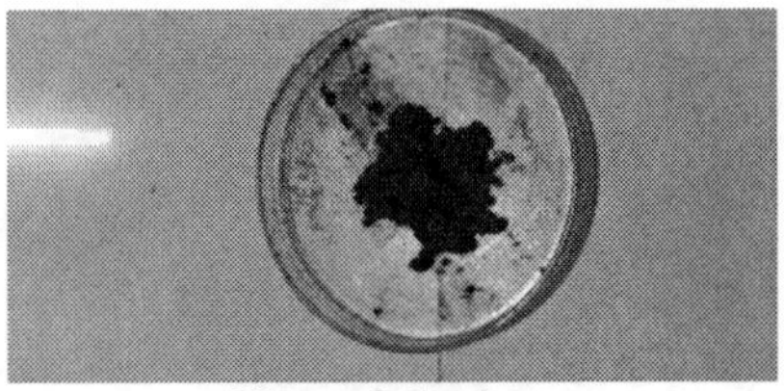

Figure. 3 Graphene

In addition to this, graphene doesn't intensify the resistance to the flow of resin in comparison to what carbon nanotubes do, empowering the consolidation of higher concentration into the arrangement. [31]. Graphene can reinforce the compound; however, this fortification stays restricted in view of the huge surface vitality that is created by the solid van der Waals bond [32]. Studies have likewise demonstrated that Graphene oxide (GO) sheets have upgraded the interfacial just as rigidity of carbon fiber fortified composites [33].

1.3.2. Graphene integrated with epoxy

In the current scenario, nano-particle-reinforced polymer composites are most promising and reliable in the field of future tech designing applications with a negligible scattering of nano-particles. Another important component of a composite is epoxy (as shown in figure 4). Right now, improved adhesive saps are widely utilized in the manufacture of natural fiber-fortified composites and in making its diverse modern items because of their predominant properties in the field of mechanical, thermal and electrical [34].

It is a versatile thermosetting polymer. However, due to its inherent hardness and brittle nature, it cannot provide good strength required for high performance applications. To overcome this, epoxy resin can be modified using potential nano-materials such as

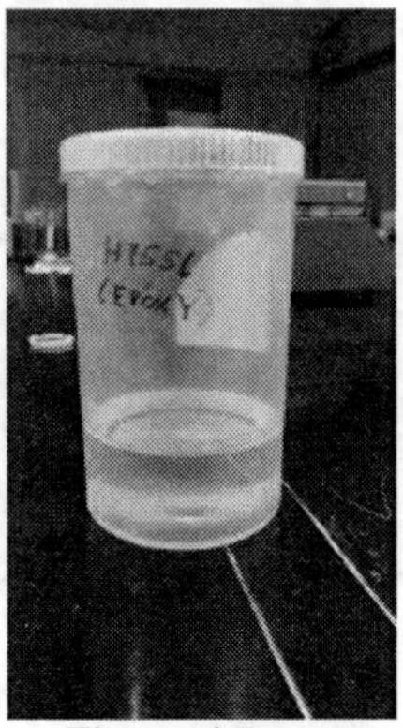

Figure. 4 Epoxy

graphene oxide (GO), graphite nano-platelet and graphene with natural or synthetic fiber as a reinforcing agent. This enhances the strength and stiffness of the composites. This process of nano modification can lead to several enhanced properties which are model dominated such as mechanical, electrical and physical. For the proper mixing and interaction of GO with epoxy resin and fibers, Cetyl trimethyl ammonium bromide (CTAB) treatment is done, which brings about upgraded mechanical properties of the matrix. From recent studies, it has been observed that 5.0 wt. % GO composites, when contrasted with flawless epoxy structures showed greater mechanical properties because of fine scattering of GO in epoxy matrix by CTAB [35-36].

1.3.3. Graphene volume content

It is also seen that graphene volume fraction has a linear relation with the toughness of the amalgamation. When the volume division is high, it results in graphene aggregation which ultimately leads to a dramatic decrement of both the modulus and strength [37]. With this new and effective approach of using GO, various opportunities can be created to develop various good potential high-performance composites for industrial applications [38]. There was an enhancement in vitality retention of the 9 layered Kevlar texture based epoxy combination on addition of graphene nano-platelets (GnPs) [39].

1.4. Spider Silk

Another potential synthetic fiber is silk. It is a protein containing fiber. It is normally created by numerous spineless creatures that incorporate web spinners (Embioptera), trichopteran hatchlings, some Hymenoptera, and all creepy crawlies (Araneidae) [40].

Figure. 5 Spider Silk

Among these, the most impressive silk toolkits are of web builders. Silks and especially spider silks are exceptionally tough. This durability is principally a result of the amino acid grouping and turning process [41]. Research on spider silk in the course of recent decades has begun to uncover the sub-atomic components that produce its mechanical properties. Spider silk proteins are comprised of many couple rehashes of explicit, exceptionally saved amino acids themes. Distinctive amino acid themes contribute explicit mechanical properties to the fiber. Major Ampullate silk, the most grounded kind of silk, is comprised of two proteins: Major ampullate Spidroin 1 (MaSp1) and Major ampullate Spidroin 2 (MaSp2). These two proteins each loan quality and flexibility to the fiber through extraordinary tandemly rehashed amino corrosive themes, which are profoundly rationed all through the sphere web weaving creepy crawly species [42]. Some important literature is shown in Table 1.

Table 1. Important conclusions from literature survey

Sr. No.	Authors	Year of publication	Main findings	Uncertainty
1.	Sean J. Blamires[1], Chung-Lin Wu[3], Todd A. Blackledge[4], and I-Min Tso[1,2]	2012	Post-emission handling causes variety in the mechanical exhibition of wild MA silk free of protein organization or turning speed crosswise over 10 types of spider.	Post discharge strands get influenced by different factors, for example, hereditary impacts, phylogenetic sign, post-emission physiological and biochemical procedures and variables acting during turning and post-turning
2.	Sean J. Blamires[1,2,*] Todd A. Blackledge[3] and I-Min Tso[1]	2016	Web-developing creepy crawlies produce to seven distinct kinds of silk, every one of which is discharged from various organs, has various capacities, and has diverse substance and physical properties.	It is basic to comprehend the characteristic procedures that reason spider silk properties to differ among species (phylogenetic inconstancy) and inside species (silk versatility).
3.	Andreas Koeppel and Chris Holland*	2017	This audit gives a diagram of the various methodologies for fake silk fiber turning and analyzes all distributed fiber properties to date which has distinguished future patterns and difficulties out and about towards reproducing elite silks.	Each approach has its very own difficulties to defeat encompassing parts of the feedstock/dope, fiber turning, and post processing: for the entrenched wet turning, apparently post processing through the enduring diminishing in fiber distance across close by an expansion in post draw proportion has seen critical upgrades in mechanical properties.
4.	Fábio de Oliveira Bragaa,b*, Lucas Tedesco Bolzana , Flávio James Humberto Tommasini Vieira Ramosa , Sergio Neves Monteiroa , Édio Pereira Lima Jr.a , Luis Carlos da Silvaa	2018	Among all the sisal fibre-fortified polyester composites considered in the present work, that with 30 vol. % displayed the best outcomes, as far as injury counteractive action and physical honesty after the effect. In this way, it was viewed as the most appropriate material to be considered as MAS second layer.	The loss of physical respectability of both the flawless polyester pitch and the 10 vol. % sisal fiber polyester composite makes them unacceptable to be utilized as a second layer of MAS in multi hit protection.
5.	R.Yahaya[1,3], S.M. Sapuan[1,2,4*] , M.Jawaid[2,5], z. Leman[1] and E.S. Zainudin[1,2]	2015	Natural fiber composites may endure when the material is presented to antagonistic situations for long periods of time.	A reduction in elastic properties of the composites was illustrated, demonstrating an extraordinary misfortune in mechanical properties of the muggy examples contrasted with the dry samples.
6.	Navaranjan N and Neitzert T	2017	Izod and Charpy techniques were at first produced for metals and	A post assessment of morphology of a NFC test after an effect test is essential to portray the material.

			the test outcomes were generally used to look at the effect reaction of the materials.	

1.5. Ballistic Test

Engineering materials possess very important mechanical property- Impact behaviour. This property is governed by various parameters such as strength, length and orientation of fibres, elastic modulus, and fibre-matrix interfacial bond strength. The type of test performed to measure the impact energy is also one important factor. Some simple methods to perform the impact test are Charpy and Izod impact test methods which are also known as pendulum methods. In these methods, the energy absorbed by the notched and the unnotched sample of a material during fracture is taken into account. This absorbed energy acts as a tool to study the ductile-brittle transition temperature. It also gives a measure of a given material's toughness. One more important test which can be performed on the composites is the ballistic test. This particular test is used to evaluate the impact resistance of the composite panels and products such as composite armors. In this test, a projectile is fired with excessive momentum on the result to be tested. The localized damage is determined after the impact [43].

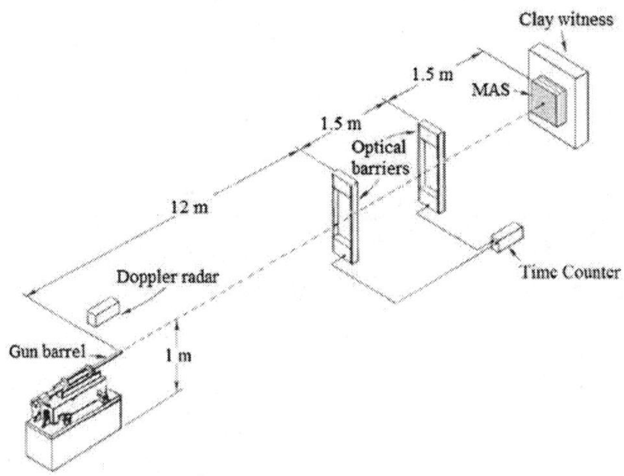

Figure. 6 Schematic view of the Ballistic Test arrangement

Because of their low density, large fracture resistance, high energy adsorption and excellent stiffness and strength, nano composites are appropriate choice for ballistic armor material. A typical Multilayer Armor System (MAS) is of ceramic from the front side further its accompanied by a layer of fiber layer composite and aluminum is present at last playing a role of back layer [44]. The first layer, which is the harder front of the ceramic plate has deforms, erodes and shatters bullet projectile. The second layer is generally made of composite material such as coir which soaks up the high speed projectile shards and the ceramic's energy. The third layer which acts as a final barrier normally involves the use of a ductile metallic layer. The second layer if substituted with natural fiber composites reinforced with graphene improves the impact absorption reduces the overall weight and cost of the system [29].

2. Conclusion

This review gives many important insights on composites. An attempt has been made to precisely mention all the necessary conclusions in the following points-

- Natural fibers are often hybridized with artificial man-made fibers in order to enhance the strength of the composites.
- Coir fibers contain the maximum amount of lignin content (48%) among all the natural fibers. Hence it is the toughest.
- Longitudinally oriented fiber composites exhibit better mechanical properties than transversally oriented fiber composites.
- Alkali treatment of fibers increases the strength of the composites in every aspect.

- The process of hybridization is environment friendly as the dependency on synthetic fibers is reduced.
- Use of reinforcing materials such as carbon fibers is known to exhibit excellent viscoelastic, thermal and mechanical properties to the composites.
- The strongest reinforcing material of carbon nano-technology which is graphene is mixed with epoxy through sonication process.
- Epoxy which is an excellent thermosetting polymer acts as a binding agent inside the composite.
- Synthetic fibers such as spider silk are potential replacements for natural fibers in the composites. These are comprised of amino acid themes which results in excellent mechanical properties.
- Nano composites are highly preferred choices for ballistic application due to their high fracture resistance, low density and high stiffness.

References:

[1]. Geethamma V, Kalaprasad G, Groeninckx G and Thomas S 2005 Dynamic mechanical behavior of short coir fiber reinforced natural rubber composites *Composites Part A: Applied Science and Manufacturing* **36** 1499–506

[2]. Thamba N B, Kuclourya T, Jain M K and Mangalaraja R V 2019 Mechanical Behaviour of Areca Nut, Sunn Hemp Natural Fibers and E-Glass Fibers Reinforced With Epoxy Composites *International Journal of Engineering and Advanced Technology* **8** 1092–101

[3]. Ferreira F, Pinheiro I, Souza S D, Mei L and Lona L 2019 Polymer Composites Reinforced with Natural Fibers and Nanocellulose in the Automotive Industry: A Short Review *Journal of Composites Science* **3** 51

[4]. Ashik K P and Sharma R S 2015 A Review on Mechanical Properties of Natural Fiber Reinforced Hybrid Polymer Composites *Journal of Minerals and Materials Characterization and Engineering* **3** 420–26

[5]. Sanjay M, Arpitha G and Yogesha B 2015 Study on Mechanical Properties of Natural - Glass Fibre Reinforced Polymer Hybrid Composites: A Review *Materials Today: Proceedings* **2** 2959–67

[6]. Pathak A K, Borah M, Gupta A, Yokozeki T and Dhakate S R 2016 Improved mechanical properties of carbon fiber/graphene oxide-epoxy hybrid composites *Composites Science and Technology* **135** 28–38

[7]. Hadden C, Klimek-Mcdonald D, Pineda E, King J, Reichanadter A, Miskioglu I, Gowtham S and Odegard G 2015 Mechanical properties of graphene nanoplatelet/carbon fiber/epoxy hybrid composites *Multiscale modeling and experiments Carbon* **95** 100–12

[8]. Anuar H, Ahmad S H, Rasid R, Ahmad A and Busu W N W 2007 Mechanical properties and dynamic mechanical analysis of thermoplastic-natural-rubber-reinforced short carbon fiber and kenaf fiber hybrid composites *Journal of Applied Polymer Science* **107** 4043–52

[9]. Chandel, Anoopsingh & Shah, Tanmay & Shah, Tarak & Varde, Dixit. (2016). A Comparative Strength Study of Coir Fibre Reinforced Concrete (CFRC) Over Plain Cement Concrete (PCC). *IOSR Journal of Mechanical and Civil Engineering* **13** 95-97

[10]. Harish S, Michael D P, Bensely A, Lal D M and Rajadurai A 2009 Mechanical property evaluation of natural fiber coir composite *Materials Characterization* **60** 44–9

[11]. Thamba N B, Kuclourya T, Jain M K and Mangalaraja R V 2019 Flexural properties of areca nut, sunn hemp and e-glass fibers reinforced with epoxy composites *Materials Research Express* **6** 105344

[12]. Naveen P and Raju T D 2013 Evaluation of Mechanical Properties of Coconut Coir Fiber Reinforced Polymer Matrix Composites *Journal of Nano Research* **24** 34–45

[13]. Das G and Biswas S 2016 Effect of fiber parameters on physical, mechanical and water absorption behaviour of coir fiber–epoxy composites *Journal of Reinforced Plastics and Composites* **35** 644–53

[14]. Quy H V and Nguyen S T T 2019 Experimental Analysis of Coir Fiber Sheet Reinforced Epoxy Resin Composite *IOP Conference Series: Materials Science and Engineering* **642** 012007

[15]. Geethamma V G, Joseph R and Thomas S 1995 Short coir fiber-reinforced natural rubber composites: Effects of fiber length, orientation, and alkali treatment *Journal of Applied Polymer Science* **55** 583–94

[16]. Saba N, Jawaid M, Alothman O Y and Paridah M 2016 A review on dynamic mechanical properties of natural fibre reinforced polymer composites *Construction and Building Materials* **106** 149–59

[17]. Pickering K, Efendy M A and Le T 2016 A review of recent developments in natural fibre composites and their mechanical performance *Composites Part A: Applied Science and Manufacturing* **83** 98–112

[18]. Hallad S A, Banapurmath N, Patil V, Ajarekar V S, Patil A, Godi M T and Shettar A S 2018 Graphene Reinforced Natural Fiber Nanocomposites for Structural Applications *IOP Conference Series: Materials Science and Engineering* **376** 012072

[19]. Ahmad M A, Güven G G and Sarıkavaklı N 2019 Some Features of Doping of Nano–Graphite in Natural Coir Fibre Epoxy Composites *European Journal of Science and Technology* **15** 491–98

[20]. Khanam P N, Khalil H P S A, Jawaid M, Reddy G R, Narayana C S and Naidu S V 2010 Sisal/Carbon Fibre Reinforced Hybrid Composites: Tensile, Flexural and Chemical Resistance Properties *Journal of Polymers and the Environment* **18** 727–33

[21]. Adeniyi A G, Onifade D V, Ighalo J O and Adeoye A S 2019 A review of coir fiber reinforced polymer composites *Composites Part B: Engineering* **176** 107305

[22]. Nam T H, Ogihara S, Tung N H and Kobayashi S 2011 Effect of alkali treatment on interfacial and mechanical properties of coir fiber reinforced poly(butylene succinate) biodegradable composites *Composites Part B: Engineering* **42** 1648–56

[23]. Kumar S C R, Shivanand H K, Vidayasagar H N and Nagabhushan V 2018 Studies on mechanical properties of graphene based hybrid composites reinforced with kenaf/glass fiber *AIP Conference Proceedings* **1943** 020115

[24]. Jesuarockiam N, Jawaid M, Zainudin E S, Sultan M T H and Yahaya R 2019 Enhanced Thermal and Dynamic Mechanical Properties of Synthetic/Natural Hybrid Composites with Graphene Nanoplateletes *Polymers* **11** 1085

[25]. Naveen J, Jawaid M, Zainudin E S, Sultan M T H and Yahaya R 2019 Improved Interlaminar Shear Behaviour of a New Hybrid Kevlar/Cocos Nucifera Sheath Composites with Graphene Nanoplatelets Modified Epoxy Matrix *Fibers and Polymers* **20** 1749–53

[26]. Hill C A S and Khalil H P S A 2000 Effect of fiber treatments on mechanical properties of coir or oil palm fiber reinforced polyester composites *Journal of Applied Polymer Science* **78** 1685–97

[27]. Zhiyuan Yu 2018 Carbon Fiber Reinforced Epoxy Resin Matrix Composites Materials *Science: Advanced Composite Materials* 1. 10.18063/msacm.v1i1.500.

[28]. Borah M, Dahiya M, Sharma S, Mathur R B and Dhakate S R 2014 Few Layer Graphene Derived from Wet Ball Milling of Expanded Graphite and Few Layer Graphene Based Polymer *Composite Materials Focus* **3** 300–9

[29]. Benzait Z and Trabzon L 2018 A review of recent research on materials used in polymer–matrix composites for body armor application *Journal of Composite Materials* **52** 3241–63

[30]. Praveen D, Kumar M E S and Pramod R 2018 Dielectric studies of Graphene and Glass Fiber reinforced composites *IOP Conference Series: Materials Science and Engineering* **310** 012153

[31]. Vázquez-Moreno J, Sánchez-Hidalgo R, Sanz-Horcajo E, Viña J, Verdejo R and López-Manchado M 2019 Preparation and Mechanical Properties of Graphene/Carbon Fiber-Reinforced Hierarchical Polymer Composites *Journal of Composites Science* **3** 30

[32]. Abdellaoui H, Raji M, Essabir H, Bouhfid R and Qaiss A E K 2019 Mechanical behavior of carbon/natural fiber-based hybrid composites In: Jawaid M., Thariq M., Saba N., editors *Mechanical and Physical Testing of Biocomposites, Fibre-Reinforced Composites and Hybrid Composites* 103–22

[33]. Zhang X, Fan X, Yan C, Li H, Zhu Y, Li X and Yu L 2012 Interfacial Microstructure and Properties of Carbon Fiber Composites Modified with Graphene Oxide *ACS Applied Materials & Interfaces* **4** 1543–52

[34]. Saba N, Jawaid M, Alothman O Y, Paridah M and Hassan A 2015 Recent advances in epoxy resin, natural fiber-reinforced epoxy composites and their applications *Journal of Reinforced Plastics and Composites* **35** 447–70

[35]. Saba N, Jawaid M, Alothman O Y, Paridah M and Hassan A 2015 Recent advances in epoxy resin, natural fiber-reinforced epoxy composites and their applications *Journal of Reinforced Plastics and Composites* **35** 447–70

[36]. Ga Koranlev 2019 Enriched Mechanical Properties of Epoxy/Coir Fiber Composites with Graphene

Oxide *Research & Development in Material Science* **10** 31031

[37]. Ji X, Xu Y, Zhang W, Cui L and Liu J 2016 Review of functionalization, structure and properties of graphene/polymer composite fibers *Composites Part A: Applied Science and Manufacturing* **87** 29–45

[38]. Chen L, Chai S, Liu K, Ning N, Gao J, Liu Q, Chen F and Fu Q 2012 Enhanced Epoxy/Silica Composites Mechanical Properties by Introducing Graphene Oxide to the Interface *ACS Applied Materials & Interfaces* **4** 4398–404

[39]. Naveen J, Jawaid M, Zainudin E, Sultan M T and Yahaya R 2019 Effect of graphene nanoplatelets on the ballistic performance of hybrid Kevlar/Cocos nucifera sheath-reinforced epoxy composites *Textile Research Journal* **89** 4349–62

[40]. Blamires S J, Blackledge T A and Tso I-M 2017 Physicochemical Property Variation in Spider Silk: Ecology, Evolution, and Synthetic Production *Annual Review of Entomology* **62** 443–60

[41]. Blamires S J, Wu C-L, Blackledge T A and Tso I-M 2012 Post-secretion processing influences spider silk performance *Journal of The Royal Society Interface* **9** 2479–87

[42]. Albertson A E, Teulé F, Weber W, Yarger J L and Lewis R V 2014 Effects of different post-spin stretching conditions on the mechanical properties of synthetic spider silk fibers *Journal of the Mechanical Behavior of Biomedical Materials* **29** 225–34

[43]. Navaranjan N and Neitzert T 2017 Impact Strength of Natural Fibre Composites Measured by Different Test Methods: A Review *MATEC Web of Conferences* **109** 01003

[44]. Braga F D O, Milanezi T L, Monteiro S N, Louro L H L, Gomes A V and Lima É P 2018 Ballistic comparison between epoxy-ramie and epoxy-aramid composites in Multilayered Armor Systems *Journal of Materials Research and Technology* **7** 541–9

Acknowledgments
The authors appreciate the support of VIT management. The authors would also like to express their sincere gratitude to all the faculties and lab technicians involved in this research work for rendering their valuable guidance and help.

Influence of CNTs grading and reinforcement on dynamic characteristics of composite Plates in thermal environment

Vasuraj Garg[1], Sanidhya Saxena[2], Gaurav Goyal[3], Sourabh Kumar Soni[4] and Benedict Thomas[5*]

[1, 2, 3, 4, 5*] School of Mechanical Engineering, Vellore Institute of Technology, Vellore - 632014, Tamil Nadu, India.

E-mail: benedict.thomas@vit.ac.in

Abstract- This analytical work deals with the dynamic analysis of the functionally graded composite plate reinforced with single walled carbon nanotubes is discussed in this article. By using the extended rule of mixtures, the material properties were computed. The dimensionless frequency parameter was discovered by generating ABAQUS-solved eigenvalues and eigenvectors. The subspace process was used to produce buckling Eigenvalues for the plate. The purpose of the current research work was to analyse the influence of varying temperatures, CNTs and CNTs grading volume fractions on dimensionless frequencies and buckling factors or parameters. From the findings available in the literature, the results of this analysis have been established. CNTs volume percentage and its distribution were found to have a substantial impact on the dimensionless fundamental frequencies. Moreover, similar effect of these parameters on plate buckling were also been observed. Results reveal that the addition of CNT enhances the stiffness of the plate.

Keywords: *Buckling, CNTs, FG, CNTRCs, Vibration, Temperature.*

1. Introduction

Carbon nanotubes (CNTs) have been replacing various reinforcing materials in nanocomposites because of its salient characteristics. It possesses high strength, extreme stiffness, low weight and corrosion resistance [1]. Functionally graded CNTs (FG-CNTRC) constitute polymer or matrix (metallic) with CNTs reinforced with dispersal pattern could be uniform (UD) or could vary (FG-X, FG-V, FG-O and FG-Λ). The distribution of volume fraction provides extremely desired materials for most of the applications. Dey et al. [2] investigated the vibration characteristics of the FG composite rotor shaft. They found that the vibration response of the FG composite rotor shaft are significantly influenced by radial thickness, material properties and power law index. In earlier literature significant amount of research work has been observed in the characterization of the functionally-graded carbon nanotube-reinforced composites (FG-CNTRCs) structures such as plates, panels, beams and shells [3]. Yas and Samadi [4] discussed the critical buckling loads and natural frequencies reduces with increasing slenderness ratio. It has also been unveiled that while 'X' type distribution has higher, CeC beams have the highest frequency values. Kamarian et al. [5] investigated the vibrational responses and concluded that parameters μ and η which are associated with agglomeration have an adverse effect on vibrational responses but not so much on the wavenumber. Mirzaei and Kiani [6] found that the panel frequencies are directly proportionate to the percentage of CNT and significantly depends on the pattern of distribution across the depth.

Keleshteri et al. [7] reveal that results of the CNTRC sector plate can be improved by variation in the percentage of CNTs and type of distribution. Vibrational redistribution is the cause of reduced vibration amplitude while nonlinear frequency increases. Fantuzzi et al. [8] have investigated the problem of arbitrary shapes and agglomeration in the present work. The former is done by Non-Uniform Rational B-splines while the latter is done by Generalized Differential Quadrature. Li and Hu [9] given emphasis on stiffness-hardening and stiffness-softening effects that are shown by FG material and conclude that the non-linearities are due to stretching effects. The asymmetric formulation is provided in order to study the vibrational responses under thermal loading which is performed on the annular FG-CNTRC plates by Torabi and Ansari [10]. Zhu et al. [11] used an independent finite element formulation method, which was very novel and the results are used as a reference in the present work. The fact that FE formulated results converge well with the software ANSYS became the basis for choosing the work as a benchmark. Lei et al. [12] tabulated the response due to buckling of FGCNT plates in different loading conditions, mesh size and aspect ratio. Thomas and Roy [13] studied the vibration damping response of FG-CNTRC shell structures. They found that the addition of nanotubes in the conventional carbon fiber matrix results in good damping and vibration response of the resulting composites structures. Thomas and Roy [14] investigated the vibration characterises of the FG-CNTRC shell structures (ellipsoidal, spherical, cylindrical and doubly curved). Their findings demonstrated that the vibration and damping characteristics of the FG-CNTRC shell structures were significantly affected by the V_{fcnt} and CNT distribution. Thomas et al. [15] investigated the thermal analysis of FG-CNTRC Timoshenko beam. By employing the power law temperature distribution they examined the influence of V_{fcnt} and CNT orientation on the stresses and strains.

The convergence of results with previous literature help in referring to the tabulated outcomes and hence forms a reference for the work presented here. The present work provides an insight by varying temperature, V_{fcnt} and CNT grading on dimensionless frequencies and buckling factors (parameters) and what effects it has on them.

2. Formulations and Techniques

2.1 Modeling of Material Properties

In the present analysis, it is supposed that the composite plate under consideration is comprised of epoxy resin and CNTs as fibers. Also, the material characteristics are smoothly distributed over the thickness of the plate.

2.2 Distribution of Material

Here, Uniform type of distribution (UD) and graded distributions (FG-X type and FG-O type) of CNTs over the thickness direction of the nano-composite plate were used to evaluate the effect of various CNTs distributions on the dynamic behaviour of the plate. The V_{fcnt} of different types of nanocomposite plate can be written as:

$$V_{fcnt} = \begin{cases} V^*_{fcnt} & (UD), \\ \frac{4|z|}{h} V^*_{fcnt} & (FG\text{-}X), \\ 2\left(1 - \frac{2|z|}{h}\right) V^*_{fcnt} & (FG\text{-}O), \end{cases} \quad (1)$$

At this point, V^*_{fcnt} is a function of CNTs volume percentage, its mass portion and its density. V_{fcnt} is the volume fraction of CNTs.

2.3 Extended rule of mixture

The material properties of proposed composite plates were found by applying the extended rule of mixture (EROM), which is expressed as:

$$\begin{aligned} E_{11} &= \eta_1 V_{fcnt} E_{11}^{CNT} + V_m E_m, \\ \frac{\eta_2}{E_{22}} &= \frac{V_{fcnt}}{E_{22}^{CNT}} + \frac{V_m}{E_m}, \\ \frac{\eta_3}{G_{12}} &= \frac{V_{fcnt}}{G_{12}^{CNT}} + \frac{V_m}{G_m} \end{aligned} \quad (2)$$

Since Poisson's ratio does not depend strongly on the position, ν_{12} can be assumed as:

$$\nu_{12} = V^*_{CNT} \nu_{12}^{CNT} + V_m \nu_m$$

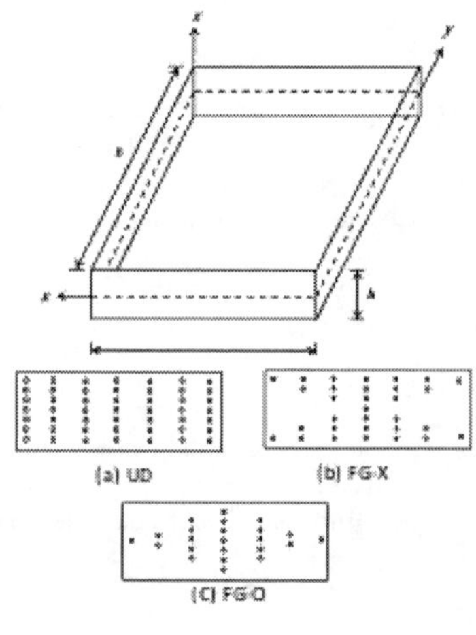

Figure 1. Distribution of CNT in various kinds of CNTRC plates. (a) UD type ; (b) FG-X type; (c) FG-O type

Figure 1(a) depicts the UD grading, where the percentage/fraction of CNTs is throughout constant. **Figure** 1(b) presented the FG-X type distribution in which has variation similar to alphabet X type fraction (i.e. highest on the top and bottom surfaces while least at the middle). **Figure** 1(c) depicts the FG-O type which is exactly opposite of X type (i.e. CNTs are plentifully present in the mid but lightly at the top regions).

2.4 Vibration Investigation

After evaluating the effective elastic properties of the proposed composite plate, finite element modelling and investigation has been performed using FEA software (ABAQUS). Finally, the vibration examination is performed using the below mentioned governing equation of motion:

$$[M_G]\{\ddot{x}\}+[K_G]\{x\}=\{F\} \tag{3}$$

here $[M_G]$ signifies the global mass matrix, $[K_G]$ represents the global stiffness matrix.

In this research work, the natural frequency of the FGCNT plate is converted to a dimensionless number (parameter) by employing the formula:

$$\bar{w}= w\left(\frac{A^2}{h}\right)\sqrt{\frac{\rho_m}{E_m}} \tag{4}$$

Where E_m = matrix elastic modulus at different temperatures.

2.5 *Investigation of Buckling*

After the vibrational analysis, buckling response of proposed composite plates are computed by solving the following equation of eigenvalue:

$$\left[K - \gamma K_g\right]X = 0$$

Where γ = critical buckling loads, X = modes of buckling , Kg = Geometric stiffness matrix, obtained using the relation:

$$K_g = \int_0^L \left[\frac{dN_w}{dx}\right]^T P \left[\frac{dN_w}{dx}\right] dx \tag{5}$$

In this work, a dimensionless critical buckling factor (parameter) is given by:

$$\overline{N}_{cr} = \frac{N_{cr} b^2}{E_m h^3} \tag{6}$$

2.6 Modelling using FEA software (ABAQUS)

Figure 2 presents the proposed model of the laminated composite plate with 0^0 orientation angle with 10 layers (plies), which explains the modeling, orientation at which the investigation has been conducted.

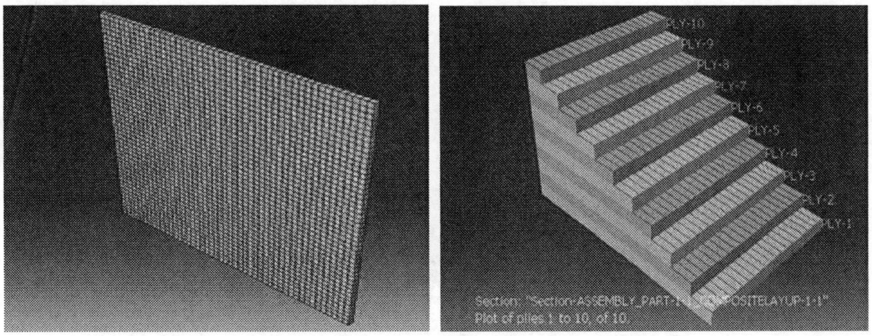

Figure 2. Proposed composite plate model in ABAQUS

Figure 2 shows the stacking sequence comprises ten layers arranged composed of same thickness (each plies relative thickness is thus t=0.1 or 10%). Red lines reflected horizontally in the figure show the orientation angle of the respective plies. Though all the lines are analogous, the angle of orientation was 0/0/0/0/0/0/0/0/0/0. The end supporting condition of C-C-C-C depicted that the all four sides of the nanocomposite plates are clamped while doing the free vibration response investigation. In this study three kind of grading patterns (UD, FG-X and FG-O) have been employed for the preparation of the plates. The grading is supposed to over the entire thickness (i.e. +h/2 to -h/2). The dimensions of the nanocomposite plate is length = a and width= b, as square plates are considered in this analysis, so a=b.

3. Results and Discussion

Table 1 depicts the validation and convergence behavior of natural frequency at 300K for a/h = 50 for UD and graded plates with different V_{fcnt}. The observed amount of the dimensionless frequency parameter agreed well with those available in the literature [11].

Table 1. Validation of dimensionless natural frequency at 300K for UD-CNT with different V_{fcnt}.

Volume percentage of CNT	Natural frequency	M-1	M-2	M-3	M-4
11	Present	43.55	49.32	63.94	88.98
	Ref. [11]	39.73	43.88	54.77	74.49
14	Present	47.94	53.06	67.04	92.21
	Ref. [11]	43.58	47.48	57.97	77.40
17	Present	51.62	56.42	69.92	94.77
	Ref. [11]	49.07	54.32	68.07	92.97

IOP Conf. Series: Materials Science and Engineering 1123 (2021) 012052 doi:10.1088/1757-899X/1123/1/012052

Here, M-1,M-2, M-3,M-4 represents the mode numbers and will be used to represent mode numbers in this article.

Table 2 represents the dimensionless frequency parameter for different V_{fcnt}, grading type for various mode numbers. It is evident from the **Table 2** that as the V_{fcnt} increases the frequency also increases. Variation of frequency parameter with different mesh sizes are shown in **Table 3**.

Table 2. The dimensionless frequency parameter for different V_{fcnt}, grading type for various mode numbers.

Temp.	V_{fcnt} %	CNT Grading	Mode Number					
			1	2	3	4	5	6
300	11	UD	43.55	49.32	63.94	88.98	109.36	111.85
		FG-O	27.55	36.19	54.74	69.86	74.48	83.24
	14	UD	47.94	53.06	67.04	92.21	119.39	120.82
		FG-O	27.96	36.52	54.99	71.08	75.61	83.45
	17	UD	51.62	56.42	69.92	94.77	127.34	128.46
		FG-O	29.60	37.74	55.78	75.25	79.38	83.98
500	11	UD	43.20	49.03	63.74	88.86	108.60	111.14
		FG-O	23.79	33.52	52.82	59.53	65.63	79.18
	14	UD	47.47	52.86	66.93	91.60	118.18	120.34
		FG-O	27.32	36.08	54.50	69.09	74.18	82.33
	17	UD	51.18	56.26	69.88	94.26	126.30	128.18
		FG-O	29.35	37.62	55.54	74.42	79.04	83.07
700	11	UD	43.02	48.86	63.61	88.76	108.17	110.71
		FG-O	24.99	34.38	53.39	62.79	68.53	81.52
	14	UD	47.28	52.69	66.80	91.52	117.77	119.94
		FG-O	27.22	36.01	54.45	68.82	73.94	82.30
	17	UD	50.98	56.08	69.75	94.17	125.88	127.78
		FG-O	29.24	37.54	55.48	74.13	78.78	83.04

Table 3. Variation of frequency parameter with different mesh sizes.

CNT %	CNT Grading	Mesh Size			
		8 × 8	16 × 16	24 × 24	32 × 32
11	UD	43.63159	43.54778	43.4659	43.458
	FG-X	50.6053	50.51882	50.4995	50.4933
	FG-O	28.1	27.96348	27.932	27.9237
14	UD	48.052	47.93973	47.9138	47.9093
	FG-X	55.2792	55.23794	55.2298	55.2241
	FG-O	27.6838	27.54566	27.5152	27.50731
17	UD	51.7019	51.6191	51.6006	51.5946
	FG-X	59.2274	59.24614	59.2464	59.2464
	FG-O	29.7428	29.59782	29.5656	29.5579

After the free vibration investigation of the nanocomposite plate, buckling investigation has been conducted and the results of buckling parameters are validated in **Table 4**.

Table 4. Validation of buckling parameter at 500K for V_{fcnt} (11%) which is UD graded subjected to uniaxial compression.

Buckling parameter	M-1	M-2	M-3	M-4
Present Study	24.23	32.976	41.988	47.198
Ref. [12]	24.95	33.4917	47.8115	50.0545

After the validation critical buckling factor (parameter) at diverse temperatures with a/h=10 and for changing V_{fcnt}, grading type under uniaxial compression are tabulated in **Table 5**.

Table 5. The critical buckling factor (parameter), N_{cr} at several temperatures for varying V_{fcnt}, grading type subjected to uni-axial compression.

Temp.	CNT %	CNT Distribution	M-1	M-2	M-3	M-4
300	11	UD	24.55	33.26	42.19	47.33
		FG-O	10.25	10.51	16.32	23.69
	14	UD	28.89	37.39	45.70	50.24
		FG-O	10.46	10.99	17.53	25.01
	17	UD	32.80	40.93	48.70	52.80
		FG-O	10.91	11.67	17.16	24.72
500	11	UD	24.23	32.98	41.99	47.20
		FG-O	9.18	9.44	15.27	22.43
	14	UD	28.46	37.04	45.46	50.09
		FG-O	10.19	10.40	16.24	23.61
	17	UD	32.41	40.68	48.60	52.81
		FG-O	10.85	11.55	17.09	24.64
700	11	UD	24.06	32.81	41.85	47.09
		FG-O	9.13	9.41	15.24	22.40
	14	UD	28.28	36.89	45.36	50.03
		FG-O	10.16	10.35	16.34	23.58
	17	UD	32.21	40.53	48.51	52.75
		FG-O	10.82	11.49	17.06	24.61

After validation, the modal shapes have been plotted and depicted in **Figure 3** for the first 6 fundamental frequencies.

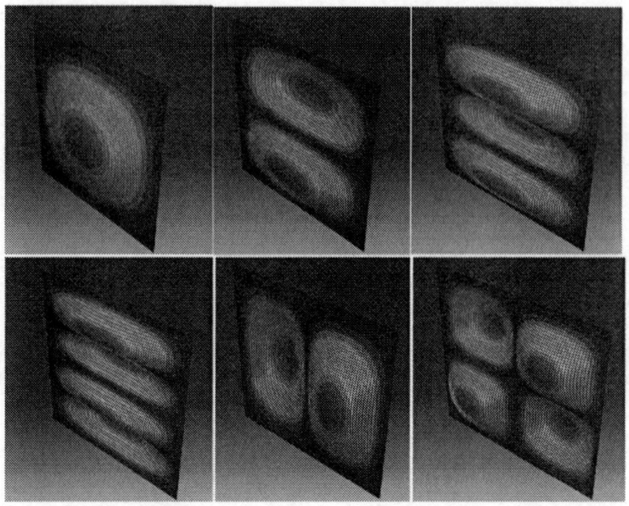

Figure 3. Mode shapes for the first six fundamental natural frequencies.

Figures 4-6 depicts the dimensionless frequency w.r.t. mode number with different CNT volume percentages and for different CNT grading (UD and FG-O) for temperature 300 K. It is evident from these figures that as the V_{fcnt} the natural frequency also increases. It is also perceived that the CNT grading profile have a noteworthy impact on the vibrational features of the nanocomposite plate.

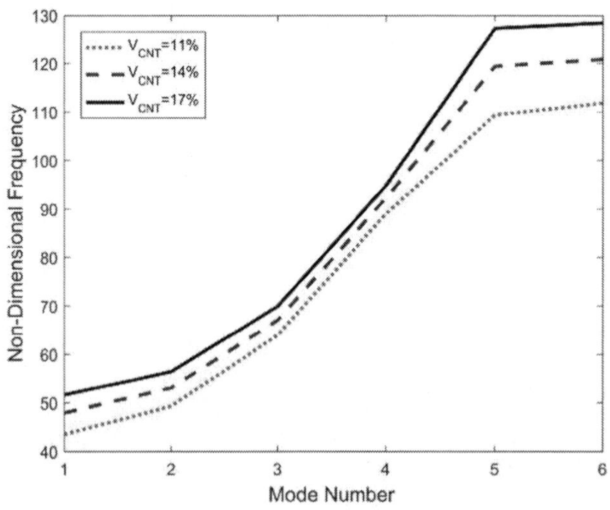

Figure 4. Dimensionless frequency variation with Mode Number of UD at 300 K

IOP Conf. Series: Materials Science and Engineering 1123 (2021) 012052 doi:10.1088/1757-899X/1123/1/012052

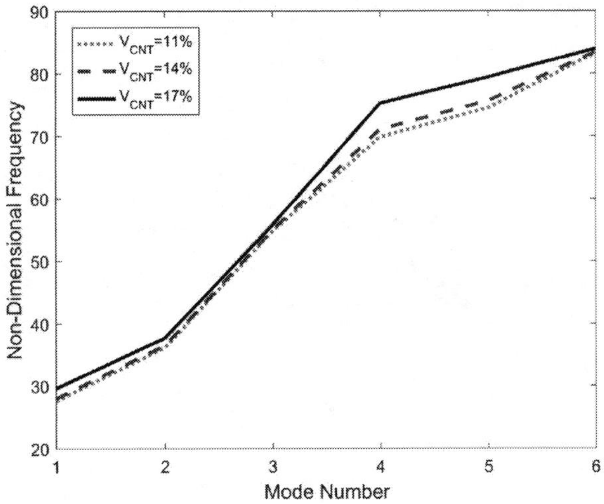

Figure 5. Dimensionless frequency variation with mode Number of FG-O at 300 K

Dimensionless frequency variation with mode number of UD at various temperatures and volume contents of CNTs is shown in **Figure 6**. Also, dimensionless frequency variation with mode number of FG-O type distribution at at various temperatures and volume contents of CNTs is shown in **Figure 7**. It is evident from both figures that the temperature and percentage of CNT have an adverse effect on dimensionless frequency. As the percentage of CNT is enhanced, the frequency also gets enhanced, whereas it reduces with the rise in temperature.

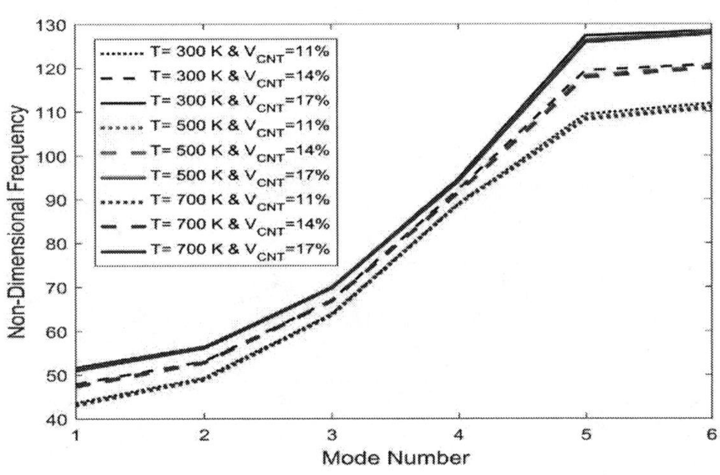

Figure 6. The dimensionless frequency vs mode number of UD type with several temperatures and volume percentages of CNTs.

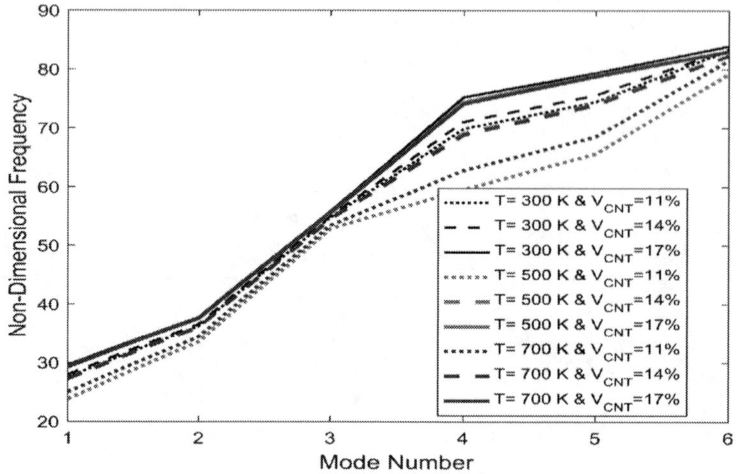

Figure 7. The dimensionless frequency vs Mode Number of FG-O type with several temperatures and volume percentages of CNTs.

Further dimensionless frequency are also plotted with respect to mesh size at 300 K, 500 K and 700 K in **Figures 8-10**, which reveals that frequency is highest for the higher volume fraction of CNT.

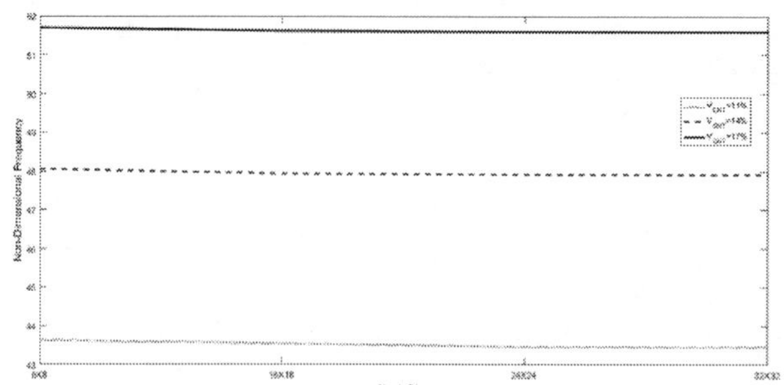

Figure 8. Dimensionless frequency vs Mesh Size at 300K

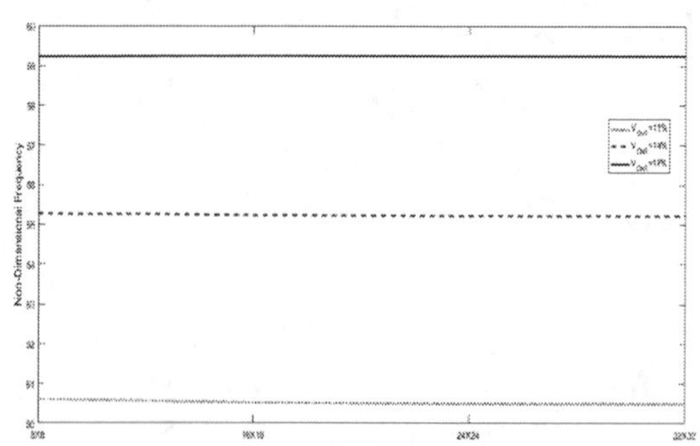

Figure 9. Dimensionless frequency vs Mesh Size at 500K

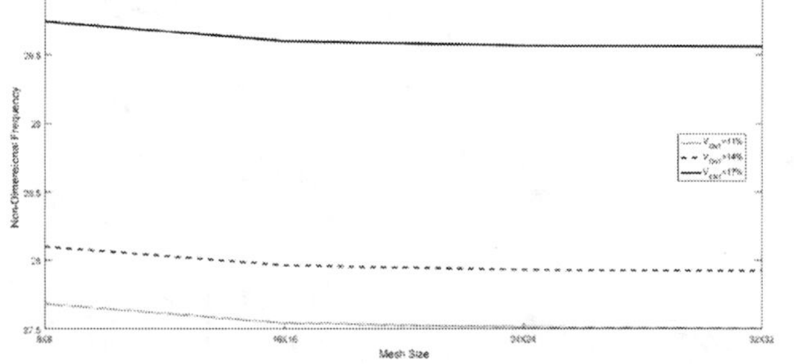

Figure 10. Dimensionless frequency vs Mesh Size at 700K

After vibration analysis for the plate under consideration, buckling analysis was also carried out and the variation of Buckling Parameter with mode number is shown in Figure 11 and Figure 12 for UD and FG-O kind of grading respectively. It is perceived from igures that for both type of grading as the V_{fcnts} buckling parameters also enhances for different temperature.

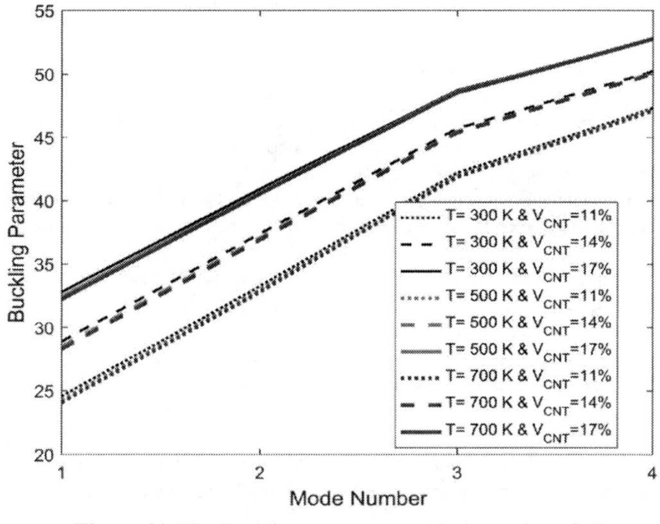

Figure 11. The Buckling parameter vs Mode number of UD

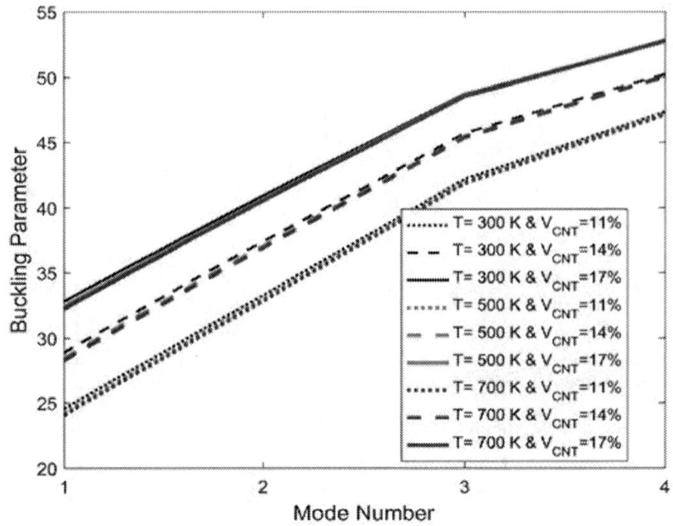

Figure 12. .Buckling Parameter vs Mode Number of FG-O

4. Conclusions

The aim of this work is to research the effect of different CNT distribution and volume fractions on the dynamic characteristics of the composite plate reinforced with CNTs. After finding the effective material properties and validation of results, vibrational analysis is conceded in order to analyse the effect of various parameters including CNTs volume (11%, 14% and 17%), distribution (UD,FG-X and FG-O) and temperatures (300 K, 500 K and 700 K) on the dynamic behaviour of the composite plate with constant a/h ratio (a/h=50). Considering the obtained results of vibrational study it is witnessed that FG-X type show the maximum values for dimensionless frequency parameters at any percentage of V_{fcnt} and temperature, followed by UD and FG-O. For the same temperature and a / h ratio, the non-dimensional frequency values increase as with higher percentage of CNT and results in a stiffer plate. For a constant V_{fcnt} and for the identical distribution type with increasing temperature, the natural frequency decreases. Meshing size does not show any prominent variation, but as mesh size decreases and mesh quantity increases, it results in reduction in frequency.

Buckling analysis on the identical plate keeping a/h =10 was performed to analyse the buckling effect. The original findings have been confirmed and the results were obtained. It is witnessed that the buckling factor for a specified CNTs fraction reduces as the temperature rises. For a given temperature, the buckling parameter value rises as the CNTs fraction rises. The buckling factor (parameter) for UD grading is found considerably greater than FG-O type at a particular temperature and CNTs fraction.

Finally, it can be inferred, based on the results obtained, that the CNTs volume and its grading show a significant role in modifying vibrational and buckling properties of resulting composite.

References

[1] Soni S K, Thomas B and Kar V R 2020 A Comprehensive Review on CNTs and CNT-Reinforced Composites: Syntheses, Characteristics and Applications *Mater. Today Commun.* **25** 101546

[2] Dey K, Verlekar J R, Soni S K and Thomas B 2019 Dynamic analysis of functionally graded shaft *FME Trans.* **47**

[3] Liew K M, Lei Z X and Zhang L W 2015 Mechanical analysis of functionally graded carbon nanotube reinforced composites: A review *Compos. Struct.* **120** 90–7

[4] Yas M H and Samadi N 2012 International Journal of Pressure Vessels and Piping Free vibrations and buckling analysis of carbon nanotube-reinforced composite Timoshenko beams on elastic foundation *Int. J. Press. Vessel. Pip.* **98** 119–28

[5] Kamarian S, Salim M, Dimitri R and Tornabene F 2016 Free vibration analysis of conical shells reinforced with agglomerated Carbon Nanotubes *Int. J. Mech. Sci.* **108–109** 157–65

[6] Mirzaei M and Kiani Y 2016 Free vibration of functionally graded carbon nanotube reinforced composite cylindrical panels *Compos. Struct.* **142** 45–56

[7] Mohammadzadeh-keleshteri M, Asadi H and Aghdam M M 2017 Geometrical nonlinear free vibration responses of FG-CNT reinforced composite annular sector plates integrated with piezoelectric layers *Compos. Struct.* **171** 100–12

[8] Fantuzzi N, Tornabene F, Bacciocchi M and Dimitri R 2017 Free vibration analysis of arbitrarily shaped Functionally Graded Carbon Nanotube-reinforced plates * *Compos. Part B* **115** 384–408

[9] Li L and Hu Y 2016 Nonlinear bending and free vibration analyses of nonlocal strain gradient beams made of functionally graded material **107** 77–97

[10] Methods C, Mech A, Torabi J and Ansari R 2017 Nonlinear free vibration analysis of thermally induced FG-CNTRC annular plates : Asymmetric versus axisymmetric study *Comput. Methods Appl. Mech. Engrg.* **324** 327–47

[11] Zhu P, Lei Z X and Liew K M 2012 Static and free vibration analyses of carbon nanotube-reinforced composite plates using finite element method with first order shear deformation plate theory *Compos. Struct.* **94** 1450–60

[12] Lei Z X, Liew K M and Yu J L 2013 Buckling analysis of functionally graded carbon nanotube-reinforced composite plates using the element-free kp-Ritz method *Compos. Struct.* **98** 160–8

[13] Thomas B and Roy T 2017 Vibration and damping analysis of functionally graded carbon nanotubes reinforced hybrid composite shell structures *JVC/Journal Vib. Control* **23** 1711–38

[14] Thomas B and Roy T 2016 Vibration analysis of functionally graded carbon nanotube-reinforced composite shell structures **599** 581–99

[15] Thomas B, Inamdar P K and Roy T 2014 Thermal analysis of randomly oriented carbon nanotube reinforced functionally graded timoshenko beam *J. Mech. Sci. Technol.* **28** 1779–88

ROBOTIC ASSEMBLY SEQUENCE GENERATION USING IMPROVED FRUIT FLY ALGORITHM BY APPLYING GROSS AND FINE MOTION AS FITNESS EQUATION

***Gunji Bala Murali* [*1], *Sridharan* [2], *BBVL Deepak*[3], *Rajashekara Reddy Muthra*[4], *BB Biswal*[5], *MVA Raju Bahubalendruni*[6]**

[1, 2, 4] Department of Design and Automation, Vellore Institute of Technology, Vellore, Tamilnadu, India-632014.
[3, 5] Department of Industrial Design, National Institute of Technology, Rourkela, India-769008.
[6] Department of Mechanical Engineering, National Institute of Technology, Pudhucherry, India- 609609.
[1]balamurali.g@vit.ac.in, [2]sridharan.k@vit.ac.in, [3]bbvl@nitrkl.ac.in, [4]rajasekhara.reddy@vit.ac.in,
[5]bbbiswal@nitrkl.ac.in, [6]bahubalindruni@gmail.com.

Abstract. Robotic assembly plays a major role in manufacturing industries to reduce the cost of overall product. As the assembly is one of the major manufacturing processes, which involves much information like contact information between the parts, tools needed during assembly, part orientation information during assembly and many more. Most of the researchers considered grippers changes, assembly time, directional changes, part orientation and energy of the part as fitness function. In this research work authors made attempt to consider gross and fine motion as fitness equation to generate optimal assembly sequences with Improved Fruit Fly Algorithm (IFFA). The proposed algorithm is implemented on many industrial products to obtain the optimal assembly sequences.

Keywords: Assembly Sequence planning, Gross, Fine motion, Fruit Fly Algorithm

1. Introduction
Robotic assembly is now replacing the traditional assembly process in the manufacturing due to high flexibility during assembly, more accurate and less to time assembly. Due to these many advantages, modern manufacturing industries replacing the traditional assembly process to robotic assembly. But robotic assembly is not as simple as traditional assembly because it requires lot of assembly information like type of tools used to assemble, type of gripper used to hold the object, contact relation between the parts and many more [1]. Most of the researchers are concentrated on contact information, gripper information, and feasibility information to obtain the optimal assembly sequences.
At the initial stages of ASP problem solving, authors like Wilson and Latombe [2] applied mathematical models to generate solution for the ASP problem. As these methods only generate feasible solutions, they are absolute by raise in demand from the costumers in terms of cost and quality.
Meanwhile, advancements in Artificial Intelligence (AI) changed the world to shift the interest towards AI for solving the engineering problems. Keeping the advantages in view regarding to the AI techniques, researchers applied Genetic Algorithm (GA), Ant Colony Optimization (ACO) algorithm, Simulated Annealing (SA) and so on to solve ASP problem [3-5].

Along with the general AI algorithms that are existed, the authors like Murali, Gunji Bala, et al.[6,7] applied Improved Cuckoo Search (ICS) and Crab Shell Search (CSS) algorithm to generate optimal assembly sequences. Out of these, CSS is a newly developed algorithm based on how crab will search the shell at the shore to survive from the foreign bodies.

As the individual algorithms are restricted to certain limits like struck at local solution and highly redundant in nature, Hybrid Algorithms (HA) are developed by the researchers to solve ASP problems. HA utilizes the advantages of the algorithms that are combining to solve ASP problem [8].

Most of the methods/techniques used to solve ASP problem uses the either mathematical approach or connector base approach to extract the assembly predicates. But Gunjia, Bala Murali, et al. applied Computer Aided Design (CAD) based method to extract the assembly predicates [9].

Along with the traditional methods Murali, G. Bala, et al. proposed assembly subset detection method using stability graph to generate optimal assembly sequences. In this method, authors try to implement the subset detection method at each stage to eliminate the higher level subsets by evaluating the fitness at each stage. By this, time of execution and search space is reduced compared to the existing methods [10].

As of now the discussed methods/algorithms to solve ASP problems uses directional changes or gripper changes or energy to carry the part or assembly tool changes as fitness function to generate optimal assembly sequences. In this paper, the authors made an attempt to generate optimal assembly sequences by considering gross and fine motion as fitness function.

To generate optimal assembly sequences using gross and fine motion as fitness equation, the authors applied Fruit Fly (FF) algorithm by improving search phenomenon of the fly.

The paper is organised as follows: section-1 deals with introduction and literature of the ASP problem. Section-2 explains about the improvement made for the fruit fly algorithm and how it is implemented to solve ASP problem. Section-3 deals with the results obtained from the proposed methodology. Section-4 explains about conclusion made from the proposed method in the paper.

2. Improved Fruit Fly Algorithm

In this section, FF algorithm is used to generate optimal assembly sequences for the considered seven part gear assembly by improving the search phenomenon of the fruit fly. FF algorithm is basically developed by the inspiration of natural behaviour of the fruit fly in search of ripened or fermenting fruits during summer season. Fruit flies are having good senses and perceptions than the other species. The individual fruit fly uses its sensitive vision to fly in the direction of the target by sampling the different scents that are present in the surroundings. As the individual fly identifies the best position that is nearer to the target, it directs the flocking location of the flies closer to the food. Through the iterative evolution, the fly swarm comes closer and closer to the target until it reaches to the fruits.

By using this phenomenon, fruit fly algorithm was developed by Pan, Wen-Tsao for the engineering problem for financial distress model. Later this algorithm was extended to many fields. In general fruit fly algorithm, initialization of the fruit fly is considered by assigning the random position in plane consisting of food as shown in the Figure 1. Due to this, the selected random position may not be in the direction of food, in that a case the solution occurs is of redundant.

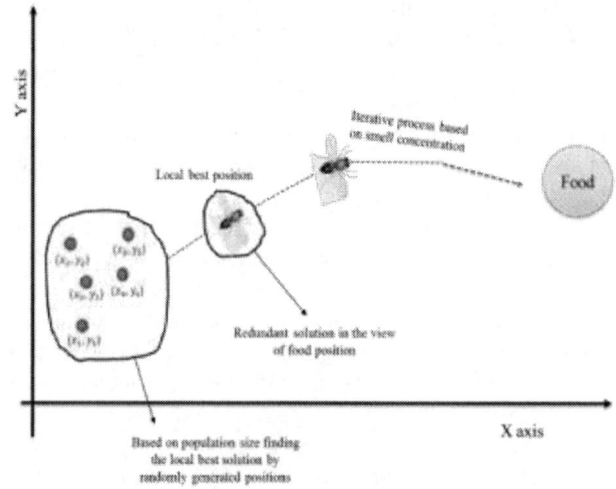

Figure 1. General Fruit Fly Algorithm

In the improved Fruit Fly algorithm instead of generating random positions to find local best, a range has been considered to generate positions of the Fruit Fly to find local best. The range is defined by a fly angle of Fruit Fly from initialization position of Fruit Fly. The range is decided by considering the food position region as the 2D plane with the coordinate system.

To avoid this phenomenon, improvement is made in the search position to find the local best solution for the fruit fly algorithm as shown in Figure 2.

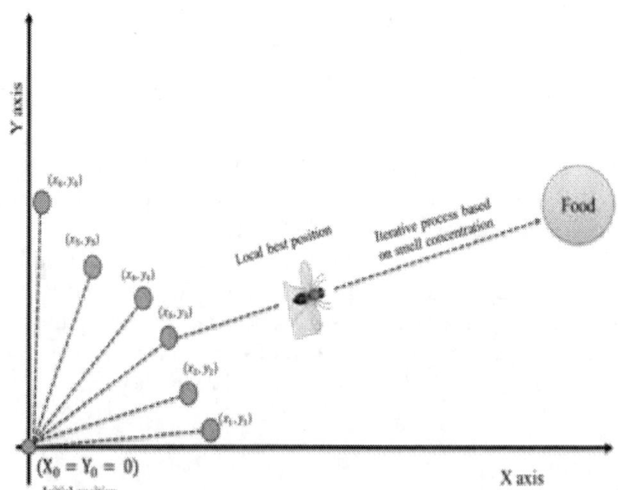

Figure 2. Improved fruit fly algorithm

In the Figure 2, if it is considered as 2D plane, then the range is given as 0-90° for the search position in the algorithm. By which location of food will compulsory in that region where the solution is not redundant.

The detailed proposed methodology is explained in Figure 3 with the flow chart to generate optimal assembly sequences. In the applied algorithm, gross and fine motions are considered as fitness equations to generate optimal assembly sequences. The generation of fitness equation in terms of gross and fine motion is shown in Eq (1).

IOP Conf. Series: Materials Science and Engineering 1123 (2021) 012053 doi:10.1088/1757-899X/1123/1/012053

$$ff = \sum_{i=1}^{n} GM_i + FM_i \quad (1)$$

Where ff= fitness function
n= Part number
GM=Gross Motion
FM=Fine Motion

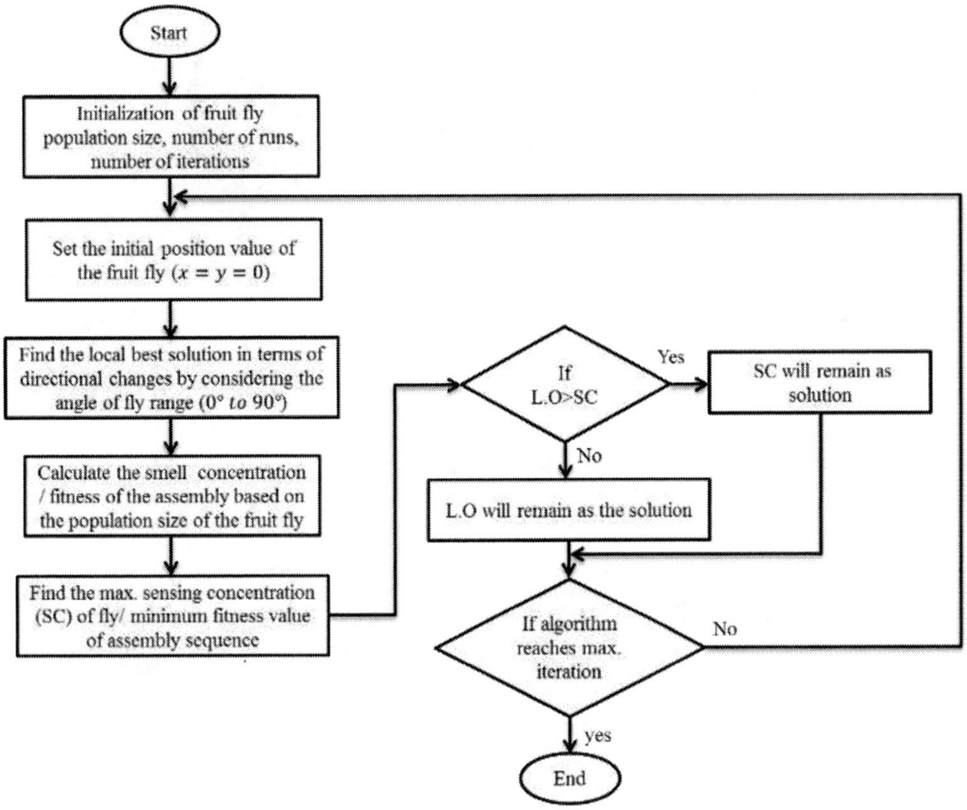

Figure 3. Represents the Improved fruit fly algorithm

2.1. Gross and Fine Motion

Gross and fine motions are two motions that robot arm will possess during assembling the parts. Gross motion is the motion carried by the robot arm with high speeds during assembly. These motions are existed till the parts come closer just before parts are in contact. Fine motion is the motion that will comes into existence once the part come contact with the other part during the assembly. These motions are very fine motions to have precision assembly and firm contact between the parts. The details about the fine and gross motion are explained with the screw jack diagram shown in the Figure 4.

Assumption made in this paper: As the robotic assembly consists of robotic work cell, where the individual parts have been kept at different places and should bring to the robotic base is where robot has been fixed by the robotic arm. During bringing of the parts to the base, the robot arm has to follow gross motion concept to reduce the time of assembly. But in this paper, this has not been considered as we are not dealing with the practical assembly.

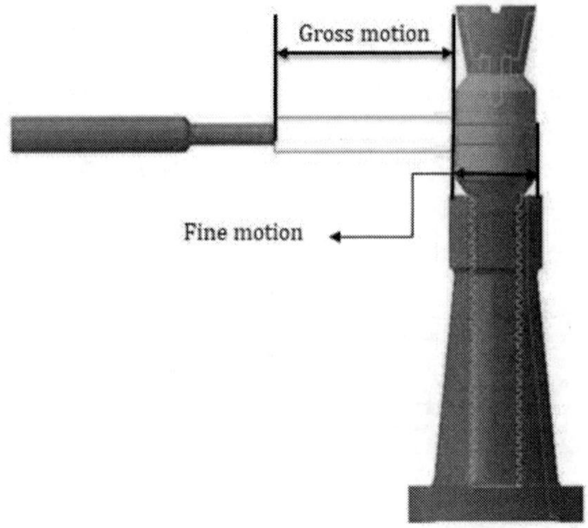

Figure 4. Represents the gross and fine motion for the screw jack assembly

3. Results and discussion

In this section, a seven part gear assembly shown in Figure 5 is considered to generate optimal assembly sequences using gross and fine motions as fitness values.

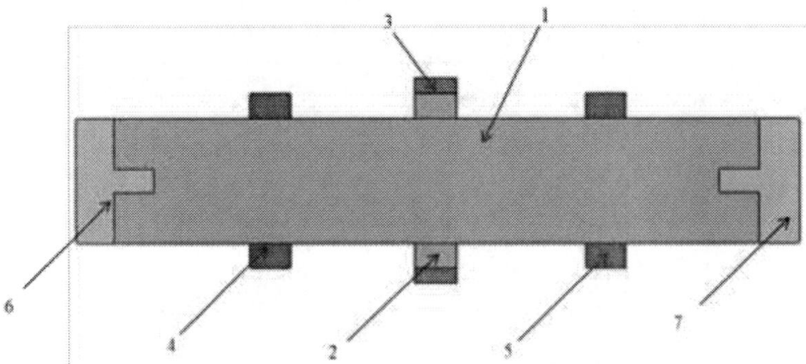

Figure 5. Represents the gear assembly

Part-1: Shaft; Part-2: Bearing; Part-3: Gear; Parts-4&5: Rotating discs; Parts-6&7: Holders.

Based on the contact relation graph, the algorithm generates feasible solution depending on the population size. After generation of the feasible solution, fitness is evaluated for the sequences to generate optimal assembly sequences. The contact relation graph is shown in Figure 6, by which the information about the contact between the parts is assigned to the algorithm in the form matrix.

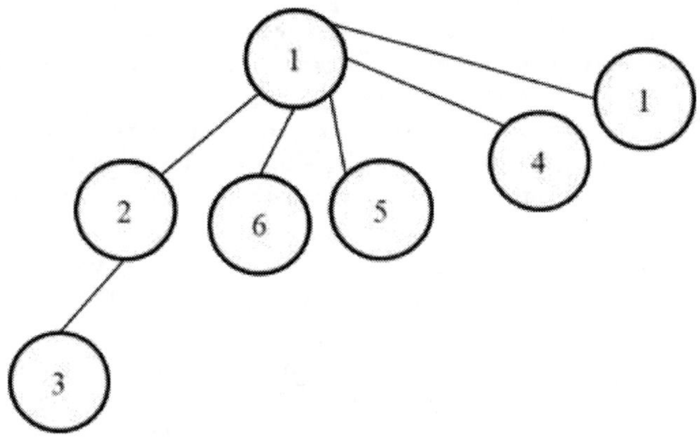

Figure 6. Contact relation graph

Contact relation matrix
This matrix summarizes the contact information that has to feed as input to the algorithm to generate feasible sequences. The detailed information about the matrix is as follows:

Contact relation matrix

	1	2	3	4	5	6	7
1	-1	0	-1	0	0	0	0
2	0	-1	0	-1	-1	-1	-1
3	-1	0	-1	-1	-1	-1	-1
4	0	-1	-1	-1	-1	-1	-1
5	0	-1	-1	-1	-1	-1	-1
6	0	-1	-1	-1	-1	-1	-1
7	0	-1	-1	-1	-1	-1	-1

Generally the matrix is represented by '0's and -1's to provide the information about the contact relation between the parts. In the matrix '0' represents the contact existed between the parts and '-1' represents the no contact between the parts.

As the assembly is not having the gross motion because as there is no part dimensions which are out of the shaft dimensions so, fine motion is only existed. The fine motion distance has been extracted from the CATIA V5 R17 Digital mock up tool bar. The fine motion distances are shown in table-1.

Table-1: Represents the fine motion distances

Part name	Fine/Gross motion distance (mm)	Direction of assembly
Bearing	90 (80 fine and 10 gross)	+Z or -Z
Gear	90 (80 fine and 10 gross)	+Z or -Z
Rotating disc	53 (43 fine and 10 gross)	+Z
Rotating disc	53 (43 fine and 10 gross)	-Z
Holder	10 fine	+Z
Holder	10 fine	-Z

By providing the contact relation information and fine motion information to the algorithm, the algorithm is made to run for 200 iterations to generate optimal assembly sequences. As the gross and fine motion are in mm, the total distance moved for any assembly sequence will be same in order to avoid this problem, the distance moved will be converted as time to show the difference in the fitness.

Assumption: The time taken to move 10mm distance in gross motion will be consider as 5sec. and the time taken to cover the 10mm distance in fine motion will be 10 sec.
A graph has been plotted between number of iterations and time taken to assembly by following the assembly sequence is shown in Figure 7. From the graph after '65' iterations, the solution is converged and the fitness value is of '456 Sec.'. The lists of assembly sequences that are generated from the algorithm, which will take 456sec to assemble the parts, are listed in table-2.

S. No.	Assembly Sequence	Time taken (sec)
1	1-2-3-4-5-7-6	456
2	1-2-3-5-4-6-7	456
3	1-4-2-3-5-7-6	456
4	1-5-2-3-4-6-7	456

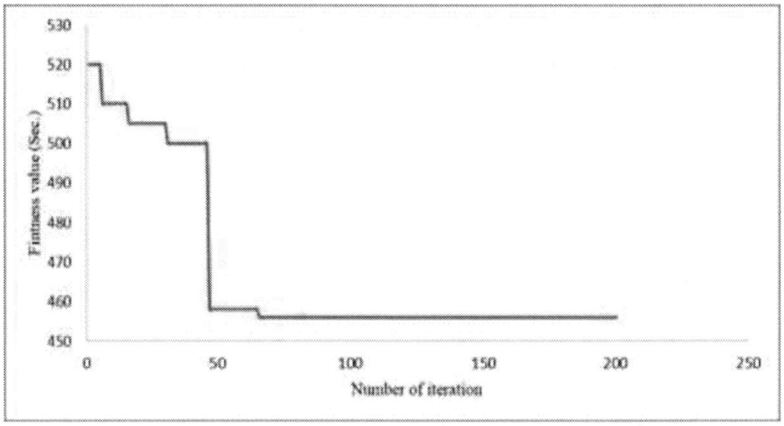

Figure 7. Represents the graph between fitness value and number of iterations

4. Conclusion
In this paper, an attempt is made to generate optimal assembly sequences by considering gross and fine motions as fitness function. A gear assembly with seven parts has been considered to generate the sequences using improved fruit fly algorithm. The results obtained are very impressive in terms of time taken to assembly the parts if a particular assembly sequence is followed.
As a future scope, robotic work cell dimensions will be taken into consideration for more practical approach of implementing the generated sequences to robot to assembly the parts.

References
[1]. Deepak, B. B. V. L., et al. "Assembly sequence planning using soft computing methods: a review." Proceedings of the Institution of Mechanical Engineers, *Part E: Journal of Process Mechanical Engineering* **233.3** (2019): 653-683.
[2]. Wilson, Randall H., and Jean-Claude Latombe. "Geometric reasoning about mechanical assembly." *Artificial Intelligence* **71.2** (1994): 371-396.
[3]. FAILLI, FRANCO, and Gino Dini. "Ant colony systems in assembly planning: a new approach to sequence detection and optimization." *2nd CIRP International Seminar on Intelligent Computation in Manufacturing Engineering-ICME 2000* **1.0 (2000)**:227-232.

[4]. Murali, G. Bala, et al. "Integrated design for assembly approach using ant colony optimization algorithm for optimal assembly sequence planning." *Computational intelligence in data mining. Springer, Singapore* **0.0** **(2019)**: 249-259.

[5]. Murali, G. Bala, et al. "Optimal assembly sequence planning towards design for assembly using simulated annealing technique." *International Conference on Research into Design. Springer, Singapore* **0.0** **(2017)**: 397-407.

[6]. Murali, Gunji Bala, et al. "Robotic Optimal Assembly Sequence Using Improved Cuckoo Search Algorithm." *Procedia computer science* **133** **(2018)**: 323-330.

[7]. Murali, Gunji Bala, et al. "A New Crab Shell Search Algorithm for Optimal Assembly Sequence Generation." *2019 9th Annual Information Technology, Electromechanical Engineering and Microelectronics Conference (IEMECON). IEEE,* **0.0** **(2019);** 108-114.

[8]. Abdel-Baset, Mohamed, and Ibrahim Hezam. "Cuckoo search and genetic algorithm hybrid schemes for optimization problems." *Appl Math* **10.3** (2016): 1185-1192.

[9]. Gunjia, Bala Murali, et al. "CAD-based automatic clash analysis for robotic assembly." *International Journal of Mathematical, Engineering and Management Sciences* **4.2** **(2019)**: 432-441.

[10]. Murali, G. Bala, et al. "Optimal robotic assembly sequence planning using stability graph through stable assembly subset identification." *Proceedings of the Institution of Mechanical Engineers, Part C: Journal of Mechanical Engineering Science* **233.15** **(2019)**: 5410-5430.

[11]. Pan, Wen-Tsao. "A new fruit fly optimization algorithm: taking the financial distress model as an example." *Knowledge-Based Systems* **26** **(2012):** 69-74.

IOP Conf. Series: Materials Science and Engineering 1123 (2021) 012054 doi:10.1088/1757-899X/1123/1/012054

Static structural analysis of hip joint with various profiles by finite element method

Abhinav Nemmani[1], Rajasekhara Reddy Mutra[2*] and Gunji Balamurali[3]

[1]School of Mechanical Engineering, Vellore Institute of Technology, Vellore, India
[2]School of Mechanical Engineering, Vellore Institute of Technology, Vellore, India

[3]School of Mechanical Engineering, Vellore Institute of Technology, Vellore, India

E-mail: rajasekhara.reddy@vit.ac.in

Abstract. The hip joint is one of the primary joints which give stability to the body. The hip anthroplasty surgery is performed due to constant wear and tear of the hip due to a person's age and various factors. Due to this surgery, the joints are replaced by the implants made up of mostly Cobalt Chromium alloy (CoCr), Titanium alloy, Stainless Steel (SS316L). A comparative study will be done amongst these materials so as to choose the better one. Also various shapes with different design profiles will be modeled and analyzed by using finite element method and the best results will be tabulated. The profiles studied are circular, ellipse and rectangular profile. The main aim of this research is to find best material as implant and also best design so that minimum amount of wear takes place during practical situations for the comfort of the person. Based on various results obtained by Finite element analysis (FEA) like Von-Mises stress, Total deformation, Equivalent Elastic Strain, it has been found that the CoCr is the best material to be used as hip implants for static loading conditions.

1. Introduction

The hip is a ball and socket joint surrounded by powerful and well balanced muscles [1]. This enables a wide range of locomotion and also exhibits remarkable stability. The hips carry forces from the ground up and also take forces from the trunk portion, head, neck and upper extremities of the human body [2]. This is the reason why this joint has become an essential part for athletic and physical activities and is mostly exposed to normal axial and torsional forces. Due to damage of hips due to various reasons and also due to old age there is increase in hip replacements in modern days. Due to this, the design for long lasting implant is required for both longevity and functionality point of view. As a result, it is driving the need for an optimized design for hip replacement [3]. This hip joint consists of a femoral head that is attached to the acetabulum. The hip anthroplasty is done to reduce the pain and induce normal movement in case of wear and tear of the joints [4]. The materials such as cobalt chromium alloy, stainless steel and titanium alloy are generally used for hip implants due to their biocompatibility property [5]. The basic necessity of the materials used is such that they should be biocompatible and produce minimal wear rate to avoid revisions due to asceptic loosening. By using the finite element analysis, it possible to predict the behaviours of the models developed. The investigation of hip joints by Anthony et al. [6] was done by considering various models of circular, ellipse, trapezoidal by using FEA. The materials used are Stainless Steel SS316L, CoCr and Titanium alloy (Ti-6Al-4V). Von-Mises stress and displacement of the stem was determined in all the cases. Zafer et al. [7] studied the static and dynamic behaviour of Ti-6Al-4V and cobalt-chromium alloy and also made a comparison to the stem developed by Charnley [8]. S.J. Hampton et al. [9] performed three dimensional stress analysis of the femoral stem of total hip prosthesis wherein stress distribution with the implanted femoral stem was studied by using finite element analysis.

In this work, three different profiles were used for hip implants viz. circular, ellipse, rectangular and each profile is investigated for three materials [10] CoCr, Ti-6Al-4V and Stainless Steel (SS316L). Finite element analysis is used to carry static analysis by using ANSYS Workbench. The boundary conditions were applied according to ASTM standards. Titanium alloy and CoCr were used due to their biocompatibility and popularity in bioimplants.

2. Materials and methods

Although several designs were popularly used in total hip anthroplasty, in this paper three profiles of circular, ellipse and rectangular were used. The geometry dimensions were taken from the previous literature Ref. [11]. The profiles were modeled by using Solid Works. The meshed part is analyzed in ANSYS. The materials used in this study are Cobalt chromium CoCr, Ti-6Al-4V and Stainless Steel (SS316L).

2.1. Meshing and boundary conditions

The models were meshed with unstructured mesh. The total number of elements and nodes obtained by modeling various implants were approximately 520000 nodes and 400000 elements. The boundary conditions are applied as per ASTM F2996-13 and loading conditions considered were according to ISO 7206-4:2010. The hip stem was constrained in all degrees of freedom at the bottom face. The fixation and load application is shown in the figure 1.

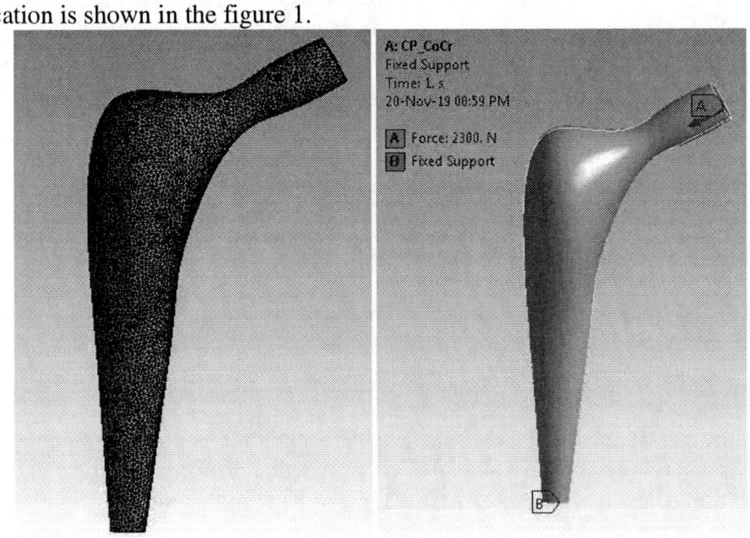

Figure 1. Meshed part of circular cross section made of CoCr and boundary conditions for the study

Static structural analysis is carried out for all profiles by giving similar bounding conditions. The mechanical properties of the materials used for the implants are shown in table 1.

Table 1. Mechanical properties of the materials used in the implant

Sr. No.	Materials	Young's modulus [GPa]	Density [gm/cm^3]	Poisson's ratio	Ultimate Tensile Strength [MPa]
1.	CoCr Alloy	200	8.5	0.30	1503
2.	Ti-6Al-4V	114	4.5	0.31	930
3.	SS316L	193	8	0.25	515

3. Results and discussion

The design is carried to preliminarily obtain results and to decide the best possible cross section for hip implant. The implants were made of commercially available materials cobalt chromium (CoCr), Titanium alloy (Ti-6Al-4V) and Stainless Steel (SS316L). The profile models were subjected to uniform loading of 2300 N. The results of CoCr are compiled in table 2 for Von-Mises stress, total deformation and equivalent elastic strain.

Table 2. Static analysis of hip implant by using CoCr

Material	Shape	Load in N	Total deformation in mm	Von-Mises stress in MPa	Equivalent elastic strain in mm/mm
CoCr	Circular	2300	6.0134	3722.2	0.018833
CoCr	Ellipse	2300	13.433	4974.9	0.025424
CoCr	Rectangular	2300	17.462	4101.7	0.021047

Table 3. Static analysis of hip implant by using Ti-6Al-4V

Material	Shape	Load in N	Total deformation in mm	Von-Mises stress in MPa	Equivalent elastic strain in mm/mm
Ti-6Al-4V	Circular	2300	11.073	3897.4	0.034544
Ti-6Al-4V	Ellipse	2300	26.027	5478.7	0.048183
Ti-6Al-4V	Rectangular	2300	35.175	4610.7	0.041623

Table 4. Static analysis of hip implant by using SS316L

Material	Shape	Load in N	Total deformation in mm	Von-Mises stress in MPa	Equivalent elastic strain in mm/mm
SS316L	Circular	2300	6.2805	3827.4	0.020041
SS316L	Ellipse	2300	14.047	4996.5	0.025938
SS316L	Rectangular	2300	18.296	4200.4	0.022145

From table 2, it can be seen that CoCr has less deformation when it is in circular cross section as compared to ellipse and rectangular. It is also observed that Von-Mises stress and elastic strain in circular cross section is less in value. From table 3, it can be seen that amongst Ti-6Al-4V material, the circular profile if offering least deformation, least Von-Mises stress and elastic strain. The rectangular cross section has maximum total deformation for SS316L as can be seen from table 4. figure 2 shows the contour plots of Von-Mises stress, deformation and equivalent elastic strain for circular profile for CoCr.

IOP Conf. Series: Materials Science and Engineering 1123 (2021) 012054 doi:10.1088/1757-899X/1123/1/012054

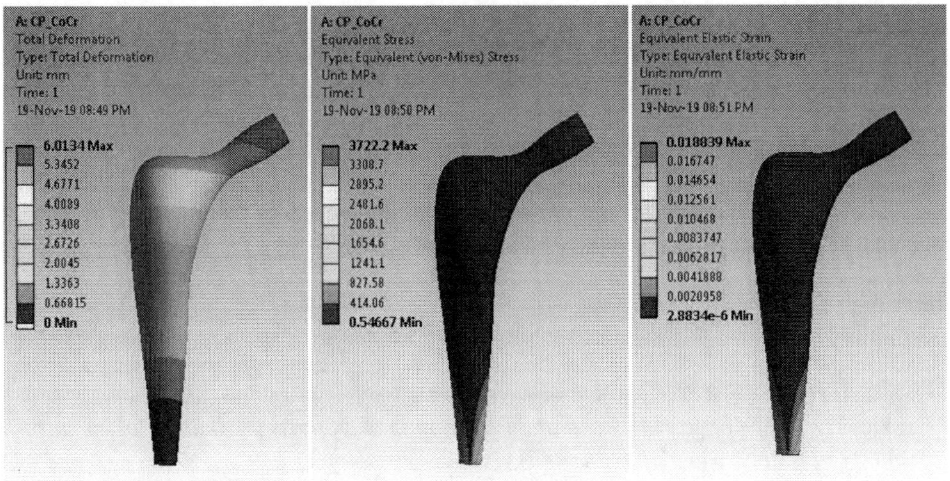

Figure 2. Circular profile for CoCr with Deformation, Von-Mises Stress and Elastic Strain

The contour plots for Ti-6Al-4V with ellipse profile were shown in figure 3. Similarly, contour plots for SS316L made of rectangular profile are shown in figure 4.

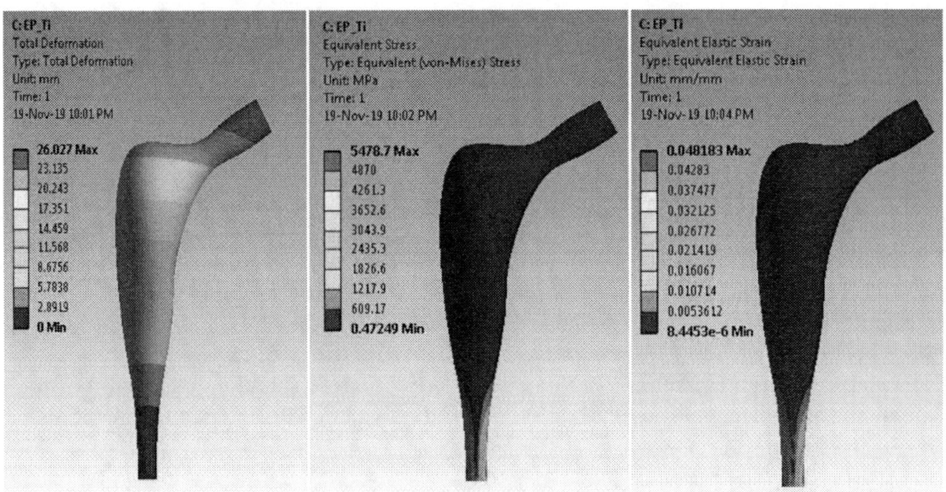

Figure 3. Elliptical profile with Ti-6Al-4V with Deformation, Von-Mises Stress and Elastic Strain

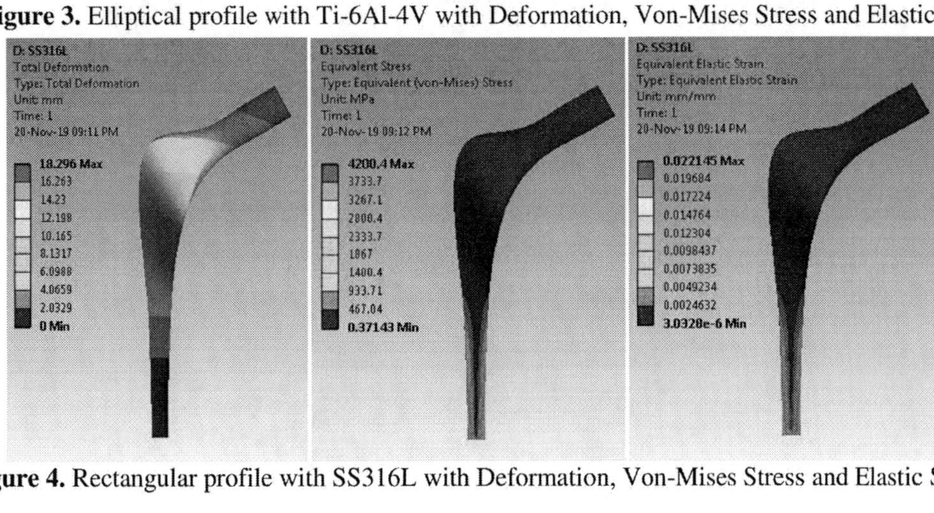

Figure 4. Rectangular profile with SS316L with Deformation, Von-Mises Stress and Elastic Strain

4. Conclusion

The present work focuses on the static structural analysis of the hip joint with various cross sections and different materials was analyzed with finite element method. From the above results, it can be seen that CoCr is offering good resistance against deformation and also inducing least Von-Mises stress. Due to this, for this material, the amount of wear taking place is less. Within CoCr, the cross section with circular profile has best possible results. It is inferred from the findings of this study that the circular profile made of CoCr material is the best one for hip joint implants. CoCr is also a popular material due to its superior mechanical properties and biocompatibility.

5. References

[1] Byrne DP, Mulhall KJ, Baker JF. Anatomy & biomechanics of the hip. *The open sports medicine Journal.* **2010**;4(1).

[2] Campbell JD, Higgs R, Wright K, Leaver-Dunn D. Pevis, hip and thigh injuries. In: Schenck RC, Guskiewicz KM, Holmes CF, Eds. *Athletic Training and Sports Medicine. Rosemount:American Academy of Orthopaedic Surgeons* **2001;** p.399.

[3] Sabatini AL, Goswami T. Hip implants VII: Finite element analysis and optimization of cross-sections. *Materials & Design. 2008 Jan 1*;**29(7**):1438-46.

[4] Izzo GM. Support for total hip replacement surgery: Structures modeling, Gait Data Analysis and Report system. *European Journal of Translational Myology*. 2012 **Mar 6;22(1-2**):69-121.

[5] Chethan KN, Zuber M, Shenoy S, Kini CR. Static structural analysis of different stem designs used in total hip arthroplasty using finite element method. *Heliyon. 2019* **Jun 1;5(6):**e01767.

[6] Sabatini AL, Goswami T. Hip implants VII: Finite element analysis and optimization of cross-sections. *Materials & Design. 2008* **Jan** 1;**29(7**):1438-46.

[7] Senalp AZ, Kayabasi O, Kurtaran H. Static, dynamic and fatigue behavior of newly designed stem shapes for hip prosthesis using finite element analysis. *Materials & design. 2007* **Jan 1;28(5):**1577-83.

[8] CAMPBELL RE, ROTHMAN RH. Charnley low-friction total hip replacement. *American Journal of Roentgenology*. 1971 **Dec;113(4**):634-41.

[9] Hampton SJ, Andriacchi TP, Galante JO. Three dimensional stress analysis of the femoral stem of a total hip prosthesis. *Journal of biomechanics. 1980* **Jan 1;13(5):**443-8.

[10] Colic K, Sedmak A, Grbovic A, Tatic U, Sedmak S, Djordjevic B. Finite element modeling of hip implant static loading. *Procedia Engineering. 2016* **Jun;149**:257-62.

[11] Darwich A, Nazha H, Abbas W. Numerical study of stress shielding evaluation of hip implant stems coated with composite (carbon/PEEK) and polymeric (PEEK) coating materials.

Performance Enhancement of Rotary Desiccant Wheel using Novel Homogenous Composite Desiccant Designs

Selvaraji Muthu[1], N Sekarapandian[1] and K Ashok[2]
[1] Department of Thermal and Energy, School of Mechanical Engineering, Vellore Institute of Technology, Vellore, India, Muthu.selji@gmail.com
[2] Department of Automotive Engineering, School of Mechanical Engineering, Vellore Institute of Technology, Vellore, India, ashok.k@vit.ac.in

Abstract. Rotary solid desiccant wheels are used as sensible and latent heat recovery wheels in the Desiccant-HVAC systems. The two major types of these wheels include enthalpy (total energy recovery) wheels which remove sensible heat and latent heat from process air and transfer them to regeneration air, and dehumidification wheels which transfer a significant amount of moisture (latent heat) at the same time minimizes heat transfer. In this work a set of novel design of hybrid rotary desiccant wheel constructed using a composite homogeneous mixture of solid desiccants (multiple types of silica gel and molecular sieves) are proposed. The transport phenomena taking place in the proposed set of novel design of hybrid rotary desiccant wheel are simulated numerically using an in house finite volume method based CFD code. The performances of these wheels are compared with conventional type of wheels made of molecular sieves and silica gel, respectively. The results show that the performance of these hybrid wheels are enhanced by up to 40 % by using these novel composite wheel designs.

1. Introduction

Dehumidification of moist air in a HVAC (Heating Ventilating, filtering and Air conditioning) system using desiccant adsorbents is one of the promising technologies to meet the demands of in-door thermal comfort regulations and norms [1]. Desiccant technology has received great attention due to its certain unique features such as overcoming the un-wanted release of adsorption heat, low level of electrical power for the system and utilization of waste heat or low grade thermal energies [2].

Rotary desiccant wheel (RDW) consists of a substrate embedded with a matrix of air passages across its thickness. The surfaces bounding these matrix structures are coated with solid desiccant materials of various types. During dehumidification process, the moisture contained in the humid air that is passing through RDW is adsorbed by the desiccant material layers and simultaneously the released adsorption heat stored in the desiccant is transported along with the processed air. When the moisture trapped in the desiccant reach its saturation level, hot regeneration air is passed through the RDW in counter flow direction so that the moisture adsorbed in the desiccant layer is removed, making it dry again, as shown in Figure 1 [3, 4]. The heat energy source required during regeneration of the wheel can be obtained from waste heat and or low grade heat resources such as other thermal equipment, solar energy, etc.

Significant theoretical researches on numerical simulation have been done to analyze the performance of these desiccant based HVAC (D-HVAC) units. A one-dimensional numerical model was built [1] to simulate the impacts of structural parameters on RDW performance under various operating conditions and obtained the heat and mass transfer correlation based on Nusselt and Sherwood numbers. The developed model is validated against the experimental data on the both air side and desiccant layer heat and mass transport properties of a RDW under various operating conditions [5, 6].

In this paper, the concepts of hybrid wheel with homogeneous mixture of different types of desiccant are presented. The purpose of this research is to develop a novel design of hybrid RDW using homogenous mixture of multiple desiccant types to predict the performance and compare the heat and mass transfer phenomena in H-RDW for the application in D-HVAC system. The aim is to predict the performance improvement potential of hybrid wheel [7-12] in comparison to conventional silica gel and molecular sieves wheels.

2. Problem Description

Figure 2 depicts the proposed novel honey comb structure of desiccant channel and the layer formation made up on homogeneous mixture of different type of desiccants. The hybrid wheel can be made using common substrate on which by coating or impregnating the non-homogeneous mixture of multiple desiccants through the depth of the wheel. The packing of desiccant in the hybrid wheel fulfills to the pore standards defined by IUPAC [13] and is reproduced in Figure 3. Accordingly, the desiccants (any type of silica gel, SG) having the adsorption capacity isotherm as per type-III (Mesopores) is used as descending concentration starting from the inlet side of the process air. The remaining layer thickness is coated with desiccants such as Zeolites, Molecular Sieves and MS (3 Å , 4 Å)) having the adsorption capacity isotherm as per type-I (Microspores).

A mathematical relation to calculate adsorption capacity of hybrid desiccant wheel with homogeneous mixture of desiccant types with fixed concentration ratios is given by [1]:

$$W = W_1\alpha + W_3\beta \tag{1}$$

Where W_1 is the adsorption capacity of desiccant having isotherm as per Type-I (Molecular sieves desiccant), W_3 is the adsorption capacity of desiccant having isotherm as per Type-III (Silica gel desiccant) , α is the concentration factor of Type-I desiccant, β is the concentration factor of Type-III desiccant. A comparison of isotherms of desiccant wheels made of conventional Silica gel and Molecular sieves with that based on the homogeneous mixtures (50:50% concentration ratios) of conventional Silica gel and Molecular sieves is shown in Figure 4. Similarly, in Fig.5 isotherm of desiccant wheel based on the homogeneous mixtures (50:50% concentration ratios) of Composite Silica gel and Molecular sieves is compared with the isotherms of wheels based on composite Silica gel and Molecular sieves. The prediction of performance of the proposed hybrid wheels are predicted using CFD simulation. In general, the desiccant wheel can be of any one particular type of desiccant and or a mixture of particles of different sizes of different type of desiccants [14]. The assumptions, nomenclatures and geometrical data adopted in the simulation are referred from Selvaraji et al. [3, 4], and [7-12].

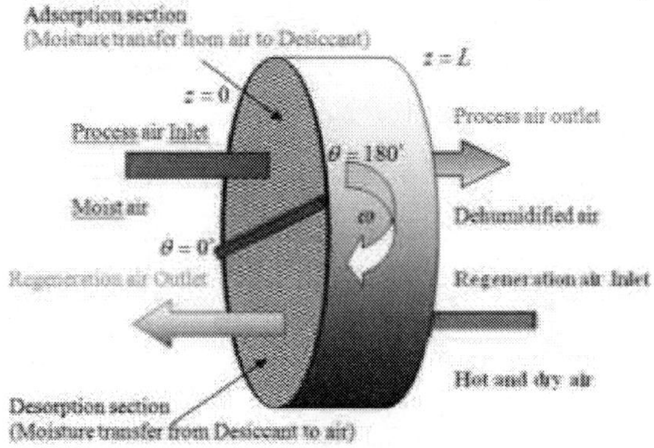

Figure 1. Schematic of rotary solid desiccant wheel (RDW)

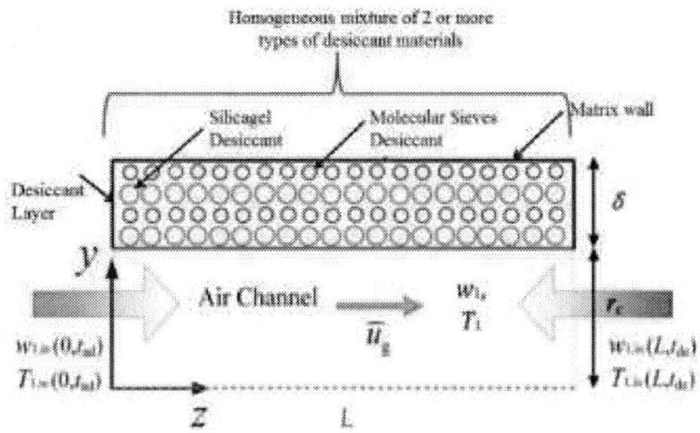

Figure 2. Novel honeycomb channel layer of hybrid solid desiccant wheel with silicagel and molecular sieves desiccant materials coated over matrix wall, a=3.5 mm, b=1.75mm

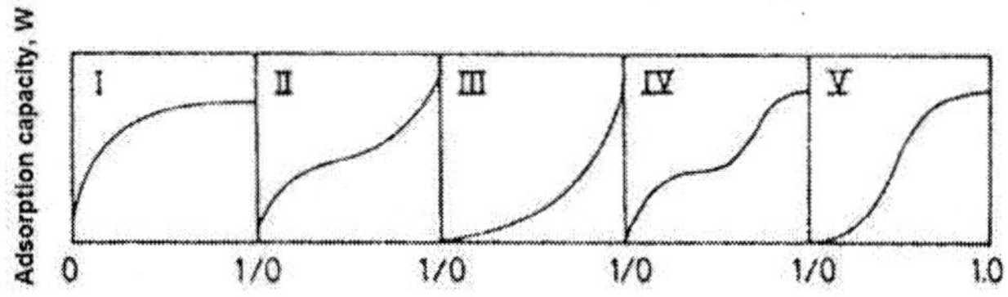

Figure 3. Classification of desiccant isotherm [8]

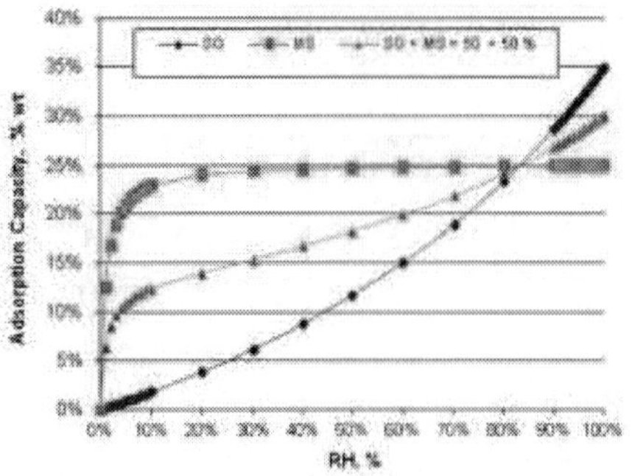

Figure 4. Isotherm of Homogeneous mixture of SG : MS

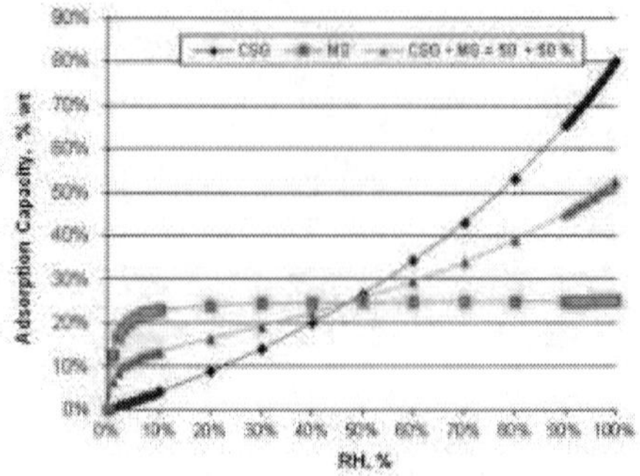

Figure 5. Isotherm of Homogeneous mixture of CSG : MS

3. Governing Equations

The governing equations are derived based on the conservation of moisture and conservation of energy for air and desiccant medium. The moisture content conservation in the process side and regeneration side air stream is given by

$$d_e\left(\frac{\partial(\rho_a Y_a)}{\partial t}+\omega\frac{\partial(\rho_a Y_a)}{\partial\theta}+u\frac{\partial(\rho_a Y_a)}{\partial z}\right)=K_y(Y_d-Y_a) \quad (2)$$

Moisture content conservation in the desiccant layer of rotary solid desiccant wheel is given by

$$\delta\rho_d\left(\frac{\partial W}{\partial t}+\omega\frac{\partial W}{\partial\theta}-\frac{\partial}{\partial z}\left(D_e\frac{\partial W}{\partial z}\right)\right)=K_y(Y_d-Y_a) \quad (3)$$

Sensible energy conservation in the process and regeneration air is given by the below equation

$$d_e c_{pa}\left(\frac{\partial(\rho_a T_a)}{\partial t}+\omega\frac{\partial(\rho_a T_a)}{\partial\theta}+u\frac{\partial(\rho_a T_a)}{\partial z}-d_e\frac{\partial}{\partial z}\left(K_a\frac{\partial T_a}{\partial z}\right)\right)=h(T_d-T_a)+c_{pv}K_y(Y_d-Y_a)(T_d-T_a) \quad (4)$$

Sensible energy and adsorption energy conservation in the desiccant layer is given by

$$c_{pd}\,\rho_d\,\delta\left(\frac{\partial(T_d)}{\partial t}+\omega\frac{\partial(T_d)}{\partial\theta}\right)-\delta\frac{\partial}{\partial z}\left(K_d\frac{\partial T_d}{\partial z}\right)=h(T_a-T_d)+c_{pv}K_y(Y_a-Y_d)(T_a-T_d)+K_y(Y_a-Y_d) \quad (5)$$

4. Results and Discussion

With variable operating parameters, the performances of the rotary solid desiccant wheels are presented for H-RDW in comparisons to conventional RDW in detail. In this work, the primary parameters are used to investigate the system performance are the moisture removal capacity (D) of process air side, relative moisture removal efficiency (η) with respect to inlet moisture content and desiccant dehumidification coefficient of performance (DCOP) and are given by the equations (6), (7) and (8), respectively.

$$\mathbf{D}=\mathbf{Y_a}|_{z=0}-\mathbf{Y_a}|_{z=L} \quad (6)$$

$$\eta=\frac{\mathbf{Y_a}|_{z=0}-\mathbf{Y_a}|_{z=L}}{\mathbf{Y_a}|_{z=0}} \quad (7)$$

$$\mathbf{DCOP} = \frac{m_p h_{fg}(Y_a|_{z=0} - Y_a|_{z=L})}{m_r(h_{r,in} - h_{r,amb})} \quad \textbf{(8)}$$

4.1 Effect of process and regeneration air inlet temperatures on D

The performance of proposed hybrid RDW system is predicted for different inlet temperature of process air, T_p from 20°C to 40°C. Fig. 6 shows the performance trend of D for conventional SG wheel and H-RDW. It can be found that MRC reduces as temperature increases, but the H-RDW is significantly having higher performance, which is about 27% higher than conventional SG wheel at 40°C.

Fig.7 shows the performance trend of D for the different inlet temperature of regeneration air T_r from 60°C to 120°C. It can be clearly seen that D increases for increase in temperature of regeneration air, as the thermal driving energy for wheel has increased and that has given a performance improvement of about 32% at 120°C for the hybrid RDW compared to conventional RDWs.

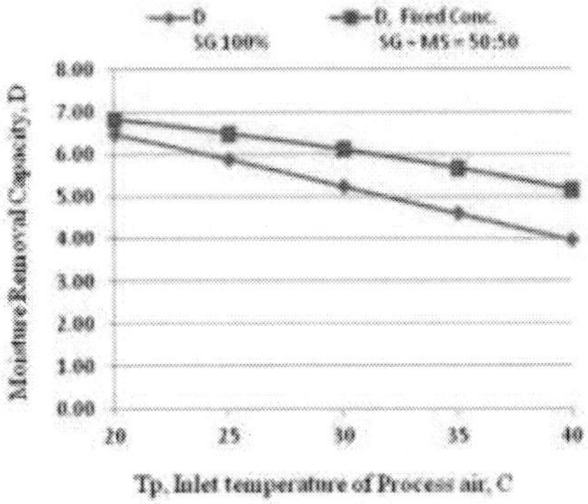

Figure 6. Performance (D) impact due to inlet temperature increase in process air

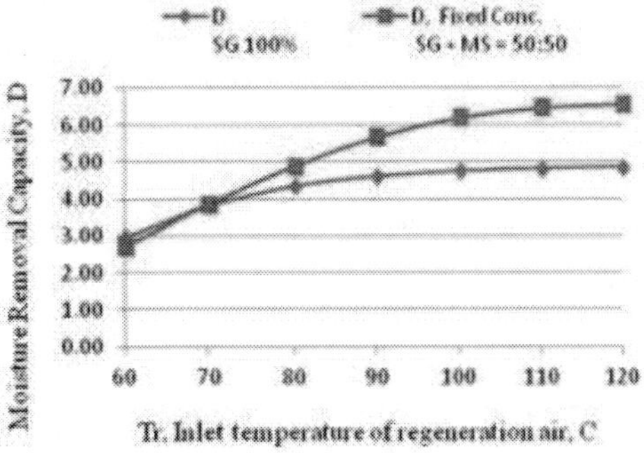

Figure 7. Performance (D) gain due to inlet temperature increase in regeneration air

4.2 Effect of moisture content of process and regeneration air-stream on D

The effect of inlet moisture content of process air on system performance is shown in Fig. 8. It is evident that, a significant increase in D is achieved using hybrid wheel for the entire range of operation. This is due to the fact that the latent heat load in the process air section has increased. This in turn enhances the moisture removal capacity by about 12% at 10 g/kg inlet moisture and about 30% increase at 30 g/kg. Fig. 9 shows the trend of Moisture Removal Capacity D (MRC) of H-RDW. It can be found that MRC reduces as temperature increases, but the H-RDW is significantly having higher performance, which is about 24% higher than conventional SG wheel at 10 g/kg inlet moisture of regeneration air-stream.

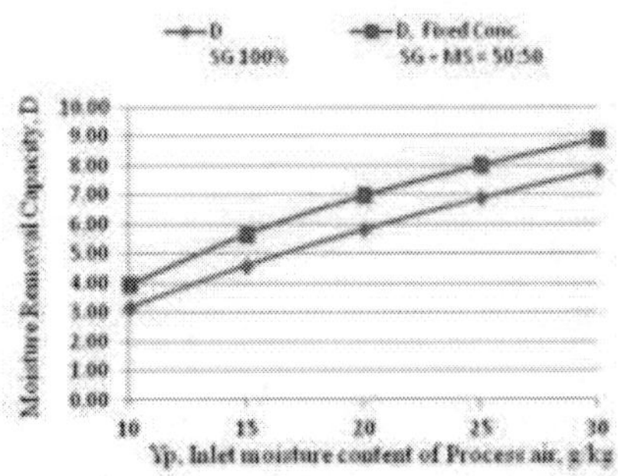

Figure 8. Performance (D) gain due to inlet moisture content increase in process air-stream

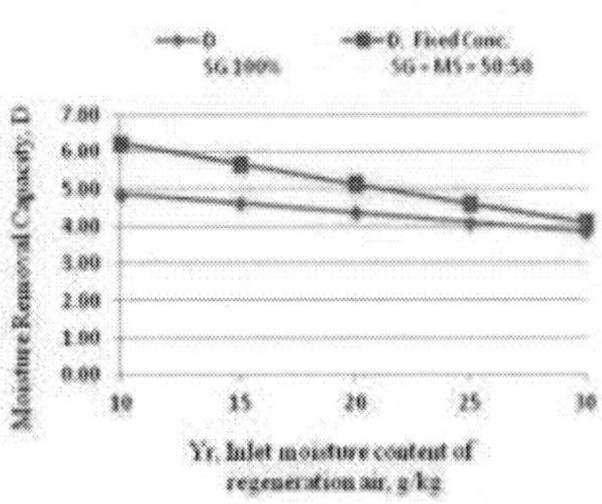

Figure 9. Performance (D) impact due to inlet moisture content to increase in regeneration air-stream

4.3 Effect of temperature of process and regeneration air-streams on RMRE & DCOP

The performance of this novel H-RDW system has been predicted for RMRE and DCOP for different inlet temperatures of process air T_p from 20°C to 40°C. Fig. 10 shows the performance trend of RMRE of conventional SG wheel and H-RDW. It can be found that RMRE reduces as temperature increases, but the H-RDW is having higher performance at 40°C. In Fig.11, it can be observed that RMRE nearly keeps constant and DCOP decreases with respect to increase in regeneration temperature T_r from 60°C to 120°C. This is due to the fact that although T_r changes, moisture content of both process air and regeneration air kept constant and the driving potential for moisture transfer almost keeps constant, therefore RMRE keeps constant. But at the same time, regeneration air with lower T_r can cool the RDW section of process air-stream better, which leads to that more sensible load of process air can be removed and DCOP increases at about 60°C - 70°C.

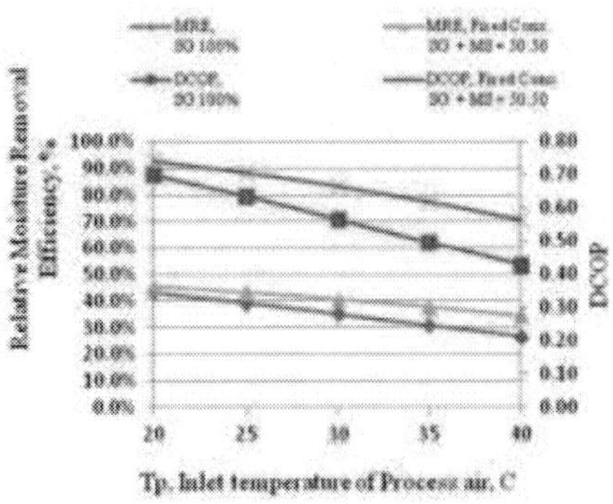

Figure 10. Performance impact (RMRE & DCOP) due to inlet temperature increase in process air-stream

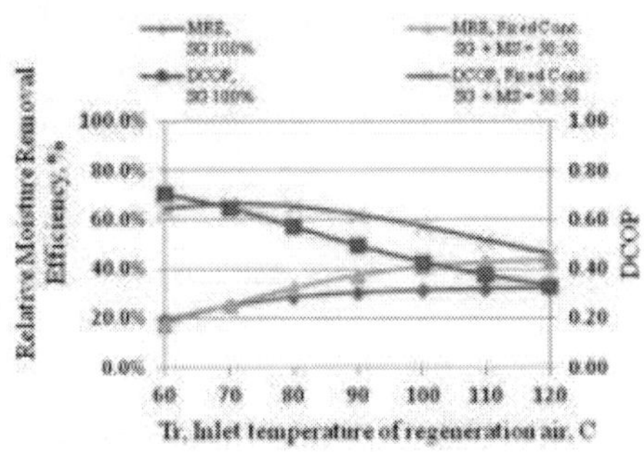

Figure 11. Performance impact (RMRE & DCOP) due to inlet temperature increase in regeneration air-stream

4.4 Effect of moisture content of process air and regeneration air-stream on RMRE & DCOP

The influence of inlet moisture content of process air-stream on system performance in terms of RMRE and DCOP are shown in Fig. 12. There is a significant increase in DCOP achieved in hybrid wheel over entire range of operation due to the fact that the latent heat load in the process air section has increased, however since the regeneration temperature of the air is fixed, then the wheel performance in terms of RMRE is slightly decreased. Fig. 13 shows the performance trend of RMRE and DCOP of H-RDW. It can be found that MRC reduces as inlet moisture of regeneration air increases, but the H-RDW is significantly having higher performance, which is about 20% higher than conventional SG wheel at 10 g/kg inlet moisture of regeneration air.

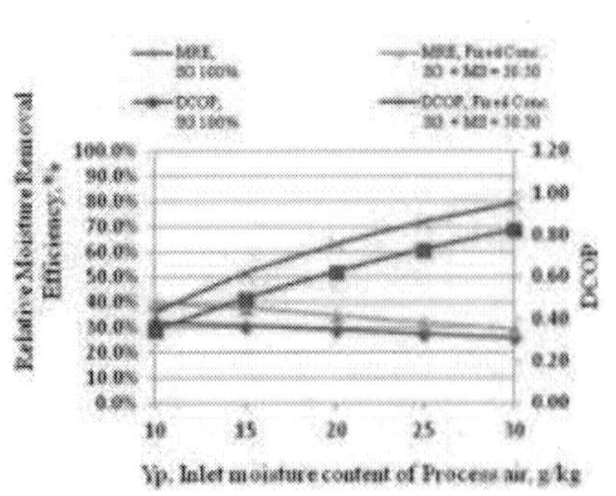

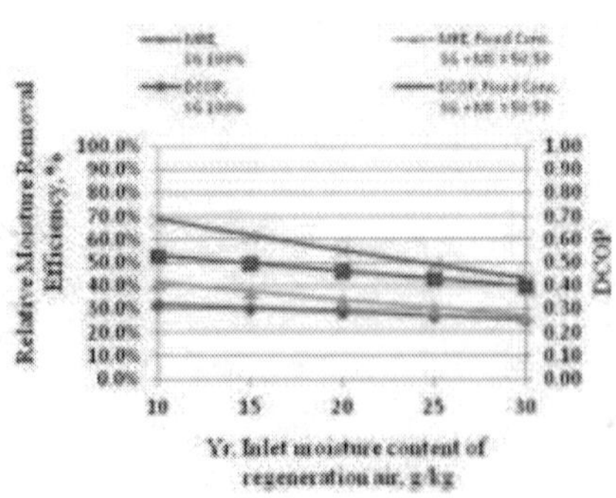

Figure 12. Performance (RMRE & DCOP) due to inlet moisture increase in process air-stream

Figure 13. Performance impact (RMRE & DCOP) due to inlet moisture increase in regeneration air-stream

4.5 Effect of temperature of regeneration air-stream on DPT & DCOP with angle of regeneration section of 180°

The potential performance improvement of the H-RDW is predicted for the 4 cases of mixture concentrations (SG : MS → 20:80%, 40:60%, 60:40%, 80:20% and together with conventional MS and SG wheels , named as 0:100 %, 100:0%, respectively and regeneration section angle fixed as 180°. In Fig.14, it can be clearly seen that the dew point temperature (DPT) achieved at the outlet air-flow of different wheel configurations and it decreases while the regeneration air-stream temperature increases. It also be seen that the optimum low value of DPT is achieved from the H-RDW of having concentration of 20:80% SG: MS). In Fig.15, it can be clearly observed that the DCOP decreases for increase in regeneration air temperature. The optimum DCOP value is achieved from H-RDW for the concentration of 40:60% SG: MS.

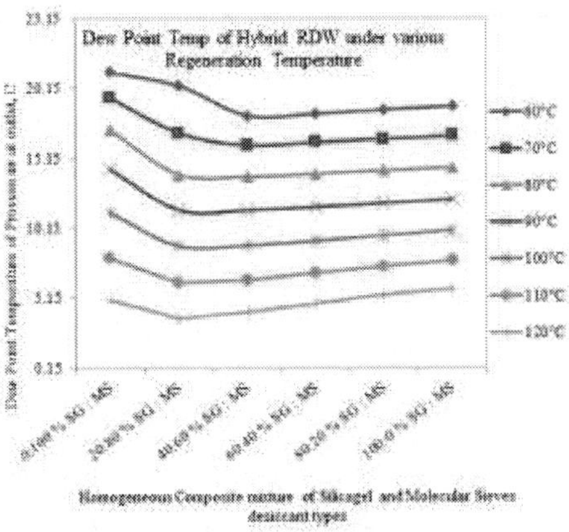

Figure 14. Dew Point Temp of Hybrid RDW under various Regeneration Temperature (regen angle 180°)

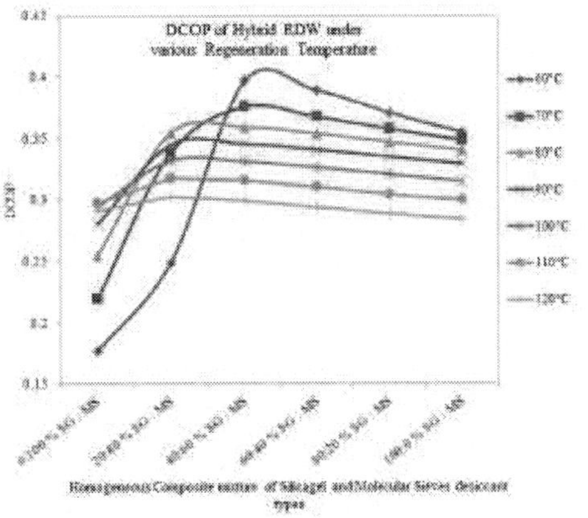

Figure 15. DCOP of Hybrid RDW under various Regeneration Temperature (regen angle 180°)

4.6 Effect of temperature of regeneration air-stream on DPT & DCOP with angle of regeneration section of 120°

Fig.16 shows the dew point temperature (DPT) achieved for the H-RDW configurations with regeneration section angle fixed of 120° and it decreases while the regeneration air-stream temperature increases. It also be observed that the optimum low value of DPT is achieved from the H-RDW of having concentration of 20:80% SG: MS, as low as 9.14°C, however, the same H-RDW has given as low as 3.7°C DPT with the regeneration angle of 180°. In Fig.17, it can be clearly noted that the DCOP decreases for increase in regeneration air temperature. The optimum DCOP value is achieved from H-RDW for the concentration of 60:40% SG: MS and as high as 0.66, which is 65% higher DCOP compared the value of 0.4 for regeneration angle of 180°.

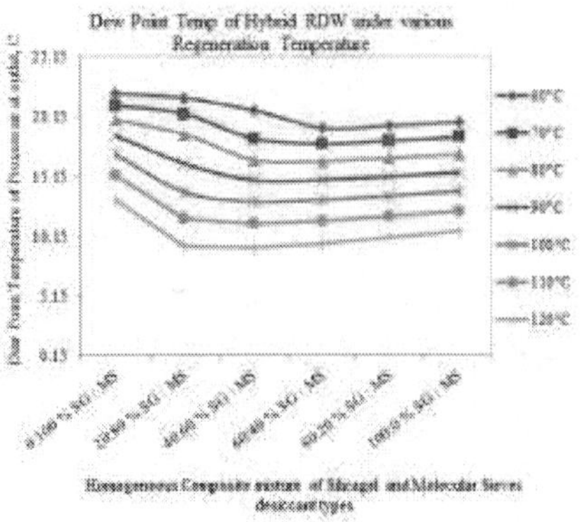

Figure 16. Dew point temperature of hybrid RDW under various regeneration temperature (regen angle 120°)

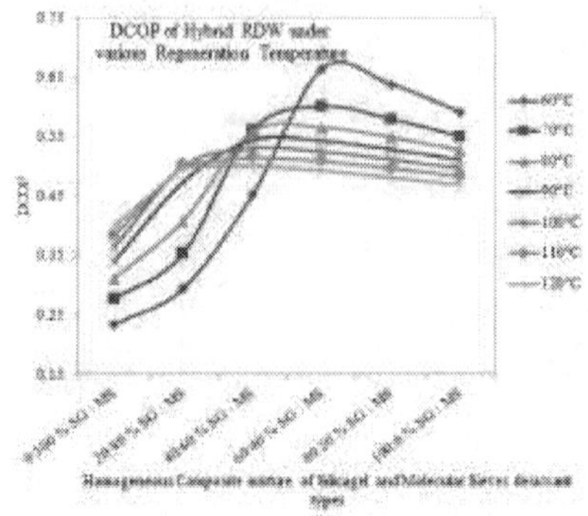

Figure 17. DCOP of hybrid RDW under various regeneration temperature (regen angle 120°)

5. Conclusion

The hybrid wheel design using homogeneous composite mixture of desiccant is explained in detail in comparison to conventional desiccant wheels. The performance improvement by using H-RDWs for a range of operating parameters is presented using different concentration ratio of mixtures. Since the hybrid rotary solid desiccant wheels are made of both SG and MS desiccant concentrations of homogeneous mixtures, the optimum performance can be achieved at wide operating conditions by bring the positive aspects of both SG and MS solid desiccants. The moisture removal capacity of the process air-stream and relative moisture removal efficiency with respect to moisture content of inlet of process air of hybrid desiccant wheels are enhanced by 17% to 24% compared to conventional desiccant wheels. The MS wheel is inferior to SG wheel at 90° and 60°C regeneration temperature in terms of MRC and RMRE; whereas the hybrid rotary solid desiccant wheels are showing the potential of superior performance than SG wheels at different regeneration section angles. The benefit of hybrid wheel design is clearly able to be seen in the " pressure dew point " achieved while comparing the conventional wheels for a wide range of regeneration air temperatures and the optimum lowest values are given by 20:80% SG : MS hybrid wheel. Optimum DCOP is achieved for hybrid wheel design of 40:60% SG : MS and 60:40% CG:MS with the regeneration section angle of 180° and 120° respectively, while comparing the conventional wheels for different regeneration air temperatures. With this hybrid rotary solid wheel design, the moisture removal capacity of process air-stream and relative moisture removal efficiency referring to process air inlet and overall DCOP can be enhanced.

REFERENCES

[1]. T.S. Ge, F. Ziegler, R.Z. Wang, 2010, "A Mathematical Model of Predicting the Performance of Compound Desiccant Wheel (A model of compound Desiccant Wheel)", *Applied Thermal Engineering*, **30**, 1005-1015.

[2]. R. Narayanan, W.Y. Saman, S.D. White, M. Goldsworthy 2011, "Comparative study of different desiccant wheel designs", *Applied Thermal Engineering* **1 -8**

[3]. Muthu, S., Talukdar, P., and Jain, S. (November 11, 2015). "Effect of Regeneration Section Angle on the Performance of a Rotary Desiccant Wheel." *ASME. J. Thermal Sci. Eng. Appl*. **March 2016; 8(1):** 011013

[4]. Selvaraji Muthu, Prabal Talukdar, Sanjeev Jain (2017) Performance Enhancement of Rotary Desiccant Wheel by Innovative Designs of Multiple Desiccant Layers. In: Saha A., Das D., Srivastava R., Panigrahi P., Muralidhar K. (eds) *Fluid Mechanics and Fluid Power – Contemporary Research. Lecture Notes in Mechanical Engineering. Springer, New Delhi,*

[5]. G. Angrisani, F. Minichiello, C. Roselli, M. Sasso, Experimental analysis on the dehumidification and thermal performance of a desiccant wheel, **Appl. Energy** *92 (2012) 563e572.*

[6]. Piero Bareschino, Giuseppe Diglio, Francesco Pepe, Giovanni Angrisani, Carlo Roselli, Maurizio Sasso, " Modelling of a rotary desiccant wheel: Numerical validation of a Variable Properties Model " *Applied Thermal Engineering* **78 (2015) 640e648**, http://dx.doi.org/10.1016/j.applthermaleng.2014.11.063

[7]. Selvaraji Muthu, Prabal Talukdar, Sanjeev Jain: Modelling and parametric simulation of coupled heat and mass transfer phenomena in a rotary desiccant wheel. In: *Proceedings of the 22th National and 11th International ISHMT-ASME Heat and Mass Transfer Conference 28–31 Dec 2013, IIT Kharagpur, India, HMTC1300857 (2013)*

[8]. Selvaraji, M., Sekarapandian, N. (2019), Periodic transient performance prediction of the rotary desiccant wheel. *Proceedings of the 46th National Conference on Fluid Mechanics and Fluid Power (FMFP) PSG College of Technology, Coimbatore, India.*

[9]. Selvaraji, M., Sekarapandian, N. (2019), Numerical Analysis of Coupled Heat and Mass Transfer in Solid Desiccant for Stationary Hybrid Desiccants Beds. *Proceedings of the 46th National Conference on Fluid Mechanics and Fluid Power (FMFP) PSG College of Technology, Coimbatore, India.*

[10]. Selvaraji, M., Sekarapandian, N. (2019), Simulation of Hybrid Rotary Desiccant Wheels having Nonhomogeneous Mixture of Multiple Desiccants Types, *Proceedings of the 25th National and 3rd International ISHMT-ASTFE Heat and Mass Transfer Conference, IIT Roorkee, Roorkee, India.*

[11]. Muthu, S., and Sekarapandian, N., 2020. "Simulation of the performance of solar driven thermoelectric based rotary desiccant wheel hvac system". *SAE Technical Paper, 2020(2020-28-0041), August.*

[12]. Muthu, S., and N, S., 2020. "Simulation of hybrid packed desiccant beds for dehumidification and drying". *SAE Technical Pape, **2020(2020-28-0019)**,* August.

[13]. D.M. Ruthven, "Principles of Adsorption and Adsorption Processes", Wiley, New York (1984).

[14]. Fischer, John C. (Marietta, GA), Mescher, Kirk T., Desiccant-coated substrate and method of manufacture, United States, Semco Incorporated (Columbia, MO), US5496397

Integration of Virtual Reality and Digital Human Model Simulations for Ergonomic Analysis of CAD Models

S Senthil Kumar[1] and S Denis Ashok[2]
School of Mechanical Engineering, Vellore Institute of Technology, Vellore, India.
Email: [1]ssenthilkumar@vit.ac.in; [2]denisashok@vit.ac.in;

Abstract. The paper deals the exercising Virtual Reality (VR) validation for ergonomic study of Computer Aided Design (CAD) virtual environment models, for the safe and comfortable operations for human. The CAD based virtual environments will subject to ergonomic analysis could reduce the costs of testing and errors in integration of sub systems compared to real-time trials. In this venture, a Digital Human Model Simulation (DHMS) was analysed in a workspace by incorporating VR simulation softwares. A transport and freight of military personnel and equipment in Armoured Personnel Carrier Vehicles (APVCs) model was employed as environment for ergonomic assessment. VR device with head-mounted display was used to perform the ergonomics analysis in virtual environment. The ergonomics study and the experience lead to changes in design features. This application could benefit in the defence, aerospace and naval sector, where ergonomically viable compact spaces for human inhabit of key importance.

1. Introduction

The fast pace modernization in research in various fields is increasing and necessary to use the IT tools for simulation and analysis. Nowadays, virtual reality to be able to virtually replicate the entire planned design and models of major projects and easily visualize them through superficially being present within. Present VR technologies supports virtual presence of user in the CAD design and feeling the entire experience of the design/object to be analysed and assists in suggesting changes associated with design issues. Contemporary Virtual Reality (VR) experience allows incorporating the dynamic models to interactively manipulate and also fit within 3-D systems environment.

In this work, a Digital Human Model (DHM) was subjected to an ergonomic analysis in an Armoured Personnel Carrier Vehicles (APVCs) workspace by incorporating VR simulation softwares. Ergonomics of human factors is the scientific discipline that is mostly concerned with understandings of the interactions among humans and also that of the other various elements of a given system, and it applies data, theories, principles and certain critical and useful methods so as to design in order to enable us to optimize the factor of human well-being and that of the overall system performance. The Digital Human Modelling (DHM) is involving CAD representation of the form of human bodies or if not, their certain parts for the purpose of virtually evaluating the ergonomics of a newly designed object's product-human compatibility.

Sougata et al [1] presented a review which informs about a solid knowledge base which is regarding the DHM's present status and its use for ergonomically evaluating the products and their workplaces in industries and emphasizes that more R&D activities are needed so as to make customized humanoid modelling designs which will be representing regional population. Mahdavian et al [2] presented various developments and the capabilities of the DHM tool IPS IMMA (Intelligent Moving Manikins) for an assembly process in VR environment and noted that in a process of product design, lack of the accuracy leads to decreased precision and confidence in the product designing processes, thus causing errors and the sequence in which components are imported does not make a difference in the simulation. Whereas, for simulation purposes, depending on the size of the component, work required for a modification and the volume, import of the point cloud, Computer aided design geometry and mannequins ease the flow of work. The precise motive of the DHM tool is to support smoother and faster decisions in the product designing process, so well as design solution for well-being of humans.

Giovanni et al [3] carried out extensive studies and introduced the Dynamic Digital Human Models, which were monitored both in force and in their acceleration. DHMs were used to simulate the experimental incorporation fitting jobs in quasi-real-time, and also used simulated times, exertions and the postures in order to calculate the index-based ergonomical assessment of the taken model. Luigi Pelliccia et al [4] developed the HuPOSE model, which was involved in the calculated simulation of an Jacobian matrix which was augmented, so as to specify the trajectories and the paths for the different points of control, considering the anthropological, bio mechanical and musco-skeletal aspects. The study was analysed based on single point analysis of multiple postures, such as that of squatting, sitting, hand straightening and a large variety of postures and positions were assessed.

Nick [5] stated that the human body and its modelling serves to the engineer concerned with human factors with a novel simulation tool which binds the various opportunities of CAD and the techniques which are encored around the user for the iterative design and also that of prototyping and focussed to create more interactive virtual environments. The very early stages of the technologies involving digital humans and their simulations in order to provide ergonomic analysis of the VEs were highlighted. Suman et al. [6] worked on the usage of digitally capturing the human motions, which will act as the source which drives the VE interaction among the DHM mannequin and its simulated VE for work. It also emphasizes on the application of dynamic ergonomical analysis of mannequins and DHMs like the Job Risking Classifications Modelling, which was rooted on the motion and interactions captured from such a virtual environmental interaction, which provides a positively encouraging way to engage in the further improvement of the current CAES systems. The study highlighted on the various shortcoming of the processes and the hindrances that exist in further developing the models and making the process more interactive and nearer to the real world. Jeffrey [7] concluded by adding that the DHM models are being taken into usage by the ergonomists increasingly over time, along with the dependency on other engineers to design virtual environments and equipments so as to help the human operators.

Faisal et al [8] presented and compared different human body joint monitoring methods for early diagnosis of musculoskeletal disorders. The current human models are limited by understandings of the ergonomics and cognitive human behavioural patterns offered by the computers for avoiding musculoskeletal disorders. Model accuracy is limited by a deficit in the current available data on 3-D human anthropometry, biomechanics, strength and postural formations.

Lack of basic information about the human postures and motions is addressed to be also one of the key problems. With the advancement of ergonomic tools, computer models will become more and more accurate over time. The CAD models are available to help designers to study and monitor fit parameters, biomechanics and many more. The current anthropometric databases have still many limitations in understandings of the human posture and motion. There is a need for the models to represent the population of the end user and thus such parameters vary accordingly, from region to region. Applying DHM has proven to be beneficial as it helps in the cause for significant reduction of project time-scale; the design costs; the manufacturing cost; the occupational hazards and it also helps in improving the quality, the productivity and the efficiency across multiple industrial sectors.

2. Digital human body model and ergonomics

The Digital Human Models (DHM) are basically computer-generated models of human beings which are used in CAD. Computer aided human body modelling is a tool to analyze and assess human factors issues. Through instilling an interactive and immersive visualization of a chosen CAD design with the representative mannequins, the arising human issues, such as fit, reach and vision can be evaluated and design changes can be recommended at the early conceptual stages of design. CAD based human body's modelling is a comparatively new form of technological system and there is a wide margin for the improvement of the commonly available tools like Android, Combiman, Poser, Safework, MannequinPro etc. The presence of the multiple degrees of freedom that do characterise the human body joints, have a significant impact on processing of the demands on the DHMs software simulations. The currently available modules support analysis on clearances and reach, visual obstructions, field of vision, biomechanical parameters, postures during a certain task/work in whole, energy expenditure metabolically, motions of various body parts, time study and anthropometric parameters.

The human body is represented as mechanical links conformed series, which are assumed to be joined at pin joints which vaguely correspond to the normal human skeleton. The complexity of such a framework may include more than 100 different links and about 150 degrees of freedom. The human body joints which are not to be modelled using rigid structure link, are represented by humanoid models. In an approach to incorporate the real-world parameters, apart from the surface geometrical parameters, most models are incorporated with different clothing aspects, which often leads to interaction of the reach, fit, range-of-motion and collision's detection etc. in a pursuit to enhance the real-world element. The various joints for computational analysis of a DHM include wrist, elbow, shoulder, stemoclavicular, ankle, knee, upper back and neck.

The biomechanical models and parameters are integrated into the simulation and DHM software systems. These parameters thus enable the estimation of the mechanical stresses, which are induced upon internal parameters of body amidst different processes and activities. The motion of human body is equal importance as of any other parameter in the design of a work environment and the DHM simulation must take enough considerations. The motion algorithms of the human body are usually included with computer packages and capable of producing lift motions, reach, and grasp motion for a model which is stationary or moving in the environment. The DHM analysis has stimulus in productions systems. Manual assembly is time and cost consuming phase in production. It is important that the design of assembly process with a consideration in ergonomics, will increase whole system's efficiency, human's well-being and quality of output at required levels.

3. Methodology

In this work, the ergonomics of a CAD generated environment of military APCV was studied and analyzed in virtual reality environment. The virtual reality tools enable the analysis of designs in early development phases and to avoid integration of any irregularities or unwanted design parameters. This will help defense sector in analyzing key vehicle design, ergonomics and features before making decisions to develop APCVs. Figure 1 shows the procedure adopted in this simulation study.

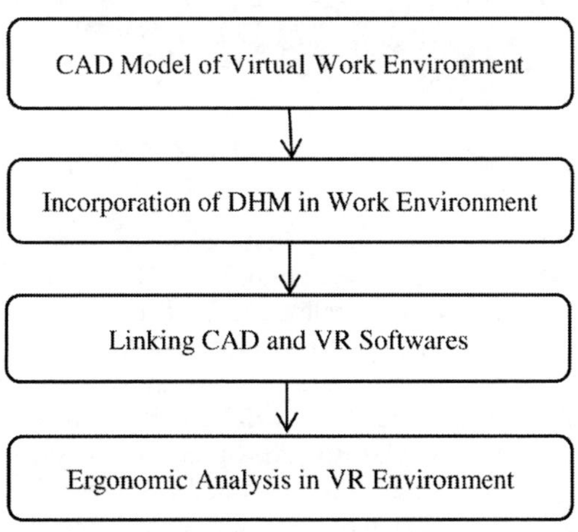

Figure 1. Methodology

The CAD softwares Unigraphics, SolidWorks and CATIA RULA were used to model VR environment and manikin (Digital Human Model). For virtual reality interaction, VR software along with head mounted VR system was employed. A high-end computing system with minimum of 8 GB RAM and graphic card support need to be used for the study.

4. CAD modelling of APCV

In this study, a CAD based virtual work environment was generated with Solid Works as a real-life scaled version of a possible Armored Personnel Carrier Vehicle (APCV) in cabin structures. The basis of the design was the Panzer IV, a World War II tank of German make. The technical specifications of the tank are listed in table 1.

Table 1. Specification of Vehicle Model

Parameter	Value
Kerb weight	23.0 tonnes
Length	6863.86 mm
Width	4371.16 mm
Height	2451.35 mm
Crew	9
Engine	12-cylinder Maybach (296 BP)
Power/weight	12 HP/tonne
Suspension	Torsion-bar suspension
Fuel capacity	310 litre
Operational range	165 km
Speed	Road: 41 km/hr Off-road: 23 km/hr

Figure 2. CAD model of APCV

A digital human model or virtual dummy (Manikin) was incorporated inside the virtual work environment of the APCV, which is shown in figure 2. A CATIA RULA analysis of the model is performed for different postures of the human male manikin. The RULA analysis system; which is based on the analysis of upper limbs assessment and is a product which is widely used in industries. It is accompanied with the CATIA software as a feature tool for the analysis of designed objects and VEs. Through the ergonomic analysis results, iterations will be performed which will thus suggest a better design ergonomically. RULA from CATIA software is used to assess the VE design and the obtained parameters are optimized for a better design. The generated CAD model was processed on SolidWorks software for VR visualization. The validation of optimal designed component is carried out by a Virtual Reality (VR) head mounted display system to assess the design parameters.

5. Ergonomic analysis in VR environment

The objective of the study is to validate the ergonomic feasibility of the workspace layout for an APCV (Armored Personnel Carrier Vehicle) in VR environment. A RULA (Rapid Upper Limb Assessment) analysis on CATIA along with the determination of reach envelope as well as binocular vision of the DHM manikin was performed in the CAD model of the Panzer IV World War II model of German make tank as the reference. The sectional top and isometric views of APCV model with seating area are depicted in figure 3. The various positions that have been assumed for RULA Analysis is marked in the sectional view of the model in the figure. Initially, two positions, at boundary seating and central seating were considered for the ergonomic analysis.

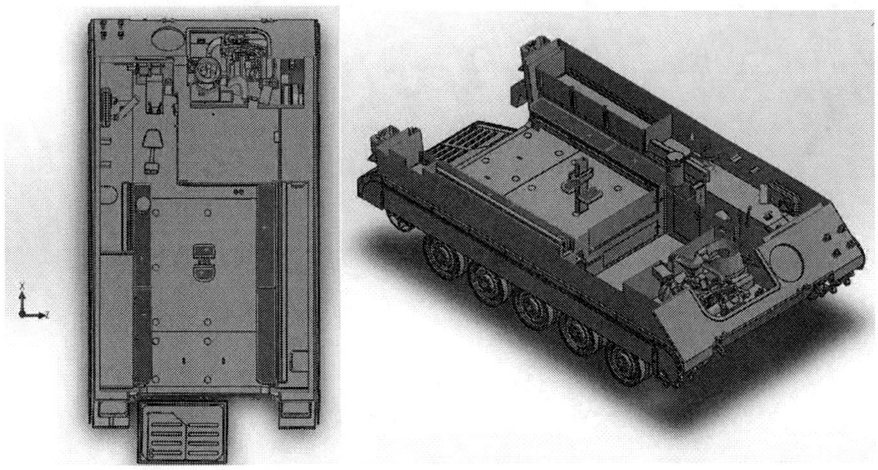

Figure 3. Sectional top and isometric views of APCV model with seating area

The SolidWorks CAD file of vehicle model was converted into *.IGS* file and imported into CATIA. A manikin DHM was added to simulate the ergonomic feasibility. For the manikin DHM, an 85 percentile male was considered as reference. During the ergonomic feasibility simulation of the model, the posture of the manikin has been changed appropriately, to depict a seated armed soldier clutching an assault rifle. The specifications of the posture are listed in the table 2.

Table 2. RULA parameters and values

Posture Parameter	Right	Left	Posture Parameter	Right	Left
LeArm	32.21°	21.113°	Thumb 2	0°	0°
Clavicular	0.495°	0°	Thumb 3	0°	18°
Foot	0°	0°	Index1	35°	0°
Forearm	77.238°	119°	Index 2	23°	64°
Full spine	0°	0°	Index 3	47°	0°
Head	0°	0°	Middle Finger 1	35°	0°
Leg	90°	90°	Middle Finger 2	23°	64°
Line of sight	0°	0°	Middle Finger 3	47°	0°
Lumbar	0°	0°	Annular 1	35°	0°
Thigh	90°	90°	Annular 2	23°	64°
Thoracic	0°	0°	Annular 3	47°	0°
Toes	0°	0°	Auricular 1	35°	0°
Hand	19.325°	-30°	Auricular 2	23°	64°
Thumb 1	0°	0°	Auricular 3	47°	0°

IOP Conf. Series: Materials Science and Engineering 1123 (2021) 012056 doi:10.1088/1757-899X/1123/1/012056

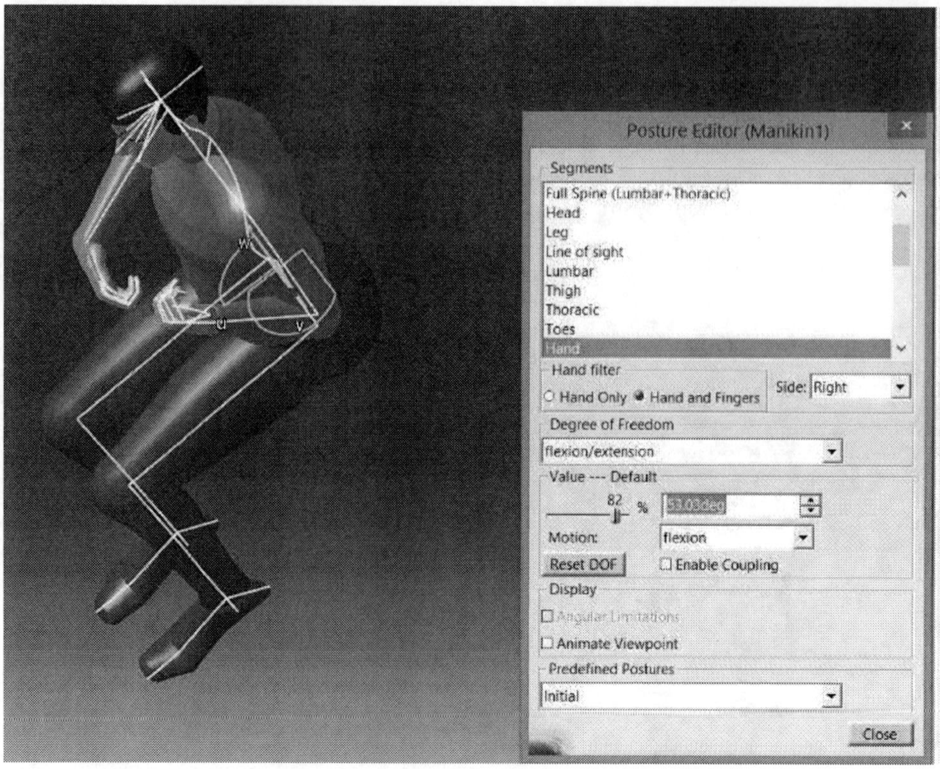

Figure 4. Defining posture of manikin DHM

The posture of the manikin was fixed according to the selected parameters as shown in figure 4 and a static load was added to the arms to simulate the weight of the rifle and subsequently, a RULA score was generated for the posture. The RULA analysis was performed for different positions within the rear area of the vehicle, for the boundary seating and central seating. The simulation results are shown in figure 5 and figure 6.

Figure 5. RULA analysis for boundary seating position

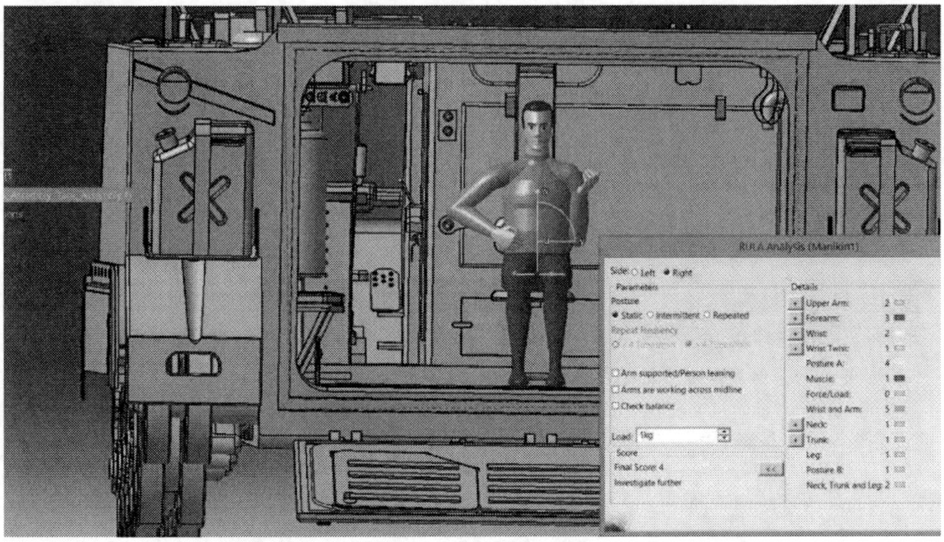

Figure 6. RULA analysis for central seating position

A RULA score of 3 was obtained for the positions considered, with individual scores being allotted for upper arm, forearm, wrist, wrist twist, muscle, force/load, wrist and arms, neck, trunk and leg. The iterations were performed multiple times by adjusting the parameters such as, the shoulder droop, wrist angle, forearm inclination, finger alignment, etc. to attain the most ergonomically sustainable posture and to reduce the load of the gun on the muscles. The posture was set as static and the number of cycles were considered based on real life scenarios. Subsequent to static posture, the reach envelope and binocular vision for the manikin was determined, the simulations were shown in figure 7, figure 8 and figure 9.

Figure 7. Binocular vision and reach envelope at boundary seating position

Figure 8. Binocular vision of manikin and upper limb reach envelope

Figure 9. Reach envelope of manikin at central seating position

6. Validation RULA model in VR environment

In order to confirm RULA results, the model was simulated in the virtual reality setup and manually checked every clearance. To validate the results of RULA analysis and reach envelope determination, the model was imported into the Virtual Reality (VR) software and scaled to the actual dimensions of the tank in VR simulation. Figure 10 shows use of VR headset and hand-held wireless sensors in experiencing APCV model clearances in the VR environment. The reach, envelope and clearances of the walls/equipment/other possible obstacles from the individual body parts were experienced in the VR environment to perceive any potential interference.

Based on the analysis, it was found that the right fore arm and left wrist were subjected to a very high static force due to the weight of the assault rifle. However, the other parameters were found to be within the desired limit. The RULA score obtained was 3, which is considered to be just acceptable. Among the individual RULA scores, none exceeded 3, which is again a desirable result and shows that the armored vehicle being considered is ergonomically sustainable. The reach envelope and binocular vision for the manikin were determined and further cemented the results of the RULA Analysis.

Figure 10. Experiencing model clearances on the VR Environment

The concept of VR analysis can further be used as demonstration to customers earlier to the decision of selling/buying a new product involving human operations.

7. Conclusion

The results of RULA analysis, reach envelope determination and the DHM model were successfully assessed in virtual reality environment with the models scaled up to actual dimensions in simulation. The experimented methodology saves vital time by evading constructions of huge and expensive physical mock-ups prior to commissioning of real setup like APCV. Integrating the VR and ergonomic analysis, crucial decisions can be made promptly, thereby saving cost, time and avoiding errors. Further research in the field of DHMs may focus Indian population, so as to get more precise and accurate results for the human workers based on local conditions. VR technology can further be developed as a fully interactive software tool for real-time solutions.

8. References

[1] Karmakar S, Sanjog J and Patel T 2014 Digital Human Modelling and Simulation in Product and Workplace Design: Indian Scenario. *Int. J. Eng. Res. Appl.* pp 6-12.

[2] Mahdavian N, Ruiz C P, Brolin E, Hogberg D and Hanson L 2017 Digital human modelling in a virtual environment of CAD parts and a point cloud. *5th Int. Digital Human Modeling Symp.*(Bonn, Germany) pp 283-91.

[3] Giovanni D M, Alain M, Jonathan S, Clarisse G and Jacques M 2015 Dynamic digital human models for ergonomic analysis based on humanoid robotics techniques *Int. J. Digital Human* **1(1)** pp 81-109.

[4] Pelliccia L 2013 HuPOSE: Human-like posture generation and biomechanical analysis for human figures *Doctoral dissertation* (University of Naples Federico II)

[5] Nick B 1997 Human Body Modelling as A Human Factors Engineering Tool, *Proceedings of The Capability of Virtual Reality to Meet Military Requirements* (Orlando, USA) pp 5-9.

[6] Suman M, Sanjib K D and Tania C 2012 Computer Aided Design in Digital Human Modeling for Human Computer Interaction in Ergonomic Assessment: A Review *Int. J. Adv. Comp. Res.* **2(4)** p 130.

[7] Jeffrey C and Woldstad 2006 Digital Human Models for Ergonomics *Int. Ency. Ergo. Hum. Fact.* (Taylor & Francis) pp 3093–96.

[8] Abu I F, Sumit M, Tapas M, David C, Sasan N and Jamal D M 2019 Monitoring Methods of Human Body Joints: State-of-the-Art and Research Challenges. *Sensors* **19 p** 2629.

Nitrogen enriched air for Nox reduction in Diesel engine

Ponnusamy Baskar [1], S Prithivi Rajan [2] and Maggi Saraf [2]
[1] Associate Professor, School of Mechanical Engineering, VIT, Vellore, India
[2] U G Scholars, School of Mechanical Engineering, VIT, Vellore, India

Abstract: This paper analyzes the impact of nitrogen enriched air on emissions, in cylinder pressure, heat released during the working cycle and fuel consumption of a diesel engine. The experiment was performed on a naturally aspirated single cylinder diesel engine by increasing the nitrogen concentration supplied through a cylinder and NOx, CO, THC emissions were all measured with the help of a five gas analyzer. It was observed that enriching the inlet air with nitrogen decreases the NOx concentration considerably with a penalty on hydrocarbon emissions and thus the fuel conversion efficiency.

1. Introduction

Diesel and gasoline engines are a major source of modern day pollution. The exhaust gases from vehicles are hazardous to the environment as well as human beings. The combination of HC and NOx in the presence of sunlight and certain atmospheric conditions also produce photochemical smog. Deplete discharges comprising of oxides of nitrogen (NOx) from inward ignition motors depict a genuine issue. Despite the fact that the issue exists for both petrol and diesel engines, the issue is more serious for the diesel engine. In diesel engines the exhaust concentrations of NOx are comparably more than that in SI engines due to lean combustion. NOx emissions depends firmly on flame temperature, and flame temperature is subordinate upon the piece of the fuel and the admission air.

A large portion of these poisonous gases begin from different imperfect methodologies amid burning, for example, incomplete ignition of fuel, high pressure and temperature ignition, greasing up oil and oil added substances and in addition burning of non-hydrocarbon parts of diesel fuel, for example, sulfur mixes and fuel added substances. Regular exhaust emissions incorporate unburned hydrocarbons (HC), carbon monoxide (CO), nitrogen oxides (NOx) or particulate matter (PM) Nitrogen oxides (NOx) are produced by the chemical reaction between nitrogen and oxygen in the combustion chamber under the high pressure and temperature conditions.

The major composition in the NOx exhaust is that of nitric oxide (NO) and a little division of nitrogen dioxide (NO2). The fuel is injected into the cylinder just before combustion starts, so throughout most of the critical parts of the cycle the fuel distribution is non-uniform. The pollutant formation process is strongly dependent on the fuel distribution and how that distribution changes with time due to mixing. Nitric oxide forms in the high temperature burnt gas regions. The critical time period is when the burned gas temperatures are at a maximum i.e. between the start of combustion and shortly after the occurrence of the peak cylinder pressure.

Mixture which burns early in the combustion process is especially important since it is compressed to a higher temperature, increasing the NO formation rate, as the combustion proceeds and cylinder pressure increases. Stringent emissions norms have been enforced by the government to restrict the emissions. Various strategies have been implemented over the years to reduce NOx emissions.

The possibility of various powertrains together with renewable vitality sources, to decrease discharges from vehicles have been broadly utilized. Enormous amount of research has been carried out on renewable energy resources such as LPG, Biodiesel etc.

2. Literature review

An exhaustive study on the nitrogen enrichment was carried out to better understand its impact on the diesel engine emissions and performance. Researchers used different techniques to reduce emissions and improve performance of a diesel engine.

Exhaust gas recirculation wherein a part of exhaust gas is recirculated into the cylinder intake charge reduces the peak cycle temperature as the inert gas acts as a heat sink [1]. There is also a reduction in the percentage of the combustion chamber oxygen content. This decreases the NOx emissions from the exhaust. Although EGR has been successfully implemented in modern day vehicles it has certain drawbacks. The NOx formation varies with the flow rate of the exhaust gases and hence the speed of the engine[2]. It also depends on the gas temperature and water content in the gases. Power and economy are significantly affected by the EGR. Smoke and particulate matter emissions are increased. Oil contamination and wear of the combustion chamber takes place due to the soot particle which are recirculated under high pressure thus engine lifespan is drastically affected [3].

Pertaining to the undesirable characteristics of EGR another way of reducing NOx is enriching the inlet with nitrogen gas [4]. As the NOx development essentially relies upon the rise in cylinder temperature and the habitation time of combustion, inclusion of nitrogen brings down peak ignition temperature with the included nitrogen contributing towards decreasing NOx instead of partaking in the ignition process. Nitrogen enhanced air effectively decreases NOx emissions without any unfavorable consequences on the engine lifespan [5].

Numerous researchers [6-10] have utilized nitrogen-enriched air to diminish NOx emissions in the diesel engine. These engine tests were led by utilizing bottled nitrogen, essentially to recreate the impacts of EGR. The flame temperature and oxygen concentration influence the NOx emissions which will be affected by nitrogen-enriched air.

Enhancing nitrogen content increases the heat capacity of the air mixture in the combustion chamber, which thus, decreases the in cylinder temperature of the combustion chamber [11]. The decreasing temperature due to mixing of high temperature gas with nitrogen gas thus freezes the NO chemistry. The impact of air oxygen content focus on the adiabatic flame temperature was analyzed by Olikara and Borman.

The adiabatic fire temperature diminishes with abatement in admitted oxygen levels at any air-fuel ratio. At the point when the oxygen content decreases from 21% to 17%, there is a diminishment in the stoichiometric fire temperature of around 250°K, which aids in reducing the NOx emissions.

This study analyzes the impact of admission of excess nitrogen on emissions, in cylinder pressure, heat released and fuel consumption. Bottled nitrogen is admitted into the engine, which displaces equal amount of oxygen.

3. Experimental setup

Tests were carried out on a Kirloskar diesel engine. The schematic diagram of the test set up is illustrated in figure 1. The engine is coupled to an eddy current dynamometer for varying the load on the engine by controlling the current supplied to the dynamometer.

Horiba exhaust gas analyzer model MEXA-584L was used to measure the THC, CO, NOx emissions. CO was measured using Non Dispersive Infrared (NDIR) method and chemiluminenscence principle was used to measure NOx. The exhaust gas was supplied to the Horiba analyzer after condensation in chilled thermally insulated container. Thermocouple sensor and display unit was used to measure the exhaust gas temperature.

AVL data acquisition system with the Indicom Mobile software user interface was connected to engine via various sensors for data measurement. Hall Effect rotary encoder (Encoders India make) was connected to the camshaft and proximity sensor to measure the TDC location was mounted near the flywheel. Piezo resistive pressure transducer (kistler make) was mounted in the combustion chamber and the signal was amplified using charge amplifier to calculate the Cylinder Pressure per degree Crankangle, Heat released per degree crank angle and MEP occurring during a cycle. Averaged results for 50 cycles were obtained via Indicom mobile software. The engine specifications are also mentioned in table 1.

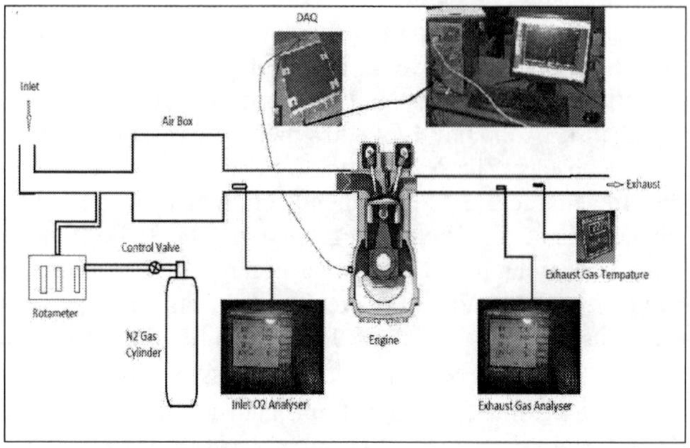

Figure 1. Schematic diagram of the experimental setup

Table 1. Engine Specification.

Engine Characteristics	Specification
Engine model and type	Kirloskar make single cylinder 4 stroke diesel engine
Cooling	Naturally aspirated
Speed	Constant
Rated power	4.4 KW at 1500 rpm
Bore	87.5mm
Stroke	110mm
Compression ratio	17.5

4. Methodology

Bottled nitrogen is introduced in the combustion chamber with the help of flow regulator attached to the nitrogen cylinder. Oxygen sensor is mounted onto the inlet supply line to monitor the percentage of oxygen. As the nitrogen is supplied to the engine it displaces equal amount of oxygen gas thus reducing the volume content of the oxygen and air becomes rich with nitrogen. The percent decrease in the oxygen concentration of air is analogous to the percent increase in the nitrogen gas. Air is enriched with nitrogen in 3 steps from 79% to 80.5%. Table 2 illustrates the oxygen concentration in the intake air measured with the help of oxygen sensor and the corresponding nitrogen gas concentration.

Table 2. Percentage of oxygen and nitrogen supplied

Oxygen %	Nitrogen %
21	79
20.5	79.5
20	80
19.5	80.5

5. Results and Discussion

Following are the results obtained from the experiments along with the discussion on the various trends observed is presented.

5.1. *Load vs Carbon Dioxide*

Figure 2 shows the CO_2 against Load when Nitrogen enriched air with various percentage of nitrogen is introduced in to the engine. CO_2 emission is almost constant as Carbon/ Hydrogen ration of fuel is not affected by the Nitrogen enriched air.

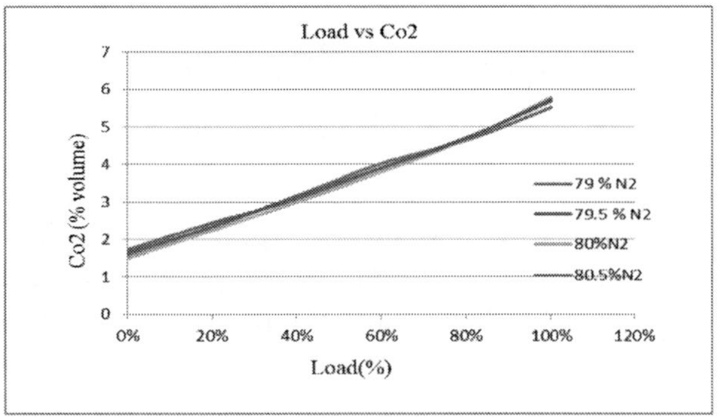

Figure. 2. Load vs CO_2

5.2. Load vs *Carbon Mono-oxide*

Figure 3 shows CO emissions for different Load when Nitrogen enriched air is admitted in to the diesel engine. As the nitrogen percentage in inlet air is increased , it is observed that variation in CO percentage of exhaust gas volume is very less and can be considerd as constant.

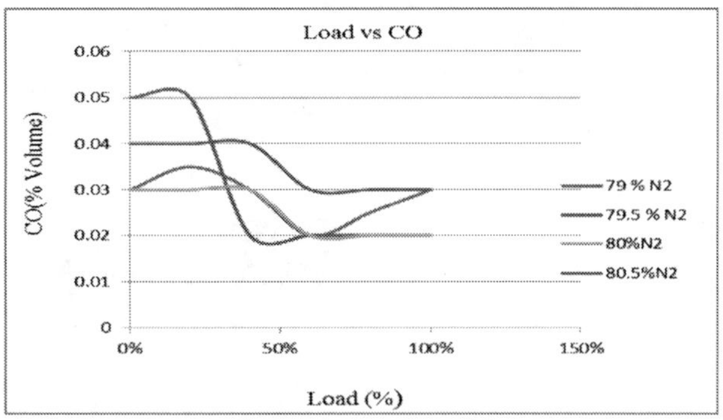

Figure 3. Load vs CO

5.3 *Load vs Hydro carbon*

Figure 4 shows the HC emissions against load with increasing concentration of Nitrogen in inlet air. It is observed that HC emissions are drastically increased, when Nitrogen concentration is increased from 79 % to 79.5 %. This increase in HC emission is due to reduction of Oxygen percentage in inlet air. When the percentage of Nitrogen is increased from 79.5 % to 80.5 %, there is slight increase in HC emissions. Increase in HC emissions with Nitrogen Enriched Air indicates that less quantity of fuel was burned during combustion.

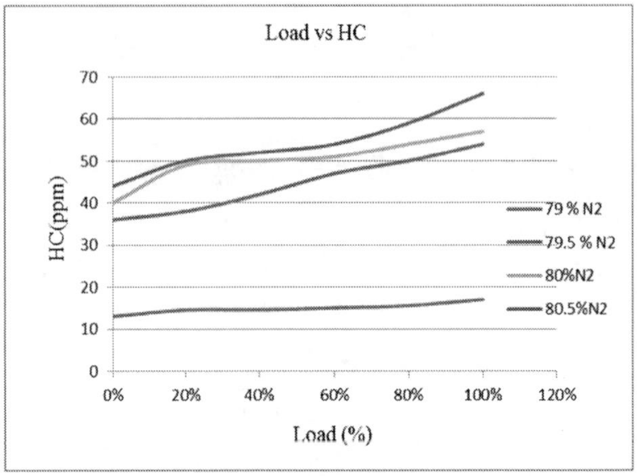

Figure 4. Load vs HC emissions

5.4 *Load vs Nitrogen Oxide*

Figure 5 shows the Load vs Oxides of Nitrogen. When the Nitrogen rich air is admitted in to the cylinder, NO_x is drastically reduced up to the 80 % of N_2 in air. This is due to less Oxygen avalibility for combustion, which in turn reduces the heat release rate, hence maximum temprature during cycle. Reduced maximun temprature resist the dissociation during cycle and Reduced values of NO_x were Observed.

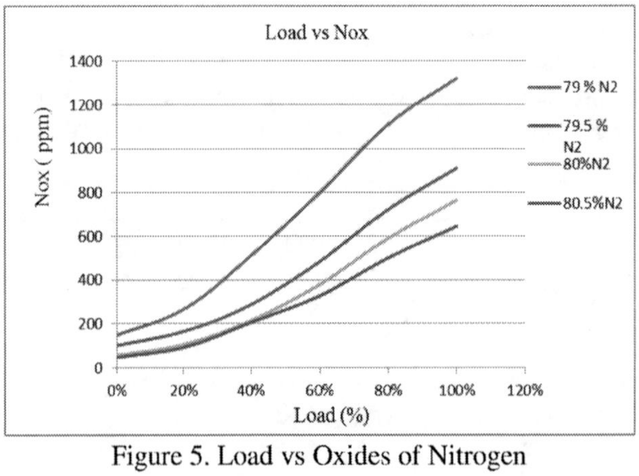

Figure 5. Load vs Oxides of Nitrogen

5.5 *Load vs Oxygen*

Figure 6 Shows trade off between Load and percentage of Oxygen in exhaust gas. Percentage of Oxygen in exhaust gas is observed as reducing with increasing with not only Load but also percentage enrichment of Nitrogen gas. With load it is reducing because as the load is increased more fule is admitted in to the cylinder in order to maintain the speed constant, more Oxygen content from inlet air is utilized to burn the fuel during combustion. It is very obivous that Oxygen content in exhaust is reduced with N_2 enrichment at inlet air which in turn reduces the Oxygen percentage at inlet and during whole cycle.

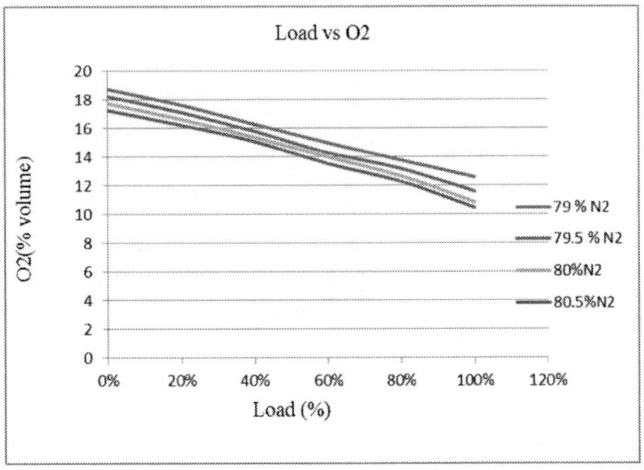

Figure 6. Load vs $O_{2\%}$ in exhaust gas

5.6 *Crank angle vs in-cylinder Pressure*

Figure 7 depicts the Crank angle against cylinder pressure for various percentage of N_2 enrichment . Pressure almost maintained constant but the maximun pressur values are delayed slightly with increasing with increasing percentage of N_2 in inlet air. It is explained as , when the N_2 is added in the inlet air, oxygen percentage in the air reduces which slowed down the combustion rate , took more time for releasing heat available in the fuel .

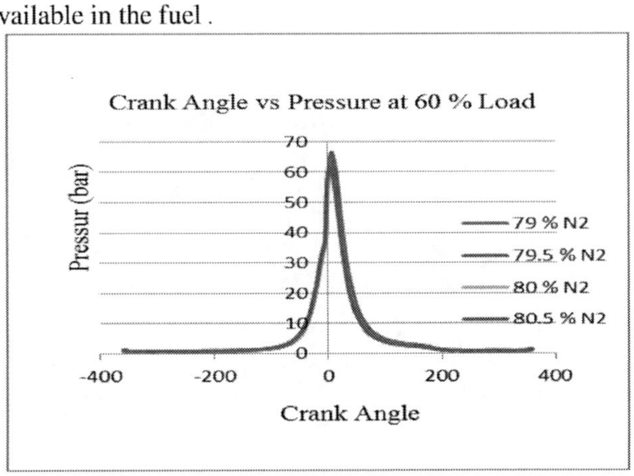

Figure 7. Crank Angle vs Pressure at 60%

5.7 *Crank Angle vs Heat release*
Figure 8 shows variation of the heat release rate for 60 % of load with crank angle . Heta release curve is obtained for different percentage of N_2 in the inlet air. As the increment in N_2 percent in the inlet air is very less , there is no significant change in the trade off between crank angle and Heat release with N_2 enrichment. From the curve it is evident that maximum heat released is shifting closer to the TDC due to slower combustion rate obtained because of less avalibility of Oxygen .

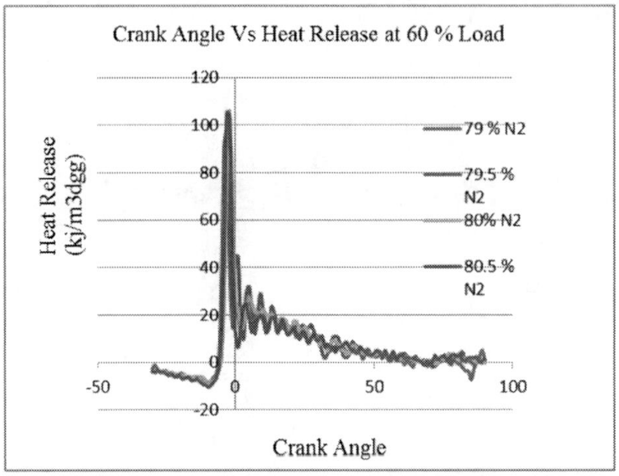

Figure 8. Variation of Heat Release at 60% load vs Crank Angle

5.8 *Load vs Indicated thermal efficiency*
Figure 9 depicts the Load versus Indicated thermal efficinency with Nitrogen enrichment. For the percentage 79.5 % to 80 % N_2 in the inlet air there was no much change was obesrved. This is due to the fact the there is very less N_2 enrichment which kept the heat release rate almost constant. But for the 80.5 % N_2 enrichment reduced indicated efficincy was observed. Hence more that 80 % N_2 enrichment is limited by the Indicated thermal efficicncy.

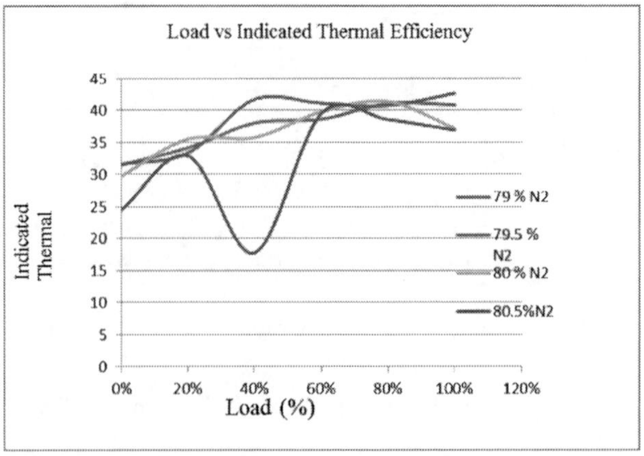

Figure 9. Load vs Indicated Thermal Efficiency

IOP Conf. Series: Materials Science and Engineering 1123 (2021) 012057 doi:10.1088/1757-899X/1123/1/012057

5.9 *Load vs BSFC*

Figure 10 shows the variation of load vs Bsfc, and it indicates that Bsfc nearly constant for various N2 enrichment because disel engiens are always operated at lean condition which ensures the availibilty of Oxygen for small increament in N_2 in inlet air.

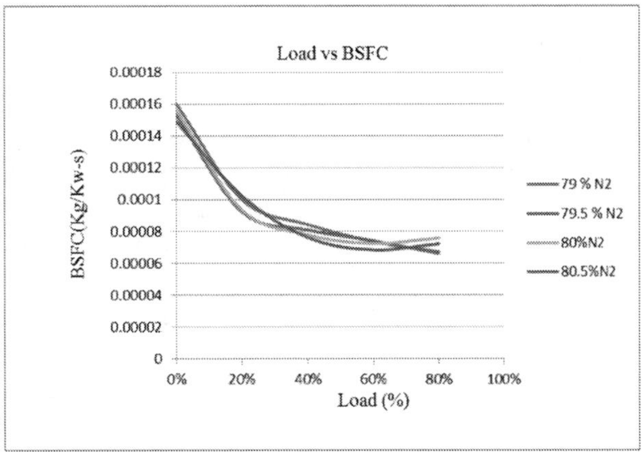

Figure 10. Load vs BSFC

6. Conclusion

Following are the conclusion drawn from the research study carried out on the diesel engine to study its impact on the nitrogen enrichment.

• Nitrogen increase from 79 % N_2 to 79.5% N_2 HC emissions increase drastically indicates more fuel is required for constant Break power. 79.5% N2 to 80.5 % N_2 increase rate of HC emission is less indicating slight increase in mass flow rate of fuel is required. Hence the N_2 enrichment slightly reduces the fuel economy.

• Nitrogen increase from the 79 % N_2 to 80.5 % N_2 , Nitrogen oxide keep on reducing but maximum value of N_2 enrichment is limited by the Indicated thermal efficiency. Optimum enrichment in order to reduce NO_x should have 80 % of N_2 in the inlet air which reduces NO_x without much increase in the break specific fuel consumption.

• For 80 percentage of N_2 enrichment in the inlet air , Indicated thermal efficiency is almost constant

7. References

[1] Ramesh B Poola, Douglas E Longman, John L Anderson, Kevin C Stork and Raj Sekar, 2000 membrane-Based Nitrogen-Enriched Air for NOx Reduction in Light-Duty Diesel Engines, *Society of Automotive Engineers, SAE technical paper no* **2000-01-0228**

[2] Mohamed Y E Selim, 2003 A Study of Some Combustion Characterstics of Dual Fuel Engine Using EGR, *Society of Automotive Engineers, SAE technical paper no* **2003-01-0776**

[3] Tie Li, Hiroyuki Izumi, Toshio Shudo and Hideyuki Ogawa, Yukihiro Okabe, 2007 Characterization of Low Temperature Diesel Combustion with Various Dilution Gases, *Society of Automotive Engineers, SAE technical paper no* **2007-01-0126**

[4] Murari Mohon Roy, 2008 Effect of Exhaust Gas Recirculation on Combustion and Odorous emissions in Direct Injection Diesel Engines, *Society of Automotive Engineers, SAE technical paper no* **2008-01-2482.**

[5] Eiji Tomita, Yuji Harada and Nobuyuki Kawahara, Atsushi Sakane, 2009 Effect of EGR on Combustion and Exhaust Emissions in Supercharged Dual-Fuel Natural Gas Engine Ignited with Diesel Fuel, *Society of Automotive Engineers*, *SAE technical paper no* **2009-01-1832.**

[6] Jianxi Zhou., Stephane Richard., Christine Mounaïm-Rousselle and Fabrice Foucher, 2013 Effects of Controlling Oxygen Concentration on the Performance, Emission and Combustion characteristics in a Downsized SI Engine, *Society of Automotive Engineers, SAE technical paper no* **2013-24-0056**.

[7] Anuj Pal, Manish V, Sahil Gupta, and Naveen Kumar, 2013 Performance and Emission Characteristics of Isobutanol-Diesel Blend in Water Cooled CI Engine Employing EGR with EGR Intercooler, *Society of Automotive Engineers*, *SAE technical paper no* **2013-24-0151.**

[8] Avolio G, Beatrice C, Del Giacomo N, Guido C, Migliaccio M , Fraioli V, 2007 Effects of Highly Cooled EGR on Modern Diesel Engine Performance at Low Temperature Combustion Condition, *Society of Automotive Engineers*, *SAE technical paper no* **2007-24-0014.**

[9] Shikha Khandelwal, Rita Yadav Chauhan, 2013 Life Cycle Assessment Of Neem And Karanja biodiesel: An Overview, *International Journal of ChemTech Research*, **Vol.5, No.2,** pp 659-65

[10] Mookan Rengasamy, Krishnasamy Anbalagan, Sundaresan Mohanraj,Velan Pugalenthi, 2014 Biodiesel Production from Pongamia pinnata Oil using Synthesized Iron Nanocatalyst, *International Journal of ChemTech Research,* **Vol.6, No.10**, pp 4511-16

[11] Mayuri Chaudhari, Akash Dhobale, 2014 Complete Utilisation of Pongamia Pinnata Preparation of Activated Carbon, Biodiesel and its purification, *International Journal of ChemTech Research*, **Vol.6, No.7,** pp 3672-76.

[12] Malhotra R K, Preeti jain, Anju Chopra and M Maheswari, 2005 Effect of bio-diesel blends on performance and exhaust emissions in passenger car, *Society of Automotive engineers*, **2005-26-032.**

[13] J Patterson, M G Hassan, A Clarke, G Shama , K Hellgardt and R. Chen, 2006 Experimental study of DI Diesel engine performance using three different biodiesel Fuels, *Society of Automotive engineers*, **2006-01-0234.**

[14] Brandon T, Tompkins, Jason Esquivel and Timothy J Jacobs, 2009 Performance parameter analysis of a Biodiesel-fuelled medium duty diesel engine, *Society of Automotive engineers*, **2009-01-0481.**

[15] P Navaneetha Krishnan, D Vasudevan., 2014 Performance Combustion and Emission Characteristics of Variable Compression Ratio Engine Fuelled with Biodiesel, *International Journal of ChemTech Research* ,**Vol.7, No.01**, pp 234-45.

Conceptual designs for solving problems

S Darius Gnanaraj

School of Mechanical Engineering (SMEC), VIT, Vellore, India.
e-mail address: dariusgnanaraj.s@vit.ac.in

Abstract: The creativity skills of students are required to be improved by making them to work on projects. Project based learning (PBL) helps students to develop their skills. VIT has introduced a project based subject known as Technical Answers for Real World Problems (TARP). This paper presents the details of the mode of handling this subject and the procedures for assigning project works to students and the method of reviewing their progress in their project component. The experience gained through handling this subject is presented in this paper along with the outcome of the project works carried out by a few groups of students.

Keywords: Creativity, TARP, PBL

1. Introduction

Industries are looking for graduates who can work in their industries without imparting training for a period of 6 months or 1 year. Every educational institute is expected to produce graduates who are 'Industry Ready'. The employability of graduates increases when they exhibit experience in handling project works successfully. Many universities around the World are concentrating on capstone projects through which every student is trained to take up a practical problem, identify a solution, design a product or develop a new process to solve the problem, demonstrate the new product or process and document it following international standards. Multidisciplinary capstone design projects are given to students to satisfy ABET student outcomes [1, 2]. Capstone projects help in bridging the gap between academics and industry practice [3]. Project based learning (PBL) improves students' creativity and outcome [4, 5]. In additions to capstone projects, VIT has introduced several subjects based on project based learning (PBL) and the subject topping this list is Technical Answers for Real World Problems (TARP). This paper presents the framework for this subject, experiences gained through handling this subject and outcomes of the project works carried out by a few group of students.

2. Frame work for TARP subject

Students are motivated to travel in and around Vellore to find real problems experienced by common man. A group of 6-10 students can work in a group to solve these problems by using the technical knowledge that they have gained through engineering subjects. One hour per week, students are required to assemble in a class room and have group discussions in the presence of the faculty and 8 hours per week, they need to work on this project along with his team members for developing conceptual designs, working models and computer simulations to demonstrate how they solve practical problems. 3 credits are given for completing this subject successfully. This subject helps students to have application oriented learning and also inculcates a sense of social responsibility in them. This subject is classified under University Core meaning that every student has to complete this subject before receiving the degree.

3. Details of conceptual design carried out by students

Students were allowed to form groups and find problems experienced by common man. They went through journal papers, magazines, newspapers, visited a few places, interacted with public, corporation, municipality office, police station to find problems experienced by common man. The responses were good and Government officials were happy to receive university students and gave them insight into real problems faced by the society and gave assurance to give support for solving those problems. The following list of problems was found and each group worked on one topic.

- Smart Pedestrian Crossing
- Oil spill collector
- Solid Waste Management
- Solar Water purifier
- An efficient stove

4. Outcomes of the project works

4.1 Smart Pedestrian Crossing

Pedestrian crossings with traffic lights are used to facilitate the time-sharing of road space between motorists and pedestrians so that pedestrians can cross the road safely. Each of these two groups of users is allocated the right-of-way alternatively with the use of traffic signals. The amount of time allocated to motorists and pedestrians is based on the principle of balancing their demands. The demand for vehicular movement is detected through the sensors placed on the road pavement. Pedestrians register their intent to cross the road by activating the push-button on the traffic light pole. The activation of the push button informs the system that someone is waiting to cross the road, and traffic will then be stopped to allow pedestrians to cross. Pedestrians can proceed to cross the road when the green signal comes on. This work is based on developing a pedestrian crossing system which is both efficient and ensures pedestrian's safety. Figure 1 shows the assembly of the traffic light post and Figure 2 shows the electronic circuit [6].

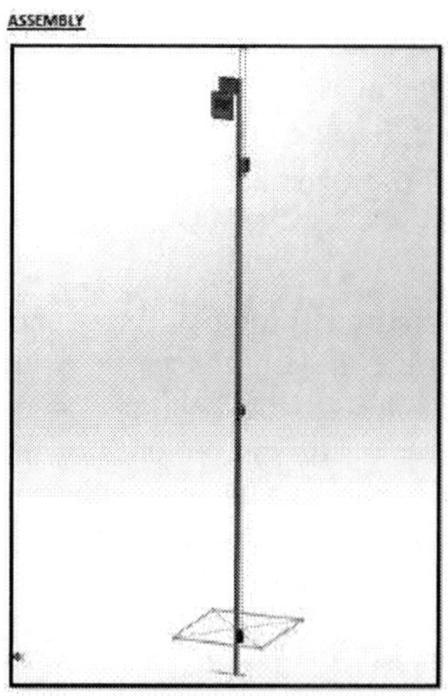

Figure 1. Traffic light post [6]

Figure 2. Electronic circuit [6]

4.2 Oil spill collector

Whenever there is an accident between two ships, oil spills on the sea. The spilled oil affects the living creatures of the sea and affects the environment close to sea shore. If the spilled oil is quickly collected by means of a collection mechanism, damage to sea and shore enviroment can be prevented. Figures 3 & 4 show two different mechanisms which can be used for collecting the spilled oil. Teflon coated on the circular discs are capable of absorbing oil and using suitable wipers, the oil is removed from the surface of the circular discs and to a floating tank through suitable path ways [7].

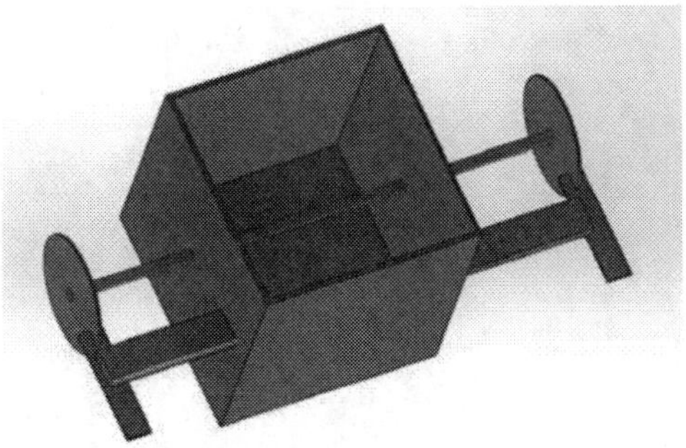

Figure 3. Oil Spill Collector Model-1 [7]

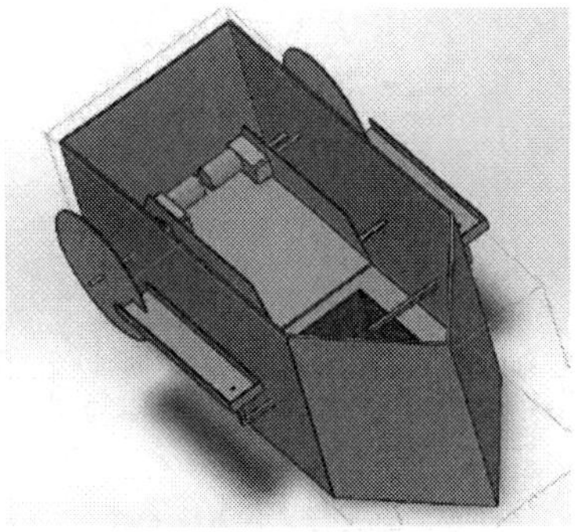

Figure 4. Oil Spill Collector Model-2 [7]

4.3 Solid Waste Management

Figure 5 shows a waste collector with two doors; one in the upper half and another in the lower half. A movable compressing plate is provided for compressing the waste placed inside the contatiner. Usually, people throw paper cups and empty paper boxes and objects of low density in the waste collector. Usually, the bin will be full within two to three hours and then the bin will start over-flowing. Sometimes, thrash is seen above the waste collector and even the thrash may spill around the bin causing unhealthy environment. A new design has been developed in which the user will open the upper door, throw the thrash and compress it using the handle. The compressing unit will have a spring which will make the plate to stay at the top of the bin always. Lower door will be used by cleaners who empty the bin. More than double of the the normal capacity can be stored in the bin apart from maintaining a healthy envirnment [8].

Figure 5. Waste collector [8]

4.4 Solar water purifier

A solar water purifier has been designed, fabricated and tested successfully. Figure 6 shows the conceptual model and Figure 7 shows the working model. 18x18x18 inches glass cubical container was fabricated having 95 liter capacity. An earthen pot with a capacity of 15 liters was procured. These two were connected by means of plastic tubes for condensed water to the pot. Black paint was applied to the glass container to absorb more solar energy. Dirty water was stored in the glass container and the working model was left in hot sun for one full day. About 5 liters of condensed water were found in the pot.

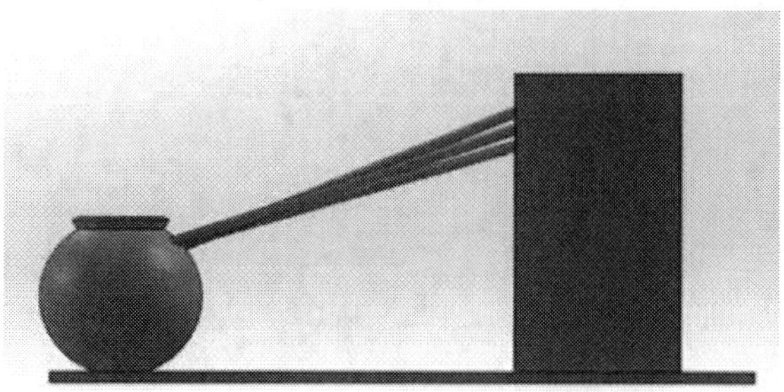

Figure 6. Conceptual model [9]

Figurc 7. Working model [9]

. pH test was conducted in the condensed water and it was found that the water was neither acidic nor basic but neutral in nature. Hardness test was carried out using lame emission photometric method and the water was found to be fit. Water was treated with lime and soda and found to be suitable for drinking [9].

4.5 An efficient stove

A few places in Tamil Nadu, India are very much backward where people are still using conventional fire wood stoves. One of such places is Thandikudi hills in Dindigul district. Places like Dharmapuri and Krishnagiri districts have 40% forest. Jawadhu and Koppam forests which share borders with Vellore district have people who use fire wood stoves. Four different shapes were considered for making the efficient stove. Square, triangle, circle, semi-circle and the square shape was selected since it was stable and easier to fabricate. The stove is made of Aluminium since it is durable, light and easier to fabricate. The stove is efficient as it utilizes the convection to warm up the vessel. The stove has two chambers and there is a gap through which wood is embedded and consumed. After it is scorched, the heat get convected through the inner chanmber and warms up the vessel kept over the stove. This stove produces less amount of smoke compared to conventional stove. The top portion of the gap between the chambers is closed and holes are made in the inner cylinder to increase the efficiency of the stove and to reduce the overall weight. Figure 8 shows the conceptual model of the stove and Figure 9 shows the working model of the stove [10].

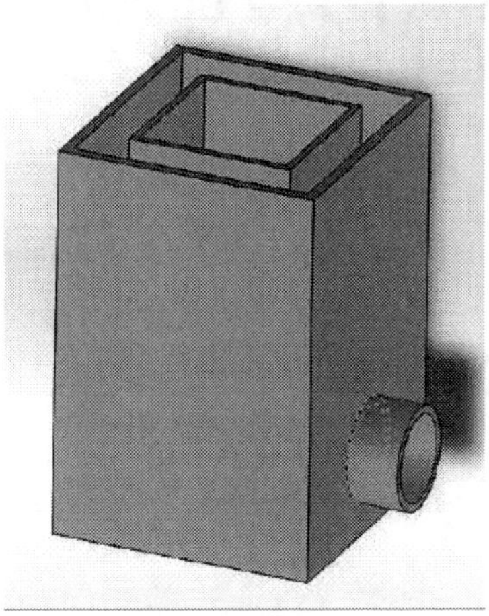

Figure 8. A model of an efficient stove [10]

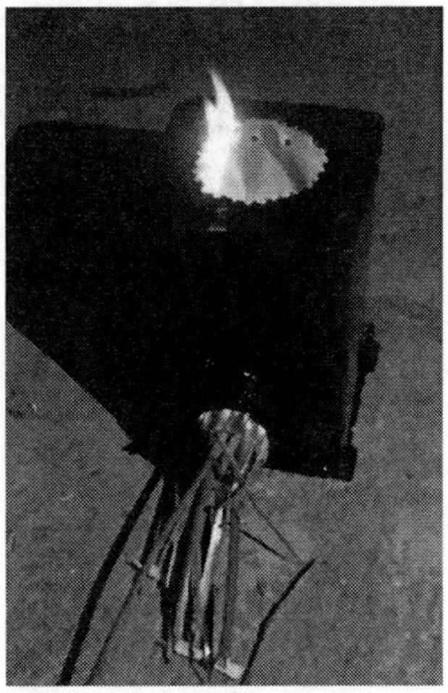

Figure 9. An efficient stove [10]

5. Conclusion

Five conceptual designs are presented in this paper: Smartt traffic light design for pedestrian crossing, oil spill collector design, design of a waste collector for solid waste management, solar water heater and an efficient stove. By working on these projects, the creativity and design skills of students improved. Students had opportunities to go into details of real problems faced by common people and to provide a solution for solving such problems. Since every student who graduate through VIT Vellore has to complete the TARP course, the student will continue to provide solutions for solving the problems faced by huminity through out his professional career.

6. References

[1]	Meah K, Hake D and Wilkerson S D 2020 *Edu. Res. Int.* **2020** DOI: 10.1155/2020/9563782
[2]	Magleby S P, Todd R H, Pugh D L and Sorensen C D 2001 *Int. J. Eng. Edu.* **17** 400
[3]	Hanna A S and Sullivan K T 2005 *J. Prof. Issues Eng. Edu & Pract.* **131** 59-62
[4]	Rodrigo F M, Lucas L C H, Pablo S and Boue, A B R 2017 IEEE Trans. Edu. **60** 229-237
[5]	Syazali M, Listiani B and Farid F 2019 *J Gifted Edu. & Creat.* **6** 209
[6]	Gulathi S 2018 *Smart Pedestrian Crossing* A report submitted to VIT Vellore India.
[7]	Sunil G 2018 *Oil Spill Collector* A report submitted to VIT Vellore India
[8]	Desai A 2018 *Solid Waste Management* A report submitted to VIT Vellore India
[9]	Mukund A 2018 *Solar Water Heater* A report submitted to VIT Vellore India
[10]	Amogh 2018 *Development of an efficient stove* A report submitted to VIT Vellore India

IOP Conf. Series: Materials Science and Engineering 1123 (2021) 012059 doi:10.1088/1757-899X/1123/1/012059

Industry 4.0 transformation initiatives in a small scale mill board unit through sustainable manufacturing: A case study

R Vezhavendhan[1], S Senthil Kumar[2], K Jayakrishna[3] and S Aravindraj[4]

[1] School of Mechanical Engineering, Vellore Institute of Technology, Vellore, India, vezhavendhan.r@vit.ac.in
[2] School of Mechanical Engineering, Vellore Institute of Technology, Vellore, India, ssenthilkumar@vit.ac.in
3 School of Mechanical Engineering, Vellore Institute of Technology, India, jayakrishna.k@vit.ac.in
4 School of Mechanical Engineering, Vellore Institute of Technology, aravindraj.s@vit.ac.in

Abstract: Small scale mill board manufacturing industries are shutting down shop at an enormous pace. Observation shows that three out of five such units close down operations within three years of its inception due to their inability to sustain in the market. These units have not been able to utilize resources effectively and have not been able to run profitably too. Further, they have been struggling to meet the norms stipulated by the state pollution control board. This article is a review of various problems identified in a mill board unit and the measures taken that made the unit self- reliant and transform into an Industry 4.0 enterprise through sustainable smart manufacturing concepts. This case study describes measures adopted to revive a struggling mill board unit and its success in terms of economic profitability and issues in incorporating Industry 4.0 concepts..

1. Introduction

Mill board is the technical name for boards made out of recycled waste paper. All small mill board industries aim at making profits and few are good at it too, while many are not. Along with fulfilling this aim lies the responsibility of every industry to fulfil their aims and goals in a responsible way by not dumping trash on this planet. Recycling industries are much prone to such activities of dumping trash as they are industries that receive lot of unwanted trash along with the ones that they are capable of handling and processing. A part of the wastes coming in are either unwanted or not capable of being processed by them. Hence these end up in landfills in surrounding areas resulting in environmental pollution. Apart from these many wastes are generated during the reprocessing of the wastes itself and there needs to be a green initiative to take care of all the wastes generated in the industries through sustainable manufacturing practices. Moreover industries are faced with challenges such as depleting resources and pollution of environment leading to heavy competition along with regulatory pressure from Government agencies to curb pollution and improve their carbon footprint points and sustainable manufacturing practices pave the way to beat all the above. To identify possible areas to implement various sustainable manufacturing practices in a Mill board manufacturing unit and transform it into a sustainable enterprise. And to present a case study of implementation of sustainable manufacturing practices to transform a mill board unit

2. Literature Review

There have been several adverse effects created by industries over the last few years leading to severe pollution of the environment and depletion of resources due to over use and abuse. Sustainability in manufacturing, which is defined as a philosophy of manufacturing by "meeting the needs of the present without compromising the ability of future generations to meet their own needs" [1] is the best way to tackle the situation. Sustainable manufacturing which can be defined as a method to develop technologies to transform materials without emitting much greenhouse gases, using non-renewable or

materials that are toxic in nature or waste generation [2] paves the way for reduced environmental pollution and effective use and reuse of materials and raw materials.
Industries need to understand that it is very important for a manufacturing organization to be economically viable for survival and the possibility to sustain the enterprise may not be bright in the long run when damages caused to the ecosystem are irreversible through the emission of greenhouse gases (GHG) and wastes that are toxic and non-renewable resources that are depleting [3]. Economic profitability is a representation of sustainability. And sustainability is also a form of duty and a responsibility towards all the environmental, social, and economic indicators, which are known as the triple bottom line of sustainability [4]. In order for sustainable manufacturing to be achieved, major changes are needed to have a thrust towards industrial processes that are more sustainable [5]. Sustainable manufacturing has a general principle to reduce materials use intensity, consumption of energy, emissions, and unwanted creation of by-products while maintaining and improving products of value to both the society as well as to the organizations [6].

- Some of objectives of a sustainable manufacturing system can be,
- Resource efficiency improvement and management of waste
- Product life cycle examination involved in manufacturing.
- Minimizing pollution of environment while ensuring economic growth
- Examining how to stimulate innovation and investment to provide cleaner technology
- Provision of employee awareness and training [7].

One of the recent trends in the segment of manufacturing is the concept of Industry 4.0 which is seen as a thing of importance among small enterprises to face competition through the use of new and innovative production and management technologies. Industry 4.0 has significant effects through the application of advanced and latest technologies, big data architecture creation and data analysis. Through the use of proper and cost effective information technology application in manufacturing, the small and medium enterprises (SMEs) can expect to have a significant increase and influence on the orders that they receive [8]. Shrouf et al. [9] have carried out a study on the concepts of Industry 4.0. in detail. The present enterprises were checked for the necessary infrastructures that are specified in the Industry 4.0 transformation process. Stock and Seliger [10] highlighted Industry 4.0 opportunities for sustainable manufacturing and production and the growth in demand for consumer goods and the limitations involved in meeting those demands.

3. Case Study

Sustainable manufacturing initiatives were carried out in a mill board unit situated in Vellore district of Tamil Nadu, India. The unit is a small scale unit with a turnover less than USD 50,000.00 per annum. The unit was established in the year 2008 and has been functioning since then catering to customers from several domains such as stationery, industrial packaging, textile manufacturers and other packaging areas etc. The unit was making low profits despite being in the field for a little over a decade. Hence the management started looking out for pertinent ways of improving its profitability some way or the other. After a lot of efforts and consultation it had decided to try its luck with the help of the concept of Sustainable Manufacturing. The plant embarked on the project of implementing sustainable manufacturing practices in the month of February 2019. The unit had picked up a few initiatives of sustainable manufacturing and only the major initiatives among them are discussed here along with the benefits reaped through them.

In order to take up sustainable manufacturing initiatives and implement it, a team that could take up and implement sustainable practices was created. There needed to be a thorough understanding of the process of manufacture of mill board. Hence the process was studied and a process map was created. The process needed to be evaluated to identify areas where sustainable manufacturing could be implemented. A brain storming session had to take place for evaluation and identification. The whole organization was asked to submit their suggestions and solutions and it was informed that the best suggestions and solutions would be awarded. This acted as a catalyst for the employees to involve themselves in the activity. Once the suggestions and solutions were collected, they were reviewed and evaluated for suitability and possibility of implementation along with that of the team. The relevant and best solutions and suggestions were implemented. Data on improvements made and wastes eliminated was collected and compared to understand the economics of implementation of sustainable manufacturing practices. The roadmap of the implementation process is given as under in Figure-1 and The process map of the mill board manufacturing process was mapped and the process map is given below in Figure 2.

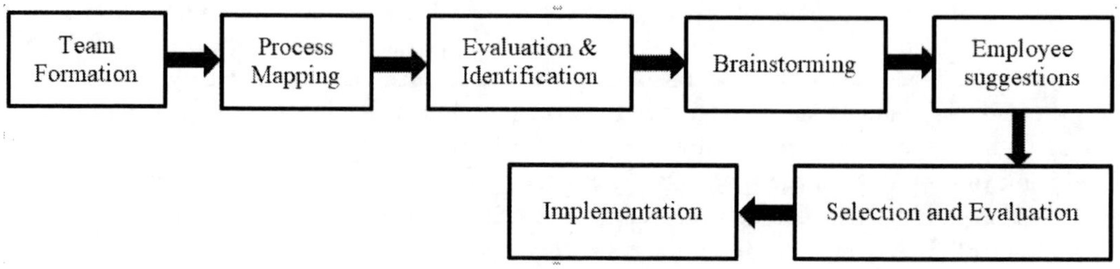

Figure 1. Implementation roadmap

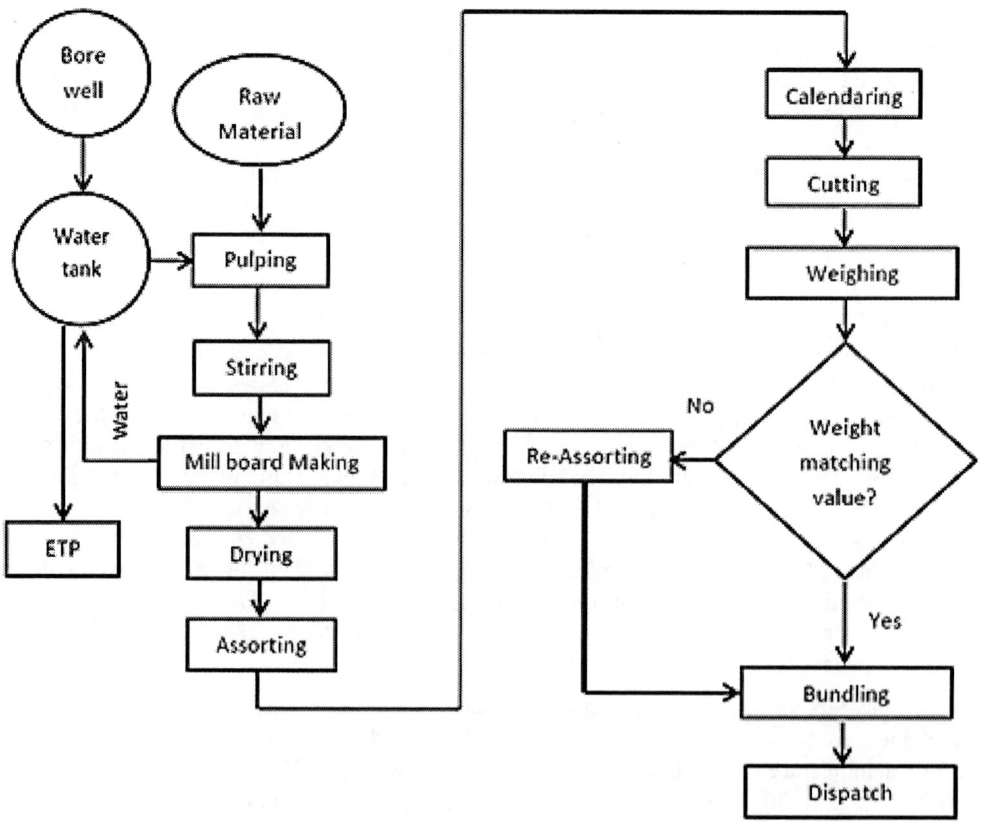

Figure 2. Process map of mill board manufacturing

4. Initiatives Taken

Given below is a list of initiatives taken

4.1 Initiative I

In due course, the team identified that there was a need of 10 kilo litres of water on a daily basis of which around 4 kilo litres was lost due to absorption during the pulping and board formation process and around 2 to 3 kilo litres of water was lost due to evaporation during the board drying process. At the end of the day, the remaining water ranging between 3 to 4 kilo litres was pumped to the Effluent Treatment Plant (ETP) for processing and the treated water was used for gardening purposes.

The process was modified wherein a tank which was called as the recycling tank was included in between so that the water coming out after the board making process could be collected in the recycling tank and reused by adding the water to waste paper (raw Material) for pulping the next day as shown in figure-3 below. This resulted in the following benefits

- Reuse of 3 to 4 kilo litres of water thereby saving water intake of 18 to 20 kilo litres per week (Approximate savings of 108,000 litres to 120,000 litres per month).
- Reduced use of ETP form daily use to usage of once in a week only to treat the same amount of water which came out in a day.
- Reduced usage of a 3 HP electric motor and connected pump only once in a week instead being used daily

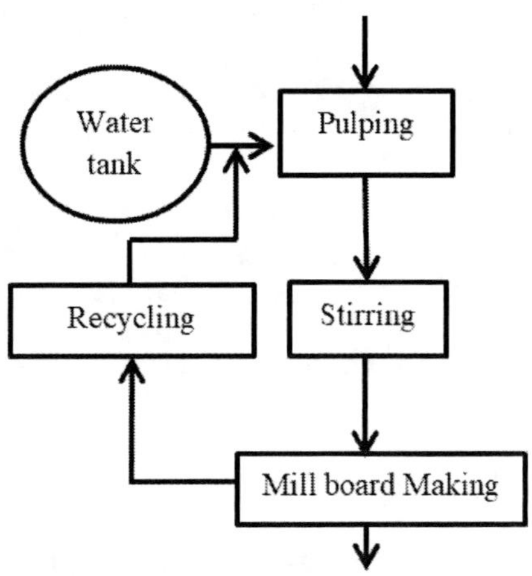

Figure 3. Modification in the process

4.2 Initiative II

It was found that a large amount of raw materials consisted of plastic laminations over them which ended up as remnants of the filtrate of the paper pulp. This usually was removed from the pulping machine and ended up in landfills causing huge environmental problems. Along with this plastic generated, raw material supplies received consisted of mixed plastic wastes accompanying the paper wastes which also found its way to landfills. Hence it was decided to check what could be done about this and how it could be done. Hence with the help of a chemical engineering expert a small pyrolysis plant with a capacity to process 200 kgs per day was set up as an initial phase. This resulted in approximately 80 litres of fuel per day, and the residue was converted into manure adding to the revenue of the enterprise.

4.3 Initiative III

The standard size of boards produced for daily calendars backs were 27 x 31 inches this resulted in 5 calendar backs from each board. The result was that there was a large scrap generation of about 600 grams on an average from each board. After a thorough study and brainstorming it was decide to

IOP Conf. Series: Materials Science and Engineering 1123 (2021) 012059 doi:10.1088/1757-899X/1123/1/012059

increase the size of the board to 30 x 34 inches and this resulted in retrieval of 6 calendar backs instead of the regular 5 numbers. Further the scrap generated from each board came down from 600 grams per board to just 150 grams per board. This also resulted in better utilization of resources, material and increased profitability to both the supplier and the customers.

4.4 Initiative IV

Analysis of the propulsion forces of Industry 4.0 to be adopted by small enterprises. By analysing the difficulties in transition process of small enterprises to Industry 4.0, it was understood that enterprise support is required in the determination of strategic steps to facilitate the transition of enterprises to Industry 4.0 and make improvement towards efficient use of resources.

For the Industry 4.0 transformation the things taken into deliberation were comparative advantages, relevance, support from management, size of the organizational, expertise available in information technology, complexity, pressure from competitors and external support . It was concluded that Cost criteria, intelligibility, distrust of benefits, transition expenses, maintenance costs, technical support, training costs, recognition and support cost criteria are to be determined according to expert opinions.

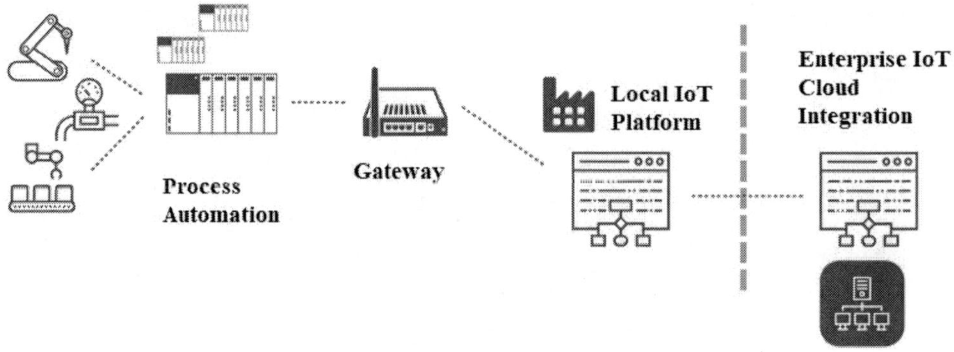

Figure 4. Architecture of Industry 4.0 transformation

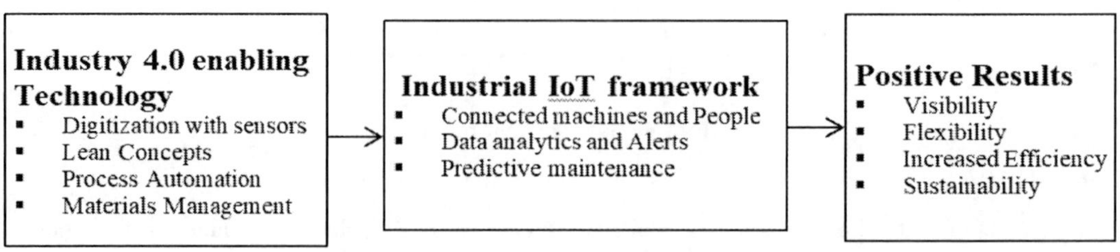

Figure 5. Framework of Industry 4.0 transformation for small enterprise

5. Conclusion

The sustainable manufacturing initiatives lent a helping hand to the mill board manufacturing unit in terms of saved resources, better utilization of resources and many more whose list is given below along with the economic savings. The benefits reaped are listed below

- Better utilization of existing resources
- Increased profitability for the unit
- Possibility of setting up a larger pyrolysis plant in the near future
- Increased customer satisfaction

- Decreased scrap production
- Savings in terms of electricity
- Water savings to a huge extent and
- Reduced (by almost 85%) environmental pollution
- Less strain on the ETP and hence longer machine/ equipment life
- Increased employee morale

After the implementation of sustainable manufacturing initiatives, today the unit stands as a profitable, satisfied and transformed unit as a result of enterprise transformation through sustainable manufacturing initiatives. The implementation of Industry 4.0 was a challenging task, which has been initiated at a lower degree and may take over the entire process in future.

.

6. References

[1] World commission on environment and development (WCED) Our Common Future 1987 *Oxford and New York: Oxford University Press*, NY.

[2] Allwood J 2009 *Sustainable Manufacturing Seminar Series* (http://www.ifm.eng.cam.ac.uk/sustainability/seminar/documents/050216lo.pdf).

[3] Sumit G, Dangayach G S, Singh A K, Meena M L and Rao P N 2018 Implementation of sustainable manufacturing practices in Indian manufacturing companies *Benchmarking: An Int. J.* (https://doi.org/10.1108/BIJ-12-2016-0186)

[4] Elita A and Annike L V 2015 Key Performance Indicators for Sustainable Manufacturing Evaluation in Cement Industry, *12th Global Conf. Sust. Manu.Procedia CIRP* **26** pp 19-23.

[5] Abdul R S H, Evans S and Longhurst P 2008 A comparison of four sustainable manufacturing strategies. *Int. J. Sust.Eng.* 1 **(3)** pp 214–29.

[6] OECD (Organization for Economic Co-operation and Development). Sustainable manufacturing and eco-innovation: towards a green economy 2009, *http://www.oecd.org.*

[7] Abdeen M O 2008 Energy, Environment and Sustainable Development *Renewable and Sustainable Energy Reviews* **12** pp 2265-300.

[8] Roblek V, Meško M and Krapež A 2014 A complex view of industry 4.0. *Sage Open* p 6.

[9] Shrouf F, Ordieres J and Miragliotta G 2014 Smart factories in Industry 4.0: A review of the concept and of energy management approached in production based on the Internet of Things paradigm. *Proc. IEEE Int. Conf. Ind. Eng. Eng. Mgmt* (Bandar Sunway, Malaysia) pp 697–701.

[10] Stock T and Seliger G 2016 Opportunities of sustainable manufacturing in industry 4.0. *Procedia CIRP* **40** 536–541.

Design and topology optimization of a rack and pinion steering system using structural and vibrational analysis

T. Narendiranath Babu, Rajkumar E, Tanmay Joshi, Vedant Patil, Wasif Mukaddam,

Associate Professor, School of Mechanical Engineering, Vellore Institute of Technology, Vellore, India, narendiranathbabu.t@vit.ac.in
Associate Professor, School of Mechanical Engineering, Vellore Institute of Technology, Vellore, India, rajkumar.e@vit.ac.in
School of Mechanical Engineering, Vellore Institute of Technology, Vellore, India, tanmayj13@gmail.com
School of Mechanical Engineering, Vellore Institute of Technology, Vellore, India, vedantpatil08.vp@gmail.com
School of Mechanical Engineering, Vellore Institute of Technology, Vellore, India, wasifmukaddam@gmail.com

Abstract. The rack and pinion steering system, in automobiles, is the most commonly used steering system. Being directly connected to the suspension system in most vehicles, it is a critical component. The main intent of this research is to design a rack and pinion steering system for an All-Terrain Vehicle (ATV) and to reduce its weight by topology optimization. The design calculations are performed using AGMA bending, and contact stress equation and modelling is done using SolidWorks. ANSYS is used for structural analysis and the stress and deformation distribution is visualized. Vibrational analysis from ANSYS is utilized to obtain the natural frequencies, which are compared with the operating frequencies of the system. Topology optimization is conducted from this analysis to obtain an optimal and lightweight design.
Keywords: Rack and Pinion, Steering, FEA, Topology Optimization

1. INTRODUCTION

Due to its simplicity in construction and compactness the rack-and-pinion steering is one of the most recurrent type of steering systems. Its implementation is mostly in, but not limited to cars, small trucks and SUVs [1]. It is a simple mechanism that directly transforms the rotational motion of the steering wheel into linear movement at the wheels [2]. It constitutes a rack-and-pinion gear set enclosed inside a metal tube, with each extremity of the rack protruding from the tube. A rod, known as tie rod, attaches to each end of the rack. The pinion gear, linked with the steering shaft, turns in tandem with the steering and consequently moves the rack. The tie rod at the extremities of the rack attaches the spindle with the steering arm

The objective of this study is to design the Rack and pinion gearbox for an All-Terrain Vehicle while reducing its weight by topology optimization. Weight is considered as one of the most important parameters in an All Terrain Vehicle [3]. Less weight leads to better acceleration, more fuel efficiency, less material costs and better handling characteristics [4].

Topology optimization is a mathematical method to optimize material arrangement contained by a given design space, directed at a particular set of boundary conditions, loads and constraints [5]. It implements Finite Element Analysis (FEM) to evaluate design performance and optimizes the component geometry [6].

The original component is designed using AGMA stress equations after assuming parameters for a typical All Terrain Vehicle [7]. Two topology optimization iterations are performed on this model to give an optimal design. All the iterations and the original model are compared on the basis of mass, maximum deformation, maximum Von Mises stress, and the range of natural frequencies.

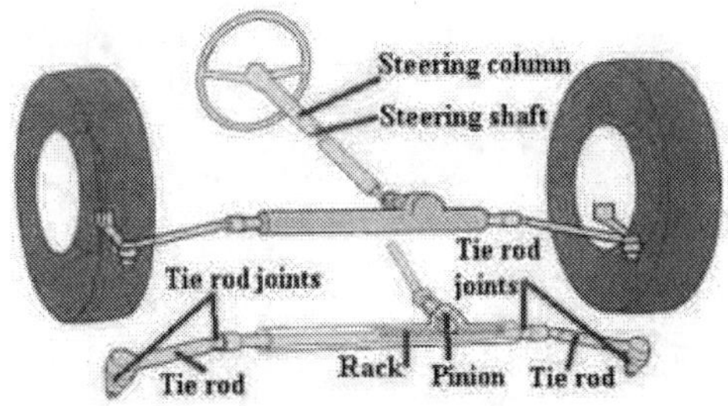

Figure 1. A typical rack and pinion steering system used in vehicles [11]

The Vibrational Analysis performed helps in determining the range of natural frequencies [8, 9]. As the steering system is directly attached to the wheels through the suspension system, it faces large vibrational loads [10]. According to previous works, the values of vibration for operating conditions are 20 Hz to 200 Hz [11 - 13]. If these vibrations are close to the natural frequencies, resonance can occur leading to large deformations [14]. Thus, this paper aims to iterate a model having a minimum range of natural frequencies, far from the operating frequencies in addition to the structural analysis for deformation and stress.

2. DESIGN

The rack and pinion steering system has been designed using the AGMA stress equation and modeled in SolidWorks CAD software [15, 16].

Calculation – AGMA Stress Equation
Material taken is alloy steel AISI 4340 with tensile yield strength = 710 MPa
Consider 1 rotation of the pinion = 4 inch displacement of the rack
Torque applied = 15Nm
N Pinion = 60 rpm
Length of teeth on rack = 6 in, Total length of rack = 14 in
$L = \pi d1 = 3.14\ d1$
$\therefore 4 = 3.14\ d1$
$\therefore d1 = 1.28$ in $= 32.6$ mm ≈ 33 mm
Assuming: module = 3 $\Rightarrow$ N1 = 33/3 = 11 teeth
Considering lock to lock steering in 1.5 turns
Rack length = $1.5 \times 4 = 6$ in, Circumference of gear equivalent = 6 in =152.4 mm
$152.4 = \pi d2$
$\therefore d2 = 48.51$ mm ≈ 48 mm
$\therefore N2 = 48/3 = 16$ teeth
Steering gear ratio = 1.45
From AGMA stress equation
Pitch line velocity V = $(\pi d1N1)/60000 = (\pi \times 33 \times 60)/60000$
V = 0.1036; KV = 1.02
Power = 94 W
Ft =94/0.1036 = 907.16N; J=0.34;
Endurance limit $\sigma e = \sigma yt/2\times K\_L\times K\_r$
Take $\sigma yt = 710$ MPa; $\sigma e = 351.12$;
Taking FOS = 2; $\sigma = 175.56$MPa
$155.81 \times 103 = (907.16\times1.02\times2.25\times1.6)/(3\times10^{\wedge}(-3)\times b\times0.34)$
$\therefore b = 20.96 \approx 21$mm (face width)

3. MODELLING AND ANALYSIS

Three models were created and analyzed in this study; one original design and two iterations. Boundary conditions used while performing FEA are

1) Fixed support for the central region (supported by the casing)

IOP Conf. Series: Materials Science and Engineering 1123 (2021) 012060 doi:10.1088/1757-899X/1123/1/012060

2) Vertical loads of 500N simulating the tie rod forces are applied to the eyes of the rack
3) Mesh parameters used are – Adaptive, Relevance center 75 and size 5mm

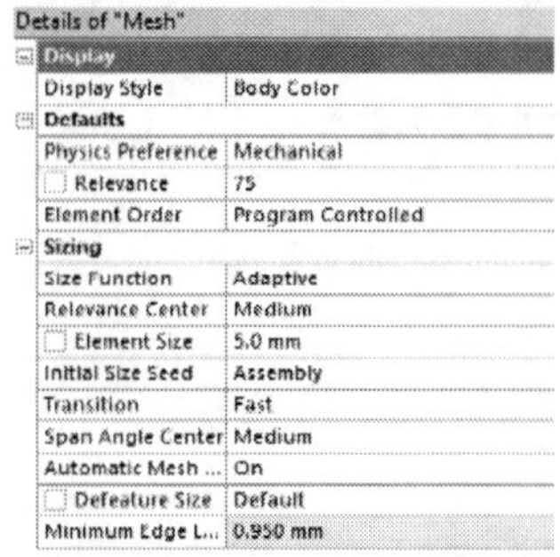

Details of "Mesh"

Display	
Display Style	Body Color
Defaults	
Physics Preference	Mechanical
Relevance	75
Element Order	Program Controlled
Sizing	
Size Function	Adaptive
Relevance Center	Medium
Element Size	5.0 mm
Initial Size Seed	Assembly
Transition	Fast
Span Angle Center	Medium
Automatic Mesh ...	On
Defeature Size	Default
Minimum Edge L...	0.950 mm

Figure 2. Mesh Parameters used

3.1 Original Design

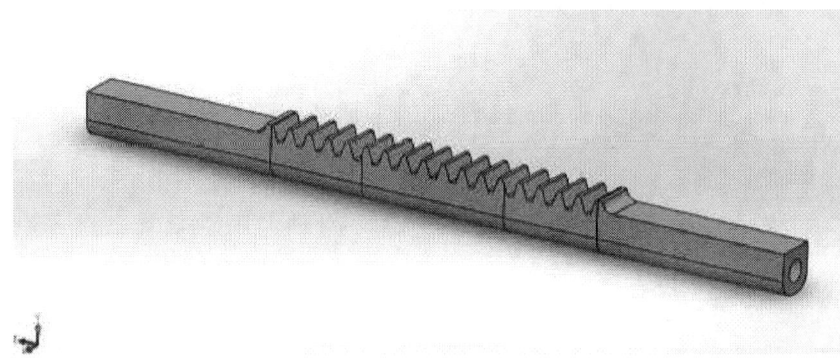

Figure 3. SolidWorks model of the original design

The dimensions obtained from the calculations are used to create a CAD model. This is referred to as the Original design in this research. From the Evaluate tab in SolidWorks, the mass of this model is obtained.
Mass = 995.34 grams

Figure 4. Total Deformation plot for Original design

The load conditions are simulated and the original model is analyzed using ANSYS software [15]. The total deformation is visualized using this contour plot. The highest value of deformation obtained is at the ends of the rack. Maximum total deformation= 0.1657mm

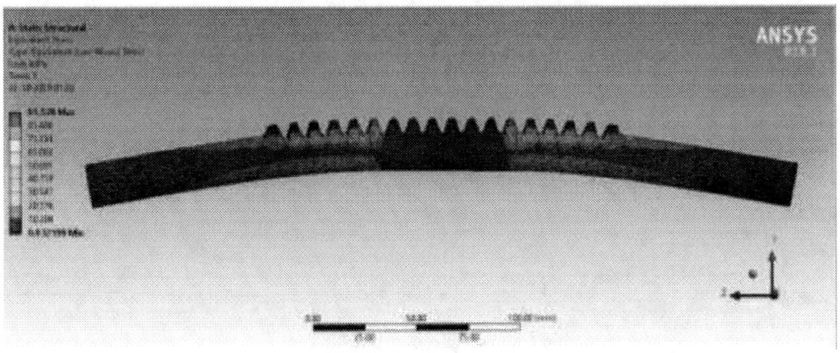

Figure 5. Equivalent stress plot for Original design

Using the same boundary conditions and loads, contour plot for equivalent Von Mises stress is obtained. The largest values obtained are close to the fixed support while the lowest values obtained are at the rack ends. Maximum stress = 91.578 Mpa

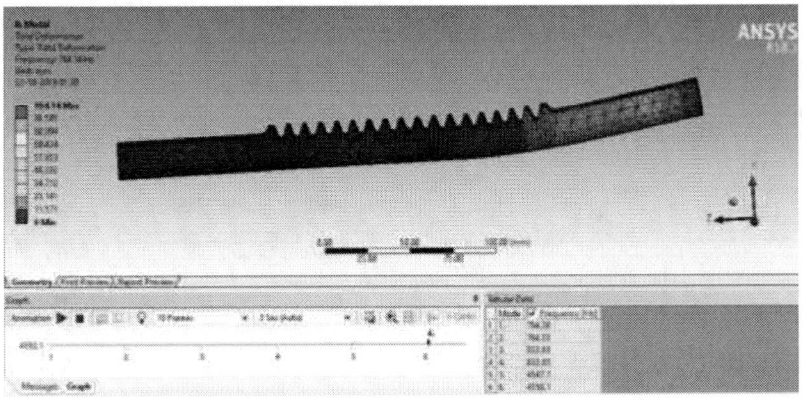

Figure 6. Vibrational analysis for Original design

Vibrational Analysis was performed using the Modal analysis system in ANSYS [16]. The range of natural frequencies and the deformation at each frequency is visualized using the contour plot. Natural frequency range = 764.36 to 4550.1 Hz

3.2 Iteration 1

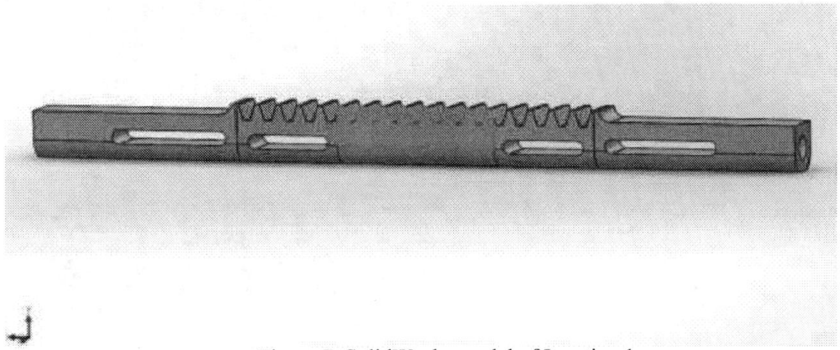

Figure 7. SolidWorks model of Iteration 1

The original model has less stress in the regions towards the ends of the rack and close to the axis. Thus, slots were added at those locations. For this iteration, all the slots are of the same thickness. Mass=836.49gram

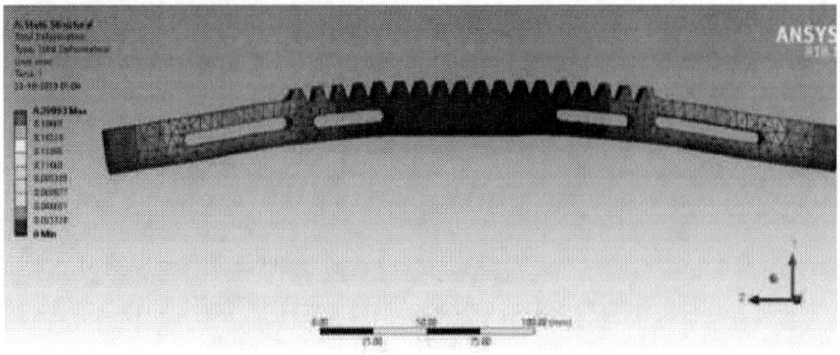

Figure 8. Total Deformation plot for Iteration 1

Similar to the earlier plot, a total deformation plot of the first iteration is obtained using ANSYS. The deformation is maximum at the ends and minimum close to the center. Maximum Deformation= 0.20993mm

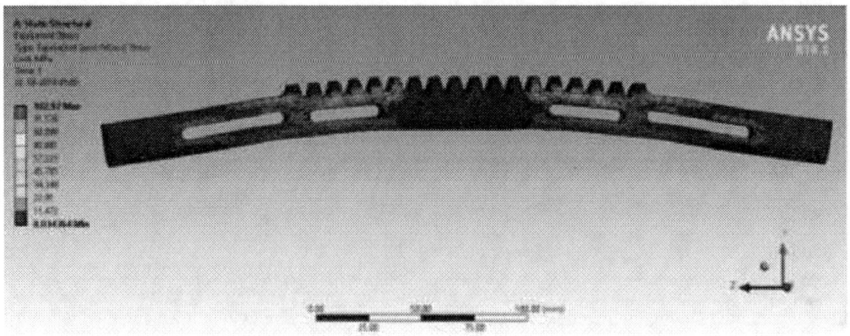

Figure 9. Equivalent stress plot for Iteration 1

The contour plot for Equivalent Von Mises stress shows better optimization than the Original design. However, there are still some regions with less stress present such as the end regions. Maximum stress is still at the teeth near the central support. Maximum Stress= 102.97 Mpa.

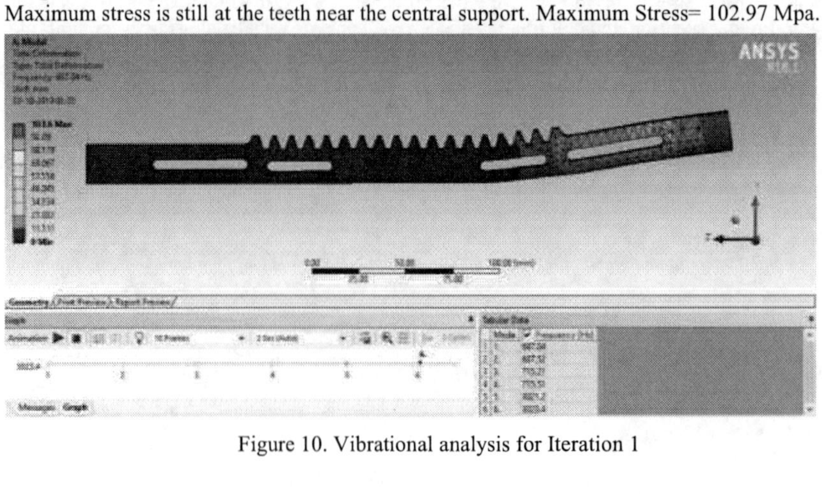

Figure 10. Vibrational analysis for Iteration 1

The characteristic deformation and the shape for vibration analysis of Iteration 1 are similar to those of the Original design. However, they have different range of frequencies. The shape of the 1st mode of vibration is visualized in this plot. Natural frequency range= 687.04 to 3023.4 Hz.

3.3 Iteration 2

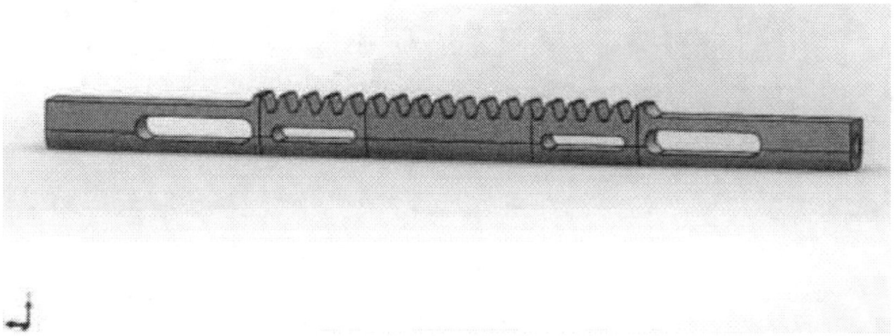

Figure 11. SolidWorks model for Iteration 2

The two slots at the ends still had regions with less stress around them and had potential for further weight reduction. Thus, the two end slots were made bigger for this iteration. Mass = 780.66 grams.

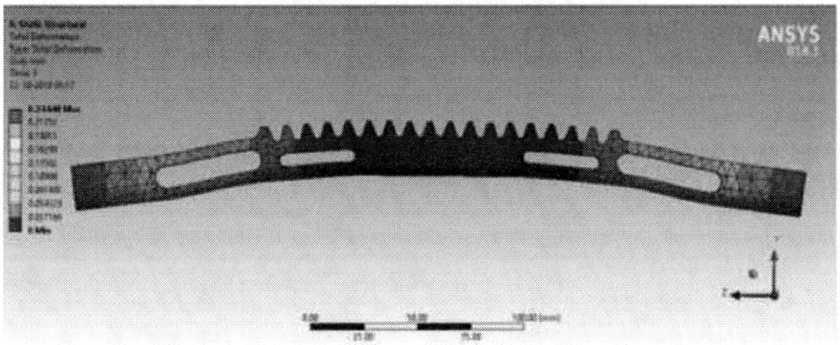

Figure 12. Total Deformation Plot for Iteration 2

The Iteration 2 is analyzed using ANSYS software to give the total deformation plot. The highest value of deformation obtained is at the ends of the rack and the lowest is near the central support. Maximum deformation= 0.2448 mm.

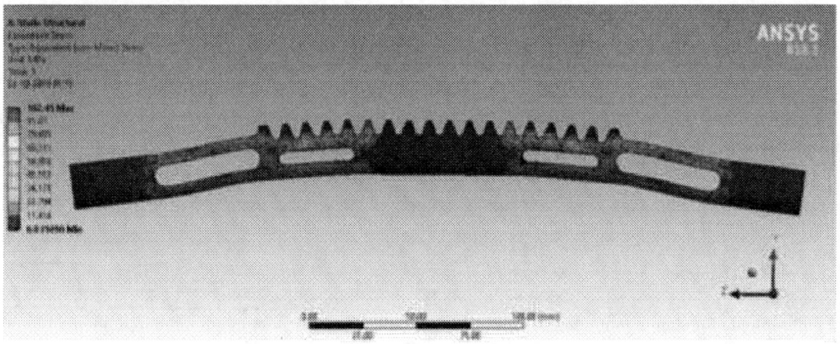

Figure 13. Equivalent Stress plot for Iteration 2

The stress plot for Iteration 2 shows a more efficient stress distribution than the earlier models. The weight reduction is optimized and there are lesser regions with low stress. Max stress= 102.45 MPa

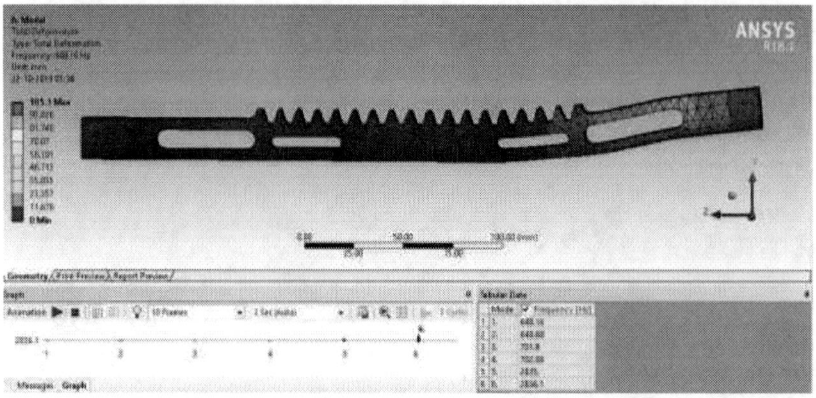

Figure 14. Vibrational analysis for Iteration 2

The natural frequency range is lesser than that of the other two models. However, the resonance deformation is larger. This visualization shows the 1st mode of vibration which is displayed using ANSYS. Natural frequency range = 648.16 to 2836.1

4. RESULTS AND DISCUSSION

A rack and pinion steering system was designed using AGMA stress equations and modelled using SolidWorks. Structural and Modal analysis was conducted using the FEA software ANSYS. Iteration based Topology Optimization was performed with an aim to reduce weight of the component while allowing a marginal increase in maximum stress. The mass, maximum deformation, maximum stress and the range of natural frequency were analyzed and compared for the original design and the two iterations [17, 18]. The results for all three models are tabulated for comparison and further analysis.

Table 1. Comparison of analyses

Analysis	Original Design	Iteration1	Iteration2
Mass (grams)	995.34	836.49	780.66
Maximum deformation (um)	165.7	209.93	244.8
Maximum stress (Mpa)	91.578	102.97	102.45
Natural frequency range (Hz)	764.36 to 4550.1	687.04 to 3023.4	648.16 to 2836.1

Iteration 1 gives a 15.96% reduction in mass from the original model. The deformation and the stress are increased by 26.69% and 12.44% respectively. The range of the natural frequency is 2336.36 Hz. Iteration 2 has 23.58% less mass than the original model. Deformation and stress increase by 47.74% and 11.87%. The natural frequency range is 2187.94 Hz.

The natural frequencies of all three models do not match with the operating frequencies of 20 Hz to 200 Hz so there is no possibility of resonance. However, there could be sudden changes in the operating frequencies in real world conditions such as due to irregular road surfaces and unbalanced or worn out tires [19]. Therefore, a model with a smaller range of natural frequencies is preferred.

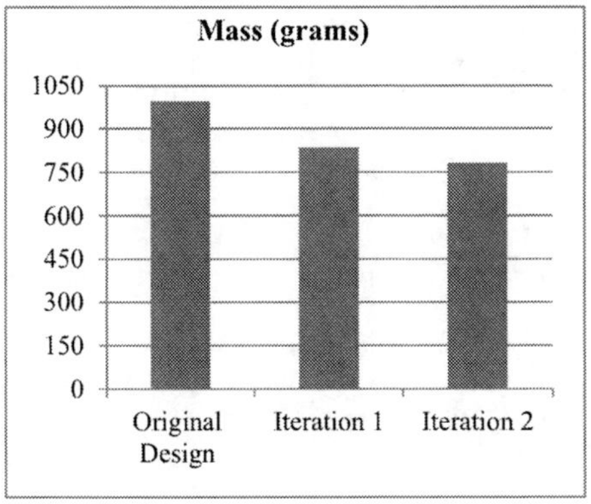

Figure 15. Comparison of mass

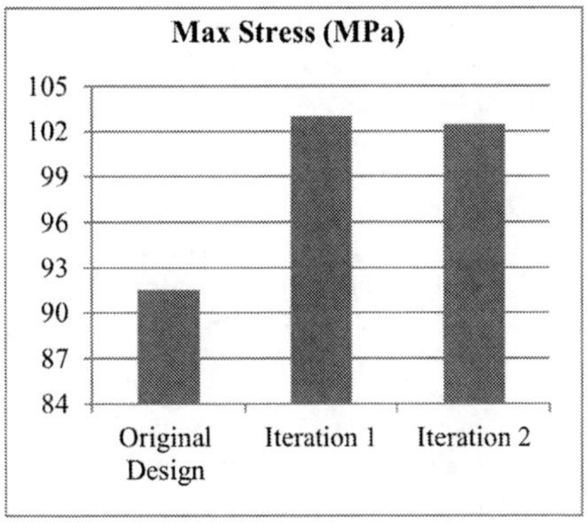

Figure 16. Comparison of max stress

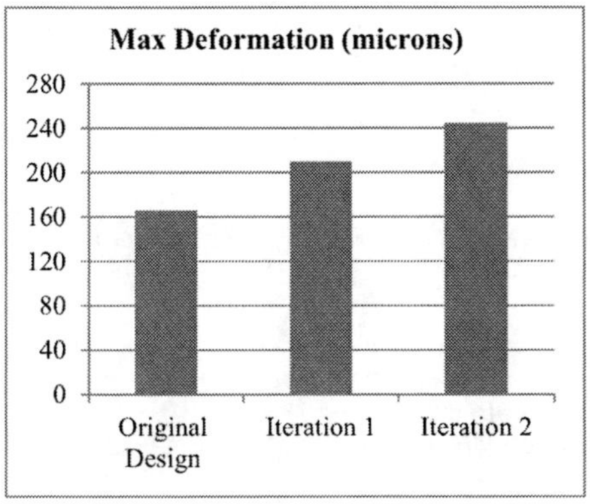

Figure 17. Comparison of max deformation

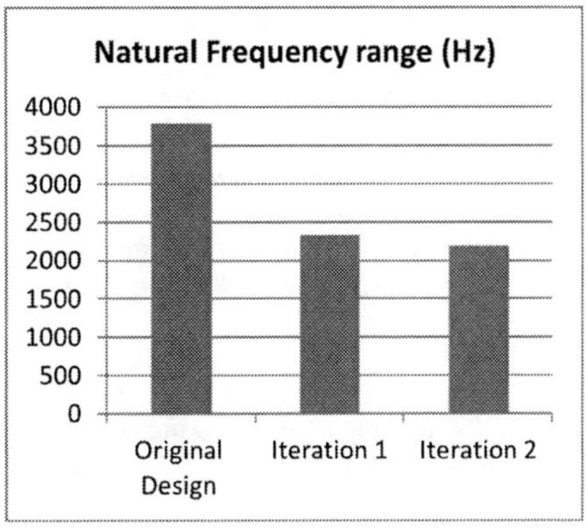

Figure 18. Comparison of natural frequency

Iteration 2 is an optimal design because it gives the most reduction in weight. It also has the smallest range of operating frequencies so it is the preferred model from Vibrational Analysis. However, it has a small increase in maximum stress over the original design and the largest deformation. If deformation is not a critical parameter, then Iteration 2 is the best model.

The proposed method used for the modeling carried out in this project can be used in investigations to solve the system faster than by using experimental methods [20]. The performance of different types of suspension and steering systems can be analyzed and compared using this procedure. Additionally, the methodology can be applied to the entire front suspension system of a car for better optimization of all components.

5. REFERENCES

[1] Hossain, Md & Hasan, Md. Rokib & Ahmed, Kazi & Shawon, Md. 2017 *Design and performance of power generation using speed breaker with the help of Rack and Pinion mechanism*. 7-11. 10.1109/ICAEE.2017.8255317.

[2] Sleesongsom, Suwin & Bureerat, Sujin. 2018. *Optimization of Steering Linkage Including the Effect of McPherson Strut Front Suspension*. 10.1007/978-3-319-93815-8_58.

[3] Thomas D. Gillespie, *Fundamentals of vehicle dynamics*, Society of Automotive Engineers, Inc. 400 commonwealth drive, Warrandale, PA 15096- 0001

[4] Bhardwaj, Saksham & Ashok, B. & Lath, Uma Shankar & Agarwal, Aman. 2018. *Design and Optimization of Steering Upright to Reduce the Weight Using FEA*. 10.4271/2018-28-0081.

[5] Sigmund, O., Maute, K. 2013, *Topology optimization approaches*. Struct Multidisc Optim 48, 1031–1055. https://doi.org/10.1007/s00158-013-0978-6

[6] Prashant L Agrawal, Sahil Shaileshbhai Patel, 2013, Shivanshu Rajeshbhai Parmar "*Design and Simulation of Manual Rack and Pinion Steering System*" International Journal for Science and Advance Research In Technology - Volume 2 Issue 7

[7] Paliwal, Shubham & Manickam, Ramachandran & Fegade, Vishal. 2018. *Numerical investigation of rack and pinion for total deformation and von mises analysis*. International Journal of Mechanical and Production Engineering Research and Development (IJMPERD) ISSN (P): 2249-6890; ISSN (E): 2249-8001

[8] Chandravanshi, M. & Mukhopadhyay, A.2013. *Modal Analysis of Structural Vibration*. ASME 2013 International Mechanical Engineering Congress and Exposition 14. 10.1115/IMECE2013-62533.

[9] Shridhar A.S., Tukkar A., Vernekar A., Badderu V., Patil A.Y., Kotturshettar B.B. 2020 *Modeling and Analysis of ATV Roll Cage*. Advances in Engineering Design and Simulation. Lecture Notes on Multidisciplinary Industrial Engineering. Springer, Singapore. https://doi.org/10.1007/978-981-13-8468-4_21.

[10] Choi, Young & Choi, E. & Jang, Sung-hyun & Park, H. & Ha, J.2005. *Vibration analysis and dynamic optimization of a rack and pinion typed feed drive system for a router machine*. ICMIT 2005: Control Systems and Robotics. 60424I-60424I. 10.1117/12.664815.

[11] Chonhenchob, Vanee & Singh, Sher & Singh, Jay & Stallings, Joseph & Grewal, Gary. 2012. *Measurement and Analysis of Vehicle Vibration for Delivering Packages in Small-Sized and Medium-Sized Trucks and Automobiles*. Packaging Technology and Science. 25. 10.1002/pts.955.

[12] Georgiev, Zlatin & Kunchev, Lilo. 2018. *Study of the vibrational behaviour of the components of a car suspension*. MATEC Web of Conferences. 234. 02005. 10.1051/matecconf/201823402005.

[13] Borowiec, M., Litak, G., Hunicz, J., Koszałka, G. and Niewczas, A. 2008, *Vertical Vibrations of a Vehicle Excited by Real Road Profiles*. Proc. Appl. Math. Mech., 8: 10713-10714. doi:10.1002/pamm.200810713

[14] Duryodhan, N. S. et al, 2015, *Life Determination by Fatigue Analysis and Modal of Intermediate Steering Shaft and Its Optimization*, International Journal of Science Technology & Engineering (IJSTE), Volume 2, Issue 1

[15] Arikan, M.A.Sahir. 2002. *Direct calculation of AGMA geometry factor J by making use of polynomial equations. Mechanics Research Communications*. 29. 257-268. 10.1016/S0093-6413(02)00249-5.

[16] Seok-Chul Hwang, Jin-Hwan Lee, Dong-Hyung Lee, Seung-Ho Han, Kwon-Hee Lee, 2013, *Contact stress analysis for a pair of mating gears*, Mathematical and Computer Modelling, Volume 57, Issues 1–2, Pages 40-49, ISSN 0895-7177, https://doi.org/10.1016/j.mcm.2011.06.055.

[17] Panchal, Nitesh & Mohanty, Akash. 2016. *A Paper Review on Finite Element Analysis with Fatigue Characteristics of Composite Multi-Leaf.* International Journal of Advance Research in Engineering, Science & Technology. 3. 226-233.

[18] Kamble, Naresh & Saha, Subir. 2007. *Developing a virtual prototype of a rack and pinion steering system*. International Journal of Vehicle Systems Modelling and Testing. 2. 10.1504/IJVSMT.2007.011426.

[19] Thin Zar Thein Hlaing, HtayHtay Win, Myint Thein 2017; *Design and Analysis of Steering Gear and Intermediate Shaft for Manual Rack and Pinion Steering System*; Int J Sci Res Publ 7(12) (ISSN: 2250-3153). http://www.ijsrp.org/research-paper-1217.php?rp=P727100

[20] K. Chen, Y. Luan, and X.H. Deng, 2014, *Transient dynamics analysis for uniform engaging process of MC nylon helical gear*, Machinery Design & Manufacture, vol. 3, pp. 110-112, 2012.

Influence of Multiwalled Carbon Nanotubes in the Structural Performance of the Curved Composite Beam

S Mageshwaran[1], S Rajeshkumar[1] and R Manoharan[1*]

[1] School of Mechanical Engineering (SMEC),
Vellore Institute of Technology (VIT), Vellore,
India – 632014.
*E-mail: rmanoharan@vit.ac.in

Abstract. In this paper, Experimental investigation on the free vibration response bheavior of multi walled carbon nanotubes (MWCNT) reinforced curved composite beams have been accomplished to evaluate the structural integrity. To obtain the uniform distribution of MWCNTs in the matrix (epoxy), the ultra-sonication technique is carried out and the curved composite beams are fabricated with the incorporation of MWCNT with the curvature radius of 0.8 m then the free vibration analysis is performed at different boundary conditions. The structural performances of the curved composite beams are explored in the relation of modal natural frequencies and damping ratio. The effective structural integrity and vibration responses are achieved with respect to the enhancement of MWCNT in the curved composite structure.

1. Introduction

Composite structures provide better strength, rigidity, and higher stiffness to the low weight ratio and it finds the place in all the industries. Curved composite structures provide a good membrane and bending energy compared to the metallic structures. Therefore it finds the place in the extensive variety of applications esspecially in automotive, aerospace, marine, astronautics, etc., due to the superior structural and mechanical properties. Numerical exploration on the structural and vibration response of the curved composite beams were performed and it has been reported in literatures. Qatu [1] presented a numerical investigation on the dynamic characteristics analysis of the thin laminated composite shallow curved composite beam based on the Kirchhoff hypothesis and it was clinched that the natural frequencies were improved with increasing the radius of curvature of the curved beam. Ascione and Fraternali [2] developed the one-dimensional curved beam model centred with the Timoshenko beam theory and in the finite element model, the penalty technique was used to solve the numerical problems. Qatu [3] performed the theoretical study on the vibration analysis of the thin and thick curved laminated composite beam based on two cases with and without considering the shear deformation. Tseng and Huang [4] examined the in-plane response of the curved composite beam based on the Timoshenko type curved beam theory with the inclusion of rotary inertia and shear distortion effects for the different curvature and good agreement of results were observed with the analytical solution. The influence of open-angle, orthotropic ratio, and the stacking sequence was also studied with the dynamic stiffness method.

Hajianmaleki and Qatu [5] investigated the static and free vibration response of the thick deep curved composite beam based on the First Order Shear Deformation Theory (FSDT) kinematics and also they developed the numerical model with the pre-defined software ANSYS for the effectiveness. From the work, they determined that the natural frequencies were reduced with an proliferation in the radius of curvature. In 0° stacking sequence they achieved effective results. Ye et al. [6] extended the procedure of series of solutions for the vibration study of the deep curved beam and the Hamilton principle was used for deriving the governing differential equation of motion based on the FSDT. The influence of boundary conditions, lamination schemes, material properties, geometric parameters, and geometry dimensions were studied. It was established that the frequencies were reduced with an rise in the orthotropic ratio (E_1/E_2) and the angle of rotation. Das and Yilmaz [7] studied the free vibration behavior of the curved composite beam through the experimentation and numerical approach. The numerical model was developed through the ABAQUS software and they found better results at fixed at both end conditions. Wang et al. [8] performed a numerical investigation based on the semi-analytical method with the FSDT kinematics. They performed a parametric investigation on varying material properties and geometry structure (elliptic, parabolic, and hyperbolic case) under different support conditions. They stated that the frequencies were reduced with an intensification in the radius of curvature (R) and the orthotropic ratio (E_1/E_2). Guo et al. [9] examined the static and dynamic response of the curved composite beam by using the domain decomposition technique with the Chebyshev orthogonal polynomials. The efficacy of the FEM is evidenced with the good agreement in results with available literature and the pre-defined software ABAQUS. It was shown that the frequencies were reduced with an rise in the included angle whereas it increases with an increase in thickness and clamped at both end condition yields the greater frequency and the clamped free yields the least frequency.

However, to induce the better strength and structural performance, the nanofillers are reinforced to the composite structures where Iijima [10] investigated the potentials of the carbon nanotubes in terms of strength and response, it gains the attention of the researchers to work on this. The dispersion of MWCNT into the polymers illustrates the importance of the vibration and mechanical features of the composite structure. Garg et al. [11] made a study on the dispersion techniques of MWCNT and they achieved better results when it was dispersed with the solvent at the 0.3 wt. % of MWCNT in the composite structure. Ayatollahi et al. [12] experimentally presented the impact of aspect ratio (L/D) and wt % of MWCNT in the composite structure and they showed that the MWCNT with lesser diameter provides the better strength at the 0.3 wt. % of MWCNT in the composite structure. Zhou et al. [13] addressed the mechanical properties of the laminated composite structure with various wt % of CNT and a 28.3 % increament in flexural modulus was noticed with 0.3 wt. % of CNT content. Khan et al. [14] experimentally evaluated the damping of the laminated composite beam and they achieved an increase in damping to the CNT content at all the modes and the better results were obtained at 1 wt. % of the CNT. Mageshwaran and Ramamoorthy [15] presented the structural analysis of composite shell panels on the free vibration mode and they performed the modal analysis on the completely free environment condition. However, from the detailed report, it was evidenced that the experimentation on the vibration response of the MWCNT embedded curved beam was not yet been reconnoitered on different support conditions.

In this work, the structural performance of the MWCNT reinforced curved composite beams are examined with the experimentation on free vibration. The ultra-sonication technique is carried out to achieve proper dispersion and the curved composite beams are fabricated with and without MWCNT reinforcement then the free vibration analysis is performed at different boundary conditions. The effectiveness of the MWCNT in structural performance is deliberated with the parameters of the modal natural frequencies and damping ratio.

2. Materials and methodology

2.1. Materials

Glass fiber reinforced polymer (GFRP) composite beams were produced with and without MWCNT reinforcement, for sample preparation a separate mold was equipped with the curvature radius (R) of 0.8 m. The uni-axis 220 GSM E-glass fiber, epoxy resin (LY556) and the hardener (HY951) were used for preparing the laminated composite beams. MWCNT with normalized length and diameter of 17 nm and 10 μm is used for this study.

2.2. Fabrication of curved composite beam

Initially, hybrid matrix was prepared, In that process, MWCNT was mixed with the epoxy through the ultra-sonic liquid processor for the proper distribution of the MWCNT in the epoxy and the 1 wt. % of the MWCNT has been used for this study. The conventional hand layup approach with the vaccum bagging setup was adopted to fabricate the curved composite beam. Initially, the curved mold with the curvature radius of 0.8 m was fabricated intended to sample preparation and over the mold, the liberating agent was applied for easy removal of the fiber reinforced composite beam then the fiber was positioned over the mold with the specified fiber angle orientations for the laminated composite beam then further the matrix was applied where the epoxy and the hardener was used in the proportion of 10:1 to achieve the better curing rate of the laminated composites. After that it was kept in the vaccum bagging process for 2 hours then it was allowed to cure at the normal environment condition for one day. Curved composite beam specimens were also fabricated with the infusion of nanofillers (MWCNT). Cross ply (0°/90°/0°/90°/0°) five layered curved composite samples were fabricated with the same weight and volume fraction of fiber and the curved beam images were presented in Figure 1.

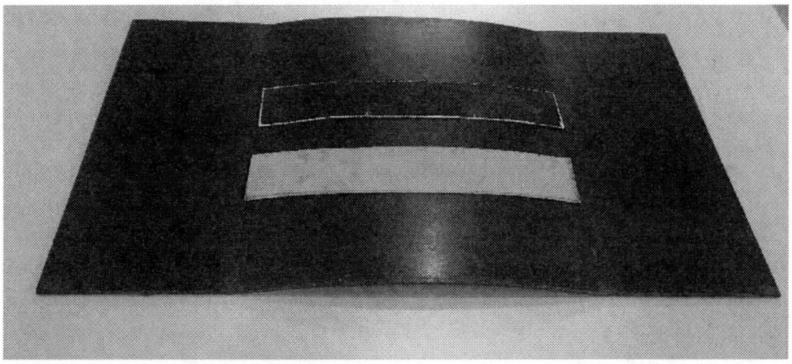

Figure 1. Curved laminated composite beam samples with and without MWCNT reinforcement.

2.3. Experimental setup for the structural analysis

Experimentations were made on the simply supported and completely free boundary condition where the experimental setup image is shown in Figure 2 for perspective understanding. The uni-axis accelerometer was placed on top surface of the specimen to collect the transverse signals and the impulse hammer was used to excite the specimens through the roving hammer technique at equal intervals then the transverse vibration responses were composed through the accelerometer from that the voltage signals were rehabilitated into frequency response function (FRF) responses over the data acquisition system and with the assist of pre-defined software Dewesoft 7.1.1. from the FRF the respective modal frequencies and modal damping ratio of the curved composite beam were evaluated. For the investigation, the samples were retained with the geometric measurements of 300 × 50 mm^2.

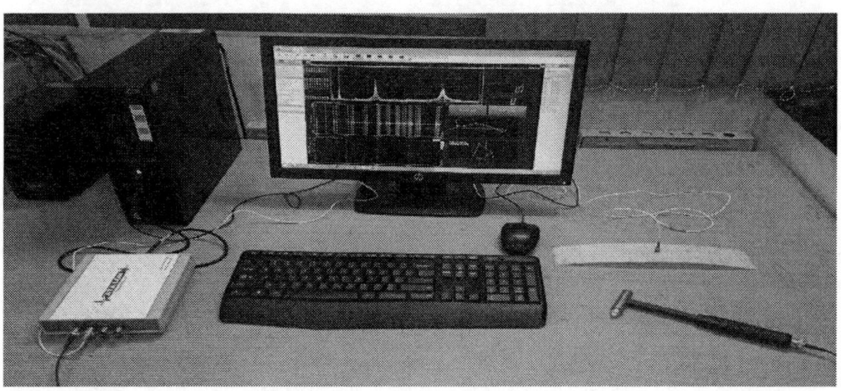

Figure 2. Experimental setup for the free vibration analysis of the curved composite beam.

3. Results and discussions

3.1. Structural examination of the GFRP curved composite beam with MWCNT infusion at simply supported (S-S) boundary Condition

The vibration behavior of the curved composite beam was studied at the simply supported (S-S) boundary conditions and this boundary condition was reported based on the real-time practical usage of applications in the surrounding environment. Experimentation on this boundary condition was performed and the corresponding first three modal natural frequencies and damping ratio are shown in Table 1.

Table 1. First three modal natural frequencies and damping ratio of the curved composite beam on the simply supported boundary condition

Modes	Natural frequencies (Hz)		Damping ratio	
	Without MWCNT	With MWCNT	Without MWCNT	With MWCNT
1	14	16	0.088	0.107
2	63	72	0.045	0.063
3	130	156	0.005	0.008

From Table 1 it can be noticed that the modal natural frequencies and modal damping ratio were improved at all the modes concerning to MWCNT infusion. The modal natural frequencies of the curved composite beams were enhanced with greater than 10 % on the incorporation of MWCNT and also the damping was improved more than 20 % in contrast to the GFRP curved composite beam. The enhancement in the structural response was due to the proper bonding between the MWCNTs, matrix, and fiber interfaces also the load transfer capability behavior was enhanced with MWCNTs.

3.2. Structural examination of the GFRP curved composite beam with MWCNT infusion at completely free (F-F) boundary Condition

In hybrid matrix preparation, the vibration behavior of the curved composite beam was also studied under the completely free (F-F) boundary conditions and the structural response are presented in Table 2. The increase in natural frequencies and damping ratio was observed by concerning the MWCNT reinforcement on the curved laminated composite beam.

Table 2. First three modal natural frequencies and damping ratio of the curved composite beam on the completely free boundary condition

Modes	Natural frequencies (Hz)		Damping ratio	
	Without MWCNT	With MWCNT	Without MWCNT	With MWCNT
1	32	40	0.03	0.035
2	97	110	0.01	0.014
3	177	205	0.004	0.006

In completely free environment conditions the sample was not constrained on the edges, so superior natural frequencies are attained in comparison with the simply supported condition where a similar trend was observed in ref [9]. In this condition also the same variation in structural response was observed as like simply supported condition. In completely free condition, the modal natural frequencies were more but the modal damping ratio is less while compare to simply supported condition due to the constrained environment.

From the discussions, it would be stated that the modal structural response of the GFRP curved laminated composite beam is enhanced at all the modes in-order of reinforcement of the MWCNT on irrespective to the boundary conditions due to the stick-slip mechanism of MWCNTs with epoxy.

4. Conclusions

Vibration behavior of the GFRP curved beam with and without MWCNT reinforcement was studied experimentally to explore the structural response. MWCNT were dispersed with the matrix through the ultra-sonication technique for the proper dispersion rate. Curved composite beams were fabricated with the infusion of MWCNT at the curvature radius (R) of 0.8 m. The free vibration analysis was accomplished at the simply supported (S-S) and completely free (F-F) boundary conditions. Better natural frequencies were obtained at the completely free boundary condition. The modal natural frequencies and modal damping ratio of the MWCNT reinforced curved composite beams were enhanced with more than 10 and 20 % at all boundary conditions. The effective stiffness and damping ratio were achieved to the enhancement of MWCNT in the curved composite structure.

References

[1] Qatu M S 1992, *J. Sound. Vib* **159** 327-338.
[2]. Ascione L, Fraternali F 1992 *Computers & Structures* **45** 985-999.
[3]. Qatu M. S 1993 *Int. J. Solids. Struct* **30**, 2743-2756.
[4]. Tseng Y. P, Huang C. S and Kao M. S 2000 *Compos. Struct* **50** 103-114.
[5]. Hajianmaleki M and Qatu M. S 2012 *Compos. Part B-Eng* **43** 1767-75.
[6]. Ye T, Jin G, Ye X and Wang X 2015 *Compos. Struct* **127** 450-65.
[7]. Das M. T and Yilmaz A 2018 *J. Sound. Vib* **436** 155-64.
[8]. Wang Q, Choe K, Tang J, Shuai C and Wang A 2019 *Mech. Adv. Mater. Struc* **24** 1-5.
[9]. Guo J, Shi D, Wang Q, Pang F and Liang Q *Mech. Adv. Mater. Struc* **26** 1390-402.
[10]. Iijima S 1991 Nature (London) **354** 6348 : 6356.
[11]. Garg P, Singh B. P, Kumar G., Gupta T, Pandey I, Seth R. K, Tandon R. P and Mathur R. B. 2011 *J. Polym. Res.* **18** 1397-1407.
[12]. Ayatollahi M. R, Shadlou S, Shokrieh M. M and Chitsazzadeh M. 2011 *Polym. Test* **30** 548-556.
[13]. Zhou Y, Pervin F, Lewis L and Jeelani S. 2008 *Mat. Science. Eng. A-Struct.,* **475** 157-165.
[14]. Khan S. U, Li C. Y, Siddiqui N. A and Kim J. K. 2011 *Compos. Sci. Technol* **71** 1486-1494.
[15]. Subramani M and Ramamoorthy M 2020 *Polym. Polym. Compos.* **28(4)** 223-232.

IOP Conf. Series: Materials Science and Engineering 1123 (2021) 012062 doi:10.1088/1757-899X/1123/1/012062

Manufacturing and Analysis of Banana- Epoxy Composite fibre

U Narendra Kumar[1], Giriraj Mannayee[1], C Thiyagu[1], G Karthik Reddy[1], P Sai Yogesh Reddy[1], V Kanishka[1] and G Venkatachalam[2]

[1] School of Mechanical Engineering, Vellore Institute of Technology, Vellore 632 014, India
[2] School of Building Sciences, Vellore Institute of Technology, Chennai 600 127, India

Abstract. The usage of natural fibre as reinforcement in polymer composites has increased in the recent years with rising concerns of environmental pollution and the heavy dependence on synthetic fibres from petroleum products of the middle east and the amount of material available in the tropical countries for such low prices. These materials can decrease our dependency on conventional materials like plastics, thereby decreasing the amount of pollution we cause on disposal and manufacturing of those. In general, the impact strength of the natural fibre composites is considered to be low and this experiment quantifies these results. This research describes the fabrication of a polymer composite, using banana fibre as the reinforcement and epoxy as the polymer matrix. A sequence of samples was prepared by varying various parameters like length, volume fraction and treatment of the fibre. The parameters were optimized using Taguchi's DOE and the combination of parameters which yields us the best result and the contribution of each parameter. The impact strength has been studied using IZOD impact tester. The work has quantified that the impact strength of the composite depends mostly on the volume fraction and also only up to a certain range and then decreases as we increase the quantity of fibre. The same goes with both alkali treatment and length of the fibre which also contribute respectively. The optimum length, volume fraction and chemical are 1mm, 25 wt% and 2% NaOH treatment shows better result.

1. Introduction

Recent years, the usage of natural fibre as reinforcement in polymer composites has increased with rising concerns of environmental pollution and the heavy dependence on synthetic fibres from petroleum products of the middle east and the amount of material available in the tropical countries for such low prices. These materials can decrease our dependency on conventional materials like plastics [1]. The manmade fibres were replaced by natural fibre to reduce the amount of pollution we cause on disposal and manufacturing of those synthetic fibre [2,3]. Natural fibre also has some advantages compare to synthetic fibre are low cost, eco-friendly nature, high strength per unit weight, nontoxicity, and biodegradability [4-6]. Some of natural fibres are banana, sisal, hemp, ramie, coir, vakka, flax, jute, and pineapple leaf fibre [7,8]. Natural fibres cell wall contains hemicellulose, cellulose and lignin. These organic components of the fibres report the stiffness and strength [9-12]. Natural fibre with thermoset matrix possesses good mechanical [13,14]. Composites materials are widely used in leisure, aerospace, sports, construction, automobile and packaging industries [15,16].

The production of banana was ~70 million metric tons in every year by subtropical and tropical regions of the world [17-19]. In 100g of Banana pulp consist of 1.15g of protein; 18.8g of carbohydrates; 73.9g of water; 0.18g of fat; 0.83g of vitamins C1 B1 B2 B6 E and 81kcal of other minerals [20]. Banana fibre has 64% of cellulose content and 11° low microfibrillar angle indicates it is suitable for reinforcement in composite [21]. The banana fibre reinforced composites are widely used in building application [22]. The dynamic mechanical property of banana fibre composite was high, if the composite consists of 40% of fibre content [23]. Cement reinforced with kraft pulped banana fibre gives better flexural strength [24]. The optimized fibre length to obtain maximum tensile and impact strength was 30mm and 40mm respectively [25].

The main draw back in natural fibre was non uniformity, moisture absorption, low mechanical properties and hydrophilic in nature [26]. Due to hydrophilic in nature, the bonding between fibre and matrix was reduced. To overcome this chemical treatment was induced. Chemical treated fibre composite gives better mechanical strength compare to untreated [27-29]. In that NaOH treatment gives better result compare to any other [30].

For optimization process the best technique was Taguchi's method [31]. In this work L9 array was incorporate for optimization. For that length of the fibre (0.1mm, 1mm, 10mm), Volume fraction (15wt%, 25wt%, 35wt%) and chemical treatment (0%, 2%, 5%) are selected as parameters. The objective of the research is to develop a polymer matrix composite from the banana fibre and epoxy polymer and test its impact strength. For the polymer to be useful in practical field especially in automobiles and aerospace they should have enough impact strength. Results can be drawn at which combination of the levels of parameters gives us the maximum toughness values. It even helps us in determining the percentage of impact by the parameters. By this an understanding can be made on how much each parameter influences the toughness. An understanding of correlation of the parameters involved in the experiment can be found out. The result was calculated by using MINITAB software.

2. Experimental

2.1. Materials
Banana fibre was bought from Bharatiya natural fibres, Coimbatore, Tamil Nadu, India. NaOH from SD fine-chemical Ltd, Mumbai, India. Epoxy (Araldite LY 556) and Hardener (HY 951) are bought from SM composites, Chennai, India.

2.2. Preparation of Composites
Initially, the fibres obtained from the supplier where nearly one meter long so they had to be trimmed down to the required dimensions. The 1mm and 10mm strands were achieved using a scissor by hand (Fig.1). With the help of Pulveriser, the fibre length was reduced from 1mm to 0.1mm (Fig. 2).

Fig 1. 1mm Banana fibre.

Fig 2. Powdered (0.1mm) Banana fibre samples.

Before the treatments, the Banana fibre was washed with normal water. Then the fibre was immersed in a NaOH solution contains 2g for 100ml distilled water for 1 hr. After 1hr the fibre was washed with running normal water to neutralize and dried for two days in oven at 60°C.

Parchment paper has been used to prepare the moulds according to ASTM 6110 the standards. Though the paper moulds have good inner surface finish they do not have enough strength to hold the composite mixture, hence wooden moulds had been prepared of plywood to support the paper moulds. The wooden samples are prepared by giving slight tolerance to fit in the paper moulds (Fig.3).

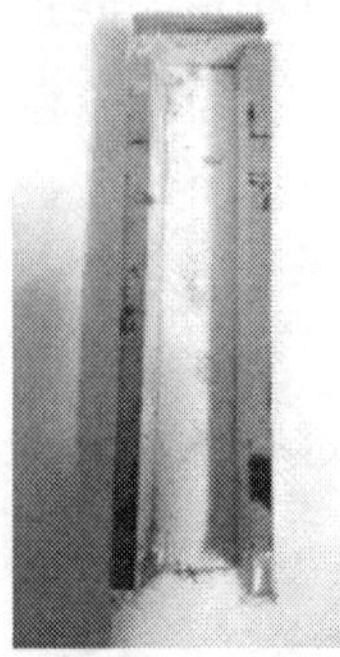

Figure 3. Wooden support.

The fibre was mixed with 10:1 ratio of epoxy and hardener by mechanical stirring. Then the mixture is slowly transfer to the mould. The composite is then allowed to settle for 48 hours in the room temperature. Later the paper moulds are scrubbed with sand paper to remove the outer paper (Fig. 4).

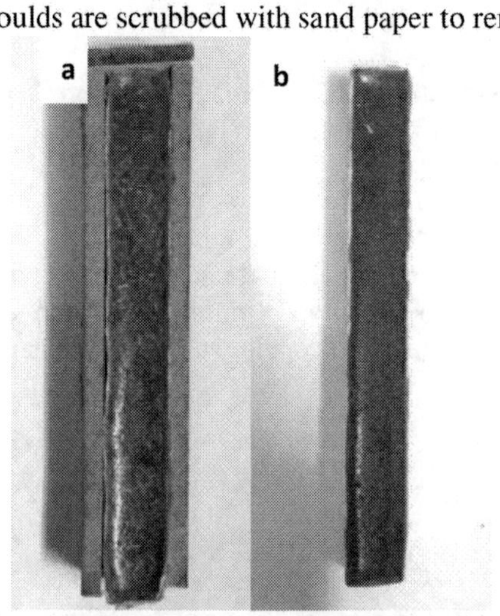

Figure 4. (a)Sample with wooden support; (b)Sample without wooden support

Sequence of samples were prepared as per L9 orthogonal array with different composition in Table 1.

Table 1. Number of experiments.

S.No	Length of the fibre (mm)	Volume Fraction (%)	Treatment with NaOH (%)
1	0.1	15	0
2	0.1	25	2

3	0.1	35	5
4	1	15	2
5	1	25	5
6	1	35	0
7	10	15	5
8	10	25	0
9	10	35	2

2.3. Characterization

The impact strength of the specimen was determined using IZOD impact tester with Scale range: 2, 4, 7.5, 15, 25 Joules (Fig.5). The specimens are cut into as per ASTM D256-88 (64mm x 12.7mm x 3.2 mm) and 45° angle notch was made at the centre of the sample for impact testing. The toughness of the samples was evaluated by charpy impact test as per ASTM D6110. It shows the amount of energy required to break the sample. From that the toughness of the specimens were calculated and the test was carried at CIPET Hyderabad.

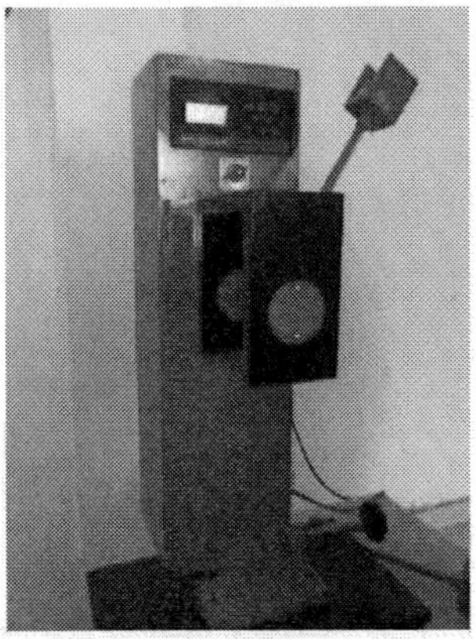

Figure 5. Impact testing machine.

3. Results and Discussion

The analyses and a mathematical model were developed by Design expert software for toughness using the input parameters (table 2)

Table 2. Experimentation table.

Length of the fibre (mm)	Volume Fraction (%)	Treatment with NaOH (%)	Toughness of sample (J/mm)
0.1	15	0	0.55
0.1	25	2	0.65
0.1	35	5	0.40
1	15	2	0.45
1	25	5	0.65
1	35	0	0.40
10	15	5	0.40

IOP Conf. Series: Materials Science and Engineering 1123 (2021) 012062 doi:10.1088/1757-899X/1123/1/012062

10	25	0	0.40
10	35	2	0.7

For the calculation purpose say the length of fibre be A, Volume fraction be B and NaOH treatment be C.

Taguchi Calculations

The optimum condition is calculated as follows:

A_1= (Sum of Toughness for samples with A factor as 1) / total no of levels

A_1= (0.55+0.65+0.40) / 3 = 0.5334

Similarly, the values are calculated for all the factors at all levels. The results come out to be,

A_1=0.5334; B_1=0.4667; C_1=0.45;

A_2=0.55; B_2=0.6166; C_2=0.60;

A_3=0.50; B_3=0.50; C_3=0.533;

The best sample is the sample which is having the highest toughness value. So, the optimum condition is A_2, B_2 & C_2.

The analysis can also be done in MINITAB software:

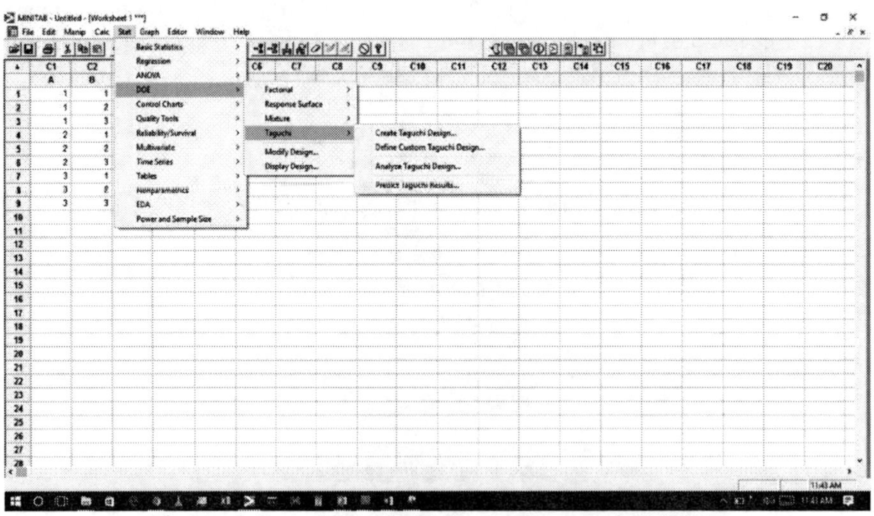

Figure 6. Taguchi Table design roadmap on MINITAB

The optimum condition can be found by analysing the Taguchi's design

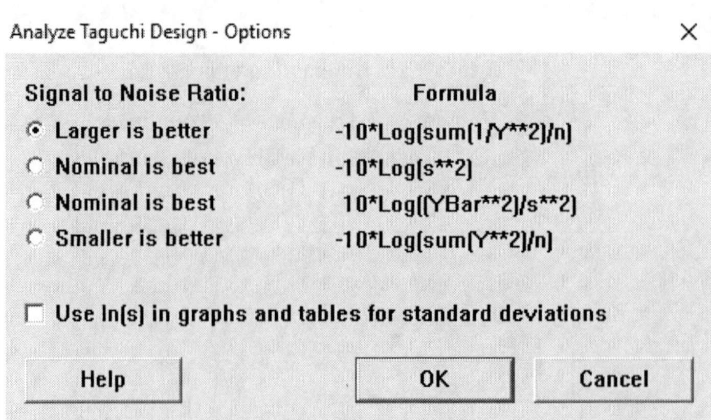

Figure 7. Option selection

When analysing we have taken the condition to be Larger is better as the sample is good only when toughness is the highest.
S/N is the Signal to Noise Ratio
The graphs for the results are as follows:

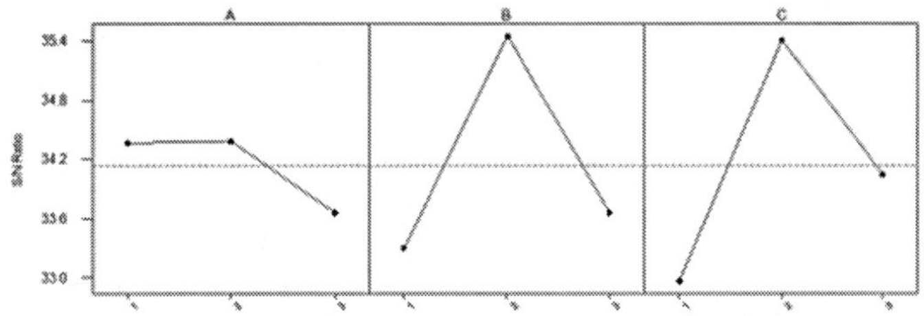

Figure 8. Main effects plot for S/N Ratios

We can see that the factor A which is the length of the fibre is having a slight increase till 2nd level which is 1mm and then from there is a sharp decrease in the toughness value. Hence the optimum value is 1mm. When it comes to factor B which is the % volume fraction you can see a massive increase in toughness when increased from 15% to 25% from there is a sharp decrease in the value. So, the optimum condition is 25% of volume fraction. For the factor C which is the % of NaOH treatment we can see the same trend as in the factor B which is increase and then decrease. So, the optimum condition is 2% NaOH treatment.
Contribution of each parameter
1) $ST = \Sigma yi^2 - T^2/N$
$T = \Sigma yi = (0.55+0.65+0.40+0.45+0.8+0.4+0.4+0.4+0.7)$
$T = 4.75$
N = no of rows in orthogonal array
$N = 9$
$\Sigma yi^2 = (0.55)^2 + (0.65)^2 + (0.4)^2 + (0.45)^2 + (0.8)^2 + (0.4)^2 + (0.4)^2 + (0.4)^2 + (0.7)^2 = 2.6975$
$T^2/N = (4.75)^2/9 = 2.5069$
$ST = 2.6975 - 2.5069 = 0.1906$
DOF = Degrees of Freedom
DOF = Levels-1
$DOF_{(Total)} = 9-1 = 8$
$DOF_{(Factor)} = level - 1 = 3-1 = 2$
$DOF_{(Error)} = DOF_{(Total)} - DOF_{(all\ factors)}$
$= (8) - (6)$
$= 2$
$SOS_A = \Sigma A^2/n - T^2/N$
$= [((0.55+0.65+0.4)2+ (0.45+0.8+0.4)2 +(0.4+0.4+0.4)2)/3] - 2.5069$
$= 2.5106 - 2.5069$
$=0.003$
Similarly, we get $SOS_B = 0.0371$
and $SOS_C = 0.0339$
$SOS_{(Error)} = ST - \text{sum (SOS)}$
$= 0.1906 - (0.003+0.0371+0.0339)$
$= 0.1906 - 0.074$
$= 0.1166$
2) Variance = SOS_A/DOF
$= 0.0037/2 = 0.00185$
Variance of Error = 0.1141/2
= 0.05705
3) PURE SUM = $SA^1 = S_A$- (variance error*F)

= 0.00185- (0*0) = 0.00185
Total sum = 0.00185+0.018+0.016965
= 0.036815
4) % of Factor = (S_1 factor/ total sum) *100
= (0.037/0.1913) *100= 19.3%

Table 3. ANOVA Table

Factor	DOF	SOS	Variance	Pure SUM	% of Factor
A	2	0.0037	0.00185	0.00185	19.3
B	2	0.0371	0.018	0.018	19.40
C	2	0.0339	0.01697	0.01697	17.72
Error DOF	2	0.1166			60.95
Total	6	0.1913			100

Similarly, we get the % of contribution by factor B i.e., % volume fraction is 19.40 and for factor C which is treatment with alkali is 17.72% and the error % which constitutes all the uncontrollable factors is almost 60%.

4. Conclusions

A sequence of banana fibre reinforced epoxy composite samples was prepared using the casting process. Different parameters, length of the fibre (0.1mm, 1mm, 10mm), Volume fraction (15wt%, 25wt%, 35wt%) and chemical treatment (0%, 2%, 5%) were incorporated for preparing the composites by casting method. Toughness testing investigation reported that the length of fibre plays a nominal role and the volume fraction plays a prominent role. However, the impact strength is not increasing continuously with increase in the volume fraction where it is reaching a maximum value and then decreasing. The same is the case with treatment with NaOH. The optimum length, volume fraction and chemical are 1mm, 25 wt% and 2% NaOH treatment shows better result.

Challenges include the homogenization of the fibre's properties and a full understanding of the crystallization and degree of polymerization, adhesion between the matrix and fibre, flame-retardant and moisture repellence properties.

References

[1] John M J and Thomas S Biofibers and biocomposites *Carbohydrate Polym* 2008; **71**: 343–364.
[2] Uma Devi L, Bhagawan S S and Thomas S Mechanical properties of pineapple leaf fiber-reinforced polyester composites *J Appl Polym Sci* 1997; **64**: 1739–1748.
[3] Murali Mohan Rao K, Mohana Rao K and Ratna Prasad A V Fabrication and testing of natural fiber composites: vakka, sisal, bamboo and banana *Mater Des 2010*; **31**: 508–813.
[4] Ramesh M, Palanikumar K, Reddy K H Influence of fiber orientation and fiber content on properties of sisal-jute-glass fiber-reinforced polyester composites *J. Appl. Polym. Sci. 2016*, **133**(1), 42968.
[5] Monteiro S N, Calado V, Rodriguez R J S, Margem F M Thermogravimetric Stability of Polymer Composites Reinforced with Less Common Lignocellulosic Fibers-an Overview *J. Mater. Res. Technol.* 2012,**1**, 117-126.
[6] Zini E, Scandola M Green composites: An overview *Polym. Compos.* 2011, **32**, 1905.
[7] Hepworth D G, Hobson R N, Bruce D M, Farrent J W The use of unretted hemp fibre in composite

manufacture. *Compos Part A: Applied Sci and Manuf.* 2000; **31**: 1279–1283.

[8] Asasutjarit C, Hirunlabh J, Khedari J, Charoenvai S, Zeghmati B, Cheul Shin U Development of coconut coir-based light weight cement board *Constr Build Mater* 2007; **21**: 277–288.

[9] Zini E, Focarete L, Noda I and Scandola M Bio-composite of bacterial poly (3-hydroxybutyrate-co-3-hydroxyhexanoate) reinforced with vegetable fibers *Compos Sci Tech.* 2007; **67**, 2085-2094.

[10] Chandramohan D, Marimuthu K Drilling of Natural Fiber Particle Reinforced Polymer Composite Material *Int. J. Adv. Eng. Res.Studies* 2011, **1**(1), 134–145.

[11] Sarasini F, Tirillo J, Puglia D, Kenny J M, Dominici F, Santulli C, Tofani M and Santis R D Effect of different lignocellulosic fibres on poly(ε-caprolactone)-based composites for potential applications in orthotics *RAC Advance,*2015; **5**(30): 23798-23809.

[12] Almeida J H S Jr, Amico S C, Botelho E C, Amado F D R *Compos. Part B Eng. 2013*, **55**, 492.

[13] Wang W and Huang G. Characterization and utilization of natural coconut fibers composites. *Mater Des* 2009; **30**: 2741–2744.

[14] Bodros E, Pillin I, Montrelay N et al Could Biopolymers have reinforced by randomly scattered flax fiber be used in structural applications. *Compos Sci Technol* 2007; **67**: 462– 469.

[15] Malkapuram R, Kumar V, Yuvraj S N Recent development in natural fibre reinforced polypropylene composites *J Reinf Plast Compos* 2008; **28**:1169–89.

[16] Wambua P, Ivens J, Verpoest I Natural fibres: can they replace glass in fibre reinforced plastics. *Compos Sci Technol* 2003; **63**:1259–64.

[17] Muller D H and Krobjilowski A (2003) New Discovery in the Properties of Composite Reinforced with Natural Fibers *Journal of Industrial Textiles*, **33**: 111130.

[18] Joshi S V, Drza L T, Mohanty A K and Arora S (2004) Are Natural Fibers Composites Environmentally Superior to Glass Fiber Reinforced Composites, *Composites Part A*, **35**: 371376.

[19] Umair S (2006). Environmental Effect of Fiber Composite Materials-Study of Life Cycle Assessment of Materials Used for Ship Structure, *M S Thesis Dissertation, Royal Institute of Technology*, Stockholm.

[20] Bilba K, Arsene M A and Quesanga A (2007) Study of Banana and Coconut Fibers: Botanical Composition, Thermal Degradation and Textural Observation, *Journal of Bioresource Technology*, **98**:5868.

[21] Satyanarayana K G, Kulkarni A G and Rohatgi P K (1981) Potential of natural fibers as a resource for industrial materials in Kerala *J. Sci. Ind. Res.,***40**:222.

[22] Coutts R S P Banana fibres as reinforcement for building products *J. Mater. Sci.*1990, **9**:1235.

[23] Pothan L A, Oommen Z and Thomas S Dynamic Mechanical Analysis of Banana Fiber Reinforced Polyester Composites *Composites Science and Technology* 2003; **63**(2): 283–293.

[24] Corbière-Nicollier T, Laban B G, Lundquist L, Leterrier Y, Manson J A E, and Jolliet O Life Cycle Assessment of Biofibers Replacing Glass Fibers as Reinforcement in Plastics *Resources, Conservation and Recycling* 2001; **33**(4): 267–287.

[25] Pothan L A, Thomas S and Neelakantan N R Short Banana Fiber Reinforced Polyester Composites: Mechanical, Failure and Aging Characteristics *J. Reinforced Plastics and Composites*. 1997; **16**(8): 744–765.

[26] Ghorbani F M, Kaffashi B, Shokrollahi P, Seyedjafari E and Ardeshirylajimi A PCL/chitosan/Zn-doped nHA electrospun nanocomposite scaffold promotes adipose derived stem cells adhesion and proliferation *Carbohydrate Polymers*.2015; **118**:133-142.

[27] Paul A, Joseph K and Thomas S Effect of surface treatments on the electrical properties of low-density polyethylene composites reinforced with short fibers *Compos Sci Technol* 1997; **57**: 67–79.

[28] Alvarez VA, Ruscekaite R A, Vazquez A. Mechanical properties and water absorption behavior of composites made from a biodegradable matrix and alkaline-treated sisal fibers *J Compos Mater* 2003; **37**: 1575–1588.

[29] Goud G and Rao R N. The effect of alkali treatment on dielectric properties of roystonea

regia/epoxy composites. *Int J Polym Anal Charact* 2011; **16**: 239–250.
[30] Rassmann S, Reid R G and Paskaramoorthy R Effects of processing conditions on the mechanical and water absorption properties of resin transfer moulded kenaf fibre reinforced polyester composite laminates *Composites: Part A*. 2010; **41**:1612-1619.
[31] Taguchi G. Introduction to quality engineering. New York, USA: *McGraw- Hill*; 1990.

IOP Conf. Series: Materials Science and Engineering 1123 (2021) 012063 doi:10.1088/1757-899X/1123/1/012063

Effects of pulsing current and filler materials on the mechanical and metallurgical properties of DSS 2205 weldments

M Sathishkumar[1,2], Sawan Nagpal[2], Harshit Gupta[2], Sanjalp Bhargava[2], S A Nithin Joseph Reddy[3], K Gokul Kumar[2], M Manikandan[2*], N Arivazhagan[2],
[1]Department of Mechanical Engineering, Amrita School of Engineering, Chennai, Amrita Vishwa Vidyapeetham, India
[2]School of Mechanical Engineering, Vellore Institute of Technology, Vellore, India
[3]Hyndai Motor India Limited, Chennai, India

*E-mail: mano.manikandan@gmail.com

Abstract. The control of balance in austenite to ferrite ratio of duplex stainless steel (2205) during welding is critical due to not choosing the optimum heat input and cooling rate and welding processes, which leads to the development of embrittlement of intermetallic precipitates in the weldment. Further, the imbalance of austenite to ferrite ratio primes the stress corrosion cracking (ferrite < 25 %), low impact toughness and corrosion resistance (ferrite > 75 %). The pulse current gas tungsten arc (PCGTA) welding has utilized to weld the DSS 2205 using ERNiCrMo-3 and ERNiCrMo-4 fillers for better control of austenite to ferrite ratio. The tensile strength of Mo-3 weldment is 4.06 % higher than Mo-4. Besides, the Mo-3 weldment also provided the enhanced impact toughness and microhardness compared to Mo-4. The macrographs show similar weld bead width. The microstructure of both fillers Mo-3 & Mo-4 confirmed columnar structure at the weld interface and the finer equiaxed structure at weld center. The scanning electron microscope along with energy dispersive spectroscopy assured the absence of Cr2N, Chi and sigma phases. The PCGTA of DSS 2205 with Mo-3 is giving the improved properties with respect to Mo-4.

1. Introduction

Duplex stainless steel (DSS) is basically a two phased alloy which has equal amount of austenitic and ferritic structures. Typical chemical composition consists Cr-25% and Ni-5%. Cr plays an important role in increasing the strength, corrosion resistance and wear resistance. Owing to its high strength, the machinability of this grade in general is quite low. Ni helps to get an equalized microstructure in DSS as it stabilizes an austenite phase. The anticorrosive properties of DSS 2205 (better than grade 316) allow it to be used in several marine applications, both onshore and offshore. DSS has played a crucial role in making of marine structures and machinery for the last decade. Its important usage lies in constructing gas pipelines, ocean-based machinery and to contain different kinds of chemicals. DSS is a very popular material to be used where Ferritic and Austenitic grade properties are needed. When compared the yields strength of DSS 2205 is around 2-3 times greater than commercial austenitic

[2] M Manikandan, School of Mechanical Engineering, Vellore Institute of Technology, Vellore, India.

stainless steel (ASS) grades. One of the economic advantages of using DSS is that, used at low thickness as compared to other grades, which in turn helps in weight reduction and the cost as well without compromising the strength of material.

The welded duplex steel has found many applications in the marine projects, heat exchangers, reactors etc. The reason behind this is that the constituents of our filler materials offer excellent strength, wear resistance and anti-corrosive properties. Vinoth Jebaraj et al. [1] studied different properties of DSS 2205 having several marine applications with reference to DSS's machinability, surfacing and weldability. Various kinds of solid-state and fusion welding processes on joining DSS 2205 were studied. The changes occurring in the microstructure during cooling process was studied along with the transformations occurring in the heat affected zone (HAZ). These results were compared between different welding processes. Also, a comparison was made between DSS 2205 and other commercial ASS grade materials on the basis of machinability. Vinoth Jebaraj et al. [2] studied the change in microstructures effect on the weldment and base metal's impact toughness. DSS 2205 was taken for this research. Gas tungsten arc welding (GTAW) was carried out on DSS 2205 and the ferrite to austenite ratio was assessed in the HAZ and the base metal. Changes in the impact toughness upon heat treatment processes were also studied. When heated at 1050 °C for about 1 hr, the austenite phases were found to be nucleated which in turn enhanced the value of impact toughness. But increasing the temperature beyond 1050 °C witnessed a high reduction in the value of impact toughness which was because of incomplete dissolution of the sigma phase in the microstructure observed.

Elsaady et al. [3] evaluated the effects of pitting corrosion on DSS 2205 which was welded using ER 2209 filler material. The welded DSS sample went through pitting corrosion tests as well as the microstructural examination. Four critical temperatures were found for the DSS weld, with reference to pitting corrosion- Severe pitting corrosion and maximum metal loss was found to be occurring at 850 °C. Very less pitting corrosion was observed at a temperature of 1050 °C since at this temperature complete dissolution occurs. The pitting corrosion was reencountered when the heating was extended to above 1050 °C, due to ferritic structures growth. Daniela Ramminger Pissanti et al. [4] studied the microstructural changes over the welded cross section of UNS S32205 Duplex stainless steel and assess the Charpy impact toughness of various regions formed for the pipeline girth friction welding. The results showed that major microstructural changes were observed at the weld interface due to very high temperature and plastic deformation. The impact toughness drop was observed in this region because of changes in austenite morphology and enormous ferrite grain growth. Emami et al. [5] examined the microstructural evolution of SAF 2205 duplex stainless steel using friction stir welding at 50 mm/min and 400 rpm using optical microscope and scanning tunneling microscope. The observations showed refined grains and plain shear texture specimens had been formed in every constituent phase of ferrite and austenite. Also, recrystallization texture components of cube were observed in austenite phase. Shuwan Cui et al. [6] inspected the microstructure, grain boundary misorientation angle distribution (GBMAD), grain size of the duplex stainless-steel plates 10.8 mm thick welded using keyhole tungsten inert gas (K-TIG) welding to characterize the impact toughness of the weld metal. Performed Charpy tests showed that the impact absorbed energy (IAE) of the weld metal improves with the heat supply. Also, the results proved that the impact energy of the weld metal improved with the increase in austenite volume fraction.

Sathishkumar and Manikandan [7–9] studied the preclusion of microsegregation occurring in the aerospace grade Hastelloy X. GTAW and pulsed gas tungsten arc welding (PCGTAW) processes were used along with the filler wires ERNiCrMo-2, ERNiCrCoMo-1, and ERNiCr-3. SEM was used to study the microstructures of the weldments. The EDS analysis showed that PCGTAW was successful in removing the microsegregation caused by Cr precipitate. The XRD analysis showed that Molybdenum rich carbide phase is present in GTAW while NiCrCoMo phase was observed in PCGTAW. The tensile results of the PCGTAW samples gave better results than that of GTAW samples. Chennaiah MB et al. [10] studied the advantages of PCGTAW welding by observing the parameters on the microstructures of Aluminium 6061 alloy. Optical metallography along with micro-hardness tests were conducted for the study. The research showed that finer grains were produced in the welding when the duration of the

pulse was short keeping pulse frequency constant. Also increasing the heat input reduces the frequency of the pulse which in turn makes grains finer. Keeping the frequency of pulse as constant, lower peak currents produced finer grains as compared to that of higher peak currents.
From the studies it is identified that very scarce work on the effect of heat supply on microstructure and tensile properties of PCGTA on DSS have been reported. Hence it is considered to be important and current study would be beneficial in gaining the performance of welded plates using various inputs. In this study, welding has to be carried out using two different filler materials such as ERNiCrMo-3 and ERNiCrMo-4 and main aim is to study the weldment's mechanical properties so that the welded material can be used accordingly. The project aims to investigate the welding properties of Duplex stainless steel (DSS 2205) by using PCGTA welding to evaluate the mechanical, metallurgical properties of the weldment.

2. Materials and Methodology

Initially the plates were cut using Wire cut EDM (Electrical discharge machine) machine with dimensions 150 mm x 100 mm x 4 mm. The elemental composition of DSS 2205 is displayed in Table 1. The plates were then cleaned using Acetone and V grooves were made at the edges at an angle of 45°. The PCGTA Welding was performed using the filler wires ERNiCrMo-3 and ERNiCrMo-4 using KEMPPI 400 machine. The process parameters for the welded joint are pulsed currents between 90-130 A. Argon was used with the supply of 13 L/min for shielding. The peak current and base current as set in welding process were 130 A and 90 A. The total heat input (Equation 1) for ERNiCrMo-3 filler is 1.906 kJ/mm and for ERNiCrMo-4 is 1.817 kJ/mm (Table 2). The microstructure sample along with the tensile and impact samples were cut using the Wire Cut EDM Machine. The microstructure ample was mounted for microstructural examination. To reveal the microstructure an etchant made of Oxalic acid was used and electrolytic etching was performed. The sample was examined using the Optical microscope, scanning electron microscope (SEM), and change in microstructure was observed at the weldment region. Tensile test (ASTM E8), microhardness (E384) were performed then to gauge the strength of the weldment and determine the place of failure. Impact test (ASTM E23) was conducted using Charpy machine to find the impact energy of different specimens and further compare them to that of the base material.

Heat Input, $$\boldsymbol{HI} = \eta \times \frac{I_m \times V}{S_w} \quad \left(\frac{kJ}{mm}\right) \qquad (1)$$

Where, I_m= Mean Current (A), V= Voltage (V), η = Efficiency, S_w= Feed rate (mm/sec)

Table 1. Chemical composition of DSS 2205 and filler materials

Materials	Cr	Ni	Mn	Mo	Fe	W	Nb	Si	N	P	S	C
DSS 2205	22.1	5.2	1.8	3.4	66.3	-	-	0.8	0.2	0.03	0.02	0.03
ERNiCrMo-3	22.2	64.1	0.5	8.8	0.2	-	3.55	0.5	-	0.02	0.015	0.1
ERNiCrMo-4	16.3	56.9	0.7	15.9	5.7	3.5	-	0.08	-	0.04	0.03	0.02

Table 2. Heat calculations of DSS 2205 weldment by PCGTAW

Weldments	Pass	Current (A)	Voltage (V)	Welding Speed (S_w) (mm/sec)	I_m (A)	Heat Input (kJ/mm)	Total Heat Input (kJ/mm)
PCGTA ERNiCrMo-3	Root	Ip = 130 Ib = 90	10.46	1.02	110	0.0789	1.906
	First	Ip = 130 Ib=90	10.46	0.721	110	0.1116	
PCGTA ERNiCrMo-4	Root	Ip = 130 Ib = 90	10.6	1.064	110	0.0767	1.817
	First	Ip = 130 Ib = 90	10.6	0.777	110	0.105	

3. Results and Discussion

3.1. Weld quality examination

The micrograph of the weldment joined by PCGTA welding technique with ERNiCrMo-3 and ERNiCrMo-4 is displayed in figure 1 (a & b). It has been confirmed from the figure that welding defects like solidification and liquation cracks, inclusion, porosity and lack of penetration are not found in weld zone as well as weld interface region. This confirmed optimum process parameters in both the filler wire condition. Further, both the weldment revealing the similar weld bead profile which is due to the larger thermal gradient and quicker cooling rate observed during welding.

3.2. Microstructure examination

The microstructure of the DSS 2205 metal is displayed in figure 2. It is clearly showing the presence of equal amount of austenite (γ) and ferrite (α) element throughout the structure. The microstructure of DSS 2205 weldment joined with PCGTA ERNiCrMo-3 is displayed in figure 3 (a & b) and PCGTA ERNiCrMo-4 is displayed in figure 3 (c & d). The type of structure observed after welding is totally different from the base metal microstructure. The use of Ni enriched filler such as ERNiCrMo-3 and ERNiCrMo-4 produced the balanced microstructure. Vinoth jebaraj et al. [1] also reported the similar structure. Both ERNiCrMo-3 and ERNiCrMo-4 weldments are confirmed the columnar structure at the weld interface and equiaxed structure at the weld centre and. It also revealed no liquation cracks in the HAZ region. The similar kind of results are also observed by others [1,11].

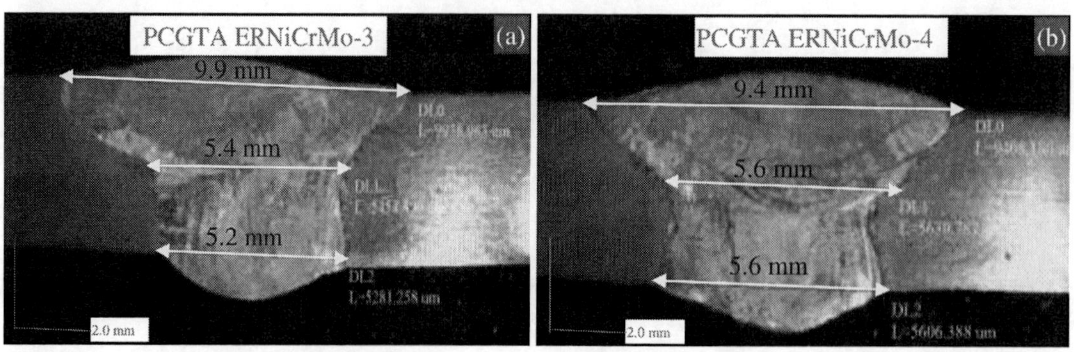

Figure 1. Macrograph of DSS 2205 weldment; (a) PCGTA ERNiCrMo-3 (b) PCGTA ERNiCrMo-4

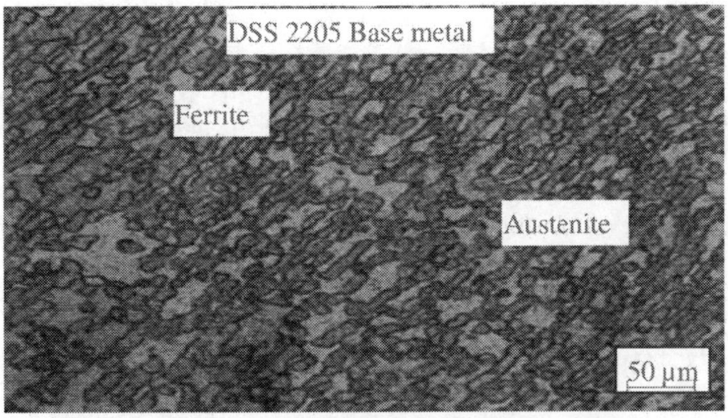

Figure 2. Microstructure of DSS 2205 (base metal)

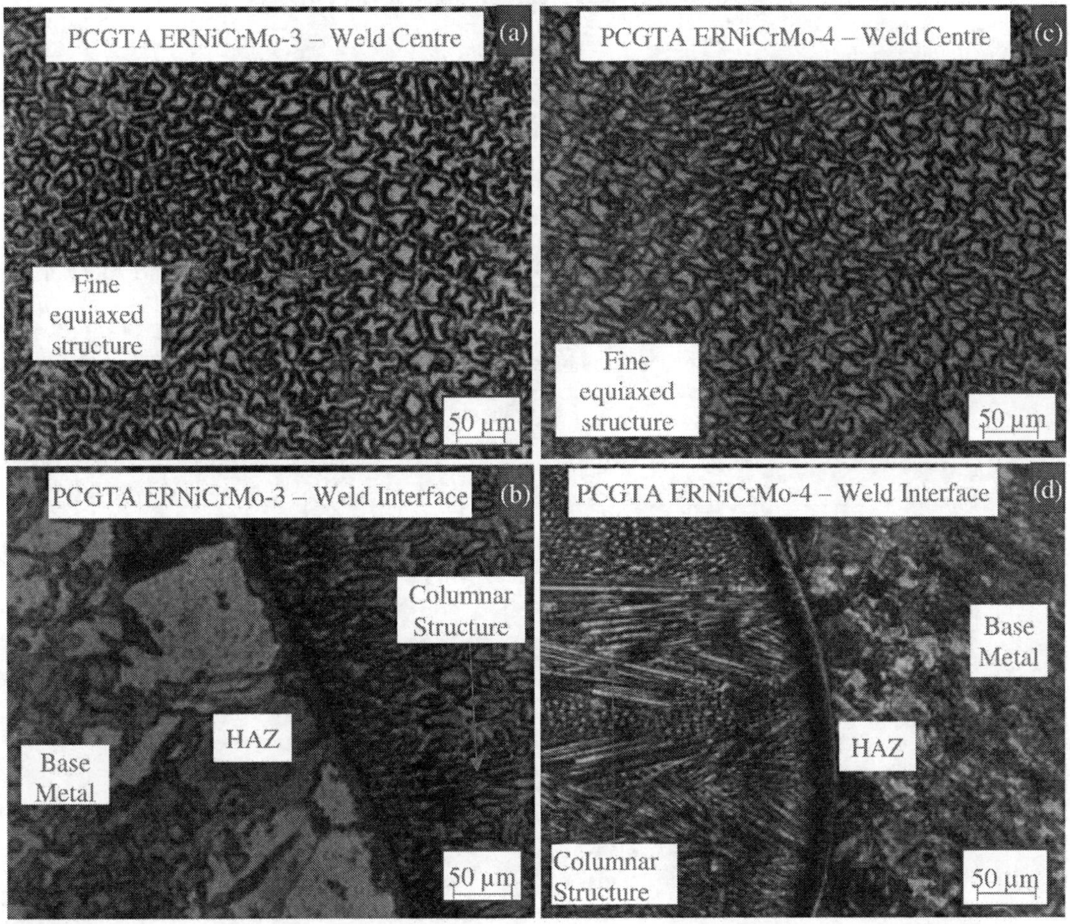

Figure 3. Micrograph of DSS 2205 weldment by; (a) PCGTA ERNiCrMo-3 – Weld centre (b) PCGTA ERNiCrMo-3 – Weld Interface (c) PCGTA ERNiCrMo-4 – Weld Centre (d) PCGTA ERNiCrMo-4 – Weld Interface

3.3. SEM/EDS analysis

The occurrence of the austenite and ferrite phases in the weldments are confirmed with SEM/EDS analysis. The results of both ERNiCrMo-3 and ERNiCrMo-4 fillers are displayed in figure 4. The occurrence of austenite phases in the grain boundaries of ferrite matrix at both weld centre and interface region. Also, the EDS composition at the dendritic core region of both ERNiCrMo-3 and ERNiCrMo-4 fillers at weld centre confirmed the more amount of Fe and less amount of Ni when compared to interdendritic region of the weldments (figure 4 (i-vi)). Further, the presence of more amount of Ni at interdendritic region confirmed the occurrence of austenite phase and similarly, the higher Fe elements in the dendritic region confirmed the ferrite phase. The sigma phase (tetragonal unit cell) occurs when Cr is more than 25 % and Mo of more than filler material and base metal but, in this study no sigma phase is found in all the specimens. The chi phases (cubic unit cell) occurs when Mo is more than 15 %, but this phase also not formed in the weldments. The similar kind of conclusion also stated by Elsaady et al. [3]. The presence of N in both filler materials are absent and this controls the development of Cr_2N phase in the weld centre and interface. This also confirmed with EDS values.

3.4. Tensile test analysis

The tensile testing has been carried out in three trails in order to obtain the repeatability and results are displayed in Table 3. PCGTA ERNiCrMo-3 is giving 3.90 % higher ultimate tensile strength value (UTS) compared to PCGTA ERNiCrMo-4. This is due to the presence of more amount of Ni and less amount of Fe in ERNiCrMo-3 filler stabilize the austenite phase than ferrite phase whereas, high Fe and low Ni in ERNiCrMo-4 stabilize the ferrite phase than austenite. The failure occurred at the weld region for all the specimens. The presence of high austenite improves the tensile properties of PCGTA ERNiCrMo-3 compared to PCGTA ERNiCrMo-4. The ductility of both the weldments are more or less similar each other. The same observation also concluded by other researches [7,8].

Table 3. Tensile strength analysis of DSS 2205 weldment by PCGTA

Weldments	Trail	UTS (MPa)	Mean UTS (MPa)	Elongation (%)	Mean Elongation (%)
Base Metal	Trail 1	815	815	49.8	45.9
	Trail 2	818		45.9	
	Trail 3	813		41.9	
PCGTA ERNiCrMo-3	Trail 1	747	743	14.5	14.1
	Trail 2	746		16.2	
	Trail 3	737		14.6	
PCGTA ERNiCrMo-4	Trail 1	714	714	13.9	15.3
	Trail 2	729		14.5	
	Trail 3	701		14.4	

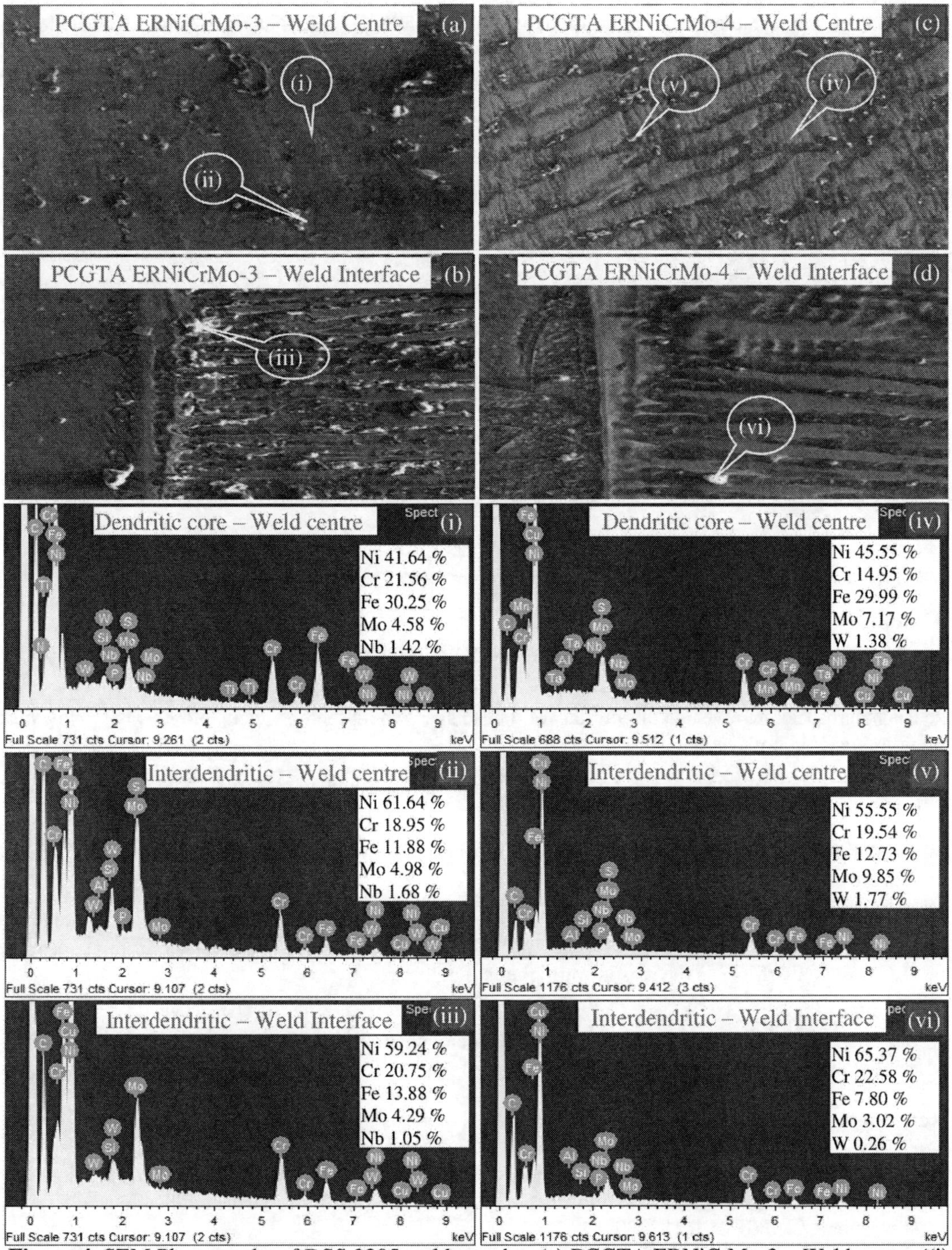

Figure 4. SEM Photographs of DSS 2205 weldment by; (a) PCGTA ERNiCrMo-3 – Weld centre ((i) EDS dendritic core, (ii) EDS interdendritic) (b) PCGTA ERNiCrMo-3 – Weld Interface ((iii) EDS interdendritic) (c) PCGTA ERNiCrMo-4 – Weld Centre ((iv) EDS dendritic core, (v) EDS interdendritic) (d) PCGTA ERNiCrMo-4 – Weld Interface ((vi) EDS interdendritic)

3.5 Impact toughness and microhardness measurement

The results of impact toughness and microhardness of both PCGTA ERNiCrMo-3 and ERNiCrMo-4 is mentioned in Table 4. It is come to know that, the PCGTA ERNiCrMo-3 is providing 24.32 % and 6.94 % higher impact toughness and microhardness than PCGTA ERNiCrMo-4 respectively. The high austenite makes the material harder and increase the impact strength where ferrite makes the weldment to more ductile and decrease the impact toughness and microhardness of the weldments [12,13]. This is the reason for higher toughness and microhardness in PCGTA ERNiCrMo-3 than PCGTA ERNiCrMo-4. Many others suggested the similar results [1–3].

Table 4. Mean microhardness and impact toughness of DSS 2205 weldment by PCGTAW

Weldments	Impact energy (J)				Microhardness (HV)
	Trail 1	Trail 2	Trail 3	Mean	Mean
Base Metal	106	107	104	106	305
PCGTA ERNiCrMo-3	35	39	37	37	288
PCGTA ERNiCrMo-4	25	31	29	28	268

4. Conclusions

1. The welding of DSS 2205 has been successfully performed using ERNiCrMo-3 and ERNiCrMo-4 filler in PCGTA welding condition.

2. The macrostructure revealed the similar weld bead size in both the weldment and it is due to low heat input provided in both weldments.

3. The microstructure confirmed the presence of ferrite and austenite structure in both weldments.

4. SEM/EDS analysis confirmed the more ferrite phase in weld dendritic core region compared to weld interdendritic region which shown higher austenite phase. Further, the presence of sigma, chi and Cr2N phases are not found in all the weldments.

5. The PCGTA ERNiCrMo-3 filler is giving 3.90 %, 24.32 % and 6.94 % higher tensile strength, impact toughness and microhardness than PCGTA ERNiCrMo-4 respectively. The presence of high austenite in PCGTA ERNiCrMo-3 is the reason for higher strength.

Acknowledgement

Author thank Vellore Institute of Technology management, Vellore for giving facilities to complete this research work.

References

[1] Vinoth Jebaraj A, Ajaykumar L, Deepak C R and Aditya K V V 2017 Weldability, machinability and surfacing of commercial duplex stainless steel AISI2205 for marine applications – A recent review *J. Adv. Res.* **8** 183–99

[2] Jebaraj A V and Ajaykumar L 2013 Influence of micro structural changes on impact toughness of weldment and base metal of duplex stainless steel AISI 2205 for low temperature applications *Procedia Eng.* **64** 456–66

[3] Elsaady M A, Khalifa W, Nabil M A and El-Mahallawi I S 2018 Effect of prolonged temperature exposure on pitting corrosion of duplex stainless steel weld joints *Ain Shams Eng. J.* **9** 1407–15

[4] Pissanti D R, Scheid A, Kanan L F, Dalpiaz G and Kwietniewski C E F 2019 Pipeline girth friction welding of the UNS S32205 duplex stainless steel *Mater. Des.* **162** 198–209
[5] Emami S, Saeid T and Khosroshahi R A 2018 Microstructural evolution of friction stir welded SAF 2205 duplex stainless steel *J. Alloys Compd.* **739** 678–89
[6] Cui S, Shi Y, Cui Y and Zhu T 2018 The impact toughness of novel keyhole TIG welded duplex stainless steel joints *Eng. Fail. Anal.* **94** 226–31
[7] Sathishkumar M and Manikandan M 2019 Preclusion of carbide precipitates in the Hastelloy X weldment using the current pulsing technique *J. Manuf. Process.* **45** 9–21
[8] Sathishkumar M and Manikandan M 2020 Influence of pulsed current arc welding to preclude the topological phases in the aerospace grade Alloy X *Proc. Inst. Mech. Eng. Part L J. Mater. Des. Appl.* **234** 637–53
[9] Sathishkumar M and Manikandan M 2020 Development of Pulsed Current Arc Welding to Preclude Carbide Precipitates in Hastelloy X Weldment Using ERNiCr-3 *J. Mater. Eng. Perform.*
[10] Kumar PN C M 2015 Effect of Pulsed TIG Welding Parameters on the Microstructure and Micro-Hardness of AA6061 Joints *J. Mater. Sci. Eng.* **04** 4–7
[11] Sathishkumar M, Naiju C D and Manikandan M 2019 Investigation of Metallurgical and Mechanical Properties of Hastelloy X by Key-Hole Plasma Arc Welding Process *SAE Tech. Pap.* 1–5
[12] Manikandan M, Raj A D, Kumar M S, Arivazhagan N, Gunachandran R, Kumar J K M, Vignesh V and Yoganathan D 2018 Investigation on Microstructure, Micro segregation and Mechanical Properties of ATIG welded Alloy C-276 *Mater. Today Proc.* **5** 6702–10
[13] Manikandan M, Sasikumar P, Arul Murugan B, Sathishkumar M and Arivazhagan N 2015 Microsegregation Studies on Pulsed Current Gas Tungsten Arc Welding of Alloy C-276 *Int. J. Sci. Eng. Res.* **6** 33–8

IOP Conf. Series: Materials Science and Engineering 1123 (2021) 012064 doi:10.1088/1757-899X/1123/1/012064

Fuzzy Logic-based Digital Hydraulics Control of Blade Pitch Angle in Wind Turbines

V Lakshmi Narayanan[1] and *R Ramakrishnan[2]

[1, 2] Department of Design and Automation, Vellore Institute of Technology, Vellore 632014, Tamil Nadu, India.

*Author to whom any correspondence should be addressed

ramakrishnan.rc@vit.ac.in

Abstract. The pitch control system is generally employed in a wind turbine to mitigate load and maintain uniform power generation at above-rated wind speed regions. The hydraulic system has more power to weight ratio and so it is incorporated in the pitch system of large scale wind turbines. Some of the issues related to the usage of conventional valves in the Hydraulic Pitch System (HPS) are: internal leakage, throttling losses, high power loss, less fault-tolerant, requires high manufacturing tolerance, and more sensitive to contamination. To overcome these issues, digital hydraulics technology should be introduced as Digital Hydraulics Pitch System (DHPS). Commercially, Proportional Integral Derivative (PID) controller is used as a pitch controller but these controller performances don't hold good when excessive disturbance or change in operating point occurs in the system. So, heuristic-based Fuzzy Logic Controller (FLC) is preferred which can surpass the PID problems. In this paper, a heuristic FLC based control strategy is proposed for a novel DHPS configuration to control the pitching action. The simulation model of DHPS is developed and system simulations are conducted. The comparative study on the effectiveness of FLC for DHPS and conventional valve controlled HPS is conducted.

1. Introduction

The necessity for renewable energy has started increasing exponentially due to the harmful effects created by non-renewable energy sources over the environment. On comparing the renewable energy sources, wind energy seems to be the most promising source of electric energy [1-3]. Since, Wind Turbine (WT) is given more priority than a conventional power plant by many countries, the technology related to WT has started to increase. To generate optimal power and also to safeguard the WT during high wind speeds, Pitch Control System (PCS) and yaw control system are mostly preferred [4-6]. Generally, PCS is employed at above-rated wind speed condition to maintain the uniform power generation and to mitigate erratic blade loads [7, 8]. The two types of PCS used in WT are: electro-mechanical and Hydraulic Pitch System (HPS). Electric motors are used as pitch actuators in case of an electro-mechanical pitch system whereas hydraulic cylinder/motor are used in the case of HPS. Since HPS are more robust to disturbance and has more power to weight ratio than its counterpart, HPS is mostly preferred in large scale WT [9, 10]. The advantage of using the hydraulic motor as the end actuator in HPS over the hydraulic cylinder is that the final pitch angle is directly proportional to the displacement of the hydraulic motor [11, 12]. Conventionally, pump control, and valve control technique are preferred in HPS [13]. The reason for considering Valve Controlled (VC) HPS to be better because it has higher bandwidth than the pump control HPS [14]. Conventional

VCHPS uses proportional or servo valves to control flow rate and uses a directional valve to control the direction of the fluid [15] as shown in figure 1.

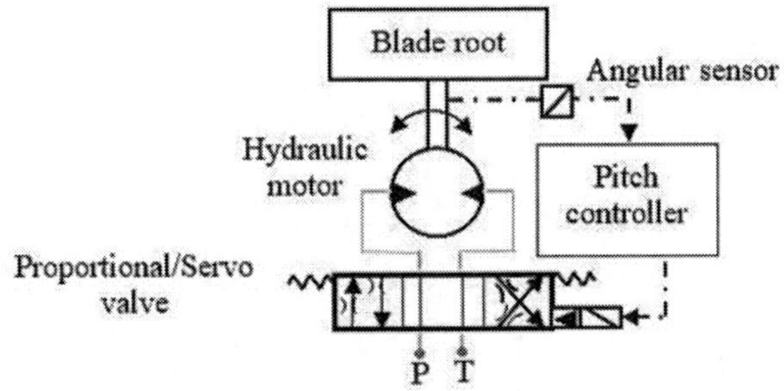

Figure 1. Conventional valve controlled hydraulic pitch system

These valves have few drawbacks like internal leakage, throttling losses, high power loss, less fault-tolerant, requires high manufacturing tolerance, high cost, and more sensitive to contaminations [16-18]. The drawback of the conventional valve controlled HPS can be overcome by implementing Digital Hydraulics (DH) technology. The DH uses a 2/2 hydraulic valve arranged in parallel combination along with various sizes of orifice known as Digital Flow Control Unit (DFCU) as shown in figure 2. Based on positioning the DFCU in the hydraulic system, DH can be categorized into few types such as: digital displacement motor and pump, switching control, and parallel valve technology [19, 20]. Here the digital motor concept is implemented. While designing the DFCU, the sizing of the orifice plays a vital role. The different methods utilized for sizing the orifice are to attain the varying flow rates are discussed in [19, 21].

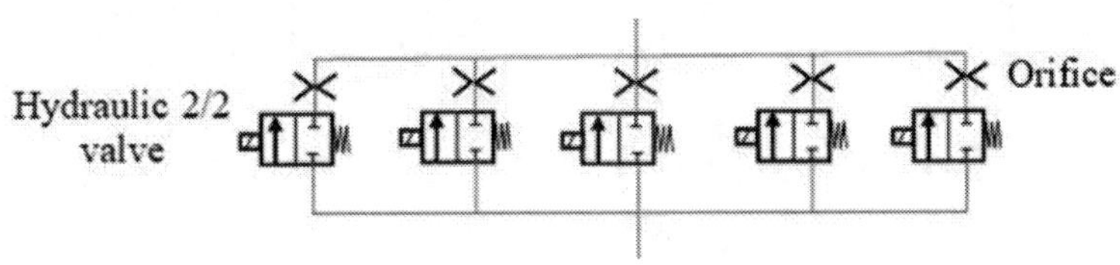

Figure 2. Digital Flow Control Unit

Though there are many advantages in the DH, the controllability of DH is still a tedious task [16, 23]. The different valve actuation methods that are implemented in the DH system are: Pulse Width Modulation (PWM), Pulse Frequency Control (PFC), Modified Pulse Width Modulation (MPWM), Pulse Number Modulation (PNM), and Pulse Code Modulation (PCM) [19, 21, 24]. P + PID controllers were incorporated in [24] to analyze the robustness in the stability of DH which is subjected to unknown load and results show that the controller has better performance. In [25] a novel fine positioning method was developed where four DFCU were used which resulted in the accurate positioning of the hydraulic system. Zero-Flowrate-Switching (ZFS) control method makes the valve to turn off when the flow rate through the valve is zero [26]. The output of ZFS was found to be effective. Though different control techniques have opted in DH, still DH control has a large scope to improve the controllability due to the extreme non-linearity of 2/2 valves. To improve the controllability by resolving the nonlinearities in the system, heuristic-based Fuzzy Logic Controller (FLC) is implemented in this paper.

The contribution of this paper involves the development of a novel DHPS for wind turbine and also FLC. The output of FLC is DFCU pair valve state selection which results in varying the flow rate and also controlling the direction of the hydraulic fluid by using DFCU so that the required pitching action takes place at the pitch actuator to achieve the appropriate pitch angle. This paper is structured as follows: Section 2 presents a description of the system. Section 3 details the modeling of the proposed system and FLC. Section 4 reveals the simulation results and discussion. Finally, section 5 discusses the conclusion.

2. System description

Figure 3 illustrates the proposed novel DHPCS which consists of FLC and DHPS. The DHPS consists of a fixed displacement pump driven by an electric motor. As discussed earlier the DFCU consist of 2/2 valves and orifice which is attached at the end of each valve.

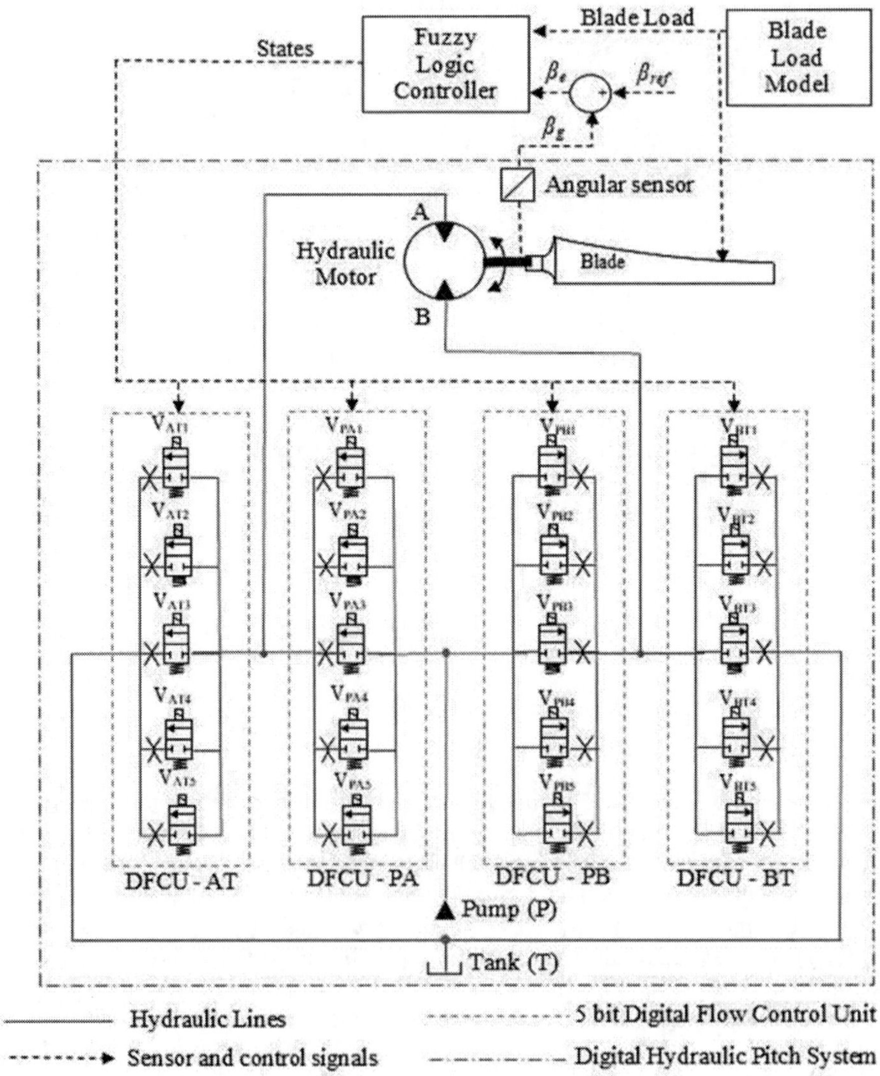

Figure 3. Proposed digital hydraulic pitch control system

The sizing of the orifice was carried out using the binary series method $[2^0, 2^1, 2^2,...,2^n]$. Since five valves are used in each DFCU, the DFCU is called 5-bit DFCU and it is possible to achieve 2^n-1 states. In this proposed configuration 2^5-1=31 states with different flowrates for each states can be achieved as shown in table 1. DFCU AT and PA are connected at A port of the hydraulic motor and the DFCU PB and BT are connected at the B port of the hydraulic motor. Based on the required direction of rotation of the hydraulic motor either DFCU PA and BT pair or DFCU PB and AT pair is triggered so that clockwise/anticlockwise rotation is achieved. The pitch angle generated at the hydraulic motor is sensed by an angular sensor and the data are fed to the FLC.

Table 1. DFCU pair valve states

V_1	V_2	V_3	V_4	V_5	States	Flow
0	0	0	0	1	1	Q_1
0	0	0	1	0	2	Q_2
0	0	0	1	1	3	Q_3
0	0	1	0	0	4	Q_4
.	.	.	.	.	.	.
.	.	.	.	.	.	.
.	.	.	.	.	.	.
1	1	1	1	1	31	Q_{31}

3. System modeling

In this section, the blade load model and FLC are modeled for the proposed DHPCS. The closed-loop control strategy of DHPCS is shown in figure 4. FLC was chosen for the proposed configuration due to its advantage over the Proportional Integral (PI) and Proportional Integral Derivative (PID) [27]. The inputs to the FLC are blade load P_l and pitch angle error β_e where β_e is obtained from equation 1.

$$\beta_e = \beta_{ref} - \beta_g \tag{1}$$

where β_{ref} is the pitch angle reference and β_g is the pitch angle generated. The output of the FLC is states

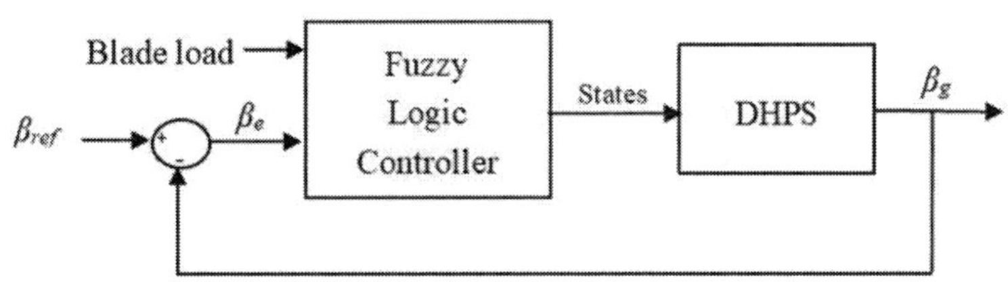

Figure 4. Block diagram of proposed digital hydraulic pitch control system

3.1 Blade load model

The load that is developed over the blades due to varying wind speed is an important parameter to be considered while designing DHPS since the DHPS should overcome the blade load P_l and has to achieve the required pitch angle. The P_l is arrived from [28] as shown in equation 2.

$$P_l = -390 + 170.9v - 92.96\beta - 0.39v^2 - 0.24v\beta - 3.73\beta^2 \tag{2}$$

The different blade loads for varying pitch angle and wind speed are shown in table 2 which is obtained by substituting the wind speed and pitch angle values in equation 2. The blade load model is developed by using equation 2 in Matlab/SIMULINK and the developed model will generate the loads which the DHPS should overcome.

Table 2. Blade load under different pitch angle and wind speed [28]

Wind speed (m/s)	Pitch angle (deg)		
	0°	5°	10°
11	1428 Nm	867 Nm	141 Nm
12	1600 Nm	1020 Nm	280 Nm
13	1765 Nm	1182 Nm	445 Nm
14	1929 Nm	1335 Nm	574 Nm
15	2114 Nm	1500 Nm	730 Nm
16	2266 Nm	1672 Nm	900 Nm
17	2420 Nm	1824 Nm	1035 Nm
18	2554 Nm	2000 Nm	1206 Nm
19	2707 Nm	2138 Nm	1355 Nm
20	2830 Nm	2307 Nm	1547 Nm

3.2 Fuzzy logic controller

FLC is based on the rules which are helpful when the dynamics of the system and also the complete nonlinearities of the system are not known. Similar to human beings making decisions, FLC applies reasoning and so the rules possess the knowledge of an expert of the system. The main advantage of FLC is that a mathematical description of the system which is to be controlled is not required. Fuzzy Logic Toolbox™ which is available in Matlab/SIMULINK. Generally, there are three stages in the FLC and they are fuzzification, fuzzy rules, and defuzzification. In the fuzzification process, the inputs are converted into fuzzy sets using linguistic terms and membership functions. Inputs and output use Triangular Membership Functions (TMF) which is shown in figure 5. TMF is highly sensitive when variables arrive at zero value [29]. Fuzzy rules are assigned as shown in table 2, where if and then statements are used to coin the rules like $IF\ x_1{=}A_1\ and\ x_2{=}A_2\ THEN\ y{=}B$. The last process in the FLC is defuzzification where the fuzzy sets are converted into precise action with real values.

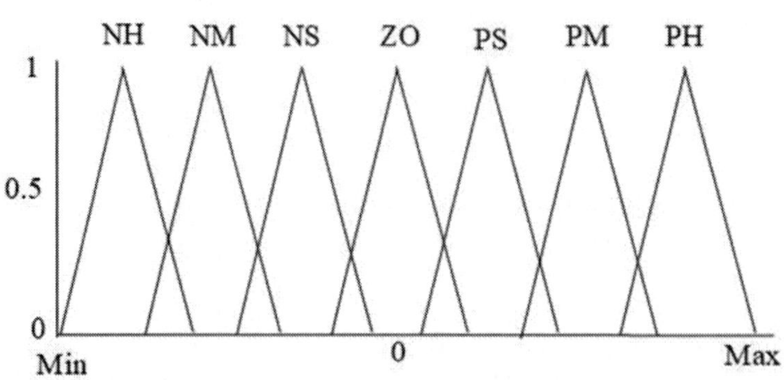

Figure 5 Triangular membership function for inputs and output

Table 3. Fuzzy rules

Blade load	Pitch angle error						
	NH	NM	NS	ZO	PS	PM	PH
NH	NH	NH	NM	ZO	PM	PH	PH
NM	NH	NM	NM	ZO	PM	PM	PH
NS	NM	NM	NS	ZO	PS	PM	PM
ZO	NM	NS	NS	ZO	PS	PS	PM
PS	NM	NM	NS	ZO	PS	PM	PM
PM	NH	NM	NM	ZO	PM	PM	PH
PH	NH	NH	NM	ZO	PM	PH	PH

4. Simulations results and discussion

The FLC performance is tested by implementing it in the DHPS. Simulations were conducted from 3 sec to 10 sec (since it took 0-3 sec for the DHPCS to initiate). Reference pitch angle data of a 2MW wind turbine was obtained from [30]. These data reveal the exact pitch angle the blade should set in the wind turbine to extract optimal power and also to mitigate blade load. Here were are not going to discuss the optimal power generated and also load mitigated. Here we are going to discuss the performance of controller tracking the reference pitch angle. A random wind profile as shown in figure 6(a) was given as an input to the reference pitch angle model to generate the required β_{ref}. Further, the states generated by the FLC is shown in figure 6(b).

The reference pitch angle generated is shown in figure 6c which was obtained by giving figure 6(a) wind profile as input. Figure 6(c) shows the comparison of β_g to β_{ref}. The β_g follows the same trend as β_{ref}. At the same time, there is a lag in terms of magnitude between β_{ref} and β_g which is due to the nonlinearities exist in the DHPS. The pitch angle error has more influence over the states, if the error is more, then higher states are assigned to DFCU so that the high flow rate is attained by DFCU to compensate the error. For more pitch error, higher states values are assigned which can be observed in figure 6(d). The maximum pitch error observed was 1.009° as indicated in figure 6(d). The minimum error shows that the FLC has a better tracking ability which was developed for DHPS. Since the wind speed is drastically changing, FLC plays a vital role in tracking the pitch angle at all conditions. Future work involves reducing the lag so that the tracking performance can be improved and also to reduce the hydraulic system initiation lag.

IOP Conf. Series: Materials Science and Engineering 1123 (2021) 012064 doi:10.1088/1757-899X/1123/1/012064

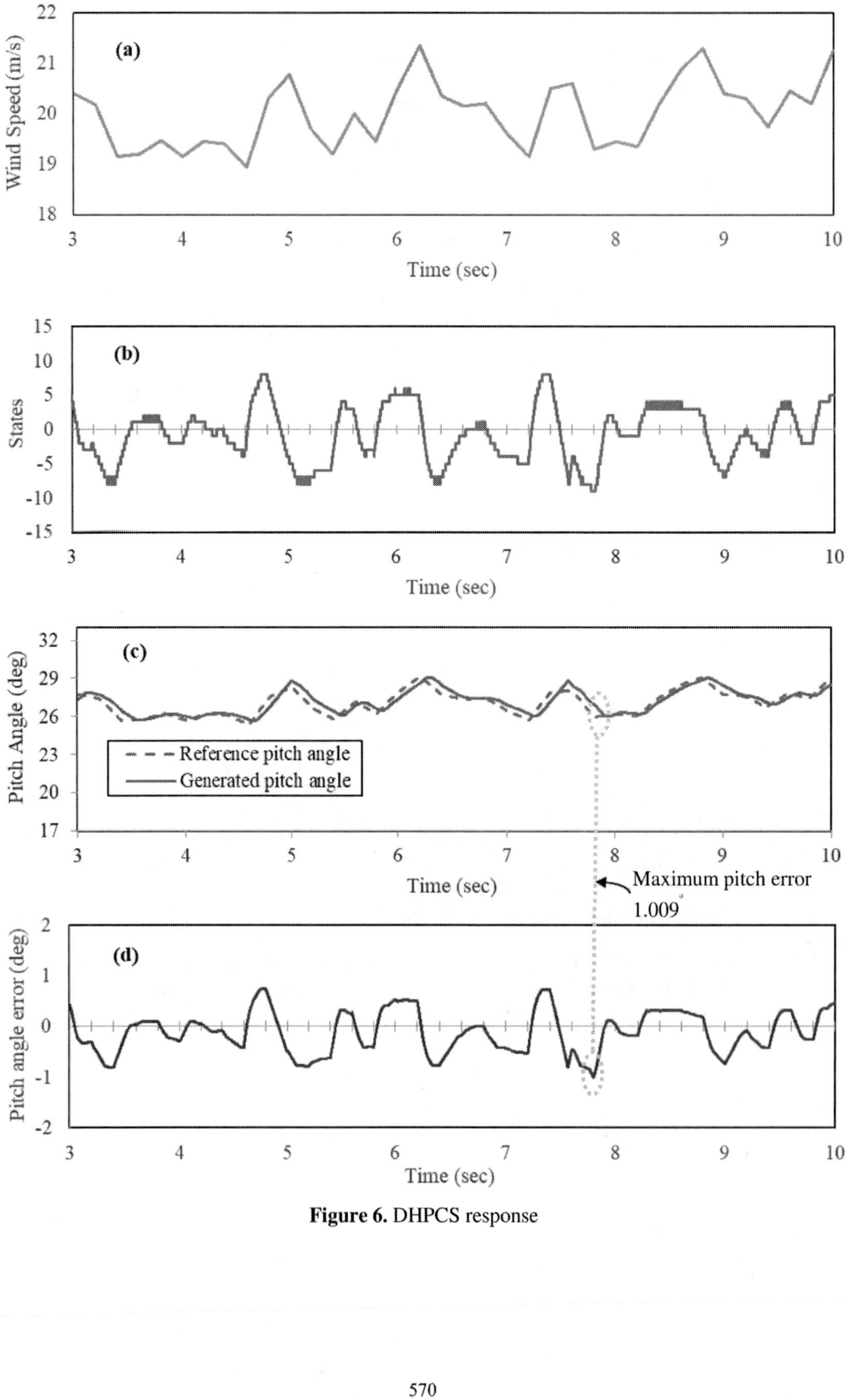

Figure 6. DHPCS response

5. Conclusion

The hydraulic pitch system delivers high power to weight ratio to mitigate load and maintain uniform power generation. In this paper, Digital hydraulics technology is implemented in a hydraulic pitch system to achieve the same performance of proportional or servo valve controlled system. The output of the fuzzy logic controller selects the DFCU pair and also the states to achieve the required pitch angle. The results show the controller has better tracking ability and the maximum pitch error observed was 1.009°. The model was tested for various wind patterns and the output was found to be effective. So by implementing the proposed DHPCS, cost-effective, highly robust, fault-tolerant, and high response pitch systems can be established in wind turbines.

References

[1] Connor B, Iyer S N, Leithead W E, and Grimble M J 1992 Control of a horizontal axis wind turbine using H infinity control. *In Proceedings of the First IEEE Conference on Control Applications*, 117-122
[2] Sholapurkar R B and Mahajan Y S 2015 Review of wind energy development and policy in India. *Energy Technology & Policy*. **2** 122-132
[3] Petersen E L 2017 In search of the wind energy potential. *Journal of Renewable and Sustainable Energy*. **9** 052301
[4] Yuan Y and Tang J 2017 On advanced control methods toward power capture and load mitigation in wind turbines. *Engineering*. **3** 494-503
[5] Tong W 2010 *Wind power generation and wind turbine design*. WIT press United kingdom
[6] Njiri J G and Soeffker D 2016 State-of-the-art in wind turbine control: Trends and challenges. *Renewable and Sustainable Energy Reviews*. **60** 377-393
[7] Fernando D B, Hernán D B and Ricardo J M 2006 Wind turbine control systems: principles, modelling and gain scheduling design
[8] Menezes E J N, Araújo A M and da Silva N S B 2018 A review on wind turbine control and its associated methods. *Journal of cleaner production*. **174** 945-953.
[9] Tong W 2010 *Fundamentals of wind energy*. WIT Press United kingdom.
[10] Yin X X, Lin Y G, Li W, Gu Y J, Lei P F and Liu H W 2015 Adaptive back-stepping pitch angle control for wind turbine based on a new electro-hydraulic pitch system. *International Journal of Control*. **88** 2316-26.
[11] Yin X X, Lin Y G, Li W, Gu Y J, Wang X J, and Lei P F 2015 Design, modeling and implementation of a novel pitch angle control system for wind turbine. *Renewable Energy*. **81** 599-608.
[12] Yin X X, Lin Y G, Li W, Liu H W and Gu Y J 2015 Adaptive sliding mode back-stepping pitch angle control of a variable-displacement pump controlled pitch system for wind turbines. *ISA transactions*. **58** 629-634.
[13] Lei X, Ni M, Li D, and Ma Y 2011 Study on simulation of digital pump-control cylinder position control system. *Procedia Engineering*. **16** 729-736.
[14] Turner C W, Babbitt G R, Balton C S, Raimao M A and Giordano D D 2004 Design and control of a two-stage electro-hydraulic valve actuation system. *SAE transactions*. 833-846.
[15] Yin X X, Lin Y G, Li W and Gu Y J 2014 Integrated pitch control for wind turbine based on a novel pitch control system. *Journal of Renewable and Sustainable Energy*. **6** 043106.
[16] Laamanen M S A and Vilenius M 2003 Is it time for digital hydraulics. *In eighth Scandinavian international conference on fluid power*.
[17] Scheidl R, Linjama M and Schmidt S 2012 Is the future of fluid power digital?. *Proceedings of the Institution of Mechanical Engineers, Part I: Journal of Systems and Control Engineering*. **226** 721-723.
[18] Laamanen A, Siivonen L, Linjama M and Vilenius M 2004 Digital flow control unit-an alternative for a proportional valve?. *In Bath Workshop on Power Transmission and Motion Control, Professional Engineering Publishing*. 297

[19] Linjama M 2011 Digital fluid power: State of the art. *In 12th Scandinavian International Conference on Fluid Power*. 18-20.
[20] Merrill K, Holland M, Batdorff M and Lumkes Jr J 2010 Comparative study of digital hydraulics and digital electronics. *International journal of fluid power*. **11** 45-51.
[21] Kalaiarassan G and Krishnamurthy K 2019 Digital hydraulic single-link trajectory tracking control through flow-based control. *Measurement and Control*
[22] Paloniitty M, and Linjama M 2018 High-linear digital hydraulic valve control by an equal coded valve system and novel switching schemes. *Proceedings of the Institution of Mechanical Engineers, Part I: Journal of Systems and Control Engineering*. **232** 258-269.
[23] Pedersen N H, Johansen P and Andersen T O 2018 Challenges with respect to control of digital displacement hydraulic units
[24] Linjama M, Huova M and Huhtala K 2016 Model-based force and position tracking control of an asymmetric cylinder with a digital hydraulic valve. *International Journal of Fluid Power*. **17** 163-172
[25] Huova M, Linjama M, Siivonen L, Deubel T, Försterling H, and Stamm E 2018 Novel Fine Positioning Method for Hydraulic Drives Utilizing On/Off-Valves. *In BATH/ASME Symposium on Fluid Power and Motion Control*. 46
[26] Peng S 2017 An Zero-Flowrate-Switching (ZFS) Control Method Applied in a Digital Hydraulic System, *In Proceedings of 15th Scandinavian International Conference on Fluid Power*. 172-177.
[27] Eker I and Torun Y 2006 Fuzzy logic control to be conventional method. *Energy conversion and management*. **47** 377-394.
[28] Wang C S and Chiang M H 2016 A novel pitch control system of a large wind turbine using two-degree-of-freedom motion control with feedback linearization control. *Energies*. **9** 791.
[29] Eltamaly A M and Farh H M 2013 Maximum power extraction from wind energy system based on fuzzy logic control. *Electric Power Systems Research*. **97** 144-150.
[30] Zhou F and Liu J 2018 Pitch controller design of wind turbine based on nonlinear PI/PD control. *Shock and Vibration*.

IOP Conf. Series: Materials Science and Engineering 1123 (2021) 012065 doi:10.1088/1757-899X/1123/1/012065

Influence of CNTs in protecting Air Plasma Sprayed Al_2O_3-3wt%TiO_2 Coated Surface

Mohammed Thalib Basha G and Venkateshwarlu B*

School of Mechanical Engg, Vellore Institute of Technology, Vellore, India-632014

*Corresponding author: venkateshwarlu.b@vit.ac.in

Abstract. The Al_2O_3-3wt%TiO_2 powders reinforced by Carbon nanotubes (CNTs) were used to prepare composite coatings by Air Plasma Spraying (APS) process. The coatings derived from APS process were varied in the percentage proportion of CNTs such as 2wt%, 4wt%, and 6wt% respectively. The coated surfaces were evaluated by means of scratch testing. The SEM micrographs of scratched samples showed the intensity of penetration during scratching gets reduced due to the presence of CNTs film over the surface. Also, the CNT reinforced coated samples showed a significant improvement in mechanical and microstructural properties such as microhardness, porosity as well as surface roughness.

1. Introduction

The coatings derived from Air Plasma Spraying process were widely functional for wear resistance applications such as surface of turbine blades, nozzle for pumps, coating of aircraft engines etc., [1-2]. Coatings derived from Plasma spraying process possess various splats to attain the requisite thickness [3]. The Alumina-Titania coatings acquired from plasma spraying process exhibits superior hardness of 1000-1200 VHN and worthy wear resistance [4, 5]. However, the ceramic coating materials still have their brittleness and moderate microhardness characterstics which makes them weak for profitable applications [6-8]. Due to this reason, only ceramic material used for coatings is not desirable. Hence, the addition of a second phase provides a solution to reduce the brittleness. In recent times, the coatings have been produced by nanostructured ceramic powders, cermet (ceramic+metal) powders and powders doped with some rare earth oxides [9]. The Multiwalled Carbon nanotubes (MWCNTs) were gaining popularity recently due to its involvement in improving mechanical properties of coatings obtained from thermal spraying process [10]. However, carbon nanotubes (CNTs) are considered as an exceptional choice for the reinforcement of ceramics with their superior properties such as physical microstructure and mechanical strength. CNTs were having a high elastic modulus of 1000 GPa, a high tensile strength of 60 GPa and thermal conductivity of 3000 W/mK [11]. It also possess unique mechanical features due to the geometry and size.

This examination is carried out to identify the crack formation phenomenon on the APS sprayed coatings obtained from conventional Al_2O_3-wt%TiO_2 powders mixed with CNTs. Also, the interpretation of toughening mechanism on the coated samples can be acquired through CNT bridging. The impact of CNT bridging on the coating microstructure and number of pores present on the as-sprayed coating while conducting scratch test, were also analyzed. The purpose of varying percentage proportion of CNTs in the coatings by 2wt%, 4wt% and 6wt%, is to evaluate the scratch resistance of coatings and to analyze the influence of CNT percentage in the scratch track.

IOP Conf. Series: Materials Science and Engineering 1123 (2021) 012065 doi:10.1088/1757-899X/1123/1/012065

2. Experimental Methods and Materials

In this present work, the thermal sprayed coating method used was Air Plasma Spraying process for the deposition of powder particles. The conventional Al_2O_3-3wt%TiO_2 powders were used for coating purpose. These powders were mixed with CNTs by varying the percentage proportion in the order of 2wt%, 4wt% and 6wt%, respectively. The scratch analysis and microstructural characterization of scratch tracks were investigated on the varied proportion of coatings mixed with CNTs. The spray parameters used for deposition of coatings is shown below in Table 1.

Table 1. Parameters for Air Plasma Spraying Process

Parameters	Air Plasma Spraying (APS)
Gun	Metco 3MB
Nozzle Type	GH
Powder Feed rate	40-50 g/min
Current	490 A
Spray Distance	(76-127 mm)
Flow rate of Primary gas (Ar)	50-60 lit/min
Flow rate of carrier gas (H_2)	8-10 lit/min

2.1. Sample Preparation

In this present work, the powder used for deposition of coatings was Al_2O_3-3wt%TiO_2. The powders were coated using Air Plasma Spraying (APS) process. The microstructural characterization and properties related to scratch analysis of the coatings were examined by varying the percentage composition of CNTs with 2wt%, 4wt% and 6wt%, respectively. The mild steel substrate of AISI 1020 (60x50x5) mm were used for the deposition of coatings. The Alex NH 500 surface grinder was used to remove excess dirt and unwanted oxide layers from the surface and also to maintain the surface roughness of sample to be 0.2 μm. The work samples were grit blasted using alumina grits of size 32 at an air pressure of 120 psi. Then the samples are to be kept in iso-propanol solution for 5-10 min to remove the foreign particles. Finally, the substrates were preheated at 200°C using Plasma Gun. Sulzer Metco 3MB plasma gun is used to deposit the coatings. Argon is used as the primary gas for generating plasma whereas hydrogen is used as the secondary and carrier gas. The mixture of Ni-5wt%CrAlY is used as the bond coating of thickness ranging from 130-150 μm were deposited first for further deposition of $Al_2O_3+TiO_2$ top coat mixed with various percentage proportions of CNTs. The thickness of ceramic top coating is ranging from 300-350 μm. The process parameters used for the deposition of coatings are shown in Table.1. For examination purpose, the specimens of 10 x 5 x 5 mm size were cut from coated samples, using a low speed diamond saw. For metallographic examinations, hot mounts with specimen inserts were prepared using Bakelite powder hot mounting process. The hot mounted samples, with coatings cross-section, were polished using Silica-Carbide abrasive papers of various grades such as 220, 400, 600, 800 and 1000. The process of polishing was carried out for a period of 15 minutes on each grade of abrasive paper and disc polishing was done for period of 20 minutes.

2.2. Characterization Procedure

Coatings microstructures were analyzed using a Zeiss Evo 60 scanning electron microscope (SEM). For a given CNT percentage, three SEM images were taken at variable magnifications of range 100X to 10,000X. X-ray diffraction studies were carried out using a PANalytical X'pertPRO PW1070 instrument with CuKα radiation, operating at 40 kV voltage and 30 mA current, scanning step of 0.01670 and step time of 0.13 s. The scratch tests were carried out using a scratch tester (Model: TR-101-M5, Made: Ducom Instruments, Bangalore, India). The scratch tester integrated with an indenter of Rockwell C type with 120° apex angle and 200 μm tip radius was used. Three scratches were made on the as-sprayed coating for microstructural investigation as well as examination of coating failure. The table 2 provides the detailed standard parameters used for scratch test [12]. Also, the most familiar technique to evaluate the porosity of a coating system i.e. preparing metallographic sample and then taking multiple micrographs for analysis was done. The image analysis was carried out by means of contrast difference between voids i.e., pores to estimate percentage of porosity [13].

Table 2. Detailed Parameters used for Scratch Test

Parameters	Scratching test
Indenter type	Rockwell C
Variable Load	30-100 N
Scratching speed	0.1 mm/s
Scratch distance	7 mm
Number of Scratches	3

3. Results and Discussion

3.1. Microstructural Characterization

The powder morphologies of Al_2O_3-3wt%TiO_2 were examined by using SEM. The characterization was carried out without CNT and after adding CNT in the proportion of 2wt%, 4wt% and 6wt%. The CNTs were consistently detached throughout the powders. It can be seen that Alumina-Titania powders are regular, wedge-shaped, and also contain pores in it whereas CNTs are looking like bundle-shaped structure (See Figure 1a and 1b) [14]. The presence of CNTs in Alumina-Titania composition can be visible at higher magnification (see Figure 1b). The XRD analysis of Al_2O_3-3wt%TiO_2 powder shows the presence of α-Alumina and Anatase Titania phases in it as shown in Figure 2a. When CNTs were added to Al_2O_3-3wt%TiO_2 powder, the carbon phase gets included. It was also found that the presence of α-alumina in the coating powder provides the resistance to high temperature and very little amount of anatase titania with tetragonal crystal structure provides stability to α-alumina.

The phases present in Al_2O_3-3wt%TiO_2 coatings after adding CNTs are shown in Figure 2b. In Alumina-Titania coatings with CNTs, the main phase observed was α-alumina which is present in the coatings without adding CNTs also. The anatase titana phase is transferred to Rutile Phase and Magneli phase after adding CNT [15]. Rutile phase is the most stable form of TiO_2 than anatase phase at all temperatures. Magneli phase is obtained due to the reduction (i.e. removal of oxygen) of titania at high temperatures. Since the phases are common in 2wt%, 4wt% and 6wt% CNT added coating systems, the XRD patterns for only 6wt% CNT coatings are shown in Figure. 2c.

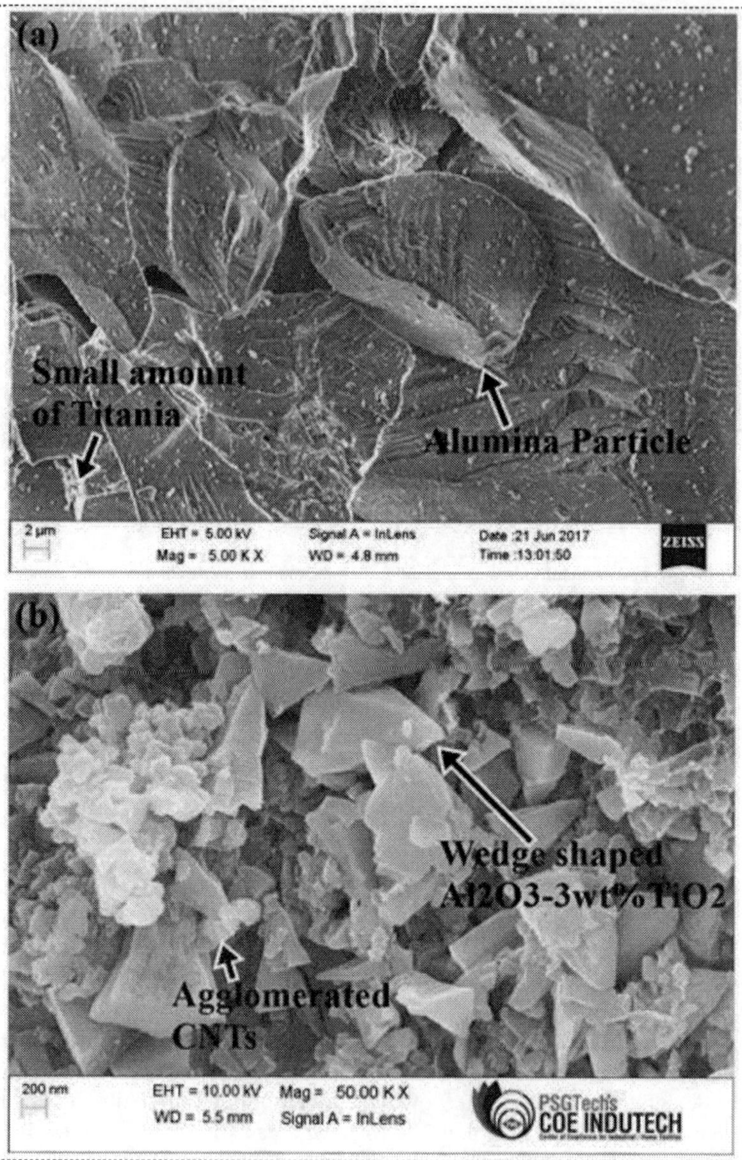

Figure 1. Powder Morphology of (a) Conventional Al_2O_3-3wt%TiO_2 Powders, (b) Conventional Al_2O_3-3wt%TiO_2 Powders with 6wt% CNTs

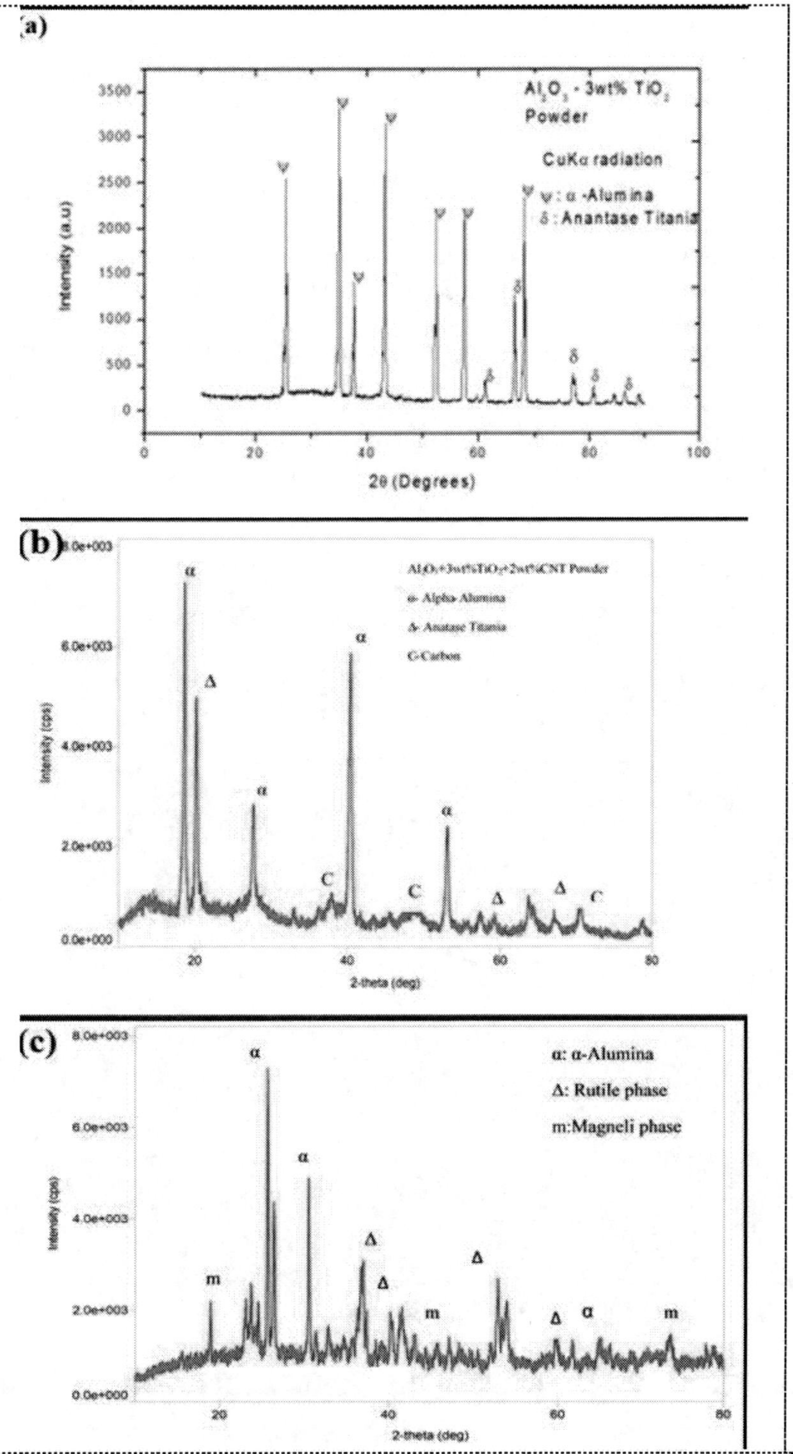

Figure 2. XRD peaks of (a) Conventional Alumina-Titania Powders (b) Powders mixed with 2wt%CNT (c) Coatings with 6wt% CNTs

3.2. Coating Scratch Characterization

The various test parameters used for performing scratch test using scratch tester have been shown in Table 2. The SEM micrographs of Alumina-Titania coating system after conducting scratch test were shown in Fig.3, Fig 4, and Fig 5. The variation of CNT percentage in the coating system were 2, 4 and 6 percent respectively. Three sets of images were taken and analyzed after the scratches are done on the coating. In that, location of failure and intensity of penetration for the percentage proportion of CNTs were examined. The back scattered electron images (BSE) were taken to interpret location of spots that initiate cracks. The brighter and darker intensity of BSE images provide meaningful information regarding greater and lower dispersion of CNTs in the coating system. In Fig. 3, the 2wt% of CNTs mixed with Al_2O_3-3wt%TiO_2 powders were coated and scratch test was performed on as-sprayed coatings. The Fig.3 (a) shows SEM image of scratch track in which an intense penetration was clearly visible due to less percentage of CNTs i.e., 2wt%. But, the intensity of penetration gets reduced with increasing percentage of CNT as shown in Fig. 4 (a) and Fig. 5 (a). The scratching failure mechanism is mainly attributed to tensile cracks. The location of failure in each percentage of CNTs can be visible in Fig. 3 (b), 4 (a) and 5 (c). The failure occurs early when the scratch starts, whereas the location of failure can be observed far from the start of scratch.

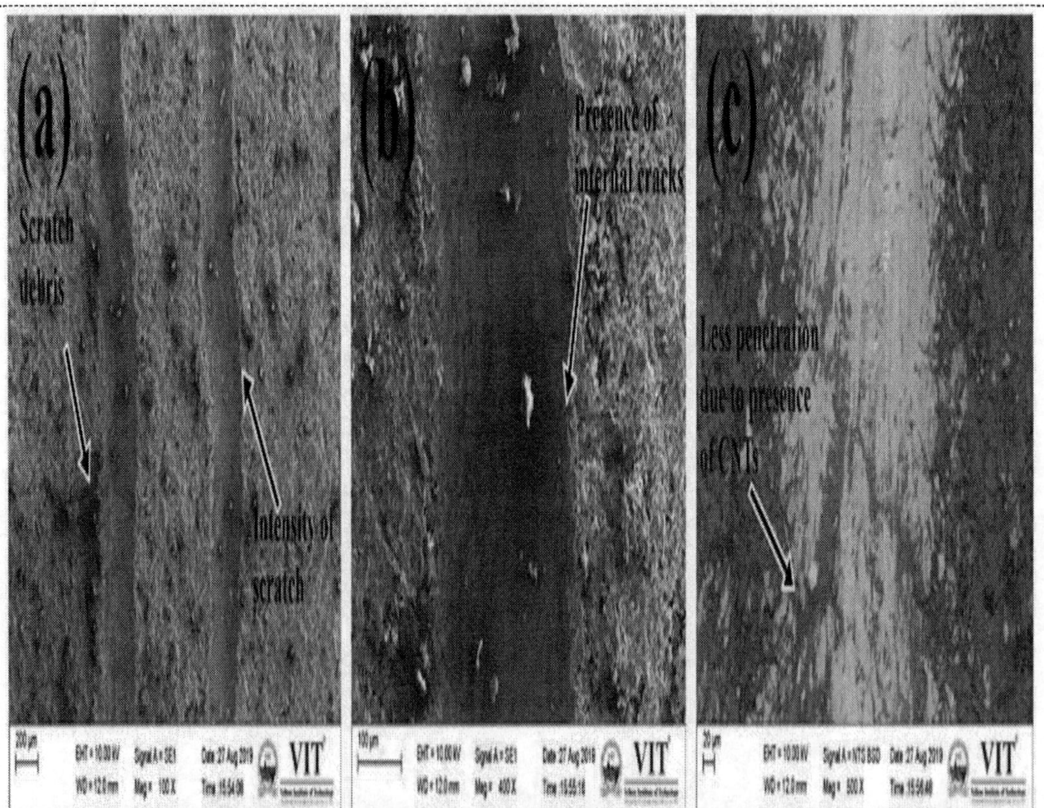

Figure 3. SEM Morphology of Alumina-Titania coatings with 2wt%CNT (a) Scratch track (b) Location of failure and (c) Back Scattered Image of scratch track

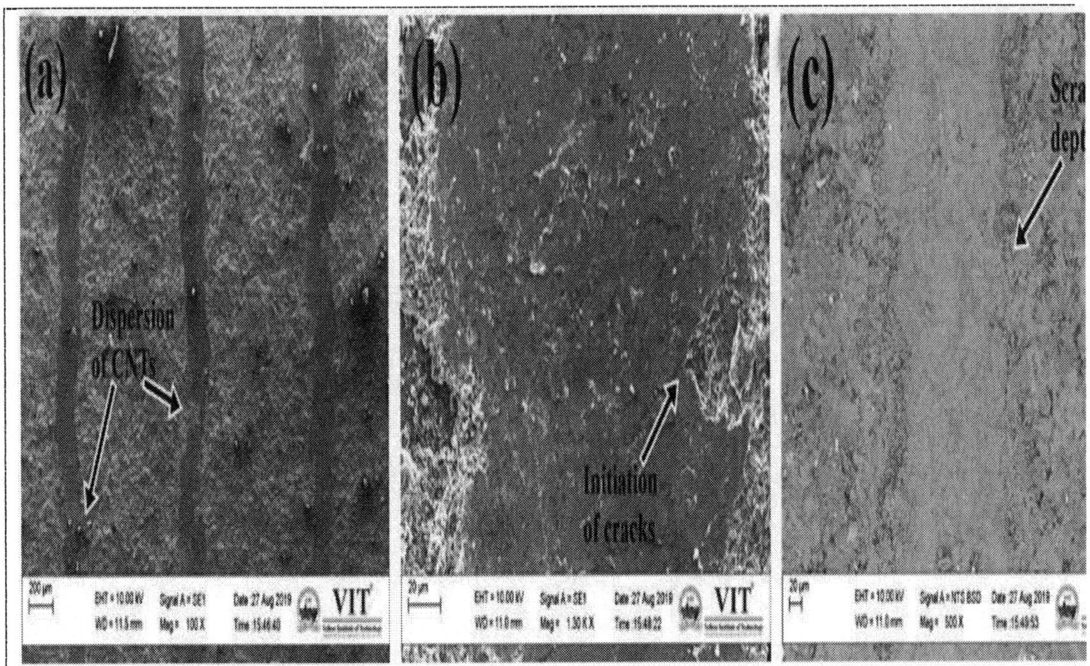

Figure 4. SEM Morphology of Alumina-Titania coatings with 4wt% CNT (a) Location of failure, (b) Scratch Track and (c) Back scattered Image of scratch track

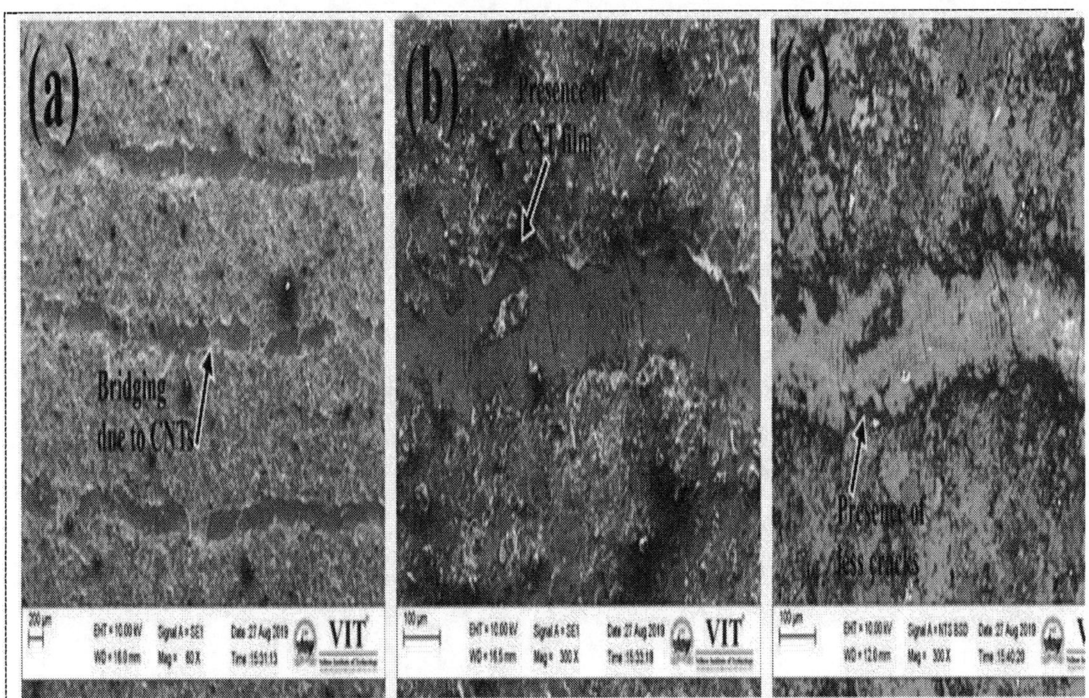

Figure 5. SEM Morphology of Alumina-Titania coatings with 6wt%CNT (a) Intensity of scratch, (b) Location of failure and (c) Back scattered Image of scratch track

3.3. Porosity Analysis

The porosity percentage has been calculated from the obtained SEM micrographs using Image J software. It can be observed from Figure 6 that the percentage of porosity decreases with the increase of CNTs percentage. It can also be noted that the porosity percentage for conventional coatings without adding CNTs was measured 7.4±0.4 percent where as for the coatings with 6wt% of CNT, the porosity was observed as 2.4±0.5 percent. This reduction in porosity can be attributed to filling of pores with CNT bundles in the coating microstructure.

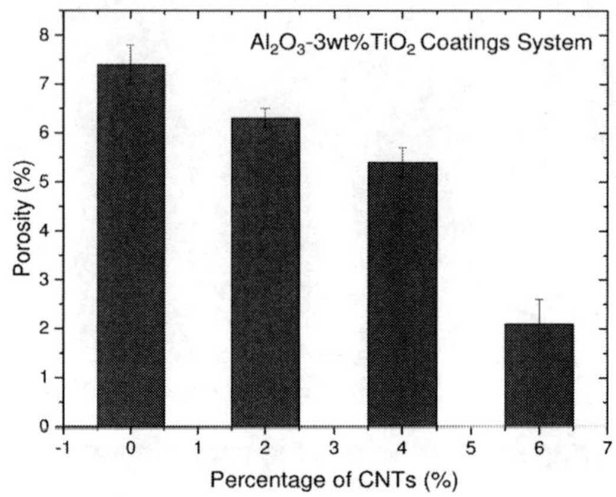

Figure 6. Porosity of Alumina-3wt%Titania coatings with different CNT percentage

4. Conclusion

The scratch tests were performed on the coatings obtained from conventional Al_2O_3-3wt%TiO_2 powders mixed with various percentage of CNTs like 2wt%, 4wt% and 6wt% respectively. The intensity of penetration during scratch test was high for 2wt% CNT mixed coating system due to the presence of maximum number of pores. The intensity of penetration was less due to the reduction in percentage of porosity, i.e., 2.4±0.5 in coating system mixed with 6wt% CNTs. Also, the location of failure which is due to tensile cracks were initiated early from starting of scratch at 2wt% CNT, whereas, the crack initiation in 6wt% CNT were delayed due to the toughening of coating microstructure. Hence, it can be concluded that by increasing the percentage of CNTs, the intensity of penetration was reduced due to CNT bridging between the splats in the coating microstructure.

5. Acknowledgement

This work was carried out with the financial support of Vellore Institute of Technology (VIT- an Institute of Eminence), Vellore, Tamilnadu, India under SEED grant scheme. The authors are thankful to VIT-Vellore for supporting this work.

6. References

[1] Balani K, Agarwal A, Process Map for Plasma Sprayed Aluminum Oxide- Carbon Nanotube Nanocomposite Coatings, *Journal of Surface & Coatings Technology* 2008; **202,** pp. 4270-4277.

[2] Wang Y, Jiang S, Wang M, Wang S, Xiao TD, Strutt PR, Abrasive Wear Characteristics of Plasma Sprayed Nanostructured Alumina/Titania Coatings, *Wear.* 2000; 237: 176-185.

[3] Bandyopadhyay P P, Hadad M, Jeaggi C, Siegmann S, Microstructural, tribological and corrosion aspects of thermally sprayed Ti-Cr-Si coatings, *Surface & Coatings Technology*, 2008; **203,**pp. 35-45.

[4] Hazra S, Sabiruddin K, Bandyopadhyay P P, Plasma and HVOF sprayed WC-Co coatings as hard chrome replacement solution, *Surface Engineering*, 2012, **28**:pp. 37-43.

[5] Sabiruddin K, Joardar, P. P. Bandyopadhyay P P, Analysis of phase transformation in plasma sprayed alumina coatings using Rietveld refinement, *Surface & Coatings Technology*, 2010, **204**: pp. 3248-3253.

[6] Lima R, Marple B, Superior Performance of high-velocity oxy fuel-sprayed nanostructured TiO_2 in comparison to air plasma-sprayed conventional Al2O3-13TiO2, *Journal of Thermal Spray Technology*, **14**(2005) pp. 397-404.

[7] Lehman RL, *Overview of Ceramic design and process engineering*, Engineered Materials Handbook **4**, 1991, pp. 29-37.

[8] Kang J, Xu B, Wang C, Influence of contact stress on rolling contact fatigue composite ceramic coatings plasma sprayed on a steel roller, *Tribology International*, 2014, **73**: pp. 47-56.

[9] Pengfie He, Guozheng Ma, Haidou Wang, Qingsong Yong, Shuying Chne, Microstructure and mechanical properties of a novel plasma-spray TiO_2 coating reinforced by CNTs, *Ceramic International*, 2016, **42**:pp. 13319-13325.

[10] Hirlekar R, Yamagar M, Gharse H, Vij M, Kadam V, Carbon nano tubes its applications: review, *Asian J Pharm Clinical Research,* 2009, **2**: pp. 17-27.

[11] B. Mukherjee, O.S. Asiq Rahman, M. Sribalaji, S.R. Bakshi, A.K. Keshri, Synergistic effect of carbon nanotube as sintering aid and toughening agent in spark plasma sintered molybdenum disilicide-hafnium carbide composite, *Mat.Sci. Eng, A-Struct*, 2016, **678**:pp. 299-307.

[12] Sudhakar C. Jambagi, Scratch adhesion strength of plasma sprayed carbon nanotube reinforced ceramic coatings, *J.Alloys and Compounds*, 2017, **728**: pp. 126-137.

[13] Fowler, D.B, Riggs, Russ, Inspecting Thermal Sprayed Coatings, *Advanced Mater Proc.* 1990; **11**: pp. 41-52.

[14] J.G. Thakare, R.S. Mulik, M.M Mahapatra, Effect of carbon nanotubes and aluminum oxide on the properties of a plasma sprayed thermal barrier coating. *Ceramics International*, 2017; http://dx.doi.org.10.1016/j.ceramint.2017.09.196.

[15] S. Aruna, N. Balaji, J. Shedthi, V.W. Grips. Effect of critical plasma spray parameters on the microstructure, microhardness and wear and corrosion resistance of plasma sprayed alumina coatings. *J. Surf. Coat and Technol*, 2012, **208**: pp. 92-100.

IOP Conf. Series: Materials Science and Engineering 1123 (2021) 012066 doi:10.1088/1757-899X/1123/1/012066

A Study on Reliance of Artificial Intelligence Embodied Robots roles in Health Applications

P Mohan Kumar
School of Information Technology and Engineering VIT University, Vellore. Tamilnadu.

Email: pmohankumar@vit.ac.in

Abstract. A ramifications of PC usage to demonstrate and/or recreate insightful conduct is called man-made consciousness. As a rule where the improvement and examination that learn and/or to perform clever conduct with insignificant human intercession. In this paper the scientist centre around the information encapsulation in mechanical framework with man-made reasoning generalizes the estimation of realities substantially more powerfully than different elements, theoretically changing the associations among individuals or in any event, moving specialists. An example investigation of a huge clinic that has connected with various Computer based intelligence computerized systems in various bits of its restorative administrations organization office was taken for sorts of data epitome, each with an obvious fixation and sees how this AI robot structure supports ability to get information, process it and give a well-described respect the end-customer. Further an investigation was made on the connection between avoidance or treatment procedures and patient's results dependent on the perplexing calculations sent in the AI automated framework. Challengingly, scientist endeavours to legitimize the focal issue could be caught by the inquiry "How does strolling identify with intuition?"

1. Introduction

In this paper, we look into the human factors consequences of fielding restorative facility organization robots that will essentially reason about which tasks to perform and when to perform them. Investigations of human-computerization connection in flying – another security basic area – have demonstrated that human directors can improperly trust in and depend upon suggestions made via robotization frameworks [1]. For instance, various aeronautics episodes have been credited to human over dependence on flawed computerization. Different investigations have inspected the impacts of changes in framework dependability, and found that it prompted problematic control assignment techniques and decreased degrees of trust in the frameworks [2, 3]. There is likewise proof that recommendations gave by encapsulated operators induce over-trust and improper dependence [4]. This worry is a basic boundary to fielding savvy clinic administration robots that step up to the plate and take an interest with medical caretakers in basic leadership. The impacts of exemplification was seen on conviction and need essentially incorporate target evaluations of need and assignment execution notwithstanding emotional appraisal of the clients trust in the framework [4, 5, 6, 7].Analysing the pervious results provide by the researchers they concentrate not only on positive aspect but also on the negative issues amid the response of robots based on the decision making systems, AI rules, and human response to the system. In general the problems were categorized as Type I and Type II errors and made progress to improve the system performance. The basics of AI role in health care will be discussed in the below passage and follows the robotic system that how this robot get embodied with AI rules and perform the applications in the real world. Further the samples with experimental proofs which concluded as advantages, disadvantages with future extension and solution the quote stated as "How does walking relate to thinking".

2. Role of AI in health care system

The essential point of wellbeing related simulated intelligence applications is to investigate connections between counteractive action or treatment systems and patient results. Mechanized thinking (man-made knowledge) is a general term that recommends the use of a PC to show just as mirror brilliant direct. Man-made awareness (man-made reasoning) in restorative supervisions is the use of complex estimations and programming to duplicate human perception in the assessment of confounded helpful data. Specifically, computerized reasoning is the limit with respect to PC counts to derived finishes without direct human information. Research in man-made intelligence centres around the improvement and investigation of calculations that learn or potentially perform insightful conduct with insignificant human mediation. Information epitome, taking a social informatics viewpoint, alludes to the change of information into a structure wherein its worth gets apparent. What perceives PC based knowledge development from standard headways in therapeutic administrations is the ability to get information, process it and give a well-described respect the end-customer. PC based insight does this through AI counts. These counts can see structures in lead and make its own one of a kind method of reasoning. In order to reduce the space for botches, PC based insight computations ought to be attempted on and on. Man-made knowledge counts continue extraordinarily as opposed to individuals in two unique manners: (1) figuring's are careful: if you set a goal, the estimation can't adjust itself and simply grasp what it has been told plainly, (2) and computations are mystery components; counts can anticipate incredibly precise, anyway not the explanation or the why. The underneath table highlights the activity of reproduced insight in the therapeutic field by then and now.

Table 1. The role of AI in the medical field

S.no	Task	Descriptions
1.	Supervision Medical Records and Original Data	Starting stage in restorative administrations is assembling and separating information (like helpful records and different past history), data the administrators is the most for the most part used utilization of man-made cognizance and modernized robotization. Robots assemble, store, re-arrangement, and pursue data to give snappier, continuously consistent access.
2.	Continuous work performance	Supplementary unremarkable endeavors should wholly remain conceivable faster than, other unequivocally by robots.
3.	Handling methodology	Fake knowledge systems have been made to dismember data – notes and reports from a patient's record, external research, and clinical capacity – to help select the right, independently revamp treatment way.
4.	Discussion based on ordinal mode	Clients report their symptoms into the application, which uses talk affirmation to investigate against a database of diseases.
5	Virtual Nurses	A propelled restorative orderly to help people with watching patient's condition and make up for lost time with medications, between master visits. The program uses AI to support patients, increasing useful involvement with endless diseases.
6	Medication Management	A wireless' webcam is united together with AI to self-overseeing attest that patients are taking their cures and causes them manage their condition. Most ordinary customers could be people with certified sicknesses, patients who will when all is said in done clash with pro admonishment, and individuals in clinical fundamentals.
7	Drug Creation	A program powered by AI was used to channel existing drugs that could be refreshed to fight the illness.

8	Precision Medicine	With the help of AI, body yields can spot harmful development and vascular contaminations early and predict the therapeutic issues people may go up against reliant on their hereditary qualities.
9	Nursing	Prosperity tracking using wearable devices.
10	Healthcare System Analysis	Utilizations AI to channel through the data to highlight messes up in drugs, work process inefficient viewpoints, and helps locale social protection systems avoid silly patient hospitalizations.

Man-made reasoning can help address huge prosperity challenges, yet might be limited by the idea of open prosperity data, and by the feebleness of AI to give some human characteristics. The usage of AI raises moral issues, including: the potential for AI to choose off base decisions. The beneath session talk about on how the AI can be exemplified with robot and sent for constant application. For our situation as partner nursing organization for basic leadership and settling tolerant needs.

3. Role of AI as embodied robot

Computerized reasoning is getting dynamically present day at doing what individuals do, anyway more beneficially, more quickly and at a lower cost. The potential for both AI and mechanical self-governance in human administrations is colossal. Much equivalent to in our reliably lives, AI and mechanical innovation are continuously a bit of our restorative administrations eco-structure. The underneath graph Fig1.shows the vision of robot encapsulated with AI for basic leadership and PC strong viewpoints.

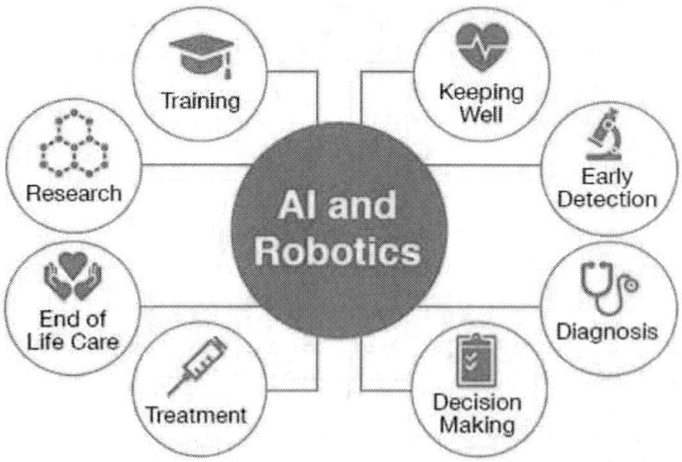

Figure 1. Vision of AI embodied robot

3.1. Role of AI embodied Robot as Associate nursing

A key test is guaranteeing that advancement of AI and using it in straightforward and good with people in general intrigue, while animating and driving development in the area. Presently we feature eight different ways that exhibit how this change is right now in progress. Bigger automated machines can be utilized to do arduous physical assignments like moving patients, and littler intelligent robots are being utilized to battle dejection and dormancy in the old populace as appeared in Figure 2.

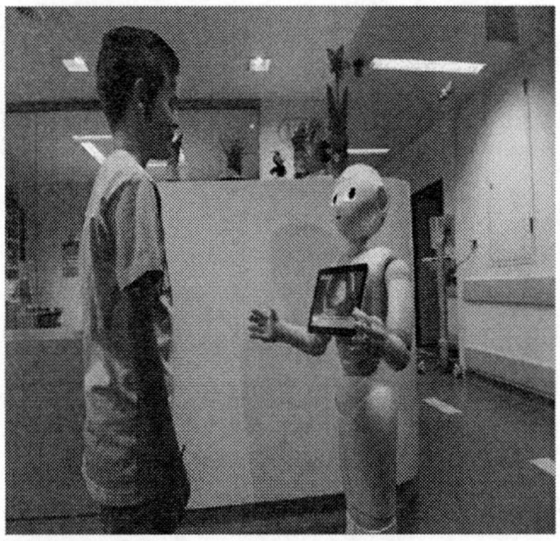

Figure 2 Robot Nurse bear

Figure 3 Nurse Robot

Meet the robot nurture bear appeared in Figure3 – Japanese apply self-rule engineers at Riken and Sumitomo Riko Labs have made a mechanized bear fit for helping care for more seasoned patients. This bear can lift a patient from standing position or from the floor, move a patient to a wheelchair, pass on a patient from show A B, and turn patients in bed.

3.2 Robots relieve some burden from Nurses

All things considered, computerized reasoning (AI) will be actualized in clinical settings quickly and on worldwide scale. Models looking like "robot medical caretakers" could have both negative and positive effect. A potential con of automated nursing is the capacity for canny online programmers to penetrate the human services databases and access delicate private data. Another conceivable downside of mechanical nursing help is the absence of "human touch." And obviously, more robots as medical caretakers may mean less interest for attendants, albeit most likely impossible. All things considered, the guide of automated nursing will simply enable attendants to satisfy increasingly significant obligations and work stressors would diminish. Besides, the potential advantages of automated nursing far exceed the disadvantages. Automated medical attendants will be able to triage patients in centers, crisis divisions, and through telehealth benefits so as to streamline mind and furnish institutionalized ways to deal with side effect the board with far less assets. With AI as a mechanical assembly to help treat and foresee sickness, the amount of facility assertions and complexities that result from nonappearance of preparing or access to prosperity organizations could lessen. Man-made knowledge will be associated with singular PDAs to help patients at home hold quickly to medication frameworks, pursue ailment the board, check vitals and blood glucose when appeared, etc.

4. Is improved Patient Care performed by the partner nursing approach?

Mechanized nursing [9] will be used as a phase for industrious guidance as well. In centers, outpatient settings, and at home, patients will have the choice to get to a few chronicles spreading out remedy utilizes, side effects, contamination the administrators, reinforce gathering, excited assistance hotlines, and all at the squint of an eye in any language. Robots are altered with estimations that diminishing calculation errors to in every way that really matters zero, so it will over the long haul be progressively secure to twofold check drugs with a robot – even chemotherapy and blood things. With less weight on medicinal orderlies and improved nature of care for patients, composed exertion with sustain robots will improve current examples of nursing inadequacies and dangerous patient extents. Robots will have the choice to help with discharge masterminding and systems. They will be joined into all center progressions and screens with the objective that we can earlier and even more unequivocally distinguish signs of patient decompensation. Thusly, patients will see improved prosperity results. Huge data surges of information set away by AI [10] will empower restorative administrations providers to have congruity of mind and depend less on the helpful history given by patients. Robot medicinal chaperons will help us with booking tasks during our nursing shift and compose our thought with the snap of a catch. Graphing will end up being to a lesser degree a weight and less dreary for therapeutic guardians considering the way that most will be therefore relegated to AI. With everything to be amped up for, there are parts of nursing that can't be supplanted by robots (or so we may think). Our capacity as medical attendants to get setting, translate concealed feelings, perceive suggestions, reflect sympathy, and follow up on instinct, are on the whole natural and humanly abilities that drive our prosperity as attendants. It is to be seen whether mechanical technology can parallel these qualities, however until the opportunity arrives, there is as yet momentous incentive in those individual traits.

5. Experimental Analysis

In order to explore the study made on how AI embodied robot helps in health care system the researcher took a case study on the real world scenario. A tertiary care hospital data set was observed for sample and analyzed. Especially human-subject experimentation planned for looking at trust in and reliance upon an encapsulated robot collaborator versus PC based choice help in a populace of doctors and enrolled medical caretakers. This mockup gave clients the chance to expect the jobs and duties of an asset nurture. An example occupation of distinguishing legitimate enlisted patient, for example, day schedule vacancy specialist in MRO segment (Medical report square) which included doling out work nurture and moving patients all through different consideration offices inside the office. The analysis utilizing an inside subjects structure that controlled two free factors: exemplification – subjects got exhortation from either a robot or a PC, and suggestion quality – subjects got high-or low quality guidance.

5. 1. Theories and Methods

Speculations and Methods T1 Rates of proper consistence with and dependence on mechanical choice help will be practically identical to or more noteworthy than those watched or PC based choice help. Target proportions of consistence and dependence were evaluated dependent on the members' "acknowledge" or "dismiss" reaction to every choice help suggestion. Examination member envisioned getting counsel from the mechanical choice help more well by clients than PC based intelligent frameworks. We hypothesized that the automated framework in our investigation would evoke this good reaction (T2), while causing suitable paces of consistence and dependence (T1). This would show a positive sign for the fruitful appropriation of a medical clinic administration robot that takes an interest in basic leadership.

5. 2. Observations prepared

Observation were prepared based on the proposed algorithm as performance analysis on monitoring the task. A sample set of work performed in the MRO(reception section and OPD)for a private registration alone for a day slot(8hrs) was taken and simulated using set of codes, A comparison was then studied among the person and robot measured the performance by the opinion of the patients arrived as satisfied/unsatisfied/may be improved. The sample of work involved is identified the patient ID, time, doctor. These data will be preloaded in the system and it will be controlled by the AI-based program. The system can also learn data from the patient record and get updated by default and response to as well as guide them for further progress such as go to concern room for prior test/pharmacy/go for scanning/come afternoon similar. These sample we tested both manually and AI embodied robot system. The observed result were plotted as graph shown in Figure 4 and Figure 5.The work comparison chart was shown in Fig 4 where the task performed by the AI based nurse and a physical nurse, based on the category of the patients taken as sample and plotted. Here it was observed that AI embodied system seems to be optimistic. But it was not satisfactory thus the patients' response was observed and a sample chart is plotted as shown in Figure 5.here it seemed to be few satisfied, un satisfied and partial as well as vice-versa. Results were shown in Table 2: and graphs relevant. These sample supports to improve the performance of the system work i.e. these data sets will be fine-tuned and utilized for learning aspects of an AI robot to perform future task in an optimized way.

Proposed Algorithm

```
Performance analysis on monitoring the task
--------------------------------------------
Input: Pi patient ID,
Po patient option in choosing the nurse
(robot_nurse (Rn)/Physical_nurse (Pn))
Output: Ps patient satisfaction based on completion of Task.
Read the patient option;
If (Patient option == Rn)
Goto 10:
  Else
Goto 100:
z=observation ();//response by the nurses.}

Observation_status(z)
{ if(number_of_task=n)
   {
    Display(task completed and patient satisfied)
   Else
    Display( pending task and patient unsatisfied)}
}
10: //Task by Robot nurse (Rn)
Robot_nurse_work()
   { While (1)
      {Check the patient Id
        If (valid)
            for ( i=0;i<n; i++)
              {  task=0;
                 Assign (room_no, time, doctor_id,
chart_relevant);
                 task++;
}return( );}

100: //Task by Physical nurse (Pn)
Physical_nurse_work()
   { While (1)
      {Check the patient Id
        If (valid)
            for ( i=0;i<m; i++)
              {  task=0;
                 Assign(room_no, time, doctor_id,
chart_relevant);
                 task++;
          }return( m);}
```

IOP Conf. Series: Materials Science and Engineering 1123 (2021) 012066 doi:10.1088/1757-899X/1123/1/012066

Table 2: Patient Status.

Sno	Nurse_type	Satisfied	Patient status Unsatisfied
1	Type I(robot)	Moderate	Moderate
2	Type II(manual)	High	Low

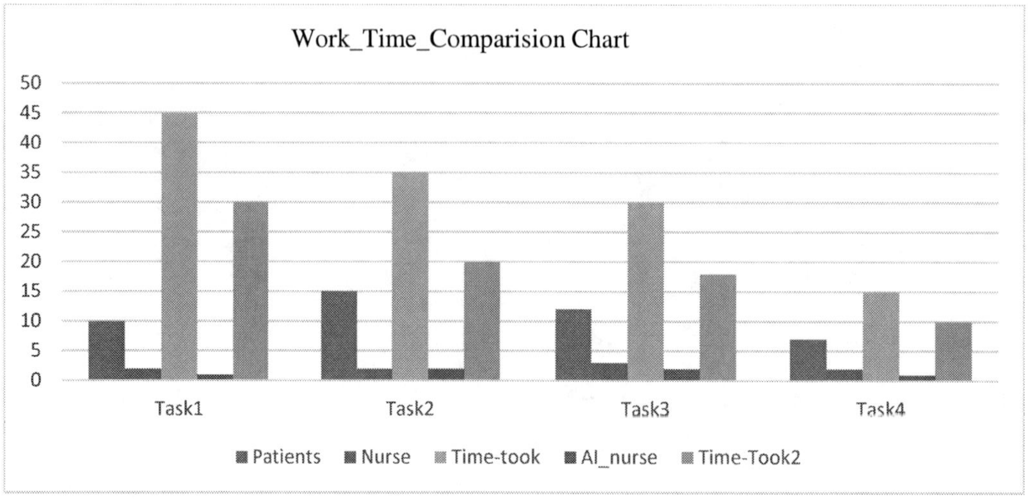

Figure 4 Work Time comparison chart

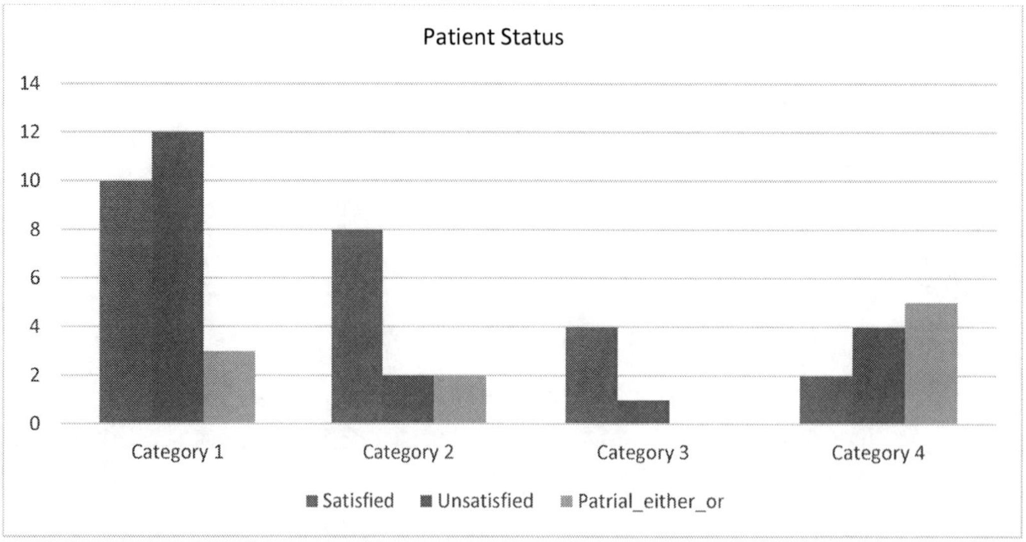

Figure 5 Patients response chart

6. Conclusion

Considering forward, numerous assignments that resemble in medicinal services will be available to rise, yet won't be completely robotized. In this paper the intensive examination was made on the human parts implications of fielding facility organization robots that will in a general sense reason about which endeavors to perform and when to perform them. Further the analyst concentrated on the information epitome in automated framework with man-made reasoning generalizes the estimation of actualities substantially more heartily than different elements, theoretically changing the associations among individuals or in any event, moving specialists. Test contextual analysis was made on constant informational index and perception were plotted in diagrams. Later Concludes as The development of mechanization will empower new bits of knowledge into social insurance forms. . So, these stages will encourage making the association between what is done, and the result accomplished, making genuine "shut circle" drug a genuine probability. Hence the arrangement the statement expressed as "How does strolling identify with intuition" so far strived might be for future augmentation.

References

[1]. R. K. Dismukes, B. A. Berman, and L. D. Loukopoulous, 2007," The Limits of Expertise: Rethinking Pilot Error and the Causes of Airline Accidents." *Ashgate Publishing.*

[2]. Munjal Desai, Mikhail Medvedev, 2012. "Effects of changing reliability on trust of robot systems. In Human-Robot Interaction (HRI)," *7th ACM/IEEE International Conference on,* pp73–80.

[3]. Aaron Steinfeld, and Holly Yanco, 2013, "Impact of robot failures and feedback on real-time trust". *In Proceedings of the 8th ACM/IEEE International Conference on Human-robot Interaction, HRI* **'13,** pp 251–258, Piscataway,

[4]. P. Robinette, A. M. Howard, and A. R. Wagner, 2016, "Over trust of robots in emergency evacuation scenarios". In Proc. HRI.

[5]. Wilma A Bainbridge, 2011,"The benefits of interactions with physically present robots over video-displayed agents*". International Journal of Social Robotics,* **3(1):**41–52,

[6]. P. McKnight, S. Scheid, 2012" The world is not enough: Trust in cognitive agents". *Proceedings of the Human Factors and Ergonomics Society Annual Meeting,* **56(1):**263–267.

[7]. Sara Kiesler, Aaron Powers, Susan R Fussell, and Cristen Torrey, 2008, "Anthropomorphic interactions with a robot and robot-like agent. Social Cognition**," 26(2):**169–181.

[8]. Brian Scassellati, 2014," Personalizing robot tutors to individuals' learning differences". *In Proceedings of the ACM/IEEE international conference on Human-robot interaction,* pages 423–430.

[9]. *European Group on Ethics in Science and New Technologies* 2018, Artificial intelligence, robotics, and 'autonomous' systems.

[10]. US National Science and Technology Council, 2016, *Preparing for the future of artificial intelligence.*

IOP Conf. Series: Materials Science and Engineering 1123 (2021) 012067 doi:10.1088/1757-899X/1123/1/012067

Design of a Sub-Surface Drilling Mechanism through Solid Modelling and Analysis

Kaustubh Mohta[1], Samarth Yelvande[2], S Venkatesan[3]

[1]UG Student, School of Mechanical Engineering, Vellore Institute of Technology, TN, India.

[2]UG Student, School of Mechanical Engineering, Vellore Institute of Technology, TN, India.

[3]Professor, School of Mechanical Engineering, Vellore Institute of Technology, TN, India.

Abstract. Current mechanisms for drilling and milling fall short in creating holes beneath the surface without disrupting bulk portion of the material. Presently, the only way to drill underground is through tunnel boring machines - each of them weighing hundreds of tons and costing millions of dollars. Clearly, they can't be used to drill small diameter holes. This paper produces a novel idea for making U-shaped holes and for drilling beneath the surface without any digging. This has wide range applications in fields of road and transport, wires and pipe installation, component manufacturing etc. and can be used for all the sub-surface drilling operations. The materials used for this purpose are HSS Steel, Carbide Steel, and Cobalt-alloy Steel. These materials are assigned to the body of the drill and FEM analysis is conducted to find the total deformation and the Von-Mises stresses induced in the drill body.

1. Introduction

Drilling is the most common type of metal cutting operation in the current industry. About 75% of the machining operations are different type of drilling. There are various methods to make a hole in a metal or other material. Depending on the requirement, the types include vertical drilling, step drilling, countersink drilling, centre drilling, etc. Among these methods, horizontal drilling is also an operation. Horizontal drilling includes extruding a horizontal hole in the workpiece by piercing the side walls of the metal by a sharp cutting edge of the cutting tool and then making the required hole. The process of making a horizontal hole in a workpiece without disturbing the side walls is very uncommon and difficult. To address this issue, a new method has been presented. In this paper, the procedure of designing the tool, working and simulation of the cutting tool (using ANSYS Workbench) is done along with the possible applications of this method. Since concepts of horizontal drilling are employed, an apparatus containing a flux sensitive coil and directional drilling control apparatus can be used as a guidance system [6]. Alternatively, manual supervision may be possible for small distances.

During drilling operation, axial thrust and cutting torque are generated. These two parameters greatly affect the viability of the drilling and the materials used for drilling operation. Relations between cutting parameters and the corresponding forces have been modelled through Artificial neural networks (ANNs) by Gurumukh Das and Padam Das [1]. Similar analyses have been performed based on ANOVA (analysis of variance) and Response Surface Methodology in the paper "Prediction of Thrust Force and Cutting Torque in Drilling Based on the Response Surface Methodology" [2]. Both these papers have been considered during design stages. Drilling requires rotational as well as translational motion. In order to transmit power for sub-surface drilling an incompressible fluid have been chosen. The is justified by the fact that fluids do not have a definite shape, making power transfer practical even over varying geometry. To facilitate analysis of fluid dynamics, moment of momentum equation has been used as described in the book "Fundamentals of Fluid Mechanics" [3]. In this paper, a horizontal, sub-surface drilling system has been designed and modelled using SolidWorks and Ansys. A load of 5KN compressive load is applied on the tool. The corresponding deformations and von-mises stresses are calculated for three different materials. Material properties are varied in order to perform structural and modal analysis on cutting tool. The different materials taken in analysis in this study are high carbon steel, cobalt-steel alloys and high speed steels (HSS).

2. Materials and Methods

As shown in Figure 1 two tubes are wound around two cylinders namely cylinder A and cylinder B. These two tubes are connected through a circular or helical pathway embedded in the drilling tool. The ends of the tubes are connected to an oil chamber. Oil flows towards the drilling cone through tube 1 and flows back to oil chamber through tube 2. A third tube is wound around another cylinder, namely cylinder C.

The drilling tool consists of two parts, a stationary cylinder and other a rotating cone. One end of the cylinder has 3 ports, two of which carry the oil used to drive the drill while the third port is connected to a tube that maintains pressure during cutting action and drives the tool forward. The interior portion of the cylinder contains a cavity which houses the rotor. The rotor arms attached to the drilling cone making the cone rotate when oil flows through the rotor. The main advantage of having a split design is that, since the cylinder doesn't rotate, the pipes attached to it won't wind up. Further design modifications to the tool are possible in case certain amount of lubrication is required during cutting. This can be done by channelling a small portion of oil to the drill cone.

The drilling tool consists of a pathway which allows flow of oil into the rotor at the base. The fluid moves through the rotor and exits into the cavity in the tool cylinder. The spent oil then moves into tube 2 from an exit port. Flute angle, cutting age, rake angle etc can be made on the conical part of the drill. Oil chamber contains the oil which is used to cool as well as drive the drilling zone. The chamber is sealed with a plunger that is driven downwards by rotating the screw using a drilling machine.

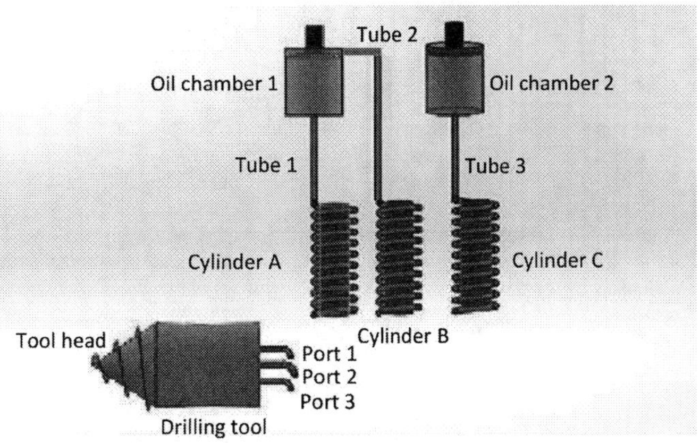

Figure 1. Nomenclature of the drilling system.

3. Results and Discussion

The tool requires rotational as well as translation motion in order to have cutting action. Translation is provided by connecting the pipe from cylinder 'C' to port 3. The other end of the pipe can be connected to a reservoir containing a pressurised fluid. The pressure force will cause the drill to move forward. However, this will also pull the pipe forward. Thus, sufficient length of pipe must be provided (At least 1.2 times the drilling length).

For rotational motion, tubes from cylinder 'A' and cylinder 'B' are connected to ports 1 and 2 respectively. As the piston moves down in the oil chamber oil enters into the drill tool through port 1. Oil then enters into the base of the rotor (as shown in Figure 2) and moves through the neck. Finally, oil exits through the rotor arms. As the exit directions of the two arms are opposite to one another, there is no net force acting on the system. Thus, application of momentum equation to the rotor arm is of no

IOP Conf. Series: Materials Science and Engineering 1123 (2021) 012067 doi:10.1088/1757-899X/1123/1/012067

significance. However, the moment of momentum equation predicts that the rotor arms will experience a couple which will cause rotation. Therefore, when oil moves out of the rotor arm, it will impart rotational motion.

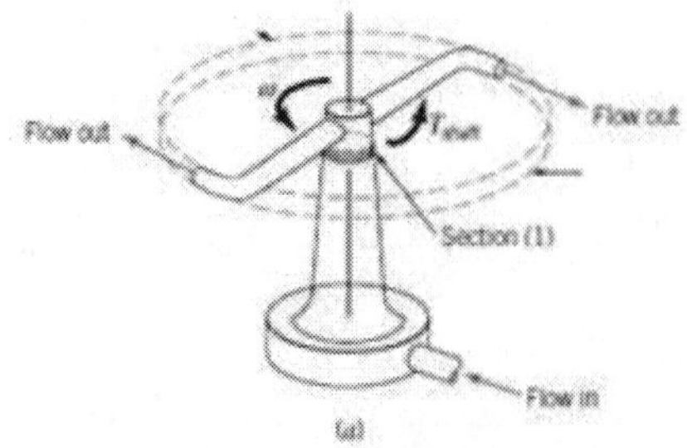

Figure 2. Rotor mechanism.

4. Design Calculations

Moment of momentum equation [3]:

$$\frac{\partial}{\partial t}\int_{cv}(r\times V)\rho d\forall+\int_{cs}(r\times V)\rho V.n\wedge dA=\sum(r\times F) \tag{1}$$

Here, r is the radius of rotor arm; V is the linear velocity of fluid leaving rotor arm; CV represents the control volume; CS corresponds to the control surface and F is force applied (refer Figure 3.).

1. The flow is assumed to be one-dimensional with an uniform distribution of average velocity at each and every section.
2. Analysis is performed for “steady in the mean” or steady cyclical flows. Therefore, at any given point of time the flow is assumed to be steady on a time-average basis (for cyclical unsteady flows).
3. Calculations are performed considering the component of equation (1) resolved along the axis of rotation.

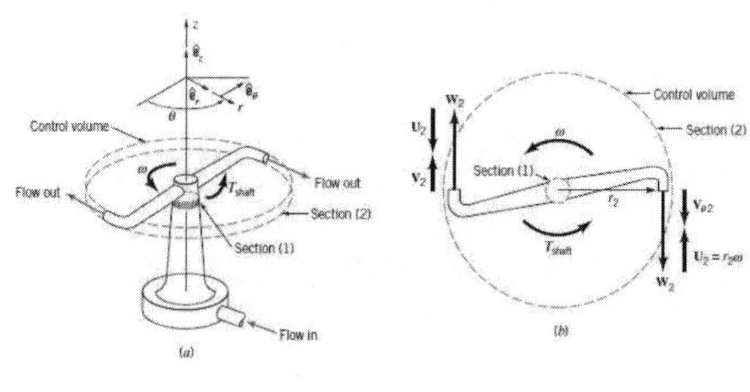

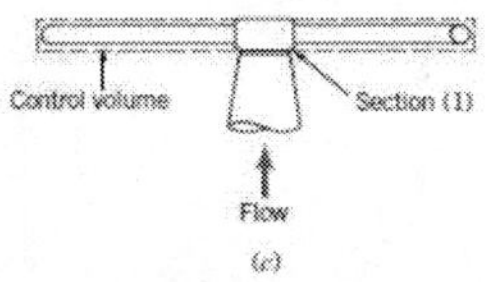

Figure 3. (a) Rotor (b) Rotor plane view (c) Rotor side view.

Considering the the rotor, as the magnitude and direction of the flow changes from rotor inlet to outlet of the rotor arm, the liquid exerts a torque on the rotor head. The causes a rotational motion or an impending rotational motion in the direction shown (similar to a turbine rotor). Applying the moment of momentum equation [3]:

$$-r_2 V_{\theta 2}\dot{m} = T_{shaft} \quad (2)$$

$$\dot{W}_{shaft} = -U_2 V_{\theta 2}\dot{m} \quad (3)$$

No load speed is [3]:

$$\omega = \frac{W_2}{r} \quad (4)$$

In the equation (4),

1. W_2 is the velocity of fluid relative to the exit nozzle
2. T_{shaft} is the external torque on shaft during cutting
3. W_{shaft} is shaft power
4. U is linear velocity of rotor arm nozzle
5. $v_{\theta 2}$ is linear velocity of fluid leaving rotor arm
6. $\dot{m}$ is mass flow rate
7. ω is angular momentum

5. Modelling of the horizontal drilling tool

SolidWorks is a computer aided design software is a with analytical tools and an easy to learn windows graphical user interface. It facilitates construction of 3-Dimensioal solid models through automatic or user-defined relations. The 3D model (as shown in Figure 4.) of the drilling system is designed using SolidWorks.[8]

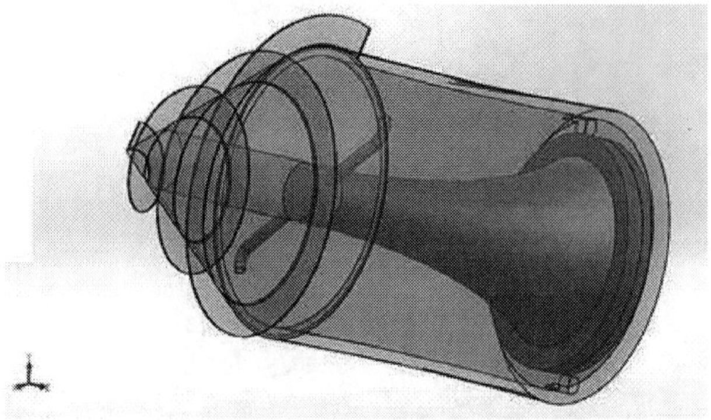

Figure 4. New Drill bit model.

a. Material Properties

The design file is imported into ANSYS workbench by converting it into IGES/STEP format. Structural analysis is performed for the drilling system. Static structural analysis is chosen and 3 different materials are used, namely Carbide steel, HSS steel and Cobalt steel alloy. The following table (Table 1.) gives the material property for the materials mentioned above.[8]

Table 1. Material properties

Parameter	Carbide steel	HSS steel	Cobalt-steel alloy
Density (kg/m^3)	15630	7972	8500
Modulus of Elasticity (GPa)	700	215	210
Poisson's Ratio	0.31	0.29	0.3

b. Analysis using ANSYS Workbench

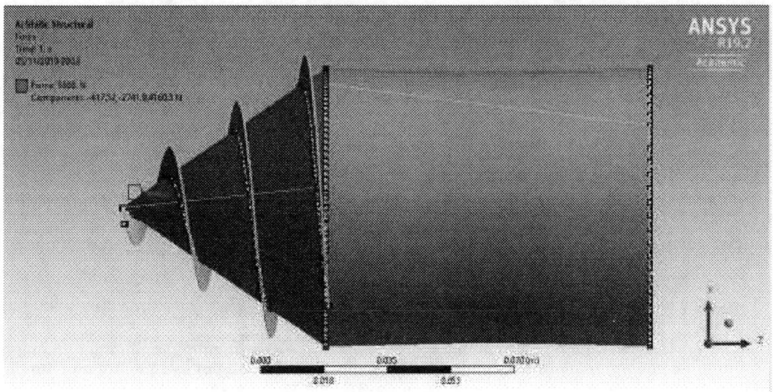

Figure 5. Load applied and Boundary conditions applied.

Once the IGES file is imported into Ansys a mesh generated. The boundary conditions and the loads are applied on the body as shown in Figure 5. and then static structural analysis is performed. The results are then simulated.

i. Carbide Steel

A mesh is generated to perform static structural analysis. A 5000N load is applied in compression, normal to the head of the drill bit. The lower end of the drilling system is defined as fixed. Material chosen is Carbide steel and the solutions for deformation and Von-Mises stress are obtained. The Figures. 6-7 show the output solutions simulated on ANSYS workbench 18.1[8].

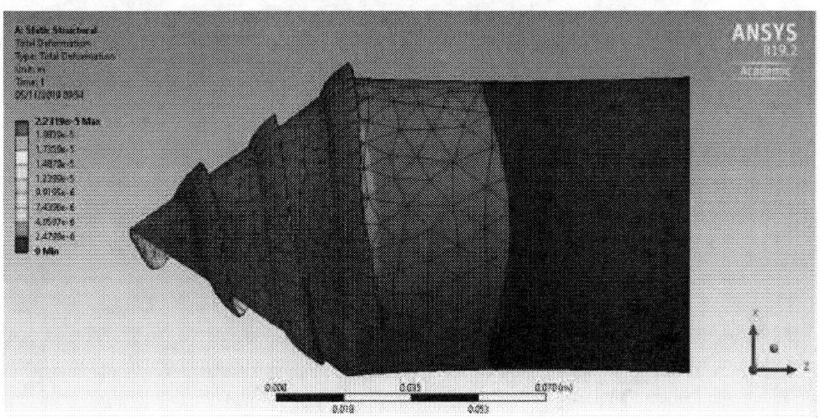

Figure 6. Total deformation of Carbide steel.

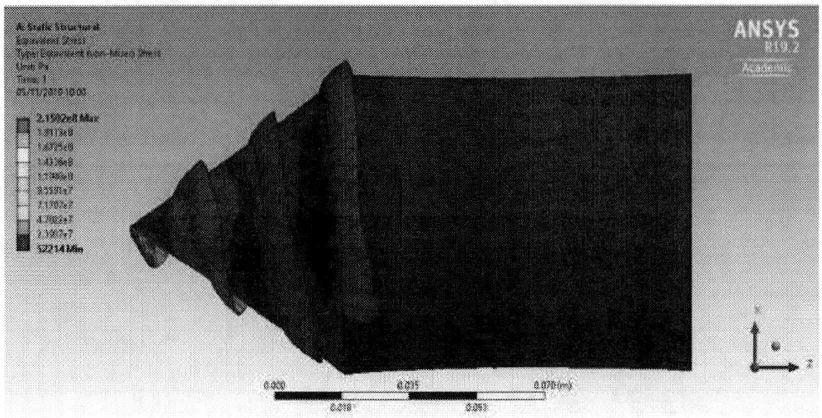

Figure 7. Von-Mises stress of Carbide steel.

ii. HSS Steel

A mesh is generated to perform static structural analysis. A 5000N load is applied in compression, normal to the head of the drill bit. The lower end of the drilling system is defined as fixed. Material chosen is HSS steel and the solutions for deformation and Von-Mises stress are obtained. The Figure. 8-9 show the output solutions simulated on ANSYS workbench 18.1[8].

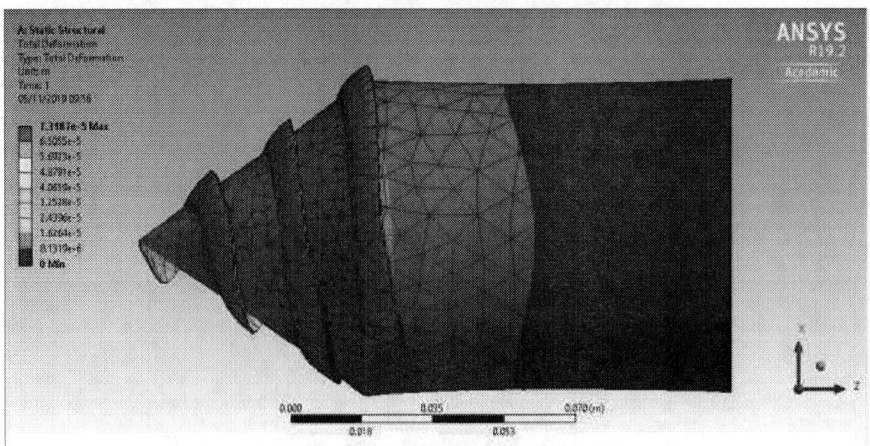

Figure 8. Total deformation of HSS steel.

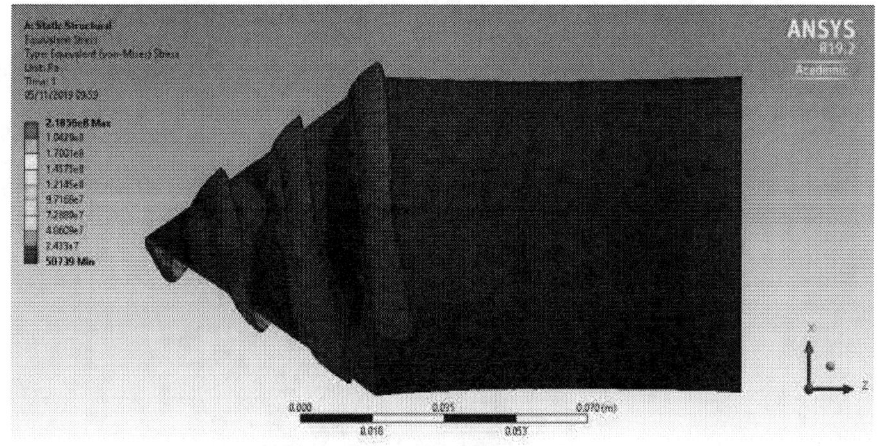

Figure 9. Von-Mises stress of HSS steel.

iii. Cobalt-Steel alloy

A mesh is generated to perform static structural analysis. A 5000N load is applied in compression, normal to the head of the drill bit. The lower end of the drilling system is defined as fixed. Material chosen is Cobalt-steel alloy and the solutions for deformation and Von-mises stress are obtained. The Figure. 8-9 show the output solutions simulated on ANSYS workbench 18.1[8].

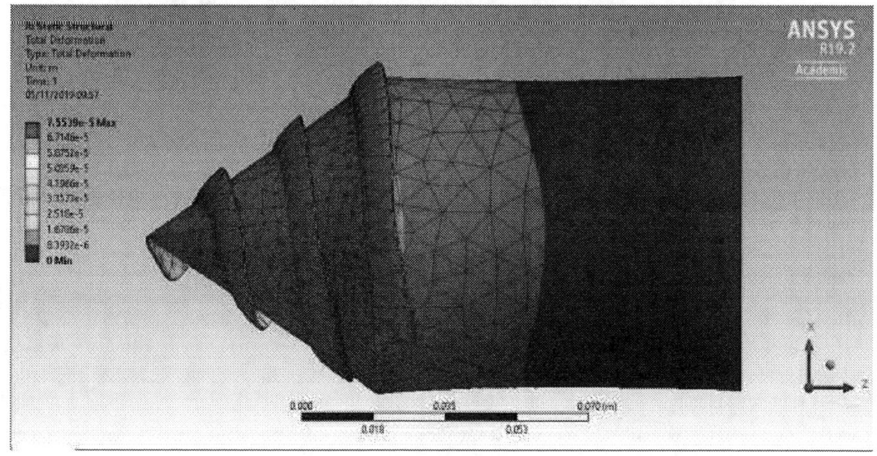

Figure. 10 Total deformation of Cobalt-steel alloy.

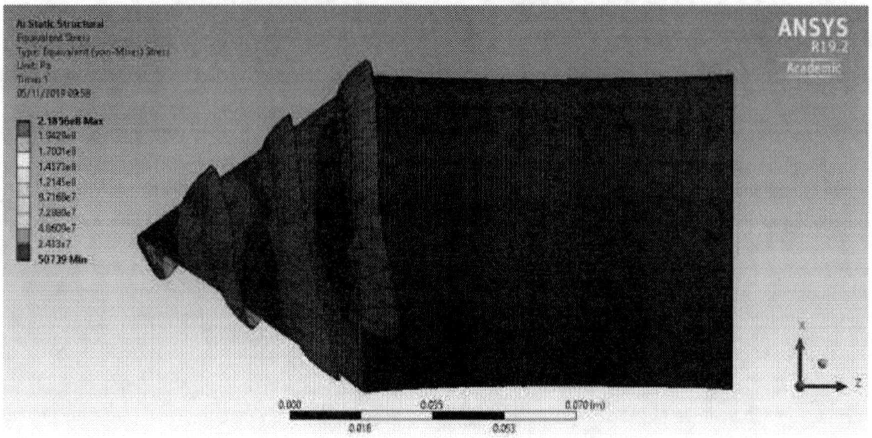

Figure 11. Von-Mises stress of Cobalt-steel alloy.

6. Conclusion

U-shaped/sub-surface drilling is a very crucial method to save material and drill in places where reaching end walls is difficult. For instance, drilling an underground hole across a road without disturbing the overall structure of the road. A practical application would be pipe laying and wire laying. Further, the concept of using fluids for power transfer can be utilised for drilling in various geometries and for making internal holes. Therefore, geometrical constraints to drilling have been addressed in this paper. From design point of view, the tool has a rotating as well as a stationary part. This prevents entanglement of tubes carrying the working fluid.

The analysis of the horizontal drilling system is done using ANSYS Workbench 18.1. The material used for the drilling system is changed for each analysis and the corresponding values are obtained. From this analysis it is inferred that the stress values of the materials used are within the acceptable range. Hence the method developed is feasible. The design can also be optimized and modified to suit the required environment. Sensors and flux sensitive coils can be integrated to serve as a guidance system for complex or large-scale applications [6]. Design modifications to the tool are possible for lubrication that might be required during cutting. This can be easily done by channelling a small portion of oil (or working fluid) to the drill cone. The method developed can serve as a basis for future developments and outcomes.

7. References

[1] Gurumukh Das and Padam Das, 2015. Cutting Forces in Drilling Operation: Measurement and Modeling for Medium-scale Manufacturing Firms. *International Journal of Computer Applications* ***121*** *8* 0975-8887

[2] Kyratsis P, Markopoulos AP, Efkolidis N, Maliagkas Vand Kakoulis K. 2018 Prediction of thrust force and cutting torque in drilling based on the response surface methodology. *Machines* **6**(2) 24

[3] Bruce R. Munson, Ted H. Okiishi, Wade W. Huebsch and Alric P. Rothmayer 2013. *Fundamentals of Fluid Mechanics seventh edition John Wiley & Sons, Inc*

[4] Maurer 1980, W.C. *Advanced drilling techniques.* United States

[5] J. Paulo Davim, 2018. *Machining: Fundamentals and Recent Advances*, Springer

[6] Dahl HD, Edmond TO, inventors; Conoco Inc, assignee. Guidance system for a horizontal drilling apparatus. United States patent US 3,907,045. 1975 Sep 23

[7] Lijo Paul, J. Babu and J. Paulo Davim 2020. Non-conventional Micro-machining Processes, Springer *InMaterials Forming, Machining and Post Processing* 109-139

[8] K. Sathishkumar, G. Dinesh, 2019. Design and Material analysis of a Suspension System in Scooter by using Finite Element Analysis Method. *International Research Journal of Multidisciplinary Technovation* **1**(1) 25-37 .

Fabrication of Bio-Scaffold by Additive manufacturing technique for Bone Synthesis using Generative Design

Renold Elsen S[1], Mahendran T[2], Piyush P Atkare[3], Amit A Bhosale[4]
[1] Associate Professor, School of Mechanical Engineering (SMEC), VIT University, Vellore.
[2] Research Scholar, School of Mechanical Engineering (SMEC), VIT University, Vellore.
[3] Masters Student, Master of Engineering, School of Mechanical Engineering (SMEC), VIT University, Vellore.
[4] Masters Student, Master of Engineering, School of Mechanical Engineering (SMEC), VIT University, Vellore.
E-mail: renoldelsen.s@vit.ac.in

Abstract. In tissue engineering, the main objective is the regeneration of damaged tissues and organs which is caused due to diseases, injuries, and inherited medical conditions. Today with the help of tissue engineering, damaged tissues and organs can be repaired with the use of artificially made structures for support, often called the scaffold. To construct a 3D structure, materials with special properties such as bioresorsablity and biocompatibility are used. This 3D structure is similar to the area of tissue implant to recover injury and regeneration of tissue. The porosity of the matrix should be maintained to promote the differentiation of cells at the control rate. Generative design is an iterative process that includes a Machine-learning program that creates a definite number of design outputs for the given loads and boundary conditions by the designer. The main objective of generative design is to increase the stiffness and to reduce the mass keeping the strength constant. In this Generative design has been used to get the optimum results using Polylactic Acid (PLA). The model of scaffold is based on the injury in femur bone thus the forces acting on the Juvenile hip joint in a one-legged stance are considered. In the last decade, intensive research had been carried out for the development of scaffold with the use of bio fabrication methods to reduce the time of manufacturing, increase precision, and decrease the cost of production.

Keywords:* - *Tissue Engineering, Bone Scaffold, Generative Design, and 3D printing

1. Introduction

Tissue engineering is using a mixture of cells, engineering, and substances with methods, as well as appropriate biochemical factors to improve or update organic tissues [1]. For the generation of new tissue in diseased/removed tissue scaffolds are used. The three key elements to be considered for the success of tissue improvement are (i) tissue creating cells (ii) supporting structure (scaffold) and (iii) cellular-matrix (scaffold) interactions that direct the tissue growth [2]. For the proper formation of bone Mesenchymal stem cells are used and it gets differentiated in the bone and starts to multiply. Generally, for the proper growth of a human cell, strong support is required which is provided by scaffold. These scaffolds are a 3D structure made of biomaterial that gives a certain surrounding for the cells in order to regenerate tissues and organs. Scaffolds provide proper directions to tissue growth and to maintain structural strength. Until the tissues are regenerated, the scaffold should be structurally stable. Scaffold biomaterials for tissue engineering may be produced in plenty of distinct strategies relying on the applications and the substances used [1]. Moreover, they are implanted in the body to regenerate different tissues and organs in the body. Irrespective of the tissue form, Physical properties, Mechanical properties, and Biological properties are imperative while designing a scaffold to be used in tissue engineering. For a 3D generation of tissue, differentiation, cell growth, proliferation and the degradation product of scaffold huge porosity and connectivity of pores are essential. The porosity is the measure of the availability of void spaces present in the scaffolds. Highly porous scaffolds are needed for the body cells to infiltrate and attach to it. Hence, the porosity of the scaffold plays a very important role in bone formation and the diffusion of nutrients within the construct. The ideal porosity of the three-dimensional structure should be more than 90%, to successfully promote cell adhesion and growth. For the better supply of oxygen and nutrients pore dimensions should be more. Generally, the scaffold pore size is required to be in the range of 300 to 500 μm [4]. Numerous scaffolds were manufactured with different biomaterials using quite a few fabrication techniques. The biomaterials used to develop scaffolds acts as a template for tissue regeneration and to support the growth of new tissue development in the affected part [5]. Among all the materials available, the most appropriate materials for generating scaffolds are the ceramics, natural and synthetic polymers. These materials are fully degradable into the body's natural metabolites providing the apt background for the rejuvenation of tissues and organs [6]. Normally replacement of bone tissues will be done by autograft or allograft. Nevertheless, harvesting autograft is costly, traumatic, and restricted with anatomical restrictions. Similarly, allografts also have some limitations in accessing the required limit of tissues for all of the patients and the transplant rejections due to the fact that the recipient's immune system does not comply with donor tissue [7]. This may also lead to the likelihood of introducing infection or disease to the patient from the donor. To overcome issues like this, highly porous biomaterials can be utilized as a permanent implant that can be synthesized according to various patients' requirements. These bio-material based scaffolds are biocompatible with the body cells and should guide the cell growth [8]. The scaffold's effectiveness is assessed by how well the cells adhere to the scaffold and the manner it interacts with proteins and other bioactive agents embedded with the scaffold. The most important consideration to be adhered to in selecting any scaffold for tissue engineering is biocompatibility. The biocompatibility is the ability of the biomaterial to adhere naturally to body cells, supports successive cellular activity, and migrate onto the surface forming a new matrix. Post implantation of the scaffold, it must act in accordance immune response to avert from developing adverse inflammatory response which might affect healing negatively or effects rejection by the body [9]. The biodegradability of bio-scaffolds is another important consideration in tissue engineering. Biodegradability is the ability of the bone substrate scaffolds to facilitate the body's tissues, over time gradually substitute the implanted scaffolds, which allows the cells to develop extracellular matrix. Also, scaffolds materials must be non-toxic and able to exit the body without being toxic or affecting the function of other organs [10]. 3D scaffold fabrication can be done with the use of 3D printing technique namely fused deposition modelling & ink-jet printing with the scaffold design developed in Computer Aided Design (CAD) software. In the above-mentioned method, a Polymeric material is melted and parallely forced out through a moving nozzle in the X and Y direction and stacked layer by layer in the Z direction. 3D printing provides the freedom to incorporate biological agents in the scaffold. Based on the above discussions the scaffold design and materials are an important factor for the regeneration of cells.

2. Materials and Methods

2.1 Scaffold Material and Design

Polylactic Acid (PLA) is used as a material for a scaffold in this work, due to its biocompatibility and biodegradable assets, which can be similar to ideal bio-scaffold properties. It is the most frequently used homogenous, isotropic, and linear elastic fabric used for 3D Printing [11]. PLA 3D-scaffolds can bear up to three compression-heating-compression cycles without delamination [12]. Lactic acid is formed when PLA is degraded which is the main reason to select PLA [13]. For Scaffold design, requirements on a macroscopic scale must duplicate human anatomy, whereas, on a microscopic level, it must accomplish short-term tissue function and in turn improves tissue regeneration. They must be balanced perfectly to have achieved impeccable scaffold design. The scaffold must be designed to have the mechanical properties consistent with the structure of the human part. It must be resilient enough for surgical handling in the course of the implantation procedure. The mechanical property needed in a scaffold will be varying in accordance with the part in which the implantation has to be done. A recent tactic to scaffold design is established on the usage of hierarchical structures formed by the replication of a unit cell of known properties and geometry [11]. This offers the opportunity to forecast the properties of the scaffolds with the known properties of the unit cell. However, the unit cell approach does not mimic the structure of the target tissue [14,15]. To overcome this limitation, generative design approach can be used to create scaffolds that can mimic the target tissue structure. A porous and interconnected microstructure, as seen in bone, can be achieved using this method.

2.2. Generative Design

The Generative Design (GD) uses the Machine learning approach by exploring the design space autonomously based on the load case provided, and then reporting again to the designer with options it considers promising for further assessment. Because a computer can process data a lot faster than a human, this sort of software permits a far deeper exploration of complex design areas. Traditionally, such an approach has been used to optimize a given model to attain maximum viable overall performance primarily based on a concrete goal. Besides exploring many format options, some other gain of the generative design technique is that we will evaluate designs at a higher degree of detail than possible with traditional processes. Once a set of interesting designs is chosen, they can be further analyzed by way of the human designer, discussed with the stakeholders, and advanced right into a final design. It is vital to note that as it follows a stochastic method based totally on sampling a restrained huge kind of designs from the design space, the overall optimal design will no longer be available through the search process [16]. In this work Porosity, Pore Diameter, Pore interconnections, and Pore Structure are the parameters considered for designing the bone scaffold. For different injuries, different scaffolds have to be modeled. The pore diameter depends on the normal bone unit (223 μm) but pore diameter ranges from 300 and 500 μm.

2.2.1Forces in femur

The scaffold design is done by assuming an injury in femur bone for which the forces in the femur during the gait cycle/walking cycle is taken for further analysis. The gait cycle discusses the whole activity from the heal which is initially placed on the ground and the following heel contact with the same feat. Thus, it considers all the static and dynamic forces acting on Femur. A two-dimensional biomechanical model of the juvenile hip joint was developed in the one-legged stance. The one-legged stance is the relevant phase during walking considering the loading of the femur [17].

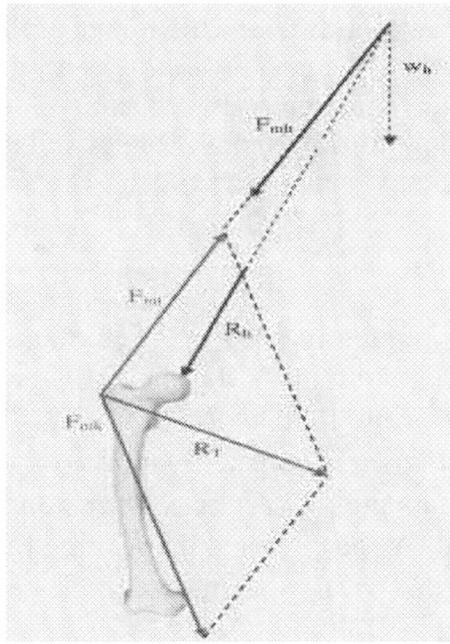

Figure 1 Forces on the femur

W_h = partial body weight, F_{mh} = force of muscle on hip, R_h = hip joint force, R_t = Trochanter resultant force F_{mk} = Knee muscle + additional muscle force

Table 1. Forces on femur Vs Inclination to the perpendicular

	Force (X bodyweight)	An inclination to the perpendicular
$\mathbf{W_h}$	53,87	0°
$\mathbf{F_{mh}}$	150,57	23°
$\mathbf{R_h}$	201,19	17°
$\mathbf{F_{mk}}$	209,63	8°
$\mathbf{R_t}$	111,63	52°

This scaffold is modeled in Fusion 360 and will serve as a reference for designing the right parameters as a starting body along with two subcomponents named obstacle and preserve. The above-mentioned model of 20x20x20mm has given the PLA material and loaded with the forces on a Femur bone for the generation of iterative models.

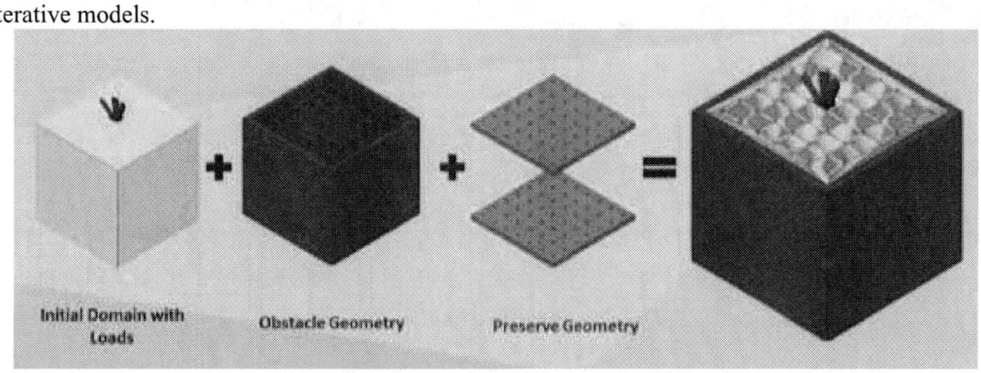

Figure 2 Load case and other constraints in Generative Design

In this study, the Factor of Safety is taken as two, and the weight of the body is taken 85 kg. The load case with the obstacle and preserve geometry is given in Figure 2. The ideal scaffold should have similar features like the extracellular matrix. Thus there is no ideal shape and size of a bone-scaffold and pores, so it is considered with a cube for the scaffold of size 20mm x 20mm x 20mm with various sized circular pores over all the surfaces and the orientation of pores is kept 90°.

2.2.2Fused deposition modeling (FDM)

The FDM method is one of the frequently used additive manufacturing techniques. The typical process flow of 3D printing begins with a CAD model, converting to an STL file the slicing for CNC code generation, and finally printing. In an FDM process, thermoplastic polymers, like ABS, PC, PA, and PLA, the material filament is constantly delivered and heated inside a moving nozzle at temperatures marginally above its melting for easy extrusion via the nozzle to form layers [18]. Through an FDM, a pore size that can be obtained by printing thermostatic polymer is around 50–1000 μm [14]. The major benefits of 3D printing are high cell density and automation than a traditional scaffold-based method [19].

Figure 33D printing of scaffolds

The developed scaffold is tested for the profile and micro features using a Digital USB microscope and the ImageJ open code is used to assess the geometric features of the scaffolds.

3. Result and Discussion

The Generative design approach gives the various design based on the design constraint provided with considering the manufacturing approach as well is done using AUTODESK FUSION 360. The design provided by the GD can be assessed and the most suitable design is selected and further analysis has been done (Figure 4).

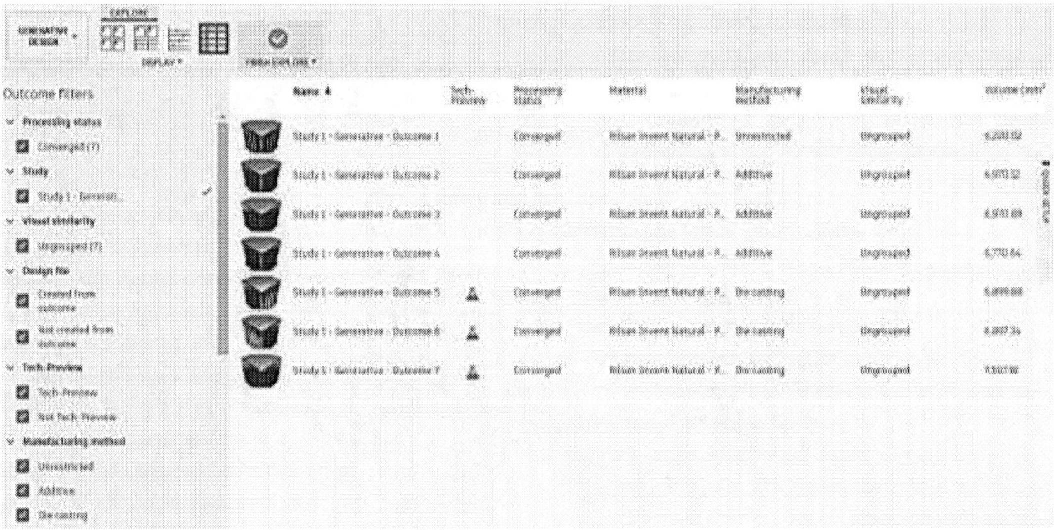

Figure 4 Design given by Generative Design

The following are the models' Iteration 1 and 2, which were obtained by Generative design with higher porosity (300 microns and 500 microns) and lower mass than the original model. In Figure 4 the design of scaffold with 20 x 20 x 20 mm^3 volume and 300 microns developed by GD with the cross-section views are given to show the interconnected pore structures as well the stress patterns based on the load applied.

Figure 5 Scaffold design with 300 micron

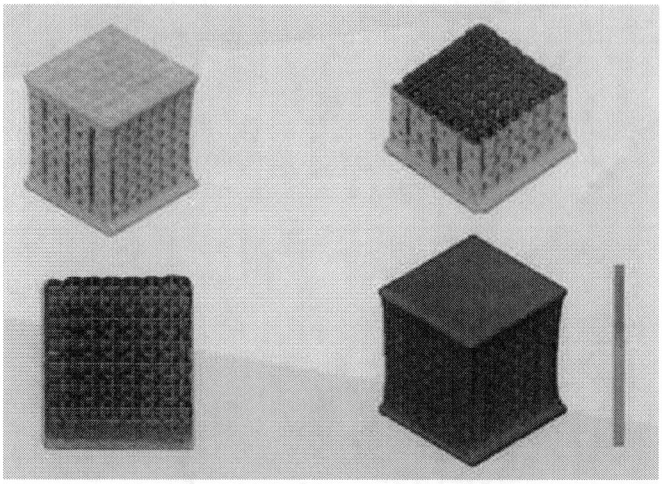

Figure 6 Scaffold design with 500 micron

The second iteration design of scaffold with 20 x 20 x 20 mm^3 volume and **5**00 microns developed by GD with the cross-section views are given to show the interconnected pore structures as well the stress patterns based on the load applied (Figure 6). The design iteration 1 and 2 had no pore opening at the top and the bottom of the structure so a third iteration was done to have pore opening at the top and bottom. The scaffold design with an opening at the top and bottom with 20 x 20 x 20 mm^3 volume and **3**00 microns is developed. The cross-section views of the final iteration with the interconnected pore structures as well as the stress patterns based on the load applied are given in Figure 7.

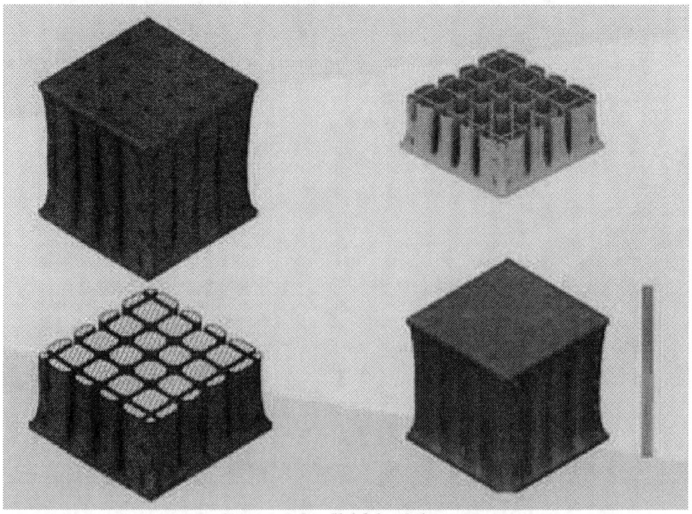

Figure 7 Scaffold final design

To assess the mechanical and biological properties of the generated design Figure 8(c) the other standard rod-type configuration Figure 8(a) and the hole type configuration Figure 8(b) are modeled in AUTODESK FUSION 360.

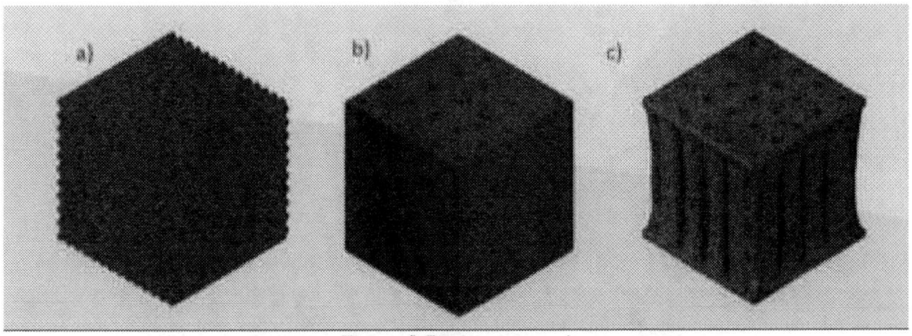

Figure 8 Design comparison

To develop the actual scaffold prototype using the FDM process a special code is required to covert the Virtual model. To achieve the above task a renowned open-source code name CURA is used and the slicing done for the GD model is given in Figure 9.

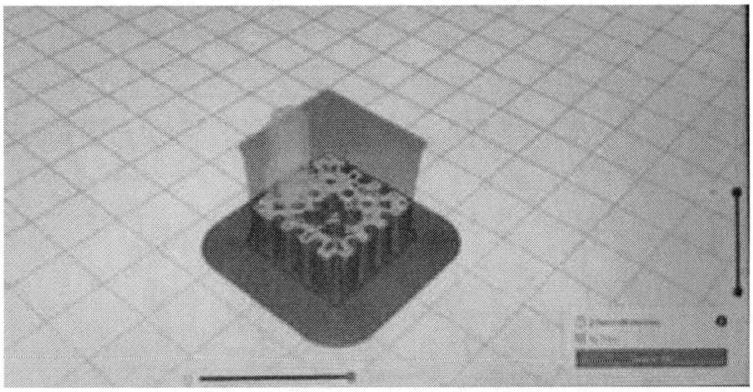

Figure 9 Layer Generation in CURA software

Figure 10 Scaffold Design Fabricated

The developed scaffold prototypes by the FDM process are shown in Figure 10 and they are used for further analysis. The various samples of scaffolds are analyzed through a USB microscope to have a microscopic view on the surface characteristics. ImageJ software was used to measure pore area, strand diameter, strand area, layer thickness, and distance between the pores both vertically and horizontally from various samples.

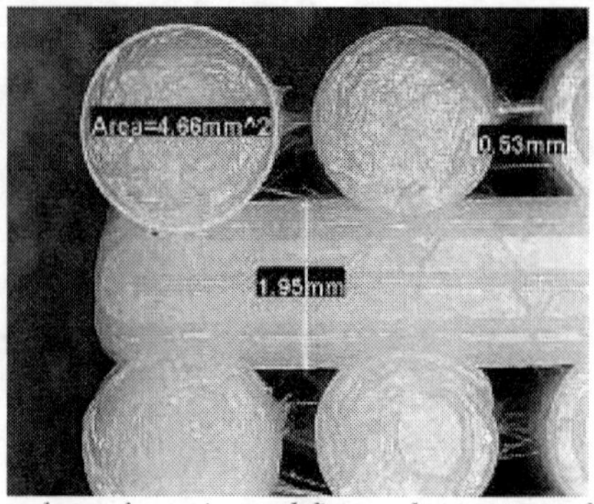

Figure 11 Strand area, layer size, and distance between strands for Rod type

The rod type scaffold configuration is given in Figure 11 in which the microscopic view of the side portion is analyzed. The area of the strand is found to be 4.66mm^2 and from that value diameter of the strand, found using the area of circle formula is 2.435mm. Then the distance between the two strands was 0.53mm. The adjacent layer to the layer of strands measures 1.95mm.

Figure 12 Pore size and distribution for Hole type

For the hole type configuration as shown in Figure 12 pore area was measured to be 23-micron pore are while the virtual model had 30-micron pore area and the distance between the pores in both vertical and horizontal direction are same and measures 3.914 mm which suggest the Pore distribution is found to the uniform.

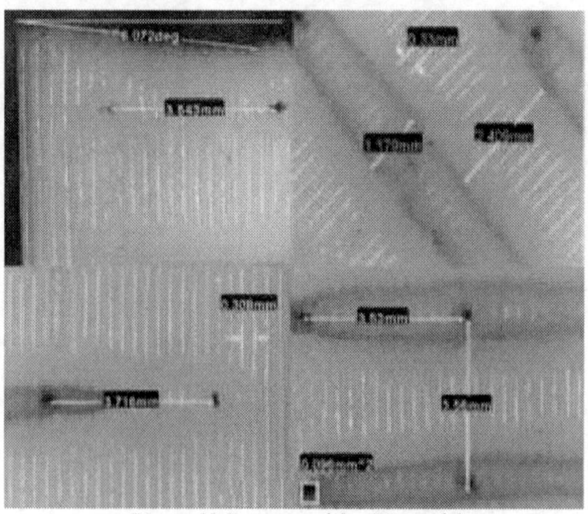

Figure 13 Features of the GD model

The GD model is fabricated and the feature generated by the FDM process is assed and reported in Figure 13. The distance between the pores along the vertical direction is measured to be 3.543mm also the angle of inclination of the side surface is 6.072^0 to the top surface. The distance between the pores along the vertical direction and the layer thickness is measured and the layer thickness is found to be 309 to 330 micron in different regions. The distance between the transverse pores measures 3.84 mm, which is slightly less than the hole type configuration. The strand length measures 2.409mm and the gap between the two parallel strands is 1.179 mm. The distance between the two pores along the horizontal direction measures 3.716 mm. The distance between the pores in both vertical and horizontal direction are almost the same from 3.716 to 3.56 mm also the area of pore over the undercut surface has a pore area 96 micro.

4. Conclusion

In this work, the Generative Design module available in Autodesk FUSION 360 was used to develop the scaffold design based on the load case and other objectives. To assess the characteristic of the design given two standard configurations namely rode type and hole type is modeled and the geometrical assessment was done using image analysis technique. The physical, mechanical, and biological characteristics are to be done in the future.

Reference

[1] Lacroix D Planell J A and Prendergast PJ 2009 *Phil. Trans. R. Soc. A* (Philos) pp 1993–2009
[2] Murugan R and Ramakrishna S 2007 *Tissue Eng.* vol 13 (New York) pp 1845–1866
[3] Chocholata P Kulda V and Babuska V 2019 *Materials* vol 12 (Basel: MDPI) p 568
[4] Mallick S Tripathi S and Srivastava P 2015 *IOSR J. Pharm. Biol. Sci.* vol 10 (Ghaziabad) pp 37–54
[5] O'Brien FJ 2011 *Mater. Today* vol 14 (Kidlington) pp 88–95
[6] Nikolova MP and Chavali MS 2019 *Bioact. Mater.* vol 4 (Beijing) pp 271–292
[7] Chen FM and Liu X 2016 *Progress in Polym. Sci.* vol 53 (Amsterdam: Elsevier) pp 86–168
[8] Eltom A Zhong G and Muhammad A 2019 *Adv. Mater. Sci. and Eng.* vol 2019 (London: Hindawi) p 1–13
[9] El-Sherbiny IM and Yacoub MH 2013 *Global Cardiology Sci. and Practice* (Doha) pp 316–342
[10] Chen CL Zeng YT Kankala RK Zhang SS and Chen AZ and Wang SB 2018 *Materials* vol 11 (Basel: MDPI) pp 1832–1852
[11] Gómez S Vlad MD López J and Fernández E 2016 *Acta Biomater.* vol 42 (Amsterdam: Elsevier) pp 341–350
[12] Senatov FS Niaza KV Zadorozhnyy MY Maksimkin AV Kaloshkin SD and Estrin YZ *J. Mech. Behav. Biomed. Mater.* vol 57 (Amsterdam: Elsevier) pp 139–148
[13] Gregor A et al 2017 *J. Biol. Eng.* vol 11 (Kentucky: IBE) pp 1–21
[14] Chua CK Leong KF Cheah CM and Chua SW 2003 *Int. J. Adv. Manuf. Technol.* vol 21 (London: Springer-Verlag London Limited) pp 302–321
[15] Johnson J Ghosh A and Lannutti J 2007 *J. App. Polym. Sci.* vol 104 (New York: Wiley–Interscience) pp 2919–2927
[16] Nagy D Lau D Locke J Stoddart J Villaggi L Wang R Zhao D and Benjamin D 2017 *Proc. Symp. Simul. Arch. Urban Design.* vol 49 (Toronto: SimAUD–2017) pp 49–56
[17] Renner S 2007 *Chair of Struc. Analy.* (Munich: Technische Universität München) pp 1–97
[18] Wittbrodt B and Pearce JM 2015 *Addit. Manuf.* vol 8 (Amsterdam: Elsevier) pp 110–116
[19] An J Teoh EM Suntornnond R and Chua CK 2015 *Eng.* vol 1 (Amsterdam: Elsevier) pp 261–268

Modelling and Numerical Simulation of Fluid Flow and Heat Transfer through Open Celled Metal Foam

K Gokul Kumar[1] and C D Naiju[2]

[1,2]School of Mechanical Engineering, Vellore Institute of Technology, Vellore, India
Email:kgokulkumar@vit.ac.in
Email:cdnaiju@vit.ac.in

Abstract - The use of open-cell metal foam in numerous technologies is increasing rapidly. Heat transfer measurements inside rectangular blocks of commercially-available aluminium foam subjected to constant heat flux at metal foam surface are presented. Cases for different pore velocities, porosities and pore densities of the foam are given. Each foam block is cooled by a confined stream of ambient air and the inlet-outlet boundaries of the computational domain were assigned. Simulations were carried out using the commercial software Fluent with a specific mass flow rate at the periodic boundaries of the domain and heat flux at the structure of the metal foam thus enabling convection. Different meshing schemes and different grid sizes were used for mesh independent to get the results with different flow velocities. Different approaches for heat transfer analysis of metal foam is considered and the most suitable approach is proposed. The temperature and the pressure drop were determined with different velocities. The heat removal rate for one unit cell was extrapolated to calculate it for the whole bed. For flow and heat transfer analysis, two fluids: air and water are used and the various results obtained are compared. Thus the variation of flow and heat transfer properties was studied with varying flow and foam parameters.

Keywords: Simulation, Open Cell Form, Heat Transfer.

1. Introduction

Only in the past twenty five years, transport phenomena in metal foam have received more attention. They are a relative new class of materials with very promising applications in which low density and other thermal, mechanical, electrical and acoustical properties make this material an excellent means of performance improvement. High-porosity metal foam continues to be applied in many new technologies, which warrants more characterization and assessment of its performance and properties. Different metal foams have been used in aerospace systems, geothermal operations and in petroleum reservoirs. Thermal management applications of foams include compact heat exchangers for airborne equipment, air-cooled condenser towers and compact heat sinks for power electronics.

Open cell metal foams are found useful for the construction of light weight structures, energy absorption devices, currently being used by some vehicle manufacturers, and for various fluid flow and thermal applications which is our interest in this work. The open high porosity (often greater than 0.9), high thermal conductivity of the solid ligaments, the large embedded surface, and the ability to generate turbulence and high-level mixing in the cooling fluid, make metal foam heat exchangers compact, efficient and light-weight. Considerable amount of work has been reported on the numerical simulation of open celled form. An analytical study of the forced convection heat transfer characteristics in high porosity open-cell metal-foam filled pipes was reported [1]. The Brinkman-extended Darcy momentum model and two-equation heat transfer model for porous media were employed. Based on the analytical solutions, the velocity and temperature distributions in metal-foam filled pipes were obtained. The use of open-celled metal foams as compact heat exchangers, exploiting convective cooling was studied and presented [2]. Based on heat transfer data, an analytical model is developed with simple cubic unit cells with heated slender cylinders. Analysis is carried out for a form chilled channel with constant wall temperatures. As a function of foam density, distribution of temperature inside channel is found. Studies of flow through an intricately structured porous medium was modelled using a new approach is reported [3]. A numerical simulation was carried out and results for mesh-independent on a flow through the form were compared with experiments and reported. The increase in pressure drop by the container walls was also compared. Using a periodic unit cell, a direct simulation of the transport in open cell metal foam is carried out assuming the pore to be spherical [4]. From a direct simulation, the effective conductivity, pressure drop, and local heat transfer coefficient is compared with existing experimental values.

Generally metal foams are categorized in two groups based on the liquid fraction and cell geometry. The liquid fraction of what is considered dry foam is relatively low, typically less than 1% and with film type geometry. This dry foam has such a small liquid fraction that the individual cells can be considered single surfaces in which soap bubbles represent its geometry very well. The differences between open cell and closed cell metal foams are mainly how the geometry of the cell is formed. In the open cell group the cells are not closed from each other and the flow of other materials through one cell occurs freely to another adjacent cell. This type of foam is generally created during the foam manufacturing by removing the inter-cellular membranes of closed cell foam. In the closed cell arrangement the surface tension and wetness scales increase and the geometry takes a very different shape compared to the open cell arrangements. It is characterized by the continuous cell walls which completely close the cells from one another with the formation of individual cell compartments. The cells nearly take a spherical shape in these types of metal foams. It is designated as wet foam when the liquid fraction exceeds 5%. The variation of flow and heat transfer properties was studied with varying flow and foam parameters. Heat transfer measurements inside rectangular blocks of commercially-available aluminum foam subjected to constant heat flux at metal foam surface are presented. Cases for different pore velocities, porosities and pore densities of the foam are given. The heat removal rate for one unit cell was extrapolated to calculate it for the whole bed. For flow and heat transfer analysis, two fluids: air and water are used and the various results obtained are compared. The proposed structure of the unit cell used in this work is a Tetrakaidecahedron model which is the most efficient packing structure proposed for the unit cells.

2. Modelling Of Cell Form

2.1 Cell Modelling

The first step in understanding the structure of a cell is obtained by considering how cells of equal volume pack themselves under pressure to most efficiently use space. From a popular experiment of peas Lord Kelvin postulated that the ideal packing cell shape was the so-called ''tetrakaidecahedron'' which is a figure consisting of six planar quadrilateral faces and eight non planar hexagons of zero net curvature. This shape also satisfies Plateau's conditions for a network of foam films. Surfaces which bound the cells meet at 1200, and the lines which are formed by their intersections meet at the tetrahedral angle cos-1 (-1/3). The structure of the metal foam used in this work is Tetrkaidecahedron which is modeled using commercial software GABIT 2.3.16. This geometric shape has been shown experimentally to be less representative than a multi cell unit cell in polymer foam; however, it is the most representative single-celled, space-filling, unit cell of the foam. This is similar to the structure proposed by Boomsma and Poulikakos [5] except with spherical nodes, which was deemed more physically accurate than cubes. The parameters used to fully describe the Tetra K unit cell are the relative density, specific surface area, average ligament diameter, ligament length, and a node size parameter. The following relationships are used to calculate modelling parameters of the model.

An elongated tetrakaidecahedron also packs to fill the space. It contains eight hexagonal faces, two horizontal square faces and four vertical diamond faces. The horizontal square faces have sides of length b and the diamond faces have sides of length L. The hexagonal faces have four sides with length L and two sides with length b. The inclination angle defines the orientation of the hexagonal faces with respect to the rise direction as well as the obtuse angle of the vertical diamond faces. When using a tetrak structure with all its sides equal then b=L=1; we obtain =45. Thus each hexagonal plane is inclined at an angle of 450.with horizontal thus forming a closed geometry.

Figure 1 shows the structure modeled in Gambit 2.3.16 as per the dimensions of ligament and the spherical nodes calculated. The TetraK model ligament cross section is circular and the diameter is constant over its length. The geometry of the structure was determined by solving the above equations using the values. The structure is similar to the structure proposed by Boomsma and Poulikakos except with spherical nodes, which was deemed more physically accurate than cubes.

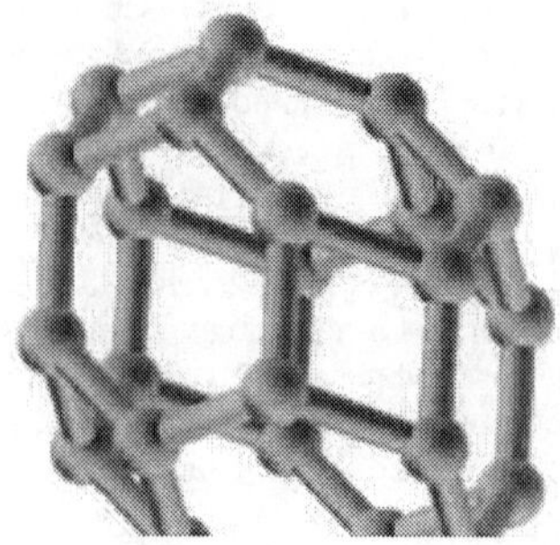

Figure. 1 Methodology of thermo-mechanical analysis

2.2 Parameters of Metal Foam Model

Diameter of ligament, df = 0.20 units.
Diameter of spherical nodes = 0.45 units.
Average Pore Diameter = 1.52 units
Side of the unit cell: 4.242 e -03 m
Surface area of the Aluminum ligament: 4.550367e-05 m^2
Volume of Al ligament: 2.678266 e -09 m^3
Volume of fluid: 7.363946 e -08 m^3
Total volume of unit cell: 7.631772 e -08 m^3
Porosity of the unit cell: 96.49 %

2.3 Computational Domain

A computational domain is to be defined for applying the boundary conditions and for limiting the volume for simulation A box is made as a computational domain enclosing the unit cell such that it is symmetric at all surfaces and is cut from the foam bed. Figure 2 illustrates the bounding box enclosing tetrakaicdecahedron as the unit cell. The bounding box, the Representative Elementary Volume (REV) which defines the symmetry is constructed and the foam is trimmed at the intersection with the bounding box so that the cell ligaments are fused with the bounding box. The dimensions of the box taken in the present work are 4.242 × 4.242 × 4.242 units. The unit cell is a squared Tetrakaidecahedron structure such that, Height = Width = 2.82 Units.

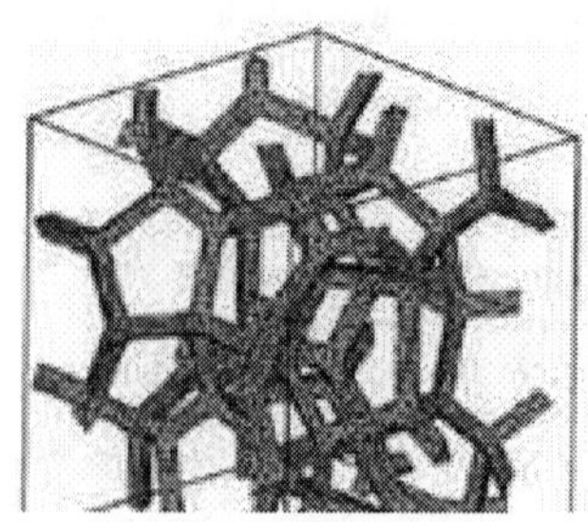

Figure. 2 Computational Domain

2.4 Meshing the Structure

The tetrak structure should be meshed as 2D triangular mesh on the surface and 3D tetrahedral mesh in the volume. The basic periodic cellular unit of the foam structure can be used as a building block to represent a much larger foam network. This is accomplished by meshing the visible void region of the structure as a fluid volume with a periodic grid at the boundaries. While meshing the geometry a bounding box or a computational domain is created and the volume enclosed by it is meshed using tetrahedral elements. The mesh generated on the surface of the structure is made to connect to the mesh generated in the void volume. Connectivity of the mesh is necessary for the flow simulation to be continuous both in the void and over the surface of the structure. The mesh generated is of size such that it contains 214194 tetrahedral elements. Figure 3 (a) depicts the meshing of the solid structure of cell and the Figure 3 (b) shows the meshing of the fluid region and solid region. With the faces of the whole structure being connected, the mesh of the fluid is generated taking the mesh on solid region as source mesh. This ensures connectivity between the nodes of fluid and solid region. A mesh check was performed throughout the volume and it was found that a minimum percentage of nodes i.e. 0.25% were found to have a skewness of 0.7 to 0.8, and none of the elements had any higher skewness. Thus the mesh generated is of good quality.

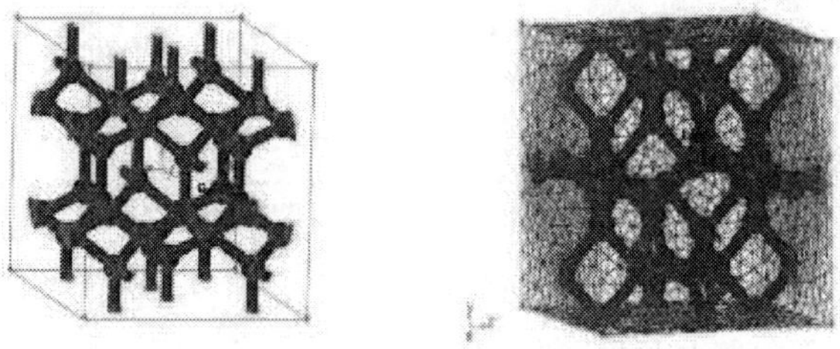

Figure. 3 (a) Solid Volume, (b) Fluid Volume

3. Fluid Flow Simulations Through Metal Foam

The porous structures have long been thought to behave as packed granular or spherical beds. But because of the structural difference between open-cell metal foams and such granular beds, it is important to characterize the particular fluid behaviour of metal foams with a different approach. This approach is different in the sense that metal foam is a rather periodic structure where the geometry and the flow pattern repeat itself. In the flow characterization of these type of material the Reynolds number is a usual non-dimensional parameter used to indicate the flow regime used in the experimentation. Although various definitions of the Reynolds number have been proposed in the past literature, the most accepted definition used at the macroscopic level is the permeability based Reynolds number, another non-dimensional term to describe the ratio of macroscopic form to viscous forces on porous media.

3.1 Boundary Conditions

Boundary conditions are given at the structure of the metal foam as a heat flux value of 1000 w/m2 on the walls of the metal foam, thus ensuring the convection of heat taking place. It is assumed that no conduction takes place through the solid wall as there is no temperature gradient. The boundary Conditions are given as follows: Inlet Face: Periodic, Outlet Face: Periodic, Other Faces: Symmetry, Structure of the Cell: Wall /Solid, Void Volume: Fluid.

As Periodic boundary conditions are assigned at the inlet and outlet faces of the computational domain to model the large foam network in the stream wise direction, hence it is obligatory to link these two faces in Gambit, this requires to give each loop on the face and a vertex to link. The faces are linked in the reverse orientation so that node to node correspondence takes place. Linking the faces ensures accurate and exact mapping of mesh nodes of one face on to another. Periodic boundary conditions are translational in nature i.e. the flow repeats itself translationally in one direction i.e. stream wise direction. A volumetric mass flow rate is also specified at the periodic boundaries as the input thus different values of velocities are used hence experimenting with different values of fluid. Symmetry boundary conditions are used when the physical geometry of interest, and the expected pattern of the flow/thermal solution, has mirror symmetry. They can also be used to model zero-shear slip walls in viscous flows.

Symmetry boundary conditions are given to define planes about which the cell can be duplicated. Symmetry boundary conditions take zero shear stress values and zero surface fluxes as q=0. As stated above, these conditions determine a zero flux across the symmetry plane, which is required by the definition of symmetry. Since the shear stress is zero at a symmetry boundary, it can also be interpreted as a "slip" wall when used in viscous flow calculations There is no diffusion flux across a symmetry plane: the normal gradients of all flow variables are thus zero at the symmetry plane. The symmetry boundary condition can therefore be summarized as follows: Zero normal velocity at a symmetry plane zero normal gradients of all variables at a symmetry plane Structure of the metal foam is given as solid wall and the void volume as the fluid. By giving a constant heat flux value of 1000 W/m2 it is assumed that there is no heat conduction in the walls of the metal foam. Only the fluid is taking heat from it by forced convection.

4. Simulations Of Metal Forms

While solving for the flow, turbulence and energy equations algorithm is used and the discretization technique used is second order upwind for better accuracy. Convergence can be controlled by using aggressive Algebraic multi grid solver and by using the under relaxation factors of 0.7 for momentum, 1 for energy. For iteration initial values of velocities, energy turbulence kinetic energy and dissipation rate are specified. Heat transfer, convergence and pressure gradient are monitored through the iteration. Following are the approaches and parametric variations carried out for simulations: Simulations are done for different flow velocities varying linearly from 1 to 5 m/s for a constant heat flux. Pressure gradient, heat transfer coefficient and temperature of fluid and solid, Nusselt's Number and Reynolds's Number are determined for every velocity and being plotted to show the graphical variation. Mesh independent results are shown using a coarse mesh of size 0.1 at the solid and 0.15 at the fluid. Heat removal rate for a sample foam bed is determined from one unit cell. Curve for the static pressure is drawn which is opposite in nature to the velocity curve thus a higher velocity results in lower pressure. Under relaxation factors are given. These factors are responsible in determining the convergence of the solution.
Algebraic Multi Grid (AMG) solver is used in this work which uses flexible cycle for all equations except energy for which it uses F cycle, the under relaxation factors can be reduced for an F cycle to 1 for convergence. Second order upwind provides a better accuracy although it takes more time. When the flow is aligned with the grid (e.g., laminar flow in a rectangular duct modeled with a quadrilateral or hexahedral grid) the first-order upwind discretization may be acceptable. When the flow is not aligned with the grid (i.e., when it crosses the grid lines obliquely), however, first-order convective discretization increases the numerical discretization error (numerical diffusion). While the first-order discretization generally yields better convergence than the second-order scheme, it generally will yield less accurate results, especially on tri/tet grids. Let the fluid's approach velocity be U. The pore velocity, u, in the pores of open-cell metal foam is also defined. This formulation has been used using fluent to calculate the net area averaged velocity at the inlet-outlet pair for a flow velocity of 1 to 5 m/s.

5. Results And Discussions

The variation of velocity across the unit cell is shown in figure 4 (a) and (b) gives the variation of static pressure across the velocity. By the nature of the two graphs we infer that the nature of pressure variation is opposite to that of velocity variation. A sudden dip in the value of the velocity is because of the momentarily rest position of the fluid particle in the when it collides with the wall surface. This results into an instant accumulation of particles there and hence increasing the pressure.

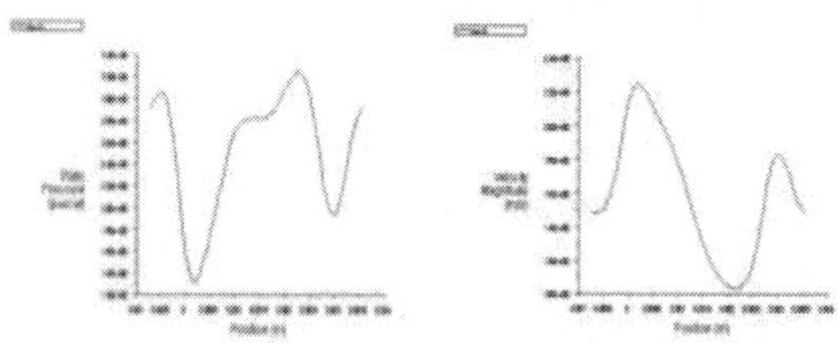

Figure. 4 (a) Velocity Magnitude Variation, (b) Pressure Static Variation

A sudden decrease in velocity around the mid plane is the result of more intricate geometry that reduces the pressure around this area. The Velocity contour in figure 5 (a) depicts the periodic boundary condition assigned at the inlet-outlet pair. The contour above suggest the exact mapping of velocity values on two periodic modules for modeling the large foam network in the stream wise direction. The Zero velocity of the fluid at the walls or the surface of the metal foam suggests "No slip Condition" at the surface hence due to wall shear stress a boundary layer is formed from pressure contour in figure 5 (b), we obtain that pressure is very low or negative at the face of the ligaments not facing the flow; this is due to the wake formation as the flow moves around the surface of structure.

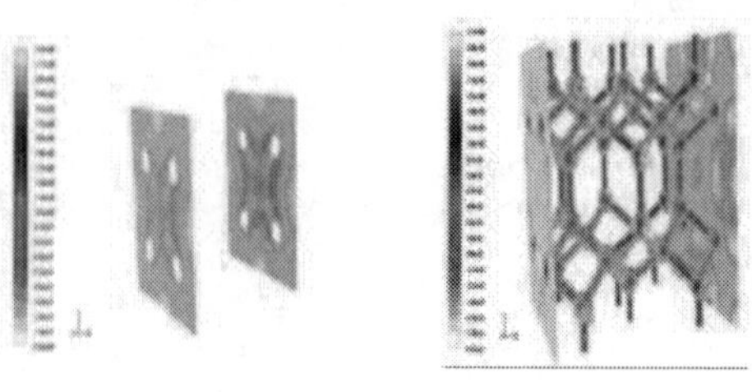

Figure. 5 (a) Velocity Contour, (b) Pressure Contour [Over inlet-outlet pair and Wall]

As the flow velocity is varied from 1 to 5 m/s, variations were observed in pressure variation profiles across the unit cell as plotted in figure 6 (a). With an increase in velocity of the flow, pressure drop across the cell increase. The porosity of the metal foam modeled is higher, hence lesser number of peaks is observed in the pressure profiles across the cell. The variation of negative pressure gradient and the graph is plotted in figure 6 (b).The graph shows a decrease in pressure gradient with the velocity. This shows the pore diameter, and not the porosity, increases the specific surface area (Ao) through the geometrical relationship involved in the scaling of the structure. This increase in the specific surface area has a larger effect on the permeability and overall flow resistance of foam than an increase in the solid fraction of the foam, since viscous drag is the dominating factor of pressure drop through a porous medium which is governed by the solid-fluid interfacial area.

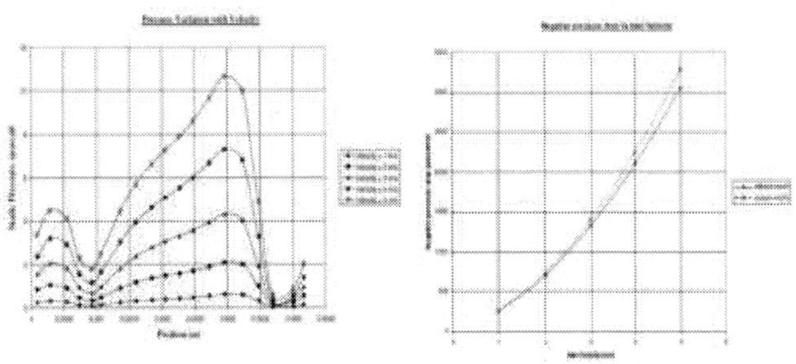

Figure. 6 (a) Variation of Pressure, (b) Variation of Inlet Velocity

A unit cell of pore diameter of 2.3 mm has a pore density of 40 ppi and the unit cell in the work is of 1.52 mm, hence we conclude its pore density to be 20 ppi with a porosity of 96%. Since viscous drag is the dominating factor of pressure drop through a porous medium, which is governed by the solid-fluid interfacial area we conclude that since the tetrakaidecahedron geometry provides a higher specific surface area thus producing more viscous drag hence giving larger pressure gradient values. From Pressure contour, we obtain that pressure is very low or negative at the face of the ligaments not facing the flow; this is due to the wake formation as the flow moves around the surface of structure.

A rise in pressure is attributed to the momentarily rest position of the fluid particle at the surface of wall, leading to increase in stagnation pressure. The pressure gradient is composed of two parts (i) The gradient of the periodic component and (ii) The gradient of linearly varying component, this linearly varying component is associated with the viscous drag. With an increase in velocity the shear stress due to viscous effects increase by Newton's Law of viscosity resulting in increasing pressure gradient. Because the value of pressure gradient is not known a priori, it must be iterated on until the mass flow rate that have been defined is achieved in the computational model. This correction of gradient term occurs in the pressure correction step of the SIMPLE, SIMPLEC, or PISO algorithm where its value is updated based on the difference between the desired mass flow rate and the actual one. In the present work, these iterations have been restricted to 2 and is monitored by providing an under relaxation factor. Area weighted average of pressure at the periodic boundaries gives periodic pressure.

6. Conclusion

In the present work the simulations are done for open celled metal foam with periodic boundary conditions on the inlet- outlet pair. Applying periodic boundary conditions helps us in simulating a unit cell and extrapolating the results for the whole of the bed, thus the periodic conditions in this work helped us to model the large foam network in the stream wise direction. The flow simulations obtained showed the occurrence of flow reversal formation of wake regions around the cell ligament thus affecting the linearly varying component of the periodic pressure, by viscous drag. Hence it is concluded about the pressure drop that the intricate geometry of Tetrakaidcahedron helped in increasing the net viscous drag thus varying the pressure and the velocity across the cell. This variation creates the necessary turbulence in the cell thus producing fluctuating velocities. Flow simulations led us to conclude that with an increase in velocity the pressure gradient is decreased as given by graph. Forced convection through metallic foams has been proved to substantially enhance heat transfer rates. Given this characteristic of the material, metal foams are well suited for the use in high performance compact heat exchangers and heat sinks for the use in electronic equipment and other industrial applications. Convective heat transfer in a fibrous medium involves the formation of complex flow and temperature fields around individual fibers. Flow separation may occur around the fibers at higher velocities than that of creep flow. Even though this leads to higher pressure drop values, the associated mixing substantially enhances the heat transfer rate. In addition when the fluid saturated medium has a larger effective thermal conductivity than the fluid alone, the heat dissipation is enhanced this in turn helps to keep the surface temperature within acceptable limits.

Acknowledgments
The authors appreciate the support of VIT management. The authors would also like to express their sincere gratitude to all the faculties and lab technicians involved in this research work for rendering their valuable guidance and help.

References:

[1] Lu, W., Zhao, C. Y., Tassou, S. A., "Thermal analysis on metal-foam filled heat exchangers Part I: Metal-foam filled pipes" International Journal of Heat and Mass Transfer, 2005, 49, p.2751-2761.

[2] Lu, T. J., Stone, H. A., Ashby, M. F., "Heat transfer in open-cell metal foams" Acta Materialia, 1998, 46, p. 3619-3635.

[3] Boomsma, K., Poulikakos, D., Ventikos, Y., "Simulations of flow through open cell metal foams using an idealized periodic cell structure" International Journal of Heat and Fluid Flow, 2003, 24, p. 825–834.

[4] Krishnan, S., Jayathi, Y. M., Suresh, V. G., "Direct simulation of transport in open-cell metal foam" Journal of Heat Transfer, 2006, 128, p.793-799.

[5] Boomsma, K., Poulikakos, D., "On the effective thermal conductivity of a three dimensionally structured fluid-saturated metal foam", International Journal of Heat and Mass Transfer, 2001, 44, p.827-836.

AUTHOR INDEX